Non-fibrillar Amyloidogenic Protein Assemblies—Common Cytotoxins Underlying Degenerative Diseases

Farid Rahimi • Gal Bitan
Editors

Non-fibrillar Amyloidogenic Protein Assemblies—Common Cytotoxins Underlying Degenerative Diseases

Springer

Editors
Farid Rahimi
Research School of Biology
Australian National University
Linnaeus Way
Canberra, ACT 0200
Australia
farid.rahimi@anu.edu.au

Gal Bitan
Neurology
University of California at Los Angeles
Charles E. Young Drive South 635
Los Angeles, CA 90095-7334
USA
gbitan@mednet.ucla.edu

ISBN 978-94-007-2773-1 e-ISBN 978-94-007-2774-8
DOI 10.1007/978-94-007-2774-8
Springer Dordrecht Heidelberg London New York

Library of Congress Control Number: 2011944432

Printed on acid-free paper

Springer is part of Springer Science+Business Media (www.springer.com)

Preface

Amyloid is a fascinating phenomenon. Proteins that have been shaped by millions of years of evolution lose their structure, if they had one, and gain a new structure, in which regardless of their amino acid sequence, they form tightly bound "one-dimensional" arrays of indefinite length and become insoluble. This new form of the protein typically is abnormal and is associated with various diseases, though in some cases, the amyloid form is functional and used in normal physiology.

The relationship between presence of amyloid and etiology of disease has been the subject of numerous studies and has caused much debate in the scientific community. An important realization that gradually has been taking hold in our understanding of this relationship is that the amyloid may be a hallmark of each disease but not necessarily the cause of the disease. Rather, oligomers of amyloid-forming proteins likely are the real perpetrators of the cytotoxicity that is characteristic of all amyloid-related diseases. Such oligomers typically are thought of as precursors of the amyloid, but in some cases have been observed to form down separate folding and assembly pathways. Therefore, we refer to them generally as non-fibrillar protein assemblies.

It has been our pleasure to serve as editors of *Non-fibrillar Amyloidogenic Protein Assemblies—Common Cytotoxins Underlying Degenerative Diseases*. We hope that this compilation of reviews will serve as a resource for researchers, students, and industry and medical professionals who are interested in the phenomenon of amyloid, the particular proteins discussed in the book, and the associated degenerative diseases.

The book begins with our historical account of the term *amyloid*, an introduction to fibrillar and non-fibrillar assemblies and their toxicity, and a discussion of often-overlooked methodological and experimental challenges in studying amyloid diseases. In Chap. 2, Vinters et al. illustrate the neuropathologic features of Alzheimer's disease and several non-Alzheimer dementias. Chapter 3 by Frydman-Marom et al. details the methods used for characterization of various oligomeric protein assemblies and reviews methods used to prepare these assemblies *in vitro*.

The following ten chapters are dedicated to specific amyloidogenic proteins and the diseases associated with them. In Chap. 4, Wilcox and coworkers cover biological targeting and activity of amyloid β-protein assemblies in Alzheimer's disease and highlight the interconnections between Alzheimer's disease and insulin signaling in the aging brain. In Chap. 5, Bhaskar and Lamb highlight different aspects of the proteins related to Alzheimer's disease and discuss studies on formation and detection of toxic assemblies of amyloid β-protein and τ protein, summarizing the current evidence on how these proteins cause neurotoxicity.

Chapter 6 by Hong et al. recounts illustrative examples of soluble, toxic or non-toxic amyloid oligomers and emphasize the roles of soluble oligomers of α-synuclein in the pathogenesis of Parkinson's disease.

Degaki et al. in Chap. 7 report various aspects of cytotoxicity of islet amyloid polypeptide in the pathogenesis of type-2 diabetes mellitus. Kerman and Chakrabartty elucidate key structural features of the misfolded superoxide dismutase 1 and its potential toxic effects in amyotrophic lateral sclerosis, in Chap. 8.

Chapters 9 and 10 by Legname et al. and Morales et al. discuss misfolding of prions and prion diseases. Murphy et al., in Chap. 11, review the roles of expanded polyglutamine proteins in neurodegeneration. In Chaps. 12 and 13, Hodkinson et al. and Saraiva et al. describe the roles of β_2-microglobulin and transthyretin in dialysis-related amyloidosis and familial amyloid polyneuropathy, respectively. In the closing chapter of the book, Lanning and Meredith comprehensively review therapeutic strategies for inhibiting abnormal protein self-assembly.

We are grateful to the scholars who dedicated their time and energy and contributed the different chapters of the book. We acknowledge and thank all the international experts who served as peer reviewers. Creation of this book would not have been possible without their contributions.

Farid Rahimi and Gal Bitan

Contents

Chapter 1
Overview of Fibrillar and Oligomeric Assemblies of Amyloidogenic Proteins

Farid Rahimi and Gal Bitan

Abstract Aberrantly folded proteins are implicated in over 40 human diseases, including neurodegenerative diseases such as Alzheimer's, Parkinson's, Huntington's, and Creutzfeldt–Jakob diseases; diseases of particular organs, including desmin-related cardiomyopathy or type-2 diabetes mellitus; and systemic diseases, such as senile systemic amyloidosis or light-chain amyloidosis. Although the proteins involved in each disease have unrelated sequences and dissimilar native structures, they all undergo conformational alterations and "misfold" to form fibrillar polymers characterized by a cross-β structure. Fibrillar assemblies build up progressively into intracellular or extracellular proteinaceous aggregates generating the pathognomonic amyloid-like lesions *in vivo*. Substantial evidence accumulated in the last decade suggest, that in many amyloid-related diseases, the lesions containing the protein aggregates are the end state of aberrant protein folding whereas the actual culprits causing the disease are soluble, non-fibrillar assemblies preceding the insoluble aggregates. The non-fibrillar protein assemblies are diverse and range from small, low-order oligomers to large assemblies, including spherical, annular, and protofibrillar species. Oligomeric species with different degrees of structural order are believed to mediate various pathogenic mechanisms that may lead to

F. Rahimi (✉)
Research School of Biology, Division of Biomedical Science and Biochemistry,
College of Medicine, Biology, and Environment, The Australian National University,
Canberra, ACT, Australia
e-mail: farid.rahimi@anu.edu.au

G. Bitan (✉)
Department of Neurology, David Geffen School of Medicine; Brain Research Institute;
and Molecular Biology Institute, University of California at Los Angeles,
635 Charles E. Young Drive South, Los Angeles, CA 90025, USA
e-mail: gbitan@mednet.ucla.edu

F. Rahimi and G. Bitan (eds.), *Non-fibrillar Amyloidogenic Protein Assemblies—Common Cytotoxins Underlying Degenerative Diseases*, DOI 10.1007/978-94-007-2774-8_1,

cellular dysfunction, cytotoxicity, and cell loss, eventuating in disease-specific degeneration. The particular pathologies thus are determined by the afflicted cell types, organs, systems, and the proteins involved. In many cases, the structure–function interrelationships amongst the various protein assemblies described *in vitro* are still elusive. Moreover, structural and mechanistic studies of amyloid proteins have been challenging due to the dynamic and metastable nature of the non-fibrillar oligomers and the non-crystalline nature of fibrillar protein aggregates. These factors have confounded the development and potential *in vivo* application of specific detection tools for non-fibrillar amyloid assemblies. Nevertheless, evidence suggests that non-fibrillar amyloid assemblies may share structural features and possibly common mechanisms of action as assessed *in vitro* or *in situ*. Deciphering these intricate structure–function correlations will help in understanding a complex array of pathogenic mechanisms, some of which may be common across different diseases albeit affecting different cell types or systems. This prefatory chapter aims to give an overview of historical definitions of amyloid along with a general discussion of fibrillar and non-fibrillar amyloid assemblies and their toxicity. The chapter also discusses some methodological challenges, which often are overlooked.

Keywords Amyloid, Cytotoxicity, Degeneration, Oligomer, Protein misfolding, Amyloid fibrils

1.1 Etymology of the Term "Amyloid"

The term "amyloid", in fact a misnomer, has been used in the context of histopathology since its neologism in 1838. Its first use then was by a German botanist, Matthias Schleiden (Schleiden 1838), who described amylaceous constituents of plant cell walls (reviewed in Kyle 2001; Steensma and Kyle 2007). Later in 1854, Rudolph Virchow, a German physician-scientist, used this term when examining the brain *corpora amylacea*, which stained pale blue upon treatment with iodine and violet when subsequently treated with sulfuric acid (Virchow 1854a, b) (reviewed in Kyle 2001; Sipe and Cohen 2000; Steensma and Kyle 2007). Because these staining characteristics are similar to those of starch, Virchow concluded that *corpora amylacea* were essentially cellulose and described the lesions as amyloid (i.e., starch-like). The term amyloid is derived from the Latin *amylum*, a transliteration of the Greek *amylon*, which was a term meaning "not ground at the mill" and referring to fine grains, especially starch (Steensma and Kyle 2007). At that time, the distinction between starch (in animals) and cellulose (in plants) was unclear (Sipe and Cohen 2000; Steensma and Kyle 2007).

In 1859, based on the high nitrogen content of amyloid lesions, Carl Friedreich and August Kekulé reported that the amyloid lesions contained albuminoid material and nothing chemically corresponding to *amylon* or cellulose (Friedreich and Kekulé 1859) (reviewed in Kyle 2001; Sipe and Cohen 2000; Steensma and Kyle 2007). They finally established the proteinaceous nature of amyloid lesions.

A century later, electron microscopic (EM) studies of human or animal amyloid lesions allowed observation of the fibrillar ultrastructure of amyloid (Cohen and Calkins 1959). Further progression of biochemical and biophysical techniques facilitated isolation of amyloid fibrils from tissue amyloid lesions in 1964 (Cohen and Calkins 1964) and their characteristic structure was determined by X-ray fiber diffraction in 1968 (Eanes and Glenner 1968). By then, the term amyloid had survived the test of time and was no longer a misnomer.

In the nineteenth and twentieth centuries, extensive studies have focused on deciphering the molecular and pathological mechanisms of protein misfolding, amyloid formation, amyloid-associated toxicity, and disease. This book provides a compendium of chapters describing these studies, with each chapter focusing on a particular protein, a particular disease, or a particular aspect of the relationship between protein misfolding and disease. Our opening chapter gives an overview of amyloids and different assemblies of amyloid proteins. The following chapter describes the pathologic lesions found in certain common amyloid diseases. All other chapters discuss structures of different amyloidogenic proteins and mechanisms mediated by these proteins involved in individual diseases. The final chapter outlines current therapeutic opportunities targeting these diseases and the associated amyloidogenic proteins.

1.2 Amyloid and Disease

To date, 27 human diseases are defined as classic amyloidoses (Westermark et al. 2007; Harrison et al. 2007; Sipe et al. 2010). These diseases are classified also as proteopathies (Walker et al. 2006), degenerative diseases (Dickson 2009), and conformational, protein-misfolding, protein-aggregation, or protein-deposition diseases (Surguchev and Surguchov 2010; Dobson 2004; Eisenberg et al. 2006). More than 40 human diseases collectively fall under the abovementioned classifications (Chiti and Dobson 2006). Of these, several neurodegenerative diseases, including Alzheimer's (AD) (Kril and Halliday 2001; Mayeux 2010; Aguzzi and O'Connor 2010; Lublin and Gandy 2010; Querfurth and LaFerla 2010), Parkinson's (PD) (Bagetta et al. 2010; Halliday and McCann 2010; Obeso et al. 2010; Pahwa and Lyons 2010; Postuma and Montplaisir 2009, Shulman 2010), Huntington's (HD) (Bauer and Nukina 2009; Cardoso 2009; Pfister and Zamore 2009; Rozas et al. 2010; Sassone et al. 2009), and prion diseases (Frost and Diamond 2010; Aguzzi and Calella 2009; Kupfer et al. 2009; Mallucci 2009; Sharma et al. 2009) are characterized pathognomonically by intracellular or extracellular microscopic lesions containing the proteinaceous amyloid aggregates. These diseases are characterized also by extensive neuron loss and atrophy in selected, vulnerable cerebral regions (Double et al. 2010), determining clinical presentations and outcomes. Amyloid-related diseases such as amyotrophic lateral sclerosis (ALS) (Cozzolino et al. 2008; Eisen 2009), non-neuropathic systemic diseases, e.g., light-chain and senile systemic amyloidoses (Comenzo 2006, 2007; Sanchorawala 2006), and other organ-specific diseases, such as dialysis-related amyloidosis (Dember and Jaber 2006; Kiss et al. 2005; Yamamoto et al. 2009), hereditary renal amyloidosis (Hawkins 2003; Kissane 1973; Eshaghian et al. 2007; McCarthy and Kasper 1998), atrial amyloidosis (Eshaghian et al. 2007; McCarthy and Kasper 1998; Goette and Rocken 2004; Rocken et al. 2002; Benvenga and Facchiano 1995; Looi 1993), and type-2 diabetes mellitus (Hayden et al. 2005; Khemtemourian et al. 2008; Li and Holscher 2007; Scheuner and Kaufman 2008) also are characterized by extracellular deposition of aberrantly folded, insoluble amyloid proteins.

Although the proteins contributing to different amyloidoses may have dissimilar sequences or unrelated native tertiary structures, they all form insoluble amyloid fibrils, ultimately lose their soluble, functional states, and deposit as amyloid, or amyloid-like lesions (Sipe and Cohen 2000). Extracellularly deposited amyloid material can be distinguished from non-amyloid deposits by: (1) characteristic straight, unbranched fibrillar morphology; (2) a typical cross-β pattern, in which β-strands run perpendicularly to the fiber axis; and (3) characteristic tinctorial properties, particularly binding of the dyes Congo red and thioflavin S. The cross-β pattern consists of two characteristic fiber-diffraction signals located on axes perpendicular to one another, a sharp, intense meridian reflection (parallel with the fiber axis) at ~4.7–4.8 Å and an equatorial signal at ~10 Å (Sunde et al. 1997). Binding of Congo red gives rise to characteristic blue-green birefringence under polarized light (Harrison et al. 2007; Merlini and Westermark 2004; Westermark et al. 2007; Frid et al. 2007), and binding of thioflavin S results in a hyperchromic shift in thioflavin-S fluorescent emission spectrum compared with free thioflavin S (Khurana et al. 2005; LeVine 1999). In addition to their major, fibrillar proteinaceous component, amyloid deposits contain metal ions, glycosaminoglycans, serum amyloid P, apolipoprotein E, collagen, nucleic acids, and other components (Hirschfield and Hawkins 2003; Alexandrescu 2005; Ginsberg et al. 1999, 1998; Marcinkiewicz 2002; Liao et al. 2004; Kahn et al. 1999).

Besides extracellularly deposited amyloid lesions in amyloidoses, many different intranuclear/intracytoplasmic amyloid-like aggresomes (inclusion bodies) also have been associated with specific diseases. The term inclusion body or "inclusion" is used frequently in the context of protein misfolding and aggregation (Kopito 2000; Cruts et al. 2006). Inclusion bodies in the latter context are distinct from bacterial inclusion bodies. Aggresomes are inclusion bodies formed by retrograde transport of aggregated proteins on microtubules (Kopito 2000). They contain a major aggregated protein and are also enriched in various molecular chaperones (Kopito 2000). Bacterial inclusion bodies typically are highly enriched for a single protein species and isolation of bacterial inclusion bodies is usually the first step in purification of heterologous proteins recombinantly expressed in bacteria. Aggresomes share certain properties with amyloid fibrils but some of them do not meet all the characteristics required by the classic definition of amyloid (Westermark et al. 2007; Sipe et al. 2010). Biochemists and biophysicists formally call the latter "amyloid-like," "amyloid-related," or "amyloidogenic" proteins. For instance, in PD and HD, protein aggregates accumulate intracellularly, generating disease-specific aggresomes—Lewy bodies and Huntington bodies, respectively. These structures have been excluded from the classification of amyloid by the Nomenclature Committee of the International Society of Amyloidosis (Westermark et al. 2007; Sipe et al. 2010) despite the fact that fibrils derived from the respective proteins, α-synuclein (Chap. 6) and polyglutamine-expanded huntingtin (Chap. 11), show all the characteristic features of amyloid (Conway et al. 2000a; Scherzinger et al. 1997; Chen et al. 2002; McGowan et al. 2000). A recent exception is the intracellular neurofibrillary tangles in AD. The main component of the neurofibrillary tangles is the microtubule-associated protein τ in a hyperphosphorylated form (Ihara et al. 1983; Joachim et al. 1987; Kosik et al. 1986; Nukina et al. 1987; Perl 2010; Steiner et al. 1990; Goedert et al. 1988). Despite their predominantly intracellular location, neurofibrillary tangles are now regarded as true amyloid (Sipe et al. 2010; Westermark et al. 2007) because of their fibrillar structure, cross-β X-ray diffraction pattern, and typical amyloid staining with Congo red (von Bergen et al. 2001; Giannetti et al. 2000; Berriman et al. 2003; Inouye et al. 2006).

1.3 Fibrillar Assemblies of Amyloid Proteins

In an Editorial in *Accounts of Chemical Research*, Ronald Wetzel once used the term "common threads" to allude to the common morphology of amyloid fibrils (Wetzel 2006). These fibrils revealed by transmission-electron microscopy (TEM) usually consist of 2–6 protofilaments each with diameter of 2–5 nm (Serpell et al. 2000). The protofilaments intertwine and form thread-like fibrils that are typically 7–13 nm wide (Serpell et al. 2000; Sunde and Blake 1997) or associate laterally to form long ribbons typically 2–5 nm thick and up to 30 nm wide (Bauer et al. 1995; Saiki et al. 2005). X-ray fiber-diffraction data indicate that in individual protofilaments the polypeptide chains are arranged in β-strands running perpendicular to the long axis of the fibril, forming the cross-β pattern (Sunde and Blake 1997).

The presence of highly organized and stable fibrillar deposits in affected organs in amyloid-related diseases long was viewed as a common causative link between aggregate formation and pathological symptoms. This had led to postulations such as the "amyloid cascade hypothesis" in the AD field. Originally, this hypothesis stated that deposition of amyloid β-protein (Aβ), the main component of amyloid plaques in AD-afflicted brains, was the cause of AD. This view was later reinforced by findings that Aβ-derived fibrils were neurotoxic (Pike et al. 1991; Lorenzo and Yankner 1994) and caused both membrane depolarization and alterations in the frequency of action potentials (Hartley et al. 1999). It was shown also that microinjection of fibrillar, but not soluble, Aβ into cerebral cortex of aged rhesus monkeys (*Macaca mulatta*) resulted in pathological events associated with AD, including profound neuronal loss, τ phosphorylation, and microglial proliferation (Geula et al. 1998). Similarly, monosialoganglioside GM1, a neuronal membrane component that is released from damaged neurons and is found in higher levels in cerebrospinal fluid from patients with AD than from age-matched controls, was found to enhance formation of Aβ fibrils with cytotoxicity and cell affinity much stronger than those of Aβ fibrils formed in phosphate-buffered saline (Okada et al. 2007).

A similar fibril-centered hypothesis was thought to apply contextually to all amyloidoses. For example, cytotoxic effects have been reported for fibrillar prion protein (Novitskaya et al. 2006) and lysozyme (Gharibyan et al. 2007). Insulin aggregation has been associated with rare injection-related amyloidosis (Swift 2002). In addition, insulin aggregation has been studied *in vitro* by multiple groups as a convenient model of protein fibrillogenesis (Murali and Jayakumar 2005; Dzwolak et al. 2007, 2006; Grudzielanek et al. 2007a). Biophysical investigations of insulin fibrillogenesis have identified oligomeric populations with conformations distinct from those of natively folded insulin dimer and hexamer (Ahmad et al. 2005). In a recent study combining structural characterization and cytotoxicity experiments, Grudzielanek et al. found no toxicity for low-order insulin oligomers whereas substantial toxicity was measured for high-order, β-sheet-rich aggregates that displayed either fibrillar or amorphous morphology (Grudzielanek et al. 2007b). Other studies using primates and transgenic murine diabetes models have shown the importance of islet amyloid in the pathogenesis of type-2 diabetes. It was thought that amyloid fibrils preceded formation of islet amyloid deposits and that fibrils derived from islet amyloid polypeptide (IAPP) were likely toxic to β-cells, thereby causing islet dysfunction (Lorenzo et al. 1994). Similarly, deposition of islet amyloid was considered an early event in type-2 diabetes and its progressive accumulation as the cause for parenchymal mass reduction and dysfunction (Westermark and Wilander 1978; Clark et al. 1988). This was thought to lead to progressively deficient insulin secretion, reduced glucose tolerance, and eventual emergence of fasting hyperglycemia (Kahn et al. 1999). Studies in mice harboring the human IAPP transgene suggested that not only hyperglycemia was associated with the development of islet amyloid, but that amyloid contributed to generation of hyperglycemia due to loss of β-cells (Hoppener et al. 2000).

One of the factors responsible for fibril-induced cytotoxicity is thought to be the physicochemical compositions of the surface of amyloid fibrils (Yoshiike et al. 2007). Significant morphological variations exist among different fibrils derived from the same peptide or protein, e.g., calcitonin (Bauer et al. 1995), SH3 domain of phosphatidylinositol-3′-kinase (Jiménez et al. 1999), insulin (Jiménez et al. 2002), Aβ (Petkova et al. 2005; Paravastu et al. 2006; Wetzel et al. 2007), and IAPP (Goldsbury et al. 1997; Radovan et al. 2008). Even single-residue alterations have been shown to affect fibril structure profoundly. For example, the substitution of D23 by N in Aβ, linked to severe cerebral amyloid angiopathy in an Iowa kindred (Van Nostrand et al. 2001), causes formation of Aβ40 fibrils considerably faster than wild-type Aβ40 (Tycko et al. 2009). At the molecular level, D23N-Aβ40 fibrils are arranged predominantly in an anti-parallel array, in contrast to the in-register, parallel β-sheet structure commonly found in wild-type Aβ40 fibrils and most other amyloid fibrils (Tycko et al. 2009). Despite these differences in molecular arrangement, the gross morphology, X-ray diffraction pattern, and dye-binding properties of amyloid fibrils appear to be universal amongst fibrillar structures of amyloid proteins.

Morphological differences in amyloid fibrils are governed also by conditions used for fibril preparation. For example, EM and nuclear magnetic resonance (NMR) enable visualization of polymorphic structure of Aβ40 fibrils prepared under agitated or quiescent conditions (Petkova et al. 2005). Seeding experiments may facilitate detailed structural characterization of amyloid fibrils developing *in vivo* and elucidate the controversial role of fibrils (Hardy and Selkoe 2002) in human amyloidoses. To study fibril polymorphism *in vivo* and based on the ability of preformed amyloid fibrils to propagate their structures through seeded growth *in vitro* (Petkova et al. 2005), fibrils extracted from AD brain tissue were used to seed growth of synthetic Aβ40 fibrils (Paravastu et al. 2009). This allowed Paravastu et al. to deduce putative structures of fibrils extracted from AD brain by recapitulating these structures using seeded fibrillar growth of synthetic Aβ40. Paravastu et al. showed that fibrils grown after being seeded with material extracted from two separate AD patients' brains had predominantly the same two fibril structures (Paravastu et al. 2009). These predominant fibril structures differed from the two previously described, purely synthetic Aβ40 fibril structures (Paravastu et al. 2008; Petkova et al. 2006), indicating that seeded growth combined with structural studies may determine the molecular structures of fibrils developing in AD brain or in fibrils involved in other amyloid diseases *in vivo*. The results described above suggest that each amyloid protein potentially forms a spectrum of structurally distinct fibrils, and that kinetic and microenvironmental factors determine which of these alternatives predominate under given circumstances, which can differ considerably *in vitro* and *in vivo*.

Direct correlation between specific molecular organization and fibril toxicity may be important where pathogenic mechanisms of sporadic and genetic forms of amyloid diseases are studied. For example, some genetic cases of AD (Taddei et al. 1998; Miyoshi 2009; Moro et al. 2010; McDonald et al. 2010) and PD (Dawson 2007; Gasser 2009; Inzelberg and Polyniki 2010; Schiesling et al. 2008; Houlden et al. 2001) have an earlier onset and a faster progression than sporadic forms of these diseases suggesting potentially different underlying molecular mechanisms. Studies similar to those of Paravastu et al. could be extended to compare fibril structures of Aβ in sporadic *versus* genetic, early-onset forms of AD. Such studies will potentially delineate correlations between protein structure and disease severity or progression at the molecular level. Aging-induced spontaneous chemical modifications, such as amino-acid racemization or amino-acid isomerization—e.g., involving aspartate and asparagine residues—may affect Aβ production, polymerization, and clearance, potentially playing a pivotal role in the pathogenesis of sporadic and genetic forms of AD (Moro et al. 2010). Therefore, studies linking fibril morphology with aging-induced posttranslational protein modifications in AD may unravel correlations between fibril structure and pathogenesis. This example is potentially applicable and relevant to other amyloidoses, for example PD.

1.4 Non-fibrillar Assemblies of Amyloid Proteins

Contrary to the original amyloid cascade hypothesis (Hardy and Higgins 1992), substantial evidence suggests that fibrillar aggregates are the end state of aberrant protein folding and eventuate as potentially protective sinks for the cytotoxic, oligomeric, non-fibrillar protein assemblies. The transient, non-fibrillar assemblies likely are the actual culprits. These assemblies are believed to initiate the pathogenic mechanisms that lead to cellular dysfunction, cell loss, loss of functional tissue, and disease-specific regional or organ-specific atrophy (Kirkitadze et al. 2002; Gurlo et al. 2010; Haataja et al. 2008; Luibl et al. 2006; Meier et al. 2006).

Amyloid β-protein (Aβ), the causative agent in AD, is considered an archetypal amyloidogenic protein. The multitude and variety of structural, functional, and pathophysiological studies of Aβ exemplify the complexity of research findings covering non-fibrillar assemblies of amyloidogenic proteins. Extensive biophysical studies in the Aβ field have led to functional and structural descriptions of non-fibrillar and pre-fibrillar Aβ assemblies. For example, the discovery of Aβ protofibrils (Walsh et al. 1997; Harper et al. 1997) and other toxic non-fibrillar Aβ assemblies, including low-order oligomers, Aβ-derived diffusible ligands, and paranuclei (reviewed in Rahimi et al. 2008) have led to a paradigm shift (Kirkitadze et al. 2002; Haass and Selkoe 2007; Glabe 2006; Glabe and Kayed 2006) in AD research, challenging the original, fibril-centered, amyloid cascade hypothesis (Hardy and Higgins 1992). An updated version of the hypothesis presented a decade after the original one (Hardy and Selkoe 2002) emphasizes that early, pre-fibrillar Aβ assemblies or Aβ assemblies unrelated to fibrils are the primary cytotoxins in AD pathogenesis leading to synaptic dysfunction and neuron loss (Sakono and Zako 2010; Gong et al. 2003; Klein 2002a; Cleary et al. 2005; Lambert et al. 1998). This paradigm shift and the centrality of non-fibrillar Aβ assemblies in AD research have led to a search for similar non-fibrillar protein assemblies in other amyloid-related diseases. To date, at least 40 different proteins have been identified as causative agents of amyloidoses (Bellotti et al. 2007; Chiti and Dobson 2006). In most cases, including prion proteins (Simoneau et al. 2007, also discussed in Chaps. 9 and 10), transthyretin (Sorgjerd et al. 2008 and Chap. 13), α-synuclein (van Rooijen et al. 2010 and Chap. 6), apolipoprotein C-II (Ryan et al. 2008), τ (Peterson et al. 2008; Sahara et al. 2008; Kayed et al. 2009 and Chap. 5), superoxide dismutase (Cozzolino et al. 2009 and Chap. 8), polyglutamine-expanded proteins (Legleiter et al. 2010 and Chap. 11), and islet amyloid polypeptide (Haataja et al. 2008 and Chap. 7), non-fibrillar protein assemblies have been found and shown to exert adverse biological effects similar to those of non-fibrillar Aβ oligomers (Kirkitadze et al. 2002; Caughey and Lansbury 2003; Ferreira et al. 2007; Glabe 2006; Jellinger 2009; Kitamura and Kubota 2010; Sakono and Zako 2010; Roychaudhuri et al. 2009 and Chaps. 3, 4, 5).

Before the focus in the amyloid field shifted from fibrils to non-fibrillar assemblies, it was known that despite sequence dissimilarity among amyloidogenic proteins, amyloid fibrils were largely similar in the core regions (Eisenberg et al. 2006; Serpell 2000). The realization that the non-fibrillar oligomeric structures may be the proximate disease-causing agents in the amyloidoses related to these proteins raised the question of whether oligomeric structures were also similar. High-resolution microscopic studies of oligomeric structures, mostly by TEM and AFM, have demonstrated that in most cases the morphologies observed were spherical, annular, or protofibrillar (worm-like). Despite morphological similarities, studies have demonstrated that small structural changes may have a large impact on the oligomer populations formed by the same protein (Bitan et al. 2003a, b, c).

Protofibrils, the penultimate precursors of fibrillar assemblies, are curvilinear, fibril-like structures of 4–8 nm diameter, ≤200 nm length (Walsh et al. 1997), and may have an axial twisting periodicity of 20 nm (Hartley et al. 1999). They have been described as spherical beads of 2–5 nm diameter arranged as beaded chains in linear, curvilinear, or annular arrangements in studies originally reporting them (Harper et al. 1997; Walsh et al. 1999, 1997). The annular protofibrils have been the predominant structures found in several studies (Caughey and Lansbury 2003; Lashuel et al. 2002a, b; Ding et al. 2002; Kayed et al. 2009). However, as discussed elsewhere (Bitan et al. 2005), it is important to note that in many cases the term protofibril has been used even though the morphologies of the assemblies under study were distinct from those originally defined as protofibrils. It is also important to distinguish between protofibrils and protofilaments, which are the constituent units of mature fibrils (Serpell et al. 2000; Teplow 1998).

One of the most-studied amyloidogenic proteins is α-synuclein (Chap. 6). α-Synuclein, first characterized in zebra finch (*Taeniopygia guttata*) (George et al. 1995) (under the UniProt accession number Q91448, the organism described is *Serinus canaria* (Island canary) or *Fringilla canaria*), was thought to be important in neural plasticity during vertebrate development. The exact function of α-synuclein still is not clear though it is thought to be part of the proteasomal system (reviewed in Layfield et al. 2003; Betarbet et al. 2005), vesicle trafficking and endocytosis (Varkey et al. 2010), and/or SNARE complex assembly (Burré et al. 2010). α-Synuclein has been shown to form α-helical structures when interacting with artificial (Jao et al. 2008; Trexler and Rhoades 2009; Georgieva et al. 2008) or biological membranes (Kim et al. 2006). As discussed earlier, α-synuclein is the predominant component in Lewy bodies, the pathological hallmarks in PD brains. It has been implicated also in other degenerative disorders (synucleinopathies), including dementia with Lewy bodies and multiple-system atrophy (Ian et al. 2001; Jellinger 2009; Chiti and Dobson 2006). Similar to Aβ, α-synuclein belongs to a growing family of "intrinsically disordered" proteins (Tompa 2002; Dyson and Wright 2005), a characteristic that perhaps renders these proteins more prone to undergoing amyloidogenic assembly because of their structural instability. Mutant α-synuclein alloforms linked to familial PD were found to oligomerize faster than the wild-type protein, whereas the rate of fibril formation did not correlate with the presence of disease-causing mutations (Conway et al. 2000b). Non-fibrillar assemblies of both wild-type and mutant α-synuclein included spherical oligomers, protofibrillar structures, and most abundantly, annular protofibrils (Ding et al. 2002; Lashuel et al. 2002b). The latter morphology suggested that the mechanism whereby α-synuclein induces toxicity is pore formation in cell membranes. In agreement with this idea, protofibrillar α-synuclein was found to permeabilize synthetic vesicles (Volles et al. 2001). Interestingly, this effect was increased by the familial PD-linked mutants A30P and A53T (Volles and Lansbury 2002), but not by the mutant E46K (Fredenburg et al. 2007). Thus, although pore formation may be involved in α-synuclein-induced toxicity, other mechanisms also have been implicated, but these are not understood well (Takeda et al. 2006).

IAPP aggregation is thought to cause type-2 diabetes. IAPP is a 37-residue peptide hormone produced in pancreatic β-cells and co-secreted with insulin. Early stages of type-2 diabetes are characterized by insulin resistance followed by increased insulin and IAPP secretion. Elevated IAPP levels lead to its assembly into toxic, soluble oligomers and insoluble aggregates (Marzban et al. 2003). Oligomeric and protofibrillar IAPP were shown to interact with synthetic membranes (Anguiano et al. 2002), a characteristic that decreases with further aggregation, providing a clue for the mechanism of IAPP toxicity (Porat et al. 2003). Similar to α-synuclein, interaction with biological membranes may induce a transient α-helical conformation in IAPP, presumably facilitating penetration of the oligomers into the membrane resulting in solute leakage across the membrane (Jayasinghe and Langen 2005; Knight et al. 2006). Strong evidence for the cytotoxic role of IAPP oligomers in type-2 diabetes was given in a study in which rifampicin, an inhibitor of IAPP fibril, but not oligomer, formation, did not protect pancreatic β-cells against apoptosis induced by either exogenous or endogenously expressed IAPP (Meier et al. 2006). More recent data have suggested that *in vivo*, toxic IAPP oligomers are formed intracellularly and therefore, oligomer-specific antibodies do not prevent cell death *in vitro* or *in vivo* (Lin et al. 2007).

1.4.1 Analytical Challenges in Studies of Amyloid Protein Oligomers

Structural studies of oligomers of amyloidogenic proteins have been challenging because these assemblies typically are metastable and comprise heterogeneous mixtures of species. Immunological insights have been obtained by Glabe and co-workers, who developed antibodies that bound specifically to oligomers but not to the monomeric or fibrillar forms of proteins of unrelated sequences (Kayed et al. 2003). The first polyclonal antibody, A11, and similar antibodies developed in follow-up studies (Kayed and Glabe 2006; Georganopoulou et al. 2005; Lafaye et al. 2009), showed remarkable ability to bind to oligomers formed by proteins as diverse as Aβ, α-synuclein, IAPP, lysozyme, insulin, polyglutamine, and prion fragments (Kayed et al. 2003). In a recent study, iterative immunization of aged beagles with an aggregated Aβ preparation (Head et al. 2008) was shown to produce antibodies specific for monomeric, non-fibrillar, or fibrillar Aβ42 preparations (Vasilevko et al. 2010). However, dot-blotting results in this study were not conclusive enough to designate the canine antibodies as purely oligomer-specific anti-Aβ antibodies because some degree of cross-reactivity (50% by densitometry) was evident and the results were not complemented by structural studies of Aβ preparations used for antibody-specificity assays.

As discussed above, recent studies by Paravastu et al. (2009) showed that the dominant structure of Aβ fibrils grown by Aβ fibril seeds derived from AD-afflicted brains differed from that in fibrils derived from purely synthetic Aβ40, suggesting that fibrillization conditions *in vitro*, and by inference, oligomerization conditions, differ from conditions *in vivo*. Findings based on the above studies by Paravastu et al. (Paravastu et al. 2009) and others (Petkova et al. 2005; Inaba et al. 2005; Lee et al. 2007; Kayed et al. 2009; Nekooki-Machida et al. 2009) argue against the idea, based on immunoreactivity data, TEM, and AFM studies, that non-fibrillar or pre-fibrillar amyloid assemblies are structurally similar. Although *in vitro* studies provide valuable insight into the structure and activity of non-fibrillar amyloid assemblies, these studies must be interpreted carefully because: (1) the conditions *in vivo* differ from those *in vitro* due to the complexity of cellular and tissue milieus; (2) mutations, amino-acid substitutions, or amino-acid modifications can result in different oligomer populations, different levels of oligomer toxicity or different fibrillar structures with different toxic properties (Bitan et al. 2003b; Yoshiike et al. 2007; Hung et al. 2008); and (3) fibrils grown in the presence of monosialoganglioside GM1 released from damaged neurons are more toxic than those prepared in buffer alone (Okada et al. 2007). Conclusively, non-fibrillar amyloid structures and compositions *in vivo* likely differ, at least to some degree, from those produced, analyzed, and studied *in vitro*. Some of the confounding factors in these cases involve post-extraction or post-analysis sample handling (e.g., freeze–thaw cycles, transportation, etc.). For examples, it was initially shown that Aβ dimers isolated from human brain tissue inhibited long-term potentiation (LTP), enhanced long-term depression, and reduced dendritic spine density in rodent hippocampal neurons (Shankar et al. 2008). However, these toxic activities were ascribed later to Aβ protofibrils, which formed readily from covalently stabilized Aβ dimers (O'Nuallain et al. 2010). These data suggest that by the time the activity of a certain Aβ preparation is measured, potentially inert Aβ species (e.g., dimers) may have converted to toxic species (e.g., protofibrils).

The same argument may apply to studies whereby non-fibrillar amyloid assemblies were extracted and studied *in vitro* (Shankar et al. 2008, 2007; Paleologou et al. 2009; Klucken et al. 2006; Sharon et al. 2003, Lesné et al. 2006; Head et al. 2010). Many extraction procedures use detergents, such as sodium dodecyl sulfate (SDS), which are known to disrupt the structure of non-fibrillar amyloid assemblies (Bitan et al. 2005; Hepler et al. 2006). Although electrophoretic separation of proteins in the presence of SDS (SDS–PAGE) generally is an excellent analytical method, the effect of SDS on all proteins is not equivalent (Gudiksen et al. 2006). Different proteins, different conformations of the same protein (Leffers et al. 2004), or truncated versions of certain proteins (Kawooya et al. 2003) may not bind stoichiometric amounts of SDS. In addition, in certain cases, SDS can induce or stabilize secondary or quaternary structures rather than denaturing them (Leffers et al. 2004; Montserret et al. 2000; Yamamoto et al. 2004). Further, SDS may cause dissociation of some protein assemblies or conversely induce protein self-association, depending on the specific protein studied (Yamamoto et al. 2004; Rangachari et al. 2007, 2006; Piening et al. 2006). For example, Aβ42-derived "globulomers" are oligomeric species produced by incubating Aβ42 in the presence of 0.2% SDS (Barghorn et al. 2005). Apparent electrophoretic fractionation of monomeric or oligomeric components in a protein mixture does not necessarily indicate existence of such components prior to SDS treatment. Examples of this shortcoming of SDS–PAGE have been reported in its applications to studies of Aβ (Bitan et al. 2005; Hepler et al. 2006) and α-synuclein (Moussa et al. 2004). A recent example is a study of Aβ40 dimers stabilized by an intermolecular disulfide bridge, which showed the same SDS–PAGE profile before and after formation of β-sheet-rich protofibrils (O'Nuallain et al. 2010).

Because of the structural instability of amyloidogenic protein oligomers and the abovementioned analytical artifacts, studies reporting on structural properties of amyloidogenic proteins based on SDS–PAGE findings must be interpreted cautiously. This is particularly relevant to those studies, which have reported characterization of antibodies specific for oligomeric assemblies of amyloidogenic proteins relying on SDS–PAGE and western blotting. Recently, an elaborate study using ultrathin array tomography and immunofluorescence showed that senile plaques in brains of a murine model of AD are surrounded by "haloes of oligomeric Aβ" (Koffie et al. 2009) based on immunoreactivity of an antibody (NAB61), which apparently was reactive to oligomeric Aβ assemblies fractionated by SDS–PAGE (Lee et al. 2006). The original paper, which described this antibody, reported that NAB61 also recognized synthetic Aβ fibrils (Lee et al. 2006). Considering these caveats, one may question the major conclusions drawn by Koffie et al. because of the use of an antibody that was claimed to be specific for SDS–PAGE-fractionated oligomeric Aβ but was also cross-reactive with fibrillar Aβ assemblies. Similar cross-reactivity was apparent in antibodies that were produced and characterized after iterative immunization of beagles (Vasilevko et al. 2010) with an aggregated Aβ preparation (Head et al. 2008). Caveats regarding binding specificity of reagents ostensibly recognizing non-fibrillar amyloid assemblies also are relevant to aptamers. Aptamers are short ribo- or single-stranded deoxyribo-oligonucleotides used as specific molecular recognition tools in research, diagnostics, and therapy. Recently, we have found that aptamers bind fibrillar assemblies of amyloid proteins avidly yet non-specifically (Rahimi et al. 2009). Despite the fact that our aptamers were selected using covalently stabilized oligomeric preparations of Aβ40, they were found to bind not only Aβ-derived fibrillar structures, but also fibrils of other amyloid proteins (Rahimi et al. 2009). Similar high affinity for fibrils was observed using aptamers selected by multiple rounds of enrichment and two non-enriched, "naïve" RNA libraries demonstrating that fibril binding was a general phenomenon rather than a characteristic of particular RNA sequences (Rahimi et al. 2009). Comparable findings were observed previously with aptamers selected against fibrillar and non-fibrillar β_2-microglobulin (Bunka et al. 2007). Moreover, nucleic acids have been shown to enhance formation of amyloid fibrils and interactions between amyloid-forming peptides and nucleic acids have been shown to cause formation of combined protein–nucleic-acid fibrils (Braun et al. 2011). These findings suggest that non-specific reactivity with oligonucleotides may be a universal property of amyloid proteins (Rahimi and Bitan 2010; Rahimi et al. 2009; Braun et al. 2011; Ylera et al. 2002; Bunka et al. 2007). Therefore, aptamers developed against non-fibrillar amyloid assemblies or those claimed to be specific for non- or pre-fibrillar amyloid assemblies must be tested for specificity for fibrillar assemblies of amyloid proteins. This is particularly applicable to studies reporting aptamers "specific" for monomeric or oligomeric Aβ (Takahashi et al. 2009) or α-synuclein (Tsukakoshi et al. 2010). Further delineation of specific mechanisms governing these interactions requires additional studies and will be important in interpretation of structure–function relationships and for designing reagents that recognize non-fibrillar amyloid assemblies specifically or potentially block amyloid-related toxicity. A relevant recent News article in *Nature* (Ledford 2010) has highlighted similar challenges researchers are facing in studying diseases with complex mechanisms and outlined some of the complexities and controversies involved in studies linking the prion protein and Aβ in AD research.

1.5 Non-fibrillar and Fibrillar Assemblies of Disease-Unrelated Proteins

The milieu of a polypeptide chain may cause it to adopt a multitude of conformations, or interconvert among many, in a wide temporal range (Dobson 2001; Dzwolak et al. 2007; Frieden 2007; Guijarro et al. 1998; Gursky and Aleshkov 2000; Stefani and Dobson 2003; Kelly 1998; Cruz et al. 2005; De Felice et al. 2004). This complexity is more relevant *in vivo* where interactions amongst proteins and interactions between proteins and other cellular components govern various cellular functional processes (Canale et al. 2006; Kitamura and Kubota 2010; Stefani and Dobson 2003; Zhang et al. 2004). Conformational heterogeneity renders the study of amyloidogenic proteins particularly difficult due to the transient nature of the adopted conformations, which populate closely related minima in the thermodynamic energy landscape (Miller et al. 2010). Besides disease-associated amyloid-forming proteins and proteins that naturally form non-pathological, functional amyloid-like fibrils (reviewed in Chiti and Dobson 2006) (see also Maji et al. 2009a), disease-unrelated proteins (Stefani and Dobson 2003) and artificially designed peptides (Fezoui et al. 2000; Wang et al. 2007; Kammerer and Steinmetz 2006) were shown to form amyloid under particular non-native conditions. The first proteins shown to form amyloid fibrils were reported by (Guijarro et al. 1998 and Litvinovich et al. 1998). The src-homology 3 (SH3) domain of bovine phosphatidyl inositol 3-kinase (PI3K), an 85-residue, β-structured protein, was shown to form amyloid fibrils slowly under acidic conditions (Guijarro et al. 1998). Thenceforth, the disease-unrelated SH3 domain has served as an excellent model system for studies examining structural properties of amyloid fibrils and molecular mechanisms of amyloid formation (Jiménez et al. 1999; Zurdo et al. 2001a, b; Carulla et al. 2005). It was found that the initial protein aggregates were relatively dynamic and flexible to allow particular interactions guiding formation of the highly ordered fibrils (Polverino de Laureto et al. 2003).

After Litvinovich et al. demonstrated formation of amyloid-like fibrils by self-association of a murine fibronectin type-III module (Litvinovich et al. 1998), others reported that similar conversions in a number of disease-unrelated proteins could be induced *in vitro* by a deliberate, rational choice of experimental conditions (Chiti et al. 2001, 1999; Stefani and Dobson 2003). Formation of fibrils from full-length proteins occurs under solution conditions that partially or completely disrupt the native structure of the protein but do not completely break hydrogen bonds (Chiti et al. 2001). On the other hand, in the aggregation of unstructured proteins, e.g., Aβ, partially structured conformers have been shown to be necessary for fibril formation (Fezoui and Teplow 2002; Kirkitadze et al. 2001; Maji et al. 2005). It was shown that proteins with as few as four residues, and amino-acid homopolymers unable to fold into stable globular structures, form fibrils readily (Stefani and Dobson 2003; Tjernberg et al. 2002; Lopez De La Paz et al. 2002). Therefore, it has been suggested that the ability to form amyloid fibrils could be a generic property of polypeptide chains (Stefani and Dobson 2003).

In contrast to the hypothesis that adoption of amyloid or amyloid-like conformation is a generic property of the polypeptide backbone with only a minor contribution by the amino-acid side-chains (Dobson 2001), Maji et al. argued that side-chain interactions are essential in the aggregation process (Maji et al. 2009b) as demonstrated in fibril-related crystal structures (Nelson et al. 2005; Nelson and Eisenberg 2006; Sawaya et al. 2007), in studies showing the sequence-specific nature of amyloid aggregation (Tjernberg et al. 2002; Margittai and Langen 2006; Zanuy and Nussinov 2003), and by the scale of amino-acid aggregation propensities determined experimentally, ranging from aggregation-prone hydrophobic residues to aggregation-interfering, charged side-chains (Fernandez-Escamilla et al. 2004; Tartaglia et al. 2008). These studies suggest that under non-physiological conditions, including acidic pH, extremes of protein concentration, or addition of aprotic solvents (Guijarro et al. 1998; Chiti et al. 1999; Polverino de Laureto et al. 2003; Marcon et al. 2005), the influence of side-chains in the aggregation process can be altered, eventually driving the protein of interest into amyloid fibrils (Maji et al. 2009b).

1.6 Studying the Toxicity of Non-fibrillar Amyloid Assemblies

One of the main pathogenic mechanisms (Jellinger 2010) underlying many neurodegenerative diseases is abnormal protein dynamics and protein misfolding (Skovronsky et al. 2006; Herczenik and Gebbink 2008) accompanied by an imbalance between protein production and degradation, proteasomal/autophagy impairment, and dysfunction or mutation of molecular chaperones (Jellinger 2009, 2010). Oxidative stress in the form of reactive oxygen/nitrogen species, free radical formation, and lipid peroxidation also is involved in protein-misfolding diseases

(Butterfield et al. 2010; Kahle et al. 2009; Sesti et al. 2010; Ahmad et al. 2009). Oxidative stress goes hand-in-hand with inflammatory mechanisms and production of cytokines and chemokines in the disease-affected tissues (Ahmad et al. 2009; Lee et al. 2009; Lucin and Wyss-Coray 2009; Sugama et al. 2009; Tansey and Goldberg 2010; Sokolova et al. 2009; Shepherd et al. 2006). Mitochondrial dysfunction, DNA damage, disruption of ion homeostasis, and impaired bioenergetics coincide with oxidative stress and inflammatory conditions (Jellinger 2009, 2010). All these pathogenic mechanisms, inter-related in complex cycles, lead to cellular dysfunction, apoptosis, and/or necrosis. In the central nervous system, depending on the cell populations affected, these pathogenic mechanisms lead to emergence of specific or mixed disease phenotypes and complex clinical presentations and outcomes (Dickson 2009; Boeve 2007; Murray et al. 2005; Pittock and Lucchinetti 2007; Lansbury and Lashuel 2006).

Numerous experimental approaches have facilitated study of cytotoxic mechanisms of non-fibrillar assemblies of amyloidogenic proteins. *In vitro* experiments using cell culture and tissue slices along with biophysical studies have been performed to examine the toxic mechanisms of non-fibrillar amyloid assemblies using recombinant, synthetic, cell-, or tissue-derived variants of the amyloidogenic proteins. As discussed above, a concern in these experimental setups is that only a small proportion of the artificial assemblies may closely resemble non-fibrillar assemblies occurring *in vivo*. Other experimental approaches include the use of animal models, such as insect (Botella et al. 2009; Cowan et al. 2010; Iijima and Iijima-Ando 2008; Iijima-Ando and Iijima 2010; Khurana 2008; Lu 2009; Lu and Vogel 2009; Park et al. 2009; van Ham et al. 2009), *Caenorhabditis elegans* (Johnson et al. 2010), *Brachydanio rerio* (Sager et al. 2010; Ingham 2009; Malaga-Trillo and Sempou 2009), murine (Ashe and Zahs 2010; Dawson et al. 2010; Elder et al. 2010; Guyenet et al. 2010; Park et al. 2010; Taylor et al. 2010), rat (Flood et al. 2009), canine (Barsoum et al. 2000; Green and Tolwani 1999; Lossi et al. 2005; Vasilevko and Head 2009; Woodruff-Pak 2008), and simian models (Yang et al. 2008; Wang and Qin 2006; Qin et al. 2006; Walker 1997) to assess various aspects of etiology and pathogenesis, including genetics, behavior, system functions, or nutritional and therapeutic applications.

Toxicity mechanisms of non-fibrillar amyloid assemblies in various diseases are discussed in detail in individual chapters of this book. Here we highlight a few examples of non-fibrillar amyloid assemblies and their associated toxicity mechanisms. In one prominent example, synthetic Aβ oligomers derived from cells transfected with amyloid precursor protein (Podlisny et al. 1995) were shown to disrupt LTP in hippocampal tissue slices and *in vivo* (Townsend et al. 2006; Walsh et al. 2002), impair the memory of a complex pre-learned behavior (Cleary et al. 2005), memory consolidation, and synaptic remodeling causing loss of functional synapses in rats (Freir et al. 2011).

Another type of oligomer studied extensively is Aβ-derived diffusible ligands (ADDLs), which are synthetic Aβ42-derived species formed in the presence of apoJ (Oda et al. 1995), in F-12 media (Klein 2002b), or in phosphate-buffered saline (De Felice et al. 2008) as small globules 3–8 nm in diameter (Chromy et al. 2003) in polydisperse mixtures of 150–1,000-kDa complexes (Hepler et al. 2006). ADDLs have been shown to be highly neurotoxic (Lambert et al. 1998; Xia et al. 1997), inhibit LTP (Lambert et al. 1998), promote oxidative stress and increased $[Ca^{2+}]_i$ (De Felice et al. 2007), induce τ phosphorylation (De Felice et al. 2008), and enhance interleukin-1β, inducible nitric oxide synthase (iNOS), nitric oxide, and tumor-necrosis-factor-α expression in astrocytes (White et al. 2005). Recently, it has been shown that ADDLs are sequestered into, and seed, new amyloid plaques in the brains of a murine AD model (Gaspar et al. 2010). However, the underlying mechanisms of this observation require further studies.

In neurodegenerative diseases characterized by intraneuronal α-synuclein deposition, even modest α-synuclein elevations can be toxic, though the precise mechanisms underlying synaptotoxicity in these diseases are unclear. Recently, a quantitative model system was used to evaluate the time-course and localization of evolving α-synuclein-induced pathologic events using cultured neurons isolated from brains of transgenic mice overexpressing fluorescently labeled human α-synuclein (Scott et al. 2010). Transgenic α-synuclein was shown to be altered pathologically over time while overexpressing neurons showed enlarged synaptic vesicles and striking deficits in neurotransmitter release (Scott et al. 2010), a phenotype characteristic of animal models lacking critical presynaptic proteins (Abeliovich et al. 2000; Chandra et al. 2004, 2005). In this model, Scott et al. showed that several endogenous presynaptic proteins were undetectable in a subset of transgenic synaptic boutons, suggesting that such diminutions triggered the overall synaptic pathology due to increased α-synuclein levels (Scott et al. 2010). Similar alterations in levels of synaptic proteins were retrospectively observed in human pathologic brains (Mukaetova-Ladinska et al. 2009; Bertrand et al. 2003), highlighting potential relevance to human disease.

Another toxic mechanism proposed for non-fibrillar assemblies of amyloid proteins is their pore- or channel-forming capacity that may lead to membrane leakage and increased $[Ca^{2+}]_i$ (Lashuel and Lansbury 2006; Lashuel et al. 2002a). In lipid bilayers *in vitro*, Aβ was shown to form uniform pore-like structures (Lin et al. 2001; Quist et al. 2005). These are thought to serve as Ca^{2+} channels and thus have been hypothesized to cause excitotoxicity and mediate Aβ-induced neurotoxicity in AD (Arispe et al. 1993b, a). Reports of various models including artificial phospholipid membrane bilayers, excised neuronal membrane patches, whole-cell patch-clamp experiments, and phospholipid vesicles support a channel-forming property of Aβ (Lin et al. 2001; Arispe et al. 1993b; Kawahara et al. 1997; Kawahara and Kuroda 2000; Sanderson et al. 1997; Rhee et al. 1998; Hirakura et al. 1999; Lin et al. 1999; Bhatia et al. 2000; Kourie et al. 2001; Kagan et al. 2002; Lin and Kagan 2002; Bahadi et al. 2003; Alarcon et al. 2006) and α-synuclein (Adamczyk and Strosznajder 2006; Di Pasquale et al. 2010; Kim et al. 2009; Tsigelny et al. 2007; Zakharov et al. 2007; Feng et al. 2010). Imaging techniques (Lin et al. 2001, 1999; Rhee et al. 1998; Bhatia et al. 2000), electrophysiological experiments (Arispe et al. 1993b; Kawahara et al. 1997; Sanderson et al. 1997; Rhee et al. 1998; Hirakura et al. 1999; Bhatia et al. 2000; Kourie et al. 2001; Bahadi et al. 2003; Alarcon et al. 2006), or cation-sensitive dyes (Bhatia et al. 2000; Jelinek and Sheynis 2010) were used to assess channel-like properties of Aβ. However, other studies have reported general disruption of the plasma membrane homeostasis without channel formation (Sokolov et al. 2006; Demuro et al. 2005; Kayed et al. 2004).

It has been shown that directed expression of the molecular chaperone, Hsp70, one of numerous molecular chaperones that guide the correct folding of polypeptides, prevented dopaminergic neuronal loss associated with α-synuclein in a *Drosophila* model of PD and that interference with endogenous chaperone activity accelerated α-synuclein toxicity (Auluck et al. 2002). This work, and similar approaches in polyglutamine-related disorders (Warrick et al. 1999; Opal and Zoghbi 2002), indicate that such diseases are indeed disorders of protein folding, suggesting that activation of chaperones and other compensatory mechanisms, such as the ubiquitin–proteasome system, potentially can decrease accumulation of misfolded proteins or enhance their clearance.

In contrast to fibrils of disease-causing amyloidogenic proteins (discussed above), those formed by disease-unrelated proteins do not cause cytotoxicity in cell-culture experiments. For example, fibrils formed by an artificially designed α-helix-turn-α-helix (αtα) peptide displayed no neurotoxicity, even though they were morphologically indistinguishable from Aβ and IAPP fibrils, which were toxic (Fezoui et al. 2000). However, the pre-fibrillar assemblies of PI3K-SH3 and HypF-N were shown to be highly toxic to PC12 cells and murine fibroblasts *in vitro* (Bucciantini et al. 2004). The extent of cellular injury caused by the cytotoxic oligomers was comparable to that by Aβ42 oligomers, whereas the corresponding fibrils of both PI3K-SH3 and HypF-N were benign.

Early pre-fibrillar HypF-N assemblies were shown to permeabilize artificial phospholipid membranes more efficiently than mature fibrils, suggesting that this disease-unrelated protein shared toxic properties with non-fibrillar assemblies of peptides and proteins involved in pathology (Relini et al. 2004). Further investigation of the cellular effects of HypF-N oligomers revealed that they entered the cytoplasm and caused an acute rise in levels of reactive oxygen species and $[Ca^{2+}]_i$, leading to cell death (Bucciantini et al. 2004). In a study in which murine fibroblasts or endothelial cells were treated with pre-fibrillar HypF-N assemblies, the two cell types underwent two different death mechanisms—fibroblasts exposed for 24 h to 10 μM HypF-N oligomers underwent necrosis, whereas endothelial cells treated similarly underwent apoptosis (Bucciantini et al. 2005). A similar study comparing cytotoxic effects of pre-fibrillar and fibrillar HypF-N assemblies using a panel of normal and pathological cell-lines showed that cells were variably affected by the same amount of pre-fibrillar aggregates, whereas mature fibrils showed little or no toxicity (Cecchi et al. 2006). Recently, it has been shown that microinjection into rat brain nucleus basalis magnocellularis of non-fibrillar assemblies of PI3K-SH3 or HypF-N, but not the corresponding mature fibrils, compromised neuronal viability dose-dependently (Baglioni et al. 2006). Taken together, these data clearly demonstrate that the non-fibrillar assemblies of disease-unrelated proteins are highly toxic whereas most of the corresponding mature fibrils are not (Baglioni et al. 2006). The toxic effects of the oligomers may arise when these assemblies assume a "misfolded" conformation, which may expose hydrophobic residues that are natively buried within the core structure. These exposed hydrophobic sequences are aggregation-prone and may interact with membranes and other cellular constituents modifying their structural/functional homeostasis. Interestingly, two types of stable, pre-fibrillar oligomers of HypF-N, which display similar morphologic and tinctorial properties, were shown to differ in their cytotoxic effects (Campioni et al. 2010). The differences in the packing of hydrophobic interactions between adjacent protein molecules in the oligomers determined the ability of the two oligomeric assemblies to cause cellular dysfunction and toxicity. Thus, a lower degree of hydrophobic packing within the oligomer core structure was found to correlate with a higher ability to penetrate the cell membrane and cause Ca^{2+} influx (Campioni et al. 2010).

1.7 Conclusions

Since the discovery and definition of amyloid lesions, intensive research has led to accumulation of data elucidating the pathogenic mechanisms of protein-misfolding diseases. Initially, pathogenic and toxic primacy was given to fibrillar forms of amyloidogenic proteins as these structures were found to be the major pathological hallmarks in neurodegenerative diseases. As discussed previously, earlier studies attributing toxicity to amyloid fibrils may have found this effect because of the inadvertent use of immature amyloid fibrils or equilibrium mixtures of oligomers and fibrils, which are cytotoxic, rather than pure preparations of mature amyloid fibrils, which often are not (Aksenov et al. 1996; Martins et al. 2008). Importantly, as our understanding of the devastating neurodegenerative and protein-misfolding diseases has been growing, an alternative paradigm has emerged. This paradigm postulates that non-fibrillar protein assemblies rather than mature amyloidogenic fibrils likely are the key neurotoxins responsible for most of the pathogenic mechanisms in protein-misfolding and neurodegenerative diseases. Accordingly, oligomeric species are thought to mediate diverse but interrelated pathogenic mechanisms that may lead to cytotoxicity and cell loss eventuating in organic and systemic involvement. This interrelation may lead to self-promoting and -propagating pathogenic cycles that worsen with age and chronicity. For instance, mechanisms associated with protein-misfolding may cause other events, such as inflammation and oxidative stress, which in turn aggravate misfolding. Overall, it is postulated that the non-fibrillar amyloidogenic proteins are "on path" to fibrillogenesis. The resulting protein fibrils are thought to be the end-stage sinks for the toxic non-fibrillar species. Fibrillar assemblies accumulate progressively into intracellular or extracellular proteinaceous amyloid aggregates generating the disease-specific lesions *in vivo*.

Global research efforts have established a framework for understanding the fundamentals of protein assembly and misfolding. A remaining challenge is to assess how these fundamental structural principles are linked to cellular and tissue microenvironments during progression of disease. Many experimental conditions have been used to study the structure and function of non-fibrillar assemblies; however, due to methodological limitations, regeneration and scrutiny of the actual *in vivo* milieus and conditions in which protein assembly, oligomerization, fibrillization, and deposition occur are difficult. Similarly, it is extremely difficult to assess all the possible interactions these assemblies may have with various cellular components and organelles in the course of pathogenesis. A multitude of detrimental mechanisms, including disruption of cellular metabolism, deregulation of synapse structure and function, membrane damage, ionic imbalance, oxidative/inflammatory stress, apoptosis, and other cytotoxic effects, have been shown to be mediated by non-fibrillar assemblies of amyloidogenic proteins, emphasizing that a single therapeutic approach likely will be insufficient to prevent or treat the progression of diseases involving protein misfolding. Involvement of complex pathogenic mechanisms in these diseases calls for multifaceted rational diagnostic and therapeutic approaches that could potentially target not only a single assembly or a single mechanism but a multitude of assemblies or mechanisms. Agents that arrest the self-assembly process at the earliest stages or divert the process into formation of non-toxic species likely have the highest chance of success preventing and treating amyloid-related diseases because they inhibit formation and/or toxicity of both initial toxic oligomers and later aggregates.

Acknowledgements We thank Drs. D. Teplow and M. Landau for reviewing this book chapter and acknowledge financial support by grants AG027818 and AG030709 from NIH/NIA and grant 07–65798 from California Department of Health Services.

References

Abeliovich A, Schmitz Y, Farinas I, Choi-Lundberg D, Ho WH, Castillo PE, Shinsky N, Verdugo JM, Armanini M, Ryan A, Hynes M, Phillips H, Sulzer D, Rosenthal A (2000) Mice lacking α-synuclein display functional deficits in the nigrostriatal dopamine system. Neuron 25:239–252

Adamczyk A, Strosznajder JB (2006) α-Synuclein potentiates Ca^{2+} influx through voltage-dependent Ca^{2+} channels. Neuroreport 17:1883–1886

Aguzzi A, Calella AM (2009) Prions: protein aggregation and infectious diseases. Physiol Rev 89:1105–1152

Aguzzi A, O'Connor T (2010) Protein aggregation diseases: pathogenicity and therapeutic perspectives. Nat Rev Drug Discov 9:237–248

Ahmad A, Uversky VN, Hong D, Fink AL (2005) Early events in the fibrillation of monomeric insulin. J Biol Chem 280:42669–42675

Ahmad R, Rasheed Z, Ahsan H (2009) Biochemical and cellular toxicology of peroxynitrite: implications in cell death and autoimmune phenomenon. Immunopharmacol Immunotoxicol 31:388–396

Aksenov MY, Aksenova MV, Butterfield DA, Hensley K, Vigo-Pelfrey C, Carney JM (1996) Glutamine synthetase-induced enhancement of β-amyloid peptide Aβ (1–40) neurotoxicity accompanied by abrogation of fibril formation and Aβ fragmentation. J Neurochem 66: 2050–2056

Alarcon JM, Brito JA, Hermosilla T, Atwater I, Mears D, Rojas E (2006) Ion channel formation by Alzheimer's disease amyloid β-peptide (Aβ40) in unilamellar liposomes is determined by anionic phospholipids. Peptides 27:95–104

Alexandrescu AT (2005) Amyloid accomplices and enforcers. Protein Sci 14:1–12

Anguiano M, Nowak RJ, Lansbury PT Jr (2002) Protofibrillar islet amyloid polypeptide permeabilizes synthetic vesicles by a pore-like mechanism that may be relevant to type II diabetes. Biochemistry 41:11338–11343

Arispe N, Pollard HB, Rojas E (1993a) Giant multilevel cation channels formed by Alzheimer disease amyloid β-protein [AβP-(1–40)] in bilayer membranes. Proc Natl Acad Sci USA 90:10573–10577

Arispe N, Rojas E, Pollard HB (1993b) Alzheimer disease amyloid β protein forms calcium channels in bilayer membranes: blockade by tromethamine and aluminum. Proc Natl Acad Sci USA 90:567–571

Ashe KH, Zahs KR (2010) Probing the biology of Alzheimer's disease in mice. Neuron 66:631–645

Auluck PK, Chan HY, Trojanowski JQ, Lee VM, Bonini NM (2002) Chaperone suppression of α-synuclein toxicity in a *Drosophila* model for Parkinson's disease. Science 295:865–868

Bagetta V, Ghiglieri V, Sgobio C, Calabresi P, Picconi B (2010) Synaptic dysfunction in Parkinson's disease. Biochem Soc Trans 38:493–497

Baglioni S, Casamenti F, Bucciantini M, Luheshi LM, Taddei N, Chiti F, Dobson CM, Stefani M (2006) Prefibrillar amyloid aggregates could be generic toxins in higher organisms. J Neurosci 26:8160–8167

Bahadi R, Farrelly PV, Kenna BL, Curtain CC, Masters CL, Cappai R, Barnham KJ, Kourie JI (2003) Cu^{2+}-induced modification of the kinetics of Aβ(1–42) channels. Am J Physiol Cell Physiol 285:C873–C880

Barghorn S, Nimmrich V, Striebinger A, Krantz C, Keller P, Janson B, Bahr M, Schmidt M, Bitner RS, Harlan J, Barlow E, Ebert U, Hillen H (2005) Globular amyloid β-peptide oligomer—a homogenous and stable neuropathological protein in Alzheimer's disease. J Neurochem 95:834–847

Barsoum SC, Callahan HM, Robinson K, Chang PL (2000) Canine models for human genetic neurodegenerative diseases. Prog Neuropsychopharmacol Biol Psychiatry 24:811–823

Bauer PO, Nukina N (2009) The pathogenic mechanisms of polyglutamine diseases and current therapeutic strategies. J Neurochem 110:1737–1765

Bauer HH, Aebi U, Haner M, Hermann R, Muller M, Merkle HP (1995) Architecture and polymorphism of fibrillar supramolecular assemblies produced by *in vitro* aggregation of human calcitonin. J Struct Biol 115:1–15

Bellotti V, Nuvolone M, Giorgetti S, Obici L, Palladini G, Russo P, Lavatelli F, Perfetti V, Merlini G (2007) The workings of the amyloid diseases. Ann Med 39:200–207

Benvenga S, Facchiano A (1995) Atrial natriuretic peptide and amyloidosis. J Intern Med 237:525–526

Berriman J, Serpell LC, Oberg KA, Fink AL, Goedert M, Crowther RA (2003) Tau filaments from human brain and from *in vitro* assembly of recombinant protein show cross-β structure. Proc Natl Acad Sci USA 100:9034–9038

Bertrand E, Lechowicz W, Lewandowska E, Szpak GM, Dymecki J, Kosno-Kruszewska E, Wierzba-Bobrowicz T (2003) Degenerative axonal changes in the hippocampus and amygdala in Parkinson's disease. Folia Neuropathol 41:197–207

Betarbet R, Sherer TB, Greenamyre JT (2005) Ubiquitin–proteasome system and Parkinson's diseases. Exp Neurol 191(Suppl 1):S17–S27

Bhatia R, Lin H, Lal R (2000) Fresh and globular amyloid β protein (1–42) induces rapid cellular degeneration: evidence for AβP channel-mediated cellular toxicity. FASEB J 14:1233–1243

Bitan G, Kirkitadze MD, Lomakin A, Vollers SS, Benedek GB, Teplow DB (2003a) Amyloid β-protein (Aβ) assembly: Aβ40 and Aβ42 oligomerize through distinct pathways. Proc Natl Acad Sci USA 100:330–335

Bitan G, Tarus B, Vollers SS, Lashuel HA, Condron MM, Straub JE, Teplow DB (2003b) A molecular switch in amyloid assembly: Met35 and amyloid β-protein oligomerization. J Am Chem Soc 125:15359–15365

Bitan G, Vollers SS, Teplow DB (2003c) Elucidation of primary structure elements controlling early amyloid β-protein oligomerization. J Biol Chem 278:34882–34889

Bitan G, Fradinger EA, Spring SM, Teplow DB (2005) Neurotoxic protein oligomers—what you see is not always what you get. Amyloid 12:88–95

Boeve BF (2007) Links between frontotemporal lobar degeneration, corticobasal degeneration, progressive supranuclear palsy, and amyotrophic lateral sclerosis. Alzheimer Dis Assoc Disord 21:S31–S38

Botella JA, Bayersdorfer F, Gmeiner F, Schneuwly S (2009) Modelling Parkinson's disease in *Drosophila*. Neuromolecular Med 11:268–280

Braun S, Humphreys C, Fraser E, Brancale A, Bochtler M, Dale TC (2011) Amyloid-associated nucleic acid hybridisation. PLoS One 6:e19125

Bucciantini M, Calloni G, Chiti F, Formigli L, Nosi D, Dobson CM, Stefani M (2004) Prefibrillar amyloid protein aggregates share common features of cytotoxicity. J Biol Chem 279: 31374–31382

Bucciantini M, Rigacci S, Berti A, Pieri L, Cecchi C, Nosi D, Formigli L, Chiti F, Stefani M (2005) Patterns of cell death triggered in two different cell lines by HypF-N prefibrillar aggregates. FASEB J 19:437–439

Bunka DH, Mantle BJ, Morten IJ, Tennent GA, Radford SE, Stockley PG (2007) Production and characterization of RNA aptamers specific for amyloid fibril epitopes. J Biol Chem 282:34500–34509

Burré J, Sharma M, Tsetsenis T, Buchman V, Etherton MR, Sudhof TC (2010) α-Synuclein promotes SNARE-complex assembly *in vivo* and *in vitro*. Science 329:1663–1667

Butterfield DA, Bader Lange ML, Sultana R (2010) Involvements of the lipid peroxidation product, HNE, in the pathogenesis and progression of Alzheimer's disease. Biochim Biophys Acta 1801:924–929

Campioni S, Mannini B, Zampagni M, Pensalfini A, Parrini C, Evangelisti E, Relini A, Stefani M, Dobson CM, Cecchi C, Chiti F (2010) A causative link between the structure of aberrant protein oligomers and their toxicity. Nat Chem Biol 6:140–147

Canale C, Torrassa S, Rispoli P, Relini A, Rolandi R, Bucciantini M, Stefani M, Gliozzi A (2006) Natively folded HypF-N and its early amyloid aggregates interact with phospholipid monolayers and destabilize supported phospholipid bilayers. Biophys J 91:4575–4588

Cardoso F (2009) Huntington disease and other choreas. Neurol Clin 27:719–736, vi

Carulla N, Caddy GL, Hall DR, Zurdo J, Gairi M, Feliz M, Giralt E, Robinson CV, Dobson CM (2005) Molecular recycling within amyloid fibrils. Nature 436:554–558

Caughey B, Lansbury PT (2003) Protofibrils, pores, fibrils, and neurodegeneration: separating the responsible protein aggregates from the innocent bystanders. Annu Rev Neurosci 26:267–298

Cecchi C, Pensalfini A, Baglioni S, Fiorillo C, Caporale R, Formigli L, Liguri G, Stefani M (2006) Differing molecular mechanisms appear to underlie early toxicity of prefibrillar HypF-N aggregates to different cell types. FEBS J 273:2206–2222

Chandra S, Fornai F, Kwon HB, Yazdani U, Atasoy D, Liu X, Hammer RE, Battaglia G, German DC, Castillo PE, Sudhof TC (2004) Double-knockout mice for α- and β-synucleins: effect on synaptic functions. Proc Natl Acad Sci USA 101:14966–14971

Chandra S, Gallardo G, Fernandez-Chacon R, Schluter OM, Sudhof TC (2005) α-Synuclein cooperates with CSPα in preventing neurodegeneration. Cell 123:383–396

Chen S, Berthelier V, Hamilton JB, O'Nuallain B, Wetzel R (2002) Amyloid-like features of polyglutamine aggregates and their assembly kinetics. Biochemistry 41:7391–7399

Chiti F, Dobson CM (2006) Protein misfolding, functional amyloid, and human disease. Annu Rev Biochem 75:333–366

Chiti F, Webster P, Taddei N, Clark A, Stefani M, Ramponi G, Dobson CM (1999) Designing conditions for *in vitro* formation of amyloid protofilaments and fibrils. Proc Natl Acad Sci USA 96:3590–3594

Chiti F, Bucciantini M, Capanni C, Taddei N, Dobson CM, Stefani M (2001) Solution conditions can promote formation of either amyloid protofilaments or mature fibrils from the HypF *N*-terminal domain. Protein Sci 10:2541–2547

Chromy BA, Nowak RJ, Lambert MP, Viola KL, Chang L, Velasco PT, Jones BW, Fernandez SJ, Lacor PN, Horowitz P, Finch CE, Krafft GA, Klein WL (2003) Self-assembly of Aβ(1–42) into globular neurotoxins. Biochemistry 42:12749–12760

Clark A, Wells CA, Buley ID, Cruickshank JK, Vanhegan RI, Matthews DR, Cooper GJ, Holman RR, Turner RC (1988) Islet amyloid, increased A-cells, reduced B-cells and exocrine fibrosis: quantitative changes in the pancreas in type 2 diabetes. Diabetes Res 9:151–159

Cleary JP, Walsh DM, Hofmeister JJ, Shankar GM, Kuskowski MA, Selkoe DJ, Ashe KH (2005) Natural oligomers of the amyloid-β protein specifically disrupt cognitive function. Nat Neurosci 8:79–84

Cohen AS, Calkins E (1959) Electron microscopic observations on a fibrous component in amyloid of diverse origins. Nature 183:1202–1203

Cohen AS, Calkins E (1964) The isolation of amyloid fibrils and a study of the effect of collagenase and hyaluronidase. J Cell Biol 21:481–486

Comenzo RL (2006) Systemic immunoglobulin light-chain amyloidosis. Clin Lymphoma Myeloma 7:182–185

Comenzo RL (2007) Managing systemic light-chain amyloidosis. J Natl Compr Canc Netw 5:179–187

Conway KA, Harper JD, Lansbury PT Jr (2000a) Fibrils formed *in vitro* from α-synuclein and two mutant forms linked to Parkinson's disease are typical amyloid. Biochemistry 39:2552–2563

Conway KA, Lee SJ, Rochet JC, Ding TT, Williamson RE, Lansbury PT Jr (2000b) Acceleration of oligomerization, not fibrillization, is a shared property of both α-synuclein mutations linked to early-onset Parkinson's disease: implications for pathogenesis and therapy. Proc Natl Acad Sci USA 97:571–576

Cowan CM, Chee F, Shepherd D, Mudher A (2010) Disruption of neuronal function by soluble hyperphosphorylated tau in a *Drosophila* model of tauopathy. Biochem Soc Trans 38:564–570

Cozzolino M, Ferri A, Carri MT (2008) Amyotrophic lateral sclerosis: from current developments in the laboratory to clinical implications. Antioxid Redox Signal 10:405–443

Cozzolino M, Pesaresi MG, Amori I, Crosio C, Ferri A, Nencini M, Carri MT (2009) Oligomerization of mutant SOD1 in mitochondria of motoneuronal cells drives mitochondrial damage and cell toxicity. Antioxid Redox Signal 11:1547–1558

Cruts M, Gijselinck I, van der Zee J, Engelborghs S, Wils H, Pirici D, Rademakers R, Vandenberghe R, Dermaut B, Martin JJ, van Duijn C, Peeters K, Sciot R, Santens P, De Pooter T, Mattheijssens M, Van den Broeck M, Cuijt I, Vennekens K, De Deyn PP, Kumar-Singh S, Van Broeckhoven C (2006) Null mutations in progranulin cause ubiquitin-positive frontotemporal dementia linked to chromosome 17q21. Nature 442:920–924

Cruz L, Urbanc B, Borreguero JM, Lazo ND, Teplow DB, Stanley HE (2005) Solvent and mutation effects on the nucleation of amyloid β-protein folding. Proc Natl Acad Sci USA 102:18258–18263

Dawson TM (2007) Unraveling the role of defective genes in Parkinson's disease. Parkinsonism Relat Disord 13(Suppl 3):S248–S249

Dawson TM, Ko HS, Dawson VL (2010) Genetic animal models of Parkinson's disease. Neuron 66:646–661

De Felice FG, Vieira MN, Meirelles MN, Morozova-Roche LA, Dobson CM, Ferreira ST (2004) Formation of amyloid aggregates from human lysozyme and its disease-associated variants using hydrostatic pressure. FASEB J 18:1099–1101

De Felice FG, Velasco PT, Lambert MP, Viola K, Fernandez SJ, Ferreira ST, Klein WL (2007) Aβ oligomers induce neuronal oxidative stress through an *N*-methyl-D-aspartate receptor-dependent mechanism that is blocked by the Alzheimer drug memantine. J Biol Chem 282:11590–11601

De Felice FG, Wu D, Lambert MP, Fernandez SJ, Velasco PT, Lacor PN, Bigio EH, Jerecic J, Acton PJ, Shughrue PJ, Chen-Dodson E, Kinney GG, Klein WL (2008) Alzheimer's disease-type neuronal tau hyperphosphorylation induced by Aβ oligomers. Neurobiol Aging 29:1334–1347

Dember LM, Jaber BL (2006) Dialysis-related amyloidosis: late finding or hidden epidemic? Semin Dial 19:105–109

Demuro A, Mina E, Kayed R, Milton SC, Parker I, Glabe CG (2005) Calcium dysregulation and membrane disruption as a ubiquitous neurotoxic mechanism of soluble amyloid oligomers. J Biol Chem 280:17294–17300

Di Pasquale E, Fantini J, Chahinian H, Maresca M, Taieb N, Yahi N (2010) Altered ion channel formation by the Parkinson's-disease-linked E46K mutant of α-synuclein is corrected by GM3 but not by GM1 gangliosides. J Mol Biol 397:202–218

Dickson DW (2009) Neuropathology of non-Alzheimer degenerative disorders. Int J Clin Exp Pathol 3:1–23

Ding TT, Lee SJ, Rochet JC, Lansbury PT Jr (2002) Annular α-synuclein protofibrils are produced when spherical protofibrils are incubated in solution or bound to brain-derived membranes. Biochemistry 41:10209–10217

Dobson CM (2001) The structural basis of protein folding and its links with human disease. Philos Trans R Soc Lond B Biol Sci 356:133–145

Dobson CM (2004) Principles of protein folding, misfolding and aggregation. Semin Cell Dev Biol 15:3–16

Double KL, Reyes S, Werry EL, Halliday GM (2010) Selective cell death in neurodegeneration: why are some neurons spared in vulnerable regions? Prog Neurobiol 92:316–329

Dyson HJ, Wright PE (2005) Intrinsically unstructured proteins and their functions. Nat Rev Mol Cell Biol 6:197–208

Dzwolak W, Loksztejn A, Smirnovas V (2006) New insights into the self-assembly of insulin amyloid fibrils: an H–D exchange FT-IR study. Biochemistry 45:8143–8151

Dzwolak W, Loksztejn A, Galinska-Rakoczy A, Adachi R, Goto Y, Rupnicki L (2007) Conformational indeterminism in protein misfolding: chiral amplification on amyloidogenic pathway of insulin. J Am Chem Soc 129:7517–7522

Eanes ED, Glenner GG (1968) X-ray diffraction studies on amyloid filaments. J Histochem Cytochem 16:673–677

Eisen A (2009) Amyotrophic lateral sclerosis: a 40-year personal perspective. J Clin Neurosci 16:505–512

Eisenberg D, Nelson R, Sawaya MR, Balbirnie M, Sambashivan S, Ivanova MI, Madsen AO, Riekel C (2006) The structural biology of protein aggregation diseases: fundamental questions and some answers. Acc Chem Res 39:568–575

Elder GA, Gama Sosa MA, De Gasperi R (2010) Transgenic mouse models of Alzheimer's disease. Mt Sinai J Med 77:69–81

Eshaghian S, Kaul S, Shah PK (2007) Cardiac amyloidosis: new insights into diagnosis and management. Rev Cardiovasc Med 8:189–199

Feng LR, Federoff HJ, Vicini S, Maguire-Zeiss KA (2010) α-Synuclein mediates alterations in membrane conductance: a potential role for α-synuclein oligomers in cell vulnerability. Eur J Neurosci 32:10–17

Fernandez-Escamilla AM, Rousseau F, Schymkowitz J, Serrano L (2004) Prediction of sequence-dependent and mutational effects on the aggregation of peptides and proteins. Nat Biotechnol 22:1302–1306

Ferreira ST, Vieira MN, De Felice FG (2007) Soluble protein oligomers as emerging toxins in Alzheimer's and other amyloid diseases. IUBMB Life 59:332–345

Fezoui Y, Teplow DB (2002) Kinetic studies of amyloid β-protein fibril assembly. Differential effects of α-helix stabilization. J Biol Chem 277:36948–36954

Fezoui Y, Hartley DM, Walsh DM, Selkoe DJ, Osterhout JJ, Teplow DB (2000) A *de novo* designed helix-turn-helix peptide forms nontoxic amyloid fibrils. Nat Struct Biol 7:1095–1099

Flood DG, Lin YG, Lang DM, Trusko SP, Hirsch JD, Savage MJ, Scott RW, Howland DS (2009) A transgenic rat model of Alzheimer's disease with extracellular Aβ deposition. Neurobiol Aging 30:1078–1090

Fredenburg RA, Rospigliosi C, Meray RK, Kessler JC, Lashuel HA, Eliezer D, Lansbury PT Jr (2007) The impact of the E46K mutation on the properties of α-synuclein in its monomeric and oligomeric states. Biochemistry 46:7107–7118

Freir DB, Fedriani R, Scully D, Smith IM, Selkoe DJ, Walsh DM, Regan CM (2011) Aβ oligomers inhibit synapse remodelling necessary for memory consolidation. Neurobiol Aging 32: 2211–2218
Frid P, Anisimov SV, Popovic N (2007) Congo red and protein aggregation in neurodegenerative diseases. Brain Res Rev 53:135–160
Frieden C (2007) Protein aggregation processes: in search of the mechanism. Protein Sci 16:2334–2344
Friedreich N, Kekulé FA (1859) Zur Amyloidfrage. Virchows Arch Pathol Anat Physiol XVI:50–65
Frost B, Diamond MI (2010) Prion-like mechanisms in neurodegenerative diseases. Nat Rev Neurosci 11:155–159
Gaspar RC, Villarreal SA, Bowles N, Hepler RW, Joyce JG, Shughrue PJ (2010) Oligomers of β-amyloid are sequestered into and seed new plaques in the brains of an AD mouse model. Exp Neurol 223:394–400
Gasser T (2009) Molecular pathogenesis of Parkinson disease: insights from genetic studies. Expert Rev Mol Med 11:e22
Georganopoulou DG, Chang L, Nam JM, Thaxton CS, Mufson EJ, Klein WL, Mirkin CA (2005) Nanoparticle-based detection in cerebral spinal fluid of a soluble pathogenic biomarker for Alzheimer's disease. Proc Natl Acad Sci USA 102:2273–2276
George JM, Jin H, Woods WS, Clayton DF (1995) Characterization of a novel protein regulated during the critical period for song learning in the zebra finch. Neuron 15:361–372
Georgieva ER, Ramlall TF, Borbat PP, Freed JH, Eliezer D (2008) Membrane-bound α-synuclein forms an extended helix: long-distance pulsed ESR measurements using vesicles, bicelles, and rodlike micelles. J Am Chem Soc 130:12856–12857
Geula C, Wu CK, Saroff D, Lorenzo A, Yuan M, Yankner BA (1998) Aging renders the brain vulnerable to amyloid β-protein neurotoxicity. Nat Med 4:827–831
Gharibyan AL, Zamotin V, Yanamandra K, Moskaleva OS, Margulis BA, Kostanyan IA, Morozova-Roche LA (2007) Lysozyme amyloid oligomers and fibrils induce cellular death *via* different apoptotic/necrotic pathways. J Mol Biol 365:1337–1349
Giannetti AM, Lindwall G, Chau MF, Radeke MJ, Feinstein SC, Kohlstaedt LA (2000) Fibers of tau fragments, but not full length tau, exhibit a cross β-structure: implications for the formation of paired helical filaments. Protein Sci 9:2427–2435
Ginsberg SD, Galvin JE, Chiu TS, Lee VM, Masliah E, Trojanowski JQ (1998) RNA sequestration to pathological lesions of neurodegenerative diseases. Acta Neuropathol 96:487–494
Ginsberg SD, Crino PB, Hemby SE, Weingarten JA, Lee VM, Eberwine JH, Trojanowski JQ (1999) Predominance of neuronal mRNAs in individual Alzheimer's disease senile plaques. Ann Neurol 45:174–181
Glabe CG (2006) Common mechanisms of amyloid oligomer pathogenesis in degenerative disease. Neurobiol Aging 27:570–575
Glabe CG, Kayed R (2006) Common structure and toxic function of amyloid oligomers implies a common mechanism of pathogenesis. Neurology 66:S74–S78
Goedert M, Wischik CM, Crowther RA, Walker JE, Klug A (1988) Cloning and sequencing of the cDNA encoding a core protein of the paired helical filament of Alzheimer disease: identification as the microtubule-associated protein tau. Proc Natl Acad Sci USA 85:4051–4055
Goette A, Rocken C (2004) Atrial amyloidosis and atrial fibrillation: a gender-dependent "arrhythmogenic substrate"? Eur Heart J 25:1185–1186
Goldsbury CS, Cooper GJ, Goldie KN, Muller SA, Saafi EL, Gruijters WT, Misur MP, Engel A, Aebi U, Kistler J (1997) Polymorphic fibrillar assembly of human amylin. J Struct Biol 119:17–27
Gong Y, Chang L, Viola KL, Lacor PN, Lambert MP, Finch CE, Krafft GA, Klein WL (2003) Alzheimer's disease-affected brain: presence of oligomeric Aβ ligands (ADDLs) suggests a molecular basis for reversible memory loss. Proc Natl Acad Sci USA 100:10417–10422
Green SL, Tolwani RJ (1999) Animal models for motor neuron disease. Lab Anim Sci 49:480–487

Grudzielanek S, Smirnovas V, Winter R (2007a) The effects of various membrane physical-chemical properties on the aggregation kinetics of insulin. Chem Phys Lipids 149:28–39

Grudzielanek S, Velkova A, Shukla A, Smirnovas V, Tatarek-Nossol M, Rehage H, Kapurniotu A, Winter R (2007b) Cytotoxicity of insulin within its self-assembly and amyloidogenic pathways. J Mol Biol 370:372–384

Gudiksen KL, Gitlin I, Whitesides GM (2006) Differentiation of proteins based on characteristic patterns of association and denaturation in solutions of SDS. Proc Natl Acad Sci USA 103:7968–7972

Guijarro JI, Sunde M, Jones JA, Campbell ID, Dobson CM (1998) Amyloid fibril formation by an SH3 domain. Proc Natl Acad Sci USA 95:4224–4228

Gurlo T, Ryazantsev S, Huang CJ, Yeh MW, Reber HA, Hines OJ, O'Brien TD, Glabe CG, Butler PC (2010) Evidence for proteotoxicity in β cells in type 2 diabetes: toxic islet amyloid polypeptide oligomers form intracellularly in the secretory pathway. Am J Pathol 176:861–869

Gursky O, Aleshkov S (2000) Temperature-dependent β-sheet formation in β-amyloid Aβ(1–40) peptide in water: uncoupling β-structure folding from aggregation. Biochim Biophys Acta 1476:93–102

Guyenet SJ, Furrer SA, Damian VM, Baughan TD, La Spada AR, Garden GA (2010) A simple composite phenotype scoring system for evaluating mouse models of cerebellar ataxia. J Vis Exp 39:1787. doi:10.3791/1787

Haass C, Selkoe DJ (2007) Soluble protein oligomers in neurodegeneration: lessons from the Alzheimer's amyloid β-peptide. Nat Rev Mol Cell Biol 8:101–112

Haataja L, Gurlo T, Huang CJ, Butler PC (2008) Islet amyloid in type 2 diabetes, and the toxic oligomer hypothesis. Endocr Rev 29:303–316

Halliday GM, McCann H (2010) The progression of pathology in Parkinson's disease. Ann N Y Acad Sci 1184:188–195

Hardy JA, Higgins GA (1992) Alzheimer's disease: the amyloid cascade hypothesis. Science 256:184–185

Hardy J, Selkoe DJ (2002) The amyloid hypothesis of Alzheimer's disease: progress and problems on the road to therapeutics. Science 297:353–356

Harper JD, Wong SS, Lieber CM, Lansbury PT (1997) Observation of metastable Aβ amyloid protofibrils by atomic force microscopy. Chem Biol 4:119–125

Harrison RS, Sharpe PC, Singh Y, Fairlie DP (2007) Amyloid peptides and proteins in review. Rev Physiol Biochem Pharmacol 159:1–77

Hartley DM, Walsh DM, Ye CP, Diehl T, Vasquez S, Vassilev PM, Teplow DB, Selkoe DJ (1999) Protofibrillar intermediates of amyloid β-protein induce acute electrophysiological changes and progressive neurotoxicity in cortical neurons. J Neurosci 19:8876–8884

Hawkins PN (2003) Hereditary systemic amyloidosis with renal involvement. J Nephrol 16:443–448

Hayden MR, Tyagi SC, Kerklo MM, Nicolls MR (2005) Type 2 diabetes mellitus as a conformational disease. JOP 6:287–302

Head E, Pop V, Vasilevko V, Hill M, Saing T, Sarsoza F, Nistor M, Christie LA, Milton S, Glabe C, Barrett E, Cribbs D (2008) A two-year study with fibrillar β-amyloid (Aβ) immunization in aged canines: effects on cognitive function and brain Aβ. J Neurosci 28:3555–3566

Head E, Pop V, Sarsoza F, Kayed R, Beckett TL, Studzinski CM, Tomic JL, Glabe CG, Murphy MP (2010) Amyloid β-peptide and oligomers in the brain and cerebrospinal fluid of aged canines. J Alzheimers Dis 20:637–646

Hepler RW, Grimm KM, Nahas DD, Breese R, Dodson EC, Acton P, Keller PM, Yeager M, Wang H, Shughrue P, Kinney G, Joyce JG (2006) Solution state characterization of amyloid β-derived diffusible ligands. Biochemistry 45:15157–15167

Herczenik E, Gebbink MF (2008) Molecular and cellular aspects of protein misfolding and disease. FASEB J 22:2115–2133

Hirakura Y, Lin MC, Kagan BL (1999) Alzheimer amyloid Aβ1–42 channels: effects of solvent, pH, and Congo Red. J Neurosci Res 57:458–466

Hirschfield GM, Hawkins PN (2003) Amyloidosis: new strategies for treatment. Int J Biochem Cell Biol 35:1608–1613
Hoppener JW, Ahren B, Lips CJ (2000) Islet amyloid and type 2 diabetes mellitus. N Engl J Med 343:411–419
Houlden H, Crook R, Dolan RJ, McLaughlin J, Revesz T, Hardy J (2001) A novel presenilin mutation (M233V) causing very early onset Alzheimer's disease with Lewy bodies. Neurosci Lett 313:93–95
Hung LW, Ciccotosto GD, Giannakis E, Tew DJ, Perez K, Masters CL, Cappai R, Wade JD, Barnham KJ (2008) Amyloid-β peptide (Aβ) neurotoxicity is modulated by the rate of peptide aggregation: Aβ dimers and trimers correlate with neurotoxicity. J Neurosci 28:11950–11958
Ian VJM, Virginia MYL, John QT (2001) Synucleinopathies: a pathological and molecular review. Clin Neurosci Res 1:445–455
Ihara Y, Abraham C, Selkoe DJ (1983) Antibodies to paired helical filaments in Alzheimer's disease do not recognize normal brain proteins. Nature 304:727–730
Iijima K, Iijima-Ando K (2008) *Drosophila* models of Alzheimer's amyloidosis: the challenge of dissecting the complex mechanisms of toxicity of amyloid-β42. J Alzheimers Dis 15:523–540
Iijima-Ando K, Iijima K (2010) Transgenic *Drosophila* models of Alzheimer's disease and tauopathies. Brain Struct Funct 214:245–262
Inaba S, Okada T, Konakahara T, Kodaka M (2005) Specific binding of amyloid-β-protein to IMR-32 neuroblastoma cell membrane. J Pept Res 65:485–490
Ingham PW (2009) The power of the zebrafish for disease analysis. Hum Mol Genet 18:R107–R112
Inouye H, Sharma D, Goux WJ, Kirschner DA (2006) Structure of core domain of fibril-forming PHF/Tau fragments. Biophys J 90:1774–1789
Inzelberg R, Polyniki A (2010) Are genetic and sporadic Parkinson's disease patients equally susceptible to develop dementia? J Neurol Sci 289:23–26
Jao CC, Hegde BG, Chen J, Haworth IS, Langen R (2008) Structure of membrane-bound α-synuclein from site-directed spin labeling and computational refinement. Proc Natl Acad Sci USA 105:19666–19671
Jayasinghe SA, Langen R (2005) Lipid membranes modulate the structure of islet amyloid polypeptide. Biochemistry 44:12113–12119
Jelinek R, Sheynis T (2010) Amyloid–membrane interactions: experimental approaches and techniques. Curr Protein Pept Sci 11:372–384
Jellinger KA (2009) Recent advances in our understanding of neurodegeneration. J Neural Transm 116:1111–1162
Jellinger KA (2010) Basic mechanisms of neurodegeneration: a critical update. J Cell Mol Med 14:457–487
Jiménez JL, Guijarro JI, Orlova E, Zurdo J, Dobson CM, Sunde M, Saibil HR (1999) Cryo-electron microscopy structure of an SH3 amyloid fibril and model of the molecular packing. EMBO J 18:815–821
Jiménez JL, Nettleton EJ, Bouchard M, Robinson CV, Dobson CM, Saibil HR (2002) The protofilament structure of insulin amyloid fibrils. Proc Natl Acad Sci USA 99:9196–9201
Joachim CL, Morris JH, Kosik KS, Selkoe DJ (1987) Tau antisera recognize neurofibrillary tangles in a range of neurodegenerative disorders. Ann Neurol 22:514–520
Johnson JR, Jenn RC, Barclay JW, Burgoyne RD, Morgan A (2010) *Caenorhabditis elegans*: a useful tool to decipher neurodegenerative pathways. Biochem Soc Trans 38:559–563
Kagan BL, Hirakura Y, Azimov R, Azimova R, Lin MC (2002) The channel hypothesis of Alzheimer's disease: current status. Peptides 23:1311–1315
Kahle PJ, Waak J, Gasser T (2009) DJ-1 and prevention of oxidative stress in Parkinson's disease and other age-related disorders. Free Radic Biol Med 47:1354–1361
Kahn SE, Andrikopoulos S, Verchere CB (1999) Islet amyloid: a long-recognized but underappreciated pathological feature of type 2 diabetes. Diabetes 48:241–253
Kammerer RA, Steinmetz MO (2006) *De novo* design of a two-stranded coiled-coil switch peptide. J Struct Biol 155:146–153

Kawahara M, Kuroda Y (2000) Molecular mechanism of neurodegeneration induced by Alzheimer's β-amyloid protein: channel formation and disruption of calcium homeostasis. Brain Res Bull 53:389–397

Kawahara M, Arispe N, Kuroda Y, Rojas E (1997) Alzheimer's disease amyloid β-protein forms Zn^{2+}-sensitive, cation-selective channels across excised membrane patches from hypothalamic neurons. Biophys J 73:67–75

Kawooya JK, Emmons TL, Gonzalez-DeWhitt PA, Camp MC, D'Andrea SC (2003) Electrophoretic mobility of Alzheimer's amyloid-β peptides in urea-sodium dodecyl sulfate–polyacrylamide gel electrophoresis. Anal Biochem 323:103–113

Kayed R, Glabe CG (2006) Conformation-dependent anti-amyloid oligomer antibodies. Methods Enzymol 413:326–344

Kayed R, Head E, Thompson JL, McIntire TM, Milton SC, Cotman CW, Glabe CG (2003) Common structure of soluble amyloid oligomers implies common mechanism of pathogenesis. Science 300:486–489

Kayed R, Sokolov Y, Edmonds B, McIntire TM, Milton SC, Hall JE, Glabe CG (2004) Permeabilization of lipid bilayers is a common conformation-dependent activity of soluble amyloid oligomers in protein misfolding diseases. J Biol Chem 279:46363–46366

Kayed R, Pensalfini A, Margol L, Sokolov Y, Sarsoza F, Head E, Hall J, Glabe C (2009) Annular protofibrils are a structurally and functionally distinct type of amyloid oligomer. J Biol Chem 284:4230–4237

Kelly JW (1998) The alternative conformations of amyloidogenic proteins and their multi-step assembly pathways. Curr Opin Struct Biol 8:101–106

Khemtemourian L, Killian JA, Hoppener JW, Engel MF (2008) Recent insights in islet amyloid polypeptide-induced membrane disruption and its role in β-cell death in type 2 diabetes mellitus. Exp Diabetes Res 2008:421287

Khurana V (2008) Modeling tauopathy in the fruit fly *Drosophila melanogaster*. J Alzheimers Dis 15:541–553

Khurana R, Coleman C, Ionescu-Zanetti C, Carter SA, Krishna V, Grover RK, Roy R, Singh S (2005) Mechanism of thioflavin T binding to amyloid fibrils. J Struct Biol 151:229–238

Kim YS, Laurine E, Woods W, Lee SJ (2006) A novel mechanism of interaction between α-synuclein and biological membranes. J Mol Biol 360:386–397

Kim HY, Cho MK, Kumar A, Maier E, Siebenhaar C, Becker S, Fernandez CO, Lashuel HA, Benz R, Lange A, Zweckstetter M (2009) Structural properties of pore-forming oligomers of α-synuclein. J Am Chem Soc 131:17482–17489

Kirkitadze MD, Condron MM, Teplow DB (2001) Identification and characterization of key kinetic intermediates in amyloid β-protein fibrillogenesis. J Mol Biol 312:1103–1119

Kirkitadze MD, Bitan G, Teplow DB (2002) Paradigm shifts in Alzheimer's disease and other neurodegenerative disorders: the emerging role of oligomeric assemblies. J Neurosci Res 69:567–577

Kiss E, Keusch G, Zanetti M, Jung T, Schwarz A, Schocke M, Jaschke W, Czermak BV (2005) Dialysis-related amyloidosis revisited. AJR Am J Roentgenol 185:1460–1467

Kissane JM (1973) Hereditary disorders of the kidney. II. Hereditary nephropathies. Perspect Pediatr Pathol 1:147–187

Kitamura A, Kubota H (2010) Amyloid oligomers: dynamics and toxicity in the cytosol and nucleus. FEBS J 277:1369–1379

Klein WL (2002a) ADDLs & protofibrils—the missing links? Neurobiol Aging 23:231–235

Klein WL (2002b) Aβ toxicity in Alzheimer's disease: globular oligomers (ADDLs) as new vaccine and drug targets. Neurochem Int 41:345–352

Klucken J, Ingelsson M, Shin Y, Irizarry MC, Hedley-Whyte ET, Frosch M, Growdon J, McLean P, Hyman BT (2006) Clinical and biochemical correlates of insoluble α-synuclein in dementia with Lewy bodies. Acta Neuropathol 111:101–108

Knight JD, Hebda JA, Miranker AD (2006) Conserved and cooperative assembly of membrane-bound α-helical states of islet amyloid polypeptide. Biochemistry 45:9496–9508

Koffie RM, Meyer-Luehmann M, Hashimoto T, Adams KW, Mielke ML, Garcia-Alloza M, Micheva KD, Smith SJ, Kim ML, Lee VM, Hyman BT, Spires-Jones TL (2009) Oligomeric amyloid β associates with postsynaptic densities and correlates with excitatory synapse loss near senile plaques. Proc Natl Acad Sci USA 106:4012–4017

Kopito RR (2000) Aggresomes, inclusion bodies and protein aggregation. Trends Cell Biol 10:524–530

Kosik KS, Joachim CL, Selkoe DJ (1986) Microtubule-associated protein tau (τ) is a major antigenic component of paired helical filaments in Alzheimer disease. Proc Natl Acad Sci USA 83:4044–4048

Kourie JI, Henry CL, Farrelly P (2001) Diversity of amyloid β protein fragment [1–40]-formed channels. Cell Mol Neurobiol 21:255–284

Kril JJ, Halliday GM (2001) Alzheimer's disease: its diagnosis and pathogenesis. Int Rev Neurobiol 48:167–217

Kupfer L, Hinrichs W, Groschup MH (2009) Prion protein misfolding. Curr Mol Med 9:826–835

Kyle RA (2001) Amyloidosis: a convoluted story. Br J Haematol 114:529–538

Lafaye P, Achour I, England P, Duyckaerts C, Rougeon F (2009) Single-domain antibodies recognize selectively small oligomeric forms of amyloid β, prevent Aβ-induced neurotoxicity and inhibit fibril formation. Mol Immunol 46:695–704

Lambert MP, Barlow AK, Chromy BA, Edwards C, Freed R, Liosatos M, Morgan TE, Rozovsky I, Trommer B, Viola KL, Wals P, Zhang C, Finch CE, Krafft GA, Klein WL (1998) Diffusible, nonfibrillar ligands derived from Aβ1–42 are potent central nervous system neurotoxins. Proc Natl Acad Sci USA 95:6448–6453

Lansbury PT, Lashuel HA (2006) A century-old debate on protein aggregation and neurodegeneration enters the clinic. Nature 443:774–779

Lashuel HA, Lansbury PT Jr (2006) Are amyloid diseases caused by protein aggregates that mimic bacterial pore-forming toxins? Q Rev Biophys 39:167–201

Lashuel HA, Hartley D, Petre BM, Walz T, Lansbury PT Jr (2002a) Neurodegenerative disease: amyloid pores from pathogenic mutations. Nature 418:291

Lashuel HA, Petre BM, Wall J, Simon M, Nowak RJ, Walz T, Lansbury PT Jr (2002b) α-Synuclein, especially the Parkinson's disease-associated mutants, forms pore-like annular and tubular protofibrils. J Mol Biol 322:1089–1102

Layfield R, Cavey JR, Lowe J (2003) Role of ubiquitin-mediated proteolysis in the pathogenesis of neurodegenerative disorders. Ageing Res Rev 2:343–356

Ledford H (2010) Key Alzheimer's findings questioned. Nature 466:1031

Lee EB, Leng LZ, Zhang B, Kwong L, Trojanowski JQ, Abel T, Lee VM (2006) Targeting amyloid-β peptide (Aβ) oligomers by passive immunization with a conformation-selective monoclonal antibody improves learning and memory in Aβ precursor protein (APP) transgenic mice. J Biol Chem 281:4292–4299

Lee S, Fernandez EJ, Good TA (2007) Role of aggregation conditions in structure, stability, and toxicity of intermediates in the Aβ fibril formation pathway. Protein Sci 16:723–732

Lee JK, Tran T, Tansey MG (2009) Neuroinflammation in Parkinson's disease. J Neuroimmune Pharmacol 4:419–429

Leffers KW, Schell J, Jansen K, Lucassen R, Kaimann T, Nagel-Steger L, Tatzelt J, Riesner D (2004) The structural transition of the prion protein into its pathogenic conformation is induced by unmasking hydrophobic sites. J Mol Biol 344:839–853

Legleiter J, Mitchell E, Lotz GP, Sapp E, Ng C, DiFiglia M, Thompson LM, Muchowski PJ (2010) Mutant huntingtin fragments form oligomers in a polyglutamine length-dependent manner *in vitro* and *in vivo*. J Biol Chem 285:14777–14790

Lesné S, Koh MT, Kotilinek L, Kayed R, Glabe CG, Yang A, Gallagher M, Ashe KH (2006) A specific amyloid-β protein assembly in the brain impairs memory. Nature 440:352–357

LeVine H 3rd (1999) Quantification of β-sheet amyloid fibril structures with thioflavin T. Methods Enzymol 309:274–284

Li L, Holscher C (2007) Common pathological processes in Alzheimer disease and type 2 diabetes: a review. Brain Res Rev 56:384–402

Liao L, Cheng D, Wang J, Duong DM, Losik TG, Gearing M, Rees HD, Lah JJ, Levey AI, Peng J (2004) Proteomic characterization of postmortem amyloid plaques isolated by laser capture microdissection. J Biol Chem 279:37061–37068

Lin MC, Kagan BL (2002) Electrophysiologic properties of channels induced by Aβ25–35 in planar lipid bilayers. Peptides 23:1215–1228

Lin H, Zhu YJ, Lal R (1999) Amyloid β protein (1–40) forms calcium-permeable, Zn^{2+}-sensitive channel in reconstituted lipid vesicles. Biochemistry 38:11189 11196

Lin H, Bhatia R, Lal R (2001) Amyloid β protein forms ion channels: implications for Alzheimer's disease pathophysiology. FASEB J 15:2433–2444

Lin CY, Gurlo T, Kayed R, Butler AE, Haataja L, Glabe CG, Butler PC (2007) Toxic human islet amyloid polypeptide (h-IAPP) oligomers are intracellular, and vaccination to induce anti-toxic oligomer antibodies does not prevent h-IAPP-induced β-cell apoptosis in h-IAPP transgenic mice. Diabetes 56:1324–1332

Litvinovich SV, Brew SA, Aota S, Akiyama SK, Haudenschild C, Ingham KC (1998) Formation of amyloid-like fibrils by self-association of a partially unfolded fibronectin type III module. J Mol Biol 280:245–258

Looi LM (1993) Isolated atrial amyloidosis: a clinicopathologic study indicating increased prevalence in chronic heart disease. Hum Pathol 24:602–607

Lopez De La Paz M, Goldie K, Zurdo J, Lacroix E, Dobson CM, Hoenger A, Serrano L (2002) *De novo* designed peptide-based amyloid fibrils. Proc Natl Acad Sci USA 99:16052–16057

Lorenzo A, Yankner BA (1994) β-Amyloid neurotoxicity requires fibril formation and is inhibited by Congo red. Proc Natl Acad Sci USA 91:12243–12247

Lorenzo A, Razzaboni B, Weir GC, Yankner BA (1994) Pancreatic islet cell toxicity of amylin associated with type-2 diabetes mellitus. Nature 368:756–760

Lossi L, Cantile C, Tamagno I, Merighi A (2005) Apoptosis in the mammalian CNS: lessons from animal models. Vet J 170:52–66

Lu B (2009) Recent advances in using *Drosophila* to model neurodegenerative diseases. Apoptosis 14:1008–1020

Lu B, Vogel H (2009) *Drosophila* models of neurodegenerative diseases. Annu Rev Pathol 4:315–342

Lublin AL, Gandy S (2010) Amyloid-β oligomers: possible roles as key neurotoxins in Alzheimer's disease. Mt Sinai J Med 77:43–49

Lucin KM, Wyss-Coray T (2009) Immune activation in brain aging and neurodegeneration: too much or too little? Neuron 64:110–122

Luibl V, Isas JM, Kayed R, Glabe CG, Langen R, Chen J (2006) Drusen deposits associated with aging and age-related macular degeneration contain nonfibrillar amyloid oligomers. J Clin Invest 116:378–385

Maji SK, Amsden JJ, Rothschild KJ, Condron MM, Teplow DB (2005) Conformational dynamics of amyloid β-protein assembly probed using intrinsic fluorescence. Biochemistry 44:13365–13376

Maji SK, Perrin MH, Sawaya MR, Jessberger S, Vadodaria K, Rissman RA, Singru PS, Nilsson KP, Simon R, Schubert D, Eisenberg D, Rivier J, Sawchenko P, Vale W, Riek R (2009a) Functional amyloids as natural storage of peptide hormones in pituitary secretory granules. Science 325:328–332

Maji SK, Wang L, Greenwald J, Riek R (2009b) Structure–activity relationship of amyloid fibrils. FEBS Lett 583:2610–2617

Malaga-Trillo E, Sempou E (2009) PrPs: proteins with a purpose: lessons from the zebrafish. Prion 3:129–133

Mallucci GR (2009) Prion neurodegeneration: starts and stops at the synapse. Prion 3:195–201

Marcinkiewicz M (2002) βAPP and furin mRNA concentrates in immature senile plaques in the brain of Alzheimer patients. J Neuropathol Exp Neurol 61:815–829

Marcon G, Plakoutsi G, Canale C, Relini A, Taddei N, Dobson CM, Ramponi G, Chiti F (2005) Amyloid formation from HypF-N under conditions in which the protein is initially in its native state. J Mol Biol 347:323 335

Margittai M, Langen R (2006) Side chain-dependent stacking modulates tau filament structure. J Biol Chem 281:37820–37827

Martins IC, Kuperstein I, Wilkinson H, Maes E, Vanbrabant M, Jonckheere W, Van Gelder P, Hartmann D, D'Hooge R, De Strooper B, Schymkowitz J, Rousseau F (2008) Lipids revert inert Aβ amyloid fibrils to neurotoxic protofibrils that affect learning in mice. EMBO J 27:224–233

Marzban L, Park K, Verchere CB (2003) Islet amyloid polypeptide and type 2 diabetes. Exp Gerontol 38:347–351

Mayeux R (2010) Clinical practice. Early Alzheimer's disease. N Engl J Med 362:2194–2201

McCarthy RE 3rd, Kasper EK (1998) A review of the amyloidoses that infiltrate the heart. Clin Cardiol 21:547–552

McDonald RJ, Craig LA, Hong NS (2010) The etiology of age-related dementia is more complicated than we think. Behav Brain Res 214:3–11

McGowan DP, van Roon-Mom W, Holloway H, Bates GP, Mangiarini L, Cooper GJS, Faull RLM, Snell RG (2000) Amyloid-like inclusions in Huntington's disease. Neuroscience 100:677–680

Meier JJ, Kayed R, Lin CY, Gurlo T, Haataja L, Jayasinghe S, Langen R, Glabe CG, Butler PC (2006) Inhibition of human IAPP fibril formation does not prevent β-cell death: evidence for distinct actions of oligomers and fibrils of human IAPP. Am J Physiol Endocrinol Metab 291:E1317–E1324

Merlini G, Westermark P (2004) The systemic amyloidoses: clearer understanding of the molecular mechanisms offers hope for more effective therapies. J Intern Med 255:159–178

Miller Y, Ma B, Nussinov R (2010) Polymorphism in Alzheimer Aβ amyloid organization reflects conformational selection in a rugged energy landscape. Chem Rev 110:4820–4838

Miyoshi K (2009) What is 'early onset dementia'? Psychogeriatrics 9:67–72

Montserret R, McLeish MJ, Bockmann A, Geourjon C, Penin F (2000) Involvement of electrostatic interactions in the mechanism of peptide folding induced by sodium dodecyl sulfate binding. Biochemistry 39:8362–8373

Moro ML, Collins MJ, Cappellini E (2010) Alzheimer's disease and amyloid β-peptide deposition in the brain: a matter of 'aging'? Biochem Soc Trans 38:539–544

Moussa CE, Wersinger C, Rusnak M, Tomita Y, Sidhu A (2004) Abnormal migration of human wild-type α-synuclein upon gel electrophoresis. Neurosci Lett 371:239–243

Mukaetova-Ladinska EB, Xuereb JH, Garcia-Sierra F, Hurt J, Gertz HJ, Hills R, Brayne C, Huppert FA, Paykel ES, McGee MA, Jakes R, Honer WG, Harrington CR, Wischik CM (2009) Lewy body variant of Alzheimer's disease: selective neocortical loss of t-SNARE proteins and loss of MAP2 and α-synuclein in medial temporal lobe. ScientificWorldJournal 9:1463–1475

Murali J, Jayakumar R (2005) Spectroscopic studies on native and protofibrillar insulin. J Struct Biol 150:180–189

Murray B, Lynch T, Farrell M (2005) Clinicopathological features of the tauopathies. Biochem Soc Trans 33:595–599

Nekooki-Machida Y, Kurosawa M, Nukina N, Ito K, Oda T, Tanaka M (2009) Distinct conformations of *in vitro* and *in vivo* amyloids of huntingtin-exon1 show different cytotoxicity. Proc Natl Acad Sci USA 106:9679–9684

Nelson R, Eisenberg D (2006) Recent atomic models of amyloid fibril structure. Curr Opin Struct Biol 16:260–265

Nelson R, Sawaya MR, Balbirnie M, Madsen AO, Riekel C, Grothe R, Eisenberg D (2005) Structure of the cross-β spine of amyloid-like fibrils. Nature 435:773–778

Novitskaya V, Bocharova OV, Bronstein I, Baskakov IV (2006) Amyloid fibrils of mammalian prion protein are highly toxic to cultured cells and primary neurons. J Biol Chem 281:13828–13836

Nukina N, Kosik KS, Selkoe DJ (1987) Recognition of Alzheimer paired helical filaments by monoclonal neurofilament antibodies is due to crossreaction with tau protein. Proc Natl Acad Sci USA 84:3415–3419

O'Nuallain B, Freir DB, Nicoll AJ, Risse E, Ferguson N, Herron CE, Collinge J, Walsh DM (2010) Amyloid β-protein dimers rapidly form stable synaptotoxic protofibrils. J Neurosci 30:14411–14419

Obeso JA, Rodriguez-Oroz MC, Goetz CG, Marin C, Kordower JH, Rodriguez M, Hirsch EC, Farrer M, Schapira AH, Halliday G (2010) Missing pieces in the Parkinson's disease puzzle. Nat Med 16:653–661

Oda T, Wals P, Osterburg HH, Johnson SA, Pasinetti GM, Morgan TE, Rozovsky I, Stine WB, Snyder SW, Holzman TF et al (1995) Clusterin (apoJ) alters the aggregation of amyloid β-peptide (Aβ1–42) and forms slowly sedimenting Aβ complexes that cause oxidative stress. Exp Neurol 136:22–31

Okada T, Wakabayashi M, Ikeda K, Matsuzaki K (2007) Formation of toxic fibrils of Alzheimer's amyloid β-protein-(1–40) by monosialoganglioside GM1, a neuronal membrane component. J Mol Biol 371:481–489

Opal P, Zoghbi HY (2002) The role of chaperones in polyglutamine disease. Trends Mol Med 8:232–236

Pahwa R, Lyons KE (2010) Early diagnosis of Parkinson's disease: recommendations from diagnostic clinical guidelines. Am J Manag Care 16(Suppl Implications):S94–S99

Paleologou KE, Kragh CL, Mann DM, Salem SA, Al-Shami R, Allsop D, Hassan AH, Jensen PH, El-Agnaf OM (2009) Detection of elevated levels of soluble α-synuclein oligomers in postmortem brain extracts from patients with dementia with Lewy bodies. Brain 132:1093–1101

Paravastu AK, Petkova AT, Tycko R (2006) Polymorphic fibril formation by residues 10–40 of the Alzheimer's β-amyloid peptide. Biophys J 90:4618–4629

Paravastu AK, Leapman RD, Yau WM, Tycko R (2008) Molecular structural basis for polymorphism in Alzheimer's β-amyloid fibrils. Proc Natl Acad Sci USA 105:18349–18354

Paravastu AK, Qahwash I, Leapman RD, Meredith SC, Tycko R (2009) Seeded growth of β-amyloid fibrils from Alzheimer's brain-derived fibrils produces a distinct fibril structure. Proc Natl Acad Sci USA 106:7443–7448

Park J, Kim Y, Chung J (2009) Mitochondrial dysfunction and Parkinson's disease genes: insights from *Drosophila*. Dis Model Mech 2:336–340

Park GH, Kariya S, Monani UR (2010) Spinal muscular atrophy: new and emerging insights from model mice. Curr Neurol Neurosci Rep 10:108–117

Perl DP (2010) Neuropathology of Alzheimer's disease. Mt Sinai J Med 77:32–42

Peterson DW, Zhou H, Dahlquist FW, Lew J (2008) A soluble oligomer of tau associated with fiber formation analyzed by NMR. Biochemistry 47:7393–7404

Petkova AT, Leapman RD, Guo Z, Yau WM, Mattson MP, Tycko R (2005) Self-propagating, molecular-level polymorphism in Alzheimer's β-amyloid fibrils. Science 307:262–265

Petkova AT, Yau WM, Tycko R (2006) Experimental constraints on quaternary structure in Alzheimer's β-amyloid fibrils. Biochemistry 45:498–512

Pfister EL, Zamore PD (2009) Huntington's disease: silencing a brutal killer. Exp Neurol 220:226–229

Piening N, Weber P, Hogen T, Beekes M, Kretzschmar H, Giese A (2006) Photo-induced cross-linking of prion protein oligomers and prions. Amyloid 13:67–77

Pike CJ, Walencewicz AJ, Glabe CG, Cotman CW (1991) *In vitro* aging of β-amyloid protein causes peptide aggregation and neurotoxicity. Brain Res 563:311–314

Pittock SJ, Lucchinetti CF (2007) The pathology of MS: new insights and potential clinical applications. Neurologist 13:45–56

Podlisny MB, Ostaszewski BL, Squazzo SL, Koo EH, Rydell RE, Teplow DB, Selkoe DJ (1995) Aggregation of secreted amyloid β-protein into sodium dodecyl sulfate-stable oligomers in cell culture. J Biol Chem 270:9564–9570

Polverino de Laureto P, Taddei N, Frare E, Capanni C, Costantini S, Zurdo J, Chiti F, Dobson CM, Fontana A (2003) Protein aggregation and amyloid fibril formation by an SH3 domain probed by limited proteolysis. J Mol Biol 334:129–141

Porat Y, Kolusheva S, Jelinek R, Gazit E (2003) The human islet amyloid polypeptide forms transient membrane-active prefibrillar assemblies. Biochemistry 42:10971–10977

Postuma RB, Montplaisir J (2009) Predicting Parkinson's disease—why, when, and how? Parkinsonism Relat Disord 15(Suppl 3):S105–S109

Qin W, Chachich M, Lane M, Roth G, Bryant M, de Cabo R, Ottinger MA, Mattison J, Ingram D, Gandy S, Pasinetti GM (2006) Calorie restriction attenuates Alzheimer's disease type brain amyloidosis in Squirrel monkeys (*Saimiri sciureus*). J Alzheimers Dis 10:417–422
Querfurth HW, LaFerla FM (2010) Alzheimer's disease. N Engl J Med 362:329–344
Quist A, Doudevski I, Lin H, Azimova R, Ng D, Frangione B, Kagan B, Ghiso J, Lal R (2005) Amyloid ion channels: a common structural link for protein-misfolding disease. Proc Natl Acad Sci USA 102:10427–10432
Radovan D, Smirnovas V, Winter R (2008) Effect of pressure on islet amyloid polypeptide aggregation: revealing the polymorphic nature of the fibrillation process. Biochemistry 47:6352–6360
Rahimi F, Bitan G (2010) Selection of aptamers for amyloid β-protein, the causative agent of Alzheimer's disease. J Vis Exp 39:1955. doi:10.3791/1955
Rahimi F, Shanmugam A, Bitan G (2008) Structure–function relationships of pre-fibrillar protein assemblies in Alzheimer's disease and related disorders. Curr Alzheimer Res 5:319–341
Rahimi F, Murakami K, Summers JL, Chen CH, Bitan G (2009) RNA aptamers generated against oligomeric Aβ40 recognize common amyloid aptatopes with low specificity but high sensitivity. PLoS One 4:e7694
Rangachari V, Reed DK, Moore BD, Rosenberry TL (2006) Secondary structure and interfacial aggregation of amyloid-β(1–40) on sodium dodecyl sulfate micelles. Biochemistry 45:8639–8648
Rangachari V, Moore BD, Reed DK, Sonoda LK, Bridges AW, Conboy E, Hartigan D, Rosenberry TL (2007) Amyloid-β(1–42) rapidly forms protofibrils and oligomers by distinct pathways in low concentrations of sodium dodecylsulfate. Biochemistry 46:12451–12462
Relini A, Torrassa S, Rolandi R, Gliozzi A, Rosano C, Canale C, Bolognesi M, Plakoutsi G, Bucciantini M, Chiti F, Stefani M (2004) Monitoring the process of HypF fibrillization and liposome permeabilization by protofibrils. J Mol Biol 338:943–957
Rhee SK, Quist AP, Lal R (1998) Amyloid β protein-(1–42) forms calcium-permeable, Zn^{2+}-sensitive channel. J Biol Chem 273:13379–13382
Rocken C, Peters B, Juenemann G, Saeger W, Klein HU, Huth C, Roessner A, Goette A (2002) Atrial amyloidosis: an arrhythmogenic substrate for persistent atrial fibrillation. Circulation 106:2091–2097
Roychaudhuri R, Yang M, Hoshi MM, Teplow DB (2009) Amyloid β-protein assembly and Alzheimer disease. J Biol Chem 284:4749–4753
Rozas JL, Gomez-Sanchez L, Tomas-Zapico C, Lucas JJ, Fernandez-Chacon R (2010) Presynaptic dysfunction in Huntington's disease. Biochem Soc Trans 38:488–492
Ryan TM, Howlett GJ, Bailey MF (2008) Fluorescence detection of a lipid-induced tetrameric intermediate in amyloid fibril formation by apolipoprotein C-II. J Biol Chem 283:35118–35128
Sager JJ, Bai Q, Burton EA (2010) Transgenic zebrafish models of neurodegenerative diseases. Brain Struct Funct 214:285–302
Sahara N, Maeda S, Takashima A (2008) Tau oligomerization: a role for tau aggregation intermediates linked to neurodegeneration. Curr Alzheimer Res 5:591–598
Saiki M, Honda S, Kawasaki K, Zhou D, Kaito A, Konakahara T, Morii H (2005) Higher-order molecular packing in amyloid-like fibrils constructed with linear arrangements of hydrophobic and hydrogen-bonding side-chains. J Mol Biol 348:983–998
Sakono M, Zako T (2010) Amyloid oligomers: formation and toxicity of Aβ oligomers. FEBS J 277:1348–1358
Sanchorawala V (2006) Light-chain (AL) amyloidosis: diagnosis and treatment. Clin J Am Soc Nephrol 1:1331–1341
Sanderson KL, Butler L, Ingram VM (1997) Aggregates of a β-amyloid peptide are required to induce calcium currents in neuron-like human teratocarcinoma cells: relation to Alzheimer's disease. Brain Res 744:7–14
Sassone J, Colciago C, Cislaghi G, Silani V, Ciammola A (2009) Huntington's disease: the current state of research with peripheral tissues. Exp Neurol 219:385–397

Sawaya MR, Sambashivan S, Nelson R, Ivanova MI, Sievers SA, Apostol MI, Thompson MJ, Balbirnie M, Wiltzius JJ, McFarlane HT, Madsen AO, Riekel C, Eisenberg D (2007) Atomic structures of amyloid cross-β spines reveal varied steric zippers. Nature 447:453–457

Scherzinger E, Lurz R, Turmaine M, Mangiarini L, Hollenbach B, Hasenbank R, Bates GP, Davies SW, Lehrach H, Wanker EE (1997) Huntingtin-encoded polyglutamine expansions form amyloid-like protein aggregates *in vitro* and *in vivo*. Cell 90:549–558

Scheuner D, Kaufman RJ (2008) The unfolded protein response: a pathway that links insulin demand with β-cell failure and diabetes. Endocr Rev 29:317–333

Schiesling C, Kieper N, Seidel K, Kruger R (2008) Review: familial Parkinson's disease—genetics, clinical phenotype and neuropathology in relation to the common sporadic form of the disease. Neuropathol Appl Neurobiol 34:255–271

Schleiden MJ (1838) Beiträge zur phytogenesis. Arch Anat Physiol Wiss Med 13:137–176

Scott DA, Tabarean I, Tang Y, Cartier A, Masliah E, Roy S (2010) A pathologic cascade leading to synaptic dysfunction in α-synuclein-induced neurodegeneration. J Neurosci 30:8083–8095

Serpell LC (2000) Alzheimer's amyloid fibrils: structure and assembly. Biochim Biophys Acta 1502:16–30

Serpell LC, Sunde M, Benson MD, Tennent GA, Pepys MB, Fraser PE (2000) The protofilament substructure of amyloid fibrils. J Mol Biol 300:1033–1039

Sesti F, Liu S, Cai SQ (2010) Oxidation of potassium channels by ROS: a general mechanism of aging and neurodegeneration? Trends Cell Biol 20:45–51

Shankar GM, Bloodgood BL, Townsend M, Walsh DM, Selkoe DJ, Sabatini BL (2007) Natural oligomers of the Alzheimer amyloid-β protein induce reversible synapse loss by modulating an NMDA-type glutamate receptor-dependent signaling pathway. J Neurosci 27:2866–2875

Shankar GM, Li S, Mehta TH, Garcia-Munoz A, Shepardson NE, Smith I, Brett FM, Farrell MA, Rowan MJ, Lemere CA, Regan CM, Walsh DM, Sabatini BL, Selkoe DJ (2008) Amyloid-β protein dimers isolated directly from Alzheimer's brains impair synaptic plasticity and memory. Nat Med 14:837–842

Sharma S, Mukherjee M, Kedage V, Muttigi MS, Rao A, Rao S (2009) Sporadic Creutzfeldt-Jakob disease—a review. Int J Neurosci 119:1981–1994

Sharon R, Bar-Joseph I, Frosch MP, Walsh DM, Hamilton JA, Selkoe DJ (2003) The formation of highly soluble oligomers of α-synuclein is regulated by fatty acids and enhanced in Parkinson's disease. Neuron 37:583–595

Shepherd CE, Goyette J, Utter V, Rahimi F, Yang Z, Geczy CL, Halliday GM (2006) Inflammatory S100A9 and S100A12 proteins in Alzheimer's disease. Neurobiol Aging 27:1554–1563

Shulman LM (2010) Understanding disability in Parkinson's disease. Mov Disord 25 (Suppl 1):S131–S135

Simoneau S, Rezaei H, Sales N, Kaiser-Schulz G, Lefebvre-Roque M, Vidal C, Fournier JG, Comte J, Wopfner F, Grosclaude J, Schatzl H, Lasmezas CI (2007) *In vitro* and *in vivo* neurotoxicity of prion protein oligomers. PLoS Pathog 3:e125

Sipe JD, Cohen AS (2000) Review: history of the amyloid fibril. J Struct Biol 130:88–98

Sipe JD, Benson MD, Buxbaum JN, Ikeda S, Merlini G, Saraiva MJ, Westermark P (2010) Amyloid fibril protein nomenclature: 2010 recommendations from the nomenclature committee of the International Society of Amyloidosis. Amyloid 17:101–104

Skovronsky DM, Lee VM, Trojanowski JQ (2006) Neurodegenerative diseases: new concepts of pathogenesis and their therapeutic implications. Annu Rev Pathol 1:151–170

Sokolov Y, Kozak JA, Kayed R, Chanturiya A, Glabe C, Hall JE (2006) Soluble amyloid oligomers increase bilayer conductance by altering dielectric structure. J Gen Physiol 128:637–647

Sokolova A, Hill MD, Rahimi F, Warden LA, Halliday GM, Shepherd CE (2009) Monocyte chemoattractant protein-1 plays a dominant role in the chronic inflammation observed in Alzheimer's disease. Brain Pathol 19:392–398

Sorgjerd K, Klingstedt T, Lindgren M, Kagedal K, Hammarstrom P (2008) Prefibrillar transthyretin oligomers and cold stored native tetrameric transthyretin are cytotoxic in cell culture. Biochem Biophys Res Commun 377:1072–1078

Steensma DP, Kyle RA (2007) A history of the kidney in plasma cell disorders. Contrib Nephrol 153:5–24
Stefani M, Dobson CM (2003) Protein aggregation and aggregate toxicity: new insights into protein folding, misfolding diseases and biological evolution. J Mol Med 81:678–699
Steiner B, Mandelkow EM, Biernat J, Gustke N, Meyer HE, Schmidt B, Mieskes G, Soling HD, Drechsel D, Kirschner MW et al (1990) Phosphorylation of microtubule-associated protein tau: identification of the site for Ca^{2+}-calmodulin dependent kinase and relationship with tau phosphorylation in Alzheimer tangles. EMBO J 9:3539–3544
Sugama S, Takenouchi T, Cho BP, Joh TH, Hashimoto M, Kitani H (2009) Possible roles of microglial cells for neurotoxicity in clinical neurodegenerative diseases and experimental animal models. Inflamm Allergy Drug Targets 8:277–284
Sunde M, Blake C (1997) The structure of amyloid fibrils by electron microscopy and X-ray diffraction. Adv Protein Chem 50:123–159
Sunde M, Serpell LC, Bartlam M, Fraser PE, Pepys MB, Blake CC (1997) Common core structure of amyloid fibrils by synchrotron X-ray diffraction. J Mol Biol 273:729–739
Surguchev A, Surguchov A (2010) Conformational diseases: looking into the eyes. Brain Res Bull 81:12–24
Swift B (2002) Examination of insulin injection sites: an unexpected finding of localized amyloidosis. Diabet Med 19:881–882
Taddei K, Kwok JB, Kril JJ, Halliday GM, Creasey H, Hallupp M, Fisher C, Brooks WS, Chung C, Andrews C, Masters CL, Schofield PR, Martins RN (1998) Two novel presenilin-1 mutations (Ser169Leu and Pro436Gln) associated with very early onset Alzheimer's disease. Neuroreport 9:3335–3339
Takahashi T, Tada K, Mihara H (2009) RNA aptamers selected against amyloid β-peptide (Aβ) inhibit the aggregation of Aβ. Mol Biosyst 5:986–991
Takeda A, Hasegawa T, Matsuzaki-Kobayashi M, Sugeno N, Kikuchi A, Itoyama Y, Furukawa K (2006) Mechanisms of neuronal death in synucleinopathy. J Biomed Biotechnol 2006:19365
Tansey MG, Goldberg MS (2010) Neuroinflammation in Parkinson's disease: its role in neuronal death and implications for therapeutic intervention. Neurobiol Dis 37:510–518
Tartaglia GG, Pawar AP, Campioni S, Dobson CM, Chiti F, Vendruscolo M (2008) Prediction of aggregation-prone regions in structured proteins. J Mol Biol 380:425–436
Taylor TN, Greene JG, Miller GW (2010) Behavioral phenotyping of mouse models of Parkinson's disease. Behav Brain Res 211:1–10
Teplow DB (1998) Structural and kinetic features of amyloid β-protein fibrillogenesis. Amyloid 5:121–142
Tjernberg L, Hosia W, Bark N, Thyberg J, Johansson J (2002) Charge attraction and β propensity are necessary for amyloid fibril formation from tetrapeptides. J Biol Chem 277:43243–43246
Tompa P (2002) Intrinsically unstructured proteins. Trends Biochem Sci 27:527–533
Townsend M, Shankar GM, Mehta T, Walsh DM, Selkoe DJ (2006) Effects of secreted oligomers of amyloid β-protein on hippocampal synaptic plasticity: a potent role for trimers. J Physiol 572:477–492
Trexler AJ, Rhoades E (2009) α-Synuclein binds large unilamellar vesicles as an extended helix. Biochemistry 48:2304–2306
Tsigelny IF, Bar-On P, Sharikov Y, Crews L, Hashimoto M, Miller MA, Keller SH, Platoshyn O, Yuan JX, Masliah E (2007) Dynamics of α-synuclein aggregation and inhibition of pore-like oligomer development by β-synuclein. FEBS J 274:1862–1877
Tsukakoshi K, Harada R, Sode K, Ikebukuro K (2010) Screening of DNA aptamer which binds to α-synuclein. Biotechnol Lett 32:643–648
Tycko R, Sciarretta KL, Orgel JP, Meredith SC (2009) Evidence for novel β-sheet structures in Iowa mutant β-amyloid fibrils. Biochemistry 48:6072–6084
van Ham TJ, Breitling R, Swertz MA, Nollen EA (2009) Neurodegenerative diseases: lessons from genome-wide screens in small model organisms. EMBO Mol Med 1:360–370
Van Nostrand WE, Melchor JP, Cho HS, Greenberg SM, Rebeck GW (2001) Pathogenic effects of D23N Iowa mutant amyloid β-protein. J Biol Chem 276:32860–32866

van Rooijen BD, Claessens MM, Subramaniam V (2010) Membrane interactions of oligomeric α-synuclein: potential role in Parkinson's disease. Curr Protein Pept Sci 11:334–342
Varkey J, Isas JM, Mizuno N, Jensen MB, Bhatia VK, Jao CC, Petrlova J, Voss JC, Stamou DG, Steven AC, Langen R (2010) Membrane curvature induction and tubulation are common features of synucleins and apolipoproteins. J Biol Chem 285:32486–32493
Vasilevko V, Head E (2009) Immunotherapy in a natural model of Aβ pathogenesis: the aging beagle. CNS Neurol Disord Drug Targets 8:98–113
Vasilevko V, Pop V, Kim HJ, Saing T, Glabe CC, Milton S, Barrett EG, Cotman CW, Cribbs DH, Head E (2010) Linear and conformation specific antibodies in aged beagles after prolonged vaccination with aggregated Aβ. Neurobiol Dis 39:301–310
Virchow R (1854a) Über den Gang der amyloiden degeneration. Archiv für pathologische Anatomie und Physiologie und für klinische Medicin 8:364–368
Virchow R (1854b) Ueber eine im Gehirn und Rückenmark des Menschen aufgefundene Substanz mit der chemsichen Reaction der Cellulose. Virchows Arch 6:135–138
Volles MJ, Lansbury PT Jr (2002) Vesicle permeabilization by protofibrillar α-synuclein is sensitive to Parkinson's disease-linked mutations and occurs by a pore-like mechanism. Biochemistry 41:4595–4602
Volles MJ, Lee SJ, Rochet JC, Shtilerman MD, Ding TT, Kessler JC, Lansbury PT Jr (2001) Vesicle permeabilization by protofibrillar α-synuclein: implications for the pathogenesis and treatment of Parkinson's disease. Biochemistry 40:7812–7819
von Bergen M, Barghorn S, Li L, Marx A, Biernat J, Mandelkow EM, Mandelkow E (2001) Mutations of tau protein in frontotemporal dementia promote aggregation of paired helical filaments by enhancing local β-structure. J Biol Chem 276:48165–48174
Walker LC (1997) Animal models of cerebral β-amyloid angiopathy. Brain Res Brain Res Rev 25:70–84
Walker LC, Levine H 3rd, Mattson MP, Jucker M (2006) Inducible proteopathies. Trends Neurosci 29:438–443
Walsh DM, Lomakin A, Benedek GB, Condron MM, Teplow DB (1997) Amyloid β-protein fibrillogenesis. Detection of a protofibrillar intermediate. J Biol Chem 272:22364–22372
Walsh DM, Hartley DM, Kusumoto Y, Fezoui Y, Condron MM, Lomakin A, Benedek GB, Selkoe DJ, Teplow DB (1999) Amyloid β-protein fibrillogenesis. Structure and biological activity of protofibrillar intermediates. J Biol Chem 274:25945–25952
Walsh DM, Klyubin I, Fadeeva JV, Cullen WK, Anwyl R, Wolfe MS, Rowan MJ, Selkoe DJ (2002) Naturally secreted oligomers of amyloid β protein potently inhibit hippocampal long-term potentiation *in vivo*. Nature 416:535–539
Wang LH, Qin ZH (2006) Animal models of Huntington's disease: implications in uncovering pathogenic mechanisms and developing therapies. Acta Pharmacol Sin 27:1287–1302
Wang C, Huang L, Wang L, Hong Y, Sha Y (2007) One-dimensional self-assembly of a rational designed β-structure peptide. Biopolymers 86:23–31
Warrick JM, Chan HY, Gray-Board GL, Chai Y, Paulson HL, Bonini NM (1999) Suppression of polyglutamine-mediated neurodegeneration in *Drosophila* by the molecular chaperone HSP70. Nat Genet 23:425–428
Westermark P, Wilander E (1978) The influence of amyloid deposits on the islet volume in maturity onset diabetes mellitus. Diabetologia 15:417–421
Westermark P, Benson MD, Buxbaum JN, Cohen AS, Frangione B, Ikeda S, Masters CL, Merlini G, Saraiva MJ, Sipe JD (2007) A primer of amyloid nomenclature. Amyloid 14:179–183
Wetzel R (2006) Amyloid fibrils—common threads in the natural history of proteins. Acc Chem Res 39:567
Wetzel R, Shivaprasad S, Williams AD (2007) Plasticity of amyloid fibrils. Biochemistry 46:1–10
White JA, Manelli AM, Holmberg KH, Van Eldik LJ, Ladu MJ (2005) Differential effects of oligomeric and fibrillar amyloid-β1–42 on astrocyte-mediated inflammation. Neurobiol Dis 18:459–465
Woodruff-Pak DS (2008) Animal models of Alzheimer's disease: therapeutic implications. J Alzheimers Dis 15:507–521

Xia W, Zhang J, Kholodenko D, Citron M, Podlisny MB, Teplow DB, Haass C, Seubert P, Koo EH, Selkoe DJ (1997) Enhanced production and oligomerization of the 42-residue amyloid β-protein by Chinese hamster ovary cells stably expressing mutant presenilins. J Biol Chem 272:7977–7982

Yamamoto S, Hasegawa K, Yamaguchi I, Tsutsumi S, Kardos J, Goto Y, Gejyo F, Naiki H (2004) Low concentrations of sodium dodecyl sulfate induce the extension of β2-microglobulin-related amyloid fibrils at a neutral pH. Biochemistry 43:11075–11082

Yamamoto S, Kazama JJ, Narita I, Naiki H, Gejyo F (2009) Recent progress in understanding dialysis-related amyloidosis. Bone 45(Suppl 1):S39–S42

Yang SH, Cheng PH, Banta H, Piotrowska-Nitsche K, Yang JJ, Cheng EC, Snyder B, Larkin K, Liu J, Orkin J, Fang ZH, Smith Y, Bachevalier J, Zola SM, Li SH, Li XJ, Chan AW (2008) Towards a transgenic model of Huntington's disease in a non-human primate. Nature 453:921–924

Ylera F, Lurz R, Erdmann VA, Furste JP (2002) Selection of RNA aptamers to the Alzheimer's disease amyloid peptide. Biochem Biophys Res Commun 290:1583–1588

Yoshiike Y, Akagi T, Takashima A (2007) Surface structure of amyloid-β fibrils contributes to cytotoxicity. Biochemistry 46:9805–9812

Zakharov SD, Hulleman JD, Dutseva EA, Antonenko YN, Rochet JC, Cramer WA (2007) Helical α-synuclein forms highly conductive ion channels. Biochemistry 46:14369–14379

Zanuy D, Nussinov R (2003) The sequence dependence of fiber organization. A comparative molecular dynamics study of the islet amyloid polypeptide segments 22–27 and 22–29. J Mol Biol 329:565–584

Zhang Q, Powers ET, Nieva J, Huff ME, Dendle MA, Bieschke J, Glabe CG, Eschenmoser A, Wentworth P Jr, Lerner RA, Kelly JW (2004) Metabolite-initiated protein misfolding may trigger Alzheimer's disease. Proc Natl Acad Sci USA 101:4752–4757

Zurdo J, Guijarro JI, Dobson CM (2001a) Preparation and characterization of purified amyloid fibrils. J Am Chem Soc 123:8141–8142

Zurdo J, Guijarro JI, Jimenez JL, Saibil HR, Dobson CM (2001b) Dependence on solution conditions of aggregation and amyloid formation by an SH3 domain. J Mol Biol 311:325–340

Chapter 2
Pathologic Lesions in Alzheimer Disease and Other Neurodegenerative Diseases—Cellular and Molecular Components

Harry V. Vinters, M.D., F.A.C.P., F.R.C.P.C, Spencer Tung, B.S., and Orestes E. Solis, M.D.

Abstract This chapter describes the neuropathologic features of Alzheimer disease [senile dementia of the Alzheimer type (SDAT)] as well as several non-Alzheimer dementias [diffuse Lewy-body disease (DLBD), frontotemporal lobar degenerations (FTLDs)], including some rare entities. These disorders are now considered 'protein-misfolding' disorders, because almost all of them are associated with abnormally folded proteins within either the nuclei or cytoplasm of neurons and/or supporting glia in the central nervous system. Diagnostic (pathologic) criteria for various disorders are discussed and illustrated. For example, Alzheimer disease is associated with abnormal deposits of amyloid β-protein (Aβ) (within senile plaques and brain parenchymal vessel walls) and phospho-tau (in neuronal cell bodies), whereas DLBD

H.V. Vinters, M.D., F.A.C.P., F.R.C.P.C (✉)
Department of Pathology and Laboratory Medicine (Neuropathology), UCLA Medical Center, CHS 18-170, 650 Charles Young Drive So, Los Angeles, CA 90095-1732, USA

Department of Neurology, UCLA Medical Center, Los Angeles, CA, USA

David Geffen School of Medicine at UCLA and Section of Neuropathology, UCLA Medical Center, Los Angeles, CA, USA
e-mail: hvinters@mednet.ucla.edu

S.Tung, B.S.
Department of Pathology and Laboratory Medicine (Neuropathology), UCLA Medical Center, CHS 18-170, 650 Charles Young Drive So, Los Angeles, CA 90095-1732, USA

O.E. Solis, M.D.
Department of Pathology and Laboratory Medicine (Neuropathology), UCLA Medical Center, CHS 18-170, 650 Charles Young Drive So, Los Angeles, CA 90095-1732, USA

David Geffen School of Medicine at UCLA and Section of Neuropathology, UCLA Medical Center, Los Angeles, CA, USA

Department of Neurology, University of Santo Tomas, Manila, Philippines

F. Rahimi and G. Bitan (eds.), *Non-fibrillar Amyloidogenic Protein Assemblies—Common Cytotoxins Underlying Degenerative Diseases*, DOI 10.1007/978-94-007-2774-8_2,

results from abnormal neuronal cytoplasmic accumulations of α-synuclein. Virtually all of the proteins implicated in pathogenesis are demonstrable within human brain by immunohistochemistry on paraffin sections, which is now a mainstay in the diagnosis of neurodegenerative diseases. A common approach to prevention or therapy of these disorders is to attempt removal of the abnormally folded proteins or modify them to such an extent that they are no longer 'toxic' to the brain. Assessing the clinical and neuropathologic effects (within the central nervous system) of these strategies will be a challenge to both clinicians interested in treating dementing disorders, and neuropathologists who study their morphologic correlates.

Keywords Alzheimer disease • Brain • Histopathology • Neurodegeneration • Neuropathology

2.1 Introduction

The problem at hand is framed well in a recent review article (Kovacs and Budka 2010): "Neurodegenerative diseases (NDDs) are traditionally defined as disorders with progressive loss of neurons in distinct anatomical distribution(s), and accordingly different clinical phenotypes. [These diseases] are also referred to as conformational diseases…, emphasizing the central pathogenic role of altered protein processing". A defining biochemical theme in the study of many neurodegenerative disorders (including the most common central nervous system (CNS) 'amyloidosis', Alzheimer disease) is that of 'protein misfolding'—the molecular nature of which is explored in depth (and from various perspectives) throughout this book. This chapter, by contrast, will discuss the practical aspects of gross, microscopic, and immunohistochemical features of some of the most common neurodegenerative disorders, how the neuropathologist approaches evaluation of an autopsy or biopsy brain specimen, and how biochemical, molecular, and genetic findings—of which there has been an explosion in recent years—inform the clinicopathologic evaluation of brains originating from afflicted individuals.

Neuropathologic examination of the brain—either at autopsy or (less commonly) biopsy—continues to be described as the 'gold standard' for the diagnosis of AD and non-AD dementias (Goedert and Ghetti 2007; Vinters et al. 1998), even as high-resolution neuroradiographic techniques are emerging that seem capable of both quantifying AD-associated cerebral atrophy, and detecting Aβ peptide or other β-pleated-sheet proteins in the brain while patients are alive and even completely asymptomatic (Mintun et al. 2006; Small et al. 1996, 2006). Yet neuropathology plays a pivotal role in illuminating the structures that are being 'imaged' by novel structural and metabolic neuroimaging methods (Vinters 2007). Careful clinicopathologic correlation, i.e., attempting to explain complex neurologic symptoms in a 'deteriorating', often end-stage CNS by autopsy examination of the brain, was a central pillar of dementia research through the 1970s, at the end of which structural imaging emerged and began to provide valuable information about the CNS *in vivo*.

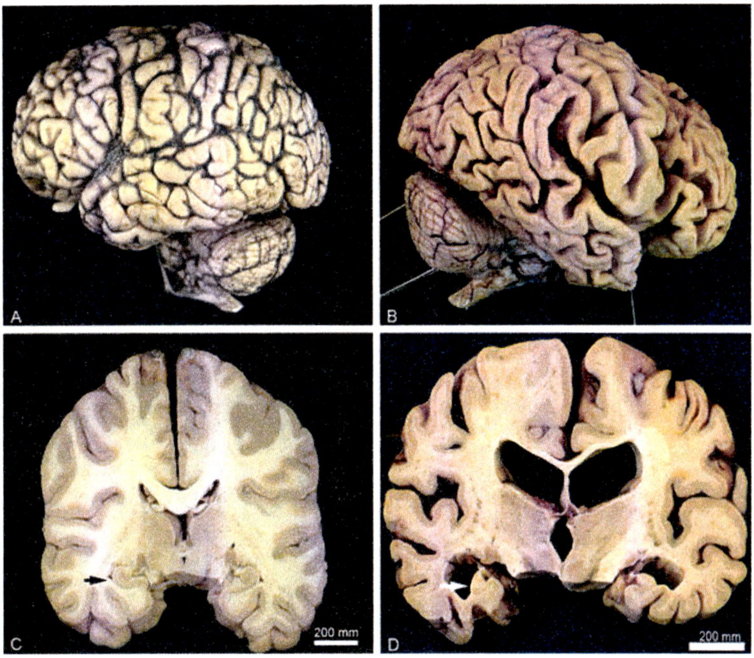

Fig. 2.1 Comparison of the appearance of normal brain (*panels A, C*, lateral view at *top*, coronal slice at *bottom*) by comparison to an AD brain (*panels B, D*, lateral view at *top*, coronal slice at *bottom*). Note profound cortical atrophy with widening of sulci, in the AD brain, which is accentuated because the leptomeninges have been removed prior to photographing the specimen. Coronal slices (*bottom* two panels) confirm cortical atrophy with thinning of the cortical ribbon (and subcortical white matter), sulcal enlargement, and pronounced enlargement of the lateral ventricles (hydrocephalus *ex vacuo*). Note especially marked atrophy of the hippocampi in AD brain, (*left* hippocampus indicated by *arrows* in each panel), with striking enlargement of temporal horns of the lateral ventricles

It bears re-emphasis that the starting point for important biochemical/molecular studies that have linked abnormally folded proteins to neurodegeneration was rapidly harvested (usually autopsy) brain tissue—neuropathologic features of which were subsequently correlated with the relevant neurochemical data (Querfurth and LaFerla 2010; Kovacs et al. 2010; Vinters et al. 1998).

The main neuropathologic feature of the AD brain on gross inspection is cortical atrophy, which is usually diffuse and fairly symmetrical throughout the cerebral hemispheres rather than being accentuated in certain lobes (as in the case of frontotemporal lobar degenerations, see below) (Vinters et al. 1998). Fresh brain weight is usually below the normal range for an adult (1,200–1,400 g), though not necessarily so, and it may be entirely normal. When the fixed brain is cut, the cortical atrophy (manifest as thinning of the cortical ribbon) is usually accompanied by enlargement of the ventricular system, or 'hydrocephalus *ex vacuo*', and sometimes shrinkage or atrophy of the subcortical white matter (Fig. 2.1). The precise etiology of the white matter change is not known—it may in part represent downstream (Wallerian) degeneration secondary to cortical atrophy with neuronal loss, or be the manifestation of an intrinsic 'leukoencephalopathy'. If the brain of a demented patient shows hydrocephalus out of proportion to the degree of cerebral cortical atrophy, the possibility of normal pressure hydrocephalus (NPH) must be considered, though microscopic lesions of AD should still be sought by the neuropathologist in such a brain and are often detected. In practice, most experienced neuropathologists (including the authors of this chapter) are struck by the variability in brain weights, cerebral cortical atrophy, and hydrocephalus *ex vacuo* among individuals who eventually have the diagnosis of AD robustly confirmed by light microscopy (Joachim et al. 1988).

2.2 Alzheimer Disease: Microscopic Confirmation of the Diagnosis of "Dementia of the Alzheimer Type (SDAT)"

The microscopic lesions that 'accumulate' in the CNS (mainly cerebral cortex) of individuals with AD can, when prominent and numerous, even be seen on routine [hematoxylin-and-eosin–stained (H-and-E-stained)] sections of the brain, but are much more easily demonstrated by the use of special stains and immunohistochemical methods. A dictum among neuropathologists is that if key lesions (senile plaques, neurofibrillary tangles, amyloid angiopathy) are identifiable on routine stains, they will be abundant using ancillary studies. Over the past 20–30 years, as the biochemical nature of AD lesions has become understood, immunohistochemistry using highly specific primary antibodies (monoclonal or polyclonal) against the components of senile plaques (SPs) and neurofibrillary tangles (NFTs) have become increasingly utilized to demonstrate these lesions in the CNS, and allow for their quantification. The special stains used to demonstrate SPs and NFTs seen in abundance in the cerebral cortex of an end-stage AD patient historically have been silver-impregnation techniques, usually the modified Bielschowsky and Bodian stains and (in more recent years) the Campbell-Switzer and Gallyas methods—the latter two used effectively in seminal studies of SP and NFT distribution by Heiko and Eva Braak and their colleagues (Braak et al. 1986, 1993; Braak and Braak 1991). Though many of these stains had an intrinsic 'beauty' and elegance, they were sometimes capricious and resulted in annoying (and inconsistent) tissue-section artifacts that tended to limit their usefulness.

2.2.1 Senile Plaques

SPs appear, on routine H-and-E-stained sections, as a coarsening of the neuropil (the 'neuritic' component of the SP) centered on an amorphous eosinophilic 'globule' of amyloid, the core of the SP (Fig. 2.2). The relationship between the amyloid core of a mature SP and its neuritic corona (both seen well on silver stains) has been debated for years and remains largely unresolved, but such

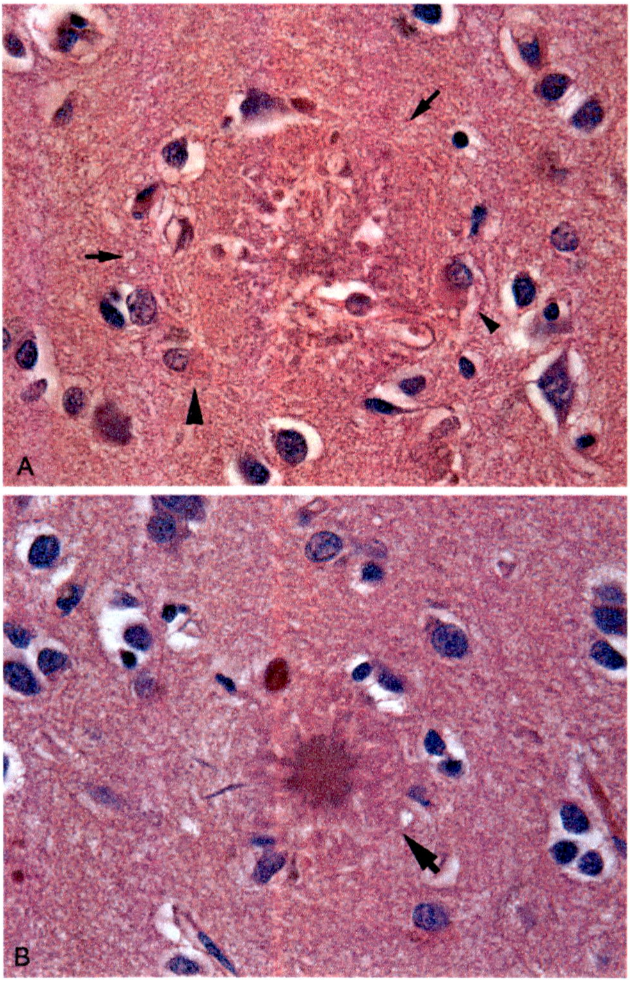

Fig. 2.2 Senile (neuritic) plaques (SP). *Panel A* shows a neuritic SP, manifest as a coarsening of the neuropil (approximate SP boundaries indicated by *arrows*). Cells surrounding the SP include reactive astrocytes (*arrowheads*). *Panel B* shows a neuritic SP with a large amyloid core (*arrow*). (Hematoxylin and eosin stain; both panels original magnification × 100)

mature 'neuritic' SPs are thought to be more representative or reflective of cortical injury (thus dysfunction) than the more diffuse SPs *lacking* a neuritic component. Excellent reviews on the hypothesized molecular pathogenesis of SPs—emphasizing the role of microglia, astrocytes, and secreted factors—have appeared (one of the best is by Dickson 1997). This author has suggested that deposition of the P3 derivative of amyloid precursor protein (Aβ amino acids 17–24) may represent a 'benign' form of brain cortical amyloid. Growth of the SP may occur through deposition of Aβ1–42 and precipitation of soluble

Aβ1–40, which leads to SPs becoming associated with activated microglia and astrocytes. Such microglia and astrocytes in the SP milieu may produce toxic molecules, e.g., reactive oxygen species, nitrogen intermediates, and proteases, as well as mediators of inflammatory cascades. Neurites bearing paired helical filaments (PHFs) may then come to surround a 'mature' SP. Though SPs (especially neuritic SPs) have a neuronal component, insofar as the neurites surrounding the amyloid core represent processes emerging from (presumably) damaged nerve cell bodies, they are substantially extraneuronal or located within the neuropil. Although SPs are often found in elderly individuals without AD, their density is in general far less than that in patients with AD (Blessed et al. 1968); however, most neuropathologists have encountered autopsy brains from cognitively intact elderly that contain abundant neuritic SPs. Recently, anecdotal reports have described all neuropathologic features of AD (abundant SPs and NFTs) in cognitively normal elderly—indeed rare individuals who had been carefully examined shortly before death (Berlau et al. 2007). In describing the neuropathologic features of an AD brain, most neuropathologists will distinguish between diffuse plaques (detectable by Aβ1–42 immunohistochemistry but barely visible on routine stains) and neuritic SPs (see above), because the latter are thought to reflect neuronal cell-process (neuritic) injury (Gearing et al. 1995). Neuritic SPs (at least their neuritic *components*) are usually immunostained by phospho-tau antibodies; while the SP cores (and a portion of its surrounding halo) are strongly immunoreactive with anti-Aβ1–42 (see below and figures).

2.2.2 *Neurofibrillary Tangles*

NFTs, the second major brain lesion of AD, are dense intraneuronal cytoplasmic aggregates of filaments that include, on ultrastructural examination, characteristic paired helical filaments (Dickson 2003; Vinters et al. 1994). NFTs are usually accompanied by neuropil threads in the adjacent brain parenchyma—these 'threads' are thought to represent processes of tangle-bearing neurons (Braak et al. 1986, Braak and Braak 1988). NFTs (Fig. 2.3) can occur with many non-AD neurodegenerative, toxic, and even inflammatory conditions, including subacute sclerosing panencephalitis, dementia pugilistica, aluminum intoxication, postencephalitic Parkinsonism, and the Parkinsonian amyotrophic lateral sclerosis (ALS)–dementia complex of Guam (Wisniewski et al. 1979). Of interest, NFT-like neuronal cytoplasmic lesions are commonly encountered within the dysmorphic and enlarged neuronal cell bodies of infants and children with epilepsy-associated cortical dysplasia or cortical tubers of tuberous sclerosis complex (TSC) (Mischel et al. 1995). These NFTs, easily demonstrable on the same silver stains (e.g., Bielschowsky) used to highlight AD lesions, are *not* composed of paired helical filaments by ultrastructural examination; rather, they show disorganized clumps of neurofilaments, straight filaments, various degenerate cytoplasmic components and neurotubules (Duong et al. 1994).

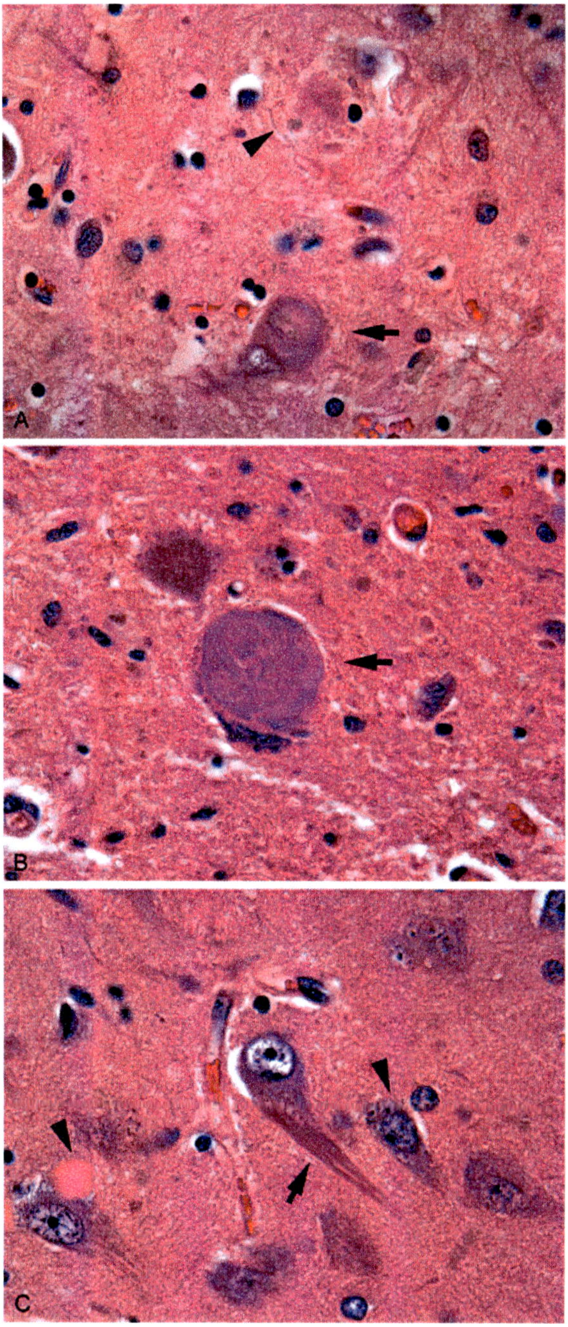

Fig. 2.3 H-and-E-stained sections showing NFTs. Tangles often take on the native shape of the neuron in which they develop—*arrows* in *panels A, B* show globose NFTs; *arrowhead* in *panel A* shows amyloid core of an SP. *Panel C* shows an NFT in a hippocampal (pyramidal layer) neuron. Arrowhead at *left* shows a Hirano body, whereas arrowhead at *right* shows GVD. (magnification × 100 all panels)

2.2.3 *Cerebral Amyloid Angiopathy*

A third important, though often underappreciated, lesion of AD is cerebral amyloid angiopathy (CAA), sometimes described as cerebrovascular amyloidosis or cerebral congophilic angiopathy (Vinters 1987). Indeed, CAA was the microscopic AD 'lesion' from which Glenner and Wong (1984) isolated A4 protein (now renamed Aβ or β amyloid). The reason that CAA is less prominently discussed (than SPs and NFTs) when considering AD neuropathologic features may be that it is extremely variable among AD patients, though when sought diligently is found (to some extent) in an estimated 90–95% of AD brains (Vinters 1987; Vinters and Gilbert 1983). CAA describes a histopathologic finding that results from a process whereby the media of parenchymal arterioles, normally composed of smooth muscle cells (SMC), undergoes progressive loss of these SMC coincident with the accumulation of an eosinophilic hyaline material (in the vessel wall) that has the staining properties of amyloid, i.e., positivity for thioflavin S or T, and congophilia (Vinters et al. 1994). When a brain with prominent CAA is stained with Congo red and polarized, the walls of affected arterioles show characteristic yellow-green birefringence. CAA may also involve cortical parenchymal venules and capillaries; some have suggested that at least a subset of SPs in the neocortex are intimately associated with capillaries and may even originate from them (Soontornniyomkij et al. 2010a). Meningeal arteries are often affected by CAA, and sometimes an amyloid-laden arteriole may be seen extending into the subarachnoid space, its wall still laden with amyloid. When CAA occurs in the subarachnoid space, the amyloid deposits are usually adventitial rather than medial in the walls of affected arteries, and have a 'chunky' appearance, suggesting they have resulted from aggregates of Aβ in the CSF. CAA almost never involves the subcortical white matter, basal ganglia, brainstem, or spinal cord, but (in severe cases) may involve the cerebellar molecular layer and meninges (Vinters and Gilbert 1983; Vinters 1987). The pathogenesis of CAA is complex, and probably involves overproduction of Aβ (from APP) in or near the vessel wall, together with abnormal/impaired clearance of Aβ, probably along perivascular adventitial pathways of brain microvessels (Weller et al. 2009).

CAA (Fig. 2.4) is also important as a cause of spontaneous (non-traumatic) intracerebral hemorrhage within the brains of elderly individuals—including many who do not manifest overt features of a dementing illness or even cognitive impairment at the time of their stroke; a small subset of these patients have predominantly severe CAA (with small loads of SPs and NFTs) as their major neuropathologic finding (Vinters 1987). CAA-related intraparenchymal hematomas are usually lobar, unlike the centrencephalic bleeds seen with hypertensive microvascular disease (Vinters 1987; Vinters et al. 1998). In some patients, multiple hematomas caused by CAA occur over months or years, leading to progressive neurologic impairment—sometimes in a stepwise or 'stuttering' fashion. These large and invariably symptomatic, sometimes fatal hematomas occur in a relatively small proportion of those with AD and severe CAA, but CAA-related *micro*bleeds (detectable on high-resolution MRI scanning using special sequences) are now accepted as a reliable biomarker for the presence of CAA

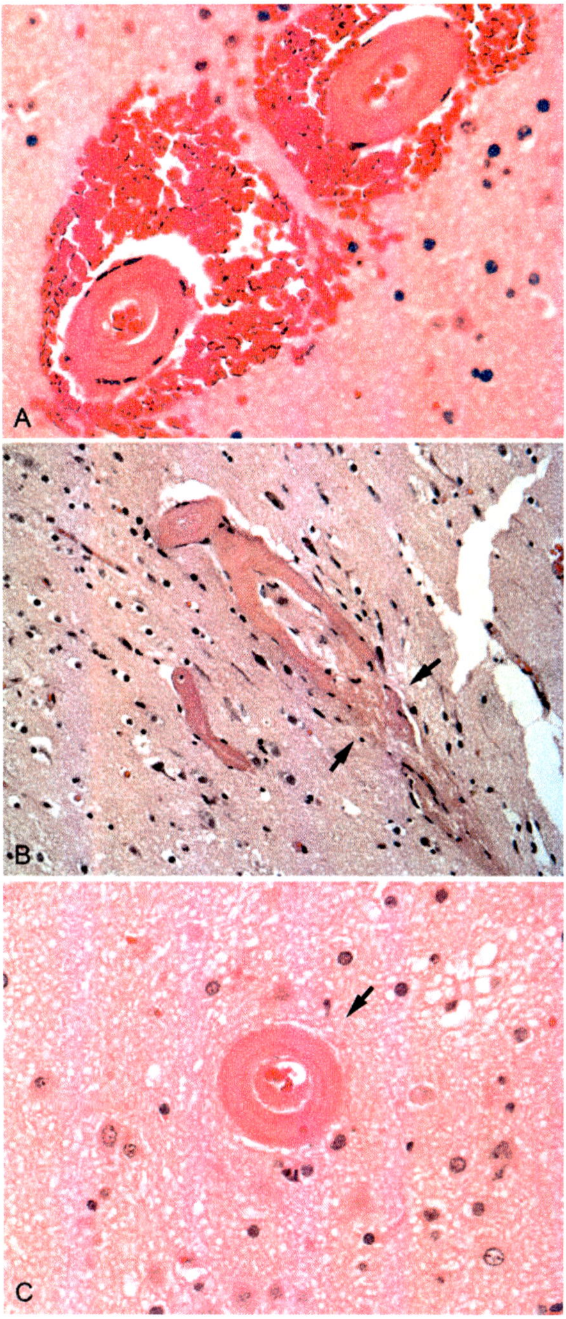

Fig. 2.4 CAA, H-and-E-stained sections, biopsy specimens. *Panel A* shows typical appearance of arterioles that have lost their smooth muscle cell media, which is replaced by fibrillar eosinophilic (amyloid) material. Note surrounding acute hemorrhage, a well-documented complication of severe CAA. *Panel B* shows a section through a similarly affected artery, cut in transverse/longitudinal section. *Arrows* indicate a possible rupture site in the artery. *Panel C* shows a severely affected arteriole within brain parenchyma (*arrow*), without surrounding reactive change or hemorrhage

within the brain (Zhang-Nunes et al. 2006). More recently, severe CAA has also been associated with the occurrence of cerebral micro-infarcts, lesions that may obviously worsen cognitive impairment in a patient already afflicted by AD parenchymal abnormalities (Soontornniyomkij et al. 2010b).

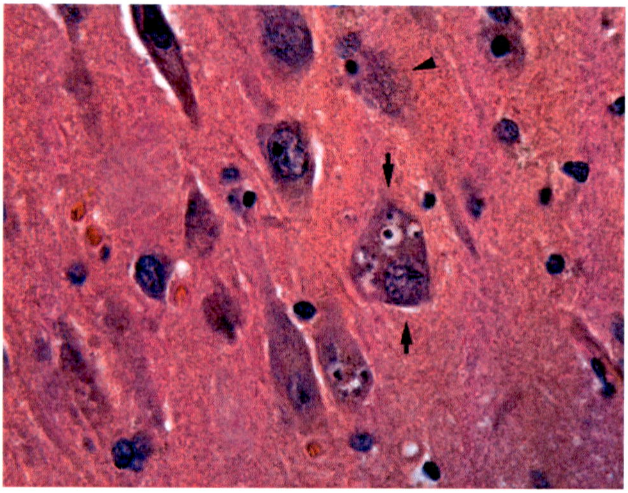

Fig. 2.5 GVD. *Arrows* indicate a large hippocampal neuron, the cytoplasm of which contains vacuoles within which are basophilic granules. *Arrowhead* indicates a nearby neuron showing less prominent GVD. (H-and-E-stained section, original magnification × 100)

2.2.4 Other Lesions

While SPs, NFTs, and CAA are the major microscopic lesions of AD and are widely distributed throughout the cortex, two others merit mention for the sake of completeness. Granulovacuolar degeneration (GVD, of Simchowicz) describes a neuronal cytoplasmic lesion in which the neuronal cytoplasm of hippocampal pyramidal cells is replaced by vacuoles containing small basophilic granules. Hippocampi showing prominent GVD (Fig. 2.5) also often show eosinophilic hyaline rod-like structures in the adjacent neuropil—these are described as Hirano bodies, structures that are composed predominantly of actin. Perhaps because of their circumscribed hippocampal distribution within the brains of AD patients and their rarity in the neocortex, GVDs and Hirano bodies have been the subject of limited study in terms of assessing their possible contributions to AD pathogenesis and progression. Instead, investigators have focused on lesions that are widely distributed within the neocortex—SPs, NFTs, and CAA (see above). Nevertheless, neurons showing GVD and Hirano bodies are a frequent finding in AD hippocampi, and are sometimes found in the hippocampi of those with non-AD dementias (e.g., progressive supranuclear palsy).

While this section has emphasized 'lesions' commonly seen in AD brains—either focally or diffusely in the cortex—one of the most important findings, demonstrable biochemically or by immunohistochemistry in AD brain, is synapse loss (Clare et al. 2010). This is shown on sections of affected cortex when they are immunostained with antibodies against a synaptic protein such as synaptophysin (Terry et al. 1991; Davidsson and Blennow 1998). Interpretation (and quantification) of the immunohistochemical signal in such cases must be done by careful comparison to brain from a cognitively normal control (and using tissue that has been comparably fixed and processed), since the loss of synaptophysin protein may be subtle. Thus immunostaining brain sections with antibodies to synaptic proteins is not usually part of the routine work-up of a dementia brain, unless a laboratory has the resources and skill set to undertake densitometric evaluation of the resultant sections. AD brains also frequently show evidence of clinical co-morbidity, not surprising given the many age-related diseases (e.g., cerebrovascular disease, neoplasms) that may impact on the aging brain (Fu et al. 2004). Coexistent Parkinson-disease changes and evidence of infarcts or hemorrhage have been seen, respectively, in as many as 1/5 and 1/4 of AD brains (Gearing et al. 1995). The theme of 'co-morbidity' between ischemic brain lesions and AD microscopic changes—both common in the elderly—features prominently in modern dementia research, possibly because it represents a more accurate and realistic scenario than considering AD or multi-infarct/ischemic vascular dementia as 'pure' entities (Vinters et al. 2000; Selnes and Vinters 2006).

2.2.5 *Immunohistochemical Features*

The amyloid cores of SPs and CAA have a major component of Aβ protein. The immunohistochemical study of AD brain lesions began as soon as the partial peptide sequence of Aβ (then called A4) was first published (Glenner and Wong 1984), several groups developed antibodies to synthetic peptides representing portions of the molecule (Vinters et al. 1988). Currently, numerous commercially available antibodies to Aβ (varied amino-acid lengths), tau, ubiquitin, α-synuclein (to detect Lewy bodies), and TDP-43 (see below) are available to facilitate accurate immunohistochemical characterization of a given necropsy brain specimen. SPs are more prominently immunoreactive for the 1–42 amino-acid length of Aβ, whereas CAA immunolabels more strongly with antibodies to Aβ1–40, though there are striking exceptions to this 'predominance' of a given Aβ length in one or the other lesion (Fig. 2.6). Diffuse SPs are shown well by anti-Aβ antibodies incorporated into appropriate immunohistochemical protocols, as are the amyloid 'cores' of mature SPs. The neuritic coronas of mature SPs, NFTs, and neuropil threads are prominently immunolabeled with antibodies to phosphorylated tau (Fig. 2.7). In cases of severe CAA, gamma-trace may also be found in affected vessel walls and a heavily infiltrated arteriole may be surrounded by tau-immunoreactive neuritis or a 'halo' of perivascular Aβ immunoreactivity (Vinters et al. 1990).

2.2.6 *'Staging' AD and Quantifying AD Lesions*

Since essentially all AD lesions described above may be encountered in the cerebral cortex of cognitively normal elderly individuals, it is useful to quantify (or *semi-quantify*) these abnormalities and assess their topographic distribution; ideally, this

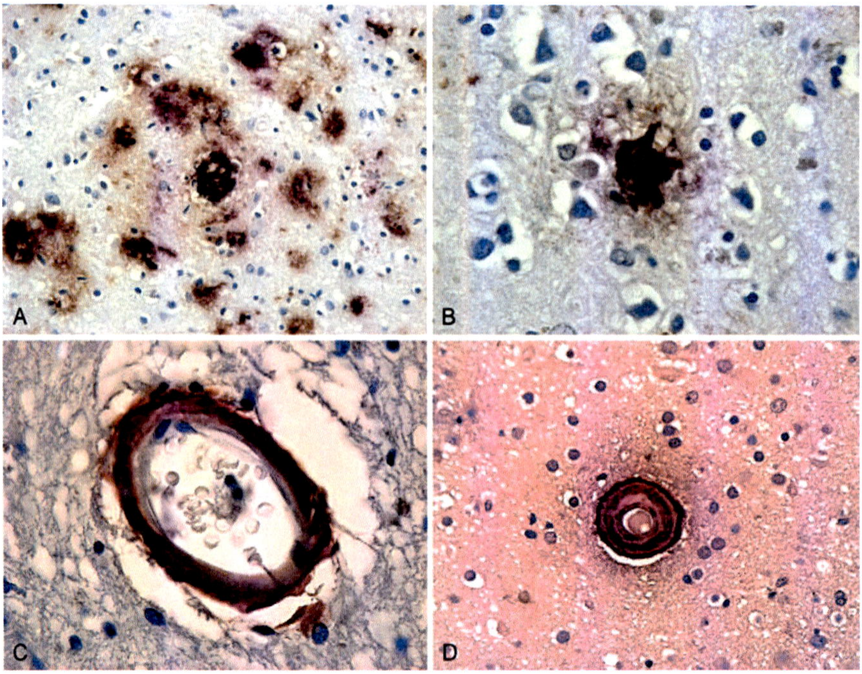

Fig. 2.6 Immunohistochemical properties of SPs and CAA. All panels are images from sections stained with anti-Aβ1–40. *Panel A* shows low-power view, highlighting abundant (mainly) diffuse SPs. *Panel B* shows a large amyloid core of a neuritic SP (compare with Fig. 2.2b). *Panels C, D* show severe amyloid angiopathy, in which arterial/arteriolar walls (normally containing smooth muscle cells) are composed almost entirely of Aβ, which has replaced the smooth muscle cells. (magnification *panel A* × 20, *panels B–D* × 40)

information can and should be incorporated into the autopsy report of a given patient. Correlations between lesion 'load', severity of neuropathologic findings and (*in vivo*) neuropsychological symptoms in a given patient are important, even essential. They become problematic, however, when the neuropathologist has the brain of an 'end-stage' patient to examine, yet that patient may have experienced his/her maximal neurologic deficit months or even years before he/she expired (Galasko et al. 1994). Rarely, biopsies are carried out to confirm the diagnosis of AD, but in that situation only a small portion of the brain is available for examination. However, small studies have used biopsy and autopsy data from one and the same patient to describe the progression of AD lesions over many years (Di Patre et al. 1999). These investigations have shown, among other things, that there can be significant AD lesions in the brain of an individual who is, as judged by a reasonably high mini-mental-state examination (MMSE), at a cognitively 'early' stage of clinical symptoms.

Many attempts have been made to standardize neuropathologic diagnostic criteria for AD/SDAT vs. normal aging (Mirra et al. 1991; Gearing et al. 1995). A paper

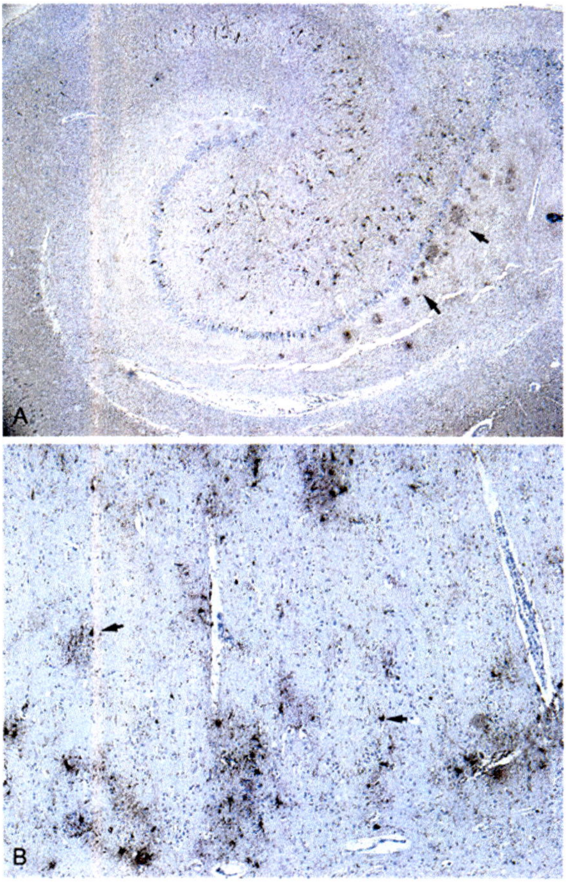

Fig. 2.7 Tau-immunoreactivity in hippocampus of a patient with severe AD. *Panel A* (photographed at low magnification) shows endplate region. Note numerous tau-immunoreactive neurons (most representing NFTs) in the endplate region, as well as many immunoreactive neurons in the granule cell layer. *Arrows* indicate tau-immunoreactive (neuritic) SPs adjacent to the granule cell layer. *Panel B* shows occipital cortex at a slightly higher magnification, indicating neuritic SPs and numerous tau-immunoreactive (NFT) neurons (*arrows*)

that resulted from a consensus conference in the 1980s resulted in the widely used 'Khachaturian criteria' for the neuropathologic diagnosis of AD (Khachaturian 1985). These were modified and updated by the Consortium to Establish a Registry for AD (Mirra et al. 1991; Gearing et al. 1995). Braak criteria for AD severity (Braak et al. 1993; Braak and Braak 1991) assume a progression of neuropathologic abnormalities (predominantly NFT and neuropil thread accumulation) from the transentorhinal cortex (stages I and II) to the hippocampus (III and IV), with ultimate widespread involvement of the neo-/isocortex (stages V and VI). It has been argued that Braak stage III and IV AD neuropathologic change is associated clinically with mild cognitive impairment (MCI) but not overt dementia. In reality, brains of subjects with amnestic MCI (aMCI) are available for examination infrequently, and show a wide range of neuropathologic lesion density as well as co-existent superimposed ischemic vascular lesions (Petersen et al. 2006; Vinters 2006).

In the late 1990s, the 'NIA-Reagan Institute Criteria for the Neuropathologic Diagnosis of AD' came into widespread use, and have been tested and operationalized by various groups (Newell et al. 1999). [These criteria assign a *high, intermediate, or low* likelihood that a given individual's dementia was due to AD neuropathologic features.] One study found not only a good correlation between a high NIA-Reagan 'probability/likelihood' of AD and clinical dementia, but ascertained that the 'older' Khachaturian and CERAD criteria correlated fairly well with those of 'NIA-Reagan', a pleasant surprise (Newell et al. 1999). Occasional cases arise—especially among the *oldest old*, e.g., nonagenarians and centenarians—where Braak stage VI AD changes and an 'NIA-Reagan' assessment of 'high likelihood of AD' are clearly present in the brain of a subject who was known to be cognitively intact until shortly before death (Berlau et al. 2007). Quantification of AD neuropathologic changes is increasingly facilitated by the ability to digitize immunostained glass slides, retain the images as a permanent electronic record of a given autopsy, and if necessary use these digital images as a starting point for further quantitative morphometry of that specimen. As well, the neuropathologic diagnosis of AD and mixed dementias will increasingly need to be reconciled with—or considered in the context of—research criteria for the clinical diagnosis of AD, which extensively incorporate biomarkers (including CSF biochemical assays of Aβ and phospho-tau) and novel neuroimaging data of the type derived from the Alzheimer Disease Neuroimaging Initiative (ADNI) (Dubois et al. 2007).

2.3 Diffuse Lewy-Body Disease (DLBD)

This disorder, which is almost always associated with dementia (it is then sometimes better described as dementia with Lewy bodies, DLB) is clinically distinct from AD/SDAT—though the overlap between DLBD and AD neuropathologic changes is striking (the authors of this chapter have rarely encountered a case of DLBD in which there was complete *absence* of some AD changes, and usually the AD changes are quite advanced). The related clinical challenge is identifying the etiology of dementia when it occurs in a patient with Parkinson's disease (i.e., Parkinson's disease dementia or PDD). There is significant debate in the literature as to the structural brain changes underling dementia in a PD patient. It may be due to concomitant AD changes, predominant LB deposition in the neocortex (rarely seen in isolation, see above), a combination of the two, or even concomitant non-AD, non-DLBD neuropathologic changes (for an excellent review, see Kalaitzakis and Pearce 2009). DLBD or DLB is characterized by fluctuating cognition with variations in attention and alertness, neuroleptic sensitivity, recurrent visual hallucinations, and Parkinsonian features (McKeith et al. 2005). LBs and associated Lewy neurites (Fig. 2.8) are demonstrated by immunohistochemistry using primary antibodies to either α-synuclein or ubiquitin—the major component of LBs is abnormally aggregated α-synuclein (Maries et al. 2003), therefore this is the optimal antibody to use; anti-ubiquitin suffers from the 'weakness' of being an antibody that will highlight AD lesions non-specifically (SPs, NFTs), which (see above) are often found in DLBD brains. Counting of LBs is problematic and yields significant inter-observer variability.

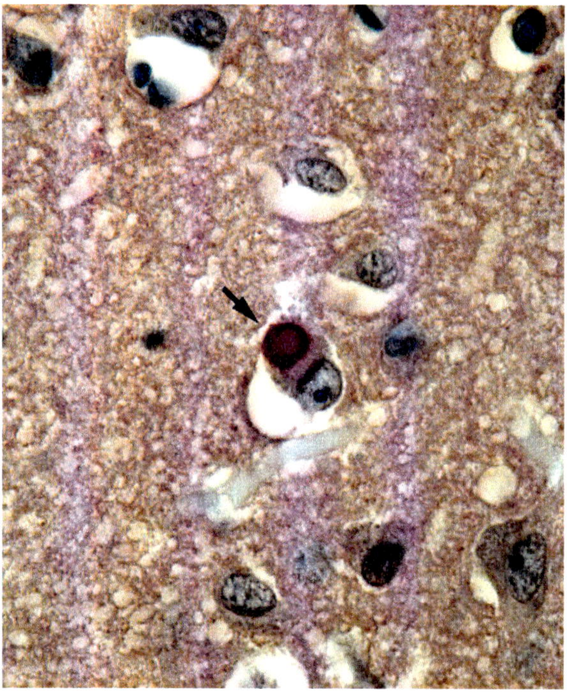

Fig. 2.8 α-Synuclein-immunoreactive Lewy body (*arrow*) in the cortex of a patient with DLBD. Apparent immunoreactivity in surrounding cortex probably represents cross-immunoreactivity of the labeling antibody with synaptic proteins (magnification × 100)

Therefore, a consensus panel recommended (in 2005) a semi-quantitative scoring system for these cytoplasmic inclusions: the scale goes from 0 (none), to 4 (abundant LBs and numerous Lewy neurites). Almost all individuals with DLBD have significant brainstem pathologic changes of the type seen in idiopathic Parkinson's disease, i.e., pigmented neuron loss with astrocytic gliosis in the substantia nigra, locus ceruleus and dorsal motor nucleus of the vagus nerve, with LBs in remaining neurons as well as non-pigmented neurons throughout the brainstem.

α-Synuclein abnormalities are also implicated in non-DLBD/non-Parkinsonian disorders, especially multiple-system atrophy (MSA). In MSA, α-synuclein-immunoreactive inclusions are often seen in non-neuronal cells, especially glia (both astrocytes and oligodendroglia) in the context of degeneration in multiple neuronal systems throughout the brain (Apostolova et al. 2006; Dickson 2003).

2.4 Frontotemporal Lobar Degeneration(s) (FTLDs)

The 'competition' to be identified as the second most common form of parenchymal dementia rages between DLBD (see above) and diseases within the FTLD spectrum—one problem being that both groups of disorders share features with AD, i.e., DLBD rarely occurs without some degree of Alzheimerization of the brain, whereas FTLDs are frequently associated with tau pathology, which is integral to AD/SDAT (see above). The morphoanatomical study of frontotemporal lobar degenerations has become one of the most challenging areas of diagnostic neuropathology. Credit is due the Lund and Manchester groups for their seminal clinicopathologic studies aimed at characterizing this interesting group of entities when their nosologic features were first recognized, and initially described as 'frontotemporal dementia (FTD)' (Snowden et al. 1992; Neary et al. 1988; Brun et al. 1994). Early studies of FTDs showed that many of the clinical entities described had neuropathologic similarities: brain weight was slightly to moderately reduced below normal; grossly the brain showed varying degrees of frontal and/or anterior temporal atrophy; neuronal loss, gliosis, and mild-to-moderate spongiform changes were found, primarily in the first 2–3 layers of the cortex, though not in the transcortical pattern characteristic of spongiform encephalopathy (Creutzfeldt–Jakob disease/CJD); gliosis (especially in regions of neuron loss) was easily visualized with immunohistochemical stains for glial fibrillary acidic protein. Cortical SPs and NFTs were not prominent. Subcortical structures, such as the substantia nigra, were sometimes abnormal, e.g., depigmented.

The larger group of redefined FTLDs now encompasses many disorders. They much more commonly have a genetic basis than does AD/SDAT, though the genes mutated vary from family to family (Kumar-Singh and Van Broeckhoven 2007). Comprehending their pathogenesis has been revolutionized by key genetic and immunohistochemical findings and observations that have accelerated and intensified over the past 10–12 years. Kumar-Singh and Van Broeckhoven (2007) have presented an illuminating synthesis of the FTLDs, integrating details of their 'core' clinical features and syndromes, preferential regions of brain involvement, distinctive neuropathologic and biochemical features, and genetic etiology. There are prominent regions of clinical (and neuropathologic) overlap among the entities, as well as many cases that are difficult to subclassify and 'pigeonhole'. This highlights the significance of detailed and careful clinicopathologic correlation in patients who come to autopsy. Many FTLD patients show evidence of aphasia and behavioral abnormalities (including disinhibited behavior), extrapyramidal and other motor disorders. Previously distinct nosologic entities incorporated into the FTLD family include: FTLD with tau abnormalities (including frontotemporal dementia and Parkinsonism linked to chromosome 17/FTDP-17, Pick disease, corticobasal ganglionic degeneration/CBGD, progressive supranuclear palsy/PSP (Dickson et al. 2007), and argyrophilic grain disease/AGD), FTLD-U with ubiquitin abnormalities, dementia lacking distinctive histology (DLDH) (Knopman et al. 1990), and FTLD associated with motor-neuron disease/MND.

Some investigators argue for the inclusion of even more entities under this umbrella, given their high frequency of tau pathology in the form of NFTs: 'NFT- or tangle-predominant' AD, dementia pugilistica, multiple-system tauopathy with dementia/MSTD, and Parkinson-dementia complex of Guam/PDG. Genes in which mutations have been found to cause some of these disorders (especially FTDP-17) include microtubule-associated protein tau (*MAPT*), hence this subgroup of the FTLDs is often described using the term 'tauopathies' (Hernandez and Avila 2007; van Swieten and Spillantini 2007). Other mutations described in FTLD families include those found in the genes charged multivesicular body protein 2b (*CHMP2B*), valosin-containing protein gene (*VCP*), and progranulin (*GRN*) (Mackenzie et al. 2009). *VCP* mutations encode a distinctive phenotype characterized by Paget's disease of the bone and inclusion-body myopathy (to be distinguished from *sporadic inclusion-body myositis*), in addition to a frontal lobe degeneration.

As the above comments imply, the neuropathologic work-up of these entities is incomplete without detailed immunohistochemical study. The proteins tau, ubiquitin, and (most recently) TDP-43 (TAR DNA-binding protein 43) must be diligently sought in brain tissue sections. (TDP-43 refers to "transactive response DNA-binding protein of molecular weight 43 kDa"). TDP-43 often co-localizes with ubiquitin, and 'immunopositivity' may be difficult to judge with absolute certainty, as it is 'counted' when positive signal is noted in the cytoplasm *rather than* the nucleus of a given neuron. Some neuropathologists have suggested subclassifying all FTLDs as either "tauopathies" or "ubiquitinopathies", depending upon which protein is detected in the brain (Bigio 2008). A recent 'position paper' has suggested a specific nomenclature for neuropathologic subtypes of FTLD, to which the interested reader is referred (Mackenzie et al. 2009). Increasingly, this group of disorders is described using (as part of a given entity's name) the predominant abnormal brain protein deposited—e.g., FTLD-tau, FTLD-ubiq(uitin), FTLD-TDP-43, FTLD-FUS (see below).

To many neuropathologists, the 'paradigmatic' FTLD remains Pick disease, characterized by severe (often 'knife edge') though often asymmetrical atrophy of the frontal and temporal lobes, characteristic affliction of the middle and inferior temporal gyri with sparing of a portion of the superior temporal gyrus, and intraneuronal 'Pick bodies' (Fig. 2.9). The latter are found in abundance in the neocortex and hippocampus (including pyramidal and granule cell layer neurons). The molecular genetic and neuropathologic investigation of FTLDs is arguably the most rapidly growing, confusing, and simultaneously challenging area of clinical neuroscience as applied to neurodegenerative diseases—a field that is certain to yield major insights in the coming years. The significance of TDP-43 translocation from the neuronal nucleus to the cytoplasm (Fig. 2.10) as a marker for neurodegenerative disease (especially FTLD) is the subject of intense investigation and significant controversy. One recent multi-center study, for instance, has found that TDP-43 expression is highly heterogeneous, with division of cases into four or five subtypes. The authors concluded that "(1) pathological variation in FTLD-TDP is best described as a 'continuum,' (2) [cortical] vacuolation was the single greatest source of [diagnostic] variation, and (3) within the FTLD-TDP 'continuum,' cases with *GRN* mutation and with coexisting motor-neuron disease or hippocampal sclerosis may have a more distinctive pathology" (Armstrong et al. 2010).

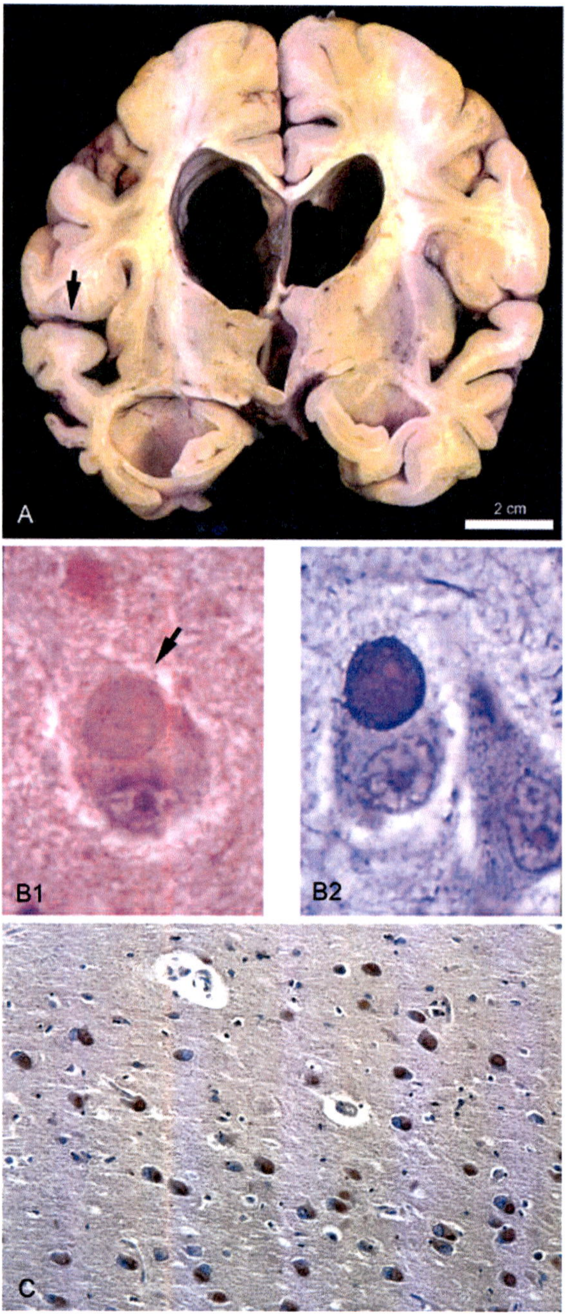

Fig. 2.9 Pick's disease (Pick variant of FTLD). *Panel A* shows coronal slice from fixed brain of an affected patient. Note pronounced and significantly asymmetrical hydrocephalus *ex vacuo* (*left* lateral ventricle significantly larger than *right*). *Arrow* indicates *left* temporal lobe, which shows selective sparing of the superior temporal gyrus but profound atrophy of middle and inferior temporal gyri; note that, by comparison, *right* temporal lobe appears relatively intact. *Panel B* shows neuronal cytoplasmic Pick bodies (H-and-E stain in B1, *arrow*, silver impregnation in *panel B2*), manifest as a skein of filamentous material in neuronal cytoplasm (magnification × 100). *Panel C* shows lower magnification view of neurons containing tau-immunoreactive Pick bodies

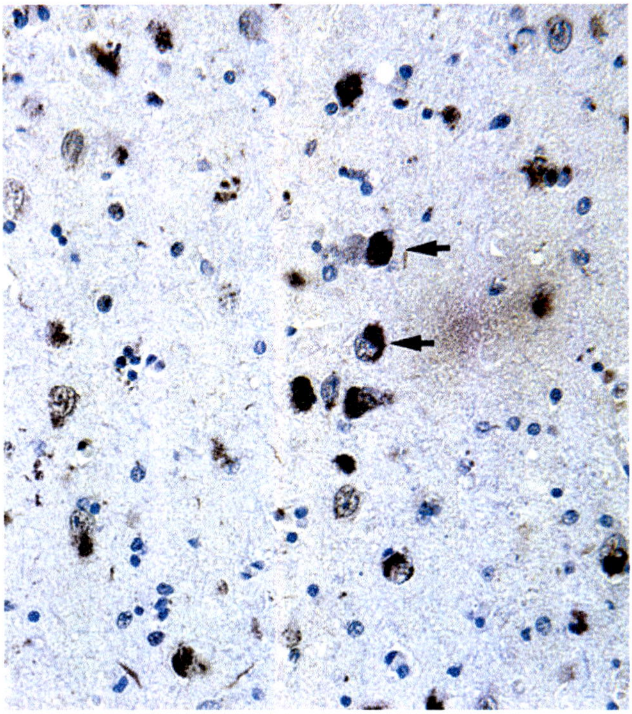

Fig. 2.10 TDP-43-immunostained section (from a patient with FTLD) shows many neurons with intense cytoplasmic staining (e.g., indicated by *arrows*)

2.5 Other (Miscellaneous) Disorders

Most of this chapter has focused on commonly encountered neurodegenerative disorders—AD/SDAT, DLBD, the FTLD spectrum—and the proteinopathies that are biochemically and immunocytochemically relevant to their pathogenesis. In a diagnostic neuropathology laboratory charged with characterizing and studying these disorders, use of antibodies to the proteins already mentioned will 'detect' at least 90–95% of relevant disorders. One important feature that has been touched on only briefly is the need to characterize cerebrovascular co-morbidity with parenchymal lesions, especially in the 'oldest old'—a major consideration given that aging is the leading risk factor for cerebrovascular disease, just as it is for SDAT (Vinters et al. 2000; Selnes and Vinters 2006). We have also not mentioned the importance of using primary antibodies to PrP on tissue sections, in confirming the diagnosis of transmissible spongiform encephalopathy (TSE), when this is suspected. When one encounters a case of suspected TSE within the USA, the resources of the National Prion Disease Pathology Surveillance Center at Case Western Reserve University in Cleveland, Ohio, are also invaluable in work-up of the necropsy brain.

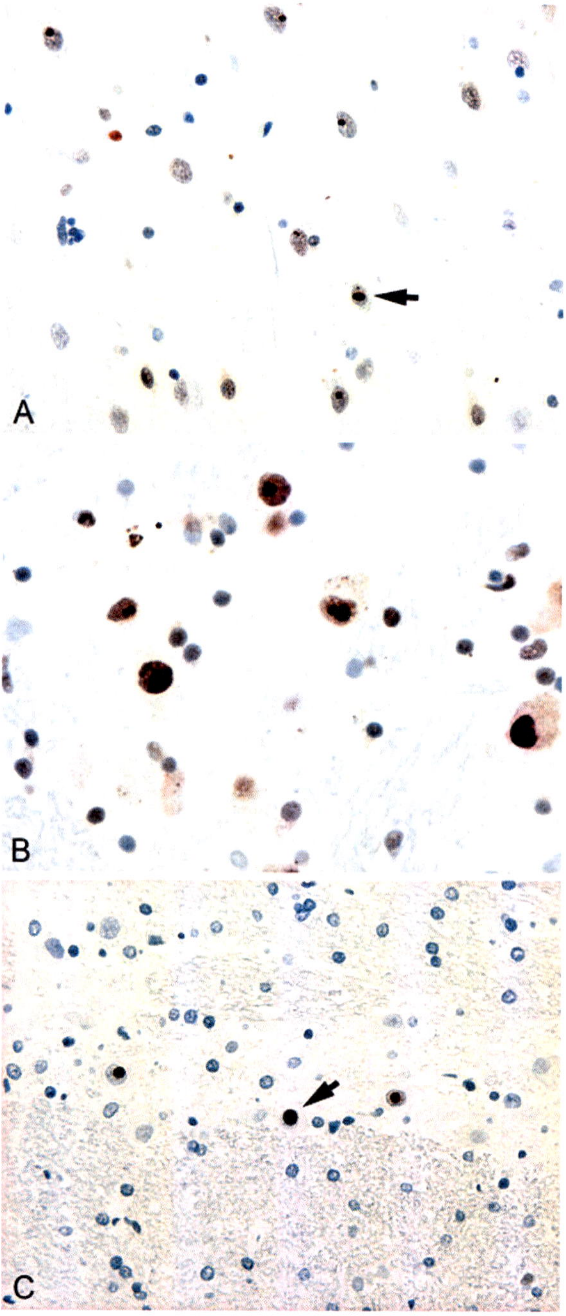

Fig. 2.11 FUS-immunoreactive neurons show various patterns of immunoreactivity of neuronal intranuclear inclusions (for details, see Woulfe et al. 2010). Some cell nuclei show curvilinear or linear inclusions (*arrow* in *panel A*), others fairly uniform reactivity throughout the nucleus (*arrow* in *panel C*), whereas others still show moderate immunoreactivity throughout the nucleus, but focally accentuated staining within it (*panel B*) (Images courtesy of Dr. John Woulfe, University of Ottawa, Canada)

Other rare neurodegenerative diseases must be considered. These include neuronal intermediate filament inclusion disease (NIFID), which some now tend to group with the FTLDs (Woulfe et al. 2010). In some disorders with abnormal neuronal intranuclear inclusions, these can be decorated with antibodies to a nuclear protein, "fused-in-sarcoma" (FUS) (Fig. 2.11). Another very rare disease (never knowingly encountered by the authors) is basophilic inclusion body disease (BIBD).

2.6 Conclusion and Future Directions

The full neuropathologic characterization of new and challenging types of neurodegenerative disease will provide 'full employment' for neuropathologists in the years to come. Not only will they be charged with characterizing abnormal 'shadows' and signals detected by neuroradiologists, but with providing feedback to clinicians on how well therapies aimed at 'clearing' abnormal proteins from the brain have worked. Finally, will clearing abnormal brain proteins lead to clinical improvement in patients? This will be the crucible in which novel therapeutic strategies will need to be tested.

Acknowledgements Work in HV Vinters' laboratory supported in part by PSH grant P50 AG16570 and the Daljit S. and Elaine Sarkaria Chair in Diagnostic Medicine. Nikki Yin assisted with preparation of some of the illustrations.

References

Apostolova LG, Klement I, Bronstein Y, Vinters HV, Cummings JL (2006) Multiple system atrophy presenting with language impairment. Neurology 67:726–727

Armstrong RA, Ellis W, Hamilton RL, Mackenzie IRA, Hedreen J, Gearing M, Montine T, Vonsattel J-P, Head E, Lieberman AP, Cairns NJ (2010) Neuropathological heterogeneity in frontotemporal lobar degeneration with TDP-43 proteinopathy: a quantitative study of 94 cases using principal components analysis. J Neural Transm 117:227–239

Berlau DJ, Kahle-Wrobleski K, Head E, Goodus M, Kim R, Kawas C (2007) Dissociation of neuropathologic findings and cognition. Case report of an apolipoprotein E ε2/ε2 genotype. Arch Neurol 64:1193–1196

Bigio EH (2008) Update on recent molecular and genetic advances in frontotemporal lobar degeneration. J Neuropathol Exp Neurol 67:635–648

Blessed G, Tomlinson BE, Roth M (1968) The association between quantitative measures of dementia and of senile change in the cerebral grey matter of elderly subjects. Br J Psychiatry 117:797–811

Braak H, Braak E (1988) Neuropil threads occur in the dendrites of tangle-bearing nerve cells. Neuropathol Appl Neurobiol 14:39–44

Braak H, Braak E (1991) Neuropathological staging of Alzheimer related changes. Acta Neuropathol 82:239–259

Braak H, Braak E, Grundke-Iqbal I, Iqbal K (1986) Occurrence of neuropil threads in the senile human brain and in Alzheimer's disease: a third location of paired helical filaments outside of neurofibrillary tangles and neuritic plaques. Neurosci Lett 65:351–355

Braak H, Duyckaerts C, Braak E, Piette F (1993) Neuropathological staging of Alzheimer-related changes with psychometrically assessed intellectual status. In: Corain B, Iqbal K, Nicolini M, Winblad B, Wisniewski H, Zatta P (eds) Alzheimer's disease: advances in clinical and basic research. Wiley, Chichester/New York, pp 131–137

Brun A, Englund B, Gustafson L, Passant U, Mann DMA, Neary D, Snowden JS (1994) Clinical and neuropathological criteria for frontotemporal dementia. J Neurol Neurosurg Psychiatry 57:416–418

Clare R, King VG, Wirenfeldt M, Vinters HV (2010) Synapse loss in dementias. J Neurosci Res 88:2083–2090

Davidsson P, Blennow K (1998) Neurochemical dissection of synaptic pathology in Alzheimer's disease. Int Psychogeriatr 10:11–23

Di Patre PL, Read SL, Cummings JL, Tomiyasu U, Vartavarian LM, Secor DL, Vinters HV (1999) Progression of clinical deterioration and pathological changes in patients with Alzheimer disease evaluated at biopsy and autopsy. Arch Neurol 56:1254–1261

Dickson DW (1997) The pathogenesis of senile plaques. J Neuropathol Exp Neurol 56:321–339

Dickson DW (ed) (2003) Neurodegeneration: the molecular pathology of dementia and movement disorders. ISN Press, Basel, 414 pp

Dickson DW, Rademakers R, Hutton ML (2007) Progressive supranuclear palsy: pathology and genetics. Brain Pathol 17:74–82

Dubois B, Feldman HH, Jacova C, DeKosky ST, Barberger-Gateau P, Cummings J, Delacourte A, Galasko D, Gauthier S, Jicha G, Meguro K, O'Brien J, Pasquier F, Robert P, Rossor M, Salloway S, Stern Y, Visser PJ, Scheltens P (2007) Research criteria for the diagnosis of Alzheimer's disease: revising the NINCDS-ADRDA criteria. Lancet Neurol 6:734–746

Duong T, DeRosa MJ, Poukens V, Vinters HV, Fisher RS (1994) Neuronal cytoskeletal abnormalities in human cerebral cortical dysplasia. Acta Neuropathol 87:493–503

Fu C, Chute DJ, Farag ES, Garakian J, Cummings JL, Vinters HV (2004) Comorbidity in dementia—An autopsy study. Arch Pathol Lab Med 128:32–38

Galasko D, Hansen LA, Katzman R, Wiederholt W, Masliah E, Terry R, Hill LR, Lessin P, Thal LJ (1994) Clinical-neuropathological correlations in Alzheimer's disease and related dementias. Arch Neurol 51:888–895

Gearing M, Mirra SS, Hedreen JC, Sumi SM, Hansen LA, Heyman A (1995) The consortium to establish a registry for Alzheimer's disease (CERAD). Part X. Neuropathology confirmation of the clinical diagnosis of Alzheimer's disease. Neurology 45:461–466

Glenner GG, Wong CW (1984) Alzheimer's disease: initial report of the purification and characterization of a novel cerebrovascular amyloid protein. Biochem Biophys Res Commun 120:885–890

Goedert M, Ghetti B (2007) Alois Alzheimer: his life and times. Brain Pathol 17:57–62

Hernandez F, Avila J (2007) Tauopathies. Cell Mol Life Sci 64:2219–2233

Joachim CL, Morris JH, Selkoe DJ (1988) Clinically diagnosed Alzheimer's disease: autopsy results in 150 cases. Ann Neurol 24:50–56

Kalaitzakis ME, Pearce RKB (2009) The morbid anatomy of dementia in Parkinson's disease. Acta Neuropathol 118:587–598

Khachaturian ZS (1985) Diagnosis of Alzheimer's disease. Arch Neurol 42:1097–1105

Knopman DS, Mastri AR, Frey WH, Sung JH, Rustan T (1990) Dementia lacking distinctive histologic features: a common non-Alzheimer degenerative dementia. Neurology 40: 251–256

Kovacs GG, Budka H (2010) Current concepts of neuropathological diagnostics in practice: neurodegenerative diseases. Clin Neuropathol 29:271–288

Kovacs GG, Botond G, Budka H (2010) Protein coding of neurodegenerative dementias: the neuropathological basis of biomarker diagnostics. Acta Neuropathol 119:389–408

Kumar-Singh S, Van Broeckhoven C (2007) Frontotemporal lobar degeneration: current concepts in the light of recent advances. Brain Pathol 17:104–113

Mackenzie IRA, Neumann M, Bigio EH, Cairns NJ, Alafuzoff I, Kril J, Kovacs GG, Ghetti B, Halliday G, Holm IE, Ince PG, Kamphorst W, Revesz T, Rozemuller AJ, Kumar-Singh S, Akiyama H, Baborie A, Spina S, Dickson DW, Trojanowski JQ, Mann DM (2009) Nomenclature for neuropathologic subtypes of frontotemporal lobar degeneration: consensus recommendations. Acta Neuropathol 117:15–18

Maries E, Dass B, Collier TJ, Kordower JH, Steece-Collier K (2003) The role of α-synuclein in Parkinsons's disease: insights from animal models. Nat Rev Neurosci 4:727–738

McKeith IG, Dickson DW, Lowe J, Emre M, O'Brien JT, Feldman H, Cummings J, Duda JE (2005) Diagnosis and management of dementia with Lewy bodies. Third report of the DLB consortium. Neurology 65:1863–1872

Mintun MA, LaRossa GN, Sheline YI, Dence CS, Lee SY, Mach RH, Klunk WE, Mathis CA, DeKosky ST, Morris JC (2006) [^{11}C]PIB in a nondemented population. Potential antecedent marker of Alzheimer disease. Neurology 67:446–452

Mirra SS, Heyman A, McKeel D, Sumi SM, Crain BJ, Brownlee LM, Vogel FS, Hughes JP, van Belle G, Berg L (1991) The consortium to establish a registry for Alzheimer's disease (CERAD). Part II. Standardization of the neuropathologic assessment of Alzheimer's disease. Neurology 41:479–486

Mischel PS, Nguyen LP, Vinters HV (1995) Cerebral cortical dysplasia associated with pediatric epilepsy. Review of neuropathologic features and proposal for a grading system. J Neuropathol Exp Neurol 54:137–153

Neary D, Snowden JS, Northen B, Goulding P (1988) Dementia of frontal lobe type. J Neurol Neurosurg Psychiatry 51:353–361

Newell KL, Hyman BT, Growdon JH, Hedley-Whyte ET (1999) Application of the National Institute on Aging (NIA)-Reagan Institute criteria for the neuropathological diagnosis of Alzheimer disease. J Neuropathol Exp Neurol 58:1147–1155

Petersen RC, Parisi JE, Dickson DW, Johnson KA, Knopman DS, Boeve BF, Jicha GA, Ivnik RJ, Smith GE, Tangalos EG, Braak H, Kokmen E (2006) Neuropathologic features of amnestic mild cognitive impairment. Arch Neurol 63:665–672

Querfurth HW, LaFerla FM (2010) Alzheimer's disease. N Engl J Med 362:329–344

Selnes OA, Vinters HV (2006) Vascular cognitive impairment. Nat Clin Pract Neurol 2:538–547

Small GW, Komo S, LaRue A, Saxena S, Phelps ME, Mazziotta JC, Saunders AM, Haines JL, Pericak-Vance MA, Roses AD (1996) Early detection of Alzheimer's disease by combining apolipoprotein E and neuroimaging. Ann N Y Acad Sci 802:70–78

Small GW, Kepe V, Ercoli LM, Siddarth P, Bookheimer SY, Miller KJ, Lavretsky H, Burggren AC, Cole GM, Vinters HV, Thompson PM, Huang SC, Satyamurthy N, Phelps ME, Barrio JR (2006) PET of brain amyloid and tau in mild cognitive impairment. N Engl J Med 355:2652–2663

Snowden JS, Neary D, Mann DMA, Goulding PJ, Testa HJ (1992) Progressive language disorder due to lobar atrophy. Ann Neurol 31:174–183

Soontornniyomkij V, Choi C, Pomakian J, Vinters HV (2010a) High-definition characterization of cerebral β-amyloid angiopathy in Alzheimer's disease. Hum Pathol 41:1601–1608

Soontornniyomkij V, Lynch MD, Mermash S, Pomakian J, Badkoobehi H, Clare R, Vinters HV (2010b) Cerebral microinfarcts associated with severe cerebral β-amyloid angiopathy. Brain Pathol 20:459–467

Terry RD, Masliah E, Salmon DP, Butters N, DeTeresa R, Hill R, Hansen LA, Katzman R (1991) Physical basis of cognitive alterations in Alzheimer's disease: synapse loss is the major correlate of cognitive impairment. Ann Neurol 30:572–580

van Swieten J, Spillantini MG (2007) Hereditary frontotemporal dementia caused by Tau gene mutations. Brain Pathol 17:63–73

Vinters HV (1987) Cerebral amyloid angiopathy. A critical review. Stroke 18:311–324

Vinters HV (2006) Neuropathology of amnestic mild cognitive impairment. Arch Neurol 63:645–646

Vinters HV (2007) Imaging cerebral microvascular amyloid. Ann Neurol 62:209–212

Vinters HV, Gilbert JJ (1983) Cerebral amyloid angiopathy: incidence and complications in the aging brain, II: the distribution of amyloid vascular changes. Stroke 14:924–928

Vinters HV, Pardridge WM, Yang J (1988) Immunohistochemical study of cerebral amyloid angiopathy. Use of an antiserum to a synthetic 28-amino-acid peptide fragment of the Alzheimer's disease amyloid precursor. Hum Pathol 19:214–222

Vinters HV, Nishimura GS, Secor DL, Pardridge WM (1990) Immunoreactive A4 and gamma-trace peptide co-localization in amyloidotic arteriolar lesions in the brains of patients with Alzheimer's disease. Am J Pathol 137:233–240

Vinters HV, Secor DL, Read SL, Frazee JG, Tomiyasu U, Stanley TM, Ferreiro JA, Akers MA (1994) Microvasculature in brain biopsy specimens from patients with Alzheimer's disease: an immunohistochemical and ultrastructural study. Ultrastruct Pathol 18:333–348

Vinters HV, Farrell MA, Mischel PS, Anders KH (1998) Diagnostic Neuropathology. Marcel Dekker, New York, pp 453–507

Vinters HV, Ellis WG, Zarow C, Zaias BW, Jagust WJ, Mack WJ, Chui HC (2000) Neuropathologic substrates of ischemic vascular dementia. J Neuropathol Exp Neurol 59:931–945

Weller RO, Boche D, Nicoll JAR (2009) Microvasculature changes and cerebral amyloid angiopathy in Alzheimer's disease and their potential impact on therapy. Acta Neuropathol 118:87–102

Wisniewski K, Jervis GA, Moretz RC, Wisniewski HM (1979) Alzheimer neurofibrillary tangles in disease other than senile and presenile dementia. Ann Neurol 5:288–294

Woulfe J, Gray DA, Mackenzie IRA (2010) FUS-immunoreactive intranuclear inclusions in neurodegenerative disease. Brain Pathol 20:589–597

Zhang-Nunes SX, Maat-Schieman MLC, van Duinen SG, Roos RAC, Frosch MP, Greenberg SM (2006) The cerebral β-amyloid angiopathies: hereditary and sporadic. Brain Pathol 16:30–39

Chapter 3
Preparation and Structural Characterization of Pre-fibrillar Assemblies of Amyloidogenic Proteins

Anat Frydman-Marom*, Yaron Bram*, and Ehud Gazit

Abstract Accumulating evidence supports the hypothesis that early, soluble, toxic oligomers, rather than the mature fibrils, relate to diverse amyloid disorders and may represent the primary cytotoxic agents in synaptic dysfunction and death in neurodegenerative diseases. Since the "amyloid cascade hypothesis" has been investigated for the amyloid β-protein (Aβ), many groups have reported toxic pre-fibrillar assemblies that are involved in diverse amyloid-related diseases. Much experimental evidence suggests that fibrils formed *in vitro* strongly resemble those in diseased tissues. For example, protofibrillar intermediates detected *in vitro* and later *in vivo* exhibit strikingly similar structural and neurotoxic properties. Taken together, these observations indicate that the structural and mechanistic evidences resulting from *in vitro* studies pertain to the role of protein fibrillogenesis in neurodegenerative diseases. Thus, extensive research has been devoted to produce *in vitro* oligomers that resemble the original species *in vivo* and to develop innovative methodologies to characterize the structure and biological activities of these oligomeric assemblies. In this chapter, we will discuss the methods used for structural characterization of oligomeric assemblies. In addition, we will review methods used for preparing different amyloid-like oligomers *in vitro*.

Keywords Amyloid • β-sheet • Fibrils • Oligomers • Structure

---

* Both authors contributed equally to this work.

A. Frydman-Marom (✉) • Y. Bram • E. Gazit
Department of Molecular Microbiology, Tel Aviv University, Green Building Room 124, Ramat Aviv, Israel

Biotechnology Department, Tel Aviv University, Green Building Room 124, Ramat Aviv, Israel
e-mail: maromana@post.tau.ac.il; bramyaron@gmail.com; ehudg@post.tau.ac.il

F. Rahimi and G. Bitan (eds.), *Non-fibrillar Amyloidogenic Protein Assemblies—Common Cytotoxins Underlying Degenerative Diseases*, DOI 10.1007/978-94-007-2774-8_3,

3.1 Pre-fibrillar Aβ Assemblies

Although it has been established that the aggregation process of amyloid β-protein (Aβ) plays a central role in Alzheimer's disease (AD), in recent years it has been debated whether the mature fibrils or rather earlier soluble intermediates in the fibrillization process represent the pathogenic species. Accumulating evidence has revealed a relatively weak correlation between neurological dysfunction and the density of fibrillar amyloid plaques (Terry et al. 1991). Moreover, cognitive impairment in transgenic mouse models of AD is observed before the appearance of amyloid plaques (Mucke et al. 2000). However, soluble Aβ levels strongly correlate with the extent of synaptic dysfunction, neuronal loss, dementia, and death (Lesné et al. 2006; Lue et al. 1999; McLean et al. 1999; Wang et al. 2002). These observations have led to the hypothesis that apparently soluble, pre-fibrillar protein assemblies, rather than mature fibrillar deposits, are the villain in AD and, by extension, in other amyloidogenic diseases (Walsh and Selkoe 2007; Kayed et al. 2003; Kirkitadze et al. 2002; Baglioni et al. 2006; Lansbury and Lashuel 2006).

In the past decade, extensive efforts have been directed toward identifying, isolating, and characterizing the oligomeric species that are present in solution prior to the appearance of fibrils, both because of their likely role in the mechanisms underlying fibril formation and because of their implication as the toxic species. Although oligomers are kinetic intermediates, it is not yet clear whether they form intermediates in the course of fibril formation (Harper et al. 1997b), or whether oligomers populate a different aggregation pathway that is distinct from the typical nucleation-dependent fibril-assembly pathway (Harper et al. 1997b; Dobson 2003; Necula et al. 2007; Gellermann et al. 2008). Therefore, the published data are ambiguous as to whether oligomers are intermediates in the pathway leading to fiber formation (Harper et al. 1997a; Serio et al. 2000) or whether they represent "off-pathway" aggregates that populate an alternative aggregation pathway (Gorman et al. 2003; Gosal et al. 2005; Morozova-Roche et al. 2004; Gellermann et al. 2008). The same debate can be extended to the aggregation of many other amyloidogenic proteins since many types of amyloids display the same type of kinetically unstable intermediates (Gosal et al. 2005; Grudzielanek et al. 2006; Morozova-Roche et al. 2004; Green et al. 2004).

Moreover, the literature describes many types of natural and synthetic assembly forms of Aβ that differ in size, shape, and biological activities (Haass and Selkoe 2007; Rahimi et al. 2008; Finder and Glockshuber 2007). In order to use the same jargon, we will define soluble Aβ oligomers as any form of Aβ that is soluble in aqueous buffer and remains in solution following high-speed centrifugation, as opposed to Aβ fibrils or aggregates that pellet following ultracentrifugation (Walsh et al. 1997; Walsh and Selkoe 2007). In the next section, we will provide an extensive account of the natural and synthetic oligomeric Aβ assemblies that were published in the literature (see Table 3.1).

Table 3.1 Natural or synthetic oligomeric Aβ assemblies reported in the literature

Oligomeric assembly	Characteristics	References
Aβ-derived diffusible ligands (ADDLs)	Globular structure Molecular mass of 17–42 kDa Diameter of ~3–8 nm Stable, do not convert to fibrils Cannot be produced from $A\beta_{1-40}$ but require $A\beta_{1-42}$	Lambert et al. (1998), Klein (2002), Chromy et al. (2003) and Younkin (1998)
Protofibril	Diameter of 6–10 nm and length up to 200 nm Soluble precursor of mature Aβ fibrils Rich in β-sheet structures Can be produced by both $A\beta_{1-40}$ and $A\beta_{1-42}$	Walsh et al. (1997), Harper et al. (1997a) and Walsh et al. (1999)
Aβ pores/annular assemblies	Channel-like structure Outer diameter of 8–12 nm and an inner diameter of 2–2.5 nm Composed of 4–8 monomers Rich in β-sheet structures $A\beta_{1-42}$ has a higher propensity to form channels than $A\beta_{1-40}$	Arispe et al. (1993a), Kawahara et al. (1997), Alarcon et al. (2006), Lashuel et al. (2003), Durell et al. (1994), Lin et al. (2001), Quist et al. (2005) and Lashuel et al. (2002a)
Paranuclei	Composed of pentamer/hexamer units Can further polymerize to form fibrils Cannot be produced from $A\beta_{1-40}$ but require $A\beta_{1-42}$	Bitan et al. (2003a)
β-Amyloid balls	Sphere morphology Diameters of ~20–200 μm Highly stable structures Produced from $A\beta_{1-40}$	Westlind-Danielsson and Arnerup (2001)
Amylospheroids	Sphere morphology Diameters of 10–15 nm Highly stable structures that do not convert into fibrils Can be produced by both $A\beta_{1-40}$ and $A\beta_{1-42}$	Hoshi et al. (2003) and Roychaudhuri et al. (2009)
Globulomers	Molecular mass of ~60 kDa, correlating to 12 $A\beta_{1-42}$ subunits Highly stable structures that do not convert into fibrils Composed of mixed parallel and antiparallel β-sheet structure Cannot be produced from $A\beta_{1-40}$ —only by $A\beta_{1-42}$	Barghorn et al. (2005) and Yu et al. (2009)

3.1.1 Pre-fibrillar Aβ Assemblies In Vitro

3.1.1.1 Aβ-Derived Diffusible Ligands (ADDLs)

Oda et al. were the first to report that clusterin (apoJ) partially blocks aggregation of synthetic $A\beta_{1-42}$, which results in Aβ complexes greater than 200 kDa (Oda et al. 1995). Later, Lambert and colleagues described clusterin-free preparation of Aβ-derived diffusible ligands (ADDLs) that have the same biochemical and neurotoxic characteristics as the species described above (Lambert et al. 1998). Clusterin-derived ADDLs, examined by atomic-force microscopy (AFM), are globular in nature, whereas, peptides prepared in the absence of clusterin tend to be more fibrillar in nature (Lambert et al. 1998; Klein 2002). The globe-shaped oligomers were shown to have a diameter of ~3–8 nm, with a molecular mass of 17–42 kDa. Further support for the low-mass oligomeric structure of ADDLs was provided by non-denaturing gel electrophoresis and SDS–PAGE, where bands were observed at predicted masses of 17 and 27 kDa (Klein 2002; Lambert et al. 1998). In addition, analyses by AFM and gel electrophoresis revealed that ADDLs are stable, independent entities rather than short-lived structures that rapidly convert into much larger assemblies such as protofibrils. Moreover, ADDLs have been shown to resist dissociation by low SDS concentrations (0.01%) (Oda et al. 1995). However, when supramicellar SDS concentrations were used, ADDLs and fibrils migrated with the same electrophoretic profile yielding monomeric, trimeric, and tetrameric moieties (Hepler et al. 2006). Chromy et al. characterized ADDLs preparations using AFM, non-denaturing electrophoresis, and size-exclusion chromatography (SEC). They identified two pre-fibrillar species with a high-mass component having a toxic effect on primary neurons, and a low-mass component around 13 kDa (Chromy et al. 2003). Another study in 2006 used SEC with multi-angle laser light scattering (SEC-MALLS) and analytical ultracentrifugation (AU) providing a much more accurate representation of the solution structure of ADDLs (Hepler et al. 2006). According to previous reports by Chromy et al., the low-molecular-mass component was composed of low-*n* oligomers. However, these components most likely represent artifacts induced by the peptide's interaction with detergent. In addition, they demonstrated that only the high-molecular-mass oligomeric components of an ADDLs preparation are capable of binding to subpopulations of primary hippocampal neurons *in vitro* (Hepler et al. 2006). Importantly, ADDLs have been shown to cause neuronal death and to block long-term potentiation (LTP) (Lambert et al. 1998; Wang et al. 2002; Dahlgren et al. 2002; Kim et al. 2003). Recently, small soluble Aβ oligomers, characterized as equivalent to synthetic ADDLs, have been found to accumulate in AD brains (Gong et al. 2003). Interestingly, ADDLs cannot be produced from $A\beta_{1-40}$ —only by $A\beta_{1-42}$ (Younkin 1998). Although $A\beta_{1-40}$ apparently can form oligomers using chemical cross-linking (Bitan et al. 2001; Klein 2002), apparently they are unstable and their formation appears to require a high peptide concentration. It is possible that formation of highly stable oligomers by $A\beta_{1-42}$ but not $A\beta_{1-40}$ is the underlying basis for the pathogenic role of $A\beta_{1-42}$.

ADDLs are prepared by dissolving Aβ_{1-42} in cold 1,1,1,3,3,3-hexafluoro-2-propanol (HFIP), incubated at room temperature for at least 1 h to establish monomerization, followed by HFIP removal by evaporation. The dry Aβ_{1-42} is then dissolved in 100% DMSO to a concentration of 5 mM and diluted into cold (4°C) phenol-red-free F12 cell-culture media to a final concentration of 100 μM. After dilution, the solution is mixed by a vortex, incubated at 4–8°C for 24 h, and centrifuged at 14,000×*g* for 10 min at 4°C. In addition, ADDLs can be generated by dissolving Aβ_{1-42} to 50 nM in clusterin-free brain-slice culture media at 37°C for 24 h (Lambert et al. 1998).

3.1.1.2 Protofibrils

Protofibrils (PFs) were first described by Walsh et al. and Harper et al. as direct, soluble precursors of mature Aβ fibrils (Walsh et al. 1997; Harper et al. 1997a, b). By using SEC, quasi-elastic light scattering (QLS), and transmission electron microscopy (TEM), Walsh et al. and Harper et al. demonstrated that both Aβ_{1-40} and Aβ_{1-42} can form PF assemblies. Structurally, PFs are characterized by regular β-sheet arrays with diameter of 6–10 nm and length of up to 200 nm (Walsh et al. 1997). An independent study by Harper et al. using AFM identified the same PFs as metastable intermediates formed during Aβ fibrillization. In addition, they revealed that their formation is strongly dependent on concentration, pH, and ionic strength (Harper et al. 1997a, b, 1999). As previously mentioned, PFs have high β-sheet content, as confirmed by circular dichroism (CD) and they can bind dyes such as Congo red and thioflavin T (Walsh et al. 1997, 1999), characteristics similar to those of mature amyloid fibrils. PFs act almost as true fibril intermediates in that they can rapidly convert into Aβ fibrils *in vitro* when seeded by small amounts of preformed fibrils (Harper and Lansbury 1997). Moreover, PFs can also dissociate to lower-molecular-weight Aβ species (Walsh et al. 1999). Another interesting fact that was determined using hydrogen–deuterium-exchange–mass spectrometry (HX-MS) is that the *C*-termini and *N*-termini in PFs are highly exposed to hydrogen exchange as opposed to their core, which is highly resistant (Kheterpal et al. 2000). An additional study revealed that Aβ fibrils are more structured in the core of the molecules in residues 22–29 as compared to the PFs (Williams et al. 2005). *In vitro* fibrillization studies showed that Aβ containing the E22G substitution related to the Arctic mutation formed PFs faster and in larger quantities than did wild-type Aβ (Nilsberth et al. 2001). By using EM, Lashuel et al. found that E22G forms annular protofibrils, which have an outer diameter of 6–9 nm, and an inner diameter of 1.5–2 nm; however, large spherical species with an average diameter of 18–24 nm were also observed (Lashuel et al. 2003). *In vitro*, PFs are potentially toxic (Hartley et al. 1999; Johansson et al. 2007) and several studies indicated that they are also neurotoxic species in AD *in vivo* (Harper and Lansbury 1997; Rochet and Lansbury 2000).

In the literature, different protocols to generate PFs were published:

1. Peptides are dissolved in 100% DMSO, then diluted with water and Tris (10 mM, pH 7.4) or phosphate buffer (10 mM NaH_2PO_4, 137 mM NaCl, and 27 mM KCl, pH 7.4) (Harper et al. 1999; Walsh et al. 1997).

2. Peptides are dissolved using NaOH because it readily dissolves the peptide, controls the pH, and reduces the aggregation rate. Briefly, peptide is dissolved at 4.2 mg/mL in 1 mM NaOH containing phenol red, followed by empirically adding 130–150 μL of 10 mM NaOH to bring the sample to pH 7–7.4 using the added phenol red as an indicator. The sample is then diluted with water and 10× phosphate-buffered saline (PBS), bringing the final peptide concentration to 500 μM in PBS (pH 7.4). Samples are then diluted to 100 μM in water (Hartley et al. 1999).
3. Peptides are dissolved in water to 88–143 μM, then mixed briefly by a vortex, and diluted with equal amounts of 50 mM Na_2HPO_4/NaH_2PO_4 (pH 7.4) containing 0.1 M NaCl (Nilsberth et al. 2001).

In all protocols the peptide solutions are centrifuged at 17,000*g* for 5 min to remove insoluble particles and then incubated for 16–60 h.

3.1.1.3 Aβ Pores/Annular Assemblies

In 1993, Arispe et al. were the first to demonstrate the ability of $A\beta_{1-40}$ to form channel-like structures in an *in vitro* lipid-bilayer system. They proposed that channel formation by Aβ is responsible for Aβ-induced toxicity in AD (Arispe et al. 1993a, b). This finding has been replicated many times, in several different laboratories, using many membrane models (Arispe et al. 1994; Kawahara et al. 1997; Alarcon et al. 2006; Lashuel et al. 2003). Based on the amino-acid sequence of Aβ peptide and experimental evidence of multilevel Aβ ion-channel conductance, Durell et al. proposed a theoretical model for the channel structure formed by multimeric Aβ peptide, with a subunit stoichiometry ranging from 4 to 8 monomers (Durell et al. 1994). Supporting results were obtained by incorporating Aβ into planar lipid bilayers and observations *via* AFM, which revealed multimeric channel-like structures with four and six apparent subunits (Fig. 3.1A) (Lin et al. 2001; Quist et al. 2005). Formation of annular, pore-like structures is promoted by a mutated form of Aβ (E22G) and this accelerates Aβ oligomerization *in vitro* (Fig. 3.1B) (Lashuel et al. 2003; Lashuel and Lansbury 2006; Nilsberth et al. 2001). It was also shown that $A\beta_{1-42}$ has a higher propensity to form channels than does $A\beta_{1-40}$ (Lashuel and Lansbury 2006). These annular assemblies of synthetic Aβ are doughnut-shaped structures having a centralized pore-like depression with an outer diameter of 8–12 nm, and an inner diameter of 2–2.5 nm (Lashuel et al. 2002b; Bitan et al. 2003a; Lin et al. 2001). This observation led to the channel hypothesis, according to which annular Aβ oligomers are toxic, causing membrane disruption, thus leading to disruption of cellular ionic homeostasis (Kagan et al. 2002; Lashuel et al. 2002a; Quist et al. 2005; Lin et al. 2001). A recent study by Jang et al. modeled the Aβ ion channels of different sizes (12- to 36-mers) in the lipid bilayer using molecular dynamics (MD) simulations. Their study indicated that the channels are formed by β-sheets that spontaneously break into loosely interacting dynamic units that associate and dissociate, leading to a toxic ionic flux (Jang et al. 2009). Recently, Kayed et al. reported that the toxicity of Aβ annular assemblies is related to their

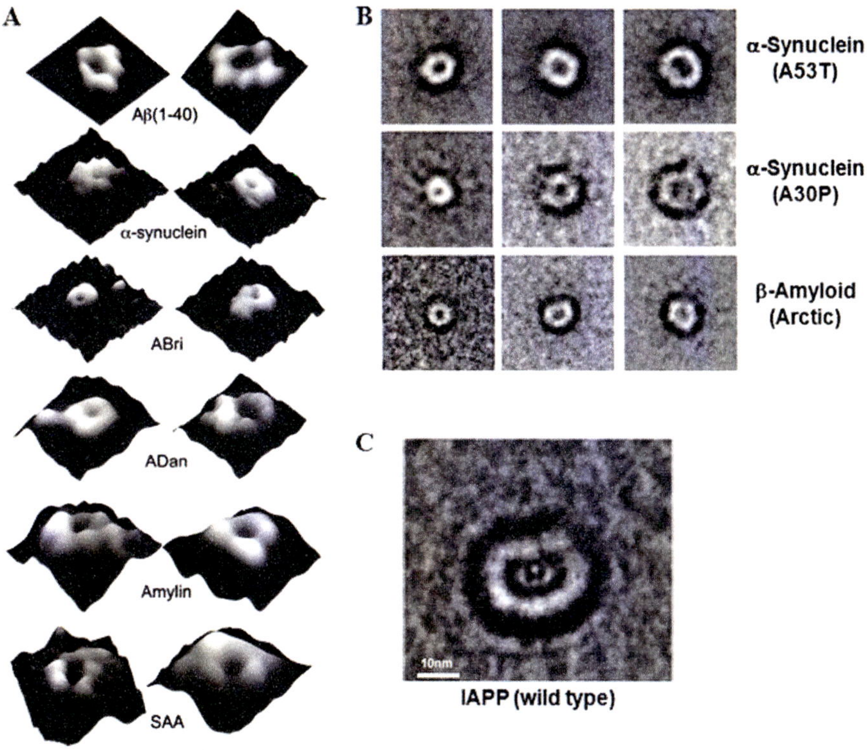

Fig. 3.1 Common annular structures formed by amyloidogenic peptides of unrelated origin. (**A**) AFM pictures of channel-like structures of different amyloid peptides: Aβ, α-synuclein, ABri, and ADan (which have been implicated in familial British and Danish dementia), amylin (IAPP), and SAA (Quist et al. 2005). (**B**) TEM pictures showing the annular assemblies of mutants of α-synuclein and Aβ (Lashuel et al. 2002b). (**C**) Field Emission gun transmission electron microscopy (FEG-TEM) allowed a clear visualization of the annular assemblies by human IAPP (Porat et al. 2003; Gazit 2004)

ability to form membrane-permeabilizing, β-barrel pores, a mode of action that resembles Gram-negative bacteria toxins (Kayed et al. 2009). In addition, different studies showed that soluble oligomers of Aβ peptides increase the conductance of planar lipid bilayers and increase calcium entry into cells, independent of discrete channel formation, pore formation, or ion selectivity. The conductance depends on the concentration of oligomers and it can be reversed by an anti-oligomer antibody (Sokolov et al. 2006; Demuro et al. 2005; Kayed et al. 2004).

Preparation of annular Aβ assemblies incorporated into lipid bilayers has been described by different protocols. First, $A\beta_{1-42}$ is dissolved in chloroform and mixed with 1,2-dioleolyl-sn-glycero-3-phosphatidyl-choline (DOPC) in chloroform at a 4:100 molar ratio. Next, the mixture is dried with argon and resuspended in 10 mM HEPES (pH 7.4) at 1 mg/mL. The lipid–$A\beta_{1-42}$ mixture is then bath sonicated for 20 min to form a liposome–lipid bilayer (Lin et al. 2001). In other studies, $A\beta_{1-40}$

was dissolved in water to form a 0.46-mM stock solution and then incorporated into an artificial membrane (Arispe et al. 1993a, b). The oligomers that do not cause channel pores are prepared by dissolving 1.0 mg Aβ in 400 μL HFIP for 10–20 min at room temperature. Next, 100 μL of the resulting seed-free Aβ solution is added to 900 μL ddH_2O in a siliconized Eppendorf tube. After 10–20 min incubation at room temperature, the samples are centrifuged for 15 min at 14,000 *g* and the supernatant fraction (pH 2.8–3.5) is transferred to a new siliconized tube and subjected to a gentle stream of N_2 for 5–10 min to evaporate the HFIP. The samples are then stirred at 500 RPM using a Teflon-coated micro stir bar for 24–48 h at 22°C (Kayed et al. 2004).

3.1.1.4 Paranuclei

Studies of fibril-formation kinetics have shown that $A\beta_{1-42}$ forms fibrils significantly faster than does $A\beta_{1-40}$ (Jarrett et al. 1993), leading to the popular view that $A\beta_{1-42}$ is more amyloidogenic than $A\beta_{1-40}$ (Teplow 1998). Bitan and co-workers, using photo-induced cross-linking of unmodified proteins (PICUP), SEC, dynamic light scatter (DLS), CD spectroscopy, and EM demonstrated that in contrast to $A\beta_{1-40}$, $A\beta_{1-42}$ preferentially forms pentamer/hexamer units, termed paranuclei, which can further polymerize to form large oligomers, protofibrils, and fibrils. The fact that at similar concentrations, paranuclei were not observed for $A\beta_{1-40}$ provides a possible explanation for the distinct biological activity of oligomeric preparations of the two Aβ alloforms. Moreover, it was found that the critical residue promoting the initial oligomerization of $A\beta_{1-42}$ is Ile-41, because addition of Ile-41 to $A\beta_{1-40}$ is sufficient to induce formation of paranuclei (Bitan et al. 2003a). Bitan et al. showed that oligomerization of $A\beta_{1-40}$ involves a rapid equilibrium among monomer, dimer, trimer, and tetramer, and that it is regulated by the peptide *N*-terminus and charged residues at positions 22–23. In the case of paranuclei formation by $A\beta_{1-42}$, the controlling elements are the *C*-terminal dipeptide and the central hydrophobic cluster (CHC). Paranuclei formation depends on the presence of a hydrophobic side-chain in amino acid 41 with a size at least as large as that of a methyl group. In addition, the self-association step requires presence of residue 42 and is substantially facilitated by replacing the *C*-terminal carboxyl group by a carboxamide (Bitan et al. 2003c). Oxidation of Met35 in $A\beta_{1-42}$ blocks paranuclei formation and produces oligomers indistinguishable in size and morphology from those produced by $A\beta_{1-40}$ (Bitan et al. 2003b). $A\beta_{1-40}$ (E22G) forms paranuclei with a propensity similar to that of $A\beta_{1-42}$ (Urbanc et al. 2010). These results reveal that specific regions and residues that control Aβ oligomerization differ between $A\beta_{1-40}$ and $A\beta_{1-42}$ and that the strong etiologic association of $A\beta_{1-42}$ with AD may thus result from assemblies formed at the earliest stages of peptide oligomerization (Bitan et al. 2003a; Kirkitadze and Kowalska 2005).

Aβ paranuclei are prepared by dissolving the peptide into a 2-mg/mL peptide solution in DMSO and sonicating for 1 min, followed by centrifugation for 10 min at 16,000 *g*. The resulting supernatant is then isolated by SEC using a 10/30 Superdex

75 HR column eluted at 0.5 mL/min with 10 mM sodium phosphate, pH 7.4 (Bitan et al. 2003a). Freshly isolated low-molecular-weight (LMW) peptides are immediately subjected to PICUP (Bitan et al. 2001), where 1 μL of 1 mM Ru(Bpy) and 1 μL of 20 mM APS in 10 mM sodium phosphate, pH 7.4, are added to 18 μL of freshly isolated LMW peptide. The mixture is then irradiated with visible light and the reaction is quenched immediately with 10 μL Tricine sample buffer containing 5% β-mercaptoethanol (β-ME).

3.1.1.5 β-Amyloid Balls

β-Amyloid balls (βamy balls) are formed spontaneously in a cell-free system and under physiological conditions by $A\beta_{1-40}$ (300–600 μM). βamy balls' spherical morphology was determined by TEM, showing diameters of ~20–200 μm. These supramolecular structures exhibit weak birefringence with Congo-red staining but have high stability with prolonged incubation times at 30°C, freezing, and dilution in H_2O. $A\beta_{1-42}$ lacks the ability to form βamy balls; however, it accelerates $A\beta_{1-40}$ βamy ball formation at low stoichiometric levels. It was shown that $A\beta_{1-40}$ (E22G) has the ability to form βamy balls as well (Westlind-Danielsson and Arnerup 2001). Interestingly, an independent study by Anderson et al. revealed *in vivo* extracellular retinal deposits, termed drusen (Aβ-containing macromolecular assemblies and are a pathologic sign of age-related macular degeneration), which have an apparent similarity to βamy balls (Anderson et al. 2004).

The βamy balls are formed by dissolving $A\beta_{1-42}$ in ddH_2O and then diluting with an equal volume of 2× PBS. The final PBS concentration is 50 mM Na_2HPO_4, NaH_2PO_4 (pH 7.4), and 0.1 M NaCl (0.02% w/v NaN_3). The final peptide concentration is 60–600 μM. Samples (150 μL) are incubated at 30°C without agitation, in the dark, for various periods of time. Incubation is stopped by centrifuging samples in a fixed-angle rotor at 14,900 *g* for 10 min at 16°C (Westlind-Danielsson and Arnerup 2001).

3.1.1.6 Amylospheroids

Amylospheroids (ASPDs) were first described by Hoshi et al. as stable and highly toxic Aβ moieties (Hoshi et al. 2003). Examination of the aggregates purified by glycerol-gradient centrifugation by AFM and TEM revealed that the toxic moiety is a perfect sphere with diameters ranging from 10 to 15 nm. ASPDs are off-pathway spheroidal structures that are formed by both $A\beta_{1-40}$ and $A\beta_{1-42}$. It was shown that the yield of isolated ASPDs is quite low, around 7.3% of toxic ASPDs > 10 nm and ~12.1% of nontoxic ASPDs < 10 nm. As mentioned before, the ASPDs are very stable, it was shown that they were stable for more than 2 months at 4°C and did not convert to fibrils even after several days with slow rotation. Toxicity was correlated with sphere size, where 10–15 nm ASPDs were highly toxic, and ASPDs < 10 nm were nontoxic. $A\beta_{1-42}$-derived ASPDs formed more rapidly, damaged neurons at lower concentrations, and exhibited 100-fold higher toxicity than did $A\beta_{1-40}$-derived ASPDs (Hoshi et al. 2003; Roychaudhuri et al. 2009). In a recent work, Noguchi et al. reported the selective immunoisolation of neurotoxic native ASPDs, 10–15-nm, spherical Aβ assemblies from patient brains with AD and dementia with Lewy bodies, using tertiary-structure-dependent antibodies against ASPDs. The native ASPDs are above 100 kDa; thus, they are larger in mass than other reported assemblies (Noguchi et al. 2009). Moreover, they are A11-negative (Kayed et al. 2003) and probably have a distinct surface tertiary structure. It was suggested that the native ASPDs have a distinct toxic surface that binds presynaptic targets on mature neurons, consequently causing neuronal death (Noguchi et al. 2009).

ASPDs are prepared by dissolving synthetic Aβ in ultra-pure water to a concentration of 700 μM. Thereafter, they are incubated at 4°C for 30 min, and then diluted with filtered Dulbecco's PBS without Ca^{2+} and Mg^{2+} to a concentration of 350 μM. Aβ solution (350 μM) is rotated slowly at 37°C for 5–7 days by using a rotating cultivator.

3.1.1.7 Globulomers

Barghorn et al. first described $A\beta_{1-42}$ globulomer (short form for globular oligomer) as a highly stable group of oligomers that are water soluble and have a mass of ~60 kDa, correlating to 12 $A\beta_{1-42}$ subunits (Barghorn et al. 2005). According to the proposed model, the hydrophobic *C*-termini of $A\beta_{1-42}$ globulomers form the inside core, whereas the hydrophilic *N*-terminal residues 19–23 are exposed at the outer surface, making the globulomer highly water soluble. $A\beta_{1-42}$ globulomer interactions are exclusively non-covalent; thermal denaturation reverted $A\beta_{1-42}$ globulomer to monomeric $A\beta_{1-42}$. Nevertheless, the non-covalently linked globulomer exhibited long-term stability at physiological temperatures without obvious disassembly or further polymerization to fibrils. In addition, globulomers were prepared with 95% purity and with only 5% cross-contamination of monomeric $A\beta_{1-42}$ (Barghorn et al. 2005). Furthermore, CD spectroscopy revealed a strong minimum at 216 nm, suggesting that a substantial fraction of the $A\beta_{1-42}$ globulomer folds in the β-structure conformation, a common characteristic of amyloid proteins in their pathological form (Rochet and Lansbury 2000). In a more recent study, using solution nuclear magnetic resonance (NMR), it was found that globulomers have a mixed parallel and antiparallel β-sheet structure that differs from fibrils containing only parallel β-sheets (Yu et al. 2009). In accordance with their results, Yu et al. proposed a model in which $A\beta_{1-42}$ globulomers populate an independent pathway of aggregation with a unique conformation of Aβ polypeptide distinct from that of Aβ fibrils (Barghorn et al. 2005; Gellermann et al. 2008). In a standard protocol, they used sodium dodecyl sulfate (SDS) to induce $A\beta_{1-42}$ globulomer formation; however, they demonstrated that globulomers with the same characteristics were produced by incubation with lauric, oleic, or arachidonic acids, a hint for their physiological role.

Globulomers bind to neurons and specifically suppress spontaneous synaptic activity resulting from a reduction of vesicular release at terminals of both

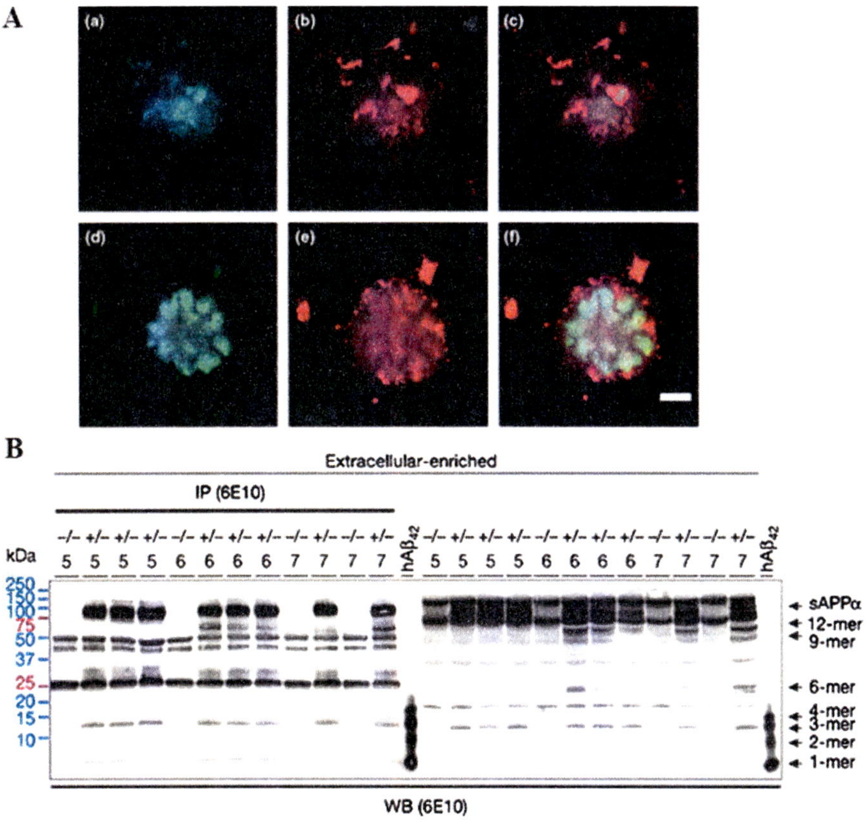

Fig. 3.2 Detection of amyloid β oligomers in *in vivo* studies. (**A**) Immunohistochemistry of cortical samples from transgenic mice and patients with Alzheimer's disease. Staining of amyloid plaques with thioflavin S (*a, d*) and specific anti-Aβ globulomer antibodies (*b, e*), double staining (*c, f*). *a–c* are cortical sections from Alzheimer's disease patients, *d–f* are transverse sections of the cortex of a 12-month-old Tg2576 mouse. Both sections show the presence of amyloid plaque and the presence of oligomers presenting globulomer epitopes both in model mice and in Alzheimer's disease patients (Barghorn et al. 2005). (**B**) SDS–PAGE analysis of Aβ oligomers in soluble, extracellular-enriched extracts of proteins from brains of 5-, 6-, and 7-month-old mice (age indicated above lanes), assessed by western blot (WB) with or without immunoprecipitation (IP), −/− indicates wild-type mice and −/+ indicates APP transgenic mice. In these models, mice less than 6 months old have normal memory and lack neuropathology, but from 6 months they develop memory deficits without neuronal loss. By using PAGE analysis and western blots, extracellular accumulation of a 56-kDa soluble Aβ assembly was detected around the time that memory deficits manifested (Lesné et al. 2006)

GABAergic and glutamatergic synapses (Barghorn et al. 2005; Nimmrich et al. 2008). Globulomers are present in the brains of AD patients and in $A\beta_{1-42}$-overproducing transgenic mice (Fig. 3.2A) (Barghorn et al. 2005). Comparable globular structures of $A\beta_{1-40}$ could be formed after an 18-h incubation in 25 mM 2-morpholinoethanesulfonic acid buffer (pH 4.5) in a "hanging-drop" environment (Moore et al. 2007).

The Aβ_{1-42} globular oligomers can be easily and reproducibly generated from synthetic Aβ. Briefly, synthetic Aβ_{1-42} is suspended in 100% HFIP at 6 mg/mL and is incubated for complete solubilization with shaking at 37°C for 1.5 h in order to eliminate pre-existing structural inhomogeneities in the Aβ. HFIP is removed by evaporation in a SpeedVac and Aβ_{1-42} is resuspended at a concentration of 5 mM in DMSO for 20 s. It is then diluted in PBS (20 mM NaH_2PO_4, 140 mM NaCl, pH 7.4) to 400 μM and 1/10 volume 2% SDS. An incubation for 6 h at 37°C results in the 16-/20-kDa Aβ_{1-42} globulomer intermediate. The 38-/48-kDa Aβ_{1-42} globulomers are generated by a further dilution with three volumes of H_2O and incubation for 18 h at 37°C. After centrifugation at 3,000 *g* for 20 min, the sample is concentrated by ultrafiltration (30-kDa cut-off), dialyzed against 5 mM NaH_2PO_4, 35 mM NaCl, pH 7.4, centrifuged at 10,000 *g* for 10 min and the supernatant containing the 38-/48-kDa Aβ_{1-42} globulomer is withdrawn. As an alternative to dialysis, the 38-/48-kDa Aβ_{1-42} globulomer can also be precipitated by a ninefold excess (v/v) of ice-cold methanol/acetic acid solution (33% methanol, 4% acetic acid) for 1 h at 4°C. The 38-/48-kDa Aβ_{1-42} globulomer is then pelleted (10 min at 16,200 *g*), resuspended in 5 mM NaH_2PO_4, 35 mM NaCl, pH 7.4, and then the pH adjusted to 7.4 (Barghorn et al. 2005).

3.1.2 Pre-fibrillar Aβ Assemblies In Vivo

3.1.2.1 Secreted Soluble Aβ Dimers and Trimers

Detection of small amounts of SDS-stable low-*n* oligomers (dimer, trimer) from culture media of APP-expressing Chinese hamster ovary (CHO) cells was described initially by Podlisny et al. (1995, 1998). Such SDS-stable oligomers have also been detected *in vivo* in human cerebrospinal fluid (CSF) (Walsh et al. 2000; Vigo-Pelfrey et al. 1993) and by western blotting in APP transgenic mouse brain and human brain (Funato et al. 1999; Roher et al. 1996; Kawarabayashi et al. 2004; Mc Donald et al. 2010). Dimers isolated from human brains appear to have a very stable conformation and have a prolate ellipsoid shape with an equatorial diameter of 3–8 nm according to AFM (Roher et al. 1996). In addition, molecular modeling showed that the dimer structure displays a hydrophobic core surrounded by hydrophilic residues creating shallow crevices into which the non-polar *C*-termini are folded (Chaney et al. 1998). *In vitro* experiments, using fluorescence resonance energy transfer (FRET) and gel-filtration chromatography, revealed formation of stable dimers of Aβ_{1-40} in low concentrations. According to Garzon-Rodriguez et al., the dimers formed in solution represent the initial event in amyloid aggregation and probably represent the fundamental building block for further fibril assembly (Garzon-Rodriguez et al. 1997). Oligomers with similar sizes have been shown to inhibit LTP *in vitro* (Walsh et al. 2002). The high toxicity of low-*n* Aβ oligomers is also supported by *in vitro* studies where Aβ dimers were shown to be threefold more toxic than monomers (Ono et al. 2009). It has also been shown that secreted

Aβ oligomers are resistant to SDS and to the Aβ-degrading protease, insulin-degrading enzyme, which degrades Aβ peptides (Walsh et al. 2002).

3.1.2.2 Aβ*56

Recently, Lensé et al. demonstrated that memory deficits in middle-aged Tg2576 mice (6–14-month-old mice that develop memory deficits without neuronal loss) are caused by extracellular accumulation of a 56-kDa dodecameric, soluble Aβ assembly, termed Aβ*56 (Fig. 3.2B). Moreover, reintroduction of the purified Aβ*56 from brains of impaired Tg2576 mice disrupts memory when administered to young rats (Lesné et al. 2006). It has also been shown that intracerebroventricular injection of Aβ*56 into rats tested under the cognition test—Alternating Lever Cyclic Ratio (ALCR)—induced concentration-dependent cognitive impairment (Reed et al. 2011). In addition, longitudinal water-maze spatial training significantly improves subsequent learning performance and reduces the amount of Aβ*56 and tau neuropathology (Billings et al. 2007). Memantine (*N*-methyl-*D*-aspartate receptor antagonist that is approved for treating moderate to severe AD) treatment in 3×Tg-AD mice decreased the concentration of dodecameric Aβ*56 by approximately 70% and improved cognition (Martinez-Coria et al. 2010). Taken together, these findings suggest that Aβ*56 impairs memory and may contribute to cognitive deficits associated with AD (Lesné et al. 2006).

3.2 Pre-fibrillar Assemblies of Disease-Associated Amyloidogenic Proteins Other Than Aβ

More than 20 human proteins form amyloid fibrils and oligomers *in vivo* (Stefani and Dobson 2003). Interestingly, different amyloid oligomers are recognized by the same antibody, suggesting that they display a common conformation-dependent structure that is unique to many oligomers regardless of sequence (Kayed et al. 2003). The shared conformational epitopes may be involved in pathogenesis and may suggest a shared mechanism (Fig. 3.1) (Lashuel and Lansbury 2006). The structures reported for different amyloidogenic oligomers resemble those described earlier for Aβ, such as annular pore-like PFs and spherical oligomers. In the next section we will describe several disease-associated amyloidogenic proteins that form oligomers *in vitro* and *in vivo*.

3.2.1 Pre-fibrillar α-Synuclein Assemblies

Similar to AD, in Parkinson's disease (PD), soluble α-synuclein oligomers rather than mature fibrils are suspected to be the pathological villains (Goldberg and Lansbury 2000; Caughey and Lansbury 2003). It was shown that PD patients have high levels of soluble α-synuclein oligomers in their plasma (El-Agnaf et al. 2006). Transgenic mice that overexpress human α-synuclein become symptomatic but do not produce fibrillar deposits (Masliah et al. 2000). α-Synuclein belongs to a class of natively unfolded proteins (Weinreb et al. 1996), a property that makes α-synuclein more prone to self-assembly (Uversky and Fink 2004; Uversky 2008). According to AFM analysis, α-synuclein oligomerization results in spherical, chain-like, and annular morphologies (Conway et al. 2000a, b; Ding et al. 2002). Mutations linked to early-onset PD (A30P, A53T) were shown to accelerate α-synuclein oligomerization (Conway et al. 2000b), where the A30P variant was observed to promote formation of annular, pore-like protofibrils. Another variant, A53T, promotes formation of annular and tubular protofibrillar structures (Ding et al. 2002; Conway et al. 2000a, b; Lashuel et al. 2002b; Goldberg and Lansbury 2000). Wild-type α-synuclein also forms annular protofibrils, but only after extended incubation. CD analysis revealed that α-synuclein protofibrils contain β-sheet structures; TEM revealed annular species with diameters of 8–12 nm and inner diameters of 2.0–2.5 nm (Lashuel et al. 2002b).

Recently it was shown that α-synuclein can form soluble, SDS-resistant oligomers in the presence of dopamine, in addition to its ability to dissociate preformed α-synuclein fibrils into soluble non-ordered aggregates (Li et al. 2004; Conway et al. 2001; Norris et al. 2005). The formed oligomers are highly stable towards SDS denaturation, which supports a covalent modification of α-synuclein (Conway et al. 2001). The oligomers formed in the presence of dopamine can form short non-ordered fragments. These aggregates lack a defined secondary structure that directs α-synuclein down a separate folding pathway (Cappai et al. 2005). However, it was shown that α-synuclein oligomerization depends on pH and that pH 4.0 promotes formation of SDS-resistant, insoluble oligomers that further associate to form sheet-like assemblies of fibrils (Pham et al. 2009). Previous work demonstrated that dopamine and its reactive intermediates oxidize all Met residues in monomeric α-synuclein, which with time associate to form α-synuclein oligomeric species (Conway et al. 2001; Li et al. 2004; Norris et al. 2005; Cappai et al. 2005; Leong et al. 2009a, b). A recent paper by Rekas et al. presented SAXS and CD data showing that Met-oxidized α-synuclein monomers are elongated worm-like structures, similar to monomeric untreated α-synuclein lacking significant secondary structure elements, whereas α-synuclein dimers, and trimers appeared to contain more β-sheet and turn structures (Rekas et al. 2010).

α-Synuclein protofibrils bind very strongly to vesicle membranes and cause leakage of small compounds entrapped within synthetic vesicles (Volles et al. 2001; Volles and Lansbury 2002). This typical pore-like behavior was consistent with the observation that addition of spherical protofibrils of α-synuclein to purified brain-derived vesicle fractions results in the formation of pore-like structures (Ding et al. 2002). Additionally, reconstitution of α-synuclein in lipid bilayers also results in the formation of pore-like structures that exhibit channel-like properties (Quist et al. 2005). Tryptophan fluorescence spectroscopy revealed that the negatively charged *C*-termini of oligomers are the most solvent-exposed part of the protein and that the *N*-termini are critical in oligomer–lipid-binding interactions (van Rooijen et al. 2009).

It was demonstrated that polyunsaturated fatty acids that interact with α-synuclein both *in vitro* and *in vivo* accelerate production of oligomers, whereas saturated fatty acids decrease it (Sharon et al. 2003). In a recent paper, Hong et al., using CD, FTIR, SEC-HPLC, and AFM, demonstrated that flavonoid baicalein inhibits α-synuclein fibrillization and induces formation of α-synuclein oligomers (Hong et al. 2008). The preformed, partially structured oligomers are composed of β-sheet and are characterized as compact globular species that are highly stable. Baicalein-stabilized oligomers have a mild effect on membrane integrity and permeability, similar to the effect of α-synuclein monomers (Hong et al. 2008). The structural features of α-synuclein annular protofibrils resemble the bacterial pore-forming toxins. This may explain the membrane-permeabilization activity of α-synuclein protofibrils and its contribution to PD pathogenesis (Ding et al. 2002; Lashuel et al. 2002a).

Preparation of α-synuclein oligomers is as follows: lyophilized recombinant α-synuclein is dissolved in PBS (0.01 M sodium phosphate buffer (pH 7.4), 150 mM NaCl) to obtain concentrations of 300–700 mM. The stock solution is then incubated on ice for 30–60 min before being centrifuged at 16,000 *g* for 5 min, and filtered through a 0.22-μm nylon spin filter to remove insoluble particles (Lashuel et al. 2002b).

In a different protocol, oligomers were prepared by dissolving lyophilized protein in 50 mM sodium-phosphate buffer, pH 7.0, containing 20% ethanol to a final concentration of 7 μM. After 4 h of shaking, the oligomers were re-lyophilized and resuspended with one-half of the starting volume in 50 mM sodium-phosphate buffer, pH 7.0, containing 10% ethanol. This was followed by shaking the oligomers for 24 h at room temperature with open lids to evaporate residual ethanol, accompanied by 6 days of incubation with closed lids (Danzer et al. 2007).

3.2.2 *Pre-fibrillar Oligomers of Prion Proteins (PrP)*

The prion diseases, which include Creutzfeldt–Jakob disease and bovine spongiform encephalopathy are characterized by the presence of an abnormal form of PrP in the brain, termed PrP^{Sc}. PrP^{Sc} fibrils, like in other amyloid diseases, may not be the neurotoxic form of the protein (Cohen and Prusiner 1998). Transgenic mouse models that overproduce a disease-associated form of PrP become symptomatic before PrP^{Sc} can be detected (Chiesa and Harris 2001). In yeast and mouse models of prion diseases, toxicity was produced in the absence of the PrP^{Sc} (Ma and Lindquist 2002). Caughey et al. suggested that those particles with the highest specific infectivity and specific converting activity are protofibrillar PrP particles with molecular mass ranging from 300 to 600 kDa (corresponding to 14–28 PrP molecules), roughly spherical to elliptical in shape, and 14–28 nm in diameter (Silveira et al. 2005; Caughey et al. 2009). PrP^{Sc} fragment (residues 89–231) oligomerization in an acidic pH results in off-pathway β-sheet-rich oligomers (Baskakov et al. 2002; Redecke et al. 2007). In addition, when FTIR and CD were used, variants of PrP (recombinant Syrian hamster prion protein) were shown to form an arranged antiparallel β-sheet structure; TEM revealed a spherical and annular structure (Sokolowski et al. 2003). It was proposed that the pathogenicity of the PrP may be related to abnormal pore formation like in other neurodegenerative diseases (Lashuel et al. 2002a). It has been shown that PrP fragments can interact with membranes and form ion channels (Kourie et al. 2003; Bahadi et al. 2003; Lin et al. 1997; Lashuel and Lansbury 2006), disrupt membranes (Pillot et al. 1997), and can induce toxicity in rat cortical neurons (Pillot et al. 2000).

3.2.3 Pre-fibrillar Oligomers of ABri in Familial British Dementia

ABri is the major component of amyloid deposits in the brains of patients with familial British dementia (Ghiso et al. 2000). El-Agnaf and co-workers demonstrated that the intramolecular disulfide bond and *C*-terminal extension are required for forming oligomeric, amyloid-like β-sheet structures and that the ABri peptide induces apoptotic cell death, whereas the wild-type ABri is non-toxic to cells (El-Agnaf et al. 2001a, b). Using AFM, Srinivasan et al. showed that in the initial step, ABri produces spherical aggregates (0.4–1.5 nm) that act as building blocks and combine into beaded chain-like protofibrils (1.5–2.3 nm) and/or annular structures (1.5–2.3 nm height). Once produced, the chain-like protofibrils can undergo further assembly to produce mature fibrils and ring-like structures (Srinivasan et al. 2004, 2003).

ABri oligomers were prepared by dissolving the peptide (5 mg/mL) in filter-sterilized 100-mM Tris-HCl (pH 9). This solution is aged at 37°C for 3 weeks. Then, in order to isolate the protofibrils, 200 μL of aged ABri solution is centrifuged at 16,000 *g* for 10 min, and an aliquot of the supernatant is fractionated by SEC (El-Agnaf et al. 2001a, b).

3.2.4 Pre-fibrillar Oligomers of Islet Amyloid Polypeptide (IAPP)

IAPP (also known as amylin) is the protein component of amyloid deposits in type-2 diabetes. IAPP is a 37-amino-acid peptide hormone, packaged and secreted with insulin by pancreatic β-cells in secretory granules (Westermark et al. 1990). In this disease as well as others described earlier, pre-fibrillar oligomers may be the pathogenic species (Gurlo et al. 2010). IAPP can generate β-sheet-rich oligomeric species *in vitro*, can act as nonselective ion channels, and disrupt membranes by a pore-like mechanism (Anguiano et al. 2002; Janson et al. 1999), a characteristic that decreases with further aggregation into larger fibrillar deposits (Porat et al. 2003). It was shown that interaction of IAPP with biological membranes may induce a transient α-helical conformation in IAPP, presumably facilitating penetration of the oligomers into the membrane, resulting in solute leakage across the membrane (Knight et al.

2006). According to high-resolution TEM, IAPP forms spheroid assemblies of 15–20 nm in diameter (Porat et al. 2003) (Fig. 3.1C).

One of the protocols for IAPP oligomerization includes dissolving IAPP in HFIP to a concentration of 0.5 mg/mL and thereafter sonicating it for 2 min. The solution is then freeze-dried overnight. The lyophilized product is then dissolved in water to 0.36 mg/mL and spun for 10 min at 20,000 *g*. Aggregation experiments were initiated by diluting the peptide stock solution into buffer and quickly filtering the mixture through a 0.22-μm filter. Final peptide concentrations were 28–34 μM (based on amino acid analysis), and the buffer composition was 10 mM Tris-HCl (pH 7). Samples were incubated at either room temperature or 37°C (Anguiano et al. 2002).

Second protocol: Synthetic hIAPP was dissolved in HFIP to 1.95 mg/mL and diluted to a final concentration of 5 μM in 10 mM sodium acetate buffer (pH 6.5), and to a final HFIP concentration of 1%. Immediately after dilution, and every 30 min, 1-mL aliquots were transferred to a microtube and centrifuged for 15 min at 20,000 *g* at 4°C. Thereafter, the supernatant fractions were transferred to another tube, and pellet fractions were gently resuspended in the remaining 0.4 mL. The most active membrane-reactive pre-fibrils assembled in the hIAPP solution after approximately 1 h (Porat et al. 2003).

3.2.5 *Pre-fibrillar Oligomers of Polyglutamine*

Huntington's disease (HD) is an inherited, neurodegenerative disorder resulting from an expanded polyglutamine [poly(Q)] region in the *N*-terminus of huntingtin. HD patients have a stretch of 36 or more glutamine residues and the disease age-of-onset inversely correlates with the length of the expanded poly(Q) region (Gusella and MacDonald 2000). The presence of fibrillar huntingtin aggregates does not correlate well with neuronal death (Sanchez et al. 2003; Lashuel and Lansbury 2006), suggesting that these aggregates may not be the toxic species. Poirier et al. demonstrated that unstructured poly(Q) adopts a β-structure *via* conformational changes and forms globular assemblies with a diameter of 4–5 nm, which over time can associate linearly to form single protofibrils (Poirier et al. 2002). Accumulating evidence suggests that membrane disruption is mediated by direct interaction with polyglutamine repeats. It was shown that poly(Q) can form cation-selective channels when incorporated into artificial planar lipid bilayer membranes where the appearance of the channel depends critically on the length of polyglutamine chains. Ion channels were observed with 40-residue stretches, whereas no significant conductance changes were detected with a 29-residue polyglutamine (Monoi et al. 2000). Kayed and co-workers reported that a polyglutamine protein forms homogeneous annular pore-like protofibrillar structures *in vitro* and specifically increases lipid-bilayer conductance (Kayed et al. 2004). Takahashi et al., using FRET in living cells, reported that detectable soluble poly(Q) oligomers are more toxic to neuronally differentiated SH-SY5Y cells than poly(Q) monomers or fibrils

(Takahashi et al. 2008). Monoi and colleagues proposed that a single chain of poly(Q) polypeptide is capable of forming cylindrical pores by forming a right-handed helix, termed μ-helix, which is further stabilized by backbone, side-chain hydrogen-bonding interactions between the amide groups and glutmaine side-chains (Monoi et al. 2000; Monoi 1995; Lashuel and Lansbury 2006).

3.3 Pre-fibrillar Assemblies of Other Non-Diseases-Associated Amyloidogenic Protein

The ability to undergo amyloid-like fibrous aggregation is not restricted to proteins associated with amyloid diseases. Many proteins are able to aggregate *in vitro* into fibrils under special excipient conditions, suggesting that the amyloid fibril is an intrinsically stable structure (Dobson 2001). One advantage of studying aggregation using proteins whose folding and unfolding pathways have been well-characterized is to understand better key early steps in protein misfolding and aggregation, which are the conformational changes that proteins undergo when partially destabilized. One well-characterized protein is insulin, which is not associated with disease, but it has been studied by multiple groups as a convenient *in vitro* model (Murali and Jayakumar 2005; Nettleton et al. 2000).

Several biophysical methods have identified at least two major populations of oligomeric intermediates of insulin between the native monomer and fibrils. Both have significantly non-native conformations (Ahmad et al. 2005). The oligomers formed are predominantly helical, and the formation of a β-sheet structure occurs simultaneously with the appearance of well-defined fibrils (Nettleton et al. 2000). The small-angle X-ray scattering (SAXS) technique was used to visualize the helical insulin oligomers as a bead-on-a-string assembly of six units, each with dimensions correlating to insulin monomers (Vestergaard et al. 2007). Another small protein (10.1 kDa), named Barstar, is a natural inhibitor of barnase, an extracellular endoribonuclease in *Bacillus amyloliquefaciens*. At low pH, Barstar partially unfolds to a molten globule-like A-form that possesses 60% of the secondary structure present in the native protein, but it is devoid of well-defined tertiary interactions. NMR spectroscopy revealed that the protein forms symmetrical aggregates containing 15 or 16 molecules. Time-resolved fluorescence anisotropy decay measurements and AFM revealed that at higher temperatures, Barstar transforms into protofibrils (with a diameter of 10 nm and length of 100–200 nm) and then slowly transforms into fibrillar aggregates (Mukhopadhyay et al. 2006).

The calcium-binding protein equine lysozyme self-assembles into protofibrils and fibrils of various morphologies under partial denaturing conditions (Malisauskas et al. 2003). The protein forms rings (with a diameter of 40–80 nm) and linear structures depending on pH and metal-ion concentrations (Malisauskas et al. 2003). Similarly, heat denaturation causes oligomerization of the core domain of the tumor-suppressor protein p53, which forms pore-like structures that are toxic to cultured cells (Ishimaru et al. 2003).

3.4 Methods Used to Study Pre-fibrillar Protein Assemblies

In the past decade substantial efforts were directed toward identifying, isolating, and characterizing the oligomeric species, both because of their likely role in the mechanism underlying fibril formation and because of their implications as primary toxic species in protein-misfolding diseases. The structure–activity relationship holds the key for better understanding the cytotoxicity of oligomeric assemblies. However, the structural characterization of amyloidogenic assemblies is not a trivial task due to their metastable nature and coexistence with constantly changing assembly products. High-resolution methods such as X-ray crystallography and NMR were not successful until now in determining detailed structural conformations of amyloidogenic oligomers. Several low-resolution methods have been used to characterize different oligomeric assemblies. Here we outline some of the methods that have been used for detection and classification.

3.4.1 Sodium Dodecyl Sulfate–Polyacrylamide-Gel Electrophoresis (SDS-PAGE)

SDS–PAGE is a widely used method enabling separation of proteins according to their mobility, which is influenced by the primary to quaternary structures of proteins. In order to perform this analysis, the protein solution is mixed with SDS, an anionic detergent that binds proteins through its dodecyl hydrophobic groups, leaving its sulfate groups exposed. It applies a negative charge to each protein in proportion to its mass; as a result, each protein has a similar charge-to-mass ratio.

A number of studies have used SDS–PAGE in order to analyze oligomer size distribution both *in vitro* and *in vivo*. One of the first studies that linked Aβ oligomers to neuronal death (Lambert et al. 1998) showed that oligomers can be formed *in vitro* by incubation of $A\beta_{1-42}$ in cold F-12 cell-culture medium. Interestingly, the same structures were present when aggregation was inhibited by clusterin (Apo J). PAGE analysis of oligomers formed by both methods revealed two major species with molecular weights of 17 and 27 kDa. Because oligomers were present in non-denaturing as well as with denaturing SDS–PAGE, their small size could not be attributed to strong detergent action.

Ashe and co-workers published several studies on the effect of Aβ oligomers on cognitive function. In one study (Cleary et al. 2005), they purified naturally secreted Aβ from Chinese hamster ovary (CHO) cells that stably express a mutated form of amyloid precursor protein (APP). Size-distribution analysis *via* SDS–PAGE revealed monomers, dimers, and trimeric assemblies. Another study (Lesné et al. 2006) investigated the connection between cognitive impairment and Aβ self-assembly in transgenic mice. In this animal model (Tg2576), mice under 6 months of age had normal memory and lacked neuropathology, whereas middle-aged mice (6–14 months old) developed memory deficits without neuronal loss, and old mice (>14 months old) formed abundant Aβ-containing neuritic plaques. Comparison of Aβ assemblies by SDS–PAGE at different ages of mice revealed an extracellular, 56-kDa oligomer termed Aβ*56. This oligomer was first observed in 6-month-old mice, when memory decline manifested. Oligomer size distribution was confirmed by SEC under more "native" conditions.

Barghorn et al. (2005) showed that it is possible to stabilize Aβ oligomers with SDS, termed by the authors, globulomers (because of their globular morphology). These oligomers were validated with fatty acids instead of SDS, which revealed the same assemblies, suggesting that SDS can simulate a membrane-like environment. PAGE analysis revealed monomers, 16-/18-kDa intermediates, and 48-/52-kDa globulomers. Although these assemblies formed *in vitro* under relatively harsh conditions, similar epitopes were also observed *in vivo* in AD patients' brains and in amyloid precursor protein transgenic mice.

Work done in our lab (Frydman-Marom et al. 2009) showed that the protocol described above, followed by PAGE analysis, can be utilized for rapid and easy screening of potential inhibitors of Aβ self-assembly (Fig. 3.3A). By following the disappearance of toxic globulomer assemblies by PAGE analysis in an initial screen, one can identify potent inhibitors. A potential inhibitor identified by this protocol was administered to a mouse model and it led to cognitive recovery, hence validating the initial screening process (Fig. 3.3B).

The use of SDS–PAGE for determining size distribution of oligomeric assemblies is an easy and straightforward method, but the effect of SDS on proteins is not equivalent. In some cases it can stabilize some assemblies but in others SDS can perform as a chaotropic agent leading to disassembly of some oligomers. A study by Teplow's group demonstrated that SDS can induce aggregation leading to higher Aβ assemblies (Bitan et al. 2005). Native gels can be applied for detection of oligomeric assemblies but one should consider that different oligomeric species may have similar mass–charge ratios, thus resulting in poor separation and low resolution.

Given the above, SDS–PAGE is an easy method but this technique cannot be used solely for size determination and size-distribution analyses and must be accompanied by complementary techniques.

3.4.2 *Size-Exclusion Chromatography (SEC)*

SEC utilizes a non-interactive mode of separation. It employs a stationary phase composed of a macromolecular gel containing a porous network, in which two liquid phases are inside the pores and between them. Proteins are separated solely according to their hydrodynamic volume. Proteins with a hydrodynamic volume larger than the largest pores of the stationary phase cannot penetrate the pores of the gel, and then pass through the spaces between the gel particles and elute first. Smaller proteins will equilibrate in both areas, resulting in retention to varying degrees and will elute later. Size can be determined by calculating the retention volume of analyzed protein to a pre-calculated standard curve.

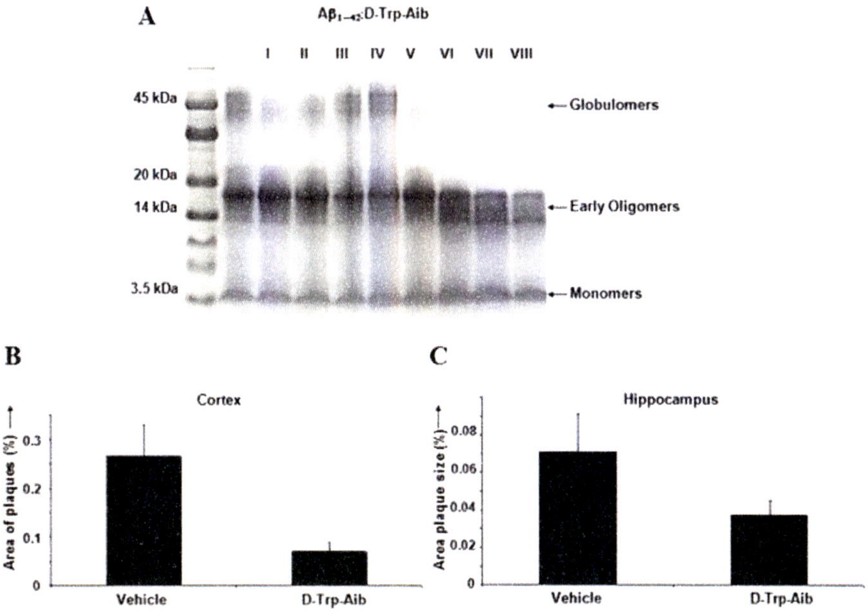

Fig. 3.3 Correlation between amyloid β oligomer inhibition *in vitro* and cognitive recovery *in vivo*. (**A**) The protocol established by Hillen and co-workers was used for screening potential inhibitors for Aβ oligomers. The chosen lead compound was a small dipeptide, D-Trp-Aib. This dipeptide combines an indole, which we identified as a potent aromatic binder of Aβ, and α-aminoisobutyric acid (Aib), which we identified as a β-breaker element. Inhibition was assessed by SDS–PAGE in the following molar ratio; (*I*) 1:1 ($A\beta_{1-42}$: D-Trp-Aib), (*II*) 1:5, (*III*) 1:10, (*IV*) 1:20,(*V*) 1:40, (*VI*) 1:60, (*VII*) 1:80, (*VIII*) 1:100. (**B–C**) The effect of D-Trp-Aib on AD model mice was studied on a total of 23 transgenic mice (aged 4.5 months) that overexpress hAPP. Animals were treated for 120 days to enable the clear-end observation of changes in plaque load and learning abilities. The relative plaque-load area of hAPP transgenic mice treated with d-Trp-Aib and vehicle-treated hAPP transgenic mice was determined by labeling with thioflavin S in the cortex area and hippocampus (Frydman-Marom et al. 2009)

This method can be utilized in order to estimate the extent of oligomerization and the size distribution of a sample under native conditions. SEC molecular-weight separation covers 10^2–10^6 Da. However, because retention volume is affected by hydrodynamic volume, changes in tertiary structure can affect retention and distort size determination. This problem can be solved by performing SEC coupled with dynamic light scattering (DLS). DLS directly measures the protein size without the need for a calibration curve.

SEC is a good method for studying the dynamics of the self-assembly processes; by calculating a peak area, it is possible to estimate the abundance of a specific oligomer in a heterogeneous population and to measure its change over time.

In a study investigating the interaction between amyloidogenic peptides and the polyphenol compound epigallocatechin gallate (EGCG) (Ehrnhoefer et al. 2008), thioflavin T (ThT)-binding assay revealed inhibition properties and lower kinetics of fibril formation. The effect of EGCG on α-synuclein aggregation was analyzed by SEC; untreated monomeric α-synuclein was eluted from the column with a molecular mass much higher than its expected mass. This was attributed to its natively unfolded structure. Additional experiments showed that EGCG remodeled the self-assembly process and promoted formation of unstructured non-toxic oligomers.

3.4.3 Dynamic Light Scattering (DLS)

DLS is a popular method for determining particle size, where the speed at which the particles are diffusing due to Brownian motion is measured (Lomakin and Teplow 2006; Lomakin et al. 2005). This is done by measuring the rate at which the intensity of the scattered light fluctuates when detected using an optical detector. The larger the particle, the slower the Brownian motion will be. Smaller particles are more influenced by solvent collisions; thus they move more rapidly. The experimental duration is rapid, ensuring very small changes in the analyzed sample; this is an important advantage when analyzing amyloidogenic peptides owing to their rapid self-assembly kinetics. Although dynamic scattering is capable of distinguishing whether a protein self-assembles to low-molecular-weight oligomers, it is much less accurate for distinguishing small oligomers than is static light scattering or analytical ultracentrifugation. The advantage of using dynamic scattering is the possibility to analyze samples containing broad distributions of species of widely differing molecular masses, and to detect very small amounts of the high-mass species. Avidan-Shpalter and Gazit have used DLS to characterize the dynamic growth process of preliminary intermediates transformed into larger structures of calcitonin (amyloidogenic peptide involved in thyroid carcinoma) (Avidan-Shpalter and Gazit 2006). Another problematic aspect in DLS or in fact any particle-sizing technique is the shape of non-spherical particles; if the shape of the particles is not identical, it will affect their diffusion, leading to incorrect size determination (Fig. 3.4).

3.4.4 Analytical Ultracentrifugation (AU)

AU is an extremely versatile and powerful technique for characterizing the solution-state behavior of macromolecules (Lebowitz et al. 2002). In AU, a constant centrifugal force is applied and a real-time observation apparatus allows examining the sedimentation and diffusion coefficients. AU can determine sample purity, protein size, conformation changes, and can monitor self-assembly process. AU is usually used in two main experiments: sedimentation velocity and sedimentation equilibrium. For a sedimentation velocity experiment, an initially uniform solution is placed in a cell and centrifuged at a high angular velocity, thus, causing rapid sedimentation of proteins towards the cell bottom. Sedimentation is followed by use of

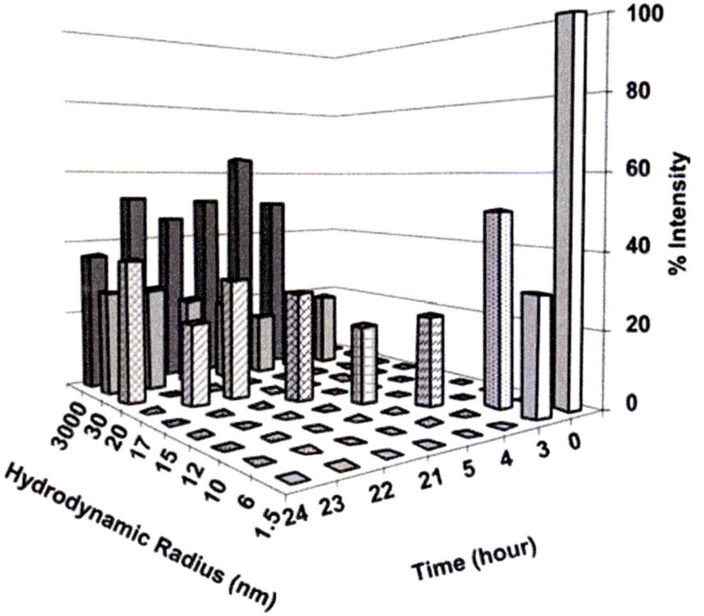

Fig. 3.4 Size-distribution analysis of calcitonin by dynamic light scattering. Hydrodynamic radius distribution of a human calcitonin sample at short time intervals. The figure shows a gradual, increased conversion of a homogeneous, low-molecular-weight population of particles to high-molecular-weight conformers. At the beginning of the experiment, a homogenous population of small particles appeared, comprising of monomers, dimers, and trimers. However, the prevalence of these particles decreased rapidly immediately after their appearance, and larger particles appeared instead, with a diverse hydrodynamic radius. The rapid assembly process and the simultaneous disappearance of the primary population support the notion that this population contained nuclei oligomers that comprise a few molecules (Avidan-Shpalter and Gazit 2006)

a spectrophotometer, which can determine the sedimentation coefficient (*s*), which is directly linked to the mass and size. In sedimentation equilibrium, a relatively lower centrifugal force than sedimentation velocity is used; thus proteins are not pelleted but reach equilibrium between two forces: diffusion and sedimentation. At the equilibrium point in the centrifugal tube, a gradient of proteins is created while separated by their molecular weight. High-molecular-mass proteins will be located at the bottom and low-molecular-mass proteins will be located higher in the cell.

As was mentioned before, ADDLs were studied by using SEC and AU in order to determine their oligomer size (Hepler et al. 2006). SEC analysis revealed two major populations. Their molecular weights were calculated using a standard curve and was correlated to 13- and 75-kDa oligomers. Analysis of fractions using a multi-angle laser light-scattering (MALLS) detector showed inconsistencies regarding the size calculated using the standard curve. The first peak correlated to 4.5 kDa and the second peak was a polydisperse mixture of oligomers ranging in size from 150 kDa at the trailing edge to nearly 1,000 kDa at the leading edge of the peak. The authors also used AU to determine oligomer size distribution; sedimentation velocity analysis performed on the oligomer-containing fraction yielded an average sedimentation coefficient of 6.7 S with an estimated mass of 223 kDa. The overall distribution was broad, ranging from about 3 to 11 S, again suggesting that a polydisperse population of oligomers was present at this peak. Velocity analysis of the low-molecular-mass components yielded a sedimentation coefficient of 0.55 S with a corresponding calculated mass of 4.5 kDa. The inconsistencies between SEC analysis and AU regarding light scattering can be explained by the fact that when using a calibration curve one assumes structural similarities between the proteins used in calibrating the analyte, and differences in molecular shape can have significant effects on migration times. It can also be explained by the nonspecific interactions between Aβ and the stationary phase in SEC. When performing SEC, the stationary phase is supposed to be inert; if not, false results could occur since the separation process is not only due to molecular weight.

3.4.5 Ion-Mobility Spectrometry–Mass Spectrometry (IMS–MS)

Mass spectrometry is a unique physical tool used for studying biological macromolecules. Coupling of ion mobility and mass spectrometry (IMS–MS) allows insight into the properties of protein assemblies (Uetrecht et al. 2010). Ion mobility separates the ions based on their ability to move through a certain medium in a gas phase under the influence of a static electric field. The movement depends on the charge and shape of certain ions. Mass spectrometry, on the other hand, is used to analyze the charge-to-mass ratio of the molecule. The homo-oligomers of a certain protein often have the same mass-to-charge ratio, and ion mobility can be used to analyze such readings by calculating the size component. When both methods are combined, mass and shape can be determined simultaneously.

The surface area of an ion which collides with the gas molecules result in a retention time (collision cross-section). Ions with compact structures move faster in comparison to wide-structured ions, with greater collision cross-sections, meaning larger ion sizes. This phenomenon can be utilized to study folding of proteins, for example, an unfolded protein will have a much greater collision cross-section than it would in its folded state.

In a study on the self-assembly process of Aβ by IMS–MS (Bernstein et al. 2009), $A\beta_{1-40}$ tested in solution displayed three conformers: monomer, dimer, and tetramer. The same analysis was performed on Aβ alloforms Pro^{19} and Met^{35} of $A\beta_{1-40}$, and Met^{35} alloform of $A\beta_{1-42}$. Those point-mutations were shown to reduce or eliminate the aggregation process (Klein et al. 2004; Bitan et al. 2003b). In all cases, the largest oligomer that was seen was a tetramer. Analysis of $A\beta_{1-42}$, a much more aggregating peptide than $A\beta_{1-40}$, revealed complex structural changes, where the largest assembly observed was a dodecamer and another multi-peak correlated to dimer, tetramer, and hexamer. By analyzing the collision cross-section, the above authors analyzed conformation changes of the monomeric unit in each oligomer. For all peptides, the monomeric size decreased as oligomerization proceeded. This indicated that conformational changes occurred in the monomeric unit during the self-assembly process. Another interesting finding was that $A\beta_{1-42}$ remained relatively large compared to the other peptides, suggesting a much more stable conformation compared with $A\beta_{1-40}$. This correlates with a previous study (Maji et al. 2005) that showed a relatively more stable center than did $A\beta_{1-40}$. The formation of dodecamers by $A\beta_{1-42}$ correlated with previous studies in transgenic mice (Lesné et al. 2006) showing a correlation between cognitive reduction and appearance of a dodecamer.

3.4.6 Single-Molecule Spectroscopy (SMS)

In the past few years, single-molecule spectroscopy (SMS) was utilized to quantify the degree of oligomerization. This quantification is based on counting the photobleaching steps of labeled oligomers. Two independent groups have used this assay in order to quantify the degree of $A\beta_{1-40}$ oligomerization in subnanomolar concentrations (Dukes et al. 2008; Ding et al. 2009). Dukes et al. used $A\beta_{1-40}$ tagged in the *N*-terminus with carboxyfluorescein (FAM) and *C*-terminus tagged with biotin. In addition, peptides were analyzed under different conditions and attached to streptavidin coverslips for reading. At basic pH values, frequent conformers found were monomers, dimers, and lower populations of trimers. At pH 5.8, dimers were the most frequent population; monomers were reduced and trimers were elevated. Zn^{2+}, which is known to bind and promote aggregation, was also used and shifted the size distribution towards the oligomeric species.

Ding et al. analyzed $A\beta_{1-40}$ tagged with Hilyte fluor 488 in neutral pH; single-molecule data that were compared to HPLC gel-filtration data showed a relatively good fit between the two methods applied. SMS data showed the presence of different conformers, ranging from monomers to hexamers, whereas the gel-filtration analysis showed that the largest oligomer was tetramers. Ding et al. explained this discrepancy by suggesting that larger oligomers were not stable in the gel-filtration process (Ding et al. 2009).

More recently, in a study by Gafni and other groups, a lipid bilayer was incorporated into the SMS apparatus. This allowed them to follow simultaneously the correlation between the oligomerization process and membrane conductivity changes (Schauerte et al. 2010). The experimental results agreed with those of the antimicrobial peptides' membrane-damage model (Huang 2000). This model describes a process whereby membrane pores are formed by a process initiated by the monomer binding to the membrane surface followed by surface diffusion and the subsequent assembly into discrete pore structures. Monomers and dimers that incorporated into the lipid bilayer did not change membrane conductivity but larger assemblies caused conductivity changes.

3.4.7 Transmission Electron Microscopy (TEM)

Transmission electron microscope (TEM) operates by emitting an electron beam. Due to the relatively short wavelength of the electrons, a 1,000 times better resolution can be obtained compared with a light microscope. Briefly, a monochromatic beam of electrons is accelerated through a potential difference of 40–100 kV to a vacuum tube. A strong electromagnetic field acts as a lens and focuses the electrons onto a very thin beam. The beam is projected on the sample and depending on the sample density, some of the electrons are scattered and disappear from the beam. At the end of the microscope the unscattered electrons hit a light-sensitive screen, which gives rise to a mirror image of the sample. A modern TEM has 0.2-nm resolution.

TEM has been used extensively to examine the morphology of amyloid oligomer assemblies. A very partial list includes the following: islet amyloid polypeptide (Porat et al. 2003) (Fig. 3.1C), ABri (El-Agnaf et al. 2001a, b), $PrP_{106-126}$ (Walsh et al. 2009), Sup35NM (Ohhashi et al. 2010), and low-molecular-weight Aβ oligomers (Bitan et al. 2003c).

3.4.8 Atomic-Force Microscopy (AFM)

Atomic-force microscopy (AFM) is high-resolution, scanning-probe microscopy. AFM allows examining surface morphology either in the air or when the sample is immersed in liquid. In AFM, a cantilever with a sharp tip is used. When a specimen is scanned, deflection forces occur between the sample and the cantilever, causing the tip to bend. Deflection is measured by a laser aimed on the cantilever or by piezoresistive elements. By scanning a surface that is under a constant force, an image of the surface area can be obtained.

An important advantage of AFM over other electron-microscopy methods is the ability to directly monitor dynamic changes in the conformation, association, or functional state of individual biomolecules in an aqueous environment, by mimicking their physiological surroundings *in situ*.

A study of the self-assembly processes of the human islet amyloid polypeptide (hIAPP) by time-lapse AFM (Green et al. 2004) showed that hIAPP elongation processes occurred in two phases: after 30 s, small round oligomeric structures predominated with average heights of 2.3 ± 1.9 nm and lengths of 23 ± 14 nm. Oligomers were also observed in the samples that had been incubated for 60 s; however, their heights were 4.6 ± 2.1 nm, and their lengths were 47 ± 28 nm. In samples that had been allowed 120 s for assembly, fibrillar structures were present, with heights and lengths of 10.6 ± 7.8 nm and 203 ± 170.0 nm, respectively. After height-*versus*-length diagrams were analyzed, it became evident that oligomers increased their heights to ~6 nm before extended in length. The authors also used phenol red, a polyphenol compound known for its aggregation-inhibition properties (Porat et al.

2004). hIAPP with phenol red in a molar ratio of 1:30, respectively, formed oligomers within 4–10 min, but elongation to mature fibrils was hardly observed. These data support the finding that hIAPP fibrillogenesis follows two distinct phases: oligomer growth and elongation into fibrils.

In a review by Lashuel and Lansbury (2006), images of different oligomers taken by AFM and TEM were found to exhibit high morphological resemblances (Fig. 3.1). The shared morphological and toxic properties of amyloid oligomers suggest that toxicity depends on shared structural features. Based on earlier studies on the ability of amyloid oligomers to damage-membrane models, Lashuel and Lansbury suggested that, similar to bacterial toxins, the annular oligomer morphology contributes to their pore-forming abilities. In a later study, a conformational antibody raised against Aβ oligomers (Kayed et al. 2003) was immunoreactive towards pore-forming proteins such as the pore-forming bacterial toxin, α-hemolysin, and human perforin from cytotoxic lymphocytes (Yoshiike et al. 2007). These findings suggest that there are structural and functional resemblances between amyloid oligomers and pore-forming proteins and that they may share a common mechanism of pathogenesis involving membrane permeabilization.

3.4.9 Circular Dichroism (CD)

CD is a useful spectroscopic method for studying conformational changes that occur during protein folding and unfolding. CD measures the interaction of a chiral molecule with polarized light, which is defined as the difference in absorption of left-handed and right-handed circularly polarized light by optically active compounds (Sreerama and Woody 2004). Different structural elements have characteristic CD spectra; for example, α-helical proteins have a negative peak at 222 and 208 nm and a positive peak at 193 nm. Proteins with well-defined β-sheets have negative bands at 218 nm and positive bands at 195 nm. By comparing a measured protein spectrum to well-defined protein-spectrum databases, secondary-structure content of proteins can be defined to some extent. The CD spectrum measurement represents the overall secondary structure present, which is why the population secondary structure can be determined and monitored, but it does not necessarily reflect the secondary structure of a specific oligomer.

A study on IAPP interaction with lipid bilayers (Knight et al. 2006) compared hIAPP and the non-amyloidogenic rat IAPP regarding their ability to interact with a liposome model. Both peptides were evaluated for secondary structure transitions in the presence of lipids. Both peptides adopt a predominantly α-helical conformation in the presence of membranes. Importantly, the CD analysis revealed no evidence of a β-sheet structure, indicative of amyloid fiber formation. Further experiments have shown that both peptides can cause membrane damage under the right conditions. The authors explained that α-helical oligomers can cause membrane damage and fiber formation not directly coupled to toxicity.

3.4.10 Fourier-Transform Infrared (FTIR) Spectroscopy

FTIR enables determination of the secondary-structure composition of proteins in solid state or in a solution (Berthomieu and Hienerwadel 2009). By measuring absorption in the vibrational spectrum of the C=O component of the amide-I bond from 1,600 to 1,700 cm^{-1}, one can evaluate a protein's secondary structure. The spectrum consists of multiple overlapping peaks for each amide. The amide peak depends on its secondary structure. By applying deconvolution to the entire spectrum, each peak can be assigned to a specific conformation; for α-helix (1,654 cm^{-1}), β-strand (1,624, 1,631, 1,637, and 1,675 cm^{-1}) (1,663, 1,670, 1,683, 1,688, and 1,684 cm^{-1}) or others (1,645 cm^{-1}), the relative amount of a specific secondary structure can be calculated from a peak integral.

Like other spectroscopic methods, these peaks are not precise and can deviate for up to 4 cm^{-1}. The main advantage of FTIR, especially when working with aggregation-prone proteins, is the long wavelength by which measurements are performed; a turbid solution can be analyzed without introducing scattering artifacts. In a recent study examining how the salt concentration influences assembly pathways of the mouse prion protein (Jain and Udgaonkar 2010), the authors showed that β-sheet-rich oligomers were formed in low- and high-salt concentrations as determined by FTIR and DLS measurements. Comparison of the secondary structures of these oligomers at pH 2 in 120 and 200 mM salt concentrations revealed different secondary-structure distributions. In the amide-I region, those oligomers formed in 120 mM NaCl solution exhibited two peaks, at ~1,620 and ~1,650 cm^{-1}. In contrast, those oligomers formed in 200 mM NaCl exhibited a single peak at ~1,628 cm^{-1}. The observation that the β-sheet-rich oligomers formed in 200 mM NaCl exhibited a peak at 1,628 cm^{-1} but did not exhibit a peak at 1,650 cm^{-1} implies that these oligomers have more β-sheet content and fewer other secondary structures, if any. Furthermore, the difference of 8 cm^{-1} in the 1,620 cm^{-1} region also suggests that those β-sheet-rich oligomers formed at low and high NaCl concentrations also differ in the internal structures of their β-sheets. Interestingly, the authors also examined the secondary structure of amyloid fibrils formed under the same conditions. The FTIR spectra differed in the amide-I region in both salt concentrations and correlated to the FTIR spectra of the β-sheet-rich oligomers at the same NaCl concentration.

3.4.11 Nuclear Magnetic Resonance (NMR)

NMR is a powerful method for studying three-dimensional structures of proteins in solution (O'Connell et al. 2009). Protein structure can be calculated by measuring nuclear Overhauser effects (NOE) that are formed due to dipolar interactions between different nuclei in a magnetic field. Magnetic nuclei are affected by each other as well as by the applied field, both through chemical bonds and over short distances through space. This can be exploited to assign resonance signals to particular nuclei in a complex structure, and derive constraints for the distances that separate them. In analyzing amyloidogenic peptides, the self-assembly process and aggregation can result in peak broadening due to slow tumbling times. Consequently, it is impossible to determine the structure and because of this phenomenon the oligomer structures are almost impossible to solve in a solution.

A study done by Abbott Laboratories on soluble Aβ oligomers (16 kDa) stabilized by SDS (Yu et al. 2009) revealed that the characteristic backbone atom chemical shifts, the protected amides, and the NOE data are all consistent with two β-strands from V18 to D23 and from K28 to V40. Residues V18–D23 form one strand of an intra-chain antiparallel β-sheet connected by a β-hairpin to the other intra-chain strand K28–G33, whereas L34–V40 forms an inter-chain in-register parallel β-sheet (Fig. 3.5A). From this finding, Yu et al. proposed a model for forming oligomers, suggesting that dimers serve as a repetitive unit in the assembly of larger oligomers. They also compared their model to the structure of two strands of mature fibrils of $A\beta_{1-42}$ previously reported by hydrogen–deuterium exchange and solid-state-NMR studies. Both structures exhibit an inter-strand parallel β-sheet for the *C*-terminal residues; however, in contrast to the fibril structure, the oligomers have an intra-strand antiparallel β-sheet connected by a β-hairpin between D23 and K28. Residues 10–16, which are part of the first β-sheet in fibrils, are disordered in oligomers.

3.4.12 X-Ray Crystallography

X-ray crystallography is used for determining the rearrangement of atoms in a crystal. Because X-rays have wavelengths similar to the size of atoms, they are useful for exploring atoms within crystals. Briefly, an X-ray beam is projected onto the crystal and is diffracted in many directions. The diffraction is caused by clouds of electrons in the molecule's crystal structure. Electron density reflects the molecule's shape; by analyzing the diffraction angles and their intensities, one can determine the protein structure.

Because of their aggregating nature, amyloidogenic proteins are difficult to analyze, and they tend to aggregate more rapidly than they crystallize. A work by (Sawaya et al. 2007) described 30 segments from different amyloidogenic proteins including Aβ, IAPP, PrP, lysozyme, myoglobin, α-synuclein, β2-microglobulin, insulin, and tau protein. By using X-ray crystallography, Sawaya et al. determined the atomic structural organization of these segments (Fig. 3.5B). Importantly, they were able to determine that the basic unit of amyloid-like fibrils is a steric zipper, formed by two tightly interlocked β-sheets, with the possibility of more complicated geometries with multiple steric zippers. Based on this discovery, they suggested that the first step in amyloid formation involves unmasking of the steric zippers, permitting them to stack into β-sheets, after which the sheets interdigitate. This also explains why nucleation seems to be a time-limiting step in amyloid formation; recruitment requires only one molecule at a time to unmask its fibril-forming sequence, but formation of the steric-zipper nucleus requires several molecules to unmask their zipper-forming segments simultaneously. A study

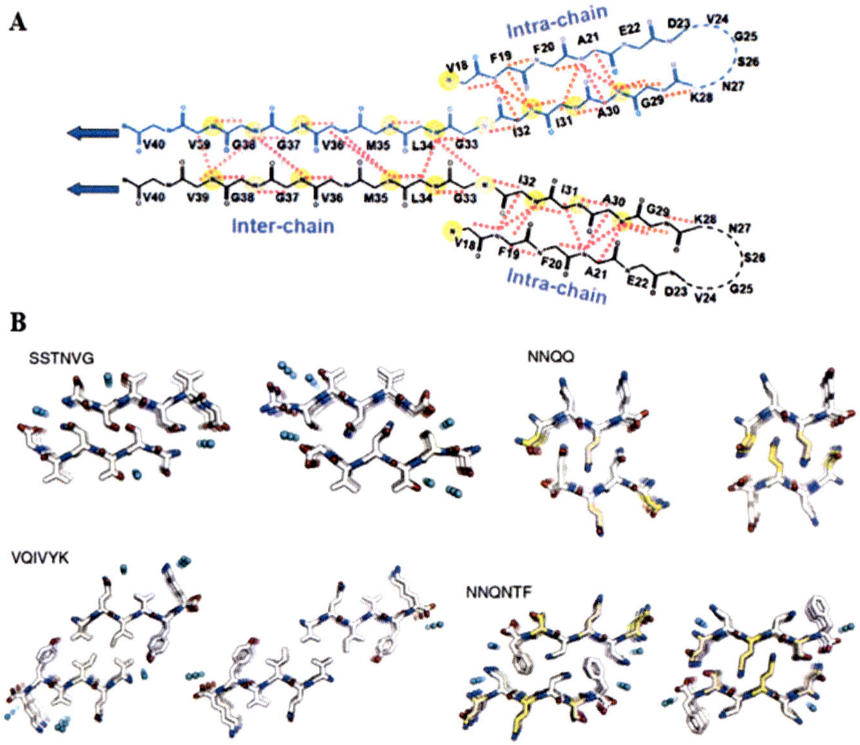

Fig. 3.5 Amyloidogenic interactions by NMR and X-ray crystallography studies. (**A**) A model of an NMR study of 16-kDa amyloid β oligomers, the model is based on Nuclear Overhauser Effect (NOE) constraints. *Dashed lines* indicate observed NOEs. *Circles* indicate the backbone amides that exhibit slow exchange in the NH/ND exchange experiments (Yu et al. 2009). (**B**). Packing polymorphism of steric zippers, determined by X-ray microcrystallography. A steric zipper is a pair of interlocked β-sheets, generally with a dry interface between them. Several amyloidogenic peptides show steric zipper interactions in several conformations: SSTNVG from IAPP, VQIVYK from tau protein, NNQQ from yeast prion Sup35, and NNQNTF from elk prion protein. The polymorphism is due to the different conditions of crystallization that took place. The polymorphism shown in this study suggests complexity of protein interactions that even a small protein can exhibit, depending on the molecular and environmental surrounding (Wiltzius et al. 2009)

published later by (Wiltzius et al. 2009) suggested that the steric-zipper packing contributes to protein-derived inheritance in prion proteins. By analyzing the structure of different segments of the same proteins, the authors discovered polymorphism in the steric-zipper packing. In considering the possible variety of packing arrangements and segmental and combined structures for steric zippers, it is clear that a substantial variety of prion strains associated with a single protein can be encoded by steric zippers. In the mechanisms proposed by the authors for prion strains, information transfer is achieved largely by the steric fit (van der Waals bonding) of short, self-complementary amino-acid sequences, with hydrogen bonding maintaining the zipper spine.

To date, no crystal structures of amyloid oligomers have been successfully determined, probably due to the fast aggregation kinetics of these peptides. Nevertheless, in a recent article, protein engineering was used to form stable non-aggregated Aβ oligomers (Sandberg et al. 2010). A disulfide bond, A21C and A30C, designed to stabilize a β-hairpin conformation, was introduced. With this addition, the hairpin is predicted to cover residues 17–23 and 30–36 as antiparallel β-strands connected by a turn between residues 25 and 29. The disulfide bond prevents formation of amyloid fibrils, resulting in formation of stable neurotoxic oligomeric forms. Addition of the reducing agent tris-2-carboxyethyl-phosphine (TCEP), used to break the disulfide bond, results in accelerating ThT binding to levels equal to those observed with wild-type Aβ40 fibrils. Better understanding the molecular assembly and structure of amyloid oligomers is extremely important in order to understand their cytotoxicity. Thus, this kind of work presents a good solution for determining the molecular structure of amyloid oligomers. In this way, by using stable non-aggregated oligomers, one can overcome limitations of crystallization of metastable conformers and can determine their structures in high resolution.

3.4.13 Immunological Classification of Amyloid Oligomers

In the past decade, several groups were able to isolate conformational antibodies, which could recognize the oligomeric state of amyloidogenic proteins without recognizing the monomeric state (Kayed et al. 2003; Barghorn et al. 2005; Acero et al. 2009; Meli et al. 2009; Masuda et al. 2009; Wang et al. 2009; Lafaye et al. 2009). In one of the earliest studies done by Kayed et al. (2003), an oligomer mimic was used to produce a conformational antibody. The oligomer mimic was composed of colloidal gold with $A\beta_{1-40}$ chains covalently bound. In order to examine the epitope that these antibodies recognized, the authors examined Aβ aggregation kinetics and antibody immunoreactivity. $A\beta_{1-42}$ immunoreactivity was observed at 6 h; however, it was maximal between 24 and 168 h, whereas after 332 h, it was not detected at all. With $A\beta_{1-40}$, the kinetics and immunoreactivity were similar but with a delay of up to 24 h. To show that the antibody recognizes oligomeric species, Aβ assemblies were fractionated and immunoreactivity was examined for every fraction. The smallest-size oligomer that is recognized by oligomer specific serum elutes at a position of 40 kDa, which corresponds to the approximate size of an octamer. The authors also examined the specificity of the antibodies to other amyloid peptides. Surprisingly, the antibody also recognizes the oligomeric assemblies of other peptides. This includes oligomeric and protofibrillar aggregates from α-synuclein, IAPP, polyglutamine, lysozyme, human insulin, and $\text{prion}_{106-126}$. These data suggest that a unique epitope is presented by oligomeric assemblies and is not present in the soluble monomer or in low-molecular-weight oligomers. Recognition of other oligomeric peptides implies that a common denominator exists between oligomers, which does not depend on amino-acid side-chains and is derived from backbone conformation.

A study by Yoshiike et al. used this antibody as a probe for detecting proteins with oligomeric conformation (Yoshiike et al. 2008). Interestingly, the antibody recognizes several proteins including those with aggregation-inhibition properties, such as heat-shock proteins. These conformational antibodies can be used as a simple direct tool for detecting oligomeric epitopes, but because of the cross-reactivity of conformational antibodies, one cannot rely solely on results using this kind of antibodies and a complementary method must be used to validate the results especially in *in vivo* or *ex vivo* studies.

3.5 Discussion

In this chapter we have reviewed various methods for isolating, preparing, and characterizing different amyloid oligomers. The nature of the actual toxic substance responsible for the pathology of amyloid diseases is under constant debate. A typical example is the controversy regarding the role of early soluble oligomers *versus* fibrils of the Aβ peptide in AD pathology. Whereas in the past most emphasis was on the fibrils, it is now clear that the appearance of the fibrils (amyloid plaques) in the brain merely indicates that they are the end product in the process of disease progression. A key role for smaller, early oligomeric forms of Aβ, and likewise in other amyloidogenic proteins, in both the cellular toxicity and final pathology of these diseases is now generally accepted. Thus, during the last several years, substantial efforts were directed toward identifying, isolating, and characterizing the oligomeric species because of their implications as the toxic species. In addition, roles of early oligomers in the process of fibril formation remain unclear. Several studies, however, have shown that oligomers may constitute an obligatory step in the process of fibril formation, i.e., 'on-pathway' (Harper et al. 1999; Serio et al. 2000), whereas other reports suggest that oligomers are formed independently of the pathway to fibril formation, i.e., 'off-pathway' (Morozova-Roche et al. 2004; Gellermann et al. 2008). Regardless of the precise roles played by oligomeric species of amyloidogenic proteins in the overall process of fibril formation, studying their structures and understanding mechanisms underlying their formation are extremely important, especially because these species could be the primary toxic agents involved in amyloid disorders.

To date, more than 55,000 articles have been published discussing the different aspects of protein aggregation. However, the vast majority of our understanding of the protein self-assembly processes, molecular recognition motifs, and cytotoxicity remains to be completed.

Aggregation is a very complex self-assembly process characterized by a vast polymorphism of different conformers that are greatly influenced by surrounding conditions. Thus, analyzing and understanding each of these self-assembling units is a difficult task. However, identifying and characterizing each of these conformers is a key step in better understanding the etiology of protein-aggregation-associated diseases and provides a much needed mechanistic insight for future therapy.

References

Acero G, Manoutcharian K, Vasilevko V, Munguia ME, Govezensky T, Coronas G, Luz-Madrigal A, Cribbs DH, Gevorkian G (2009) Immunodominant epitope and properties of pyroglutamate-modified Aβ-specific antibodies produced in rabbits. J Neuroimmunol 213:39–46

Ahmad A, Uversky VN, Hong D, Fink AL (2005) Early events in the fibrillation of monomeric insulin. J Biol Chem 280:42669–42675

Alarcon JM, Brito JA, Hermosilla T, Atwater I, Mears D, Rojas E (2006) Ion channel formation by Alzheimer's disease amyloid β-peptide (Aβ40) in unilamellar liposomes is determined by anionic phospholipids. Peptides 27:95–104

Anderson DH, Talaga KC, Rivest AJ, Barron E, Hageman GS, Johnson LV (2004) Characterization of β amyloid assemblies in drusen: the deposits associated with aging and age-related macular degeneration. Exp Eye Res 78:243–256

Anguiano M, Nowak RJ, Lansbury PT Jr (2002) Protofibrillar islet amyloid polypeptide permeabilizes synthetic vesicles by a pore-like mechanism that may be relevant to type II diabetes. Biochemistry 41:11338–11343

Arispe N, Pollard HB, Rojas E (1993) Giant multilevel cation channels formed by Alzheimer disease amyloid β-protein [AβP-(1–40)] in bilayer membranes. Proc Natl Acad Sci USA 90:10573–10577

Arispe N, Rojas E, Pollard HB (1993) Alzheimer disease amyloid-β protein forms calcium channels in bilayer membranes: blockade by tromethamine and aluminium. Proc Natl Acad Sci USA 90:567–571

Arispe N, Pollard HB, Rojas E (1994) The ability of amyloid β-protein [AβP(1–40)] to form Ca^{2+} channels provides a mechanism for neuronal death in Alzheimer's disease. Ann NY Acad Sci 747:256–266

Avidan-Shpalter C, Gazit E (2006) The early stages of amyloid formation: biophysical and structural characterization of human calcitonin pre-fibrillar assemblies. Amyloid 13:216–225

Baglioni S, Casamenti F, Bucciantini M, Luheshi LM, Taddei N, Chiti F, Dobson CM, Stefani M (2006) Prefibrillar amyloid aggregates could be generic toxins in higher organisms. J Neurosci 26:8160–8167

Bahadi R, Farrelly PV, Kenna BL, Kourie JI, Tagliavini F, Forloni G, Salmona M (2003) Channels formed with a mutant prion protein PrP(82–146) homologous to a 7-kDa fragment in diseased brain of GSS patients. Am J Physiol Cell Physiol 285:C862–C872

Barghorn S, Nimmrich V, Striebinger A, Krantz C, Keller P, Janson B, Bahr M, Schmidt M, Bitner RS, Harlan J, Barlow E, Ebert U, Hillen H (2005) Globular amyloid β-peptide oligomer—a homogenous and stable neuropathological protein in Alzheimer's disease. J Neurochem 95:834–847

Baskakov IV, Legname G, Baldwin MA, Prusiner SB, Cohen FE (2002) Pathway complexity of prion protein assembly into amyloid. J Biol Chem 277:21140–21148

Bernstein SL, Dupuis NF, Lazo ND, Wyttenbach T, Condron MM, Bitan G, Teplow DB, Shea J-E, Ruotolo BT, Robinson CV, Bowers MT (2009) Amyloid β protein oligomerization and the importance of tetramers and dodecamers in the etiology of Alzheimer's disease. Nat Chem 1:326–331

Berthomieu C, Hienerwadel R (2009) Fourier transform infrared (FTIR) spectroscopy. Photosynth Res 101:157–170

Billings LM, Green KN, Mcgaugh JL, Laferla FM (2007) Learning decreases Aβ*56 and tau pathology and ameliorates behavioral decline in 3×Tg-AD mice. J Neurosci 27:751–761

Bitan G, Lomakin A, Teplow DB (2001) Amyloid β-protein oligomerization: prenucleation interactions revealed by photo-induced cross-linking of unmodified proteins. J Biol Chem 276: 35176–35184

Bitan G, Kirkitadze MD, Lomakin A, Vollers SS, Benedek GB, Teplow DB (2003a) Amyloid β-protein (Aβ) assembly: Aβ40 and Aβ42 oligomerize through distinct pathways. Proc Natl Acad Sci USA 100:330–335

Bitan G, Tarus B, Vollers SS, Lashuel HA, Condron MM, Straub JE, Teplow DB (2003b) A molecular switch in amyloid assembly: Met35 and amyloid β-protein oligomerization. J Am Chem Soc 125:15359–15365

Bitan G, Vollers SS, Teplow DB (2003c) Elucidation of primary structure elements controlling early amyloid β-protein oligomerization. J Biol Chem 278:34882–34889

Bitan G, Fradinger EA, Spring SM, Teplow DB (2005) Neurotoxic protein oligomers what you see is not always what you get. Amyloid 12:88–95

Cappai R, Leck SL, Tew DJ, Williamson NA, Smith DP, Galatis D, Sharples RA, Curtain CC, Ali FE, Cherny RA, Culvenor JG, Bottomley SP, Masters CL, Barnham KJ, Hill AF (2005) Dopamine promotes α-synuclein aggregation into SDS-resistant soluble oligomers *via* a distinct folding pathway. FASEB J 19:1377–1379

Caughey B, Lansbury PT (2003) Protofibrils, pores, fibrils, and neurodegeneration: separating the responsible protein aggregates from the innocent bystanders. Annu Rev Neurosci 26:267–298

Caughey B, Baron GS, Chesebro B, Jeffrey M (2009) Getting a grip on prions: oligomers, amyloids, and pathological membrane interactions. Annu Rev Biochem 78:177–204

Chaney MO, Webster SD, Kuo YM, Roher AE (1998) Molecular modeling of the Aβ1–42 peptide from Alzheimer's disease. Protein Eng 11:761–767

Chiesa R, Harris DA (2001) Prion diseases: what is the neurotoxic molecule? Neurobiol Dis 8:743–763

Chromy BA, Nowak RJ, Lambert MP, Viola KL, Chang L, Velasco PT, Jones BW, Fernandez SJ, Lacor PN, Horowitz P, Finch CE, Krafft GA, Klein WL (2003) Self-assembly of Aβ(1–42) into globular neurotoxins. Biochemistry 42:12749–12760

Cleary JP, Walsh DM, Hofmeister JJ, Shankar GM, Kuskowski MA, Selkoe DJ, Ashe KH (2005) Natural oligomers of the amyloid-β protein specifically disrupt cognitive function. Nat Neurosci 8:79–84

Cohen FE, Prusiner SB (1998) Pathologic conformations of prion proteins. Annu Rev Biochem 67:793–819

Conway KA, Harper JD, Lansbury PT Jr (2000a) Fibrils formed *in vitro* from α-synuclein and two mutant forms linked to Parkinson's disease are typical amyloid. Biochemistry 39:2552–2563

Conway KA, Lee SJ, Rochet JC, Ding TT, Williamson RE, Lansbury PT Jr (2000b) Acceleration of oligomerization, not fibrillization, is a shared property of both α-synuclein mutations linked to early-onset Parkinson's disease: implications for pathogenesis and therapy. Proc Natl Acad Sci USA 97:571–576

Conway KA, Rochet JC, Bieganski RM, Lansbury PT Jr (2001) Kinetic stabilization of the α-synuclein protofibril by a dopamine-α-synuclein adduct. Science 294:1346–1349

Dahlgren KN, Manelli AM, Stine WB Jr, Baker LK, Krafft GA, Ladu MJ (2002) Oligomeric and fibrillar species of amyloid-β peptides differentially affect neuronal viability. J Biol Chem 277:32046–32053

Danzer KM, Haasen D, Karow AR, Moussaud S, Habeck M, Giese A, Kretzschmar H, Hengerer B, Kostka M (2007) Different species of α-synuclein oligomers induce calcium influx and seeding. J Neurosci 27:9220–9232

Demuro A, Mina E, Kayed R, Milton SC, Parker I, Glabe CG (2005) Calcium dysregulation and membrane disruption as a ubiquitous neurotoxic mechanism of soluble amyloid oligomers. J Biol Chem 280:17294–17300

Ding TT, Lee SJ, Rochet JC, Lansbury PT Jr (2002) Annular α-synuclein protofibrils are produced when spherical protofibrils are incubated in solution or bound to brain-derived membranes. Biochemistry 41:10209–10217

Ding H, Wong PT, Lee EL, Gafni A, Steel DG (2009) Determination of the oligomer size of amyloidogenic protein β-amyloid(1–40) by single-molecule spectroscopy. Biophys J 97:912–921

Dobson CM (2001) The structural basis of protein folding and its links with human disease. Philos Trans R Soc Lond B Biol Sci 356:133–144

Dobson CM (2003) Protein folding and misfolding. Nature 426:884–890

Dukes KD, Rodenberg CF, Lammi RK (2008) Monitoring the earliest amyloid-β oligomers *via* quantized photobleaching of dye-labeled peptides. Anal Biochem 382:29–34

Durell SR, Guy HR, Arispe N, Rojas E, Pollard HB (1994) Theoretical models of the ion channel structure of amyloid β-protein. Biophys J 67:2137–2145

Ehrnhoefer DE, Bieschke J, Boeddrich A, Herbst M, Masino L, Lurz R, Engemann S, Pastore A, Wanker EE (2008) EGCG redirects amyloidogenic polypeptides into unstructured, off-pathway oligomers. Nat Struct Mol Biol 15:558–566

El-Agnaf OM, Nagala S, Patel BP, Austen BM (2001a) Non-fibrillar oligomeric species of the amyloid ABri peptide, implicated in familial British dementia, are more potent at inducing apoptotic cell death than protofibrils or mature fibrils. J Mol Biol 310:157–168

El-Agnaf OM, Sheridan JM, Sidera C, Siligardi G, Hussain R, Haris PI, Austen BM (2001b) Effect of the disulfide bridge and the *C*-terminal extension on the oligomerization of the amyloid peptide ABri implicated in familial British dementia. Biochemistry 40:3449–3457

El-Agnaf OM, Salem SA, Paleologou KE, Curran MD, Gibson MJ, Court JA, Schlossmacher MG, Allsop D (2006) Detection of oligomeric forms of α-synuclein protein in human plasma as a potential biomarker for Parkinson's disease. FASEB J 20:419–425

Finder VH, Glockshuber R (2007) Amyloid-β aggregation. Neurodegener Dis 4:13–27

Frydman-Marom A, Rechter M, Shefler I, Bram Y, Shalev DE, Gazit E (2009) Cognitive-performance recovery of Alzheimer's disease model mice by modulation of early soluble amyloidal assemblies. Angew Chem Int Ed Engl 48:1981–1986

Funato H, Enya M, Yoshimura M, Morishima-Kawashima M, Ihara Y (1999) Presence of sodium dodecyl sulfate-stable amyloid β-protein dimers in the hippocampus CA1 not exhibiting neurofibrillary tangle formation. Am J Pathol 155:23–28

Garzon-Rodriguez W, Sepulveda-Becerra M, Milton S, Glabe CG (1997) Soluble amyloid Aβ-(1–40) exists as a stable dimer at low concentrations. J Biol Chem 272:21037–21044

Gazit E (2004) The role of prefibrillar assemblies in the pathogenesis of amyloid diseases. Drugs Future 29:613–619

Gellermann GP, Byrnes H, Striebinger A, Ullrich K, Mueller R, Hillen H, Barghorn S (2008) Aβ-globulomers are formed independently of the fibril pathway. Neurobiol Dis 30:212–220

Ghiso J, Vidal R, Rostagno A, Mead S, Revesz T, Plant G, Frangione B (2000) A newly formed amyloidogenic fragment due to a stop codon mutation causes familial British dementia. Ann NY Acad Sci 903:129–137

Goldberg MS, Lansbury PT Jr (2000) Is there a cause-and-effect relationship between α-synuclein fibrillization and Parkinson's disease? Nat Cell Biol 2:E115–E119

Gong Y, Chang L, Viola KL, Lacor PN, Lambert MP, Finch CE, Krafft GA, Klein WL (2003) Alzheimer's disease-affected brain: presence of oligomeric Aβ ligands (ADDLs) suggests a molecular basis for reversible memory loss. Proc Natl Acad Sci USA 100:10417–10422

Gorman PM, Yip CM, Fraser PE, Chakrabartty A (2003) Alternate aggregation pathways of the Alzheimer β-amyloid peptide: Aβ association kinetics at endosomal pH. J Mol Biol 325:743–757

Gosal WS, Morten IJ, Hewitt EW, Smith DA, Thomson NH, Radford SE (2005) Competing pathways determine fibril morphology in the self-assembly of β2-microglobulin into amyloid. J Mol Biol 351:850–864

Green JD, Goldsbury C, Kistler J, Cooper GJ, Aebi U (2004) Human amylin oligomer growth and fibril elongation define two distinct phases in amyloid formation. J Biol Chem 279: 12206–12212

Grudzielanek S, Smirnovas V, Winter R (2006) Solvation-assisted pressure tuning of insulin fibrillation: from novel aggregation pathways to biotechnological applications. J Mol Biol 356: 497–509

Gurlo T, Ryazantsev S, Huang CJ, Yeh MW, Reber HA, Hines OJ, O'brien TD, Glabe CG, Butler PC (2010) Evidence for proteotoxicity in β cells in type 2 diabetes: toxic islet amyloid polypeptide oligomers form intracellularly in the secretory pathway. Am J Pathol 176:861–869

Gusella JF, Macdonald ME (2000) Molecular genetics: unmasking polyglutamine triggers in neurodegenerative disease. Nat Rev Neurosci 1:109–115

Haass C, Selkoe DJ (2007) Soluble protein oligomers in neurodegeneration: lessons from the Alzheimer's amyloid β-peptide. Nat Rev Mol Cell Biol 8:101–112

Harper JD, Lansbury PT Jr (1997) Models of amyloid seeding in Alzheimer's disease and scrapie: mechanistic truths and physiological consequences of the time-dependent solubility of amyloid proteins. Annu Rev Biochem 66:385–407

Harper JD, Lieber CM, Lansbury PT Jr (1997a) Atomic force microscopic imaging of seeded fibril formation and fibril branching by the Alzheimer's disease amyloid-β protein. Chem Biol 4:951–959

Harper JD, Wong SS, Lieber CM, Lansbury PT (1997b) Observation of metastable Aβ amyloid protofibrils by atomic force microscopy. Chem Biol 4:119–125

Harper JD, Wong SS, Lieber CM, Lansbury PT Jr (1999) Assembly of Aβ amyloid protofibrils: an *in vitro* model for a possible early event in Alzheimer's disease. Biochemistry 38:8972–8980

Hartley DM, Walsh DM, Ye CP, Diehl T, Vasquez S, Vassilev PM, Teplow DB, Selkoe DJ (1999) Protofibrillar intermediates of amyloid β-protein induce acute electrophysiological changes and progressive neurotoxicity in cortical neurons. J Neurosci 19:8876–8884

Hepler RW, Grimm KM, Nahas DD, Breese R, Dodson EC, Acton P, Keller PM, Yeager M, Wang H, Shughrue P, Kinney G, Joyce JG (2006) Solution state characterization of amyloid β-derived diffusible ligands. Biochemistry 45:15157–15167

Hong DP, Fink AL, Uversky VN (2008) Structural characteristics of α-synuclein oligomers stabilized by the flavonoid baicalein. J Mol Biol 383:214–223

Hoshi M, Sato M, Matsumoto S, Noguchi A, Yasutake K, Yoshida N, Sato K (2003) Spherical aggregates of β-amyloid (amylospheroid) show high neurotoxicity and activate tau protein kinase I/glycogen synthase kinase-3β. Proc Natl Acad Sci USA 100:6370–6375

Huang HW (2000) Action of antimicrobial peptides: two-state model. Biochemistry 39:8347–8352

Ishimaru D, Andrade LR, Teixeira LS, Quesado PA, Maiolino LM, Lopez PM, Cordeiro Y, Costa LT, Heckl WM, Weissmuller G, Foguel D, Silva JL (2003) Fibrillar aggregates of the tumor suppressor p53 core domain. Biochemistry 42:9022–9027

Jain S, Udgaonkar JB (2010) Salt-induced modulation of the pathway of amyloid fibril formation by the mouse prion protein. Biochemistry 49:7615–7624

Jang H, Arce FT, Capone R, Ramachandran S, Lal R, Nussinov R (2009) Misfolded amyloid ion channels present mobile β-sheet subunits in contrast to conventional ion channels. Biophys J 97:3029–3037

Janson J, Ashley RH, Harrison D, Mcintyre S, Butler PC (1999) The mechanism of islet amyloid polypeptide toxicity is membrane disruption by intermediate-sized toxic amyloid particles. Diabetes 48:491–498

Jarrett JT, Berger EP, Lansbury PT Jr (1993) The carboxy terminus of the β amyloid protein is critical for the seeding of amyloid formation: implications for the pathogenesis of Alzheimer's disease. Biochemistry 32:4693–4697

Johansson AS, Garlind A, Berglind-Dehlin F, Karlsson G, Edwards K, Gellerfors P, Ekholm-Pettersson F, Palmblad J, Lannfelt L (2007) Docosahexaenoic acid stabilizes soluble amyloid-β protofibrils and sustains amyloid-β-induced neurotoxicity *in vitro*. FEBS J 274:990–1000

Kagan BL, Hirakura Y, Azimov R, Azimova R, Lin MC (2002) The channel hypothesis of Alzheimer's disease: current status. Peptides 23:1311–1315

Kawahara M, Arispe N, Kuroda Y, Rojas E (1997) Alzheimer's disease amyloid β-protein forms Zn^{2+}-sensitive, cation-selective channels across excised membrane patches from hypothalamic neurons. Biophys J 73:67–75

Kawarabayashi T, Shoji M, Younkin LH, Wen-Lang L, Dickson DW, Murakami T, Matsubara E, Abe K, Ashe KH, Younkin SG (2004) Dimeric amyloid β protein rapidly accumulates in lipid rafts followed by apolipoprotein E and phosphorylated tau accumulation in the Tg2576 mouse model of Alzheimer's disease. J Neurosci 24:3801–3809

Kayed R, Head E, Thompson JL, Mcintire TM, Milton SC, Cotman CW, Glabe CG (2003) Common structure of soluble amyloid oligomers implies common mechanism of pathogenesis. Science 300:486–489

Kayed R, Sokolov Y, Edmonds B, Mcintire TM, Milton SC, Hall JE, Glabe CG (2004) Permeabilization of lipid bilayers is a common conformation-dependent activity of soluble amyloid oligomers in protein misfolding diseases. J Biol Chem 279:46363–46366

Kayed R, Pensalfini A, Margol L, Sokolov Y, Sarsoza F, Head E, Hall J, Glabe C (2009) Annular protofibrils are a structurally and functionally distinct type of amyloid oligomer. J Biol Chem 284:4230–4237

Kheterpal I, Zhou S, Cook KD, Wetzel R (2000) Aβ amyloid fibrils possess a core structure highly resistant to hydrogen exchange. Proc Natl Acad Sci USA 97:13597–13601

Kim HJ, Chae SC, Lee DK, Chromy B, Lee SC, Park YC, Klein WL, Krafft GA, Hong ST (2003) Selective neuronal degeneration induced by soluble oligomeric amyloid β protein. FASEB J 17:118–120

Kirkitadze MD, Kowalska A (2005) Molecular mechanisms initiating amyloid β-fibril formation in Alzheimer's disease. Acta Biochim Pol 52:417–423

Kirkitadze MD, Bitan G, Teplow DB (2002) Paradigm shifts in Alzheimer's disease and other neurodegenerative disorders: the emerging role of oligomeric assemblies. J Neurosci Res 69:567–577

Klein WL (2002) Aβ toxicity in Alzheimer's disease: globular oligomers (ADDLs) as new vaccine and drug targets. Neurochem Int 41:345–352

Klein WL, Stine WB Jr, Teplow DB (2004) Small assemblies of unmodified amyloid β-protein are the proximate neurotoxin in Alzheimer's disease. Neurobiol Aging 25:569–580

Knight JD, Hebda JA, Miranker AD (2006) Conserved and cooperative assembly of membrane-bound α-helical states of islet amyloid polypeptide. Biochemistry 45:9496–9508

Kourie JI, Kenna BL, Tew D, Jobling MF, Curtain CC, Masters CL, Barnham KJ, Cappai R (2003) Copper modulation of ion channels of PrP[106–126] mutant prion peptide fragments. J Membr Biol 193:35–45

Lafaye P, Achour I, England P, Duyckaerts C, Rougeon F (2009) Single-domain antibodies recognize selectively small oligomeric forms of amyloid β, prevent Aβ-induced neurotoxicity and inhibit fibril formation. Mol Immunol 46:695–704

Lambert MP, Barlow AK, Chromy BA, Edwards C, Freed R, Liosatos M, Morgan TE, Rozovsky I, Trommer B, Viola KL, Wals P, Zhang C, Finch CE, Krafft GA, Klein WL (1998) Diffusible, nonfibrillar ligands derived from Aβ1-42 are potent central nervous system neurotoxins. Proc Natl Acad Sci USA 95:6448–6453

Lansbury PT, Lashuel HA (2006) A century-old debate on protein aggregation and neurodegeneration enters the clinic. Nature 443:774–779

Lashuel HA, Lansbury PT Jr (2006) Are amyloid diseases caused by protein aggregates that mimic bacterial pore-forming toxins? Q Rev Biophys 39:167–201

Lashuel HA, Hartley D, Petre BM, Walz T, Lansbury PT Jr (2002a) Neurodegenerative disease: amyloid pores from pathogenic mutations. Nature 418:291

Lashuel HA, Petre BM, Wall J, Simon M, Nowak RJ, Walz T, Lansbury PT Jr (2002b) α-Synuclein, especially the Parkinson's disease-associated mutants, forms pore-like annular and tubular protofibrils. J Mol Biol 322:1089–1102

Lashuel HA, Hartley DM, Petre BM, Wall JS, Simon MN, Walz T, Lansbury PT Jr (2003) Mixtures of wild-type and a pathogenic (E22G) form of Aβ40 *in vitro* accumulate protofibrils, including amyloid pores. J Mol Biol 332:795–808

Lebowitz J, Lewis MS, Schuck P (2002) Modern analytical ultracentrifugation in protein science: a tutorial review. Protein Sci 11:2067–2079

Leong SL, Cappai R, Barnham KJ, Pham CL (2009a) Modulation of α-synuclein aggregation by dopamine: a review. Neurochem Res 34:1838–1846

Leong SL, Pham CL, Galatis D, Fodero-Tavoletti MT, Perez K, Hill AF, Masters CL, Ali FE, Barnham KJ, Cappai R (2009b) Formation of dopamine-mediated α-synuclein soluble oligomers requires methionine oxidation. Free Radic Biol Med 46:1328–1337

Lesné S, Koh MT, Kotilinek L, Kayed R, Glabe CG, Yang A, Gallagher M, Ashe KH (2006) A specific amyloid-β protein assembly in the brain impairs memory. Nature 440:352–357

Li J, Zhu M, Manning-Bog AB, Di Monte DA, Fink AL (2004) Dopamine and L-DOPA disaggregate amyloid fibrils: implications for Parkinson's and Alzheimer's disease. FASEB J 18:962–964

Lin MX, Mizabekov T, Kagan BL (1997) Channel formation by a neurotoxic prion protein fragment. J Biol Chem 272:44–47

Lin H, Bhatia R, Lal R (2001) Amyloid β protein forms ion channels: implications for Alzheimer's disease pathophysiology. FASEB J 15:2433–2444
Lomakin A, Teplow DB (2006) Quasielastic light scattering study of amyloid β-protein fibril formation. Protein Pept Lett 13:247–254
Lomakin A, Teplow DB, Benedek GB (2005) Quasielastic light scattering for protein assembly studies. Methods Mol Biol 299:153–174
Lue LF, Kuo YM, Roher AE, Brachova L, Shen Y, Sue L, Beach T, Kurth JH, Rydel RE, Rogers J (1999) Soluble amyloid β peptide concentration as a predictor of synaptic change in Alzheimer's disease. Am J Pathol 155:853–862
Ma J, Lindquist S (2002) Conversion of PrP to a self-perpetuating PrP^{Sc}-like conformation in the cytosol. Science 298:1785–1788
Maji SK, Amsden JJ, Rothschild KJ, Condron MM, Teplow DB (2005) Conformational dynamics of amyloid β-protein assembly probed using intrinsic fluorescence. Biochemistry 44:13365–13376
Malisauskas M, Zamotin V, Jass J, Noppe W, Dobson CM, Morozova-Roche LA (2003) Amyloid protofilaments from the calcium-binding protein equine lysozyme: formation of ring and linear structures depends on pH and metal ion concentration. J Mol Biol 330:879–890
Martinez-Coria H, Green KN, Billings LM, Kitazawa M, Albrecht M, Rammes G, Parsons CG, Gupta S, Banerjee P, Laferla FM (2010) Memantine improves cognition and reduces Alzheimer's-like neuropathology in transgenic mice. Am J Pathol 176:870–880
Masliah E, Rockenstein E, Veinbergs I, Mallory M, Hashimoto M, Takeda A, Sagara Y, Sisk A, Mucke L (2000) Dopaminergic loss and inclusion body formation in α-synuclein mice: implications for neurodegenerative disorders. Science 287:1265–1269
Masuda M, Hasegawa M, Nonaka T, Oikawa T, Yonetani M, Yamaguchi Y, Kato K, Hisanaga S, Goedert M (2009) Inhibition of α-synuclein fibril assembly by small molecules: analysis using epitope-specific antibodies. FEBS Lett 583:787–791
Mc Donald JM, Savva GM, Brayne C, Welzel AT, Forster G, Shankar GM, Selkoe DJ, Ince PG, Walsh DM (2010) The presence of sodium dodecyl sulphate-stable Aβ dimers is strongly associated with Alzheimer-type dementia. Brain 133:1328–1341
Mclean CA, Cherny RA, Fraser FW, Fuller SJ, Smith MJ, Beyreuther K, Bush AI, Masters CL (1999) Soluble pool of Aβ amyloid as a determinant of severity of neurodegeneration in Alzheimer's disease. Ann Neurol 46:860–866
Meli G, Visintin M, Cannistraci I, Cattaneo A (2009) Direct *in vivo* intracellular selection of conformation-sensitive antibody domains targeting Alzheimer's amyloid-β oligomers. J Mol Biol 387:584–606
Monoi H (1995) New tubular single-stranded helix of poly-L-amino acids suggested by molecular mechanics calculations: I. Homopolypeptides in isolated environments. Biophys J 69:1130–1141
Monoi H, Futaki S, Kugimiya S, Minakata H, Yoshihara K (2000) Poly-L-glutamine forms cation channels: relevance to the pathogenesis of the polyglutamine diseases. Biophys J 78:2892–2899
Moore RA, Hayes SF, Fischer ER, Priola SA (2007) Amyloid formation *via* supramolecular peptide assemblies. Biochemistry 46:7079–7087
Morozova-Roche LA, Zamotin V, Malisauskas M, Ohman A, Chertkova R, Lavrikova MA, Kostanyan IA, Dolgikh DA, Kirpichnikov MP (2004) Fibrillation of carrier protein albebetin and its biologically active constructs. Multiple oligomeric intermediates and pathways. Biochemistry 43:9610–9619
Mucke L, Masliah E, Yu GQ, Mallory M, Rockenstein EM, Tatsuno G, Hu K, Kholodenko D, Johnson-Wood K, Mcconlogue L (2000) High-level neuronal expression of Aβ1–42 in wild-type human amyloid protein precursor transgenic mice: synaptotoxicity without plaque formation. J Neurosci 20:4050–4058
Mukhopadhyay S, Nayak PK, Udgaonkar JB, Krishnamoorthy G (2006) Characterization of the formation of amyloid protofibrils from barstar by mapping residue-specific fluorescence dynamics. J Mol Biol 358:935–942
Murali J, Jayakumar R (2005) Spectroscopic studies on native and protofibrillar insulin. J Struct Biol 150:180–189

Necula M, Kayed R, Milton S, Glabe CG (2007) Small molecule inhibitors of aggregation indicate that amyloid β oligomerization and fibrillization pathways are independent and distinct. J Biol Chem 282:10311–10324

Nettleton EJ, Tito P, Sunde M, Bouchard M, Dobson CM, Robinson CV (2000) Characterization of the oligomeric states of insulin in self-assembly and amyloid fibril formation by mass spectrometry. Biophys J 79:1053–1065

Nilsberth C, Westlind-Danielsson A, Eckman CB, Condron MM, Axelman K, Forsell C, Stenh C, Luthman J, Teplow DB, Younkin SG, Naslund J, Lannfelt L (2001) The 'Arctic' APP mutation (E693G) causes Alzheimer's disease by enhanced Aβ protofibril formation. Nat Neurosci 4:887–893

Nimmrich V, Grimm C, Draguhn A, Barghorn S, Lehmann A, Schoemaker H, Hillen H, Gross G, Ebert U, Bruehl C (2008) Amyloid β oligomers ($A\beta_{1-42}$ globulomer) suppress spontaneous synaptic activity by inhibition of P/Q-type calcium currents. J Neurosci 28:788–797

Noguchi A, Matsumura S, Dezawa M, Tada M, Yanazawa M, Ito A, Akioka M, Kikuchi S, Sato M, Ideno S, Noda M, Fukunari A, Muramatsu S, Itokazu Y, Sato K, Takahashi H, Teplow DB, Nabeshima Y, Kakita A, Imahori K, Hoshi M (2009) Isolation and characterization of patient-derived, toxic, high mass amyloid β-protein (Aβ) assembly from Alzheimer disease brains. J Biol Chem 284:32895–32905

Norris EH, Giasson BI, Hodara R, Xu S, Trojanowski JQ, Ischiropoulos H, Lee VM (2005) Reversible inhibition of α-synuclein fibrillization by dopaminochrome-mediated conformational alterations. J Biol Chem 280:21212–21219

O'connell MR, Gamsjaeger R, Mackay JP (2009) The structural analysis of protein-protein interactions by NMR spectroscopy. Proteomics 9:5224–5232

Oda T, Wals P, Osterburg HH, Johnson SA, Pasinetti GM, Morgan TE, Rozovsky I, Stine WB, Snyder SW, Holzman TF, Krafft GA, Finch CE (1995) Clusterin (apoJ) alters the aggregation of amyloid β-peptide ($A\beta_{1-42}$) and forms slowly sedimenting Aβ complexes that cause oxidative stress. Exp Neurol 136:22–31

Ohhashi Y, Ito K, Toyama BH, Weissman JS, Tanaka M (2010) Differences in prion strain conformations result from non-native interactions in a nucleus. Nat Chem Biol 6:225–230

Ono K, Condron MM, Teplow DB (2009) Structure–neurotoxicity relationships of amyloid β-protein oligomers. Proc Natl Acad Sci USA 106:14745–14750

Pham CL, Leong SL, Ali FE, Kenche VB, Hill AF, Gras SL, Barnham KJ, Cappai R (2009) Dopamine and the dopamine oxidation product 5,6-dihydroxylindole promote distinct on-pathway and off-pathway aggregation of α-synuclein in a pH-dependent manner. J Mol Biol 387:771–785

Pillot T, Lins L, Goethals M, Vanloo B, Baert J, Vandekerckhove J, Rosseneu M, Brasseur R (1997) The 118–135 peptide of the human prion protein forms amyloid fibrils and induces liposome fusion. J Mol Biol 274:381–393

Pillot T, Drouet B, Pincon-Raymond M, Vandekerckhove J, Rosseneu M, Chambaz J (2000) A nonfibrillar form of the fusogenic prion protein fragment [118–135] induces apoptotic cell death in rat cortical neurons. J Neurochem 75:2298–2308

Podlisny MB, Ostaszewski BL, Squazzo SL, Koo EH, Rydell RE, Teplow DB, Selkoe DJ (1995) Aggregation of secreted amyloid β-protein into sodium dodecyl sulfate-stable oligomers in cell culture. J Biol Chem 270:9564–9570

Podlisny MB, Walsh DM, Amarante P, Ostaszewski BL, Stimson ER, Maggio JE, Teplow DB, Selkoe DJ (1998) Oligomerization of endogenous and synthetic amyloid β-protein at nanomolar levels in cell culture and stabilization of monomer by Congo red. Biochemistry 37:3602–3611

Poirier MA, Li H, Macosko J, Cai S, Amzel M, Ross CA (2002) Huntingtin spheroids and protofibrils as precursors in polyglutamine fibrilization. J Biol Chem 277:41032–41037

Porat Y, Kolusheva S, Jelinek R, Gazit E (2003) The human islet amyloid polypeptide forms transient membrane-active prefibrillar assemblies. Biochemistry 42:10971–10977

Porat Y, Mazor Y, Efrat S, Gazit E (2004) Inhibition of islet amyloid polypeptide fibril formation: a potential role for heteroaromatic interactions. Biochemistry 43:14454–14462

Quist A, Doudevski I, Lin H, Azimova R, Ng D, Frangione B, Kagan B, Ghiso J, Lal R (2005) Amyloid ion channels: a common structural link for protein-misfolding disease. Proc Natl Acad Sci USA 102:10427–10432

Rahimi F, Shanmugam A, Bitan G (2008) Structure–function relationships of pre-fibrillar protein assemblies in Alzheimer's disease and related disorders. Curr Alzheimer Res 5:319–341

Redecke L, Von Bergen M, Clos J, Konarev PV, Svergun DI, Fittschen UE, Broekaert JA, Bruns O, Georgieva D, Mandelkow E, Genov N, Betzel C (2007) Structural characterization of β-sheeted oligomers formed on the pathway of oxidative prion protein aggregation *in vitro*. J Struct Biol 157:308–320

Reed MN, Hofmeister JJ, Jungbauer L, Welzel AT, Yu C, Sherman MA, Lesné S, Ladu MJ, Walsh DM, Ashe KH, Cleary JP (2011) Cognitive effects of cell-derived and synthetically derived Aβ oligomers. Neurobiol Aging 32:1784–1794

Rekas A, Knott RB, Sokolova A, Barnham KJ, Perez KA, Masters CL, Drew SC, Cappai R, Curtain CC, Pham CL (2010) The structure of dopamine induced α-synuclein oligomers. Eur Biophys J 39:1407–1419

Rochet JC, Lansbury PT (2000) Amyloid fibrillogenesis: themes and variations. Curr Opin Struct Biol 10:60–68

Roher AE, Chaney MO, Kuo YM, Webster SD, Stine WB, Haverkamp LJ, Woods AS, Cotter RJ, Tuohy JM, Krafft GA, Bonnell BS, Emmerling MR (1996) Morphology and toxicity of Aβ-(1–42) dimer derived from neuritic and vascular amyloid deposits of Alzheimer's disease. J Biol Chem 271:20631–20635

Roychaudhuri R, Yang M, Hoshi MM, Teplow DB (2009) Amyloid β-protein assembly and Alzheimer disease. J Biol Chem 284:4749–4753

Sanchez I, Mahlke C, Yuan J (2003) Pivotal role of oligomerization in expanded polyglutamine neurodegenerative disorders. Nature 421:373–379

Sandberg A, Luheshi LM, Sollvander S, Pereira De Barros T, Macao B, Knowles TP, Biverstal H, Lendel C, Ekholm-Petterson F, Dubnovitsky A, Lannfelt L, Dobson CM, Hard T (2010) Stabilization of neurotoxic Alzheimer amyloid-β oligomers by protein engineering. Proc Natl Acad Sci USA 107:15595–15600

Sawaya MR, Sambashivan S, Nelson R, Ivanova MI, Sievers SA, Apostol MI, Thompson MJ, Balbirnie M, Wiltzius JJ, Mcfarlane HT, Madsen AO, Riekel C, Eisenberg D (2007) Atomic structures of amyloid cross-β spines reveal varied steric zippers. Nature 447:453–457

Schauerte JA, Wong PT, Wisser KC, Ding H, Steel DG, Gafni A (2010) Simultaneous single-molecule fluorescence and conductivity studies reveal distinct classes of Aβ species on lipid bilayers. Biochemistry 49:3031–3039

Serio TR, Cashikar AG, Kowal AS, Sawicki GJ, Moslehi JJ, Serpell L, Arnsdorf MF, Lindquist SL (2000) Nucleated conformational conversion and the replication of conformational information by a prion determinant. Science 289:1317–1321

Sharon R, Bar-Joseph I, Frosch MP, Walsh DM, Hamilton JA, Selkoe DJ (2003) The formation of highly soluble oligomers of α-synuclein is regulated by fatty acids and enhanced in Parkinson's disease. Neuron 37:583–595

Silveira JR, Raymond GJ, Hughson AG, Race RE, Sim VL, Hayes SF, Caughey B (2005) The most infectious prion protein particles. Nature 437:257–261

Sokolov Y, Kozak JA, Kayed R, Chanturiya A, Glabe C, Hall JE (2006) Soluble amyloid oligomers increase bilayer conductance by altering dielectric structure. J Gen Physiol 128:637–647

Sokolowski F, Modler AJ, Masuch R, Zirwer D, Baier M, Lutsch G, Moss DA, Gast K, Naumann D (2003) Formation of critical oligomers is a key event during conformational transition of recombinant syrian hamster prion protein. J Biol Chem 278:40481–40492

Sreerama N, Woody RW (2004) Computation and analysis of protein circular dichroism spectra. Methods Enzymol 383:318–351

Srinivasan R, Jones EM, Liu K, Ghiso J, Marchant RE, Zagorski MG (2003) pH-dependent amyloid and protofibril formation by the ABri peptide of familial British dementia. J Mol Biol 333:1003–1023

Srinivasan R, Marchant RE, Zagorski MG (2004) ABri peptide associated with familial British dementia forms annular and ring-like protofibrillar structures. Amyloid 11:10–13
Stefani M, Dobson CM (2003) Protein aggregation and aggregate toxicity: new insights into protein folding, misfolding diseases and biological evolution. J Mol Med 81:678–699
Takahashi T, Kikuchi S, Katada S, Nagai Y, Nishizawa M, Onodera O (2008) Soluble polyglutamine oligomers formed prior to inclusion body formation are cytotoxic. Hum Mol Genet 17:345–356
Teplow DB (1998) Structural and kinetic features of amyloid β-protein fibrillogenesis. Amyloid 5:121–142
Terry RD, Masliah E, Salmon DP, Butters N, Deteresa R, Hill R, Hansen LA, Katzman R (1991) Physical basis of cognitive alterations in Alzheimer's disease: synapse loss is the major correlate of cognitive impairment. Ann Neurol 30:572–580
Uetrecht C, Rose RJ, Van Duijn E, Lorenzen K, Heck AJ (2010) Ion mobility mass spectrometry of proteins and protein assemblies. Chem Soc Rev 39:1633–1655
Urbanc B, Betnel M, Cruz L, Bitan G, Teplow DB (2010) Elucidation of amyloid β-protein oligomerization mechanisms: discrete molecular dynamics study. J Am Chem Soc 132: 4266–4280
Uversky VN (2008) Amyloidogenesis of natively unfolded proteins. Curr Alzheimer Res 5:260–287
Uversky VN, Fink AL (2004) Conformational constraints for amyloid fibrillation: the importance of being unfolded. Biochim Biophys Acta 1698:131–153
Van Rooijen BD, Van Leijenhorst-Groener KA, Claessens MM, Subramaniam V (2009) Tryptophan fluorescence reveals structural features of α-synuclein oligomers. J Mol Biol 394:826–833
Vestergaard B, Groenning M, Roessle M, Kastrup JS, Van De Weert M, Flink JM, Frokjaer S, Gajhede M, Svergun DI (2007) A helical structural nucleus is the primary elongating unit of insulin amyloid fibrils. PLoS Biol 5:e134
Vigo-Pelfrey C, Lee D, Keim P, Lieberburg I, Schenk DB (1993) Characterization of β-amyloid peptide from human cerebrospinal fluid. J Neurochem 61:1965–1968
Volles MJ, Lansbury PT Jr (2002) Vesicle permeabilization by protofibrillar α-synuclein is sensitive to Parkinson's disease-linked mutations and occurs by a pore-like mechanism. Biochemistry 41:4595–4602
Volles MJ, Lee SJ, Rochet JC, Shtilerman MD, Ding TT, Kessler JC, Lansbury PT Jr (2001) Vesicle permeabilization by protofibrillar α-synuclein: implications for the pathogenesis and treatment of Parkinson's disease. Biochemistry 40:7812–7819
Walsh DM, Selkoe DJ (2007) Aβ oligomers—a decade of discovery. J Neurochem 101:1172–1184
Walsh DM, Lomakin A, Benedek GB, Condron MM, Teplow DB (1997) Amyloid β-protein fibrillogenesis. Detection of a protofibrillar intermediate. J Biol Chem 272:22364–22372
Walsh DM, Hartley DM, Kusumoto Y, Fezoui Y, Condron MM, Lomakin A, Benedek GB, Selkoe DJ, Teplow DB (1999) Amyloid β-protein fibrillogenesis. Structure and biological activity of protofibrillar intermediates. J Biol Chem 274:25945–25952
Walsh DM, Tseng BP, Rydel RE, Podlisny MB, Selkoe DJ (2000) The oligomerization of amyloid β-protein begins intracellularly in cells derived from human brain. Biochemistry 39:10831–10839
Walsh DM, Klyubin I, Fadeeva JV, Cullen WK, Anwyl R, Wolfe MS, Rowan MJ, Selkoe DJ (2002) Naturally secreted oligomers of amyloid β protein potently inhibit hippocampal long-term potentiation *in vivo*. Nature 416:535–539
Walsh P, Yau J, Simonetti K, Sharpe S (2009) Morphology and secondary structure of stable β-oligomers formed by amyloid peptide PrP(106–126). Biochemistry 48:5779–5781
Wang HW, Pasternak JF, Kuo H, Ristic H, Lambert MP, Chromy B, Viola KL, Klein WL, Stine WB, Krafft GA, Trommer BL (2002) Soluble oligomers of β amyloid (1–42) inhibit long-term potentiation but not long-term depression in rat dentate gyrus. Brain Res 924:133–140
Wang XP, Zhang JH, Wang YJ, Feng Y, Zhang X, Sun XX, Li JL, Du XT, Lambert MP, Yang SG, Zhao M, Klein WL, Liu RT (2009) Conformation-dependent single-chain variable fragment antibodies specifically recognize β-amyloid oligomers. FEBS Lett 583:579–584

Weinreb PH, Zhen W, Poon AW, Conway KA, Lansbury PT Jr (1996) NACP, a protein implicated in Alzheimer's disease and learning, is natively unfolded. Biochemistry 35:13709–13715

Westermark P, Engstrom U, Johnson KH, Westermark GT, Betsholtz C (1990) Islet amyloid polypeptide: pinpointing amino acid residues linked to amyloid fibril formation. Proc Natl Acad Sci USA 87:5036–5040

Westlind-Danielsson A, Arnerup G (2001) Spontaneous *in vitro* formation of supramolecular β-amyloid structures, "βamy balls", by β-amyloid 1–40 peptide. Biochemistry 40:14736–14743

Williams AD, Sega M, Chen M, Kheterpal I, Geva M, Berthelier V, Kaleta DT, Cook KD, Wetzel R (2005) Structural properties of Aβ protofibrils stabilized by a small molecule. Proc Natl Acad Sci USA 102:7115–7120

Wiltzius JJ, Landau M, Nelson R, Sawaya MR, Apostol MI, Goldschmidt L, Soriaga AB, Cascio D, Rajashankar K, Eisenberg D (2009) Molecular mechanisms for protein-encoded inheritance. Nat Struct Mol Biol 16:973–978

Yoshiike Y, Kayed R, Milton SC, Takashima A, Glabe CG (2007) Pore-forming proteins share structural and functional homology with amyloid oligomers. Neuromolecular Med 9:270–275

Yoshiike Y, Minai R, Matsuo Y, Chen YR, Kimura T, Takashima A (2008) Amyloid oligomer conformation in a group of natively folded proteins. PLoS One 3:e3235

Younkin SG (1998) The role of Aβ42 in Alzheimer's disease. J Physiol Paris 92:289–292

Yu L, Edalji R, Harlan JE, Holzman TF, Lopez AP, Labkovsky B, Hillen H, Barghorn S, Ebert U, Richardson PL, Miesbauer L, Solomon L, Bartley D, Walter K, Johnson RW, Hajduk PJ, Olejniczak ET (2009) Structural characterization of a soluble amyloid β-peptide oligomer. Biochemistry 48:1870–1877

Chapter 4
Biological Targeting and Activity of Pre-fibrillar Aβ Assemblies

Kyle C. Wilcox, Jason Pitt, Adriano Sebollela, Helen Martirosova, Pascale N. Lacor, and William L. Klein

Abstract The body of work from the past decade has advanced our understanding of how toxic oligomers of Aβ are capable of eliciting the spectrum of pathological and behavioral hallmarks of Alzheimer's disease. These potent neurotoxins now provide a molecular basis for the cause of this disease as well as a basis for identifying and evaluating diagnostic and therapeutic strategies. Oligomer toxicity is mediated by a number of factors—both in the targeting of these toxins to the neuronal synapses and in the transduction of this targeting into intracellular signals resulting in synapse loss and, eventually, cell death. Recent investigations have focused on defining the mechanisms of binding of toxic Aβ oligomers, the pathways modulated by these events, and strategies to treat Alzheimer's disease by targeting both aspects. One promising facet of recent research highlighted in this chapter, and in which Aβ oligomers play a central role, is the unfolding of connection between Alzheimer's disease and insulin signaling in the aging brain.

Keywords Alzheimer's disease • Aβ oligomers • Insulin signaling • Alzheimer's therapeutics

4.1 Introduction: Aβ Oligomers as a New Class of Neurotoxins

4.1.1 Pre-Oligomer Era

The original formulation of the "amyloid cascade hypothesis" predicted that reducing the buildup of amyloid plaques should reduce the memory impairment observed in Alzheimer's disease (AD). However, early efforts to correlate amyloid pathology

K.C. Wilcox • J. Pitt • A. Sebollela • H. Martirosova • P.N. Lacor • W.L. Klein (✉)
Department of Neurobiology, Northwestern University, Evanston, IL, USA
e-mail: wklein@northwestern.edu

F. Rahimi and G. Bitan (eds.), *Non-fibrillar Amyloidogenic Protein Assemblies—Common Cytotoxins Underlying Degenerative Diseases*, DOI 10.1007/978-94-007-2774-8_4,

to memory impairment in animal models found that plaque load does not dictate AD severity (Haass and Selkoe 2007). Various reports began to emerge in the early 1990s, beginning with that by Roher et al., which described soluble oligomeric forms of Aβ (Roher et al. 1991), but without explicit testing of neuronal responses to these species (Frackowiak et al. 1994; Kuo et al. 1996; Podlisny et al. 1995; Roher et al. 1993; Vigo-Pelfrey et al. 1993). An initial clue in defining the oligomeric basis of Aβ toxicity was the finding that co-incubation of Aβ peptides with clusterin increased the toxicity of the resulting species (Oda et al. 1994). This treatment, which prevents the assembly of amyloid fibrils, was expected to reduce toxicity, but contrary to this expectation, an increase in oxidative stress was induced in PC12 cells upon exposure to the clusterin–Aβ complexes.

4.1.2 Oligomer Era

On the heels of this finding, Lambert et al. (1998) provided an explicit demonstration that soluble Aβ oligomers act as potent neurotoxins that initiate alterations in cell signaling that lead to rapid inhibition of long-term potentiation (LTP) and, ultimately, to selective neuronal death. ADDLs, short for Aβ-derived diffusible ligands, were so termed to emphasize the ability of these species to act as specific toxins and the term comprises only those Aβ oligomers with dementing activity. A dual effect of ADDLs was reported, such that a rapid reduction of LTP preceded a slower phase featuring cell death induced by aberrant cell signaling. A prediction presented in that work—that if soluble Aβ oligomers proved to be important in AD pathogenesis, it is theoretically possible to halt or reverse AD progression at the early stages—has provided the basis for some of the current efforts to design therapeutics for early-stage AD. Though Aβ oligomers provided the initial statement of the oligomer hypothesis, disease-related oligomers of diverse fibrillogenic proteins have been catalogued.

4.1.3 Aβ Oligomer Species

Soluble Aβ species in various oligomeric states possess neurotoxic characteristics, making it difficult to label a single species as the most relevant to AD progression, if such a single species exists. One consistently observed species *in vitro* and *in vivo* has turned out to be approximately a dodecamer. Early comparative experiments based on centrifugal filtration to separate small and large oligomers illustrated that Aβ oligomers comprising roughly 12–24-mers bind to cultured neurons and exhibit toxicity (Chromy et al. 2003). Barghorn et al. also demonstrated that dodecamers exhibit postsynaptic binding and block LTP (Barghorn et al. 2005). In addition to the collection of dodecameric Aβ oligomers shown to be toxic in various studies, there have been demonstrations of toxicity in larger (Noguchi et al. 2009) and smaller (Shankar et al. 2008) soluble Aβ species. Protofibrils (Hartley et al. 1999) and fibrils (Lorenzo and Yankner 1994) are able to induce neuronal toxicity, but we do not rule out the possibility that smaller Aβ oligomers are responsible for the toxicity credited to fibrillar species, in the light of recent results suggesting that insoluble Aβ deposits may serve as a reservoir of toxic, soluble species (Koffie et al. 2009). Also of interest is a recent report that Aβ42 monomers are neurotrophic rather than neurotoxic (Giuffrida et al. 2009), an effect that is relevant to our following discussion of neuronal signaling mechanisms as it is mediated through the insulin-like growth factor and PI3K pathways.

Another crucial question is to what extent toxic Aβ oligomers formed *in vitro* are structurally similar to the soluble Aβ species found in AD-diseased human brain. Although there has not been a conclusive characterization of brain-derived Aβ oligomers, AD-dependent antibody detection of soluble oligomers using conformation-sensitive, anti-ADDL antibodies in human samples provides compelling evidence that ADDLs prepared *in vitro* possess structural similarities with brain-derived oligomers (Gong et al. 2003; Lambert et al. 2007; Lacor et al. 2004). These antibodies were successfully used to detect low levels of ADDLs in both brain extracts from postmortem human tissue (Gong et al. 2003) or cerebrospinal fluid (CSF) (Georganopoulou et al. 2005; Lacor et al. 2004), revealing up to a 70-fold increase in ADDLs in AD *versus* control tissues. A predominant Aβ species in detergent-free, soluble extracts of human AD brain was a 12-mer, consistent with ADDLs prepared *in vitro* (Gong et al. 2003). Other laboratories have also reported 12-mers in either AD transgenic mice or human CSF (Fukumoto et al. 2010; Lesné et al. 2006). In contrast, Shankar et al. detected Aβ dimers under denaturing conditions in human AD brain extracts (Shankar et al. 2008). It is possible that the discrepancy in the oligomer size reported in these studies comes from differences in the protocols used either to prepare or analyze the extracts, such as the presence of detergents. For instance, it was observed that in the presence of SDS high-molecular-weight Aβ oligomers formed *in vitro* dissociate into smaller species (Bitan et al. 2003). Therefore, it is reasonable to speculate that high-molecular-weight oligomers are SDS-labile, but may become SDS-resistant after interacting with endogenous compounds *in vivo*, a hypothesis supported by the finding that high-molecular-weight species are stabilized *in vitro* by interacting with prostaglandins (Boutaud et al. 2006).

4.2 Linking Aβ Oligomers to Major Facets of AD Neuropathology

4.2.1 Synaptic Targeting

Abundant binding of Aβ oligomers occurs on dendritic arbors of select neurons in hippocampal cultures. This pattern is consistently observed with synthetic ADDLs, AD brain- or CSF-derived soluble oligomers, consistent with the idea that Aβ oligomers bind to synapses (Lacor et al. 2004), as would be expected of a molecule that disrupts LTP and long-term depression (LTD). Electron microscopic immunolocalization of oligomers to synapses has not been described; however, biochemical evidence from detergent-resistant fractions of AD brains (Tomic et al. 2009) points to the possibility of synaptic oligomer enrichment.

Evidence that oligomers do, in fact, bind at synapses is the finding that ADDL hot spots co-locate with PSD-95 as visualized by high-resolution confocal fluorescence microscopy. Other synaptic markers that are highly enriched at ADDL hot spots are calcium/calmodulin-dependent kinase II (CaMKII), Arc, spinophilin, drebrin, and *N*-methyl-$_{\mathrm{D}}$-aspartate receptor (NMDAR), while the presynaptic marker, synaptophysin is opposite the ADDL hot spots (Deshpande et al. 2009; Gong et al. 2003; Lacor et al. 2004, 2007). PSD-95 co-localization is also observed in brain sections from a transgenic mouse model of AD (Koffie et al. 2009). Using oligomer-selective antibodies, a diffuse immunostaining can be visualized surrounding neuronal cell bodies in postmortem sections of early stages of AD (Lacor et al. 2004), reminiscent of the diffuse synaptic deposits observed in prion-associated diseases (Kovacs et al. 2002) and consistent with apparent oligomer binding within dendritic arbors. ADDLs preferentially associate with excitatory synapses (Lambert et al. 2007; Renner et al. 2010), as they co-locate with Homer1b/c, a scaffolding protein concentrated at excitatory synapses that interacts with metabotropic glutamate receptors and members of the Shank family (Tu et al. 1999). Association of ADDLs with gephyrin, a scaffolding protein of inhibitory synapses, is not evident (Renner et al. 2010).

Aβ oligomers appear to bind post-synaptic sites. Detergent extraction of ADDL-treated cortical synaptosomes yields an ADDL-binding complex that is, in essence, postsynaptic as demonstrated by the presence of PSD-95 and the absence of syntaxin, a presynaptic active-zone protein. Further extraction by sarkosyl and sodium dodecyl sulfate releases PSD-95 as well as subunits of the NMDARs (Lacor 2007). The immunoisolation of an ADDL-binding complex from synaptosomes has established the presence of several prominent ionotropic and metabotropic glutamate receptors (NR1, NR2, GluR1, mGluR5) and neuroligin (Renner et al. 2010). Nicotinic acetylcholine receptors and glycine receptors were not detected, again demonstrating selectivity for excitatory synapses.

4.2.2 Synaptic Damage

Synapse loss is the most reliable correlate of cognitive deficits in AD (Terry et al. 1991), and a loss of specific synaptic proteins has been observed that is correlated with AD severity and regionally specific neurodegeneration (Proctor et al. 2010). Aβ oligomers are capable of triggering loss of synapses by binding to dendritic spines (Lacor et al. 2004), whereupon they exert a collection of effects on synapse size, shape, composition, and abundance (Lacor 2007). ADDL exposure leads to a loss of stubby and mushroom spines typical of healthy neurons and appearance of elongated, fillopodia-like spines and large, branched spines (Lacor et al. 2004, 2007;

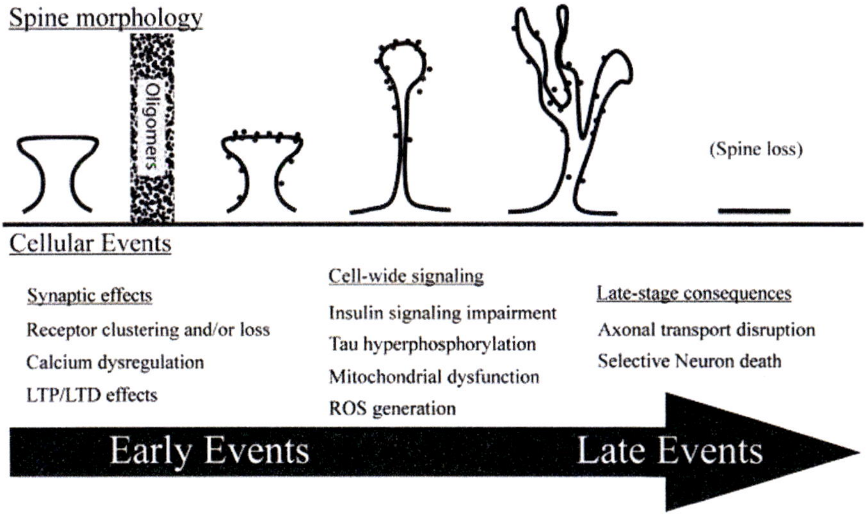

Fig. 4.1 Timeline of events triggered by Aβ oligomers. *Top*: Schematic depicting morphological changes of the dendritic spines following oligomer treatment. *Bottom*: Selected cellular events associated with toxic Aβ oligomers

Shrestha et al. 2006), perhaps due to compensatory mechanisms at targeted synapses. This is an example of an apparent multi-stage pathological process in which oligomer treatment causes a buildup of synaptic proteins in the short term (i.e., minutes to several hours) while ultimately resulting in the loss of the same proteins as dendritic spines degenerate (hours to days). This phenomenon, with respect to both spine morphology and cellular events, is illustrated in Fig. 4.1.

In the early timescale leading up to wholesale synapse loss, treatment with oligomeric toxins induces significant changes in the makeup of synaptic membranes. This receptor reorganization is manifested as a loss of surface glutamate receptors—both NMDA (Lacor et al. 2004, 2007; Snyder et al. 2005) and α-amino-3-hydroxyl-5-methyl-4-isoxazole-propionate [AMPA; (Zhao et al.)] subtypes—as well as EphB2 receptors (Lacor 2007; Lacor et al. 2007) and insulin receptors (De Felice et al. 2009) as discussed below. This alteration of synaptic receptor composition prior to structural changes and spine loss is consistent with the postulated loss of synaptic plasticity without neurodegeneration in early stages of AD (Klein et al. 2007; Lambert et al. 1998).

4.2.3 LTP/LTD

The early discovery that synthetic soluble Aβ oligomers cause severe impairment of LTP (Lambert et al. 1998), an electrophysiological correlate of learning and memory, was confirmed using oligomers produced by transfected cells (Walsh et al. 2002) and with soluble extracts from AD subjects and AD transgenic mice brains (Lesné et al. 2006; Shankar et al. 2008). Oligomers also impair LTD reversal (Walsh et al. 2002; Wang et al. 2002, 2004a), indicative of a net shift in synaptic activity toward inhibition and strongly suggestive of instability in synaptic composition and morphology (Klein et al. 2001; Lacor et al. 2004). Impairment of LTP in AD transgenic mice occurs before the development of Aβ deposits (Larson et al. 1999; Oddo et al. 2006) but correlates with accumulation of soluble oligomers (Chang et al. 2003), similar to behavioral testing for memory (Cleary et al. 2005; Lesné et al. 2006). Maintenance of LTP and LTD require activation of NMDAR and/or metabotropic glutamate receptors (mGluRs) (Citri and Malenka 2008; Kemp and Bashir 2001). Selkoe initially reported that the NMDAR–p38/MAPK pathway was affected by Aβ oligomers and, more recently, reported facilitation of LTD in hippocampal slices in the presence of buffer-soluble Aβ extracts from AD brains (mostly composed of Aβ dimers and trimers) (Shankar et al. 2008). This LTD induction leading to faulty glutamate recycling at synapses appears to depend on mGluRs rather than NMDARs.

4.2.4 Other AD Pathologies

Aβ oligomers additionally cause a variety of other specific neuronal pathologies linked to AD, including tau phosphorylation, formation of reactive oxygen species, mitochondrial dysfunction, endoplasmic reticulum stress, and selective cell death.

ADDLs and oligomer-containing AD brain extracts stimulate phosphorylation of AD-associated epitopes in tau, while similar extracts from non-AD control brains do not (De Felice et al. 2008). This phosphorylation is prevented by incubation of the brain-derived oligomers with conformation-dependent antibodies against ADDLs and also by pharmacological inhibitors of either Src-family tyrosine kinases or phosphatidylinositol-3 kinase (PI3K). These data support earlier studies showing a moderating effect of the A11 pan-oligomer-specific antibody (Kayed et al. 2003) on tau pathology upon injection into triple transgenic mice (Oddo et al. 2006).

Oxidative stress resulting from ADDL-induced production of reactive oxygen species is another salient aspect of AD pathology attributable to oligomeric Aβ. Production of reactive oxygen species at low levels is a component of normal neuronal function and integral to maintenance of LTP (Serrano and Klann 2004). ADDLs trigger a rise in reactive oxygen species through the activity of NMDARs, and this effect is abrogated by NMDAR antagonist, memantine (De Felice et al. 2007), which is currently used to treat AD patients. Aβ oligomers also cause mitochondrial dysfunction (reviewed in Bayer and Wirths 2010), possibly through direct interaction with mitochondria, and the endoplasmic reticulum also exhibits oligomer-induced disruptions in calcium homeostasis (Resende et al. 2008).

Axonal transport is affected in AD and is thought to be a consequence of tau pathology. Oligomeric Aβ disrupts axonal transport through modulation of glycogen synthase kinase β (GSK3β), a kinase responsible for tau hyperphosphorylation (Decker et al. 2010). This pathological effect occurs 18 h after exposure to oligomers, reinforcing the idea that defective axonal transport is a consequence of oligomer-induced tau phosphorylation.

Selective killing of subgroups of neurons in the hippocampus is one of the most striking features of AD, wherein the CA1 region shows the most loss (West et al. 1994). At the other end of the spectrum, the cerebellum appears to be unaffected in the AD brain. From the earliest descriptions of oligomeric Aβ toxins, it has been clear that these species are capable of producing patterns of cell toxicity consistent with AD pathology. The first experiments testing the toxicity of oligomers in slice cultures from rodent brains showed hippocampal cell death (Lambert et al. 1998). Later experiments confirmed the findings of hippocampal ADDL toxicity and further explored the subregional selectivity of their neurotoxic action—finding that the CA1 region of the hippocampus is preferentially targeted by ADDLs while the CA3 region is largely unaffected (Kim et al. 2003).

Non-neuronal populations are also affected by Aβ oligomers, which may contribute to AD progression through inflammation. Astrocytes in culture are susceptible to oligomeric Aβ, which induces release of inflammatory cytokines in a manner temporally distinct from the action of amyloid fibrils, as well as resulting in increased levels of other inflammatory markers (White et al. 2005). Demyelination surrounding amyloid deposits in the brains of AD patients and transgenic mice suggests that soluble Aβ oligomers may also exert toxic effects on oligodendrocytes (Mitew et al. 2010), while microglia may actually facilitate Aβ oligomerization through an intracellular mechanism involving CCL2 (Kiyota et al. 2009). Moreover, interaction of Aβ oligomers with microglia appears to be conformation-dependent (Heurtaux et al. 2010).

4.2.5 *Animal Models*

Animal models have been useful in demonstrating the biological presence and activity of toxic Aβ oligomers and their effects on behavior. Oligomers were identified in transgenic animals in 2003 (Chang et al. 2003), and the dependence of AD on oligomeric Aβ rather than plaques is now supported by transgenic mice that produce soluble oligomers but no plaques, even at advanced age (Tomiyama et al. 2010). These mice exhibit clear defects in hippocampal synaptic plasticity and memory that accompany loss of synapses. Similar mice were described that also lack fibrils while at the same time producing oligomeric Aβ species (Gandy et al. 2010). These two lines of transgenic animals were created by altering the same amino acid in the sequence of the amyloid precursor protein. In the mice from the former study, glutamic acid 693 was deleted based on a mutation found in members of a Japanese family that develop a form of AD with decreased Aβ levels. Studies of the mutated peptide revealed a propensity to form oligomers and no fibrillization (Tomiyama et al. 2008). In the latter mouse model, the same glutamic acid residue was changed to a glutamine (E693Q). Despite the findings that transgenic mice expressing the E693Q variant lack amyloid plaques, the mutation responsible for this form of Aβ was identified as underlying a form of hereditary cerebral hemorrhage with amyloidosis in humans (Levy et al. 1990), which is accompanied by the deposition of Aβ.

4.3 Basis of Aβ Oligomer Attachment to Synapses

4.3.1 Mode of Attachment

Three competing schools of thought dominate the study of how Aβ oligomers attack neurons. One suggested possibility is that oligomeric Aβ interacts directly with membranes to form toxic pores. This hypothesis is supported by findings that peptides and oligomers can insert into model membranes of varying compositions, perhaps forming "amyloid pores" (Lashuel and Lansbury 2006) and has been hypothesized to constitute a general property underlying the toxicity of multiple amyloid-forming proteins involved in neurodegenerative diseases (Kayed et al. 2003). Multiple studies using model membranes as well as intact cells have implicated negatively charged phospholipids in the interaction of Aβ peptides with neural membranes (Alarcon et al. 2006; Hertel et al. 1997; McLaurin and Chakrabartty 1997; Wong et al. 2009), perhaps catalyzing the conversion to oligomers.

A second hypothesis supported by the literature is that toxic Aβ oligomers exert their pathological effects from within the cell (Takahashi et al. 2004; Walsh et al. 2000). Intracellular Aβ, though not necessarily in an oligomeric form, is observed in AD brains (Gouras et al. 2000) and there is evidence that Aβ is generated intracellularly (reviewed in LaFerla et al. 2007). Recent transgenic rat models of AD featuring intracellular oligomers as detected by the oligomer-specific NU-1 monoclonal antibody suggest that oligomeric Aβ is also present inside neurons (Leon et al. 2010; Tomiyama et al. 2010). There is disagreement in the literature regarding whether oligomers form extracellularly or intracellularly. While oligomeric Aβ can form within neurons prior to export (Walsh et al. 2000), a report that Aβ monomer levels in the interstitial fluid of the brain undergo a circadian cycle in living mice highlights the likelihood that concentration of Aβ monomer in interstitial fluid can be sufficient for oligomer formation (Kang et al. 2009). Furthermore, intracellular accumulation of monomeric Aβ might not correlate with toxicity. Rather, an inverse correlation between intracellular Aβ monomers and nucleic acid oxidation might exist, constituting a possible protective mechanism against oxidative stress (Nunomura et al. 2010).

Finally, there is evidence that loss of synaptic activity associated with AD is caused by binding of soluble Aβ oligomers to specific sites on the neuronal surface. It has long been recognized that all regions of the AD brain are not equally affected upon autopsy (Braak and Braak 1991). At the molecular level, the hippocampal neurons are highly targeted by toxic oligomers in culture whereas neurons from the cerebellum are not recognized (Klein 2002; Klein et al. 2001; Gong et al. 2003). Similarly, treatment with soluble oligomers induces mitochondrial dysfunction in cortical, but not cerebellar preparations (Eckert et al. 2008), indicating specific susceptibility of certain neuronal populations. Additional evidence from primary hippocampal cultures shows that neighboring neurons exhibit dramatic cell-to-cell differences in ADDL binding such that highly labeled cells have overlapping processes with a cell that is unlabeled by ADDLs (Lacor et al. 2007). More striking than these local differences is the ADDL specificity for synapses within a single cell. ADDLs show approximately 90% colocalization with synaptic markers. Overall, however, only half of the excitatory synapses within a population show ADDL binding (Lacor et al. 2004). This fractional synaptic targeting by oligomers increases upon neuronal activation (Deshpande et al. 2009), possibly reflecting alterations in synaptic receptor content. Additional evidence for a receptor-mediated mechanism includes the trypsin sensitivity of ADDL binding and observations that ADDLs and other toxic oligomers preferentially affect mature neurons, gaining the ability to bind toxic oligomers only after 7–14 days *in vitro*, which suggests the presence of a developmentally regulated toxin receptor (Lacor et al. 2007; Lambert et al. 1998; Shughrue et al. 2010).

4.3.2 *Potential Oligomer-Binding Sites*

While there have been many studies showing that soluble Aβ oligomers accumulate at synapses acting as gain-of-function pathogenic ligands of high affinity (Lacor et al. 2004, 2007; Renner et al. 2010), their specific binding sites are still the subject of investigation. Too many receptors for Aβ oligomers have been proposed in the literature for a full accounting in this chapter. Therefore, we shall only highlight several of the most recent studies of the receptor-mediated nature of ADDL binding—these having resulted in new receptor candidates, proteins intimately involved in ADDL-induced synaptic pathology, and mechanisms for receptor-mediated ADDL clustering at synapses. As yet, no single protein seems to recapitulate all of the necessary characteristics of a true ADDL receptor.

The most notable recent example is the prion protein, PrP, which was identified in a high-throughput gene-expression screen of proteins capable of mediating ADDL binding to a non-neuronal cell line [i.e., one without endogenous ADDL-binding capability (Laurén et al. 2009)]. This finding has already been the subject of numerous follow-up studies, with subsequent reports of PrP-independent ADDL toxicity (Calella et al. 2010; Kessels et al. 2010) and ADDL-induced memory deficits (Balducci et al. 2010), suggesting a possible mechanism in which PrP may participate in ADDL clustering, allowing a neuronal response to lower ADDL concentrations (Laurén et al. 2010).

The β-2 adrenergic receptor, when heterologously expressed in human embryonic kidney cells, appears to bind Aβ, although the aggregation state of the peptide was not characterized (Wang et al. 2010). NMDARs have also been implicated in ADDL binding by findings that memantine, an NMDA antagonist, reduces ADDL-induced toxicity in neuronal cultures (De Felice et al. 2007; Lacor et al. 2007). In the same work, pretreatment with antibodies against the NMDAR subunit NR1 reduced ADDL binding by half while reducing ADDL-induced generation of reactive oxygen species to background levels. These results suggest that ADDLs are binding a site near NMDARs.

In another screen to identify mediators of oligomer binding, Zhao et al. recently reported that GluR2, a subunit of AMPA receptors (AMPARs), is involved in synaptic ADDL binding. Similar to the aforementioned antibody pretreatment against NR1 to reduce ADDL binding, pharmacological reduction in surface AMPAR expression results in an incomplete reduction in ADDL labeling of synapses (Zhao et al. 2010), further indicating the existence of other receptors or more complex factors at play.

Experiments using a knockout mouse lacking mGluR5 implicate this receptor in ADDL binding and ADDL-induced synaptic pathology (Renner et al. 2010). While this study does not address the issue of a direct mGluR5–ADDL interaction, it provides evidence for the essential role of mGluR5 in ADDL-induced synaptotoxicity, as described in Sect. 4.3.3 below.

Gangliosides and the lipid rafts they define have also been implicated in several aspects of AD pathophysiology, including amyloidogenic APP processing (Fonseca et al. 2010) and ADDL binding (Gong et al. 2003; Zampagni et al. 2010). In fact, lipid-raft GM1 gangliosides were recently reported to mediate directly neuronal toxic calcitonin oligomer binding (Malchiodi-Albedi et al. 2010). In the same work, a complete elimination of calcium influx associated with oligomer toxicity following ganglioside removal by neuraminidase was reported. While compelling, this study does not rule out a protein-based receptor, as neuraminidase enzymes are not selective for gangliosides and will nonspecifically deglycosylate lipids and proteins alike to reduce oligomer toxicity. Additionally, antibodies against GM1 gangliosides prevent oligomer toxicity, but ganglioside-specific antibodies would occlude binding to other lipid-raft components (such as PrP) as well as to gangliosides.

How can so many receptors be involved in oligomer binding? One hypothesis is that because Aβ oligomers typically comprise a distribution of states, they are capable of binding to multiple receptors. Another is that a single oligomer can bind to different, low-affinity receptors. Figure 4.2 (from Sakono and Zako 2010) effectively illustrates this conundrum. If either of these hypotheses is true, then the sum of the contributions from the various receptors should comprise ADDL binding. In fact, when each of the proposed receptors is ablated through antibody blockade, pharmacological inhibitors, or regulation of expression levels, only a fractional decrease in ADDL binding is typically observed (De Felice et al. 2007; Laurén et al. 2009; Zhao et al. 2010). However, a combination of PrP, mGluR5, and NR1 antibodies applied simultaneously to cultured hippocampal neurons does not improve this fractional reduction in ADDL binding observed for any of the individual receptor antibodies (Renner et al. 2010), and this non-additivity suggests that a more complex mechanism dictates ADDL association with synapses.

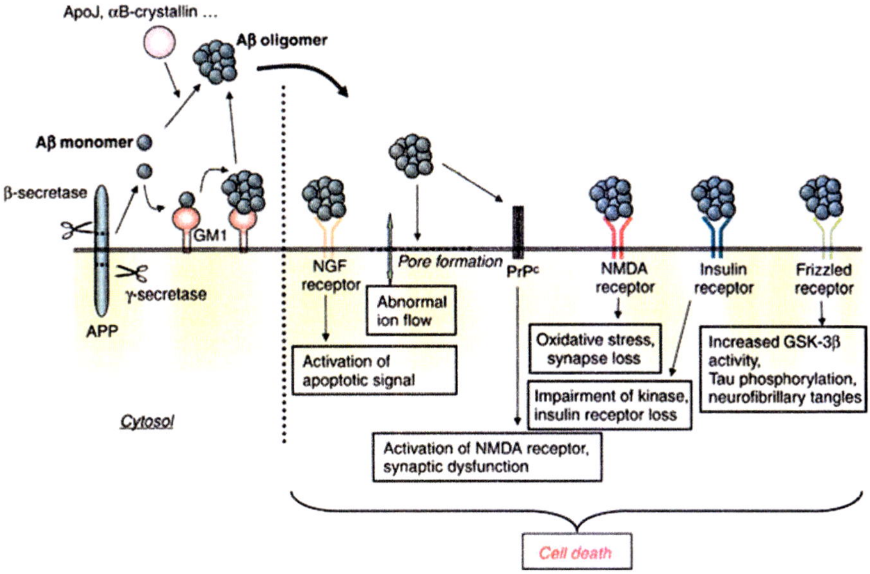

Fig. 4.2 Oligomer-binding sites. Multiple binding sites have been hypothesized for extracellular Aβ oligomers, transducing a spectrum of intracellular signals (Image credit: Sakono and Zako 2010—Fig. 4.1)

4.3.3 A New Hypothesis for ADDL-Induced Synaptotoxicity: Receptor Clustering

The ectopic clustering of mGluR5 by ADDLs acting as an extracellular scaffold represents a new hypothesis for ADDL-induced synaptotoxicity. Real-time single-particle tracking of quantum-dot-labeled ADDLs bound to the surface of living neurons reveals that upon initial binding, ADDLs exhibit diffusion typical of freely moving membrane proteins (Renner et al. 2010). Shortly thereafter, ADDLs became essentially immobile as lateral diffusion progressively decreased between 5 and 60 min (see Fig. 4.3). Both single-particle tracking and confocal imaging data imply that membrane-bound ADDLs exhibit decreased mobility as they accumulate within growing clusters. Similarly, reduced diffusion accompanies the recruitment of trans-membrane proteins to specific sites (Douglass and Vale 2005; Geng et al. 2009), suggesting that ADDL clustering and immobilization may be receptor-dependent. Suggestive of receptor involvement in the clustering of ADDLs, Renner et al. showed that ADDL treatment alters the dynamics of mGluR5 diffusion, producing a similar clustering behavior for this receptor. Furthermore, artificial clustering of mGluR5 using antibodies to an extracellular epitope reproduced ADDL-initiated dynamics changes and mGluR5 clustering, suggesting that ADDLs are acting as an extracellular scaffold for cell-surface molecules. There is currently no evidence that mGluR5 is the receptor responsible for this clustering, though its signaling activity

Fig. 4.3 Dynamics of ADDLs at the neuronal membrane. Time-resolved, single-particle tracking of ADDLs bound to live neurons 5 min (*lower left panel*) and 1 h (*lower right panel*) after exposure to ADDLs (*red*). *White lines* represent maximum projections from 5-s trajectories of surface-bound ADDLs at each time-point. The initial state of each trajectory is shown in the *upper panels*. ADDLs undergo an initial diffusive regime followed by immobilization at synaptic sites (*green*)

upon artificial clustering may be responsible for the surface withdrawal of NR1 and rise in intracellular calcium levels upon ADDL treatment.

The synapse-directed shift in mGluR5 distributions triggered by ADDLs is significant in that this receptor plays a role in synaptic plasticity mechanisms underlying learning and memory (Simonyi et al. 2005). This finding is consistent with other studies showing that mGluR5 contributes to oligomer-induced synaptotoxicity (Hsieh et al. 2006; Li et al. 2009a; Wang et al. 2004a). Interestingly, lateral diffusion of AMPA-type glutamate and GABA receptors was unaffected, suggesting the specificity of ADDLs for mGluR5-associated binding sites (Renner et al. 2010). The link to mGluR5 implicates a mechanism for the effect of ADDLs on LTP/LTD and calcium homeostasis, and indeed, mGluR5 antagonists successfully prevent oligomer-induced interference with LTP/LTD and intracellular calcium levels (Renner et al. 2010; Shankar et al. 2008; Townsend et al. 2007; Wang et al. 2004a).

In summary, this scaffolding-like action of membrane-bound ADDLs to cause redistribution of critical plasma-membrane receptors represents a new type of molecular mechanism to explain calcium dysregulation and impairment of synaptic plasticity by ADDLs.

4.4 Intracellular Signaling Mechanisms

Downstream cell-signaling alterations that occur as a consequence of ADDL binding to excitatory synapses can be usefully divided into two broad categories, one encompassing the cell-wide changes that occur as a result of ADDL-induced changes of intracellular calcium levels, and the other regarding the specific pathways vulnerable to ADDL-induced changes.

4.4.1 Calcium Signaling

Aβ oligomers exert a global effect on the regulation of intraneuronal calcium levels (Small et al. 2009). This can have a number of deleterious effects, as proper levels of calcium are crucial for synaptic plasticity and numerous neuronal receptors, kinases, and other components of signaling cascades are under the direct control of calcium.

As mentioned above, one mechanism whereby oligomers could be altering intracellular calcium levels is the ectopic clustering of mGluR5 receptors, which leads to an increase in cellular calcium that is prevented by mGluR5 antagonists (Renner et al. 2010). The increased influx of calcium through NMDAR in response to ADDLs has likewise been shown, where a nearly fourfold increase in intracellular calcium levels induced by ADDLs was prevented by memantine, an antagonist specific to open-state NMDAR, and an antibody to NR1 (De Felice et al. 2007).

A full accounting of the calcium-regulated pathways impacted by Aβ oligomers is beyond the scope of this chapter. One particularly interesting example, though, is that of calcineurin—a calcium-dependent phosphatase whose inhibitory effect on LTP can be reversed by pharmacological inhibition (Dineley et al. 2010). ADDL-induced AMPAR GluR2/3 removal from spines depends on calcineurin-induced endocytosis in clathrin-coated pits (Zhao et al. 2010). One important target of calcineurin and other pathways targeted by ADDLs, including the insulin and excitatory signaling, is the activity of the transcription factor, cAMP-response-binding-element protein (CREB). Of particular note is a study showing that a low dose of ADDLs, insufficient to produce neurodegeneration, reduces CREB activation and CREB-dependent expression of brain-derived neurotrophic factor (Tong et al. 2001). The convergence of disparate ADDL-induced signaling modalities on this transcription factor has the effect of suppressing its activation (Krafft and Klein 2010). Because CREB regulates transcription of a number of genes thought to be important for establishing late LTP, as well as preventing apoptosis (Benito and Barco 2010), the loss of CREB function may help explain the mechanism of LTP inhibition elicited by Aβ oligomers.

4.4.2 Specific Pathways Vulnerable to Toxic Aβ Oligomers

A major pathological AD hallmark is the appearance of cytosolic neurofibrillary tangles composed of hyperphosphorylated tau (Grundke-Iqbal et al. 1986). Evidence that the presence of Aβ precedes and leads to the appearance of tau pathology (Oddo et al. 2003, 2006) can be reproduced in cultured hippocampal neurons upon ADDL treatment (De Felice et al. 2008), leading to the question of which signaling pathways mediate this process.

The implication of GSK3β as a kinase responsible for tau phosphorylation in AD (Mandelkow et al. 1992) led to the investigation of the kinase's upstream regulators—notably the PI3K/Akt pathway, which negatively regulates GSK3β activity through phosphorylation (Erol 2009). However, despite ongoing investigations of the influence of ADDLs on Akt activity, whether Aβ oligomers positively or negatively regulate this pathway is unclear (Krafft and Klein 2010). Evidence that ADDLs down-regulate PI3K comes primarily from studies using upstream PI3K stimulation—namely, that stimulation of PI3K/Akt *via* α7 nicotinic acetylcholine receptor activation is neuroprotective against ADDL-induced toxicity (Kihara et al. 2001). Analogously, PI3K stimulation using the cytokine, erythropoietin, is protective against oligomer-induced tau hyperphosphorylation in SHSY5Y cells (Sun et al. 2008b). Lee et al. demonstrated that ADDL-induced Akt activation is attenuated by specific inhibition of the Akt activator, phosphoinositide-dependent kinase 1, in SHSY5Y and myotube cells, while also reporting decreased levels of total and activated Akt in AD brains (Lee et al. 2009). In opposition to these data are findings that ADDLs up-regulate PI3K/Akt. De Felice et al. demonstrated that PI3K antagonists prevent ADDL-induced tau hyperphosphorylation in hippocampal neurons (De Felice et al. 2008). Furthermore, direct quantification of activated Akt and mammalian target of rapamycin (mTOR), an Akt substrate, in mouse primary cultures reveals an increase in both proteins following ADDL exposure (Bhaskar et al. 2009). Increased levels of Akt activity have also been reported in AD brains by multiple groups (Griffin et al. 2005; Rickle et al. 2004). These apparently conflicting data may be reconciled by considering that ADDL-induced changes to PI3K/Akt activity may depend on the timing of specific events measured under the differing conditions of each experiment (e.g., short *versus* long exposures to ADDLs as discussed in Sect. 4.2.2) or the use of different cellular models.

The importance of PI3K/Akt in maintaining general cell health, metabolism, and survival have been well-established and it is not surprising that this pathway has become a major focal point of research for reasons beyond those concerning regulation of GSK3β phosphorylation. Further research is certainly necessary to define the exact role of PI3K/Akt to ADDL-induced synaptic signaling. This is especially true when taking into account recent findings regarding the effects of ADDL binding on insulin-receptor signaling discussed below.

Tau has also been shown to associate with the src-kinase family member Fyn, a tyrosine kinase up-regulated in the AD brain (Shirazi and Wood 1993). Fyn-knockout animals are protected against ADDL-induced cell death (Lambert et al. 1998), indicating a role for Fyn signaling in ADDL-induced AD progression. Fyn is activated by focal adhesion kinase in response to aggregated Aβ (Grace and Busciglio 2003; Zhang et al. 1996b) and mediates synaptic toxicity and memory impairment in transgenic mice (Chin et al. 2004, 2005). The connected roles of tau and Fyn were recently highlighted in a study demonstrating tau-dependent trafficking of Fyn to dendrites that is disrupted by engineering a tau construct lacking a microtubule-binding domain (Ittner et al. 2010). The somatic mislocalization of Fyn led to reduced synaptotoxicity in a transgenic model of AD through uncoupling of NMDAR-mediated excitotoxicity.

4.4.3 Oligomers as a Link Between Diabetes and AD

Factors that contribute to the development and progression of AD are of great interest for both mechanistic and therapeutic reasons. Though outside of the purview of this chapter, much work has gone into defining a variety of contributing factors such as hypertension (Skoog et al. 1996), inflammation (Akiyama et al. 2000), and even linguistic acuity throughout life (Snowdon et al. 1996). Germane to the present discussion of the oligomeric basis of AD, however, is accumulating evidence that at least one putative risk factor—insulin dysfunction—may impact AD pathogenesis specifically by rendering the brain more susceptible to oligomeric toxins.

Until the 1970s, the brain was considered as an insulin-insensitive organ. This view changed following several key discoveries demonstrating the presence of insulin (Havrankova et al. 1978b) and its receptor (Havrankova et al. 1978a) in the rat central nervous system, specifically at postsynaptic sites (Abbott et al. 1999), where it plays a critical role in neuronal survival (Aizenman et al. 1986). Insulin is a hormone now known to have diverse involvements in the brain, including glucose homeostasis (Clarke et al. 1984), synaptic function (Chiu et al. 2008; van der Heide et al. 2006), neuronal survival (Ryu et al. 1999; Tanaka et al. 1995), and short- and long-term memory (Marks et al. 2009).

Given the positive effects of insulin signaling on brain function, particularly memory, there is growing interest in a possible role for insulin in dementia, and several lines of study have indicated that aberrant insulin signaling is in fact a risk factor. Epidemiological studies report an increased prevalence of dementia in individuals with diabetes mellitus, including an approximately twofold increase in the risk for developing AD (Leibson et al. 1997; Ott et al. 1999). Independent of diabetes, abnormally high or low insulin levels also increase the risk of dementia (Peila et al. 2004). Further evidence connecting AD and insulin dysregulation comes from the age-related decrease in insulin signaling in the rat central nervous system (Fernandes et al. 2001), as well as the high degree of comorbidity (Heron et al. 2009).

The epidemiological connection between diabetes and AD is supported and expanded by experiments using transgenic mice. Mouse models of diabetes exhibit AD-like pathology, e.g., elevated Aβ levels and tau phosphorylation, which are partially reduced by insulin treatments (Jolivalt et al. 2008). AD transgenic mice with induced type-1 or type-2 diabetes show exacerbated AD and diabetic phenotypes (Jolivalt et al. 2010; Ke et al. 2009; Plaschke et al. 2010; Takeda et al. 2010), indicating that common underlying mechanisms (i.e., insulin dysfunction) may be involved in the progression of diabetes and AD.

The relationship between ADDL and insulin signaling at the molecular and cellular level has been investigated *in vitro*. ADDL binding markedly antagonizes insulin signaling, as ADDL-bound neurons show inhibited insulin-receptor activity as well as a reduction in synaptic insulin receptors (De Felice et al. 2009; Zhao et al. 2008). ADDLs induce the removal of insulin receptors in a calcium-dependent mechanism, involving activity of the NMDAR, casein kinase II (CKII), and CaMKII. It is likely

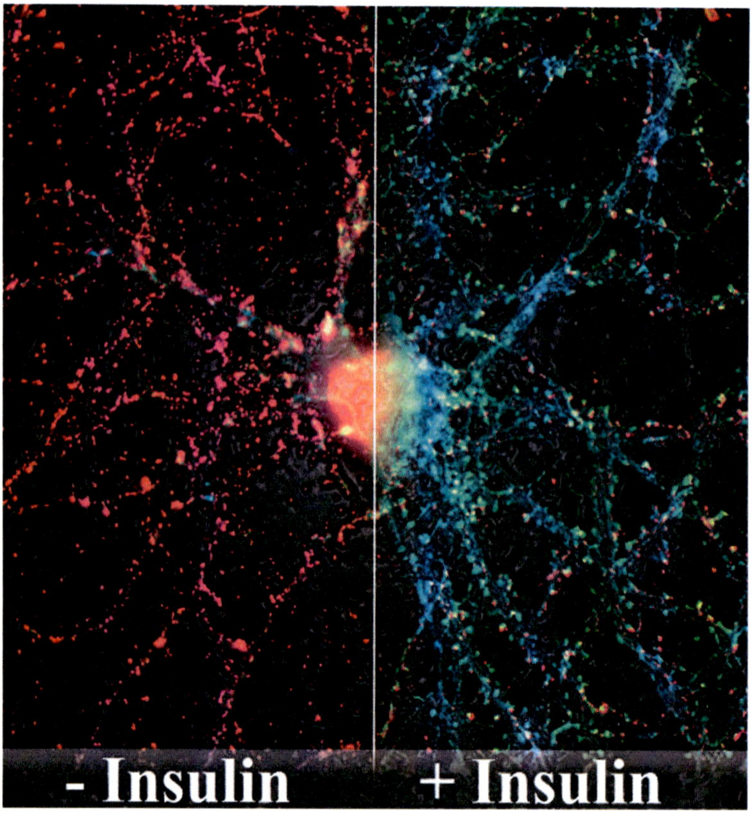

Fig. 4.4 Insulin signaling protects against ADDL-induced dendritic spine loss. A composite image showing two primary hippocampal neurons exposed to ADDLs (*red immunofluorescence*). Insulin pretreatment (*right panel*) protects against ADDL binding and loss of dendritic spines (*green drebrin immunofluorescence*)

that ADDLs act upstream of CKII, as CKII potentiates NMDAR activity (Lieberman and Mody 1999), which leads to CaMKII activation (Bayer et al. 2001; Tan et al. 1994). ADDLs and insulin appear to share common signaling elements, particularly the PI3K–Akt–mTOR signaling pathway. Further evidence that elevated mTOR signaling may underlie the observed cognitive defects in AD is that treatment with the mTOR inhibitor rapamycin protects against cognitive impairments in an AD mouse model (Caccamo et al. 2010). Insulin activation of this signaling pathway mediates several processes integral to synapse function, including receptor trafficking (Huang et al. 2004), synaptic plasticity (van der Heide et al. 2006), and protein synthesis (Lee et al. 2005). These findings provide evidence that aberrant insulin signaling resulting from ADDL treatment eventually leads to a state of insulin resistance.

Surprisingly, whereas ADDL treatment has an inhibitory effect on insulin signaling, the reverse is also true—pretreatment of neurons with insulin protects against ADDL binding and toxicity (Fig. 4.4; De Felice et al. 2009; Zhao et al. 2009). Signaling cascades downstream from the insulin receptor are required for this

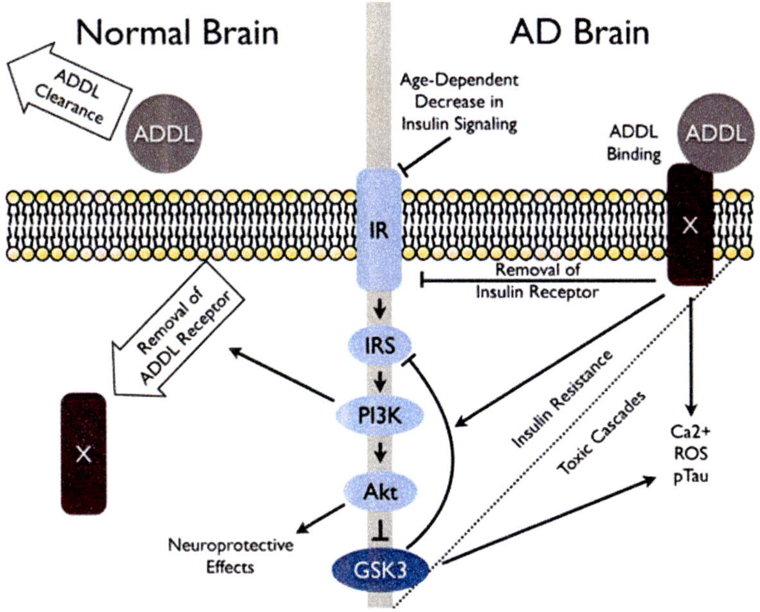

Fig. 4.5 Cross-talk between insulin- and ADDL-induced signaling pathways may underlie the development of sporadic AD. Under normal conditions, insulin signaling protects neurons from the toxic attack of ADDLs by several possible protective mechanisms: clearance of ADDLs, removal of ADDL receptors, or the disruption of ADDL-binding sites. Insulin signaling decreases as a function of age, making neurons vulnerable to the toxic attack by ADDLs. Upon ADDL binding, insulin resistance is reinforced through removal of insulin receptors from synapses and activation of GSK3β, which has an inhibitory effect on insulin-receptor substrate-1 (IRS). Once in a state of insulin resistance, neurons are further susceptible to ADDL binding, and subsequent toxicity

protection, as inhibition of insulin receptor activity can elevate the degree to which healthy neurons are targeted by ADDLs (De Felice et al. 2009). The co-antagonistic relationship between insulin and ADDL signaling provides a mechanism that could explain the development of sporadic AD, in which the age-dependent decline in insulin signaling (Fernandes et al. 2001) increases the susceptibility of neurons to the synaptotoxic attack of ADDLs (Fig. 4.5). Upon binding, ADDLs further hinder protective insulin signaling cascades, ultimately resulting in synapse loss and neuronal death. This hypothesis for the induction of sporadic AD predicts that methods of strengthening insulin signaling could potentially stave off AD pathogenesis.

4.5 AD Diagnosis and Therapy Based on the Toxic-Oligomer Hypothesis

The concept of oligomers as distinct and biologically active species is enabling improved diagnostic and therapeutic strategies to detect and treat AD.

4.5.1 ADDLs as a Biomarker for Early Detection of AD

AD is characterized by a "pre-clinical" phase during which the molecular events (e.g., ADDL binding) leading to the development of symptoms are not diagnosable using current tools such as neurological interviews (Arendt 2009). Due to the lack of curative treatments for late-stage AD, identifying biomarkers for early AD diagnosis will be a critical milestone for applying and developing AD treatments that target a phase when disease symptoms are still likely to be reversible. Proteomic analyses of AD and control tissues have been used to catalogue proteins in brain, CSF, and plasma that can be used as biomarkers of AD (Craig-Schapiro et al. 2009; Korolainen et al. 2010). However, adoption of a single biomarker has proven elusive due to the spectrum of pathologies that often accompany AD. Currently, reduction in total Aβ42 combined with increase in total and phosphorylated tau in CSF is believed to be a hallmark of the disease (Blennow and Hampel 2003). There is also evidence that this pattern of biomarkers can be used to predict progression from mild cognitive impairment to AD (Herukka et al. 2005).

In contrast to the decreased levels of total Aβ in CSF, ADDLs show an increase in the CSF of AD patients, as measured by the ultrasensitive biobarcode assay (Georganopoulou et al. 2005). Because ADDLs are elevated in the CSF of AD patients and manifest prior to other AD pathologies (Lacor et al. 2004), they may serve as a useful indicator of the initial onset of AD.

4.5.2 Therapeutic Potential of Conformation-Sensitive Antibodies

Conformation-dependent antibodies obtained after animal vaccination with ADDLs have demonstrated a disease-dependent presence of oligomers in human brain (Gong et al. 2003; Lacor et al. 2004; Lambert et al. 2001, 2007). The therapeutic exploitation of such antibodies for treating AD is also underway. After early failures targeting amyloid plaques and total soluble Aβ (including monomers), there are now several dozen therapeutic antibodies, including ADDL-specific antibodies, under evaluation in clinical trials (Krafft and Klein 2010). The ineffectiveness of non-conformation-sensitive antibodies as AD therapeutics is not surprising, since the small fraction of administered antibody that reaches the brain is likely to be further depleted by Aβ monomers and insoluble Aβ deposits (Lambert et al. 2009). Benefits of oligomer-selective antibodies as therapeutic agents include their capacity to interact specifically with oligomers in a primarily monomeric milieu (Kayed et al. 2003; Lambert et al. 2001), a regionally specific detection *in vivo* consistent with AD pathology [high level of oligomers in the cortex *versus* background detection in cerebellum (Lacor et al. 2004), and the efficient blockade of both synthetic and human-brain-derived oligomers binding to neuronal surfaces and the consequent prevention of oligomer-induced pathological responses (Lambert et al. 2001); reviewed in

(Lambert et al. 2009)]. These antibodies may also remove bound ADDLs from synapses (Pitt et al. 2009). Immunotherapy using conformation-sensitive antibodies has been successfully tested *in vivo* using animal models, showing reversal of both tau and plaque pathologies (Oddo et al. 2004) in addition to improved performance in behavioral tests (Hillen et al. 2010; Lee et al. 2006). One potential concern regarding AD immunotherapy is increased intracerebral hemorrhage, as reported in APP transgenic mice treated with anti-Aβ antibodies (Wilcock et al. 2004). To this end, oligomer-specific, single-chain variable-domain (scFv) antibodies have already been generated (Wang et al. 2009), providing another option to neutralize toxic oligomers *in vivo* without triggering Fc-mediated inflammation.

4.5.3 Modulation of Insulin Signaling

Though the precise mechanism of insulin protection against ADDL targeting remains unclear, insulin has been hypothesized to protect neurons against ADDL binding and toxicity through various endocytic mechanisms. One such mechanism is insulin-initiated endocytosis of ADDLs by neurons and glia (Zhao et al. 2009). Alternatively, insulin could promote removal of ADDL-binding sites from the plasma membrane. As discussed above, a possible ADDL-binding site is the AMPAR (Zhao et al. 2010), which is endocytosed following treatment with insulin in a clathrin-mediated mechanism (Man et al. 2000) involving PI3K/PKC (Huang et al. 2004) and phosphorylation of tyrosine residues within the GluR2 *C*-terminus (Ahmadian et al. 2004).

The protective effect of insulin against ADDLs recommends the use of compounds that enhance insulin function as AD therapeutics. Rosiglitazone is an antidiabetic compound that acts as a potent activator of peroxisome proliferator-activated receptor γ (PPARγ) (Lehmann et al. 1995). PPARγ plays an important role in adipocyte generation and insulin signaling (Zhang et al. 1996a). The protective effects of PPARγ agonists are likely the result of enhanced insulin signaling, although their regulation of calcium influx through voltage-gated calcium channels and NMDARs may also provide neuroprotective effects (Pancani et al. 2009). Rosiglitazone treatment augments protective effects of submaximal insulin doses against ADDL toxicity *in vitro* (De Felice et al. 2009). Additionally, experiments *in vivo* have shown positive effects of rosiglitazone against memory impairment, amyloid burden, and tau neuropathology in AD mouse models (Escribano et al. 2010, 2009). NSAIDs, another class of potential PPARγ agonists, have also shown therapeutic effects against tau and amyloid pathology *in vivo* (McKee et al. 2008). Despite the above evidence of protective effects of PPARγ agonists against ADDL toxicity, clinical trials with rosiglitazone have failed to show a significant benefit in AD patients.

Why was rosiglitazone unsuccessful in clinical trials? A possible explanation is that an effective dose was not achieved. Reports of cardiac failures resulting from rosiglitazone administration (Graham et al. 2010; Risner et al. 2006) necessitated use of relatively low doses, likely resulting in inadequate levels of the drug reaching the central nervous system. In addition, because a period of chronic oligomer activity is expected to lead to inactivation of neuroprotective insulin-signaling pathways, rosiglitazone might be ineffective when introduced late in the course of the disease. According to this framework, a drug such as rosiglitazone should be applied either alongside a complementary strategy to remove the oligomers responsible for the dampening of insulin signaling or, more preferably, before the presence and activity of oligomers is widespread. Improved AD biomarkers should therefore enable a more timely application of similar strategies to boost protective neuronal signaling pathways.

4.5.4 Natural Compounds Affecting Oligomer Structure

As Aβ oligomerization appears to be a critical step in the initiation of AD pathology, considerable research has focused on preventing the formation of toxic oligomeric species. Many natural compounds, e.g., *Ginkgo biloba* and curcumin, have been found to suppress pathology associated with ADDLs (Tchantchou et al. 2009), likely by altering Aβ polymerization (Wu et al. 2006; Yang et al. 2005). Studies using several phenolic compounds have shown their ability to prevent pre-fibrillar oligomerization, fibril formation, and cytotoxicity in PC12 cells and in Tg2576 mice as well as cognitive decline typically seen in Tg2576 mice (De Felice et al. 2004; Ono et al. 2008; Wang et al. 2008, 2004b; Yu et al. 2009). Salvianolic acid B, a polyphenolic compound derived from the root of *Salvia miltiorrhiza*, disrupts aggregation of Aβ into fibrils and protects against cytotoxic effects of high Aβ doses (Durairajan et al. 2008). Oleuropein, a compound extracted from olive leaves, has exhibited a non-covalent interaction with Aβ40 peptide, although structural or functional consequences of this interaction were not reported (Bazoti et al. 2008, 2006).

We recently reported the anti-ADDL activity of oleocanthal (OC) (Pitt et al. 2009), an olive-oil-derived, phenol-containing NSAID (Beauchamp et al. 2005; Smith et al. 2005) capable of disrupting tau fibrillation (Li et al. 2009b). Although OC, like other phenol compounds, does alter Aβ polymerization, the effect is unique—Aβ aggregation appears to be potentiated. OC increases the size of Aβ oligomers while protecting neurons from their synaptopathological effects (Pitt et al. 2009). Similar results have been reported with the polyphenolic ellagic acid, which promotes fibril formation and prevents cellular toxicity of Aβ (Feng et al. 2009), making these findings at odds with the current dogma that therapeutic compounds must cause Aβ disaggregation. Still, if reducing toxic oligomer burden is truly therapeutic, one strategy is to promote the transformation of soluble species to more benign insoluble species. There is evolutionary precedent for this idea in the case of Pmel17. A so-called "functional amyloid," the sequence of Pmel17 was surmised to have evolved to allow sufficiently rapid aggregation so as to reduce the time spent as potentially toxic soluble oligomers (Fowler et al. 2006).

Another natural compound capable of altering ADDL structure is scyllo-inositol. Scyllo-inositol rescues LTP in mouse hippocampi treated with ADDLs and restores cognitive abilities in rats injected with ADDLs (Townsend et al. 2006). Scyllo-inositol likely interacts directly with Aβ through hydrogen bonding to inhibit fibril formation (Sun et al. 2008a) and, as a consequence, it increases soluble Aβ42 levels (Hawkes et al. 2010). Given that ADDLs (Georganopoulou et al. 2005; Gong et al. 2003) and Aβ42 (Zhuo et al. 2008) appear to be more abundant in AD patients and transgenic mouse models, and the common view of Aβ clearance as a therapeutic endpoint, this rise in Aβ42 after therapeutic application of scyllo-inositol is unexpected. The use of scyllo-inositol to treat AD patients has been further complicated by a phase-II clinical trial reporting nine deaths in groups receiving the two highest doses (2,000 and 1,000 mg twice daily). It is possible that these deaths and other serious adverse events were caused by the breakdown of less toxic Aβ aggregates into highly toxic oligomers. However, it is yet to be established that these effects are the result of Aβ disaggregation, or are even related to scyllo-inositol treatments at all.

4.6 Summary and Future Prospects

A search of the literature currently reveals over 1,000 articles discussing Aβ oligomers. These species are present in AD brain and CSF and their activity at the cellular level is consistent with the clinical and pathological hallmarks of AD. With the retasking of the "amyloid cascade hypothesis" to feature the toxic action of oligomeric Aβ, these species have become an important tool for determining the molecular mechanisms underlying AD, as well as for diagnostics and therapeutics. While many previous, current, and proposed strategies focus on detecting and removing toxins from the brain—now including oligomers—effort is still needed to determine why they form at all. The future of this effort will rely on determining what conditions lead to an unhealthy brain that produces (or overproduces) these species. Insulin signaling in the aging brain is one such area that could provide a link between a healthy brain and one that develops AD. If this is true, we expect new models of AD focusing on the susceptibility of the diabetic brain to sporadic AD to shed light on this connection in the human population.

References

Abbott MA, Wells DG, Fallon JR (1999) The insulin receptor tyrosine kinase substrate p58/53 and the insulin receptor are components of CNS synapses. J Neurosci 19:7300–7308

Ahmadian G, Ju W, Liu L, Wyszynski M, Lee SH, Dunah AW, Taghibiglou C, Wang Y, Lu J, Wong TP, Sheng M, Wang YT (2004) Tyrosine phosphorylation of GluR2 is required for insulin-stimulated AMPA receptor endocytosis and LTD. EMBO J 23:1040–1050

Aizenman Y, Weichsel ME Jr, De Vellis J (1986) Changes in insulin and transferrin requirements of pure brain neuronal cultures during embryonic development. Proc Natl Acad Sci USA 83:2263–2266

Akiyama H, Barger S, Barnum S, Bradt B, Bauer J, Cole GM, Cooper NR, Eikelenboom P, Emmerling M, Fiebich BL, Finch CE, Frautschy S, Griffin WS, Hampel H, Hull M, Landreth G, Lue L, Mrak R, Mackenzie IR, Mcgeer PL, O'banion MK, Pachter J, Pasinetti G, Plata-Salaman C, Rogers J, Rydel R, Shen Y, Streit W, Strohmeyer R, Tooyoma I, Van Muiswinkel FL, Veerhuis R, Walker D, Webster S, Wegrzyniak B, Wenk G, Wyss-Coray T (2000) Inflammation and Alzheimer's disease. Neurobiol Aging 21:383–421

Alarcon JM, Brito JA, Hermosilla T, Atwater I, Mears D, Rojas E (2006) Ion channel formation by Alzheimer's disease amyloid β-peptide (Aβ40) in unilamellar liposomes is determined by anionic phospholipids. Peptides 27:95–104

Arendt T (2009) Synaptic degeneration in Alzheimer's disease. Acta Neuropathol 118:167–179

Balducci C, Beeg M, Stravalaci M, Bastone A, Sclip A, Biasini E, Tapella L, Colombo L, Manzoni C, Borsello T, Chiesa R, Gobbi M, Salmona M, Forloni G (2010) Synthetic amyloid-β oligomers impair long-term memory independently of cellular prion protein. Proc Natl Acad Sci USA 107:2295–2300

Barghorn S, Nimmrich V, Striebinger A, Krantz C, Keller P, Janson B, Bahr M, Schmidt M, Bitner RS, Harlan J, Barlow E, Ebert U, Hillen H (2005) Globular amyloid β-peptide oligomer—a homogenous and stable neuropathological protein in Alzheimer's disease. J Neurochem 95:834–847

Bayer TA, Wirths O (2010) Intracellular accumulation of amyloid-β—a predictor for synaptic dysfunction and neuron loss in Alzheimer's disease. Front Aging Neurosci 2:8

Bayer KU, De Koninck P, Leonard AS, Hell JW, Schulman H (2001) Interaction with the NMDA receptor locks CaMKII in an active conformation. Nature 411:801–805

Bazoti FN, Bergquist J, Markides KE, Tsarbopoulos A (2006) Noncovalent interaction between amyloid-β-peptide (1–40) and oleuropein studied by electrospray ionization mass spectrometry. J Am Soc Mass Spectrom 17:568–575

Bazoti FN, Bergquist J, Markides K, Tsarbopoulos A (2008) Localization of the noncovalent binding site between amyloid-β-peptide and oleuropein using electrospray ionization FT-ICR mass spectrometry. J Am Soc Mass Spectrom 19:1078–1085

Beauchamp GK, Keast RS, Morel D, Lin J, Pika J, Han Q, Lee CH, Smith AB, Breslin PA (2005) Phytochemistry: ibuprofen-like activity in extra-virgin olive oil. Nature 437:45–46

Benito E, Barco A (2010) CREB's control of intrinsic and synaptic plasticity: Implications for CREB-dependent memory models. Trends Neurosci 33:230–240

Bhaskar K, Miller M, Chludzinski A, Herrup K, Zagorski M, Lamb BT (2009) The PI3K-Akt-mTOR pathway regulates Aβ oligomer induced neuronal cell cycle events. Mol Neurodegener 4:14

Bitan G, Kirkitadze MD, Lomakin A, Vollers SS, Benedek GB, Teplow DB (2003) Amyloid β-protein (Aβ) assembly: Aβ40 and Aβ42 oligomerize through distinct pathways. Proc Natl Acad Sci USA 100:330–335

Blennow K, Hampel H (2003) CSF markers for incipient Alzheimer's disease. Lancet Neurol 2:605–613

Boutaud O, Montine TJ, Chang L, Klein WL, Oates JA (2006) PGH2-derived levuglandin adducts increase the neurotoxicity of amyloid β1–42. J Neurochem 96:917–923

Braak H, Braak E (1991) Neuropathological stageing of Alzheimer-related changes. Acta Neuropathol 82:239–259

Caccamo A, Majumder S, Richardson A, Strong R, Oddo S (2010) Molecular interplay between mammalian target of rapamycin (mTOR), amyloid-β, and Tau: effects on cognitive impairments. J Biol Chem 285:13107–13120

Calella AM, Farinelli M, Nuvolone M, Mirante O, Moos R, Falsig J, Mansuy IM, Aguzzi A (2010) Prion protein and Aβ-related synaptic toxicity impairment. EMBO Mol Med 2:306–314

Chang L, Bakhos L, Wang Z, Venton DL, Klein WL (2003) Femtomole immunodetection of synthetic and endogenous amyloid-β oligomers and its application to Alzheimer's disease drug candidate screening. J Mol Neurosci 20:305–313

Chin J, Palop JJ, Yu GQ, Kojima N, Masliah E, Mucke L (2004) Fyn kinase modulates synaptotoxicity, but not aberrant sprouting, in human amyloid precursor protein transgenic mice. J Neurosci 24:4692–4697

Chin J, Palop JJ, Puolivali J, Massaro C, Bien-Ly N, Gerstein H, Scearce-Levie K, Masliah E, Mucke L (2005) Fyn kinase induces synaptic and cognitive impairments in a transgenic mouse model of Alzheimer's disease. J Neurosci 25:9694–9703

Chiu SL, Chen CM, Cline HT (2008) Insulin receptor signaling regulates synapse number, dendritic plasticity, and circuit function *in vivo*. Neuron 58:708–719

Chromy BA, Nowak RJ, Lambert MP, Viola KL, Chang L, Velasco PT, Jones BW, Fernandez SJ, Lacor PN, Horowitz P, Finch CE, Krafft GA, Klein WL (2003) Self-assembly of Aβ(1–42) into globular neurotoxins. Biochemistry 42:12749–12760

Citri A, Malenka RC (2008) Synaptic plasticity: multiple forms, functions, and mechanisms. Neuropsychopharmacology 33:18–41

Clarke DW, Boyd FT Jr, Kappy MS, Raizada MK (1984) Insulin binds to specific receptors and stimulates 2-deoxy-$_D$-glucose uptake in cultured glial cells from rat brain. J Biol Chem 259:11672–11675

Cleary JP, Walsh DM, Hofmeister JJ, Shankar GM, Kuskowski MA, Selkoe DJ, Ashe KH (2005) Natural oligomers of the amyloid-β protein specifically disrupt cognitive function. Nat Neurosci 8:79–84

Craig-Schapiro R, Fagan AM, Holtzman DM (2009) Biomarkers of Alzheimer's disease. Neurobiol Dis 35:128–140

De Felice FG, Vieira MN, Saraiva LM, Figueroa-Villar JD, Garcia-Abreu J, Liu R, Chang L, Klein WL, Ferreira ST (2004) Targeting the neurotoxic species in Alzheimer's disease: inhibitors of Aβ oligomerization. FASEB J 18:1366–1372

De Felice FG, Velasco PT, Lambert MP, Viola K, Fernandez SJ, Ferreira ST, Klein WL (2007) Aβ oligomers induce neuronal oxidative stress through an *N*-methyl-$_D$-aspartate receptor-dependent mechanism that is blocked by the Alzheimer drug memantine. J Biol Chem 282:11590–11601

De Felice FG, Wu D, Lambert MP, Fernandez SJ, Velasco PT, Lacor PN, Bigio EH, Jerecic J, Acton PJ, Shughrue PJ, Chen-Dodson E, Kinney GG, Klein WL (2008) Alzheimer's disease-type neuronal tau hyperphosphorylation induced by Aβ oligomers. Neurobiol Aging 29:1334–1347

De Felice FG, Vieira MN, Bomfim TR, Decker H, Velasco PT, Lambert MP, Viola KL, Zhao WQ, Ferreira ST, Klein WL (2009) Protection of synapses against Alzheimer's-linked toxins: insulin signaling prevents the pathogenic binding of Aβ oligomers. Proc Natl Acad Sci USA 106:1971–1976

Decker H, Lo KY, Unger SM, Ferreira ST, Silverman MA (2010) Amyloid-β peptide oligomers disrupt axonal transport through an NMDA receptor-dependent mechanism that is mediated by glycogen synthase kinase 3β in primary cultured hippocampal neurons. J Neurosci 30:9166–9171

Deshpande A, Kawai H, Metherate R, Glabe CG, Busciglio J (2009) A role for synaptic zinc in activity-dependent Aβ oligomer formation and accumulation at excitatory synapses. J Neurosci 29:4004–4015

Dineley KT, Kayed R, Neugebauer V, Fu Y, Zhang W, Reese LC, Taglialatela G (2010) Amyloid-β oligomers impair fear conditioned memory in a calcineurin-dependent fashion in mice. J Neurosci Res 88:2923–2932

Douglass AD, Vale RD (2005) Single-molecule microscopy reveals plasma membrane microdomains created by protein-protein networks that exclude or trap signaling molecules in T cells. Cell 121:937–950

Durairajan SS, Yuan Q, Xie L, Chan WS, Kum WF, Koo I, Liu C, Song Y, Huang JD, Klein WL, Li M (2008) Salvianolic acid B inhibits Aβ fibril formation and disaggregates preformed fibrils and protects against Aβ-induced cytotoxicity. Neurochem Int 52:741–750

Eckert A, Hauptmann S, Scherping I, Meinhardt J, Rhein V, Drose S, Brandt U, Fandrich M, Muller WE, Gotz J (2008) Oligomeric and fibrillar species of β-amyloid (Aβ42) both impair mitochondrial function in P301L tau transgenic mice. J Mol Med 86:1255–1267

Erol A (2009) Unraveling the molecular mechanisms behind the metabolic basis of sporadic Alzheimer's disease. J Alzheimers Dis 17:267–276

Escribano L, Simon AM, Perez-Mediavilla A, Salazar-Colocho P, Del Rio J, Frechilla D (2009) Rosiglitazone reverses memory decline and hippocampal glucocorticoid receptor down-regulation in an Alzheimer's disease mouse model. Biochem Biophys Res Commun 379:406–410

Escribano L, Simon AM, Gimeno E, Cuadrado-Tejedor M, Lopez De Maturana R, Garcia-Osta A, Ricobaraza A, Perez-Mediavilla A, Del Rio J, Frechilla D (2010) Rosiglitazone rescues memory impairment in Alzheimer's transgenic mice: mechanisms involving a reduced amyloid and tau pathology. Neuropsychopharmacology 35:1593–1604

Feng Y, Yang SG, Du XT, Zhang X, Sun XX, Zhao M, Sun GY, Liu RT (2009) Ellagic acid promotes Aβ42 fibrillization and inhibits Aβ42-induced neurotoxicity. Biochem Biophys Res Commun 390:1250–1254

Fernandes ML, Saad MJ, Velloso LA (2001) Effects of age on elements of insulin-signaling pathway in central nervous system of rats. Endocrine 16:227–234

Fonseca AC, Resende R, Oliveira CR, Pereira CM (2010) Cholesterol and statins in Alzheimer's disease: current controversies. Exp Neurol 223:282–293

Fowler DM, Koulov AV, Alory-Jost C, Marks MS, Balch WE, Kelly JW (2006) Functional amyloid formation within mammalian tissue. PLoS Biol 4:e6

Frackowiak J, Zoltowska A, Wisniewski HM (1994) Non-fibrillar β-amyloid protein is associated with smooth muscle cells of vessel walls in Alzheimer disease. J Neuropathol Exp Neurol 53:637–645

Fukumoto H, Tokuda T, Kasai T, Ishigami N, Hidaka H, Kondo M, Allsop D, Nakagawa M (2010) High-molecular-weight β-amyloid oligomers are elevated in cerebrospinal fluid of Alzheimer patients. FASEB J 24:2716–2726

Gandy S, Simon AJ, Steele JW, Lublin AL, Lah JJ, Walker LC, Levey AI, Krafft GA, Levy E, Checler F, Glabe C, Bilker WB, Abel T, Schmeidler J, Ehrlich ME (2010) Days to criterion as an indicator of toxicity associated with human Alzheimer amyloid-β oligomers. Ann Neurol 68:220–230

Geng L, Zhang HL, Peng HB (2009) The formation of acetylcholine receptor clusters visualized with quantum dots. BMC Neurosci 10:80

Georganopoulou DG, Chang L, Nam JM, Thaxton CS, Mufson EJ, Klein WL, Mirkin CA (2005) Nanoparticle-based detection in cerebral spinal fluid of a soluble pathogenic biomarker for Alzheimer's disease. Proc Natl Acad Sci USA 102:2273–2276

Giuffrida ML, Caraci F, Pignataro B, Cataldo S, De Bona P, Bruno V, Molinaro G, Pappalardo G, Messina A, Palmigiano A, Garozzo D, Nicoletti F, Rizzarelli E, Copani A (2009) β-Amyloid monomers are neuroprotective. J Neurosci 29:10582–10587

Gong Y, Chang L, Viola KL, Lacor PN, Lambert MP, Finch CE, Krafft GA, Klein WL (2003) Alzheimer's disease-affected brain: presence of oligomeric Aβ ligands (ADDLs) suggests a molecular basis for reversible memory loss. Proc Natl Acad Sci USA 100:10417–10422

Gouras GK, Tsai J, Naslund J, Vincent B, Edgar M, Checler F, Greenfield JP, Haroutunian V, Buxbaum JD, Xu H, Greengard P, Relkin NR (2000) Intraneuronal Aβ42 accumulation in human brain. Am J Pathol 156:15–20

Grace EA, Busciglio J (2003) Aberrant activation of focal adhesion proteins mediates fibrillar amyloid β-induced neuronal dystrophy. J Neurosci 23:493–502

Graham DJ, Ouellet-Hellstrom R, Macurdy TE, Ali F, Sholley C, Worrall C, Kelman JA (2010) Risk of acute myocardial infarction, stroke, heart failure, and death in elderly Medicare patients treated with rosiglitazone or pioglitazone. JAMA 304:411–418

Griffin RJ, Moloney A, Kelliher M, Johnston JA, Ravid R, Dockery P, O'connor R, O'neill C (2005) Activation of Akt/PKB, increased phosphorylation of Akt substrates and loss and altered distribution of Akt and PTEN are features of Alzheimer's disease pathology. J Neurochem 93:105–117

Grundke-Iqbal I, Iqbal K, Quinlan M, Tung YC, Zaidi MS, Wisniewski HM (1986) Microtubule-associated protein tau. A component of Alzheimer paired helical filaments. J Biol Chem 261:6084–6089

Haass C, Selkoe DJ (2007) Soluble protein oligomers in neurodegeneration: lessons from the Alzheimer's amyloid β-peptide. Nat Rev Mol Cell Biol 8:101–112

Hartley DM, Walsh DM, Ye CP, Diehl T, Vasquez S, Vassilev PM, Teplow DB, Selkoe DJ (1999) Protofibrillar intermediates of amyloid β-protein induce acute electrophysiological changes and progressive neurotoxicity in cortical neurons. J Neurosci 19:8876–8884

Havrankova J, Roth J, Brownstein M (1978a) Insulin receptors are widely distributed in the central nervous system of the rat. Nature 272:827–829

Havrankova J, Schmechel D, Roth J, Brownstein M (1978b) Identification of insulin in rat brain. Proc Natl Acad Sci USA 75:5737–5741

Hawkes CA, Deng LH, Shaw JE, Nitz M, Mclaurin J (2010) Small molecule β-amyloid inhibitors that stabilize protofibrillar structures *in vitro* improve cognition and pathology in a mouse model of Alzheimer's disease. Eur J Neurosci 31:203–213

Heron M, Hoyert DL, Murphy SL, Xu J, Kochanek KD, Tejada-Vera B (2009) Deaths: final data for 2006. Natl Vital Stat Rep 57:1–134

Hertel C, Terzi E, Hauser N, Jakob-Rotne R, Seelig J, Kemp JA (1997) Inhibition of the electrostatic interaction between β-amyloid peptide and membranes prevents β-amyloid-induced toxicity. Proc Natl Acad Sci USA 94:9412–9416

Herukka SK, Hallikainen M, Soininen H, Pirttila T (2005) CSF Aβ42 and tau or phosphorylated tau and prediction of progressive mild cognitive impairment. Neurology 64:1294–1297

Heurtaux T, Michelucci A, Losciuto S, Gallotti C, Felten P, Dorban G, Grandbarbe L, Morga E, Heuschling P (2010) Microglial activation depends on β-amyloid conformation: role of the formylpeptide receptor 2. J Neurochem 114:576–586

Hillen H, Barghorn S, Striebinger A, Labkovsky B, Muller R, Nimmrich V, Nolte MW, Perez-Cruz C, Van Der Auwera I, Van Leuven F, Van Gaalen M, Bespalov AY, Schoemaker H, Sullivan JP, Ebert U (2010) Generation and therapeutic efficacy of highly oligomer-specific β-amyloid antibodies. J Neurosci 30:10369–10379

Hsieh H, Boehm J, Sato C, Iwatsubo T, Tomita T, Sisodia S, Malinow R (2006) AMPAR removal underlies Aβ-induced synaptic depression and dendritic spine loss. Neuron 52:831–843

Huang CC, Lee CC, Hsu KS (2004) An investigation into signal transduction mechanisms involved in insulin-induced long-term depression in the CA1 region of the hippocampus. J Neurochem 89:217–231

Ittner LM, Ke YD, Delerue F, Bi M, Gladbach A, Van Eersel J, Wolfing H, Chieng BC, Christie MJ, Napier IA, Eckert A, Staufenbiel M, Hardeman E, Gotz J (2010) Dendritic function of tau mediates amyloid-β toxicity in Alzheimer's disease mouse models. Cell 142:387–397

Jolivalt CG, Lee CA, Beiswenger KK, Smith JL, Orlov M, Torrance MA, Masliah E (2008) Defective insulin signaling pathway and increased glycogen synthase kinase-3 activity in the brain of diabetic mice: parallels with Alzheimer's disease and correction by insulin. J Neurosci Res 86:3265–3274

Jolivalt CG, Hurford R, Lee CA, Dumaop W, Rockenstein E, Masliah E (2010) Type 1 diabetes exaggerates features of Alzheimer's disease in APP transgenic mice. Exp Neurol 223:422–431

Kang JE, Lim MM, Bateman RJ, Lee JJ, Smyth LP, Cirrito JR, Fujiki N, Nishino S, Holtzman DM (2009) Amyloid-β dynamics are regulated by orexin and the sleep-wake cycle. Science 326:1005–1007

Kayed R, Head E, Thompson JL, Mcintire TM, Milton SC, Cotman CW, Glabe CG (2003) Common structure of soluble amyloid oligomers implies common mechanism of pathogenesis. Science 300:486–489

Ke YD, Delerue F, Gladbach A, Gotz J, Ittner LM (2009) Experimental diabetes mellitus exacerbates tau pathology in a transgenic mouse model of Alzheimer's disease. PLoS One 4:e7917

Kemp N, Bashir ZI (2001) Long-term depression: a cascade of induction and expression mechanisms. Prog Neurobiol 65:339–365

Kessels HW, Nguyen LN, Nabavi S, Malinow R (2010) The prion protein as a receptor for amyloid-β. Nature 466:E3–E4, discussion E4–5

Kihara T, Shimohama S, Sawada H, Honda K, Nakamizo T, Shibasaki H, Kume T, Akaike A (2001) α7 nicotinic receptor transduces signals to phosphatidylinositol 3-kinase to block A β-amyloid-induced neurotoxicity. J Biol Chem 276:13541–13546

Kim HJ, Chae SC, Lee DK, Chromy B, Lee SC, Park YC, Klein WL, Krafft GA, Hong ST (2003) Selective neuronal degeneration induced by soluble oligomeric amyloid β protein. FASEB J 17:118–120

Kiyota T, Yamamoto M, Xiong H, Lambert MP, Klein WL, Gendelman HE, Ransohoff RM, Ikezu T (2009) CCL2 accelerates microglia-mediated Aβ oligomer formation and progression of neurocognitive dysfunction. PLoS One 4:e6197

Klein WL (2002) Aβ toxicity in Alzheimer's disease: globular oligomers (ADDLs) as new vaccine and drug targets. Neurochem Int 41:345–352

Klein WL, Krafft GA, Finch CE (2001) Targeting small Aβ oligomers: the solution to an Alzheimer's disease conundrum? Trends Neurosci 24:219–224

Klein WL, Lacor PN, De Felice FG, Ferreira ST (2007) Molecules that disrupt memory circuits in Alzheimer's disease: the attack on synapses by Aβ oligomers (ADDLs). In: Bontempi B, Silva AJ, Christen Y (eds) Memories: molecules and Circuits. Springer, Berlin/Heidelberg

Koffie RM, Meyer-Luehmann M, Hashimoto T, Adams KW, Mielke ML, Garcia-Alloza M, Micheva KD, Smith SJ, Kim ML, Lee VM, Hyman BT, Spires-Jones TL (2009) Oligomeric amyloid β associates with postsynaptic densities and correlates with excitatory synapse loss near senile plaques. Proc Natl Acad Sci USA 106:4012–4017

Korolainen MA, Nyman TA, Aittokallio T, Pirttila T (2010) An update on clinical proteomics in Alzheimer's research. J Neurochem 112:1386–1414

Kovacs GG, Head MW, Hegyi I, Bunn TJ, Flicker H, Hainfellner JA, Mccardle L, Laszlo L, Jarius C, Ironside JW, Budka H (2002) Immunohistochemistry for the prion protein: comparison of different monoclonal antibodies in human prion disease subtypes. Brain Pathol 12:1–11

Krafft GA, Klein WL (2010) ADDLs and the signaling web that leads to Alzheimer's disease. Neuropharmacology 59:230–242

Kuo YM, Emmerling MR, Vigo-Pelfrey C, Kasunic TC, Kirkpatrick JB, Murdoch GH, Ball MJ, Roher AE (1996) Water-soluble Aβ (N-40, N-42) oligomers in normal and Alzheimer disease brains. J Biol Chem 271:4077–4081

Lacor PN (2007) Advances on the understanding of the origins of synaptic pathology in AD. Curr Genomics 8:486–508

Lacor PN, Buniel MC, Chang L, Fernandez SJ, Gong Y, Viola KL, Lambert MP, Velasco PT, Bigio EH, Finch CE, Krafft GA, Klein WL (2004) Synaptic targeting by Alzheimer's-related amyloid β oligomers. J Neurosci 24:10191–10200

Lacor PN, Buniel MC, Furlow PW, Clemente AS, Velasco PT, Wood M, Viola KL, Klein WL (2007) Aβ oligomer-induced aberrations in synapse composition, shape, and density provide a molecular basis for loss of connectivity in Alzheimer's disease. J Neurosci 27:796–807

Laferla FM, Green KN, Oddo S (2007) Intracellular amyloid-β in Alzheimer's disease. Nat Rev Neurosci 8:499–509

Lambert MP, Barlow AK, Chromy BA, Edwards C, Freed R, Liosatos M, Morgan TE, Rozovsky I, Trommer B, Viola KL, Wals P, Zhang C, Finch CE, Krafft GA, Klein WL (1998) Diffusible, nonfibrillar ligands derived from Aβ1–42 are potent central nervous system neurotoxins. Proc Natl Acad Sci USA 95:6448–6453

Lambert MP, Viola KL, Chromy BA, Chang L, Morgan TE, Yu J, Venton DL, Krafft GA, Finch CE, Klein WL (2001) Vaccination with soluble Aβ oligomers generates toxicity-neutralizing antibodies. J Neurochem 79:595–605

Lambert MP, Velasco PT, Chang L, Viola KL, Fernandez S, Lacor PN, Khuon D, Gong Y, Bigio EH, Shaw P, De Felice FG, Krafft GA, Klein WL (2007) Monoclonal antibodies that target pathological assemblies of Aβ. J Neurochem 100:23–35

Lambert MP, Velasco PT, Viola KL, Klein WL (2009) Targeting generation of antibodies specific to conformational epitopes of amyloid β-derived neurotoxins. CNS Neurol Disord Drug Targets 8:65–81

Larson J, Lynch G, Games D, Seubert P (1999) Alterations in synaptic transmission and long-term potentiation in hippocampal slices from young and aged PDAPP mice. Brain Res 840:23–35

Lashuel HA, Lansbury PT Jr (2006) Are amyloid diseases caused by protein aggregates that mimic bacterial pore-forming toxins? Q Rev Biophys 39:167–201

Laurén J, Gimbel DA, Nygaard HB, Gilbert JW, Strittmatter SM (2009) Cellular prion protein mediates impairment of synaptic plasticity by amyloid-β oligomers. Nature 457:1128–1132
Laurén J, Gimbel DA, Nygaard HB, Gilbert JW, Strittmatter SM (2010) Nature 466:E4–E5, Laurén et al. reply
Lee CC, Huang CC, Wu MY, Hsu KS (2005) Insulin stimulates postsynaptic density-95 protein translation via the phosphoinositide 3-kinase-Akt-mammalian target of rapamycin signaling pathway. J Biol Chem 280:18543–18550
Lee EB, Leng LZ, Zhang B, Kwong L, Trojanowski JQ, Abel T, Lee VM (2006) Targeting amyloid-β peptide (Aβ) oligomers by passive immunization with a conformation-selective monoclonal antibody improves learning and memory in Aβ precursor protein (APP) transgenic mice. J Biol Chem 281:4292–4299
Lee HK, Kumar P, Fu Q, Rosen KM, Querfurth HW (2009) The insulin/Akt signaling pathway is targeted by intracellular β-amyloid. Mol Biol Cell 20:1533–1544
Lehmann JM, Moore LB, Smith-Oliver TA, Wilkison WO, Willson TM, Kliewer SA (1995) An antidiabetic thiazolidinedione is a high affinity ligand for peroxisome proliferator-activated receptor γ (PPARγ). J Biol Chem 270:12953–12956
Leibson CL, Rocca WA, Hanson VA, Cha R, Kokmen E, O'brien PC, Palumbo PJ (1997) The risk of dementia among persons with diabetes mellitus: a population-based cohort study. Ann NY Acad Sci 826:422–427
Leon WC, Canneva F, Partridge V, Allard S, Ferretti MT, Dewilde A, Vercauteren F, Atifeh R, Ducatenzeiler A, Klein W, Szyf M, Alhonen L, Cuello AC (2010) A novel transgenic rat model with a full Alzheimer's-like amyloid pathology displays pre-plaque intracellular amyloid-β-associated cognitive impairment. J Alzheimers Dis 20:113–126
Lesné S, Koh MT, Kotilinek L, Kayed R, Glabe CG, Yang A, Gallagher M, Ashe KH (2006) A specific amyloid-β protein assembly in the brain impairs memory. Nature 440:352–357
Levy E, Carman MD, Fernandez-Madrid IJ, Power MD, Lieberburg I, Van Duinen SG, Bots GT, Luyendijk W, Frangione B (1990) Mutation of the Alzheimer's disease amyloid gene in hereditary cerebral hemorrhage, Dutch type. Science 248:1124–1126
Li S, Hong S, Shepardson NE, Walsh DM, Shankar GM, Selkoe D (2009a) Soluble oligomers of amyloid β protein facilitate hippocampal long-term depression by disrupting neuronal glutamate uptake. Neuron 62:788–801
Li W, Sperry JB, Crowe A, Trojanowski JQ, Smith AB 3rd, Lee VM (2009b) Inhibition of tau fibrillization by oleocanthal via reaction with the amino groups of tau. J Neurochem 110:1339–1351
Lieberman DN, Mody I (1999) Casein kinase-II regulates NMDA channel function in hippocampal neurons. Nat Neurosci 2:125–132
Lorenzo A, Yankner BA (1994) β-Amyloid neurotoxicity requires fibril formation and is inhibited by congo red. Proc Natl Acad Sci USA 91:12243–12247
Malchiodi-Albedi F, Contrusciere V, Raggi C, Fecchi K, Rainaldi G, Paradisi S, Matteucci A, Santini MT, Sargiacomo M, Frank C, Gaudiano MC, Diociaiuti M (2010) Lipid raft disruption protects mature neurons against amyloid oligomer toxicity. Biochim Biophys Acta 1802:406–415
Man HY, Lin JW, Ju WH, Ahmadian G, Liu L, Becker LE, Sheng M, Wang YT (2000) Regulation of AMPA receptor-mediated synaptic transmission by clathrin-dependent receptor internalization. Neuron 25:649–662
Mandelkow EM, Drewes G, Biernat J, Gustke N, Van Lint J, Vandenheede JR, Mandelkow E (1992) Glycogen synthase kinase-3 and the Alzheimer-like state of microtubule-associated protein tau. FEBS Lett 314:315–321
Marks DR, Tucker K, Cavallin MA, Mast TG, Fadool DA (2009) Awake intranasal insulin delivery modifies protein complexes and alters memory, anxiety, and olfactory behaviors. J Neurosci 29:6734–6751
Mckee AC, Carreras I, Hossain L, Ryu H, Klein WL, Oddo S, Laferla FM, Jenkins BG, Kowall NW, Dedeoglu A (2008) Ibuprofen reduces Aβ, hyperphosphorylated tau and memory deficits in Alzheimer mice. Brain Res 1207:225–236

Mclaurin J, Chakrabartty A (1997) Characterization of the interactions of Alzheimer β-amyloid peptides with phospholipid membranes. Eur J Biochem 245:355–363
Mitew S, Kirkcaldie MT, Halliday GM, Shepherd CE, Vickers JC, Dickson TC (2010) Focal demyelination in Alzheimer's disease and transgenic mouse models. Acta Neuropathol 119:567–577
Noguchi A, Matsumura S, Dezawa M, Tada M, Yanazawa M, Ito A, Akioka M, Kikuchi S, Sato M, Ideno S, Noda M, Fukunari A, Muramatsu S, Itokazu Y, Sato K, Takahashi H, Teplow DB, Nabeshima Y, Kakita A, Imahori K, Hoshi M (2009) Isolation and characterization of patient-derived, toxic, high mass amyloid β-protein (Aβ) assembly from Alzheimer disease brains. J Biol Chem 284:32895–32905
Nunomura A, Tamaoki T, Tanaka K, Motohashi N, Nakamura M, Hayashi T, Yamaguchi H, Shimohama S, Lee HG, Zhu X, Smith MA, Perry G (2010) Intraneuronal amyloid β accumulation and oxidative damage to nucleic acids in Alzheimer disease. Neurobiol Dis 37:731–737
Oda T, Pasinetti GM, Osterburg HH, Anderson C, Johnson SA, Finch CE (1994) Purification and characterization of brain clusterin. Biochem Biophys Res Commun 204:1131–1136
Oddo S, Caccamo A, Kitazawa M, Tseng BP, Laferla FM (2003) Amyloid deposition precedes tangle formation in a triple transgenic model of Alzheimer's disease. Neurobiol Aging 24:1063–1070
Oddo S, Billings L, Kesslak JP, Cribbs DH, Laferla FM (2004) Aβ immunotherapy leads to clearance of early, but not late, hyperphosphorylated tau aggregates via the proteasome. Neuron 43:321–332
Oddo S, Caccamo A, Tran L, Lambert MP, Glabe CG, Klein WL, Laferla FM (2006) Temporal profile of amyloid-β (Aβ) oligomerization in an *in vivo* model of Alzheimer disease. A link between Aβ and tau pathology. J Biol Chem 281:1599–1604
Ono K, Condron MM, Ho L, Wang J, Zhao W, Pasinetti GM, Teplow DB (2008) Effects of grape seed-derived polyphenols on amyloid β-protein self-assembly and cytotoxicity. J Biol Chem 283:32176–32187
Ott A, Stolk RP, Van Harskamp F, Pols HA, Hofman A, Breteler MM (1999) Diabetes mellitus and the risk of dementia: the Rotterdam study. Neurology 53:1937–1942
Pancani T, Phelps JT, Searcy JL, Kilgore MW, Chen KC, Porter NM, Thibault O (2009) Distinct modulation of voltage-gated and ligand-gated Ca2+ currents by PPAR-gamma agonists in cultured hippocampal neurons. J Neurochem 109:1800–1811
Peila R, Rodriguez BL, White LR, Launer LJ (2004) Fasting insulin and incident dementia in an elderly population of Japanese-American men. Neurology 63:228–233
Pitt J, Roth W, Lacor P, Smith AB 3rd, Blankenship M, Velasco P, De Felice F, Breslin P, Klein WL (2009) Alzheimer's-associated Aβ oligomers show altered structure, immunoreactivity and synaptotoxicity with low doses of oleocanthal. Toxicol Appl Pharmacol 240:189–197
Plaschke K, Kopitz J, Siegelin M, Schliebs R, Salkovic-Petrisic M, Riederer P, Hoyer S (2010) Insulin-resistant brain state after intracerebroventricular streptozotocin injection exacerbates Alzheimer-like changes in Tg2576 AβPP-overexpressing mice. J Alzheimers Dis 19:691–704
Podlisny MB, Ostaszewski BL, Squazzo SL, Koo EH, Rydell RE, Teplow DB, Selkoe DJ (1995) Aggregation of secreted amyloid β-protein into sodium dodecyl sulfate-stable oligomers in cell culture. J Biol Chem 270:9564–9570
Proctor DT, Coulson EJ, Dodd PR (2010) Reduction in post-synaptic scaffolding PSD-95 and SAP-102 protein levels in the Alzheimer inferior temporal cortex is correlated with disease pathology. J Alzheimers Dis 21:795–811
Renner M, Lacor PN, Velasco PT, Xu J, Contractor A, Klein WL, Triller A (2010) Deleterious effects of amyloid β oligomers acting as an extracellular scaffold for mGluR5. Neuron 66:739–754
Resende R, Ferreiro E, Pereira C, Resende De Oliveira C (2008) Neurotoxic effect of oligomeric and fibrillar species of amyloid-β peptide 1–42: involvement of endoplasmic reticulum calcium release in oligomer-induced cell death. Neuroscience 155:725–737
Rickle A, Bogdanovic N, Volkman I, Winblad B, Ravid R, Cowburn RF (2004) Akt activity in Alzheimer's disease and other neurodegenerative disorders. Neuroreport 15:955–959

Risner ME, Saunders AM, Altman JF, Ormandy GC, Craft S, Foley IM, Zvartau-Hind ME, Hosford DA, Roses AD (2006) Efficacy of rosiglitazone in a genetically defined population with mild-to-moderate Alzheimer's disease. Pharmacogenomics J 6:246–254

Roher AE, Ball MJ, Bhave SV, Wakade AR (1991) β-amyloid from Alzheimer disease brains inhibits sprouting and survival of sympathetic neurons. Biochem Biophys Res Commun 174:572–579

Roher AE, Palmer KC, Yurewicz EC, Ball MJ, Greenberg BD (1993) Morphological and biochemical analyses of amyloid plaque core proteins purified from Alzheimer disease brain tissue. J Neurochem 61:1916–1926

Ryu BR, Ko HW, Jou I, Noh JS, Gwag BJ (1999) Phosphatidylinositol 3-kinase-mediated regulation of neuronal apoptosis and necrosis by insulin and IGF-I. J Neurobiol 39:536–546

Sakono M, Zako T (2010) Amyloid oligomers: formation and toxicity of Aβ oligomers. FEBS J 277:1348–1358

Serrano F, Klann E (2004) Reactive oxygen species and synaptic plasticity in the aging hippocampus. Ageing Res Rev 3:431–443

Shankar GM, Li S, Mehta TH, Garcia-Munoz A, Shepardson NE, Smith I, Brett FM, Farrell MA, Rowan MJ, Lemere CA, Regan CM, Walsh DM, Sabatini BL, Selkoe DJ (2008) Amyloid-β protein dimers isolated directly from Alzheimer's brains impair synaptic plasticity and memory. Nat Med 14:837–842

Shirazi SK, Wood JG (1993) The protein tyrosine kinase, fyn, in Alzheimer's disease pathology. Neuroreport 4:435–437

Shrestha BR, Vitolo OV, Joshi P, Lordkipanidze T, Shelanski M, Dunaevsky A (2006) Amyloid β peptide adversely affects spine number and motility in hippocampal neurons. Mol Cell Neurosci 33:274–282

Shughrue PJ, Acton PJ, Breese RS, Zhao WQ, Chen-Dodson E, Hepler RW, Wolfe AL, Matthews M, Heidecker GJ, Joyce JG, Villarreal SA, Kinney GG (2010) Anti-ADDL antibodies differentially block oligomer binding to hippocampal neurons. Neurobiol Aging 31:189–202

Simonyi A, Schachtman TR, Christoffersen GR (2005) The role of metabotropic glutamate receptor 5 in learning and memory processes. Drug News Perspect 18:353–361

Skoog I, Lernfelt B, Landahl S, Palmertz B, Andreasson LA, Nilsson L, Persson G, Oden A, Svanborg A (1996) 15-year longitudinal study of blood pressure and dementia. Lancet 347:1141–1145

Small DH, Gasperini R, Vincent AJ, Hung AC, Foa L (2009) The role of Aβ-induced calcium dysregulation in the pathogenesis of Alzheimer's disease. J Alzheimers Dis 16:225–233

Smith AB 3rd, Han Q, Breslin PA, Beauchamp GK (2005) Synthesis and assignment of absolute configuration of (−)-oleocanthal: a potent, naturally occurring non-steroidal anti-inflammatory and anti-oxidant agent derived from extra virgin olive oils. Org Lett 7:5075–5078

Snowdon DA, Kemper SJ, Mortimer JA, Greiner LH, Wekstein DR, Markesbery WR (1996) Linguistic ability in early life and cognitive function and Alzheimer's disease in late life. Findings from the Nun Study. JAMA 275:528–532

Snyder EM, Nong Y, Almeida CG, Paul S, Moran T, Choi EY, Nairn AC, Salter MW, Lombroso PJ, Gouras GK, Greengard P (2005) Regulation of NMDA receptor trafficking by amyloid-β. Nat Neurosci 8:1051–1058

Sun Y, Zhang G, Hawkes CA, Shaw JE, Mclaurin J, Nitz M (2008a) Synthesis of scyllo-inositol derivatives and their effects on amyloid β peptide aggregation. Bioorg Med Chem 16:7177–7184

Sun ZK, Yang HQ, Pan J, Zhen H, Wang ZQ, Chen SD, Ding JQ (2008b) Protective effects of erythropoietin on tau phosphorylation induced by β-amyloid. J Neurosci Res 86:3018–3027

Takahashi RH, Almeida CG, Kearney PF, Yu F, Lin MT, Milner TA, Gouras GK (2004) Oligomerization of Alzheimer's β-amyloid within processes and synapses of cultured neurons and brain. J Neurosci 24:3592–3599

Takeda S, Sato N, Uchio-Yamada K, Sawada K, Kunieda T, Takeuchi D, Kurinami H, Shinohara M, Rakugi H, Morishita R (2010) Diabetes-accelerated memory dysfunction via cerebrovascular inflammation and Aβ deposition in an Alzheimer mouse model with diabetes. Proc Natl Acad Sci USA 107:7036–7041

Tan SE, Wenthold RJ, Soderling TR (1994) Phosphorylation of AMPA-type glutamate receptors by calcium/calmodulin-dependent protein kinase II and protein kinase C in cultured hippocampal neurons. J Neurosci 14:1123–1129

Tanaka M, Sawada M, Yoshida S, Hanaoka F, Marunouchi T (1995) Insulin prevents apoptosis of external granular layer neurons in rat cerebellar slice cultures. Neurosci Lett 199:37–40

Tchantchou F, Lacor PN, Cao Z, Lao L, Hou Y, Cui C, Klein WL, Luo Y (2009) Stimulation of neurogenesis and synaptogenesis by bilobalide and quercetin via common final pathway in hippocampal neurons. J Alzheimers Dis 18:787–798

Terry RD, Masliah E, Salmon DP, Butters N, Deteresa R, Hill R, Hansen LA, Katzman R (1991) Physical basis of cognitive alterations in Alzheimer's disease: synapse loss is the major correlate of cognitive impairment. Ann Neurol 30:572–580

Tomic JL, Pensalfini A, Head E, Glabe CG (2009) Soluble fibrillar oligomer levels are elevated in Alzheimer's disease brain and correlate with cognitive dysfunction. Neurobiol Dis 35:352–358

Tomiyama T, Nagata T, Shimada H, Teraoka R, Fukushima A, Kanemitsu H, Takuma H, Kuwano R, Imagawa M, Ataka S, Wada Y, Yoshioka E, Nishizaki T, Watanabe Y, Mori H (2008) A new amyloid β variant favoring oligomerization in Alzheimer's-type dementia. Ann Neurol 63:377–387

Tomiyama T, Matsuyama S, Iso H, Umeda T, Takuma H, Ohnishi K, Ishibashi K, Teraoka R, Sakama N, Yamashita T, Nishitsuji K, Ito K, Shimada H, Lambert MP, Klein WL, Mori H (2010) A mouse model of amyloid β oligomers: their contribution to synaptic alteration, abnormal tau phosphorylation, glial activation, and neuronal loss *in vivo*. J Neurosci 30:4845–4856

Tong L, Thornton PL, Balazs R, Cotman CW (2001) β-amyloid-(1–42) impairs activity-dependent cAMP-response element-binding protein signaling in neurons at concentrations in which cell survival is not compromised. J Biol Chem 276:17301–17306

Townsend M, Cleary JP, Mehta T, Hofmeister J, Lesné S, O'hare E, Walsh DM, Selkoe DJ (2006) Orally available compound prevents deficits in memory caused by the Alzheimer amyloid-β oligomers. Ann Neurol 60:668–676

Townsend M, Mehta T, Selkoe DJ (2007) Soluble Aβ inhibits specific signal transduction cascades common to the insulin receptor pathway. J Biol Chem 282:33305–33312

Tu JC, Xiao B, Naisbitt S, Yuan JP, Petralia RS, Brakeman P, Doan A, Aakalu VK, Lanahan AA, Sheng M, Worley PF (1999) Coupling of mGluR/Homer and PSD-95 complexes by the Shank family of postsynaptic density proteins. Neuron 23:583–592

Van Der Heide LP, Ramakers GM, Smidt MP (2006) Insulin signaling in the central nervous system: learning to survive. Prog Neurobiol 79:205–221

Vigo-Pelfrey C, Lee D, Keim P, Lieberburg I, Schenk DB (1993) Characterization of β-amyloid peptide from human cerebrospinal fluid. J Neurochem 61:1965–1968

Walsh DM, Tseng BP, Rydel RE, Podlisny MB, Selkoe DJ (2000) The oligomerization of amyloid β-protein begins intracellularly in cells derived from human brain. Biochemistry 39:10831–10839

Walsh DM, Klyubin I, Fadeeva JV, Cullen WK, Anwyl R, Wolfe MS, Rowan MJ, Selkoe DJ (2002) Naturally secreted oligomers of amyloid β protein potently inhibit hippocampal long-term potentiation *in vivo*. Nature 416:535–539

Wang HW, Pasternak JF, Kuo H, Ristic H, Lambert MP, Chromy B, Viola KL, Klein WL, Stine WB, Krafft GA, Trommer BL (2002) Soluble oligomers of β amyloid (1–42) inhibit long-term potentiation but not long-term depression in rat dentate gyrus. Brain Res 924:133–140

Wang Q, Walsh DM, Rowan MJ, Selkoe DJ, Anwyl R (2004a) Block of long-term potentiation by naturally secreted and synthetic amyloid β-peptide in hippocampal slices is mediated via activation of the kinases c-Jun *N*-terminal kinase, cyclin-dependent kinase 5, and p38 mitogen-activated protein kinase as well as metabotropic glutamate receptor type 5. J Neurosci 24:3370–3378

Wang Z, Chang L, Klein WL, Thatcher GR, Venton DL (2004b) Per-6-substituted-per-6-deoxy β-cyclodextrins inhibit the formation of β-amyloid peptide derived soluble oligomers. J Med Chem 47:3329–3333

Wang J, Ho L, Zhao W, Ono K, Rosensweig C, Chen L, Humala N, Teplow DB, Pasinetti GM (2008) Grape-derived polyphenolics prevent Aβ oligomerization and attenuate cognitive deterioration in a mouse model of Alzheimer's disease. J Neurosci 28:6388–6392
Wang XP, Zhang JH, Wang YJ, Feng Y, Zhang X, Sun XX, Li JL, Du XT, Lambert MP, Yang SG, Zhao M, Klein WL, Liu RT (2009) Conformation-dependent single-chain variable fragment antibodies specifically recognize β-amyloid oligomers. FEBS Lett 583:579–584
Wang D, Govindaiah G, Liu R, Dc Arcangelis V, Cox CL, Xiang YK (2010) Binding of amyloid β peptide to β2 adrenergic receptor induces PKA-dependent AMPA receptor hyperactivity. FASEB J 24:3511–3521
West MJ, Coleman PD, Flood DG, Troncoso JC (1994) Differences in the pattern of hippocampal neuronal loss in normal ageing and Alzheimer's disease. Lancet 344:769–772
White JA, Manelli AM, Holmberg KH, Van Eldik LJ, Ladu MJ (2005) Differential effects of oligomeric and fibrillar amyloid-β 1–42 on astrocyte-mediated inflammation. Neurobiol Dis 18:459–465
Wilcock DM, Rojiani A, Rosenthal A, Subbarao S, Freeman MJ, Gordon MN, Morgan D (2004) Passive immunotherapy against Aβ in aged APP-transgenic mice reverses cognitive deficits and depletes parenchymal amyloid deposits in spite of increased vascular amyloid and microhemorrhage. J Neuroinflammation 1:24
Wong PT, Schauerte JA, Wisser KC, Ding H, Lee EL, Steel DG, Gafni A (2009) Amyloid-β membrane binding and permeabilization are distinct processes influenced separately by membrane charge and fluidity. J Mol Biol 386:81–96
Wu Y, Wu Z, Butko P, Christen Y, Lambert MP, Klein WL, Link CD, Luo Y (2006) Amyloid-β-induced pathological behaviors are suppressed by *Ginkgo biloba* extract EGb 761 and ginkgolides in transgenic Caenorhabditis elegans. J Neurosci 26:13102–13113
Yang F, Lim GP, Begum AN, Ubeda OJ, Simmons MR, Ambegaokar SS, Chen PP, Kayed R, Glabe CG, Frautschy SA, Cole GM (2005) Curcumin inhibits formation of amyloid β oligomers and fibrils, binds plaques, and reduces amyloid *in vivo*. J Biol Chem 280:5892–5901
Yu L, Edalji R, Harlan JE, Holzman TF, Lopez AP, Labkovsky B, Hillen H, Barghorn S, Ebert U, Richardson PL, Miesbauer L, Solomon L, Bartley D, Walter K, Johnson RW, Hajduk PJ, Olejniczak ET (2009) Structural characterization of a soluble amyloid β-peptide oligomer. Biochemistry 48:1870–1877
Zampagni M, Evangelisti E, Cascella R, Liguri G, Becatti M, Pensalfini A, Uberti D, Cenini G, Memo M, Bagnoli S, Nacmias B, Sorbi S, Cecchi C (2010) Lipid rafts are primary mediators of amyloid oxidative attack on plasma membrane. J Mol Med 88:597–608
Zhang B, Berger J, Zhou G, Elbrecht A, Biswas S, White-Carrington S, Szalkowski D, Moller DE (1996a) Insulin- and mitogen-activated protein kinase-mediated phosphorylation and activation of peroxisome proliferator-activated receptor gamma. J Biol Chem 271:31771–31774
Zhang C, Qiu HE, Krafft GA, Klein WL (1996b) Aβ peptide enhances focal adhesion kinase/Fyn association in a rat CNS nerve cell line. Neurosci Lett 211:187–190
Zhao WQ, De Felice FG, Fernandez S, Chen H, Lambert MP, Quon MJ, Krafft GA, Klein WL (2008) Amyloid β oligomers induce impairment of neuronal insulin receptors. FASEB J 22:246–260
Zhao WQ, Lacor PN, Chen H, Lambert MP, Quon MJ, Krafft GA, Klein WL (2009) Insulin receptor dysfunction impairs cellular clearance of neurotoxic oligomeric Aβ. J Biol Chem 284:18742–18753
Zhao WQ, Santini F, Breese R, Ross D, Zhang XD, Stone DJ, Ferrer M, Townsend M, Wolfe AL, Seager MA, Kinney GG, Shughrue PJ, Ray WJ (2010) Inhibition of calcineurin-mediated endocytosis and α-amino-3-hydroxy-5-methyl-4-isoxazolepropionic acid (AMPA) receptors prevents amyloid β oligomer-induced synaptic disruption. J Biol Chem 285:7619–7632
Zhuo JM, Prakasam A, Murray ME, Zhang HY, Baxter MG, Sambamurti K, Nicolle MM (2008) An increase in Aβ42 in the prefrontal cortex is associated with a reversal-learning impairment in Alzheimer's disease model Tg2576 APPsw mice. Curr Alzheimer Res 5:385–391

Chapter 5
The Role of Aβ and Tau Oligomers in the Pathogenesis of Alzheimer's Disease

Kiran Bhaskar and Bruce T. Lamb, Ph.D.

Abstract Dense extracellular aggregates of amyloid β-protein (Aβ) in senile plaques (SPs) and intracellular aggregates of the hyperphosphorylated, microtubule-associated protein tau (MAPT) in neurofibrillary tangles (NFTs) within the brain are the key diagnostic hallmarks of Alzheimer's disease (AD). While initial studies focused on SPs and NFTs as the key pathogenic proteinaceous species that could account for the clinical features of AD, increasing evidence suggests that the fibrils of Aβ and MAPT are unlikely to be the unique neurotoxic entities responsible for AD pathogenesis. Instead, more recent studies have implicated small, soluble oligomeric species of both Aβ and MAPT. Indeed, a wide variety of Aβ and tau oligomers have been described in both *in vitro* and *in vivo* systems that possess a diverse set of biological properties, including substantial synapto- and neurotoxicity. While many of these oligomers have been extensively characterized by novel biophysical, biochemical, and immunological techniques, attributing particular functions and dysfunctions to particular oligomer structures *in vivo* has proven enormously difficult, as the different proteinaceous species are present in equilibrium and likely are contained within unique intracellular and extracellular environments. Nevertheless, a number of therapeutic strategies have been developed that seek to target Aβ and tau oligomerization for AD.

K. Bhaskar
Department of Neurosciences, NC30, Lerner Research Institute, The Cleveland Clinic, NC30, 9500 Euclid Avenue, Cleveland, OH 44195, USA
e-mail: bhaskak@ccf.org

B.T. Lamb, Ph.D. (✉)
Department of Neurosciences, NC30, Lerner Research Institute, The Cleveland Clinic, NC30, 9500 Euclid Avenue, Cleveland, OH 44195, USA

Departments of Neurosciences and Genetics, Case Western Reserve University School of Medicine, Cleveland, OH, USA
e-mail: lambb@ccf.org

F. Rahimi and G. Bitan (eds.), *Non-fibrillar Amyloidogenic Protein Assemblies—Common Cytotoxins Underlying Degenerative Diseases*, DOI 10.1007/978-94-007-2774-8_5,

Keywords Alzheimer's disease • Aβ oligomers • Tau oligomers • Neurofibrillary tangles • Neuropathology

5.1 Introduction

Alzheimer's disease (AD) is the most common form of dementia in the elderly and is one of the most prominent neurodegenerative diseases with progressive cognitive decline and memory impairment (Khachaturian 1985). In the absence of reliable biomarkers, a definitive diagnosis of AD can only be made *via* direct pathological examination of the brain tissue derived from either biopsy or autopsy samples (Corey-Bloom 2000). Macroscopically, the end-stage AD brain shows gross cortical atrophy with enlargement of the ventricles. Microscopically, there is widespread cellular degeneration and cortical neuronal loss, more so in temporal and frontal cortices subserving cognition than the parietal and occipital cortices. Neuronal atrophy is also accompanied by reactive gliosis, diffuse synaptic and neuronal loss as well as the presence of the two prominent pathological hallmarks of the disease, extracellular deposits of aggregated amyloid β-protein (Aβ) in senile plaques (SP) and intracellular aggregates of microtubule-associated protein tau (tau or MAPT) in neurofibrillary tangles (NFTs) (Jellinger 1990; Selkoe 1997). Aβ plaques vary in size and composition, but are generally 50–100 μm in diameter and intimately associated with swollen dystrophic axons and dendrites, reactive astrocytes, and activated microglia. NFTs are intracellular bundles of paired helical filaments of tau in their highly phosphorylated forms. Similar to Aβ plaques, NFTs are often observed in neurons of the hippocampus as well as temporal and frontal cortices that are relevant to cognition. Although SPs are specific to AD, NFTs are found in a variety of other neurodegenerative diseases, including frontotemporal dementia and parkinsonism linked to chromosome 17 (FTDP-17), progressive supranuclear palsy (PSP), corticobasal degeneration (CBD), and Pick's disease (PiD), which are collectively known as tauopathies. A detailed description of the generation of SPs and NFTs are discussed in the subsequent sections of this chapter.

Since their original description by Dr. Alzheimer over 100 years ago (Alzheimer 1907), SPs and NFTs have been recognized as the key pathological hallmarks of AD. However, the exact role of SPs and NFTs in AD pathogenesis has come into question due to lack of correlation between the presence of SPs and NFTs in postmortem brain and corresponding clinical symptoms (Ashe and Zahs 2010; Duff and Planel 2005). Notably, both postmortem pathological analysis of brain tissue as well as *in vivo* imaging analysis of SPs has revealed that upwards of 30% of aged individuals have abundant SPs in the absence of detectable cognitive deficits (Price and Morris 1999). Furthermore, a recent drug trial targeting SPs revealed robust reductions in SPs with no detectable improvements in cognitive functioning at the end stage prior to death (Holmes et al. 2008). Similarly, recent reports utilizing mouse models of tauopathies have also established a lack of correlation between NFT pathology and cognitive impairment (Berger et al. 2007; de Calignon et al. 2010).

While the reason for the lack of correlation between SPs and NFTs and clinical disease is hotly debated, there is increasing evidence that this may in part be due to the presence of neurotoxic species of Aβ and tau that form different intermediate structures termed Aβ oligomers, Aβ protofibrils, Aβ paranucleus, and tau multimers. Many of these structures are often more potent synapto- and neurotoxins than either Aβ or tau fibrils. Significant advancements in the generation of antibody reagents that can detect unique Aβ or tau species, along with enzyme-linked immuno-sorbent assays (ELISAs), cellular extraction techniques, and western blotting have enabled a more precise identification and characterization of these toxic intermediates. Notably, the levels of oligomeric Aβ, as measured by biochemical extraction and detection with these antibody reagents, correlates with the presence and degree of cognitive deficits (Lue et al. 1999; McLean et al. 1999; Näslund et al. 2000; Wang et al. 1999). Furthermore, suppression of a human tau transgene in an inducible mouse model of tauopathy (rTg4510) prevented neuronal loss while NFTs continued to exist (Spires et al. 2006), suggesting presence of other toxic tau intermediates. In a subsequent study, formation of early-stage aggregated tau species, before formation of NFTs, was detected and demonstrated to correlate strongly with cognitive impairment in rTg4510 mice (Berger et al. 2007). In the current review we discuss formation and detection of toxic intermediates of Aβ (Sect. 5.4) and tau (Sect. 5.5), and current evidence as to how these species induce neurotoxicity.

5.2 Genetics of AD and the "Amyloid Cascade Hypothesis"

Bavarian psychiatrist and neuropathologist Alois Alzheimer first made the discovery of the unique association between SPs and AD in 1906. The brain autopsy of a single patient (Auguste D.) demonstrated silver-stained "miliary foci" (SPs) and the "tangled bundle of fibrils" (NFTs) (Alzheimer 1907) that have come to characterize the disease. However, the next significant advance in understanding AD did not come until the early 1980s when the biochemists Glenner and Wong purified microvascular amyloid from the meninges of AD brains and provided a partial sequence of an ~4-kDa protein subunit that they named amyloid β (Aβ) peptide (Glenner and Wong 1984). The following year, Masters et al., similarly characterized amyloid from SPs in the brains of AD and from patients with Down syndrome (DS), who have trisomy for human chromosome 21 (Masters et al. 1985).

Perpetual occurrence of AD in multi-generational families as well as in DS prompted a race to identify the gene(s) involved in AD. The first candidate gene was localized to chromosome 21, although this gene failed to segregate in certain European families with AD (St George-Hyslop et al. 1987; Van Broeckhoven et al. 1987). Subsequent work in the same year identified the entire sequence of the gene localized to chromosome 21 that encodes the precursor to Aβ, termed the amyloid β-protein precursor (APP) (Kang et al. 1987; Robakis et al. 1987; Tanzi et al. 1987). Additional work thereafter, identified multiple APP gene mutations (Goate et al. 1991; Levy et al. 1990; Murrell et al. 1991; Van Broeckhoven et al. 1990) that caused early-onset familial AD and were also linked to altered production of Aβ (Citron et al. 1992). Finally, while mice normally do not develop either SPs or NFTs, generation of transgenic animals that overexpress APP carrying a familial AD mutation, resulted in the production of mice with robust, age-related accumulation of SPs, dystrophic neurites, and gliosis (Games et al. 1995; Hsiao et al. 1996).

Additional evidence that APP is causally related to AD came from the study of DS individuals. Virtually 100% of these individuals developed SPs very early in life as well as formation of NFTs (Prasher et al. 1998). Notably, characterization of a DS individual with partial trisomy for human chromosome 21, revealed that trisomy for human APP was necessary for the development of AD-like neuropathology (Prasher et al. 1998). Furthermore, recent identification of early-onset familial AD that is caused by gene duplication of human APP, confirmed that trisomy for APP was also sufficient for development of both SPs and NFTs (Rovelet-Lecrux et al. 2006). Taken together, these observations have led to the so-called "amyloid cascade hypothesis", which stipulates that it is the absolute levels of specific Aβ peptides (Hardy and Higgins 1992; Selkoe 1989) that dictate AD risk. The amyloid cascade hypothesis was further strengthened by the identification of mutations in the presenilin genes (*PSEN1* and *PSEN2*) that cause early-onset familial Alzheimer's disease (FAD) (Sherrington et al. 1995) and were found to alter the metabolism of Aβ (Borchelt et al. 1996; Citron et al. 1997).

Aβ is derived from sequential enzymatic cleavage of APP, a 695–770-residue, type-I integral transmembrane protein expressed in both neuronal and non-neuronal tissues. Initially, APP is cleaved either by α-secretase or β-secretase (BACE, β-site APP cleavage enzyme) competing pathways, which generates α- and β-*C*-terminal fragments (CTFs) of APP, respectively. Subsequent proteolytic cleavage of α-CTF by γ-secretase precludes the formation of Aβ peptide, instead generating an ~3-kDa peptide called p3 (Nunan and Small 2000), while cleavage of β-CTF by γ-secretase results in the production of full-length Aβ (Vassar et al. 1999; Yan et al. 1999). Many studies have demonstrated that γ-secretase is a multi-subunit protease that includes PSEN1 or PSEN2 at the catalytic core that can cleave APP β-CTF at different sites generating the predominant $A\beta_{1-40}$ and $A\beta_{1-42}$ fragments as well as $A\beta_{1-39}$ and $A\beta_{1-43}$ (Ida et al. 1996). A preponderance of evidence suggests that the $A\beta_{1-42}$ peptide is most likely to form aggregates *in vitro* and is also laid down early in SPs in human AD. Notably, most of the PSEN1 and PSEN2 familial AD mutations seem to favor the production of the $A\beta_{1-42}$ species over the $A\beta_{1-40}$ species (Scheuner et al. 1996).

5.3 Aβ-Mediated Neurotoxicity Is Dependent upon Different Aβ Conformers

Aβ is produced as a natural product of cellular metabolism and can be detected in numerous biological milieus, including plasma, cerebrospinal fluid (CSF), and brain (Haass et al. 1992; Ida et al. 1996; Seubert et al. 1992; Vigo-Pelfrey et al. 1993; Walsh et al. 2000). Therefore, generation of Aβ itself does not induce neurodegeneration. However, the unique hydrophobic properties of Aβ peptides (particularly the $A\beta_{1-42}$ species that contains two additional hydrophobic residues) render them more likely to self-aggregate (Busciglio et al. 1992; Geula et al. 1998; Pike et al. 1991). Aβ peptides exist as monomers, dimers, and higher-order oligomers, with aggregation-producing protofibrils and eventually fibrils, in a β-pleated-sheet conformation (Table 5.1). Mature SPs from the AD brain are organized into insoluble fibrils of 6–10 nm in diameter and *in vitro* recombinant Aβ peptides can self-assemble into fibrils that resemble those observed in the human brain.

Initial *in vitro* studies suggested that mature Aβ fibrils were neurotoxic, although the studies were not always reproducible and usually required a concentration of 1–50 μM in order to observe substantial toxicity. Furthermore, subsequent pathological analyses of human AD tissue and more recent *in vivo* imaging demonstrated a poor correlation between SPs and cognitive deficits (Berg et al. 1998; Cummings et al. 1996; Dickson et al. 1995; Hyman and Tanzi 1992; McKee et al. 1991, 1993; Morris and Rubin 1991; Terry et al. 1991). However, subsequent studies have observed more robust correlations between the levels of soluble Aβ (forms of Aβ remaining in the aqueous phase following high-speed centrifugation >100,000 × *g* for >1 h) and the extent of synaptic loss and severity of cognitive impairment (Lue et al. 1999; McLean et al. 1999). Finally, more recent studies have detected oligomeric Aβ assemblies in soluble extracts from the AD brain, but not in age-matched controls, *via* immunological techniques utilizing a specific antibody against Aβ oligomers and western-blot analysis (Georganopoulou et al. 2005; Kayed et al. 2003; Kokubo et al. 2005; Lambert et al. 2007; Takahashi et al. 2004; Walsh et al. 2000; Xia et al. 2009). Overall, the focus on Aβ as playing a central role in AD pathogenesis has evolved from Aβ monomers, to SPs and now to Aβ oligomers. However, the exact roles specific Aβ oligomers play in the neurotoxicity observed in human AD remain to be determined.

Following the discovery of soluble Aβ oligomers as potential neurotoxic entities, numerous recent studies have explored the types of different Aβ oligomers, their biophysical and neurotoxic properties in cell-culture systems, in animal models as well as their presence in human AD brain tissue (reviewed in Roychaudhuri et al. 2009; Rauk 2008, 2009). Overall, these studies have demonstrated that Aβ belongs to the class of "natively disordered" proteins, existing in the monomer state as an equilibrium mixture of many different conformers. On-pathway assembly requires formation of a partially folded monomer that self-associates to form a nucleus for fibril elongation, termed a paranucleus (containing six monomers). Paranuclei can subsequently self-associate to form protofibrils and then to classical amyloid-type fibrils. Other assembly pathways produce annular pore-like structures, globular dodecameric (and higher order) structures, and amylospheroids (Table 5.1). Annuli and amylospheroids appear to be off-pathway assemblies (Table 5.1 and also reviewed in Roychaudhuri et al. 2009).

Although many studies have attempted to identify and attribute unique biological activities to specific Aβ conformers, it has been extremely difficult to confidently determine which is the "most toxic" form of Aβ that is most relevant to AD pathogenesis due to variability in the results reported amongst laboratories and the

Table 5.1 Multiple assemblies of Aβ peptide and their biological activities [Adopted and modified from Roychaudhuri et al. (2009)]

Types of Aβ assembly	Schematic	Physical properties	Detection in animal model	Detection in humans	Biological effects	References
$A\beta_{1-x\ (x=39-44)}$, $A\beta_{x-28\ (x=12\ or\ 15)}$ Monomers		~4–4.5 kDa, natively "unstructured"	Yes (examples: *J20, Tg2576, 3×Tg, R1.40)*	Yes	Non-toxic at normal physiological concentrations in CSF and serum	Pike et al. (1991), Walsh and Selkoe (2007), and Shankar et al. (2007)
Aβ dimers Aβ trimers	 	~8–12 kDa, altered *C*-terminus; co-exist with β-sheeted higher oligomers	Yes (J20, *Tg2576, APP23)*	Yes	LTP inhibition, glutamate-receptor toxicity (mGluR and NMDA), synaptic dysfunction, cognitive deficits. Induction of neuronal cell-cycle events	Shankar et al. (2008), Klyubin et al. (2008), Horn et al. (2010), Wei et al. (2010), Kawarabayashi et al. (2004), Kuo et al. (2001), Townsend et al. (2006), and Varvel et al. (2008)
Aβ paranucleus (penta/hexamers)		5 nm diameter, spheroidal, $A\beta_{42}$ only	ND	ND	Neuronal death, redox effects	Bitan et al. (2003a, b)
Aβ dodecamer (Aβ*56)		~56 kDa (12 mer), 1 nm height (AFM), prolate ellipsoid	Yes (*Tg2576, J20, 3xTg, APP23*)	Yes	Synaptic dysfunction, memory disruption and cognitive deficits	Lesné et al. (2006), Cheng et al. (2007), Billings et al. (2007), and Lefterov et al. (2009)
Aβ-derived diffusible ligands (ADDL)[b]		~53 kDa, 5–6 nm height (AFM), $A\beta_{42}$ only	Yes (*Tg2576*)	Yes	Neuronal death at nanomolar concentration, NMDAR-dependent toxicity, LTP inhibition, channel formation, increase ionic conductance, memory loss	Lambert et al. (1998), and Sokolov et al. (2006)

AβO		~90 kDa (15–20 monomers), spherical vesicle with 2–5 nm diameter	ND	A11[a] + staining observed	Increase in membrane conductance, calcium influx, apoptosis	Kayed et al. (2003)
Aβ amylo-spheroids[c]		~150–700 kDa, ~10–15 nm diameter, formed by both $A\beta_{40}$ or $A\beta_{42}$	ND	ND	Neuronal death	Hoshi et al. (2003)
Aβ protofibril		~5 nm diameter, beaded, curvilinear structure, <150 nm long, formed by $A\beta_{40}$ or $A\beta_{42}$ (24–700 monomers)	Yes (tg-ArcSwe)	ND	Neuronal death, increase EPSC, NMDAR-dependent toxicity, ROS production, ion channel formation, calcium homeostasis disruption Spatial learning deficit in tg-ArcSwe mice	Walsh et al. (1997, 1999), Harper et al. (1997), and Lord et al. (2009)
Aβ annulus/annular assembly		Annular structure with 7–10-nm outer diameter, 1.5–2.0-nm inner diameter, 150–250 kDa	ND	ND	Pore formation within the plasma membrane by protofibrils	Caughey and Lansbury (2003)

(continued)

Table 5.1 (continued)

Types of Aβ assembly	Schematic	Physical properties	Detection in animal model	Detection in humans	Biological effects	References
Aβ fibrils		~6–10 nm in diameter of varying lengths	Yes (all depositing AD models)	Yes	Earlier studies have suggested that fibrillar Aβ is neurotoxic, synaptotoxic and induces neuronal cell-cycle events Recent studies implicate that genetic means of acceleration of fibril formation reduce levels of Aβ oligomers and prevent functional deficit *via* specifically reducing Aβ*56 levels in ARC6 hAPP transgenic mice (J20)	Lorenzo et al. (1994) and Cheng et al. (2007)

Abbreviations: *AβO* "Aβ Oligomer", a term defined by Deshpande et al. (2006); *AFM* atomic-force microscopy, *EPSC* excitatory post-synaptic currents; *NMDAR* *N*-methyl-$_D$-aspartate receptor, *LTP* long-term potentiation, *ND* not determined, *ROS* reactive oxygen species, *CSF* cerebrospinal fluid

[a]A11 antibody has been reported to detect 6-, 9-, and 12-mers in western blots of Tg2576 mouse brain extracts from 7- to 25-month-old animals (Lesné et al. 2006)

[b]One study reported that ADDLs, identical to Aβ annulus, can alter channel conductance (Sokolov et al. 2006). Another report suggests that ADDL are dodecamer with heights of 5–6 nm (similar to Aβ*56) (Lambert et al. 1998)

[c]Aβ amylospheroid also called ASPD, β-amyball, or β-amyloid ball

different methods used for the identification and/or purification of Aβ conformers. While it is expected that the biological effects of different types of Aβ assemblies may be closely related, it also seems that specific Aβ assemblies likely have unique biological effects that may have relevance to AD. Since the goal of this chapter is to review the biological effects of Aβ oligomers that likely are relevant to AD, we have detailed the findings reported in the literature below according to broad biological categories and also summarized these in Table 5.1.

5.3.1 Membrane Effects and Oxidative Stress

A unique property of Aβ is that it is an amphipathic peptide and can acquire a micellar structure *via* oligomerization. The side-chains of the first 16 residues out of 28 are polar and 12 are charged at neutral pH. The remaining 12 ($A\beta_{40}$) or 14 ($A\beta_{42}$) side-chains are non-polar. Such molecular arrangements render them to form micelles (Lomakin et al. 1996), and thus Aβ can interact with a variety of biomolecules, including lipids, proteins, and proteoglycans. Aβ can directly bind and interact with a vast array of phospholipids, lipoproteins, receptors, and lipid molecules and induce cellular damage (reviewed in Verdier and Penke 2004). These membrane molecules include apolipoproteins, high-density lipoproteins, lipid rafts, gangliosides, G-protein-coupled receptors, sphingolipids, and cholesterol as well as numerous membrane receptors (reviewed in Verdier and Penke 2004).

Direct membrane insertion of Aβ has also been documented (Lin et al. 2001) and thus could potentially perturb plasma membrane integrity. For example, structural analysis of the *C*-terminal domain of Aβ (residues 29–X) has been shown to have properties similar to fusion peptides of viral origin. Insertion of these fragments in a tilted manner in the membrane is thought to disrupt the parallel symmetry of the fatty acyl chains, altering the curvature of the membrane surface and destabilizing the membrane. Consistent with this hypothesis, $A\beta_{22-42}$ can also induce membrane fusion and permeabilize lipid vesicles (Arispe et al. 2007).

Several studies have demonstrated that by virtue of forming membrane pores, Aβ oligomers can act as ion channels, increasing the ionic conductance across the lipid bilayer and thereby affecting membrane potentials, although there is considerable debate in the field about the biological relevance of this finding (Sokolov et al. 2006). Aβ–membrane interactions also can result in the release of cholesterol, phospholipids, and monosialogangliosides, which in turn can indirectly lead to tau hyperphosphorylation and neurodegeneration (Tashima et al. 2004; Verdier et al. 2004). Arispe et al., have shown that $A\beta_{40}$ channel activity in planar lipid bilayer results in spontaneous transitions to higher conductance, while atomic-force-microscopy (AFM) images of Aβ-treated, reconstituted bilayers have revealed disk-like structures with pore-like concavities of 8–12-nm outer diameter and 1–2-nm inner diameter (Arispe et al. 2007). However, other authors suggest that these data may be due to non-specific interactions. For example, Aβ-oligomer-mediated interference with the surface packing of lipid head-groups effectively thins the membrane, reduces effective membrane conductance, and may produce the appearance of pores, without really forming discrete pores (Green et al. 2004; Sokolov et al. 2006). Binding of Aβ to the membrane could also have consequences on the structure of Aβ itself. For example, non-specific adhesion of Aβ to certain membrane proteins facilitates lateral aggregation of preformed fibrils and Aβ nucleation at the earliest stages of fibril formation (McLaurin et al. 1999).

Other potential targets of Aβ oligomers that several investigators have explored include a variety of membrane receptors. Multiple different studies have suggested that conversion of Aβ monomers into soluble oligomeric forms renders them highly reactive to a variety of membrane proteins and receptors, some examples of which are presented below.

5.3.1.1 Insulin Receptors

Binding assays have revealed that $A\beta_{1-40}$ and $A\beta_{1-42}$ compete with insulin for binding to the insulin receptors and that this binding involves residues 16–25 of Aβ (Xie et al. 2002; Zhao et al. 2008). Notably this stretch of amino acids within Aβ has a recognition motif that is also observed in amino acids 21–30 of insulin. This binding has been investigated with regard to the overlapping biological activities of insulin and Aβ. First, as described above, insulin and Aβ share a common sequence-recognition motif that enables them to interact with insulin receptors. Second, Aβ and insulin are substrates for insulin-degrading enzyme. Third, impaired glucose metabolism is a characteristic feature in AD (Xie et al. 2002). Finally, many studies have suggested that interaction between Aβ and insulin receptor in turn affects insulin receptor signaling, including inhibition of autophosphorylation of the insulin receptor.

Interestingly, a series of recent studies from Klein and colleagues have suggested that a specific form of Aβ assembly called “ADDLs” (Aβ derived diffusible ligands) (Table 5.1) binds to insulin receptors on specific hippocampal neuronal populations *in vitro* and induces ligand–receptor internalization. Notably, these effects seem to precede the observed neurotoxicity, namely loss of dendritic spines (De Felice et al. 2009). Furthermore, these studies demonstrated that exogenous addition of insulin can prevent the loss of surface insulin receptors, ADDL-induced oxidative stress and spine loss, suggesting that enhancing insulin signaling can counteract ADDL-induced synaptotoxicity. Similarly, we have recently demonstrated that Aβ oligomers, rich in dimers, trimers, and other low-n assemblies but not monomers, cause an imbalance in the phosphoinositide 3-kinase (PI3K)-Akt–mammalian target-of-rapamycin (mTOR) pathway, which is the downstream target of the insulin receptor pathway. In our hands, these Aβ-oligomers-induced alterations in the PI3K–mTOR pathway is upstream of another form of neurotoxicity, namely induction of neuronal cell-cycle events, which is also observed in human AD (Bhaskar et al. 2009; Herrup and Yang 2007; Varvel et al. 2008; Yang et al. 2003, 2006).

5.3.1.2 Acetylcholine Receptors

Cholinergic neurons in the basal forebrain are severely affected in human AD (Coyle et al. 1983; Herholz et al. 2004; Kar and Quirion 2004; Mufson et al. 2003; Wu et al. 2005). The binding of Aβ to α7-nicotinic acetylcholine receptor (α7nAChR) has been reported by multiple groups and the binding has been demonstrated to be dependent upon the sequence and conformation of Aβ [for example, α7nAChR binds to $A\beta_{1-40}$ with a lower affinity than to $A\beta_{1-42}$ (Wang et al. 2000)]. Notably, oligomeric Aβ induces a different physiological response with respect to acetylcholine release and Ca^{2+} influx than monomeric Aβ (Wang et al. 2000). Additional studies have suggested that the $A\beta_{1-42}$–α7nAChR interaction results in internalization of the ligand–receptor complex within neurons (Nagele et al. 2002). Tau phosphorylation, a hallmark of AD, has also been reported as a downstream pathological effect of the interaction between $A\beta_{1-42}$ and α7nAChR (Wang et al. 2003).

In contrast to the detrimental effects of $A\beta_{1-42}$–α7nAChR interactions, several recent reports suggest that $A\beta_{1-42}$–α7nAChR interactions may also be protective to neurons and that the loss of α7nAChR in a mouse model of AD (Tg2576) exacerbates cognitive and synaptic deficits (Hernandez et al. 2010). Notably, biochemical analysis on 5-month-old Tg2576 mice lacking α7nAChR (A7KO-APP) revealed significant reduction in the hippocampal and basal forebrain choline-acetyltransferase activity and profound loss of hippocampal neurons in the A7KO-APP mice compared to Tg2576 mice (Hernandez et al. 2010). Interestingly, these observations coincided with the appearance of ~56-kDa soluble Aβ oligomers (Aβ*56) in the hippocampus of A7KO-APP mice. However, other groups have found that deletion of α7nAChR in a different AD mouse model (PDAPP) results in improved cognitive outcomes (Dziewczapolski et al. 2009). Whether this apparent discrepancy is due to the animal model utilized, age of animals, or other experimental measures is unclear. Nonetheless, these results suggest that Aβ oligomers can have a direct impact on acetylcholine receptor function.

5.3.1.3 *N*-Methyl-$_D$-Aspartate Receptor (NMDAR)

Recent studies have reported that ADDLs can directly interact with NMDARs. Using highly differentiated cultures of hippocampal neurons, Klein and colleagues found that ADDLs bound directly to neurons attaching to presumed excitatory pyramidal neurons but not GABAergic neurons (Lacor et al. 2007). Furthermore, ADDLs specifically bound to spines rich in NMDARs and induced a rapid decrease in surface receptor density *via* internalization (Lacor et al. 2007). Continued exposure of ADDLs to neurons resulted in abnormal spine morphology, with induction of long thin spines reminiscent of the morphology found in mental retardation, deafferentiation, and prionoses (Lacor et al. 2007). Furthermore, the NMDAR-antagonist, memantine, prevented ADDL-induced loss of the dendritic cytoskeleton protein drebrin (Lacor et al. 2004; Sekino et al. 2006), while enhancing NMDAR activity has been linked to synaptic targeting of Aβ oligomers, which can be blocked *via* the

NMDAR antagonists 2-amino-5-phosphonovaleric acid (APV) and memantine (Deshpande et al. 2009).

Recently, NMDAR-dependent Ca^{2+} overload in rat cortical neurons *in vitro* was observed upon ADDL treatment (Alberdi et al. 2010). This massive ADDL-induced entry of Ca^{2+} through NMDA and α-amino-3-hydroxy-5-methyl-4-isoxazolepropionic acid (AMPA) receptors induced mitochondrial dysfunction as indicated by mitochondrial Ca^{2+} overload, oxidative stress, and mitochondrial membrane depolarization (Alberdi et al. 2010). In a separate study, Aβ promoted endocytosis of NMDAR from cortical neurons *in vitro* (Snyder et al. 2005). Notably, reduced levels of NMDA receptor 1 (NR1) and reduced NMDA-evoked currents were observed in the cortical neurons derived from APP_{swe} mouse model of AD, which was blocked by reducing the Aβ level *via* inhibition of γ-secretase (Snyder et al. 2005).

In addition to the effects of ADDLs, Aβ protofibrils (Table 5.1) have also been observed to bind to NMDA receptors. Aβ-protofibril-induced activity was specifically attenuated by the NMDAR antagonist, APV, while the non-NMDA glutamate receptor antagonist, 2,3-dihydroxy-6-nitro-7-sulfamoyl-benzo[f]quinoxaline-2,3-dione (NBQX), preferentially reduced Aβ-fibril-induced activity (Ye et al. 2004). Taken together, these studies suggest that different Aβ assemblies may target NMDA receptors.

In addition to insulin, α7nAChR, and NMDA receptors, different assemblies of Aβ have also been shown to interact with several other membrane receptors, including the serpin complex, metabotrophic glutamate receptors (mGluRs), AMPA receptors, integrins, complement receptors, APP, receptor for advanced-glycosylation end-products (RAGE), collagen-like Alzheimer amyloid plaque component precursor—CLAC-P/Col XXV, p75 neurotrophin receptor, scavenger receptor on microglial cells/mononuclear phagocytes, CD36 and CD47, glycosaminoglycans, prion protein, and others. A detailed description of these interactions and their biological effects are beyond the scope of this chapter but have been reviewed elsewhere (Verdier et al. 2004; Nygaard and Strittmatter 2009).

5.3.1.4 Oxidative Stress

Several studies have reported that Aβ triggers oxidative stress (reviewed in Smith et al. 2007 and Rauk 2009). Aβ has been suggested to be an inducer of lipid peroxidation (Lauderback et al. 2001; Mark et al. 1997) *via* utilization of a Met (at residue 35) of $Aβ_{1-42}$ as free radicals inserted into lipid bilayers (Curtain et al. 2001; Kanski et al. 2002a, b). Notably, these events are mediated by Cu(II)- and Fe(III)-catalyzed oxidation of Aβ (Schoneich and Williams 2002). Lipid peroxidation leads to production of reactive alkenals such as 4-hydroxy-2-nonenal (HNE) and 2-propen-1-al (acrolein), both of which are increased in AD brain (Lauderback et al. 2001; Lovell et al. 2001; Markesbery and Lovell 1998). Consistent with this observation, oxidative modification of methionine at residue 35 to methionine sulfoxide constitutes a major component of various Aβ species isolated from AD brain (Dong et al. 2003; Kuo et al. 2001; Näslund et al. 1994). Furthermore, if the methionine in Aβ is already oxidized, then it significantly reduces the lipotoxic properties of Aβ (Barnham et al. 2003; Varadarajan et al. 2001). However, a more recent study has questioned the involvement of Met35 in induction of oxidative stress, as it does not occur in the $A\beta_{1-42}$-Cu(II) complex (Jiang et al. 2007). Notably, a separate study implicated histidine 13 and histidine 14 of Aβ in the binding to Cu(II) (Hewitt and Rauk 2009). Other studies invoke Cu(II) reduction and subsequent H_2O_2 formation in the oxidative stress and neurotoxic properties of $A\beta_{1-42}$. A recent study demonstrated that Aβ oligomers and fibrils can promote generation of H_2O_2 when the concentration of co-incubated Cu(II) is below a critical level in a model of redox metal-induced oxidative stress (Fang et al. 2010). This observation is also consistent with another study, where both oligomeric and fibrous Aβ aggregates were demonstrated to induce H_2O_2 generation (Jiang et al. 2010).

In summary, numerous studies, in addition to those discussed above, suggest that different assemblies of Aβ can directly impact lipid and membrane homeostasis, cause receptor internalization and/or directly affect receptor-mediated signaling as well as elicit lipid peroxidation and promote formation of other reactive oxygen species, all of which have the potential to induce considerable neurotoxicity in the human AD brain.

5.3.2 *Mitochondrial Effects*

In recent years, a link between mitochondrial dysfunction and AD pathogenesis has been postulated. Notably, intracellular oligomeric $A\beta_{1-42}$ has been detected in mitochondria from a transgenic mouse model of AD (J-20) as well as mitochondria from human AD patients (Caspersen et al. 2005). It has been hypothesized that mitochondrial Aβ is derived either *via* the transfer of Aβ from the endoplasmic reticulum or *via* the processing of APP by mitochondria (Anandatheerthavarada et al. 2003); although the molecular pathways responsible have yet to be identified. Nonetheless, it has been reported that accumulation of mitochondrial Aβ diminishes the enzymatic activity of respiratory chain complexes (III and IV) and reduces the rate of oxygen consumption. In a separate study, a 20% decrease in cytochrome-C oxidase activity was observed in 2-month-old Tg2576 mouse brain compared to age-matched controls (Manczak et al. 2006), suggesting that soluble Aβ species may impair mitochondrial functions.

In more recent studies, specific Aβ conformers have been linked to mitochondrial dysfunction. To determine the contributions of tau pathology in Aβ-oligomer-mediated mitochondrial dysfunction, mixed cortical brain cultures from the P301L transgenic mouse model of tauopathy were exposed to different Aβ preparations. Notably, incubation of the culture with oligomeric or fibrillar $A\beta_{1-42}$, but not disaggregated (monomer-rich) $A\beta_{1-42}$ resulted in a reduced mitochondrial membrane potential (Eckert et al. 2008a, b). Furthermore, isolation and treatment of mitochondria derived from P301L mice with oligomeric or fibrillar $A\beta_{1-42}$ reduced state three respiration, respiratory control ratio, and uncoupled respiration (Eckert et al. 2008a, b).

Finally, several studies have examined the downstream effects of Aβ-oligomer-mediated mitochondrial dysfunction. For example, ADDL treatment was observed to induce mitochondrial Ca^{2+} overload within neurons (Alberdi et al. 2010). In addition, time-lapse imaging of primary hippocampal neurons following ADDL exposure suggested that a sublethal dose of ADDLs impaired axonal transport of mitochondria (Wang et al. 2010). Notably, a recent study found that Aβ oligomers could induce an acute impairment of fast mitochondrial transport, mitochondrial translocation into dendritic spines, and accumulation of AMPA receptors on the post-synaptic surface in response to repetitive membrane depolarization (Rui et al. 2010). Taken together, these studies suggest that Aβ oligomers can impact mitochondrial structure, function, and/or sub-cellular localization and this is an active area of investigation.

5.3.3 Endosomal/Lysosomal Effects

Several recent studies have suggested that soluble Aβ targeted to endosomes/lysosomes can result in altered formation of Aβ fibrils and/or altered organelle functioning. For example, when primary cortical neurons or SH-SY5Y cells were incubated with subnanomolar concentrations of soluble Aβ there was a hundredfold increase in the concentration of Aβ within late endosomes/lysosomes compared to the concentration in the extracellular media (Hu et al. 2009). Furthermore, prolonged incubation of SH-SY5Y cells with micromolar concentrations of $A\beta_{1-42}$, but not $A\beta_{1-40}$, resulted in time-dependent increases in the formation of Aβ fibrils, suggesting that late endosomes/lysosomes may promote formation of higher-order Aβ assemblies (Hu et al. 2009). Similarly, internalization of Aβ into lysosomes/endosomes and multivesicular bodies (MVB) of THP-1 monocytes resulted in the nucleation and intracellular aggregation of Aβ and subsequent formation of extracellular plaques *in vitro* (Friedrich et al. 2010). Other studies support the view that oligomerization of Aβ can occur within subcellular organelles such as multivesicular bodies and endosomal vesicles (Takahashi et al. 2004). Finally, formation of altered Aβ assemblies within endosomes/lysosomes could help explain the presence of lysosomal/endosomal abnormalities observed both in human AD and in mouse models of AD, although at present there remains little direct evidence linking the two (Cataldo et al. 2004). Nonetheless, these studies provide evidence for the existence of intracellular Aβ aggregates within endosomes/lysosomes that could alter higher-order structures of Aβ and possibly impact organelle function.

5.3.4 Effects on Synaptotoxicity, Cognitive Function, Neuronal Cell-Cycle Events, and Neurodegeneration

There is considerable evidence to support the view that AD is a disorder of synaptic dysfunction. For example, quantification of a wide variety of synaptic markers *via* electron microscopy, biochemistry, or immunohistochemical staining has revealed a significant decrease in synaptic density in the association cortices and hippocampi of human AD brains that appears to correlate with cognitive decline (Bertoni-Freddari et al. 1989; Davies et al. 1987; DeKosky and Scheff 1990; Masliah et al. 1990, 2001; Reddy et al. 2005; Sze et al. 1997; Terry et al. 1991). Extensive experiments in animal models of aging and AD have suggested that defective communication in the perforant pathway, which connects the hippocampi to other limbic and cortical structures, and failed synaptic plasticity likely underlie the progressive memory loss in AD (Geinisman 1999; Geinisman et al. 1992; Hyman et al. 1986). Additional functional studies have also established that long-term potentiation (LTP) and long-term depression (LTD), which are used as models of learning and memory and believed to play important roles in neural circuits (Lynch 2004; Morris 2003), are profoundly affected in AD brains (Mesulam 1999). Mechanistically, the imbalances in cell-surface glutamate receptors as well as downstream signaling cascades that form the basis for LTP/LTD are also altered in AD brains (Greenamyre and Young 1989). Induction of LTP is associated with spine formation and increased spine volume, whereas induction of LTD results in decreased spine volume and spine elimination (Bastrikova et al. 2008; Matsuzaki et al. 2004; Zhou and Poo 2004).

5.3.4.1 Naturally Produced Aβ Oligomers

The most compelling evidence that soluble Aβ oligomers can induce both synaptic and cognitive dysfunction has come from transgenic mouse models overexpressing human APP. First, a single intraperitoneal injection of an anti-Aβ antibody reversed Aβ-mediated memory deficits in the PDAPP mice (Dodart et al. 2002). In this acute (<24 h) experiment, brain SPs were not decreased, suggesting that the antibody was acting on soluble, diffusible species of Aβ and that neutralization or clearing the soluble Aβ species allowed rapid improvement in performance in an object-recognition test of memory. Second, in an elegant series of experiments, Lesné and colleagues reported the presence of both ~42-kDa (nonamer) and ~56-kDa (dodecamer or Aβ*56) Aβ species in the extracellular-enriched soluble fraction of brain extracts from 6-month-old Tg2576 mice. Notably, presence of these Aβ species, but not Aβ monomer, trimer, and other oligomeric forms, correlated with cognitive deficits observed in Tg2576 mice. Furthermore, Lesné et al. purified the 56-kDa Aβ species from the brain extracts and demonstrated that direct intracerebroventricular injections of purified Aβ*56/dodecamer into normal pretrained wild-type rats perturbed the memory of a learned behavior (Lesné et al. 2006; Reed et al. 2011), suggesting that Aβ dodecamers can directly induce cognitive dysfunction *in vivo*.

However, additional studies have suggested that in addition to Aβ*56, other Aβ oligomers may also impact learning and memory in rodents. First, Tg2576 mice exhibit poor performance in the hippocampus-dependent contextual fear-conditioning tests of memory, decreased spine density in the dentate gyrus, and LTP impairments at ages several months prior to the first detection of Aβ*56 (Dineley et al. 2002; Jacobsen et al. 2006). Second, Kawarabayashi and colleagues reported the appearance of SDS-stable dimers present in lipid rafts of Tg2576 mice that coincided with the onset of impairments in spatial reference memory (Kawarabayashi et al. 2004). These contrasting findings suggest that multiple different Aβ oligomers likely exist in rodent models of AD and that each oligomeric Aβ species may have unique effects on synaptic structure and function. Indeed, recent reports suggest multiple biologically active species of Aβ in APP transgenic mouse models of AD (Lesné et al. 2006; Shankar et al. 2009). Consistent with this view, several different Aβ oligomers (Table 5.1) were detected in the cortical brain extracts from the J20 mouse model of AD at different stages of disease progression (Li et al. 2009), suggesting that different oligomeric Aβ species may play unique roles in inducing cognitive dysfunction.

In an attempt to identify Aβ species that account for memory dysfunction in human AD, several investigators have examined human AD brain tissue for the presence of Aβ oligomers. Notably, low-n Aβ oligomers (dimers and trimers; see Table 5.1) have been detected in both mouse models of AD as well as in human AD brain samples (Poling et al. 2008; Rosen et al. 2010; Townsend et al. 2006; Walsh et al. 2000; Xia et al. 2009). To understand the toxicity of these low-n Aβ oligomers, Selkoe and colleagues have utilized human mutant APP-overexpressing cell lines (Podlisny et al. 1995; Walsh et al. 2002; Xia et al. 1997) that constitutively secrete Aβ oligomers into the conditioned media (CM). For example, Chinese hamster ovary (CHO) cells engineered to express mutant (V717F) human APP (referred to as 7PA2 cells) generate and secrete low nanomolar amounts of Aβ species that migrate in denaturing SDS–PAGE with molecular weights consistent with Aβ monomers, dimers, and trimers (Podlisny et al. 1995; Walsh et al. 2002). Subsequent functional studies have demonstrated that CM from 7PA2 cells, which are rich in low-n Aβ oligomers, exhibit neurotoxicity and promote cognitive dysfunction. First, 7PA2 CM can block LTP both *in vivo* (Cleary et al. 2005; Klyubin et al. 2008; Walsh et al. 2002) and *in vitro* (Walsh et al. 2005; Wang et al. 2004). Second, 7PA2 CM reduce spine density and synaptic plasticity in cultured neurons (Calabrese et al. 2007; Freir et al. 2010; Shankar et al. 2007). Finally, purification of Aβ oligomers from 7PA2 CM and injection into the ventricles of non-transgenic rats resulted in impairments of both working memory and consolidation of learned avoidance behavior (Freir et al. 2010; Poling et al. 2008).

In addition to low-n oligomers, Aβ protofibrils (PFs) also exhibit neurotoxicity that has been observed to result in either enhancement of electrical activity of neurons (Hartley et al. 1999) or reduced LTP (Hartley et al. 2008). Furthermore, when Aβ PFs were injected into the ventricles of mice, they significantly impaired learning (Martins et al. 2008). Finally, using an Aβ PF-specific antibody, termed mAb158, in an ELISA, Lord and colleagues observed elevated levels of Aβ PFs that correlated with impaired spatial learning and memory in transgenic mice with Arctic and Swedish Alzheimer's mutations (tg-ArcSwe) (Lord et al. 2009).

In an effort to generate an animal model that is enriched in Aβ oligomers, Tomiyama and colleagues generated APP transgenic mice containing the E693Δ familial AD deletion (Tomiyama et al. 2008). The E693Δ deletion in APP was first discovered in a Japanese pedigree exhibiting AD and Alzheimer's disease-like dementia (Tomiyama et al. 2008). Notably, this deletion is located within the Aβ sequence and produces a variant of Aβ lacking Glu22 (E22Δ). This deletion promotes formation of Aβ oligomers, but not fibrils (Tomiyama et al. 2008). Notably, E693Δ transgenic mice displayed age-dependent accumulation of intraneuronal Aβ oligomers, but no SPs (Nishitsuji et al. 2009; Tomiyama et al. 2010). However, other features of AD were observed, including impaired hippocampal synaptic plasticity, abnormal tau phosphorylation, microglial activation, astrocytosis, and neuronal cell loss. Together, these findings suggest that promotion of the formation of Aβ oligomers induces not only synaptic alterations but also other features of AD pathology in the absence of SPs. Thus, taken altogether, there is evidence implicating a wide variety of naturally occurring Aβ species, from low-n Aβ oligomers to Aβ*56 to Aβ PFs in inducing neurotoxicity and cognitive dysfunction.

5.3.4.2 Synthetic Preparations of Aβ Oligomers

Klein and colleagues generated ADDLs from synthetic preparations of Aβ (Lambert et al. 1998) and demonstrated that ADDLs could both block LTP and induce neuronal death (Lambert et al. 1998; Wang et al. 2002). ADDLs preferentially bind to neuronal synapses (Lacor et al. 2004) and evidence suggests that these interactions are dependent upon the neuronal populations examined, as ADDLs seem to target and induce synaptic aberrations only in a subset of susceptible hippocampal neurons *in vitro* (Lacor et al. 2004, 2007). Evidence for the functional role of ADDLs came from inhibition studies, in which ADDL-specific antibodies (M93 and M94) prevented Aβ-induced neurotoxicity (Lambert et al. 2001). Furthermore, dot-blot analysis of soluble extracts from aged Tg2576 mice revealed the presence of substantial anti-ADDL immunoreactivity (Chang et al. 2003). In addition to causing synaptotoxicity and/or deficits in LTP, ADDLs have also been shown to induce neurodegeneration. For example, low concentrations of ADDLs (5 nM) added for a 24-h incubation resulted in 20% loss of neurons in organotypic mouse brain slice cultures, whereas at higher concentrations of ADDLs (500 nM) and brief incubation periods (45–60 min) neuronal loss was not evident, but a robust inhibition of LTP was observed (Lambert et al. 1998; Wang et al. 2004).

In addition to ADDLs, a number of other studies have utilized slightly different methods to prepare non-fibrillar oligomers (Table 5.1) (Hoshi et al. 2003; Kayed et al. 2003; Kelly and Ferreira 2006; Maloney et al. 2005; Whalen et al. 2005). For example, in a study by Deshpande and colleagues, three separate preparations of synthetic Aβ oligomers (oligomeric or AβO, fibrillar Aβ, or Aβf, and ADDLs) independently exerted neurotoxicity at different levels (Deshpande et al. 2006). Notably, AβOs (Demuro et al. 2005) were neurotoxic within 24 h at 5 μM, while ADDLs took 5-times longer to induce similar neurotoxicity and Aβf took 10 days at fourfold higher concentrations to induce only modest cell death (Deshpande et al. 2006). Additional data suggested that the neurotoxic effects of high-n oligomers and ADDLs were likely mediated *via* synaptic targeting in physiologically viable neurons (Deshpande et al. 2006). By contrast, another study reported that both soluble and fibrillar Aβ are required over prolonged incubation times for significant overt neurotoxicity (Wogulis et al. 2005). However, this apparent discrepancy may be due to alterations in the Aβ species present in the cultures throughout the incubation period, as the various Aβ species exist in an equilibrium and thus Aβ oligomers can self-aggregate into higher-ordered structures and Aβ fibrils can dissociate into monomers. Unfortunately, most studies have not examined the structure of the synthetic preparations of Aβ oligomers at both the beginning and end of the experiment and thus it is extremely difficult to confidently identify and assign biological effects to specific Aβ species (Kelly and Ferreira 2006; Maloney et al. 2005; Snyder et al. 2005; Tamagno et al. 2006; White et al. 1999; Zhao et al. 2006). Future development of novel immunological and biochemical techniques is essential to address this issue and more clearly define the toxicity associated with different assemblies of Aβ.

Finally, the Lamb and Herrup laboratories have provided unique data that link particular Aβ species to induction of aberrant neuronal cell-cycle events (CCEs) in AD. Notably, neuronal populations within the AD brain, which are destined to degenerate, express cell-cycle proteins and also exhibit evidence of DNA synthesis (Yang et al. 2003, 2006), whereas these neuronal CCEs are not observed in age-matched control brains. Notably, neuronal CCEs are observed in mild cognitive impairment (MCI), the clinical predecessor to AD, suggesting that neuronal CCEs are an early marker of neuronal distress (Herrup and Yang 2007). Earlier cell-culture studies suggested that full-length $A\beta_{1-40/42}$ and the fragment $A\beta_{25-35}$ could induce neuronal CCEs (Copani et al. 1999). However, the Aβ preparations used were fibrillar and were at high micromolar concentrations over prolonged incubation periods (Copani et al. 1999).

To gain further insight into the factors that induce neuronal CCEs in AD, we have utilized transgenic mouse models of AD. Notably, the R1.40 transgenic mouse model of AD also displays neuronal CCEs in a spatial and temporal pattern that resembles that which occurs in human AD (Varvel et al. 2008). Furthermore, the first neuronal CCEs occur at least 6 months prior to the first signs of Aβ deposition in the R1.40 model (Varvel et al. 2008) and ablating β-secretase prevents appearance of neuronal CCEs altogether, implicating the role of soluble Aβ species in the induction of neuronal CCEs. To examine this in greater detail, neurons were exposed to size-exclusion chromatography-fractionated Aβ oligomers and monomers, revealing that Aβ oligomers at subnanomolar concentrations, but not Aβ monomers induced neuronal CCEs in a concentration-dependent manner (Varvel et al. 2008). In a subsequent study, it was observed that primary neurons derived from R1.40 transgenic mice secrete Aβ oligomers (Bhaskar et al. 2009) and that induction of neuronal CCEs *via* Aβ oligomers depends on the PI3K-Akt–mTOR signaling pathway (Bhaskar et al. 2009). These studies have provided clear evidence that soluble Aβ oligomers can induce neuronal CCEs, an early marker of neuronal distress in AD.

5.3.5 Neuroinflammation

A substantial body of evidence demonstrates that extracellular deposition of fibrillar Aβ in SPs is one of the key triggers that induce neuroinflammation within the AD brain (reviewed in Eikelenboom et al. 2006). In particular, fibrillar Aβ activates microglia by engaging RAGE and other scavenger receptors (see above). Furthermore, chronic, long-term engagement of these microglial receptors results in the generation of microglia that are incapable of phagocytosing fibrillar Aβ (McDonald et al. 1998, reviewed in Cameron and Landreth 2010).

However, several recent studies have suggested that other soluble Aβ assemblies can also alter microglial activation and neuroinflammation. First, alterations in microglia, astrocytes, and cytokine production have been observed in several different transgenic mouse models of AD (Arelin et al. 2002; Frautschy et al. 1998; Kitazawa et al. 2005), before the first appearance of fibrillar Aβ. This includes the R1.40 transgenic mouse model of AD, in which alterations in cortical microglia are observed at 6 months of age, a time coincident with the first appearance of neuronal CCEs but at least 6–8 months before the first appearance of SPs (Varvel et al. 2009). Notably, removal of β-secretase blocked both alterations in microglial activation and induction of neuronal CCEs, implicating soluble Aβ species in the early alterations in neuroinflammation observed in the R1.40 mouse model (Varvel et al. 2008). In addition to these *in vivo* correlational studies, Sondag and colleagues determined that primary microglia exposed to synthetic low-n Aβ oligomers produced elevated levels of active Lyn and Syk tyrosine kinases as well as p38 mitogen-activated protein kinase (p38 MAPK) compared to Aβ fibrils (Sondag et al. 2009). Aβ oligomers also resulted in an altered chemokine/cytokine profile when compared to Aβ fibrils, including elevated levels of interleukin-6, interleukin-8 and CCL2 (Kiyota et al. 2009). Finally, Aβ oligomers stimulated enhanced neurotoxicity when primary microglia were co-cultured with primary neurons, suggesting that Aβ-oligomer-stimulated microglia released soluble factors that were toxic to neurons (Sondag et al. 2009).

5.3.6 *Tau Phosphorylation*

The amyloid cascade hypothesis of AD posits that Aβ promotes downstream NFT pathology (Hardy and Selkoe 2002; Hardy and Higgins 1992; Selkoe 1989). In support of this, Delacourte and colleagues demonstrated a synergistic interaction between Aβ and tau pathologies, although with unique spatiotemporal distribution patterns (Delacourte et al. 1999, 2002). Evidence for the interaction of Aβ SPs and NFTs came from several different studies. First, direct intracerebral injection of SP-equivalent concentrations of fibrillar, but not soluble, Aβ resulted in profound neuronal loss, tau phosphorylation, and microglial proliferation in aged rhesus monkeys (*Macaca mulatta*) (Geula et al. 1998). Second, stereotaxic injections of fibrillar $A\beta_{1-42}$ resulted in elevated tau phosphorylation and a fivefold increase in NFT levels in the amygdala of P301L tau transgenic mice compared to those injected with the non-fibrillar reverse peptide, $A\beta_{42-1}$ (Gotz et al. 2001). Third, when the Tg2576 APP transgenic mouse model of AD was crossed to the P301L tau-mutant transgenic mouse, the number of NFTs in the olfactory bulb, entorhinal cortex, and the amygdala increased ~7-fold when compared to P301L transgenic mice alone

(Lewis et al. 2001). Fourth, to develop a mouse model with both Aβ and tau pathologies, LaFerla and colleagues generated the triple transgenic (3×Tg) mouse model of AD that contained mutant PSEN1, APP, and tau transgenes. The 3×Tg mice exhibit fibrillar Aβ pathology that precedes the NFT pathology (Oddo et al. 2003). Notably, the SPs first appear in the cortex, while tau pathology is first observed in the hippocampi. In a subsequent immunization study, intracerebral administration of an anti-Aβ antibody to the 3×Tg mice resulted in the disappearance of somatodendritic tau staining (Oddo et al. 2004). This observation is also consistent with another study by Cole and colleagues where, antibodies against Aβ reduced the levels of Aβ oligomers and tau phosphorylation in the Tg2576 mouse model of AD as well as in SH-SY5Y cells *in vitro* (Ma et al. 2006). Together, these studies suggest that extracellular, fibrillar Aβ deposits can exacerbate intraneuronal tau pathology (Blurton-Jones and LaFerla 2006).

However, in addition to the potential role Aβ fibrils play in the induction of tau pathology, there is also increasing evidence that Aβ oligomers may promote tau pathology (Caughey and Lansbury 2003). For example, recent studies by De Felice and colleagues suggested that Aβ oligomers, and not Aβ fibrils, are potent inducers of tau phosphorylation (De Felice et al. 2008). Treatment of mature cultures of hippocampal neurons and neuroblastoma cells with sub-nanomolar concentrations of ADDLs induced significant tau phosphorylation that could be blocked by antibodies against ADDLs. In addition, a study by Cole and colleagues demonstrated that Aβ oligomers activated c-Jun *N*-terminal kinase (JNK) and induced tau hyperphosphorylation in cultured hippocampal neurons and that adding a specific JNK inhibitor reversed this effect (Ma et al. 2009). Finally, transgenic mice with the E693Δ APP mutation, which promotes formation of Aβ oligomers, also exhibited elevated tau phosphorylation (Tomiyama et al. 2010). Similarly, other studies suggest that Aβ protofibrils can also induce tau phosphorylation (Martins et al. 2008).

In summary, studies using synthetic Aβ peptides, Aβ-containing CM, APP transgenic mice, and human AD brain tissue demonstrate that Aβ toxicity is a complex and multifaceted phenomenon that may be induced by multiple, different assemblies of Aβ, which can result in a wide variety of AD-related pathologies ranging from reversible changes in synaptic structure and function all the way to neuronal cell-cycle changes and cell death.

5.4 Tau Aggregates in AD-Related Pathologies

5.4.1 General Structure and Function of Tau Protein

In 1975, Mark Kirchner's laboratory discovered that tau is a microtubule-associated protein in the nervous system (Weingarten et al. 1975). They also demonstrated that as microtubules are a highly dynamic cytoskeletal structures, the extent of microtubule dynamicity depended on the association–dissociation of the tau protein (Weingarten et al. 1975). A large number of subsequent studies have demonstrated

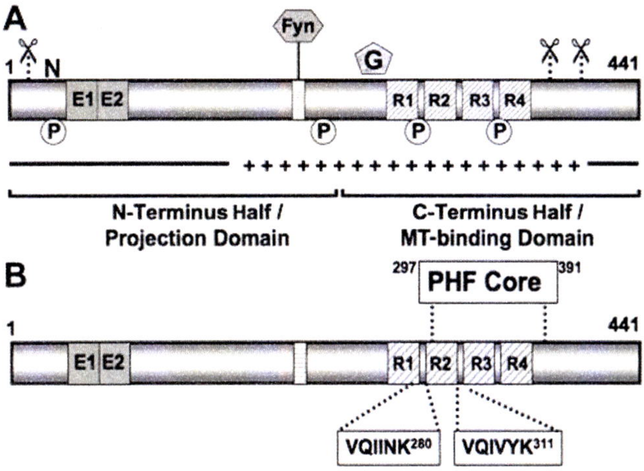

Fig. 5.1 *Tau structure*. (**A**) A schematic showing tau with four microtubule-binding repeats (R1–R4)—4R-Tau. Note that all the microtubule-binding repeats (MTBR) are present in the *C*-terminal half of the molecule. The *N*-terminal of tau has two inserts generated by exons 1 and 2 (E1 and E2). The *N*-terminal half also has proline-rich regions, one of which is a PXXP motif that interacts with the SH3 domain of Fyn. Note the charge distribution across the tau molecule. *C*-terminal MTBR is positively charged whereas the rest of the molecule is negatively charged. Some of the important sites of phosphorylation (*P*), glycosylation or glycation (*G*), nitration (*N*) and truncation (✂) are also shown. (**B**) Oligomerization of tau occurs through the PHF-core region, which comprises two hexapeptide motifs in the MTBR region (not to scale)

that tau was predominantly localized to axons, but was also detectable in both astrocytes and oligodendrocytes within the CNS (Binder et al. 1985; Cleveland et al. 1977; Couchie et al. 1992; LoPresti et al. 1995; Shin et al. 1991).

Genetic analyses confirmed that the human tau gene (*microtubule associated protein tau—MAPT*) is about 100 kb in size and located on chromosome 17q21 (Andreadis et al. 1992; Goedert et al. 1988; Neve et al. 1986). *MAPT* contains 16 exons resulting in the expression of six different isoforms of tau in the CNS *via* alternative splicing of exons 2, 3, and 10 of *MAPT* mRNA (Andreadis et al. 1992; Goedert et al. 1989). Exons 2 and 3 generate two amino-terminal inserts, while exon 10 generates a second microtubule-binding repeat (Fig. 5.1a). The shortest tau isoform contains 352 amino acids, lacks exons 2, 3, and 10 and contains three microtubule-binding repeats (3R-Tau), while the longest tau isoform contains 441 amino acids, includes all the exons and contains four microtubule-binding repeats (4R-Tau) (Fig. 5.1a). In the adult human brain, the ratio of 3R-Tau to 4R-Tau isoforms is ~1 (Goedert and Jakes 1990). In addition, the alternative splicing of tau is developmentally regulated such that only the shortest tau isoform (3R-Tau with no amino-terminal inserts) is expressed in fetal brain, while all six isoforms are expressed in the adult human brain (Goedert et al. 1989).

After the initial description of the primary structure of tau, several groups began to examine its microtubule-binding abilities. It is clear that tau directly binds to microtubules, stabilizing its own structure and promoting microtubule polymerization (Cleveland et al. 1977; Weingarten et al. 1975). Studies have also demonstrated that the microtubule-binding repeats (MTBR) are localized to the *C*-terminal half of the molecule (Fig. 5.1a). These motifs are composed of highly conserved 18-amino-acid-long-binding elements separated by flexible, but less conserved inter-repeats that are 13–14-amino-acid long (Butner and Kirschner 1991; Himmler et al. 1989; Lee et al. 1989). The binding of tau to microtubules is a complex process mediated in part by electrostatic attraction between negatively charged *C*-terminal regions of tubulin and positively charged MTBR in tau that is specifically mediated *via* a flexible array of weak microtubule-binding sites that are distributed throughout the MTBR/inter-repeats (Buee et al. 2000; Butner and Kirschner 1991; Lee et al. 1989; Rosenberg et al. 2008) (Fig. 5.1a). Because of the presence of an extra MTBR, the 4R-Tau isoforms bind to microtubules with higher affinity than 3R-Tau (Butner and Kirschner 1991; Goedert and Jakes 1990). Interestingly, studies have suggested that within 4R-Tau, the first inter-repeat, which is specifically present only in 4R-Tau (because of the inclusion of exon 10/MTBR 2) has the highest affinity for microtubules than any other repeat–inter-repeat sequence (Goode and Feinstein 1994) and thus 4R-Tau serves as a better microtubule stabilizer than 3R-Tau (Goedert and Jakes 1990).

Although numerous studies have examined the role of MTBR of tau, very little is known about the projection domain in the *N*-terminal half of the tau protein (Fig. 5.1a). The projection domain has a negatively charged acidic region and positively charged proline-rich region. The PXXP motif present in the proline-rich region of tau is responsible for tau's interaction with the SH3 domains of Src-family of non-receptor tyrosine kinases (SFKs) such as Fyn (Lee et al. 1998) (Fig. 5.1), which then regulates tau's interaction with the plasma membrane and other cytoskeletal proteins. Furthermore, interaction between the PXXP motif of tau and SH3 domain of SFKs results in phosphorylation of tau at tyrosine 18 (Lee et al. 2004) and is critical in regulating actin cytoskeletal dynamics (Sharma et al. 2007) and membrane association of tau (Ittner et al. 2010).

5.4.2 Disease Association of Tau Assemblies

Tau was first discovered to be a principal component of paired helical filaments (PHFs) within the AD brain in 1985 (Brion et al. 1985; Grundke-Iqbal et al. 1986). Both PHFs and straight filaments isolated from AD brain were later demonstrated to be composed of predominantly hyperphosphorylated tau proteins (Goedert 1998; Kondo et al. 1988; Kosik et al. 1988; Lee et al. 1991; Wischik et al. 1988a).

Compelling evidence in support of the pathologic role of the tau protein in neurodegeneration was provided by the discovery of tau-positive filamentous aggregation in Aβ-independent, sporadic tauopathies (reviewed in Lee et al. 2001). The most prominent and well-studied non-AD tauopathies include PSP, CBD, and PiD (Constantinidis et al. 1974; Rebeiz et al. 1967, 1968; Steele et al. 1964). Clinically, PSP is characterized by supranuclear gaze palsy as well as postural instability, while

CBD and PiD are characterized by Parkinsonism and general behavior/language impairments respectively (Constantinidis et al. 1974; Rebeiz et al. 1967, 1968).

Neuropathologically, PSP is characterized by abundant tau pathology and NFTs as well as atrophy of the basal ganglia, sub-thalamic nuclei, and brainstem with corresponding neuronal loss and gliosis (Hauw et al. 1994, 1994; Litvan et al. 1996; Pollock et al. 1986). Similarly, in the case of CBD, the neuropathological characteristics, including depigmentation of the substantia nigra and asymmetric frontoparietal atrophy are associated with spongiosis, gliosis, and intra-glial as well as intra-neuronal accumulation of filamentous tau (Iwatsubo et al. 1994; Mori et al. 1994). Another distinct neuropathological feature of CBD is the accumulation of tau-positive neuropil threads throughout the gray and white matter (Feany and Dickson 1995; Feany et al. 1996). Spongiosis and gliosis are also common neuropathological features of PiD (Constantinidis et al. 1974; Feany et al. 1996; Pollock et al. 1986). Atrophy of the frontotemporal lobes as well as limbic structures and presence of ballooned neurons with tau immunoreactive Pick's bodies are the prominent features of PiD (Dickson 1998; Lund and Manchester-Groups 1994). Together, identification of PSP, CBD, and PiD demonstrated that tau pathology, independent of amyloid pathology is sufficient to cause neurodegeneration.

To understand the structural properties of aggregated tau present in the PHFs/NFTs derived from the brains of AD and non-AD tauopathies, numerous biochemical techniques and immunological reagents were developed (Brion 2006; Dickson et al. 1987; Greenberg and Davies 1990; Ksiezak-Reding et al. 1988, 1994b; Ksiezak-Reding and Yen 1987, 1991; Tabaton et al. 1988; Wolozin et al. 1986). Analysis of PHFs purified from AD brains by SDS–PAGE revealed the presence of three major tau bands of approximately 68, 64, and 60 kDa, as well as a minor band of approximately 72 kDa (Greenberg and Davies 1990). Following dephosphorylation, six tau bands were resolved that corresponded to the six different isoforms of tau found in adult human brain (Goedert et al. 1992; Greenberg et al. 1992; Lee et al. 1991). By contrast, certain studies found only two major bands at 68 and 64 kDa in AD and PSP samples (Flament et al. 1991; Vermersch et al. 1994). Similarly, the insoluble tau from CBD brains also consisted of two major bands at 64 and 68 kDa and a variable minor band at 72 kDa (Buee-Scherrer et al. 1996; Ksiezak-Reding et al. 1994a).

A distinct pattern of tau migration on SDS–PAGE was observed for PiD. Insoluble tau from PiD was composed of two major bands at 60 and 64 kDa and a variable minor band at 68 kDa (Buee-Scherrer et al. 1996, Delacourte et al. 1996; Lieberman et al. 1998). Subsequent analysis to determine the molecular basis for these differences in the mobility of tau revealed that in AD, all six isoforms of tau participate in the aggregation process. On the other hand, 4R-Tau and 3R-Tau isoforms contribute to filamentous aggregation in PSP/CBD and PiD respectively (Chambers et al. 1999; Di Maria et al. 2000; Goedert et al. 1995; Mailliot et al. 1998; Morishima-Kawashima et al. 1995; Sergeant et al. 1997; Trojanowski and Lee 1994). Thus, the different tauopathies result in widely different neuropathological and clinic phenotypes that depend on the type of tau isoform contributing to NFT formation. Taken together, these studies suggest that tau self-assembly is a prominent feature and biochemical basis for the neuropathology observed in both the AD and non-AD tauopathies.

Recent studies document the presence of tau oligomers composed of about 40 tau molecules during early stages of AD (Maeda et al. 2006) and suggest that these oligomeric structures precede PHF formation (Maeda et al. 2007). In another recent study, Kayed and colleagues utilized a specific antibody against β-pleated annular protofibrils (Officer or T2286) along with a tau-specific antibody and discovered unique protofibrils that contained tau in P301L and rTg4510 mouse models of tauopathy as well as in human AD brain (Kayed and Jackson 2009). Thus, similar to the case for Aβ, increasing evidence has suggested that smaller soluble assemblies of tau may be relevant to neurotoxicity and cognitive dysfunction in tauopathies.

5.4.3 Factors Affecting Tau Self-assembly

5.4.3.1 Role of Post-translational Modification and Truncation of Tau

In order to gain insight into the potential triggers for the self-assembly of tau, multiple studies focused on various post-translational modifications that might play a direct role in the formation of PHFs (Fig. 5.1). For example, phosphorylation glycosylation, ubiquitination, deamidation, oxidation, tyrosine nitration, cross-linking, and glycation were all observed in tau (Avila et al. 2004; Hernandez and Avila 2007; Wang et al. 2007a) (Fig. 5.1). Among these, current evidence suggests that phosphorylation, nitration, ubiquitination, and *N*- or *C*-terminal truncation are the most relevant to tau aggregation and toxicity (reviewed in Arnaud et al. 2006; Garcia-Sierra et al. 2008; Hampel et al. 2010; Reynolds et al. 2007).

Although, the exact biochemical and biophysical mechanisms linking the post-translational modifications of tau to the formation of NFTs remain elusive, multiple studies have provided compelling evidence that these modifications play a unique role in tau aggregation. For example, hyperphosphorylation of tau appears to be an early event in the pathway leading from soluble to insoluble tau (Braak et al. 1994). The prevailing hypothesis is that hyperphosphorylation of tau disengages the protein from microtubules, thereby increasing the pool of unbound tau (Schneider et al. 1999). This unbound tau is subsequently prone to self-aggregation resulting in the production of protease-resistant tau. Notably, in support of this hypothesis, the binding of tau to microtubules was observed to generate ordered structures, indicating that microtubules can induce conformational changes in tau rendering them to be less prone to self-aggregation (Woody et al. 1983).

5.4.3.2 Tau Phosphorylation

Numerous studies have demonstrated that tau phosphorylation is one of the first steps during the formation of tau aggregates and that hyperphosphorylated tau is the principal component of PHFs in the AD brain (reviewed in Hampel et al. 2010).

Disease-associated phosphorylation of tau occurs on multiple serines and threonines that are distributed in the flanking regions of MTBRs and affect microtubule stability. Interestingly, a total of 45 phosphorylation sites have been identified in insoluble tau preparations isolated from AD brain *via* mass-spectrometric and immunological analyses (reviewed in Hanger et al. 2009; http://cnr.iop.kcl.ac.uk/hangerlab/tautable). Notably, phosphorylation of tau at S262 and S356 within the MTBRs has been suggested to detach tau from microtubules (Buee et al. 2000; Schneider et al. 1999). Phosphorylation of tau at sites away from the MTBR might also be involved in regulation of cytoskeletal stability since S214 and T231 in the proline-rich domain of tau also reduce its ability to bind to microtubules (Cho and Johnson 2004) (Fig. 5.1). While these studies suggest that phosphorylation of tau at specific serines and threonines is directly involved in the formation of tau aggregates, tau also undergoes phosphorylation of tyrosines. Notably, tau tyrosine phosphorylation correlates with an increased propensity of tau to self-aggregate (Derkinderen et al. 2005; Lee et al. 2004; Vega et al. 2005) as well as with disease progression in two different mouse models of tauopathy (Bhaskar et al. 2010a). Based on the fact that tau phosphorylation is one of the earliest markers of NFT formation and disease progression, the measurement of tau phosphorylation in CSF samples has been proposed as a unique biomarker for the early detection and progression of AD (reviewed in Fagan and Holtzman 2010).

5.4.3.3 Tau Nitration

Nitration of tau has also been proposed as a post-translational modification that could promote tau aggregation (Reynolds et al. 2006) (Fig. 5.1). Tau nitration likely occurs as a consequence of increased oxidative damage within the brain. In support of this, elevated levels of 3-nitrotyrosine and dityrosines as well as abnormally nitrated tau have been observed in AD brain tissue (Reynolds et al. 2006). Furthermore, *in vitro* studies demonstrated that incubation of recombinant tau with peroxynitrite resulted in the nitration and oligomerization of tau as well as microtubule instability in a dose-dependent manner (Zhang et al. 2005). Among five tyrosines present in the full-length 4R-Tau, nitration of Y29 and Y197 increased the average filament length without changing the steady-state polymer mass, while the nitration of Y18 and Y394 decreased the average filament length (Reynolds et al. 2005). In another recent report, tau nitrated at Y18 localizes to SPs and astrocytes within the AD brain (Reyes et al. 2008). In summary, while nitration of tau seems to exert disparate effects on tau polymerization and aggregation *in vitro*, it is conceivable that their cumulative effect *in vivo* could promote self-association and NFT formation (reviewed in Reynolds et al. 2007).

5.4.3.4 Tau Ubiquitination and Acetylation

Ubiquitination of tau may also be involved in its self-assembly and aggregation (Arnaud et al. 2006). Ubiquitin is a highly conserved protein with 76 amino acids that plays a regulatory role in protein degradation *via* the proteasome (Lehman 2009). Multiple studies have demonstrated that tau undergoes ubiquitination and is degraded by the proteasome *in vitro* (David et al. 2002), in cell culture (David et al. 2002; Dickey et al. 2006; Goldbaum et al. 2003; Goldbaum and Richter-Landsberg 2004), and in the 3×Tg-AD transgenic mouse model of AD (Oddo et al. 2004). Furthermore, recent studies have suggested that defects in the proteasomal pathway contribute to accumulation of tau in selective neuronal populations within AD brains (Liu et al. 2009). Interestingly, a positive correlation between molecular chaperones such as heat-shock proteins (Hsp90, Hsp40, Hsp27, α-crystallin, and CHIP) and soluble tau was documented in AD brain tissue (Sahara et al. 2007b). By contrast, the levels of granular tau oligomers was inversely correlated with the levels of heat-shock proteins, suggesting that proteasomal degradation of tau assisted by heat-shock proteins is critical to prevent tau self-aggregation into oligomers (Jinwal et al. 2010; Sahara et al. 2007b). Indeed, additional studies suggest that an imbalance in Hsp-mediated tau degradation can result in NFT formation and neuronal death (Dickey et al. 2006).

Studies have suggested that ubiquitination of lysines can be precluded by lysine acetylation, which inhibits proteasome-mediated degradation of numerous proteins (Gronroos et al. 2002; Ito et al. 2002; Jin et al. 2004). A recent study demonstrated that tau is acetylated and that tau acetylation prevents degradation of phosphorylated tau (Min et al. 2010). Notably, tau acetylation is elevated in patients at early and moderate Braak stages of tauopathies (Min et al. 2010). Interestingly, acetylation of tau appears to be mediated *via* histone acetyltransferase p300 and a member of mammalian sirtuins (SIRT1) deacetylates tau and mediates proteasomal degradation (Min et al. 2010). The study by Min et al. suggests that acetylation of tau also contributes to pathological aggregation in tauopathies.

5.4.3.5 Tau Truncation

Finally, several pieces of evidence suggest that tau truncation is also linked to its self-assembly (Reviewed in Kovacech and Novak 2010) (Fig. 5.1a). First, tau is truncated at N421 by caspase cleavage, generating $tau_{1–421}$, which co-localizes with NFTs. Notably, Aβ treatment can also induce generation of $tau_{1–421}$ in cultured neurons (Gamblin et al. 2003; Rissman et al. 2004). Second, tau is also a substrate for calpain cleavage, which results in the generation of $tau_{45–230}$ with a molecular mass of 17 kDa (Park and Ferreira 2005). Third, tau truncation accelerates its assembly into fibrils *in vitro* (Abraha et al. 2000; Gamblin et al. 2003; Yin and Kuret 2006).

While these findings demonstrate that tau truncation can certainly occur within the brain, there is still considerable debate as to whether tau truncation is beneficial, detrimental, or both during AD progression. While some studies suggest that tau truncation induces apoptosis and/or mitochondrial dysfunction in cell culture (Fasulo et al. 2005; Quintanilla et al. 2009), others using two-photon *in vivo* imaging of NFT formation and caspase activation in mouse models of tauopathies observe that most neurons that contain activated caspases also contain NFTs and yet do not undergo apoptotic cell death (Spires-Jones et al. 2008, 2009). Instead, caspase activation in this setting may promote tau cleavage and aggregation, rather than promoting cell death by conventional apoptotic pathways (Spires-Jones et al. 2009). Based on these studies, it has been proposed that caspase activation precedes formation of NFTs in mouse models of tauopathy (de Calignon et al. 2010). Finally, in addition to the cleavage of tau by caspase and calpain, other studies have demonstrated that tau can also undergo cleavage by thrombin (Akiyama et al. 1992; Arai et al. 2006), cathespsins (Bednarski and Lynch 1996; Kenessey et al. 1997; Khurana et al. 2010), puromycin-sensitive aminopeptidase (Karsten et al. 2006), unknown proteases (Wischik et al. 1988b), or *via* non-enzymatic mechanisms (Watanabe et al. 2004). Taken together, the proteolytic processing of tau has potential relevance to self-assembly and neurotoxicity of tau.

5.4.3.6 Role of Familial Mutations in MAPT Gene

The disease relevance of tau in neurodegeneration was confirmed based on the identification of familial mutations in the *MAPT* gene in FTDP-17. FTDP-17 is a group of autosomal dominant, non-AD tauopathies that are clinically characterized by extensive cognitive and motor disturbances (reviewed in Foster et al. 1997). Neuropathologically, FTDP-17 is characterized by frontotemporal lobar atrophy, severe neuronal loss, gliosis, and spongiosis (reviewed in Foster et al. 1997; Lee et al. 2001). Biochemical and ultrastructural studies of FTDP-17 have demonstrated abundant filamentous tau aggregates in both neurons and glial cells (Foster et al. 1997). Subsequent studies have demonstrated that FTDP-17 mutations in *MAPT* alter the ability of tau to interact with microtubules and promote tau self-assembly and aggregation (Arrasate et al. 1999; Barghorn et al. 2000; Goedert et al. 1999; Grover et al. 2003; Nacharaju et al. 1999; Neumann et al. 2001; Rizzini et al. 2000; Yen et al. 1999).

One particularly notable FTDP-17 mutation in *MAPT* is ΔK280, which lacks the lysine at position 280 in the hexapeptide motif ^{275}VQIINK^{280} of tau. This hexapeptide motif, and another (^{306}VQIVYK^{311}) present in the MTBR region of tau, exhibit a tendency to aggregate from random coil into β-pleated fibrous structures (Barghorn et al. 2000; Sahara et al. 2007a; von Bergen et al. 2000) (Fig. 5.1b). Direct evidence for the role of these hexapeptides in tau oligomerization and/or nucleation was obtained *via* generation of inducible transgenic mice that express tau with the ΔK280 FTDP-17 mutation (TauRDΔK280) and promoted tau aggregation within the brain (Eckermann et al. 2007). By contrast, generation of inducible transgenic mice, in which the isoleucines at positions 277 and 308 were replaced with prolines in ^{275}VQIINK^{280} and ^{306}VQIVYK^{311} (TauRDΔK280I277PI308P), respectively (Fig. 5.1b), disrupted the β-pleated-sheet structure and abrogated aggregation (Mocanu et al. 2008). Taken together, these studies demonstrate that specific stretches of amino acids within the tau MTBR seed nucleation for tau oligomerization and aggregation.

Finally, our group has demonstrated that FTDP-17 mutations in *MAPT* also affect non-microtubule-binding functions of tau (Bhaskar et al. 2005). Most notably, tau with P301L, V337M, R406W, G272V, and R5H FTDP-17 mutations significantly enhances interaction between the PXXP motif of 4R-Tau and the SH3 domains of Fyn and thereby upregulates Y18 phosphorylation of tau (Bhaskar et al. 2005). In summary, evidence from the characterization of the effects of the FTDP-17 mutations in *MAPT* suggest that single amino-acid mutations in tau can alter tau conformation, function, and oligomerization in the absence of Aβ pathology.

5.4.3.7 RNA, DNA, Glycosaminoglycans, Polyanions, and Dyes

There is increasing evidence suggesting that cofactors such as polyanions, anionic micelles, and planar anionic aromatic dyes can promote aggregation of both full-length tau and the tau MTBR (reviewed in Kuret et al. 2005a, b). Several studies have demonstrated that these cofactors can promote tau aggregation and stabilize protein assembly. Current biochemical models suggest that these cofactors act to reduce repulsion between cationic molecules, such as tau, resulting in self-aggregation (Alonso et al. 2001; Chirita et al. 2005; Konno et al. 2004; Kuret et al. 2005b). For example, RNA can induce the tau MTBR to self-aggregate, a reaction that depends on the intramolecular disulfide bridges involving C332 in the third repeat of tau (Kampers et al. 1996). Notably, these potential RNA–tau interactions are distinct in 3R-Tau compared to 4R-Tau because the latter can form an intramolecular disulfide bridge due to presence of two cysteines compared to one cysteines in 3R-Tau. Thus, current evidence suggests that 4R-Tau offers resistance to formation of aggregates in the absence of cofactors (Hikosou et al. 2007; Hua and He 2003; Wei et al. 2008). Similarly, a recent study demonstrated that either single- or double-stranded DNA with poly-dA/dT and poly-dG/dC, respectively, promoted self-aggregation of MTBRs of tau (Hikosou et al. 2007). Finally, sulfated glcosaminoglycans such as heparin or heparan sulfate promote formation of a β-pleated structure of tau *via* involvement of the hexapeptide, ^{306}VQIVYK^{311}, present in the MTBR (Goedert et al. 1996; Perez et al. 2001; von Bergen et al. 2000) (Figs. 5.1b and 5.2). Notably, heparan sulfate and hyperphosphorylated tau co-localize within the AD brain (Goedert et al. 1996).

In addition to those listed above, polyanions such as taurine have also been shown to promote tau polymerization and aggregation (Santa-Maria et al. 2007). While taurine decreased aggregation of Aβ at millimolar concentrations, it promoted tau assembly, into fibrillar polymers at similar concentrations *via* electrostatic interactions (Santa-Maria et al. 2007) (Figs. 5.1 and 5.2). Furthermore, anionic micelles such as arachidonic acid also accelerated induction of β-pleated-sheet conformers of tau monomers. The nucleation kinetics of recombinant tau observed with arachidonic acid were identical to that of tau isolated from the human AD brain (Abraha et al. 2000; King et al. 1999; Ksiezak-Reding and Wall 2005).

Transmission-electron microscopy and spectroscopic studies have demonstrated that planar aromatic dyes such as thiazine red, Congo red, and thioflavin S have the capacity to bind tau and trigger fibril formation (Chirita et al. 2005). Notably,

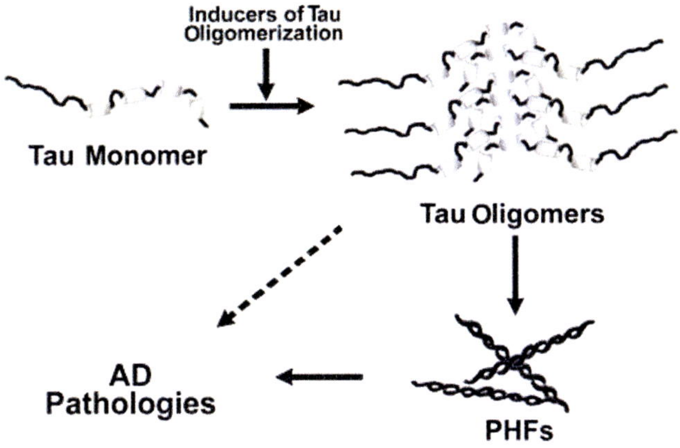

Fig. 5.2 *Structural implications of tau toxicity.* Monomeric tau assembles into tau oligomers or small aggregates in disease condition. This is mediated *via* several inducers of tau oligomerization, including oxidative stress, polyanions, or FTDP-17-related mutations. Oligomeric tau assembles into paired helical filaments (PHFs) and then to mature neurofibrillary tangles. Recent evidence shows that oligomeric tau is more potent in inducing Alzheimer's disease (AD) pathology than the mature PHFs

dye-mediated tau aggregation occurred *via* an increase in the nucleation rate of monomers (Chirita et al. 2005). By contrast, certain cyanine dyes (such as 3,3′-bis(β-hydroxyethyl)-9-ethyl-5,5′-dimethoxythiacarbocyanine iodide or N744) were observed to reduce the filament length of full-length 4R-Tau in a concentration-dependent manner (Necula et al. 2005). Interestingly, many of these dyes (and derivatives thereof) have been utilized for detection of NFTs. In addition, dyes with inhibitory properties may have therapeutic potential for the treatment of tauopathies (Chirita et al. 2005).

Finally, several studies have implicated oxidative stress pathways in tau aggregation. First, exposure of cultured rat hippocampal or human cortical neurons to the lipid-peroxidation product 4-hydroxynonenal results in tau hyperphosphorylation, which is induced by covalent modification of tau by this aldehyde (Mattson et al. 1997). 4-Hydroxynonenal levels are increased in the brains of AD patients, particularly in neurons with altered tau immunoreactivity (Lovell et al. 1997; Montine et al. 1998), suggesting a potential role of lipid peroxidation in tauopathies. Second, the dopamine oxidation product p-benzoquinone induces tau polymerization in differentiated SH-SY5Y in the presence of iron (Santa-Maria et al. 2005). Third, markers of oxidative stress precede formation of tau aggregates in cell culture (Gomez-Ramos et al. 2003), in a transgenic mouse model (Nakashima et al. 2004), as well as in Down syndrome (Nunomura et al. 2000). Finally, postmortem analysis of AD brain tissue revealed that oxidative damage to nucleic acids within neurons is *decreased* when the neuron contains an NFT (Nunomura et al. 2000). Together, these results suggest that oxidative stress may play a significant role in tau aggregation.

5.4.4 Neuropathological Effects of Tau Oligomers and Fibrils

An increasing number of studies suggest that, similar to the case for Aβ oligomers, tau oligomers may be more cytotoxic than mature NFTs, as NFT-bearing neurons appear to survive for decades within the AD brain (Morsch et al. 1999). However, thus far a specific oligomeric structure of tau that could account for specific AD phenotypes has not been identified. While tau pathologies are often viewed as downstream pathological events in AD pathogenesis, the identification of discrete stages of AD at which CSF phospho-tau is elevated and of non-AD tauopathies in which there is no Aβ pathology, suggests that generation of various toxic tau assemblies is likely critical in pathogenesis as outlined below.

5.4.4.1 Lysosomal Dysfunction, Membrane Effects, and Oxidative Stress

Several lines of evidence support a role of tau in lysosomal function and membrane integrity (Lim et al. 2001). First, neurons containing pre-tangles display morphologically aberrant lysosomes associated with increased levels of the lysosomal marker, acid phosphatase (Lim et al. 2001). Second, *N*-terminal truncation of tau occurs in the cytosol *via* an unknown protease, followed by *C*-terminal cleavage of tau within lysosomes by cathepsin L (Wang et al. 2007b). In addition, cytosolic tau enters the lysosome *via* chaperone-mediated autophagy (Wang et al. 2009). However, a subcellular analysis of tau revealed that *N*-terminally truncated tau never enters into the lumen of the lysosome and instead remains associated with the lysosomal membrane. Finally, inhibition of autophagy or expression of FTDP-17-mutant tau blocks translocation of tau fragments across the lysosomal membrane and instead promotes tau oligomerization *via* disruption of the lysosomal membrane (Wang et al. 2009). Together with data demonstrating that lysosomal dysfunction precedes tau cleavage *via* cathepsin D and tau-induced neurotoxicity (Khurana et al. 2010), these results suggest that lysosomal defects and tau oligomerization may play a role in tauopathies.

In addition to the effects of tau on lysosome function, several studies support a role for tau in promotion of oxidative stress. For example, neurons containing glycated NFTs within the AD brain exhibit evidence of oxidative stress, with elevated levels of malondialdehyde and heme oxygenase (Yan et al. 1994). Furthermore, overexpression of tau induced oxidative stress *via* inhibition of axonal transport of peroxisomes (Stamer et al. 2002). Together, these studies suggest that there may be a direct interaction between tau and markers of oxidative stress.

5.4.4.2 Microtubule Destabilization and Defects in Axonal Transport

Hundreds of published studies demonstrate that hyperphosphorylation and truncation of tau lead to microtubule instability and axonal defects in AD and non-AD tauopathies (reviewed in Cowan et al. 2010; Iqbal et al. 2010). For example, hyperphosphorylated tau derived from AD brains sequesters nonphosphorylated tau from microtubules (Alonso et al. 2001) and may result in subsequent microtubule instability (reviewed in Cowan et al. 2010). In addition, the KXGS motif in the MTBR region of tau has been identified as being responsible for detaching tau from the microtubules following phosphorylation by microtubule-affinity-regulating kinase (MARK) (Biernat et al. 1993; Mandelkow et al. 2004; Schneider et al. 1999; Stoothoff and Johnson 2005).

Numerous experiments in neuronal and non-neuronal cells have shown that tau is capable of reducing net anterograde transport of vesicles and cell organelles *via* binding to kinesin on the microtubule tracks (Seitz et al. 2002). Since tau also cargoes *via* microtubules reducing the pace of anterograde transport leads to missorting of tau itself and also starvation of cytoskeletal requirement at the synapses (Konzack et al. 2007; Mandelkow et al. 2003; Thies and Mandelkow 2007). Notably, these effects occur long before tau detaches from microtubules and aggregates into NFTs (Mandelkow et al. 2003). Finally, elevated levels of tau can block cellular trafficking of organelles, neurofilaments, and APP vesicles in neurons and enhances oxidative stress (Stamer et al. 2002). Together these studies demonstrate that tau can contribute to microtubule destabilization and axonal defects by virtue of both loss of function and/or gain-of-toxic function *via* oligomerization.

5.4.4.3 Effects on Synaptotoxicity, Cognitive Function, and Neurodegeneration

To examine more precisely the various neurotoxic effects of tau *in vivo*, numerous invertebrate and vertebrate models of tauopathies have been generated (reviewed in Lee et al. 2005). This includes *Drosophila melanogaster*, *Caenorhabditis elegans*, zebra fish, mice, and rats that overexpress either wild-type or mutant, human *MAPT* transgenes. The most common pathological features observed in these models include hyperphosphorylation and/or aggregation of tau, neurotoxicity, and/or synaptotoxicity, as well as defective axonal transport/axonopathy (reviewed in Lee et al. 2005).

Several recent transgenic mouse models of tauopathies, including rTg4510, JNPL3 and Wtau-Tg mice, have implicated tau oligomers in the observed neurotoxcity (Kimura et al. 2007; Santacruz et al. 2005). Notably, Ashe and colleagues utilized a regulatable mouse model of tauopathy (rTg4510) to examine the role of soluble *versus* aggregated tau in disease phenotypes. The rTg4510 mice exhibit age-related appearance of hyperphosphorylation and aggregation of tau as well as memory dysfunction. However, switching off the *MAPT* transgene at older ages, resulted in a dramatic improvement in memory with no significant alterations in the presence of NFTs, suggesting that soluble tau species were likely responsible for memory dysfunction observed in these mice (Santacruz et al. 2005). Berger and colleagues subsequently identified two forms of tau multimers (140 and 170 kDa) that accumulated during early stages of pathology development in rTg4510 and JNPL3 transgenic mice as well as in the brains of patients with AD and FTDP-17 (Berger et al. 2007).

Notably, levels of these tau multimers correlated with memory loss in the rTg4510 transgenic mice. Furthermore, the presence of granular tau oligomers correlated with synaptic loss in aged wild-type tau (Wtau-Tg) transgenic mice (Kimura et al. 2007). Finally, the hTau mouse model of tauopathy, which expresses all six isoforms of non-mutant human MAPT in the endogenous mouse MAPT background (Andorfer et al. 2005, 2003; Polydoro et al. 2009), exhibits age-related appearance of NFTs and considerable neuronal loss. However, neuronal death was not observed coincident with the appearance of NFTs, suggesting that the cell death was triggered by an NFT-independent pathway. Together these studies suggest that before formation of NFTs, soluble tau oligomers are associated with specific functional and anatomical deficits. Based on these and other studies, tau oligomers have been the focus of therapeutic strategies in AD and related tauopathies (Kayed and Jackson 2009).

5.4.4.4 Amyloid Toxicity

Several recent studies have provided evidence that Aβ neurotoxicity depends on the presence of the wild-type tau protein both in cell-culture and in animal models (Gomez de Barreda et al. 2010; Ittner et al. 2010; Roberson et al. 2007). Notably, tau deficiency prevented behavioral impairments in the J20 mouse model of AD and this effect was independent of alterations in Aβ levels (Roberson et al. 2007). By contrast, other reports suggest that deficiency of endogenous tau can elicit axonal degeneration and tau aggregation, and impact nuclear function in animal models of tauopathies expressing either human APP or *MAPT* with FTDP-17 mutation (Ando et al. 2010; Dawson et al. 2010; de Barreda et al. 2010). It is therefore unclear whether or not conversion of soluble tau into truncated fragments, aggregates, or oligomers may be directly involved in Aβ-induced neurotoxicity, but clearly future studies are warranted.

5.4.4.5 Neuroinflammation

Increasing evidence suggests that neuroinflammation is a common feature of tauopathies. First, activated microglia are found in the postmortem brain tissues of various human tauopathies, including AD, FTDP-17, PSP, and CBD (Gebicke-Haerter 2001; Gerhard et al. 2006; Ishizawa and Dickson 2001). Second, in an animal model of FTDP-17 (P301S), overexpression of mutant tau induced microglial activation, which preceded tangle pathology. Furthermore, an immunosuppressant drug, FK506, attenuated microglial activation, decreased tau hyperphosphorylation, and extended the lifespan in the P301S mouse model of tauopathy (Yoshiyama et al. 2007). Finally, induction of systemic inflammation *via* lipopolysaccharide significantly induced tau hyperphosphorylation in 3×Tg mouse model of AD (Kitazawa et al. 2005). Together these studies suggested that mutation or overexpression of human *MAPT* could cause neuroinflammation, which in turn exacerbates tau phosphorylation in a feed-forward manner.

To provide more direct evidence of a role of neuronal–microglial signaling in tau phosphorylation and aggregation, we recently examined mice that were deficient for the chemokine (fractalkine) receptor CX3CR1, which is solely expressed by microglia and responds to the neuronally derived ligand, CX3CL1. Notably, CX3CR1 deficiency promoted tau phosphorylation, aggregation, and behavioral impairments in the hTau mouse model of tauopathy (Bhaskar et al. 2010b). Additional mechanistic studies demonstrated that induction of tau hyperphosphorylation upon CX3CR1 deficiency was dependent on the secretion of IL-1 from microglia and activation of the p38 mitogen-activated protein-kinase pathway within neurons (Bhaskar et al. 2010b). Taken together, these studies demonstrated that altered signaling between neurons and microglia can promote tau phosphorylation and aggregation.

5.5 Conclusions

The first 5 years of the past decade provided increasing evidence that Aβ oligomers are present in human AD and in mouse models of the disease, and that various forms of these Aβ oligomers exhibit substantial neurotoxic effects both *in vitro* and *in vivo*. Similarly, recent studies examining tau pathologies are now providing evidence for the role of tau oligomers in AD-relevant phenotypes as well. Indeed, a number of recent studies implicate a subgroup of soluble tau species (tau oligomers/immature filaments), rather than NFTs *per se*, in playing critical roles in the specific AD phenotypes (Bretteville and Planel 2008; Congdon and Duff 2008) (Fig. 5.2). While the biological effects of many of the Aβ and/or tau oligomers have been extensively characterized, attributing particular functions or dysfunctions to particular oligomeric structures *in vivo* has proven enormously difficult. Since different conformers are present in dynamic equilibria and likely contained within unique intracellular and extracellular environments, this has proven an enormous technical challenge. To dissect out these complex structure–function relationships, additional studies utilizing highly sophisticated biochemical, immunological, and microscopic techniques to identify and characterize specific oligomers are required as well as methods to stabilize unique oligomeric structures so that they can be more precisely studied in isolation. However, there remain a number of critical technical hurdles that need to be overcome, namely in the reliable detection, quantification, and purification of the wide variety of specific Aβ and tau oligomers, in order to attribute accurately particular functions to particular oligomeric structures and to develop effective therapeutic strategies to target these oligomers. It is hoped that the coming decade will bring a clearer picture of the roles of these protein structures in neurodegenerative diseases.

References

Abraha A, Ghoshal N, Gamblin TC, Cryns V, Berry RW, Kuret J, Binder LI (2000) *C*-terminal inhibition of tau assembly *in vitro* and in Alzheimer's disease. J Cell Sci 113(Pt 21):3737–3745

Akiyama H, Ikeda K, Kondo H, McGeer PL (1992) Thrombin accumulation in brains of patients with Alzheimer's disease. Neurosci Lett 146:152–154

Alberdi E, Sanchez-Gomez MV, Cavaliere F, Perez-Samartin A, Zugaza JL, Trullas R, Domercq M, Matute C (2010) Amyloid β oligomers induce Ca^{2+} dysregulation and neuronal death through activation of ionotropic glutamate receptors. Cell Calcium 47:264–272

Alonso AD, Zaidi T, Novak M, Barra HS, Grundke-Iqbal I, Iqbal K (2001) Interaction of tau isoforms with Alzheimer's disease abnormally hyperphosphorylated tau and *in vitro* phosphorylation into the disease-like protein. J Biol Chem 276:37967–37973

Alzheimer A (1907) Über eine eigenartige Erkrankung der Hirnrinde. Allg. Z. Psychiatrie Psychisch-GerichtlMed 64:146–148

Anandatheerthavarada HK, Biswas G, Robin MA, Avadhani NG (2003) Mitochondrial targeting and a novel transmembrane arrest of Alzheimer's amyloid precursor protein impairs mitochondrial function in neuronal cells. J Cell Biol 161:41–54

Ando K, Leroy K, Heraud C, Kabova A, Yilmaz Z, Authelet M, Suain V, De Decker R, Brion JP (2010) Deletion of murine tau gene increases tau aggregation in a human mutant tau transgenic mouse model. Biochem Soc Trans 38:1001–1005

Andorfer C, Kress Y, Espinoza M, De Silva R, Tucker KL, Barde YA, Duff K, Davies P (2003) Hyperphosphorylation and aggregation of tau in mice expressing normal human tau isoforms. J Neurochem 86:582–590

Andorfer C, Acker CM, Kress Y, Hof PR, Duff K, Davies P (2005) Cell-cycle reentry and cell death in transgenic mice expressing nonmutant human tau isoforms. J Neurosci 25:5446–5454

Andreadis A, Brown WM, Kosik KS (1992) Structure and novel exons of the human tau gene. Biochemistry 31:10626–10633

Arai T, Miklossy J, Klegeris A, Guo JP, McGeer PL (2006) Thrombin and prothrombin are expressed by neurons and glial cells and accumulate in neurofibrillary tangles in Alzheimer disease brain. J Neuropathol Exp Neurol 65:19–25

Arelin K, Kinoshita A, Whelan CM, Irizarry MC, Rebeck GW, Strickland DK, Hyman BT (2002) LRP and senile plaques in Alzheimer's disease: colocalization with apolipoprotein E and with activated astrocytes. Brain Res Mol Brain Res 104:38–46

Arispe N, Diaz JC, Simakova O (2007) Aβ ion channels. Prospects for treating Alzheimer's disease with Aβ channel blockers. Biochim Biophys Acta 1768:1952–1965

Arnaud L, Robakis NK, Figueiredo-Pereira ME (2006) It may take inflammation, phosphorylation and ubiquitination to 'tangle' in Alzheimer's disease. Neurodegener Dis 3:313–319

Arrasate M, Perez M, Armas-Portela R, Avila J (1999) Polymerization of tau peptides into fibrillar structures. The effect of FTDP-17 mutations. FEBS Lett 446:199–202

Ashe KH, Zahs KR (2010) Probing the biology of Alzheimer's disease in mice. Neuron 66:631–645

Avila J, Perez M, Lucas JJ, Gomez-Ramos A, Santa Maria I, Moreno F, Smith M, Perry G, Hernandez F (2004) Assembly *in vitro* of tau protein and its implications in Alzheimer's disease. Curr Alzheimer Res 1:97–101

Barghorn S, Zheng-Fischhofer Q, Ackmann M, Biernat J, Von Bergen M, Mandelkow EM, Mandelkow E (2000) Structure, microtubule interactions, and paired helical filament aggregation by tau mutants of frontotemporal dementias. Biochemistry 39:11714–11721

Barnham KJ, Ciccotosto GD, Tickler AK, Ali FE, Smith DG, Williamson NA, Lam YH, Carrington D, Tew D, Kocak G, Volitakis I, Separovic F, Barrow CJ, Wade JD, Masters CL, Cherny RA, Curtain CC, Bush AI, Cappai R (2003) Neurotoxic, redox-competent Alzheimer's β-amyloid is released from lipid membrane by methionine oxidation. J Biol Chem 278:42959–42965

Bastrikova N, Gardner GA, Reece JM, Jeromin A, Dudek SM (2008) Synapse elimination accompanies functional plasticity in hippocampal neurons. Proc Natl Acad Sci USA 105:3123–3127

Bednarski E, Lynch G (1996) Cytosolic proteolysis of tau by cathepsin D in hippocampus following suppression of cathepsins B and L. J Neurochem 67:1846–1855

Berg L, Mckeel DW Jr, Miller JP, Storandt M, Rubin EH, Morris JC, Baty J, Coats M, Norton J, Goate AM, Price JL, Gearing M, Mirra SS, Saunders AM (1998) Clinicopathologic studies in cognitively healthy aging and Alzheimer's disease: relation of histologic markers to dementia severity, age, sex, and apolipoprotein E genotype. Arch Neurol 55:326–335

Berger Z, Roder H, Hanna A, Carlson A, Rangachari V, Yue M, Wszolek Z, Ashe K, Knight J, Dickson D, Andorfer C, Rosenberry TL, Lewis J, Hutton M, Janus C (2007) Accumulation of pathological tau species and memory loss in a conditional model of tauopathy. J Neurosci 27:3650–3662

Bertoni-Freddari C, Fattoretti P, Meier-Ruge W, Ulrich J (1989) Computer-assisted morphometry of synaptic plasticity during aging and dementia. Pathol Res Pract 185:799–802

Bhaskar K, Yen SH, Lee G (2005) Disease-related modifications in tau affect the interaction between Fyn and Tau. J Biol Chem 280:35119–35125

Bhaskar K, Miller M, Chludzinski A, Herrup K, Zagorski M, Lamb BT (2009) The PI3K-Akt–mTOR pathway regulates Aβ oligomer induced neuronal cell cycle events. Mol Neurodegener 4:14

Bhaskar K, Hobbs GA, Yen SH, Lee G (2010a) Tyrosine phosphorylation of tau accompanies disease progression in transgenic mouse models of tauopathy. Neuropathol Appl Neurobiol 36:462–477

Bhaskar K, Konerth M, Kokiko-Cochran ON, Cardona A, Ransohoff RM, Lamb BT (2010b) Regulation of tau pathology by the microglial fractalkine receptor. Neuron 68:19–31

Biernat J, Gustke N, Drewes G, Mandelkow EM, Mandelkow E (1993) Phosphorylation of Ser262 strongly reduces binding of tau to microtubules: distinction between PHF-like immunoreactivity and microtubule binding. Neuron 11:153–163

Billings LM, Green KN, Mcgaugh JL, Laferla FM (2007) Learning decreases Aβ*56 and tau pathology and ameliorates behavioral decline in 3×Tg-AD mice. J Neurosci 27:751–761

Binder LI, Frankfurter A, Rebhun LI (1985) The distribution of tau in the mammalian central nervous system. J Cell Biol 101:1371–1378

Bitan G, Kirkitadze MD, Lomakin A, Vollers SS, Benedek GB, Teplow DB (2003a) Amyloid β-protein (Aβ) assembly: Aβ40 and Aβ42 oligomerize through distinct pathways. Proc Natl Acad Sci USA 100:330–335

Bitan G, Vollers SS, Teplow DB (2003b) Elucidation of primary structure elements controlling early amyloid β-protein oligomerization. J Biol Chem 278:34882–34889

Blurton-Jones M, Laferla FM (2006) Pathways by which Aβ facilitates tau pathology. Curr Alzheimer Res 3:437–448

Borchelt DR, Thinakaran G, Eckman CB, Lee MK, Davenport F, Ratovitsky T, Prada CM, Kim G, Seekins S, Yager D, Slunt HH, Wang R, Seeger M, Levey AI, Gandy SE, Copeland NG, Jenkins NA, Price DL, Younkin SG, Sisodia SS (1996) Familial Alzheimer's disease-linked presenilin 1 variants elevate Aβ1–42/1–40 ratio *in vitro* and *in vivo*. Neuron 17:1005–1013

Braak E, Braak H, Mandelkow EM (1994) A sequence of cytoskeleton changes related to the formation of neurofibrillary tangles and neuropil threads. Acta Neuropathol 87:554–567

Bretteville A, Planel E (2008) Tau aggregates: toxic, inert, or protective species? J Alzheimers Dis 14:431–436

Brion JP (2006) Immunological demonstration of tau protein in neurofibrillary tangles of Alzheimer's disease. J Alzheimers Dis 9:177–185

Brion JP, Couck AM, Passareiro E, Flament-Durand J (1985) Neurofibrillary tangles of Alzheimer's disease: an immunohistochemical study. J Submicrosc Cytol 17:89–96

Buee L, Bussiere T, Buee-Scherrer V, Delacourte A, Hof PR (2000) Tau protein isoforms, phosphorylation and role in neurodegenerative disorders. Brain Res Brain Res Rev 33:95–130

Buee-Scherrer V, Condamines O, Mourton-Gilles C, Jakes R, Goedert M, Pau B, Delacourte A (1996) AD2, a phosphorylation-dependent monoclonal antibody directed against tau proteins found in Alzheimer's disease. Brain Res Mol Brain Res 39:79–88

Busciglio J, Lorenzo A, Yankner BA (1992) Methodological variables in the assessment of β amyloid neurotoxicity. Neurobiol Aging 13:609–612

Butner KA, Kirschner MW (1991) Tau protein binds to microtubules through a flexible array of distributed weak sites. J Cell Biol 115:717–730
Calabrese B, Shaked GM, Tabarean IV, Braga J, Koo EH, Halpain S (2007) Rapid, concurrent alterations in pre- and postsynaptic structure induced by naturally secreted amyloid-β protein. Mol Cell Neurosci 35:183–193
Cameron B, Landreth GE (2010) Inflammation, microglia, and Alzheimer's disease. Neurobiol Dis 37:503–509
Caspersen C, Wang N, Yao J, Sosunov A, Chen X, Lustbader JW, Xu HW, Stern D, Mckhann G, Yan SD (2005) Mitochondrial Aβ: a potential focal point for neuronal metabolic dysfunction in Alzheimer's disease. FASEB J 19:2040–2041
Cataldo AM, Petanceska S, Terio NB, Peterhoff CM, Durham R, Mercken M, Mehta PD, Buxbaum J, Haroutunian V, Nixon RA (2004) Aβ localization in abnormal endosomes: association with earliest Aβ elevations in AD and Down syndrome. Neurobiol Aging 25:1263–1272
Caughey B, Lansbury PT (2003) Protofibrils, pores, fibrils, and neurodegeneration: separating the responsible protein aggregates from the innocent bystanders. Annu Rev Neurosci 26:267–298
Chambers CB, Lee JM, Troncoso JC, Reich S, Muma NA (1999) Overexpression of four-repeat tau mRNA isoforms in progressive supranuclear palsy but not in Alzheimer's disease. Ann Neurol 46:325–332
Chang L, Bakhos L, Wang Z, Venton DL, Klein WL (2003) Femtomole immunodetection of synthetic and endogenous amyloid-β oligomers and its application to Alzheimer's disease drug candidate screening. J Mol Neurosci 20:305–313
Cheng IH, Scearce-Levie K, Legleiter J, Palop JJ, Gerstein H, Bien-Ly N, Puolivali J, Lesne S, Ashe KH, Muchowski PJ, Mucke L (2007) Accelerating amyloid-β fibrillization reduces oligomer levels and functional deficits in Alzheimer disease mouse models. J Biol Chem 282:23818–23828
Chirita CN, Congdon EE, Yin H, Kuret J (2005) Triggers of full-length tau aggregation: a role for partially folded intermediates. Biochemistry 44:5862–5872
Cho JH, Johnson GV (2004) Primed phosphorylation of tau at Thr231 by glycogen synthase kinase 3β (GSK3β) plays a critical role in regulating tau's ability to bind and stabilize microtubules. J Neurochem 88:349–358
Citron M, Oltersdorf T, Haass C, Mcconlogue L, Hung AY, Seubert P, Vigo-Pelfrey C, Lieberburg I, Selkoe DJ (1992) Mutation of the β-amyloid precursor protein in familial Alzheimer's disease increases β-protein production. Nature 360:672–674
Citron M, Westaway D, Xia W, Carlson G, Diehl T, Levesque G, Johnson-Wood K, Lee M, Seubert P, Davis A, Kholodenko D, Motter R, Sherrington R, Perry B, Yao H, Strome R, Lieberburg I, Rommens J, Kim S, Schenk D, Fraser P, St George Hyslop P, Selkoe DJ (1997) Mutant presenilins of Alzheimer's disease increase production of 42-residue amyloid β-protein in both transfected cells and transgenic mice. Nat Med 3:67–72
Cleary JP, Walsh DM, Hofmeister JJ, Shankar GM, Kuskowski MA, Selkoe DJ, Ashe KH (2005) Natural oligomers of the amyloid-β protein specifically disrupt cognitive function. Nat Neurosci 8:79–84
Cleveland DW, Hwo SY, Kirschner MW (1977) Purification of tau, a microtubule-associated protein that induces assembly of microtubules from purified tubulin. J Mol Biol 116:207–225
Congdon EE, Duff KE (2008) Is tau aggregation toxic or protective? J Alzheimers Dis 14:453–457
Constantinidis J, Richard J, Tissot R (1974) Pick's disease. Histological and clinical correlations. Eur Neurol 11:208–217
Copani A, Condorelli F, Caruso A, Vancheri C, Sala A, Giuffrida Stella AM, Canonico PL, Nicoletti F, Sortino MA (1999) Mitotic signaling by β-amyloid causes neuronal death. FASEB J 13:2225–2234
Corey-Bloom J (2000) The natural history of Alzheimer's disease. Arnold, London
Couchie D, Mavilia C, Georgieff IS, Liem RK, Shelanski ML, Nunez J (1992) Primary structure of high molecular weight tau present in the peripheral nervous system. Proc Natl Acad Sci USA 89:4378–4381

Cowan CM, Bossing T, Page A, Shepherd D, Mudher A (2010) Soluble hyper-phosphorylated tau causes microtubule breakdown and functionally compromises normal tau *in vivo*. Acta Neuropathol 120:593–604
Coyle JT, Price DL, Delong MR (1983) Alzheimer's disease: a disorder of cortical cholinergic innervation. Science 219:1184–1190
Cummings BJ, Pike CJ, Shankle R, Cotman CW (1996) β-Amyloid deposition and other measures of neuropathology predict cognitive status in Alzheimer's disease. Neurobiol Aging 17:921–933
Curtain CC, Ali F, Volitakis I, Cherny RA, Norton RS, Beyreuther K, Barrow CJ, Masters CL, Bush AI, Barnham KJ (2001) Alzheimer's disease amyloid-β binds copper and zinc to generate an allosterically ordered membrane-penetrating structure containing superoxide dismutase-like subunits. J Biol Chem 276:20466–20473
David DC, Layfield R, Serpell L, Narain Y, Goedert M, Spillantini MG (2002) Proteasomal degradation of tau protein. J Neurochem 83:176–185
Davies CA, Mann DM, Sumpter PQ, Yates PO (1987) A quantitative morphometric analysis of the neuronal and synaptic content of the frontal and temporal cortex in patients with Alzheimer's disease. J Neurol Sci 78:151–164
Dawson HN, Cantillana V, Jansen M, Wang H, Vitek MP, Wilcock DM, Lynch JR, Laskowitz DT (2010) Loss of tau elicits axonal degeneration in a mouse model of Alzheimer's disease. Neuroscience 169:516–531
De Barreda EG, Dawson HN, Vitek MP, Avila J (2010) Tau deficiency leads to the upregulation of BAF-57, a protein involved in neuron-specific gene repression. FEBS Lett 584:2265–2270
De Calignon A, Fox LM, Pitstick R, Carlson GA, Bacskai BJ, Spires-Jones TL, Hyman BT (2010) Caspase activation precedes and leads to tangles. Nature 464:1201–1204
De Felice FG, Wu D, Lambert MP, Fernandez SJ, Velasco PT, Lacor PN, Bigio EH, Jerecic J, Acton PJ, Shughrue PJ, Chen-Dodson E, Kinney GG, Klein WL (2008) Alzheimer's disease-type neuronal tau hyperphosphorylation induced by Aβ oligomers. Neurobiol Aging 29: 1334–1347
De Felice FG, Vieira MN, Bomfim TR, Decker H, Velasco PT, Lambert MP, Viola KL, Zhao WQ, Ferreira ST, Klein WL (2009) Protection of synapses against Alzheimer's-linked toxins: Insulin signaling prevents the pathogenic binding of Aβ oligomers. Proc Natl Acad Sci USA 106:1971–1976
Dekosky ST, Scheff SW (1990) Synapse loss in frontal cortex biopsies in Alzheimer's disease: correlation with cognitive severity. Ann Neurol 27:457–464
Delacourte A, Robitaille Y, Sergeant N, Buee L, Hof PR, Wattez A, Laroche-Cholette A, Mathieu J, Chagnon P, Gauvreau D (1996) Specific pathological Tau protein variants characterize Pick's disease. J Neuropathol Exp Neurol 55:159–168
Delacourte A, David JP, Sergeant N, Buee L, Wattez A, Vermersch P, Ghozali F, Fallet-Bianco C, Pasquier F, Lebert F, Petit H, Di Menza C (1999) The biochemical pathway of neurofibrillary degeneration in aging and Alzheimer's disease. Neurology 52:1158–1165
Delacourte A, Sergeant N, Champain D, Wattez A, Maurage CA, Lebert F, Pasquier F, David JP (2002) Nonoverlapping but synergetic tau and APP pathologies in sporadic Alzheimer's disease. Neurology 59:398–407
Demuro A, Mina E, Kayed R, Milton SC, Parker I, Glabe CG (2005) Calcium dysregulation and membrane disruption as a ubiquitous neurotoxic mechanism of soluble amyloid oligomers. J Biol Chem 280:17294–17300
Derkinderen P, Scales TM, Hanger DP, Leung KY, Byers HL, Ward MA, Lenz C, Price C, Bird IN, Perera T, Kellie S, Williamson R, Noble W, Van Etten RA, Leroy K, Brion JP, Reynolds CH, Anderton BH (2005) Tyrosine 394 is phosphorylated in Alzheimer's paired helical filament tau and in fetal tau with c-Abl as the candidate tyrosine kinase. J Neurosci 25:6584–6593
Deshpande A, Mina E, Glabe C, Busciglio J (2006) Different conformations of amyloid β induce neurotoxicity by distinct mechanisms in human cortical neurons. J Neurosci 26:6011–6018
Deshpande A, Kawai H, Metherate R, Glabe CG, Busciglio J (2009) A role for synaptic zinc in activity-dependent Aβ oligomer formation and accumulation at excitatory synapses. J Neurosci 29:4004–4015

Di Maria E, Tabaton M, Vigo T, Abbruzzese G, Bellone E, Donati C, Frasson E, Marchese R, Montagna P, Munoz DG, Pramstaller PP, Zanusso G, Ajmar F, Mandich P (2000) Corticobasal degeneration shares a common genetic background with progressive supranuclear palsy. Ann Neurol 47:374–377

Dickey CA, Yue M, Lin WL, Dickson DW, Dunmore JH, Lee WC, Zehr C, West G, Cao S, Clark AM, Caldwell GA, Caldwell KA, Eckman C, Patterson C, Hutton M, Petrucelli L (2006) Deletion of the ubiquitin ligase CHIP leads to the accumulation, but not the aggregation, of both endogenous phospho- and caspase-3-cleaved tau species. J Neurosci 26:6985–6996

Dickson DW (1998) Pick's disease: a modern approach. Brain Pathol 8:339–354

Dickson DW, Ksiezak-Reding H, Davies P, Yen SH (1987) A monoclonal antibody that recognizes a phosphorylated epitope in Alzheimer neurofibrillary tangles, neurofilaments and tau proteins immunostains granulovacuolar degeneration. Acta Neuropathol 73:254–258

Dickson DW, Crystal HA, Bevona C, Honer W, Vincent I, Davies P (1995) Correlations of synaptic and pathological markers with cognition of the elderly. Neurobiol Aging 16:285–298, discussion 98–304

Dineley KT, Xia X, Bui D, Sweatt JD, Zheng H (2002) Accelerated plaque accumulation, associative learning deficits, and up-regulation of α7 nicotinic receptor protein in transgenic mice co–expressing mutant human presenilin 1 and amyloid precursor proteins. J Biol Chem 277:22768–22780

Dodart JC, Bales KR, Gannon KS, Greene SJ, Demattos RB, Mathis C, Delong CA, Wu S, Wu X, Holtzman DM, Paul SM (2002) Immunization reverses memory deficits without reducing brain Aβ burden in Alzheimer's disease model. Nat Neurosci 5:452–457

Dong J, Atwood CS, Anderson VE, Siedlak SL, Smith MA, Perry G, Carey PR (2003) Metal binding and oxidation of amyloid-β within isolated senile plaque cores: Raman microscopic evidence. Biochemistry 42:2768–2773

Duff K, Planel E (2005) Untangling memory deficits. Nat Med 11:826–827

Dziewczapolski G, Glogowski CM, Masliah E, Heinemann SF (2009) Deletion of the α7 nicotinic acetylcholine receptor gene improves cognitive deficits and synaptic pathology in a mouse model of Alzheimer's disease. J Neurosci 29:8805–8815

Eckermann K, Mocanu MM, Khlistunova I, Biernat J, Nissen A, Hofmann A, Schonig K, Bujard H, Haemisch A, Mandelkow E, Zhou L, Rune G, Mandelkow EM (2007) The β-propensity of Tau determines aggregation and synaptic loss in inducible mouse models of tauopathy. J Biol Chem 282:31755–31765

Eckert A, Hauptmann S, Scherping I, Meinhardt J, Rhein V, Drose S, Brandt U, Fandrich M, Muller WE, Gotz J (2008a) Oligomeric and fibrillar species of β-amyloid (Aβ42) both impair mitochondrial function in P301L tau transgenic mice. J Mol Med 86:1255–1267

Eckert A, Hauptmann S, Scherping I, Rhein V, Muller-Spahn F, Gotz J, Muller WE (2008b) Soluble β-amyloid leads to mitochondrial defects in amyloid precursor protein and tau transgenic mice. Neurodegener Dis 5:157–159

Eikelenboom P, Veerhuis R, Scheper W, Rozemuller AJ, Van Gool WA, Hoozemans JJ (2006) The significance of neuroinflammation in understanding Alzheimer's disease. J Neural Transm 113:1685–1695

Fagan AM, Holtzman DM (2010) Cerebrospinal fluid biomarkers of Alzheimer's disease. Biomark Med 4:51–63

Fang CL, Wu WH, Liu Q, Sun X, Ma Y, Zhao YF, Li YM (2010) Dual functions of β-amyloid oligomer and fibril in Cu(II)-induced H_2O_2 production. Regul Pept 163:1–6

Fasulo L, Ugolini G, Cattaneo A (2005) Apoptotic effect of caspase-3 cleaved tau in hippocampal neurons and its potentiation by tau FTDP-mutation N279K. J Alzheimers Dis 7:3–13

Feany MB, Dickson DW (1995) Widespread cytoskeletal pathology characterizes corticobasal degeneration. Am J Pathol 146:1388–1396

Feany MB, Mattiace LA, Dickson DW (1996) Neuropathologic overlap of progressive supranuclear palsy, Pick's disease and corticobasal degeneration. J Neuropathol Exp Neurol 55:53–67

Flament S, Delacourte A, Verny M, Hauw JJ, Javoy-Agid F (1991) Abnormal Tau proteins in progressive supranuclear palsy. Similarities and differences with the neurofibrillary degeneration of the Alzheimer type. Acta Neuropathol 81:591–596

Foster NL, Wilhelmsen K, Sima AA, Jones MZ, D'amato CJ, Gilman S (1997) Frontotemporal dementia and parkinsonism linked to chromosome 17: a consensus conference. Conference Participants. Ann Neurol 41:706–715

Frautschy SA, Yang F, Irrizarry M, Hyman B, Saido TC, Hsiao K, Cole GM (1998) Microglial response to amyloid plaques in APPsw transgenic mice. Am J Pathol 152:307–317

Freir DB, Fedriani R, Scully D, Smith IM, Selkoe DJ, Walsh DM, Regan CM (2010) Aβ oligomers inhibit synapse remodelling necessary for memory consolidation. Neurobiol Aging, 32: 2211–2218

Friedrich RP, Tepper K, Ronicke R, Soom M, Westermann M, Reymann K, Kaether C, Fandrich M (2010) Mechanism of amyloid plaque formation suggests an intracellular basis of Aβ pathogenicity. Proc Natl Acad Sci USA 107:1942–1947

Gamblin TC, Chen F, Zambrano A, Abraha A, Lagalwar S, Guillozet AL, Lu M, Fu Y, Garcia-Sierra F, Lapointe N, Miller R, Berry RW, Binder LI, Cryns VL (2003) Caspase cleavage of tau: linking amyloid and neurofibrillary tangles in Alzheimer's disease. Proc Natl Acad Sci USA 100:10032–10037

Games D, Adams D, Alessandrini R, Barbour R, Berthelette P, Blackwell C, Carr T, Clemens J, Donaldson T, Gillespie F et al (1995) Alzheimer-type neuropathology in transgenic mice overexpressing V717F β-amyloid precursor protein. Nature 373:523–527

Garcia-Sierra F, Mondragon-Rodriguez S, Basurto-Islas G (2008) Truncation of tau protein and its pathological significance in Alzheimer's disease. J Alzheimers Dis 14:401–409

Gebicke-Haerter PJ (2001) Microglia in neurodegeneration: molecular aspects. Microsc Res Tech 54:47–58

Geinisman Y (1999) Age-related decline in memory function: is it associated with a loss of synapses? Neurobiol Aging 20:353–356

Geinisman Y, Detoledo-Morrell L, Morrell F, Persina IS, Rossi M (1992) Structural synaptic plasticity associated with the induction of long-term potentiation is preserved in the dentate gyrus of aged rats. Hippocampus 2:445–456

Georganopoulou DG, Chang L, Nam JM, Thaxton CS, Mufson EJ, Klein WL, Mirkin CA (2005) Nanoparticle-based detection in cerebral spinal fluid of a soluble pathogenic biomarker for Alzheimer's disease. Proc Natl Acad Sci USA 102:2273–2276

Gerhard A, Trender-Gerhard I, Turkheimer F, Quinn NP, Bhatia KP, Brooks DJ (2006) *In vivo* imaging of microglial activation with [^{11}C](R)-PK11195 PET in progressive supranuclear palsy. Mov Disord 21:89–93

Geula C, Wu CK, Saroff D, Lorenzo A, Yuan M, Yankner BA (1998) Aging renders the brain vulnerable to amyloid β-protein neurotoxicity. Nat Med 4:827–831

Glenner GG, Wong CW (1984) Alzheimer's disease: initial report of the purification and characterization of a novel cerebrovascular amyloid protein. Biochem Biophys Res Commun 120:885–890

Goate A, Chartier-Harlin MC, Mullan M, Brown J, Crawford F, Fidani L, Giuffra L, Haynes A, Irving N, James L et al (1991) Segregation of a missense mutation in the amyloid precursor protein gene with familial Alzheimer's disease. Nature 349:704–706

Goedert M (1998) Neurofibrillary pathology of Alzheimer's disease and other tauopathies. Prog Brain Res 117:287–306

Goedert M, Jakes R (1990) Expression of separate isoforms of human tau protein: correlation with the tau pattern in brain and effects on tubulin polymerization. EMBO J 9:4225–4230

Goedert M, Wischik CM, Crowther RA, Walker JE, Klug A (1988) Cloning and sequencing of the cDNA encoding a core protein of the paired helical filament of Alzheimer disease: identification as the microtubule-associated protein tau. Proc Natl Acad Sci USA 85:4051–4055

Goedert M, Spillantini MG, Jakes R, Rutherford D, Crowther RA (1989) Multiple isoforms of human microtubule-associated protein tau: sequences and localization in neurofibrillary tangles of Alzheimer's disease. Neuron 3:519–526

Goedert M, Spillantini MG, Cairns NJ, Crowther RA (1992) Tau proteins of Alzheimer paired helical filaments: abnormal phosphorylation of all six brain isoforms. Neuron 8:159–168

Goedert M, Spillantini MG, Jakes R, Crowther RA, Vanmechelen E, Probst A, Gotz J, Burki K, Cohen P (1995) Molecular dissection of the paired helical filament. Neurobiol Aging 16:325–334

Goedert M, Jakes R, Spillantini MG, Hasegawa M, Smith MJ, Crowther RA (1996) Assembly of microtubule-associated protein tau into Alzheimer-like filaments induced by sulphated glycosaminoglycans. Nature 383:550–553

Goedert M, Jakes R, Crowther RA (1999) Effects of frontotemporal dementia FTDP-17 mutations on heparin-induced assembly of tau filaments. FEBS Lett 450:306–311

Goldbaum O, Richter-Landsberg C (2004) Proteolytic stress causes heat shock protein induction, tau ubiquitination, and the recruitment of ubiquitin to tau-positive aggregates in oligodendrocytes in culture. J Neurosci 24:5748–5757

Goldbaum O, Oppermann M, Handschuh M, Dabir D, Zhang B, Forman MS, Trojanowski JQ, Lee VM, Richter-Landsberg C (2003) Proteasome inhibition stabilizes tau inclusions in oligodendroglial cells that occur after treatment with okadaic acid. J Neurosci 23:8872–8880

Gomez De Barreda E, Perez M, Gomez Ramos P, De Cristobal J, Martin-Maestro P, Moran A, Dawson HN, Vitek MP, Lucas JJ, Hernandez F, Avila J (2010) Tau-knockout mice show reduced GSK3-induced hippocampal degeneration and learning deficits. Neurobiol Dis 37:622–629

Gomez-Ramos A, Diaz-Nido J, Smith MA, Perry G, Avila J (2003) Effect of the lipid peroxidation product acrolein on tau phosphorylation in neural cells. J Neurosci Res 71:863–870

Goode BL, Feinstein SC (1994) Identification of a novel microtubule binding and assembly domain in the developmentally regulated inter-repeat region of tau. J Cell Biol 124:769–782

Gotz J, Chen F, Van Dorpe J, Nitsch RM (2001) Formation of neurofibrillary tangles in P301l tau transgenic mice induced by Aβ 42 fibrils. Science 293:1491–1495

Green JD, Kreplak L, Goldsbury C, Li Blatter X, Stolz M, Cooper GS, Seelig A, Kistler J, Aebi U (2004) Atomic force microscopy reveals defects within mica supported lipid bilayers induced by the amyloidogenic human amylin peptide. J Mol Biol 342:877–887

Greenamyre JT, Young AB (1989) Excitatory amino acids and Alzheimer's disease. Neurobiol Aging 10:593–602

Greenberg SG, Davies P (1990) A preparation of Alzheimer paired helical filaments that displays distinct tau proteins by polyacrylamide gel electrophoresis. Proc Natl Acad Sci USA 87: 5827–5831

Greenberg SG, Davies P, Schein JD, Binder LI (1992) Hydrofluoric acid-treated tau PHF proteins display the same biochemical properties as normal tau. J Biol Chem 267:564–569

Gronroos E, Hellman U, Heldin CH, Ericsson J (2002) Control of Smad7 stability by competition between acetylation and ubiquitination. Mol Cell 10:483–493

Grover A, England E, Baker M, Sahara N, Adamson J, Granger B, Houlden H, Passant U, Yen SH, Deture M, Hutton M (2003) A novel tau mutation in exon 9 (1260V) causes a four-repeat tauopathy. Exp Neurol 184:131–140

Grundke-Iqbal I, Iqbal K, Tung YC, Quinlan M, Wisniewski HM, Binder LI (1986) Abnormal phosphorylation of the microtubule-associated protein tau (tau) in Alzheimer cytoskeletal pathology. Proc Natl Acad Sci USA 83:4913–4917

Haass C, Koo EH, Mellon A, Hung AY, Selkoe DJ (1992) Targeting of cell-surface β-amyloid precursor protein to lysosomes: alternative processing into amyloid-bearing fragments. Nature 357:500–503

Hampel H, Blennow K, Shaw LM, Hoessler YC, Zetterberg H, Trojanowski JQ (2010) Total and phosphorylated tau protein as biological markers of Alzheimer's disease. Exp Gerontol 45:30–40

Hanger DP, Anderton BH, Noble W (2009) Tau phosphorylation: the therapeutic challenge for neurodegenerative disease. Trends Mol Med 15:112–119

Hardy JA, Higgins GA (1992) Alzheimer's disease: the amyloid cascade hypothesis. Science 256:184–185

Hardy J, Selkoe DJ (2002) The amyloid hypothesis of Alzheimer's disease: progress and problems on the road to therapeutics. Science 297:353–356

Harper JD, Lieber CM, Lansbury PT Jr (1997) Atomic force microscopic imaging of seeded fibril formation and fibril branching by the Alzheimer's disease amyloid-β protein. Chem Biol 4:951–959

Hartley DM, Walsh DM, Ye CP, Diehl T, Vasquez S, Vassilev PM, Teplow DB, Selkoe DJ (1999) Protofibrillar intermediates of amyloid β-protein induce acute electrophysiological changes and progressive neurotoxicity in cortical neurons. J Neurosci 19:8876–8884

Hartley DM, Zhao C, Speier AC, Woodard GA, Li S, Li Z, Walz T (2008) Transglutaminase induces protofibril-like amyloid β-protein assemblies that are protease-resistant and inhibit long-term potentiation. J Biol Chem 283:16790–16800

Hauw JJ, Daniel SE, Dickson D, Horoupian DS, Jellinger K, Lantos PL, Mckee A, Tabaton M, Litvan I (1994) Preliminary NINDS neuropathologic criteria for Steele-Richardson-Olszewski syndrome (progressive supranuclear palsy). Neurology 44:2015–2019

Herholz K, Weisenbach S, Zundorf G, Lenz O, Schroder H, Bauer B, Kalbe E, Heiss WD (2004) *In vivo* study of acetylcholine esterase in basal forebrain, amygdala, and cortex in mild to moderate Alzheimer disease. Neuroimage 21:136–143

Hernandez F, Avila J (2007) Tauopathies. Cell Mol Life Sci 64:2219–2233

Hernandez CM, Kayed R, Zheng H, Sweatt JD, Dineley KT (2010) Loss of α7 nicotinic receptors enhances β-amyloid oligomer accumulation, exacerbating early-stage cognitive decline and septohippocampal pathology in a mouse model of Alzheimer's disease. J Neurosci 30: 2442–2453

Herrup K, Yang Y (2007) Cell cycle regulation in the postmitotic neuron: oxymoron or new biology? Nat Rev Neurosci 8:368–378

Hewitt N, Rauk A (2009) Mechanism of hydrogen peroxide production by copper-bound amyloid β peptide: a theoretical study. J Phys Chem B 113:1202–1209

Hikosou R, Kurabayashi Y, Doumoto M, Hoshitoku K, Mizushima F, Minoura K, Tomoo K, Ishida T (2007) Effect of DNA on filament formation of tau microtubule-binding domain: structural dependence of DNA. Chem Pharm Bull (Tokyo) 55:1030–1033

Himmler A, Drechsel D, Kirschner MW, Martin DW Jr (1989) Tau consists of a set of proteins with repeated *C*-terminal microtubule-binding domains and variable *N*-terminal domains. Mol Cell Biol 9:1381–1388

Holmes C, Boche D, Wilkinson D, Yadegarfar G, Hopkins V, Bayer A, Jones RW, Bullock R, Love S, Neal JW, Zotova E, Nicoll JA (2008) Long-term effects of Aβ42 immunisation in Alzheimer's disease: follow-up of a randomised, placebo-controlled phase I trial. Lancet 372:216–223

Horn AH, Sticht H (2010) Amyloid-β42 oligomer structures from fibrils: a systematic molecular dynamics study. J Phys Chem B 114:2219–2226

Hoshi M, Sato M, Matsumoto S, Noguchi A, Yasutake K, Yoshida N, Sato K (2003) Spherical aggregates of β-amyloid (amylospheroid) show high neurotoxicity and activate tau protein kinase I/glycogen synthase kinase-3β. Proc Natl Acad Sci USA 100:6370–6375

Hsiao K, Chapman P, Nilsen S, Eckman C, Harigaya Y, Younkin S, Yang F, Cole G (1996) Correlative memory deficits, Aβ elevation, and amyloid plaques in transgenic mice. Science 274:99–102

Hu X, Crick SL, Bu G, Frieden C, Pappu RV, Lee JM (2009) Amyloid seeds formed by cellular uptake, concentration, and aggregation of the amyloid-β peptide. Proc Natl Acad Sci USA 106:20324–20329

Hua Q, He RQ (2003) Tau could protect DNA double helix structure. Biochim Biophys Acta 1645:205–211

Hyman BT, Tanzi RE (1992) Amyloid, dementia and Alzheimer's disease. Curr Opin Neurol Neurosurg 5:88–93

Hyman BT, Van Hoesen GW, Kromer LJ, Damasio AR (1986) Perforant pathway changes and the memory impairment of Alzheimer's disease. Ann Neurol 20:472–481

Ida N, Masters CL, Beyreuther K (1996) Rapid cellular uptake of Alzheimer amyloid βA4 peptide by cultured human neuroblastoma cells. FEBS Lett 394:174–178

Iqbal K, Liu F, Gong CX, Grundke-Iqbal I (2010) Tau in Alzheimer disease and related tauopathies. Curr Alzheimer Res 7:656–664

Ishizawa K, Dickson DW (2001) Microglial activation parallels system degeneration in progressive supranuclear palsy and corticobasal degeneration. J Neuropathol Exp Neurol 60:647–657

Ito A, Kawaguchi Y, Lai CH, Kovacs JJ, Higashimoto Y, Appella E, Yao TP (2002) MDM2-HDAC1-mediated deacetylation of p53 is required for its degradation. EMBO J 21:6236–6245
Ittner LM, Ke YD, Delerue F, Bi M, Gladbach A, Van Eersel J, Wolfing H, Chieng BC, Christie MJ, Napier IA, Eckert A, Staufenbiel M, Hardeman E, Gotz J (2010) Dendritic function of tau mediates amyloid-β toxicity in Alzheimer's disease mouse models. Cell 142:387–397
Iwatsubo T, Hasegawa M, Ihara Y (1994) Neuronal and glial tau-positive inclusions in diverse neurologic diseases share common phosphorylation characteristics. Acta Neuropathol 88:129–136
Jacobsen JS, Wu CC, Redwine JM, Comery TA, Arias R, Bowlby M, Martone R, Morrison JH, Pangalos MN, Reinhart PH, Bloom FE (2006) Early-onset behavioral and synaptic deficits in a mouse model of Alzheimer's disease. Proc Natl Acad Sci USA 103:5161–5166
Jellinger K (1990) Morphology of Alzheimer's disease and related disorders. Springer, Berlin
Jiang D, Men L, Wang J, Zhang Y, Chickenyen S, Wang Y, Zhou F (2007) Redox reactions of copper complexes formed with different β-amyloid peptides and their neuropathological [correction of neuropathalogical] relevance. Biochemistry 46:9270–9282
Jiang D, Li X, Liu L, Yagnik GB, Zhou F (2010) Reaction rates and mechanism of the ascorbic acid oxidation by molecular oxygen facilitated by Cu(II)-containing amyloid-β complexes and aggregates. J Phys Chem B 114:4896–4903
Jin YH, Jeon EJ, Li QL, Lee YH, Choi JK, Kim WJ, Lee KY, Bae SC (2004) Transforming growth factor-β stimulates p300-dependent RUNX3 acetylation, which inhibits ubiquitination-mediated degradation. J Biol Chem 279:29409–29417
Jinwal UK, Koren J 3rd, Borysov SI, Schmid AB, Abisambra JF, Blair LJ, Johnson AG, Jones JR, Shults CL, O'leary JC 3rd, Jin Y, Buchner J, Cox MB, Dickey CA (2010) The Hsp90 cochaperone, FKBP51, increases Tau stability and polymerizes microtubules. J Neurosci 30:591–599
Kampers T, Friedhoff P, Biernat J, Mandelkow EM, Mandelkow E (1996) RNA stimulates aggregation of microtubule-associated protein tau into Alzheimer-like paired helical filaments. FEBS Lett 399:344–349
Kang J, Lemaire HG, Unterbeck A, Salbaum JM, Masters CL, Grzeschik KH, Multhaup G, Beyreuther K, Muller-Hill B (1987) The precursor of Alzheimer's disease amyloid A4 protein resembles a cell-surface receptor. Nature 325:733–736
Kanski J, Aksenova M, Schoneich C, Butterfield DA (2002a) Substitution of isoleucine-31 by helical-breaking proline abolishes oxidative stress and neurotoxic properties of Alzheimer's amyloid β-peptide. Free Radic Biol Med 32:1205–1211
Kanski J, Varadarajan S, Aksenova M, Butterfield DA (2002b) Role of glycine-33 and methionine-35 in Alzheimer's amyloid β-peptide 1–42-associated oxidative stress and neurotoxicity. Biochim Biophys Acta 1586:190–198
Kar S, Quirion R (2004) Amyloid β peptides and central cholinergic neurons: functional interrelationship and relevance to Alzheimer's disease pathology. Prog Brain Res 145:261–274
Karsten SL, Sang TK, Gehman LT, Chatterjee S, Liu J, Lawless GM, Sengupta S, Berry RW, Pomakian J, Oh HS, Schulz C, Hui KS, Wiedau-Pazos M, Vinters HV, Binder LI, Geschwind DH, Jackson GR (2006) A genomic screen for modifiers of tauopathy identifies puromycin-sensitive aminopeptidase as an inhibitor of tau-induced neurodegeneration. Neuron 51:549–560
Kawarabayashi T, Shoji M, Younkin LH, Wen-Lang L, Dickson DW, Murakami T, Matsubara E, Abe K, Ashe KH, Younkin SG (2004) Dimeric amyloid β protein rapidly accumulates in lipid rafts followed by apolipoprotein E and phosphorylated tau accumulation in the Tg2576 mouse model of Alzheimer's disease. J Neurosci 24:3801–3809
Kayed R, Jackson GR (2009) Prefilament tau species as potential targets for immunotherapy for Alzheimer disease and related disorders. Curr Opin Immunol 21:359–363
Kayed R, Head E, Thompson JL, Mcintire TM, Milton SC, Cotman CW, Glabe CG (2003) Common structure of soluble amyloid oligomers implies common mechanism of pathogenesis. Science 300:486–489
Kelly BL, Ferreira A (2006) β-Amyloid-induced dynamin 1 degradation is mediated by *N*-methyl-D-aspartate receptors in hippocampal neurons. J Biol Chem 281:28079–28089

Kenessey A, Nacharaju P, Ko LW, Yen SH (1997) Degradation of tau by lysosomal enzyme cathepsin D: implication for Alzheimer neurofibrillary degeneration. J Neurochem 69: 2026–2038
Khachaturian ZS (1985) Diagnosis of Alzheimer's disease. Arch Neurol 42:1097–1105
Khurana V, Elson-Schwab I, Fulga TA, Sharp KA, Loewen CA, Mulkearns E, Tyynela J, Scherzer CR, Feany MB (2010) Lysosomal dysfunction promotes cleavage and neurotoxicity of tau *in vivo*. PLoS Genet 6:e1001026
Kimura T, Yamashita S, Fukuda T, Park JM, Murayama M, Mizoroki T, Yoshiike Y, Sahara N, Takashima A (2007) Hyperphosphorylated tau in parahippocampal cortex impairs place learning in aged mice expressing wild-type human tau. EMBO J 26:5143–5152
King ME, Ahuja V, Binder LI, Kuret J (1999) Ligand-dependent tau filament formation: implications for Alzheimer's disease progression. Biochemistry 38:14851–14859
Kitazawa M, Oddo S, Yamasaki TR, Green KN, Laferla FM (2005) Lipopolysaccharide-induced inflammation exacerbates tau pathology by a cyclin-dependent kinase 5-mediated pathway in a transgenic model of Alzheimer's disease. J Neurosci 25:8843–8853
Kiyota T, Yamamoto M, Xiong H, Lambert MP, Klein WL, Gendelman HE, Ransohoff RM, Ikezu T (2009) CCL2 accelerates microglia-mediated Aβ oligomer formation and progression of neurocognitive dysfunction. PLoS One 4:e6197
Klyubin I, Betts V, Welzel AT, Blennow K, Zetterberg H, Wallin A, Lemere CA, Cullen WK, Peng Y, Wisniewski T, Selkoe DJ, Anwyl R, Walsh DM, Rowan MJ (2008) Amyloid β protein dimer-containing human CSF disrupts synaptic plasticity: prevention by systemic passive immunization. J Neurosci 28:4231–4237
Kokubo H, Kayed R, Glabe CG, Yamaguchi H (2005) Soluble Aβ oligomers ultrastructurally localize to cell processes and might be related to synaptic dysfunction in Alzheimer's disease brain. Brain Res 1031:222–228
Kondo J, Honda T, Mori H, Hamada Y, Miura R, Ogawara M, Ihara Y (1988) The carboxyl third of tau is tightly bound to paired helical filaments. Neuron 1:827–834
Konno T, Oiki S, Hasegawa K, Naiki H (2004) Anionic contribution for fibrous maturation of protofibrillar assemblies of the human tau repeat domain in a fluoroalcohol solution. Biochemistry 43:13613–13620
Konzack S, Thies E, Marx A, Mandelkow EM, Mandelkow E (2007) Swimming against the tide: mobility of the microtubule-associated protein tau in neurons. J Neurosci 27:9916–9927
Kosik KS, Orecchio LD, Binder L, Trojanowski JQ, Lee VM, Lee G (1988) Epitopes that span the tau molecule are shared with paired helical filaments. Neuron 1:817–825
Kovacech B, Novak M (2010) Tau truncation is a productive posttranslational modification of neurofibrillary degeneration in Alzheimer's disease. Curr Alzheimer Res 7:708–716
Ksiezak-Reding H, Wall JS (2005) Characterization of paired helical filaments by scanning transmission electron microscopy. Microsc Res Tech 67:126–140
Ksiezak-Reding H, Yen SH (1987) Two monoclonal antibodies recognize Alzheimer's neurofibrillary tangles, neurofilament, and microtubule-associated proteins. J Neurochem 48:455–462
Ksiezak-Reding H, Yen SH (1991) Structural stability of paired helical filaments requires microtubule-binding domains of tau: a model for self-association. Neuron 6:717–728
Ksiezak-Reding H, Binder LI, Yen SH (1988) Immunochemical and biochemical characterization of tau proteins in normal and Alzheimer's disease brains with Alz 50 and Tau-1. J Biol Chem 263:7948–7953
Ksiezak-Reding H, Morgan K, Dickson DW (1994a) Tau immunoreactivity and SDS solubility of two populations of paired helical filaments that differ in morphology. Brain Res 649:185–196
Ksiezak-Reding H, Morgan K, Mattiace LA, Davies P, Liu WK, Yen SH, Weidenheim K, Dickson DW (1994b) Ultrastructure and biochemical composition of paired helical filaments in corticobasal degeneration. Am J Pathol 145:1496–1508
Kuo YM, Kokjohn TA, Beach TG, Sue LI, Brune D, Lopez JC, Kalback WM, Abramowski D, Sturchler-Pierrat C, Staufenbiel M, Roher AE (2001) Comparative analysis of amyloid-β chemical structure and amyloid plaque morphology of transgenic mouse and Alzheimer's disease brains. J Biol Chem 276:12991–12998

Kuret J, Chirita CN, Congdon EE, Kannanayakal T, Li G, Necula M, Yin H, Zhong Q (2005a) Pathways of tau fibrillization. Biochim Biophys Acta 1739:167–178
Kuret J, Congdon EE, Li G, Yin H, Yu X, Zhong Q (2005b) Evaluating triggers and enhancers of tau fibrillization. Microsc Res Tech 67:141–155
Lacor PN, Buniel MC, Chang L, Fernandez SJ, Gong Y, Viola KL, Lambert MP, Velasco PT, Bigio EH, Finch CE, Krafft GA, Klein WL (2004) Synaptic targeting by Alzheimer's-related amyloid β oligomers. J Neurosci 24:10191–10200
Lacor PN, Buniel MC, Furlow PW, Clemente AS, Velasco PT, Wood M, Viola KL, Klein WL (2007) Aβ oligomer-induced aberrations in synapse composition, shape, and density provide a molecular basis for loss of connectivity in Alzheimer's disease. J Neurosci 27:796–807
Lambert MP, Barlow AK, Chromy BA, Edwards C, Freed R, Liosatos M, Morgan TE, Rozovsky I, Trommer B, Viola KL, Wals P, Zhang C, Finch CE, Krafft GA, Klein WL (1998) Diffusible, nonfibrillar ligands derived from Aβ1–42 are potent central nervous system neurotoxins. Proc Natl Acad Sci USA 95:6448–6453
Lambert MP, Viola KL, Chromy BA, Chang L, Morgan TE, Yu J, Venton DL, Krafft GA, Finch CE, Klein WL (2001) Vaccination with soluble Aβ oligomers generates toxicity-neutralizing antibodies. J Neurochem 79:595–605
Lambert MP, Velasco PT, Chang L, Viola KL, Fernandez S, Lacor PN, Khuon D, Gong Y, Bigio EH, Shaw P, De Felice FG, Krafft GA, Klein WL (2007) Monoclonal antibodies that target pathological assemblies of Aβ. J Neurochem 100:23–35
Lauderback CM, Hackett JM, Huang FF, Keller JN, Szweda LI, Markesbery WR, Butterfield DA (2001) The glial glutamate transporter, GLT-1, is oxidatively modified by 4-hydroxy-2-nonenal in the Alzheimer's disease brain: the role of Aβ1–42. J Neurochem 78:413–416
Lee G, Neve RL, Kosik KS (1989) The microtubule binding domain of tau protein. Neuron 2:1615–1624
Lee VM, Balin BJ, Otvos L Jr, Trojanowski JQ (1991) A68: a major subunit of paired helical filaments and derivatized forms of normal Tau. Science 251:675–678
Lee G, Newman ST, Gard DL, Band H, Panchamoorthy G (1998) Tau interacts with src-family non-receptor tyrosine kinases. J Cell Sci 111(Pt 21):3167–3177
Lee VM, Goedert M, Trojanowski JQ (2001) Neurodegenerative tauopathies. Annu Rev Neurosci 24:1121–1159
Lee G, Thangavel R, Sharma VM, Litersky JM, Bhaskar K, Fang SM, Do LH, Andreadis A, Van Hoesen G, Ksiezak-Reding H (2004) Phosphorylation of tau by fyn: implications for Alzheimer's disease. J Neurosci 24:2304–2312
Lee VM, Kenyon TK, Trojanowski JQ (2005) Transgenic animal models of tauopathies. Biochim Biophys Acta 1739:251–259
Lefterov I, Fitz NF, Cronican A, Lefterov P, Staufenbiel M, Koldamova R (2009) Memory deficits in APP23/ABCA1$^{+/-}$ mice correlate with the level of Aβ oligomers. ASN Neuro 1:pii: e00006
Lehman NL (2009) The ubiquitin proteasome system in neuropathology. Acta Neuropathol 118:329–347
Lesné S, Koh MT, Kotilinek L, Kayed R, Glabe CG, Yang A, Gallagher M, Ashe KH (2006) A specific amyloid-β protein assembly in the brain impairs memory. Nature 440:352–357
Levy E, Carman MD, Fernandez-Madrid IJ, Power MD, Lieberburg I, Van Duinen SG, Bots GT, Luyendijk W, Frangione B (1990) Mutation of the Alzheimer's disease amyloid gene in hereditary cerebral hemorrhage, Dutch type. Science 248:1124–1126
Lewis J, Dickson DW, Lin WL, Chisholm L, Corral A, Jones G, Yen SH, Sahara N, Skipper L, Yager D, Eckman C, Hardy J, Hutton M, Mcgowan E (2001) Enhanced neurofibrillary degeneration in transgenic mice expressing mutant tau and APP. Science 293:1487–1491
Li S, Hong S, Shepardson NE, Walsh DM, Shankar GM, Selkoe DJ (2009) Soluble oligomers of amyloid β protein facilitate hippocampal long-term depression by disrupting neuronal glutamate uptake. Neuron 62:788–801
Lieberman AP, Trojanowski JQ, Lee VM, Balin BJ, Ding XS, Greenberg J, Morrison D, Reivich M, Grossman M (1998) Cognitive, neuroimaging, and pathological studies in a patient with Pick's disease. Ann Neurol 43:259–265

Lim F, Hernandez F, Lucas JJ, Gomez-Ramos P, Moran MA, Avila J (2001) FTDP-17 mutations in tau transgenic mice provoke lysosomal abnormalities and Tau filaments in forebrain. Mol Cell Neurosci 18:702–714

Lin H, Bhatia R, Lal R (2001) Amyloid β protein forms ion channels: implications for Alzheimer's disease pathophysiology. FASEB J 15:2433–2444

Litvan I, Agid Y, Calne D, Campbell G, Dubois B, Duvoisin RC, Goetz CG, Golbe LI, Grafman J, Growdon JH, Hallett M, Jankovic J, Quinn NP, Tolosa E, Zee DS (1996) Clinical research criteria for the diagnosis of progressive supranuclear palsy (Steele-Richardson-Olszewski syndrome): report of the NINDS-SPSP international workshop. Neurology 47:1–9

Liu YH, Wei W, Yin J, Liu GP, Wang Q, Cao FY, Wang JZ (2009) Proteasome inhibition increases tau accumulation independent of phosphorylation. Neurobiol Aging 30:1949–1961

Lomakin A, Chung DS, Benedek GB, Kirschner DA, Teplow DB (1996) On the nucleation and growth of amyloid β-protein fibrils: detection of nuclei and quantitation of rate constants. Proc Natl Acad Sci USA 93:1125–1129

Lopresti P, Szuchet S, Papasozomenos SC, Zinkowski RP, Binder LI (1995) Functional implications for the microtubule-associated protein tau: localization in oligodendrocytes. Proc Natl Acad Sci USA 92:10369–10373

Lord A, Englund H, Soderberg L, Tucker S, Clausen F, Hillered L, Gordon M, Morgan D, Lannfelt L, Pettersson FE, Nilsson LN (2009) Amyloid-β protofibril levels correlate with spatial learning in Arctic Alzheimer's disease transgenic mice. FEBS J 276:995–1006

Lorenzo A, Yankner BA (1994) β-Amyloid neurotoxicity requires fibril formation and is inhibited by congo red. Proc Natl Acad Sci USA 91:12243–12247

Lovell MA, Ehmann WD, Mattson MP, Markesbery WR (1997) Elevated 4-hydroxynonenal in ventricular fluid in Alzheimer's disease. Neurobiol Aging 18:457–461

Lovell MA, Xie C, Markesbery WR (2001) Acrolein is increased in Alzheimer's disease brain and is toxic to primary hippocampal cultures. Neurobiol Aging 22:187–194

Lue LF, Kuo YM, Roher AE, Brachova L, Shen Y, Sue L, Beach T, Kurth JH, Rydel RE, Rogers J (1999) Soluble amyloid β peptide concentration as a predictor of synaptic change in Alzheimer's disease. Am J Pathol 155:853–862

Lund, Manchester-Groups (1994) Clinical and neuropathological criteria for frontotemporal dementia. The Lund and Manchester Groups. J Neurol Neurosurg Psychiatry 57:416–418

Lynch MA (2004) Long-term potentiation and memory. Physiol Rev 84:87–136

Ma QL, Lim GP, Harris-White ME, Yang F, Ambegaokar SS, Ubeda OJ, Glabe CG, Teter B, Frautschy SA, Cole GM (2006) Antibodies against β-amyloid reduce Aβ oligomers, glycogen synthase kinase-3β activation and tau phosphorylation *in vivo* and *in vitro*. J Neurosci Res 83:374–384

Ma QL, Yang F, Rosario ER, Ubeda OJ, Beech W, Gant DJ, Chen PP, Hudspeth B, Chen C, Zhao Y, Vinters HV, Frautschy SA, Cole GM (2009) β-Amyloid oligomers induce phosphorylation of tau and inactivation of insulin receptor substrate *via* c-Jun *N*-terminal kinase signaling: suppression by omega-3 fatty acids and curcumin. J Neurosci 29:9078–9089

Maeda S, Sahara N, Saito Y, Murayama S, Ikai A, Takashima A (2006) Increased levels of granular tau oligomers: an early sign of brain aging and Alzheimer's disease. Neurosci Res 54:197–201

Maeda S, Sahara N, Saito Y, Murayama M, Yoshiike Y, Kim H, Miyasaka T, Murayama S, Ikai A, Takashima A (2007) Granular tau oligomers as intermediates of tau filaments. Biochemistry 46:3856–3861

Mailliot C, Sergeant N, Bussiere T, Caillet-Boudin ML, Delacourte A, Buee L (1998) Phosphorylation of specific sets of tau isoforms reflects different neurofibrillary degeneration processes. FEBS Lett 433:201–204

Maloney MT, Minamide LS, Kinley AW, Boyle JA, Bamburg JR (2005) β-Secretase-cleaved amyloid precursor protein accumulates at actin inclusions induced in neurons by stress or amyloid β: a feedforward mechanism for Alzheimer's disease. J Neurosci 25:11313–11321

Manczak M, Anekonda TS, Henson E, Park BS, Quinn J, Reddy PH (2006) Mitochondria are a direct site of Aβ accumulation in Alzheimer's disease neurons: implications for free radical generation and oxidative damage in disease progression. Hum Mol Genet 15:1437–1449

Mandelkow EM, Stamer K, Vogel R, Thies E, Mandelkow E (2003) Clogging of axons by tau, inhibition of axonal traffic and starvation of synapses. Neurobiol Aging 24:1079–1085
Mandelkow EM, Thies E, Trinczek B, Biernat J, Mandelkow E (2004) MARK/PAR1 kinase is a regulator of microtubule-dependent transport in axons. J Cell Biol 167:99–110
Mark RJ, Pang Z, Geddes JW, Uchida K, Mattson MP (1997) Amyloid β-peptide impairs glucose transport in hippocampal and cortical neurons: involvement of membrane lipid peroxidation. J Neurosci 17:1046–1054
Markesbery WR, Lovell MA (1998) Four-hydroxynonenal, a product of lipid peroxidation, is increased in the brain in Alzheimer's disease. Neurobiol Aging 19:33–36
Martins IC, Kuperstein I, Wilkinson H, Maes E, Vanbrabant M, Jonckheere W, Van Gelder P, Hartmann D, D'hooge R, De Strooper B, Schymkowitz J, Rousseau F (2008) Lipids revert inert Aβ amyloid fibrils to neurotoxic protofibrils that affect learning in mice. EMBO J 27:224–233
Masliah E, Iimoto DS, Saitoh T, Hansen LA, Terry RD (1990) Increased immunoreactivity of brain spectrin in Alzheimer disease: a marker for synapse loss? Brain Res 531:36–44
Masliah E, Mallory M, Alford M, Deteresa R, Hansen LA, Mckeel DW Jr, Morris JC (2001) Altered expression of synaptic proteins occurs early during progression of Alzheimer's disease. Neurology 56:127–129
Masters CL, Simms G, Weinman NA, Multhaup G, Mcdonald BL, Beyreuther K (1985) Amyloid plaque core protein in Alzheimer disease and Down syndrome. Proc Natl Acad Sci USA 82:4245–4249
Matsuzaki M, Honkura N, Ellis-Davies GC, Kasai H (2004) Structural basis of long-term potentiation in single dendritic spines. Nature 429:761–766
Mattson MP, Fu W, Waeg G, Uchida K (1997) 4-Hydroxynonenal, a product of lipid peroxidation, inhibits dephosphorylation of the microtubule-associated protein tau. Neuroreport 8:2275–2281
Mcdonald DR, Bamberger ME, Combs CK, Landreth GE (1998) β-Amyloid fibrils activate parallel mitogen-activated protein kinase pathways in microglia and THP1 monocytes. J Neurosci 18:4451–4460
Mckee AC, Kosik KS, Kowall NW (1991) Neuritic pathology and dementia in Alzheimer's disease. Ann Neurol 30:156–165
Mckeel DW Jr, Ball MJ, Price JL, Smith DS, Miller JP, Berg L, Morris JC (1993) Interlaboratory histopathologic assessment of Alzheimer neuropathology: different methodologies yield comparable diagnostic results. Alzheimer Dis Assoc Disord 7:136–151
Mclaurin J, Franklin T, Zhang X, Deng J, Fraser PE (1999) Interactions of Alzheimer amyloid-β peptides with glycosaminoglycans effects on fibril nucleation and growth. Eur J Biochem 266:1101–1110
Mclean CA, Cherny RA, Fraser FW, Fuller SJ, Smith MJ, Beyreuther K, Bush AI, Masters CL (1999) Soluble pool of Aβ amyloid as a determinant of severity of neurodegeneration in Alzheimer's disease. Ann Neurol 46:860–866
Mesulam MM (1999) Spatial attention and neglect: parietal, frontal and cingulate contributions to the mental representation and attentional targeting of salient extrapersonal events. Philos Trans R Soc Lond B Biol Sci 354:1325–1346
Min SW, Cho SH, Zhou Y, Schroeder S, Haroutunian V, Seeley WW, Huang EJ, Shen Y, Masliah E, Mukherjee C, Meyers D, Cole PA, Ott M, Gan L (2010) Acetylation of tau inhibits its degradation and contributes to tauopathy. Neuron 67:953–966
Mocanu MM, Nissen A, Eckermann K, Khlistunova I, Biernat J, Drexler D, Petrova O, Schonig K, Bujard H, Mandelkow E, Zhou L, Rune G, Mandelkow EM (2008) The potential for β-structure in the repeat domain of tau protein determines aggregation, synaptic decay, neuronal loss, and coassembly with endogenous Tau in inducible mouse models of tauopathy. J Neurosci 28:737–748
Montine KS, Reich E, Neely MD, Sidell KR, Olson SJ, Markesbery WR, Montine TJ (1998) Distribution of reducible 4-hydroxynonenal adduct immunoreactivity in Alzheimer disease is associated with APOE genotype. J Neuropathol Exp Neurol 57:415–425

Mori H, Nishimura M, Namba Y, Oda M (1994) Corticobasal degeneration: a disease with widespread appearance of abnormal tau and neurofibrillary tangles, and its relation to progressive supranuclear palsy. Acta Neuropathol 88:113–121

Morishima-Kawashima M, Hasegawa M, Takio K, Suzuki M, Yoshida H, Watanabe A, Titani K, Ihara Y (1995) Hyperphosphorylation of tau in PHF. Neurobiol Aging 16:365–371

Morris RG (2003) Long-term potentiation and memory. Philos Trans R Soc Lond B Biol Sci 358:643–647

Morris JC, Rubin EH (1991) Clinical diagnosis and course of Alzheimer's disease. Psychiatr Clin North Am 14:223–236

Morsch R, Simon W, Coleman PD (1999) Neurons may live for decades with neurofibrillary tangles. J Neuropathol Exp Neurol 58:188–197

Mufson EJ, Ginsberg SD, Ikonomovic MD, Dekosky ST (2003) Human cholinergic basal forebrain: chemoanatomy and neurologic dysfunction. J Chem Neuroanat 26:233–242

Murrell J, Farlow M, Ghetti B, Benson MD (1991) A mutation in the amyloid precursor protein associated with hereditary Alzheimer's disease. Science 254:97–99

Nacharaju P, Lewis J, Easson C, Yen S, Hackett J, Hutton M, Yen SH (1999) Accelerated filament formation from tau protein with specific FTDP-17 missense mutations. FEBS Lett 447:195–199

Nagele RG, D'andrea MR, Anderson WJ, Wang HY (2002) Intracellular accumulation of β-amyloid(1–42) in neurons is facilitated by the α7 nicotinic acetylcholine receptor in Alzheimer's disease. Neuroscience 110:199–211

Nakashima H, Ishihara T, Yokota O, Terada S, Trojanowski JQ, Lee VM, Kuroda S (2004) Effects of α-tocopherol on an animal model of tauopathies. Free Radic Biol Med 37:176–186

Näslund J, Schierhorn A, Hellman U, Lannfelt L, Roses AD, Tjernberg LO, Silberring J, Gandy SE, Winblad B, Greengard P et al (1994) Relative abundance of Alzheimer Aβ amyloid peptide variants in Alzheimer disease and normal aging. Proc Natl Acad Sci USA 91:8378–8382

Näslund J, Haroutunian V, Mohs R, Davis KL, Davies P, Greengard P, Buxbaum JD (2000) Correlation between elevated levels of amyloid β-peptide in the brain and cognitive decline. JAMA 283:1571–1577

Necula M, Chirita CN, Kuret J (2005) Cyanine dye N744 inhibits tau fibrillization by blocking filament extension: implications for the treatment of tauopathic neurodegenerative diseases. Biochemistry 44:10227–10237

Neumann M, Schulz-Schaeffer W, Crowther RA, Smith MJ, Spillantini MG, Goedert M, Kretzschmar HA (2001) Pick's disease associated with the novel Tau gene mutation K369I. Ann Neurol 50:503–513

Neve RL, Harris P, Kosik KS, Kurnit DM, Donlon TA (1986) Identification of cDNA clones for the human microtubule-associated protein tau and chromosomal localization of the genes for tau and microtubule-associated protein 2. Brain Res 387:271–280

Nishitsuji K, Tomiyama T, Ishibashi K, Ito K, Teraoka R, Lambert MP, Klein WL, Mori H (2009) The E693Δ mutation in amyloid precursor protein increases intracellular accumulation of amyloid β oligomers and causes endoplasmic reticulum stress-induced apoptosis in cultured cells. Am J Pathol 174:957–969

Nunan J, Small DH (2000) Regulation of APP cleavage by α-, β- and γ-secretases. FEBS Lett 483:6–10

Nunomura A, Perry G, Pappolla MA, Friedland RP, Hirai K, Chiba S, Smith MA (2000) Neuronal oxidative stress precedes amyloid-β deposition in Down syndrome. J Neuropathol Exp Neurol 59:1011–1017

Nygaard HB, Strittmatter SM (2009) Cellular prion protein mediates the toxicity of β-amyloid oligomers: implications for Alzheimer disease. Arch Neurol 66:1325–1328

Oddo S, Caccamo A, Shepherd JD, Murphy MP, Golde TE, Kayed R, Metherate R, Mattson MP, Akbari Y, Laferla FM (2003) Triple-transgenic model of Alzheimer's disease with plaques and tangles: intracellular Aβ and synaptic dysfunction. Neuron 39:409–421

Oddo S, Billings L, Kesslak JP, Cribbs DH, Laferla FM (2004) Aβ immunotherapy leads to clearance of early, but not late, hyperphosphorylated tau aggregates *via* the proteasome. Neuron 43:321–332

Park SY, Ferreira A (2005) The generation of a 17 kDa neurotoxic fragment: an alternative mechanism by which tau mediates β-amyloid-induced neurodegeneration. J Neurosci 25:5365–5375
Perez M, Arrasate M, Montejo De Garcini E, Munoz V, Avila J (2001) *In vitro* assembly of tau protein: mapping the regions involved in filament formation. Biochemistry 40:5983–5991
Pike CJ, Walencewicz AJ, Glabe CG, Cotman CW (1991) *In vitro* aging of β-amyloid protein causes peptide aggregation and neurotoxicity. Brain Res 563:311–314
Podlisny MB, Ostaszewski BL, Squazzo SL, Koo EH, Rydell RE, Teplow DB, Selkoe DJ (1995) Aggregation of secreted amyloid β-protein into sodium dodecyl sulfate-stable oligomers in cell culture. J Biol Chem 270:9564–9570
Poling A, Morgan-Paisley K, Panos JJ, Kim EM, O'hare E, Cleary JP, Lesné S, Ashe KH, Porritt M, Baker LE (2008) Oligomers of the amyloid-β protein disrupt working memory: confirmation with two behavioral procedures. Behav Brain Res 193:230–234
Pollock NJ, Mirra SS, Binder LI, Hansen LA, Wood JG (1986) Filamentous aggregates in Pick's disease, progressive supranuclear palsy, and Alzheimer's disease share antigenic determinants with microtubule-associated protein, tau. Lancet 2:1211
Polydoro M, Acker CM, Duff K, Castillo PE, Davies P (2009) Age-dependent impairment of cognitive and synaptic function in the htau mouse model of tau pathology. J Neurosci 29:10741–10749
Prasher VP, Farrer MJ, Kessling AM, Fisher EM, West RJ, Barber PC, Butler AC (1998) Molecular mapping of Alzheimer-type dementia in Down's syndrome. Ann Neurol 43:380–383
Price JL, Morris JC (1999) Tangles and plaques in nondemented aging and "preclinical" Alzheimer's disease. Ann Neurol 45:358–368
Quintanilla RA, Matthews-Roberson TA, Dolan PJ, Johnson GV (2009) Caspase-cleaved tau expression induces mitochondrial dysfunction in immortalized cortical neurons: implications for the pathogenesis of Alzheimer disease. J Biol Chem 284:18754–18766
Rauk A (2008) Why is the amyloid β peptide of Alzheimer's disease neurotoxic? Dalton Trans 1273–1282
Rauk A (2009) The chemistry of Alzheimer's disease. Chem Soc Rev 38:2698–2715
Rebeiz JJ, Kolodny EH, Richardson EP Jr (1967) Corticodentatonigral degeneration with neuronal achromasia: a progressive disorder of late adult life. Trans Am Neurol Assoc 92:23–26
Rebeiz JJ, Kolodny EH, Richardson EP Jr (1968) Corticodentatonigral degeneration with neuronal achromasia. Arch Neurol 18:20–33
Reddy PH, Mani G, Park BS, Jacques J, Murdoch G, Whetsell W Jr, Kaye J, Manczak M (2005) Differential loss of synaptic proteins in Alzheimer's disease: implications for synaptic dysfunction. J Alzheimers Dis 7:103–117, discussion 73–80
Reed MN, Hofmeister JJ, Jungbauer L, Welzel AT, Yu C, Sherman MA, Lesne S, Ladu MJ, Walsh DM, Ashe KH, Cleary JP (2011) Cognitive effects of cell-derived and synthetically derived Aβ oligomers. Neurobiol Aging 32(10):1784–1794
Reyes JF, Reynolds MR, Horowitz PM, Fu Y, Guillozet-Bongaarts AL, Berry R, Binder LI (2008) A possible link between astrocyte activation and tau nitration in Alzheimer's disease. Neurobiol Dis 31:198–208
Reynolds MR, Berry RW, Binder LI (2005) Site-specific nitration and oxidative dityrosine bridging of the tau protein by peroxynitrite: implications for Alzheimer's disease. Biochemistry 44:1690–1700
Reynolds MR, Reyes JF, Fu Y, Bigio EH, Guillozet-Bongaarts AL, Berry RW, Binder LI (2006) Tau nitration occurs at tyrosine 29 in the fibrillar lesions of Alzheimer's disease and other tauopathies. J Neurosci 26:10636–10645
Reynolds MR, Berry RW, Binder LI (2007) Nitration in neurodegeneration: deciphering the "Hows" "nYs". Biochemistry 46:7325–7336
Rissman RA, Poon WW, Blurton-Jones M, Oddo S, Torp R, Vitek MP, Laferla FM, Rohn TT, Cotman CW (2004) Caspase-cleavage of tau is an early event in Alzheimer disease tangle pathology. J Clin Invest 114:121–130
Rizzini C, Goedert M, Hodges JR, Smith MJ, Jakes R, Hills R, Xuereb JH, Crowther RA, Spillantini MG (2000) Tau gene mutation K257T causes a tauopathy similar to Pick's disease. J Neuropathol Exp Neurol 59:990–1001

Robakis NK, Ramakrishna N, Wolfe G, Wisniewski HM (1987) Molecular cloning and characterization of a cDNA encoding the cerebrovascular and the neuritic plaque amyloid peptides. Proc Natl Acad Sci USA 84:4190–4194

Roberson ED, Scearce-Levie K, Palop JJ, Yan F, Cheng IH, Wu T, Gerstein H, Yu GQ, Mucke L (2007) Reducing endogenous tau ameliorates amyloid β-induced deficits in an Alzheimer's disease mouse model. Science 316:750–754

Rosen RF, Tomidokoro Y, Ghiso JA, Walker LC (2010) SDS-PAGE/immunoblot detection of Aβ multimers in human cortical tissue homogenates using antigen-epitope retrieval. J Vis Exp. 38, pii: 1916. doi:10.3791/1916

Rosenberg KJ, Ross JL, Feinstein HE, Feinstein SC, Israelachvili J (2008) Complementary dimerization of microtubule-associated tau protein: implications for microtubule bundling and tau-mediated pathogenesis. Proc Natl Acad Sci USA 105:7445–7450

Rovelet-Lecrux A, Hannequin D, Raux G, Le Meur N, Laquerriere A, Vital A, Dumanchin C, Feuillette S, Brice A, Vercelletto M, Dubas F, Frebourg T, Campion D (2006) APP locus duplication causes autosomal dominant early-onset Alzheimer disease with cerebral amyloid angiopathy. Nat Genet 38:24–26

Roychaudhuri R, Yang M, Hoshi MM, Teplow DB (2009) Amyloid β-protein assembly and Alzheimer disease. J Biol Chem 284:4749–4753

Rui Y, Gu J, Yu K, Hartzell HC, Zheng JQ (2010) Inhibition of AMPA receptor trafficking at hippocampal synapses by β-amyloid oligomers: the mitochondrial contribution. Mol Brain 3:10

Sahara N, Maeda S, Murayama M, Suzuki T, Dohmae N, Yen SH, Takashima A (2007a) Assembly of two distinct dimers and higher-order oligomers from full-length tau. Eur J Neurosci 25:3020–3029

Sahara N, Maeda S, Yoshiike Y, Mizoroki T, Yamashita S, Murayama M, Park JM, Saito Y, Murayama S, Takashima A (2007b) Molecular chaperone-mediated tau protein metabolism counteracts the formation of granular tau oligomers in human brain. J Neurosci Res 85:3098–3108

Santacruz K, Lewis J, Spires T, Paulson J, Kotilinek L, Ingelsson M, Guimaraes A, Deture M, Ramsden M, Mcgowan E, Forster C, Yue M, Orne J, Janus C, Mariash A, Kuskowski M, Hyman B, Hutton M, Ashe KH (2005) Tau suppression in a neurodegenerative mouse model improves memory function. Science 309:476–481

Santa-Maria I, Smith MA, Perry G, Hernandez F, Avila J, Moreno FJ (2005) Effect of quinones on microtubule polymerization: a link between oxidative stress and cytoskeletal alterations in Alzheimer's disease. Biochim Biophys Acta 1740:472–480

Santa-Maria I, Hernandez F, Moreno FJ, Avila J (2007) Taurine, an inducer for tau polymerization and a weak inhibitor for amyloid-β-peptide aggregation. Neurosci Lett 429:91–94

Scheuner D, Eckman C, Jensen M, Song X, Citron M, Suzuki N, Bird TD, Hardy J, Hutton M, Kukull W, Larson E, Levy-Lahad E, Viitanen M, Peskind E, Poorkaj P, Schellenberg G, Tanzi R, Wasco W, Lannfelt L, Selkoe D, Younkin S (1996) Secreted amyloid β-protein similar to that in the senile plaques of Alzheimer's disease is increased *in vivo* by the presenilin 1 and 2 and APP mutations linked to familial Alzheimer's disease. Nat Med 2:864–870

Schneider A, Biernat J, Von Bergen M, Mandelkow E, Mandelkow EM (1999) Phosphorylation that detaches tau protein from microtubules (Ser262, Ser214) also protects it against aggregation into Alzheimer paired helical filaments. Biochemistry 38:3549–3558

Schoneich C, Williams TD (2002) Cu(II)-catalyzed oxidation of β-amyloid peptide targets His13 and His14 over His6: detection of 2-Oxo-histidine by HPLC-MS/MS. Chem Res Toxicol 15:717–722

Seitz A, Kojima H, Oiwa K, Mandelkow EM, Song YH, Mandelkow E (2002) Single-molecule investigation of the interference between kinesin, tau and MAP2c. EMBO J 21:4896–4905

Sekino Y, Tanaka S, Hanamura K, Yamazaki H, Sasagawa Y, Xue Y, Hayashi K, Shirao T (2006) Activation of *N*-methyl-$_D$-aspartate receptor induces a shift of drebrin distribution: disappearance from dendritic spines and appearance in dendritic shafts. Mol Cell Neurosci 31:493–504

Selkoe DJ (1989) Amyloid β protein precursor and the pathogenesis of Alzheimer's disease. Cell 58:611–612

Selkoe DJ (1997) Alzheimer's disease: genotypes, phenotypes, and treatments. Science 275:630–631

Sergeant N, David JP, Lefranc D, Vermersch P, Wattez A, Delacourte A (1997) Different distribution of phosphorylated tau protein isoforms in Alzheimer's and Pick's diseases. FEBS Lett 412:578–582

Seubert P, Vigo-Pelfrey C, Esch F, Lee M, Dovey H, Davis D, Sinha S, Schlossmacher M, Whaley J, Swindlehurst C, McCormack R, Wolfert R, Selkoe DJ, Lieberburg I, Schenk D (1992) Isolation and quantification of soluble Alzheimer's β-peptide from biological fluids. Nature 359:325–327

Shankar GM, Bloodgood BL, Townsend M, Walsh DM, Selkoe DJ, Sabatini BL (2007) Natural oligomers of the Alzheimer amyloid-β protein induce reversible synapse loss by modulating an NMDA-type glutamate receptor-dependent signaling pathway. J Neurosci 27:2866–2875

Shankar GM, Li S, Mehta TH, Garcia-Munoz A, Shepardson NE, Smith I, Brett FM, Farrell MA, Rowan MJ, Lemere CA, Regan CM, Walsh DM, Sabatini BL, Selkoe DJ (2008) Amyloid-β protein dimers isolated directly from Alzheimer's brains impair synaptic plasticity and memory. Nat Med 14:837–842

Shankar GM, Leissring MA, Adame A, Sun X, Spooner E, Masliah E, Selkoe DJ, Lemere CA, Walsh DM (2009) Biochemical and immunohistochemical analysis of an Alzheimer's disease mouse model reveals the presence of multiple cerebral Aβ assembly forms throughout life. Neurobiol Dis 36:293–302

Sharma VM, Litersky JM, Bhaskar K, Lee G (2007) Tau impacts on growth-factor-stimulated actin remodeling. J Cell Sci 120:748–757

Sherrington R, Rogaev EI, Liang Y, Rogaeva EA, Levesque G, Ikeda M, Chi H, Lin C, Li G, Holman K et al (1995) Cloning of a gene bearing missense mutations in early-onset familial Alzheimer's disease. Nature 375:754–760

Shin RW, Iwaki T, Kitamoto T, Tateishi J (1991) Hydrated autoclave pretreatment enhances tau immunoreactivity in formalin-fixed normal and Alzheimer's disease brain tissues. Lab Invest 64:693–702

Smith DG, Cappai R, Barnham KJ (2007) The redox chemistry of the Alzheimer's disease amyloid β peptide. Biochim Biophys Acta 1768:1976–1990

Snyder EM, Nong Y, Almeida CG, Paul S, Moran T, Choi EY, Nairn AC, Salter MW, Lombroso PJ, Gouras GK, Greengard P (2005) Regulation of NMDA receptor trafficking by amyloid-β. Nat Neurosci 8:1051–1058

Sokolov Y, Kozak JA, Kayed R, Chanturiya A, Glabe C, Hall JE (2006) Soluble amyloid oligomers increase bilayer conductance by altering dielectric structure. J Gen Physiol 128:637–647

Sondag CM, Dhawan G, Combs CK (2009) β Amyloid oligomers and fibrils stimulate differential activation of primary microglia. J Neuroinflammation 6:1

Spires TL, Orne JD, Santacruz K, Pitstick R, Carlson GA, Ashe KH, Hyman BT (2006) Region-specific dissociation of neuronal loss and neurofibrillary pathology in a mouse model of tauopathy. Am J Pathol 168:1598–1607

Spires-Jones TL, De Calignon A, Matsui T, Zehr C, Pitstick R, Wu HY, Osetek JD, Jones PB, Bacskai BJ, Feany MB, Carlson GA, Ashe KH, Lewis J, Hyman BT (2008) *In vivo* imaging reveals dissociation between caspase activation and acute neuronal death in tangle-bearing neurons. J Neurosci 28:862–867

Spires-Jones TL, Stoothoff WH, De Calignon A, Jones PB, Hyman BT (2009) Tau pathophysiology in neurodegeneration: a tangled issue. Trends Neurosci 32:150–159

St George-Hyslop PH, Tanzi RE, Polinsky RJ, Haines JL, Nee L, Watkins PC, Myers RH, Feldman RG, Pollen D, Drachman D et al (1987) The genetic defect causing familial Alzheimer's disease maps on chromosome 21. Science 235:885–890

Stamer K, Vogel R, Thies E, Mandelkow E, Mandelkow EM (2002) Tau blocks traffic of organelles, neurofilaments, and APP vesicles in neurons and enhances oxidative stress. J Cell Biol 156:1051–1063

Steele JC, Richardson JC, Olszewski J (1964) Progressive supranuclear palsy. A heterogeneous degeneration involving the brain stem, basal ganglia and cerebellum with vertical gaze and pseudobulbar palsy, nuchal dystonia and dementia. Arch Neurol 10:333–359

Stoothoff WH, Johnson GV (2005) Tau phosphorylation: physiological and pathological consequences. Biochim Biophys Acta 1739:280–297

Sze CI, Troncoso JC, Kawas C, Mouton P, Price DL, Martin LJ (1997) Loss of the presynaptic vesicle protein synaptophysin in hippocampus correlates with cognitive decline in Alzheimer disease. J Neuropathol Exp Neurol 56:933–944

Tabaton M, Whitehouse PJ, Perry G, Davies P, Autilio-Gambetti L, Gambetti P (1988) Alz 50 recognizes abnormal filaments in Alzheimer's disease and progressive supranuclear palsy. Ann Neurol 24:407–413

Takahashi RH, Almeida CG, Kearney PF, Yu F, Lin MT, Milner TA, Gouras GK (2004) Oligomerization of Alzheimer's β-amyloid within processes and synapses of cultured neurons and brain. J Neurosci 24:3592–3599

Tamagno E, Bardini P, Guglielmotto M, Danni O, Tabaton M (2006) The various aggregation states of β-amyloid 1–42 mediate different effects on oxidative stress, neurodegeneration, and BACE-1 expression. Free Radic Biol Med 41:202–212

Tanzi RE, Gusella JF, Watkins PC, Bruns GA, St George-Hyslop P, Van Keuren ML, Patterson D, Pagan S, Kurnit DM, Neve RL (1987) Amyloid β protein gene: cDNA, mRNA distribution, and genetic linkage near the Alzheimer locus. Science 235:880–884

Tashima Y, Oe R, Lee S, Sugihara G, Chambers EJ, Takahashi M, Yamada T (2004) The effect of cholesterol and monosialoganglioside (GM1) on the release and aggregation of amyloid β-peptide from liposomes prepared from brain membrane-like lipids. J Biol Chem 279:17587–17595

Terry RD, Masliah E, Salmon DP, Butters N, Deteresa R, Hill R, Hansen LA, Katzman R (1991) Physical basis of cognitive alterations in Alzheimer's disease: synapse loss is the major correlate of cognitive impairment. Ann Neurol 30:572–580

Thies E, Mandelkow EM (2007) Missorting of tau in neurons causes degeneration of synapses that can be rescued by the kinase MARK2/Par-1. J Neurosci 27:2896–2907

Tomiyama T, Nagata T, Shimada H, Teraoka R, Fukushima A, Kanemitsu H, Takuma H, Kuwano R, Imagawa M, Ataka S, Wada Y, Yoshioka E, Nishizaki T, Watanabe Y, Mori H (2008) A new amyloid β variant favoring oligomerization in Alzheimer's-type dementia. Ann Neurol 63:377–387

Tomiyama T, Matsuyama S, Iso H, Umeda T, Takuma H, Ohnishi K, Ishibashi K, Teraoka R, Sakama N, Yamashita T, Nishitsuji K, Ito K, Shimada H, Lambert MP, Klein WL, Mori H (2010) A mouse model of amyloid β oligomers: their contribution to synaptic alteration, abnormal tau phosphorylation, glial activation, and neuronal loss *in vivo*. J Neurosci 30:4845–4856

Townsend M, Shankar GM, Mehta T, Walsh DM, Selkoe DJ (2006) Effects of secreted oligomers of amyloid β-protein on hippocampal synaptic plasticity: a potent role for trimers. J Physiol 572:477–492

Trojanowski JQ, Lee VM (1994) Phosphorylation of neuronal cytoskeletal proteins in Alzheimer's disease and Lewy body dementias. Ann N Y Acad Sci 747:92–109

Van Broeckhoven C, Genthe AM, Vandenberghe A, Horsthemke B, Backhovens H, Raeymaekers P, Van Hul W, Wehnert A, Gheuens J, Cras P et al (1987) Failure of familial Alzheimer's disease to segregate with the A4-amyloid gene in several European families. Nature 329:153–155

Van Broeckhoven C, Haan J, Bakker E, Hardy JA, Van Hul W, Wehnert A, Vegter-Van Der Vlis M, Roos RA (1990) Amyloid β protein precursor gene and hereditary cerebral hemorrhage with amyloidosis (Dutch). Science 248:1120–1122

Varadarajan S, Kanski J, Aksenova M, Lauderback C, Butterfield DA (2001) Different mechanisms of oxidative stress and neurotoxicity for Alzheimer's Aβ(1–42) and Aβ(25–35). J Am Chem Soc 123:5625–5631

Varvel NH, Bhaskar K, Patil AR, Pimplikar SW, Herrup K, Lamb BT (2008) Aβ oligomers induce neuronal cell cycle events in Alzheimer's disease. J Neurosci 28:10786–10793

Varvel NH, Bhaskar K, Kounnas MZ, Wagner SL, Yang Y, Lamb BT, Herrup K (2009) NSAIDs prevent, but do not reverse, neuronal cell cycle reentry in a mouse model of Alzheimer disease. J Clin Invest 119:3692–3702

Vassar R, Bennett BD, Babu-Khan S, Kahn S, Mendiaz EA, Denis P, Teplow DB, Ross S, Amarante P, Loeloff R, Luo Y, Fisher S, Fuller J, Edenson S, Lile J, Jarosinski MA, Biere AL, Curran E, Burgess T, Louis JC, Collins F, Treanor J, Rogers G, Citron M (1999) β-Secretase cleavage of Alzheimer's amyloid precursor protein by the transmembrane aspartic protease BACE. Science 286:735–741

Vega IE, Cui L, Propst JA, Hutton ML, Lee G, Yen SH (2005) Increase in tau tyrosine phosphorylation correlates with the formation of tau aggregates. Brain Res Mol Brain Res 138:135–144

Verdier Y, Penke B (2004) Binding sites of amyloid β-peptide in cell plasma membrane and implications for Alzheimer's disease. Curr Protein Pept Sci 5:19–31

Verdier Y, Zarandi M, Penke B (2004) Amyloid β-peptide interactions with neuronal and glial cell plasma membrane: binding sites and implications for Alzheimer's disease. J Pept Sci 10:229–248

Vermersch P, Scheltens P, Barkhof F, Steinling M, Leys D (1994) Evidence for atrophy of the corpus callosum in Alzheimer's disease. Eur Neurol 34:83–86

Vigo-Pelfrey C, Lee D, Keim P, Lieberburg I, Schenk DB (1993) Characterization of β-amyloid peptide from human cerebrospinal fluid. J Neurochem 61:1965–1968

Von Bergen M, Friedhoff P, Biernat J, Heberle J, Mandelkow EM, Mandelkow E (2000) Assembly of tau protein into Alzheimer paired helical filaments depends on a local sequence motif (^{306}VQIVYK^{311}) forming β structure. Proc Natl Acad Sci USA 97:5129–5134

Walsh DM, Selkoe DJ (2007) Aβ oligomers—a decade of discovery. J Neurochem 101:1172–1184

Walsh DM, Lomakin A, Benedek GB, Condron MM, Teplow DB (1997) Amyloid β-protein fibrillogenesis. Detection of a protofibrillar intermediate. J Biol Chem 272:22364–22372

Walsh DM, Hartley DM, Kusumoto Y, Fezoui Y, Condron MM, Lomakin A, Benedek GB, Selkoe DJ, Teplow DB (1999) Amyloid β-protein fibrillogenesis. Structure and biological activity of protofibrillar intermediates. J Biol Chem 274:25945–25952

Walsh DM, Tseng BP, Rydel RE, Podlisny MB, Selkoe DJ (2000) The oligomerization of amyloid β-protein begins intracellularly in cells derived from human brain. Biochemistry 39: 10831–10839

Walsh DM, Klyubin I, Fadeeva JV, Cullen WK, Anwyl R, Wolfe MS, Rowan MJ, Selkoe DJ (2002) Naturally secreted oligomers of amyloid β protein potently inhibit hippocampal long-term potentiation *in vivo*. Nature 416:535–539

Walsh DM, Klyubin I, Shankar GM, Townsend M, Fadeeva JV, Betts V, Podlisny MB, Cleary JP, Ashe KH, Rowan MJ, Selkoe DJ (2005) The role of cell-derived oligomers of Aβ in Alzheimer's disease and avenues for therapeutic intervention. Biochem Soc Trans 33:1087–1090

Wang J, Dickson DW, Trojanowski JQ, Lee VM (1999) The levels of soluble *versus* insoluble brain Aβ distinguish Alzheimer's disease from normal and pathologic aging. Exp Neurol 158:328–337

Wang HY, Lee DH, D'andrea MR, Peterson PA, Shank RP, Reitz AB (2000) β-Amyloid(1–42) binds to α7 nicotinic acetylcholine receptor with high affinity. Implications for Alzheimer's disease pathology. J Biol Chem 275:5626–5632

Wang HW, Pasternak JF, Kuo H, Ristic H, Lambert MP, Chromy B, Viola KL, Klein WL, Stine WB, Krafft GA, Trommer BL (2002) Soluble oligomers of β amyloid (1–42) inhibit long-term potentiation but not long-term depression in rat dentate gyrus. Brain Res 924:133–140

Wang HY, Li W, Benedetti NJ, Lee DH (2003) α7 Nicotinic acetylcholine receptors mediate β-amyloid peptide-induced tau protein phosphorylation. J Biol Chem 278:31547–31553

Wang Q, Walsh DM, Rowan MJ, Selkoe DJ, Anwyl R (2004) Block of long-term potentiation by naturally secreted and synthetic amyloid β-peptide in hippocampal slices is mediated *via* activation of the kinases c-Jun *N*-terminal kinase, cyclin-dependent kinase 5, and p38 mitogen-activated protein kinase as well as metabotropic glutamate receptor type 5. J Neurosci 24:3370–3378

Wang JZ, Grundke-Iqbal I, Iqbal K (2007a) Kinases and phosphatases and tau sites involved in Alzheimer neurofibrillary degeneration. Eur J Neurosci 25:59–68

Wang YP, Biernat J, Pickhardt M, Mandelkow E, Mandelkow EM (2007b) Stepwise proteolysis liberates tau fragments that nucleate the Alzheimer-like aggregation of full-length tau in a neuronal cell model. Proc Natl Acad Sci U S A 104:10252–10257

Wang Y, Martinez-Vicente M, Kruger U, Kaushik S, Wong E, Mandelkow EM, Cuervo AM, Mandelkow E (2009) Tau fragmentation, aggregation and clearance: the dual role of lysosomal processing. Hum Mol Genet 18:4153–4170

Wang X, Perry G, Smith MA, Zhu X (2010) Amyloid-β-derived diffusible ligands cause impaired axonal transport of mitochondria in neurons. Neurodegener Dis 7:56–59

Watanabe A, Hong WK, Dohmae N, Takio K, Morishima-Kawashima M, Ihara Y (2004) Molecular aging of tau: disulfide-independent aggregation and non-enzymatic degradation *in vitro* and *in vivo*. J Neurochem 90:1302–1311

Wei Y, Qu MH, Wang XS, Chen L, Wang DL, Liu Y, Hua Q, He RQ (2008) Binding to the minor groove of the double-strand, tau protein prevents DNA from damage by peroxidation. PLoS One 3:e2600

Wei G, Jewett AI, Shea JE (2010) Structural diversity of dimers of the Alzheimer amyloid-β(25–35) peptide and polymorphism of the resulting fibrils. Phys Chem Chem Phys 12:3622–3629

Weingarten MD, Lockwood AH, Hwo SY, Kirschner MW (1975) A protein factor essential for microtubule assembly. Proc Natl Acad Sci USA 72:1858–1862

Whalen BM, Selkoe DJ, Hartley DM (2005) Small non-fibrillar assemblies of amyloid β-protein bearing the Arctic mutation induce rapid neuritic degeneration. Neurobiol Dis 20:254–266

White AR, Multhaup G, Maher F, Bellingham S, Camakaris J, Zheng H, Bush AI, Beyreuther K, Masters CL, Cappai R (1999) The Alzheimer's disease amyloid precursor protein modulates copper-induced toxicity and oxidative stress in primary neuronal cultures. J Neurosci 19:9170–9179

Wischik CM, Novak M, Edwards PC, Klug A, Tichelaar W, Crowther RA (1988a) Structural characterization of the core of the paired helical filament of Alzheimer disease. Proc Natl Acad Sci USA 85:4884–4888

Wischik CM, Novak M, Thogersen HC, Edwards PC, Runswick MJ, Jakes R, Walker JE, Milstein C, Roth M, Klug A (1988b) Isolation of a fragment of tau derived from the core of the paired helical filament of Alzheimer disease. Proc Natl Acad Sci USA 85:4506–4510

Wogulis M, Wright S, Cunningham D, Chilcote T, Powell K, Rydel RE (2005) Nucleation-dependent polymerization is an essential component of amyloid-mediated neuronal cell death. J Neurosci 25:1071–1080

Wolozin BL, Pruchnicki A, Dickson DW, Davies P (1986) A neuronal antigen in the brains of Alzheimer patients. Science 232:648–650

Woody RW, Clark DC, Roberts GC, Martin SR, Bayley PM (1983) Molecular flexibility in microtubule proteins: proton nuclear magnetic resonance characterization. Biochemistry 22:2186–2192

Wu CK, Thal L, Pizzo D, Hansen L, Masliah E, Geula C (2005) Apoptotic signals within the basal forebrain cholinergic neurons in Alzheimer's disease. Exp Neurol 195:484–496

Xia W, Zhang J, Kholodenko D, Citron M, Podlisny MB, Teplow DB, Haass C, Seubert P, Koo EH, Selkoe DJ (1997) Enhanced production and oligomerization of the 42-residue amyloid β-protein by Chinese hamster ovary cells stably expressing mutant presenilins. J Biol Chem 272:7977–7982

Xia W, Yang T, Shankar G, Smith IM, Shen Y, Walsh DM, Selkoe DJ (2009) A specific enzyme linked immunosorbent assay for measuring β-amyloid protein oligomers in human plasma and brain tissue of patients with Alzheimer disease. Arch Neurol 66:190–199

Xie L, Helmerhorst E, Taddei K, Plewright B, Van Bronswijk W, Martins R (2002) Alzheimer's β-amyloid peptides compete for insulin binding to the insulin receptor. J Neurosci 22:RC221

Yan SD, Chen X, Schmidt AM, Brett J, Godman G, Zou YS, Scott CW, Caputo C, Frappier T, Smith MA et al (1994) Glycated tau protein in Alzheimer disease: a mechanism for induction of oxidant stress. Proc Natl Acad Sci USA 91:7787–7791

Yan R, Bienkowski MJ, Shuck ME, Miao H, Tory MC, Pauley AM, Brashier JR, Stratman NC, Mathews WR, Buhl AE, Carter DB, Tomasselli AG, Parodi LA, Heinrikson RL, Gurney ME (1999) Membrane-anchored aspartyl protease with Alzheimer's disease β-secretase activity. Nature 402:533–537

Yang Y, Mufson EJ, Herrup K (2003) Neuronal cell death is preceded by cell cycle events at all stages of Alzheimer's disease. J Neurosci 23:2557–2563

Yang Y, Varvel NH, Lamb BT, Herrup K (2006) Ectopic cell cycle events link human Alzheimer's disease and amyloid precursor protein transgenic mouse models. J Neurosci 26:775–784

Ye C, Walsh DM, Selkoe DJ, Hartley DM (2004) Amyloid β-protein induced electrophysiological changes are dependent on aggregation state: *N*-methyl-$_D$-aspartate (NMDA) *versus* non-NMDA receptor/channel activation. Neurosci Lett 366:320–325

Yen S, Easson C, Nacharaju P, Hutton M, Yen SH (1999) FTDP-17 tau mutations decrease the susceptibility of tau to calpain I digestion. FEBS Lett 461:91–95

Yin H, Kuret J (2006) *C*-terminal truncation modulates both nucleation and extension phases of tau fibrillization. FEBS Lett 580:211–215

Yoshiyama Y, Higuchi M, Zhang B, Huang SM, Iwata N, Saido TC, Maeda J, Suhara T, Trojanowski JQ, Lee VM (2007) Synapse loss and microglial activation precede tangles in a P301S tauopathy mouse model. Neuron 53:337–351

Zhang YJ, Xu YF, Chen XQ, Wang XC, Wang JZ (2005) Nitration and oligomerization of tau induced by peroxynitrite inhibit its microtubule-binding activity. FEBS Lett 579:2421–2427

Zhao L, Ma QL, Calon F, Harris-White ME, Yang F, Lim GP, Morihara T, Ubeda OJ, Ambegaokar S, Hansen JE, Weisbart RH, Teter B, Frautschy SA, Cole GM (2006) Role of p21-activated kinase pathway defects in the cognitive deficits of Alzheimer disease. Nat Neurosci 9:234–242

Zhao WQ, De Felice FG, Fernandez S, Chen H, Lambert MP, Quon MJ, Krafft GA, Klein WL (2008) Amyloid β oligomers induce impairment of neuronal insulin receptors. FASEB J 22:246–260

Zhou Q, Poo MM (2004) Reversal and consolidation of activity-induced synaptic modifications. Trends Neurosci 27:378–383

Chapter 6
Oligomers of α-Synuclein in the Pathogenesis of Parkinson's Disease

Dong-Pyo Hong, Wenbo Zhou, Aaron Santner, and Vladimir N. Uversky

Abstract Misfolding and subsequent self-assembly of proteins into various aggregates is a common molecular mechanism in a number of important human diseases. Understanding the peculiarities of the protein-misfolding processes is essential for the design of successful drugs that inhibit or reverse protein aggregation, leading to protein-misfolding pathologies. Protein aggregation is a complex process characterized by remarkable polymorphism, where soluble amyloid oligomers, amyloid fibrils, and amorphous aggregates are found as the final products. This polymorphism is associated with existence of multiple, independent, and competing assembly

---

D.-P. Hong
Department of Chemistry and Biochemistry, University of California, Santa Cruz, CA 95064, USA

Research Center for Protein Chemistry, Brown Foundation Institute of Molecular Medicine and the Department of Biochemistry and Molecular Biology, The University of Texas, Houston, TX 77030, USA

W. Zhou
Department of Chemistry and Biochemistry, University of California, Santa Cruz, CA 95064, USA

A. Santner
Molecular Kinetics, Inc., 6201 La Pas Trail, Suite 160, Indianapolis, IN 46268, USA

V.N. Uversky (✉)
Department of Molecular Medicine, University of South Florida, Tampa, FL 33612, USA

Institute for Intrinsically Disordered Protein Research, Center for Computational Biology and Bioinformatics, Department of Biochemistry and Molecular Biology, Indiana University School of Medicine, 410 W. 10th Street, HS 5009, Indianapolis, IN 46202, USA

Institute for Biological Instrumentation, Russian Academy of Sciences, 142290, Pushchino, Moscow Region, Russia
e-mail: vuversky@iupui.edu

F. Rahimi and G. Bitan (eds.), *Non-fibrillar Amyloidogenic Protein Assemblies—Common Cytotoxins Underlying Degenerative Diseases*, DOI 10.1007/978-94-007-2774-8_6,

pathways leading to aggregation. Irrespectively of aggregation mechanisms, soluble oligomers inevitably form during the self-association processes. Some of these oligomers are now considered major initiators of the pathogenic neurodegenerative cascades in the corresponding diseases. However, not all oligomers are equally harmful, and several amyloidogenic proteins form non-toxic oligomers, some of which are efficient fibrillation inhibitors. Unfortunately, information on the structural properties of soluble oligomers and mechanisms of their formation, inter-conversion, and toxicity is sparse. This chapter provides an overview of some topics related to soluble oligomers and several illustrative examples of toxic, non-toxic, productive, and off-pathway amyloid oligomers. The peculiarities of soluble oligomers of α-synuclein and its relation to the pathogenesis of Parkinson's disease are also discussed.

Keywords α-synuclein • Soluble oligomer • Amyloid fibril • Protein deposition diseases • Aggregation mechanism

Abbreviations

AD	Alzheimer disease
ADDL	Aβ-derived diffusible ligand
AFM	Atomic-force microscopy
AL amyloidosis	Light-chain-associated amyloidosis
ANS	8-anilinonaphthalenesulfonic acid
APP	β-amyloid precursor protein
CD	Circular dichroism
CJD	Creutzfeldt–Jakob disease
DLB	Dementia with Lewy bodies
DLBD	Diffuse Lewy-body disease
EM	Electron microscopy
FFI	Fatal familial insomnia
FTIR	Fourier-transform infrared spectroscopy
GSS	Gerstmann–Sträussler–Scheinker syndrome
HD	Huntington disease
LCDD	Light-chain-deposition disease
LBVAD	Lewy-body variant of Alzheimer's disease
LTP	Long-term potentiation
MSA	Multiple-system atrophy
MW	Molecular weight
NFT	Neurofibrillary tangle
NIID	Neuronal intranuclear inclusion disease
PD	Parkinson's disease
PHF	Paired helical filaments
PrP^{C}	Cellular form of the prion protein
PrP^{Sc}	Scrapie form of the prion protein
SBMA	Spinal and bulbar muscular atrophy
SCA	Spinocerebellar ataxia
ThT	Thioflavin T

6.1 Introduction

Many biologically active proteins act as specific oligomers. Proteins assemble into sophisticated supramolecular complexes that play various roles intracellularly. Formation of such functional oligomers and supramolecular complexes is tightly controlled and regulated. On the other hand, protein misfolding and subsequent uncontrolled (or unwanted) self-aggregation are now considered potential driving forces for the development of a number of human diseases (Bellotti et al. 1999; Dobson 1999; Kelly 1998; Rochet and Lansbury 2000; Uversky et al. 1999a, b). In fact, pathogenic proteinaceous deposits are at the heart of several so-called conformational diseases, such as Alzheimer's disease (AD), Parkinson's disease (PD), diffuse Lewy-body disease (DLBD), Lewy-body variant of Alzheimer's disease (LBVAD), dementia with Lewy bodies (DLB), multiple-system atrophy (MSA), Hallervorden–Spatz disease, light-chain-associated amyloidosis (AL amyloidosis), light-chain-deposition disease (LCDD), amyloidosis associated with hemodialysis, Huntington disease (HD), spinal and bulbar muscular atrophy (SBMA), spinocerebellar ataxia (SCA), neuronal intranuclear inclusion disease (NIID), Creutzfeldt–Jakob disease (CJD), Gerstmann-Sträussler–Scheinker syndrome (GSS), fatal familial insomnia (FFI), and Kuru to name a few. These and many other diseases originate from conversion of harmless soluble protein moieties into stable, ordered, filamentous protein aggregates, commonly referred to as amyloid (or amyloid-like) fibrils, which can accumulate in a variety of organs and tissues. At least 21 different proteins have been recognized as causative agents of these conformational diseases (Westermark et al. 2002). Protein aggregation in general, and amyloid fibrillation in particular, is a highly selective molecular self-assembly process. As a result, proteinaceous deposits found in different diseases predominantly contain aggregated forms of a specific causative protein, unique for a given disorder. What does drive the transformation of a biologically active soluble protein into a pathogenic, misfolded conformation with high self-aggregating potential? Some of the possible mechanisms of amyloidosis were recently reviewed (Merlini and Bellotti 2003). These mechanisms include the intrinsic propensity of some proteins to assume a pathologic conformation, which becomes evident with aging [e.g., normal α-synuclein in sporadic forms of PD and other synucleinopathies (Uversky 2007), and normal transthyretin in patients with senile systemic amyloidosis (Saraiva 2001)]. Pathological protein conformations may also result from unnaturally and persistently high cellular or plasma concentrations of a certain protein [e.g., triplication of a normal α-synuclein gene in some familial forms of PD (Singleton et al. 2004, 2003; Farrer et al. 2004), accumulation of β_2-microglobulin in patients undergoing long-term hemodialysis (Verdone et al. 2002), and locally high insulin concentrations at injection sites due to slow insulin release from the injection site (Shikama et al. 2010)]. Amyloidosis can also be triggered by genetic alterations, including amino-acid point-mutations in causative proteins (e.g., familial forms of AD and PD, various hereditary amyloidoses) or the genetic expansion of a CAG repeat in the open reading frames of genes encoding the proteins implicated in HD, SBMA, and SCA. Abnormal post-translational modifications of the causative proteins can result in aggregation. Examples include hyperphosphorylation of tau protein in AD (Medeiros et al. 2010; Chen et al. 2004) and proteolytic cleavage of a certain precursor protein [e.g., β-amyloid precursor protein (APP) in AD (Lichtenthaler 2010), and tau protein in AD (Kovacech and Novak 2010)]. Exposure to some environmental agents is correlated with PD pathogenesis (Di Monte et al. 2002; McCormack et al. 2002). Some of these environmental factors can affect protein conformation and modulate fibrillation of α-synuclein *in vitro* (Uversky and Eliezer 2009; Uversky 2008, 2007, 2003; Fink 2006; Dev et al. 2003). In fact, certain pesticides, herbicides, fungicides, heavy metals, polycations, polyanions, glycosaminoglycans, lipids, membranes, and macromolecular crowding are the conditions that induce changes in α-synuclein conformation due to increased concentrations of partially folded intermediates (Uversky et al. 2001b, a; Martinez et al. 2007; Zhu et al. 2003). Some factors, such as nitration and oxidation of α-synuclein or interaction of this protein with flavonoids (e.g., baicalein) or some other chemical reagents (e.g., compounds such as nicotine and hydroquinone) can inhibit α-synuclein fibrillation (Zhu et al. 2004; Hong et al. 2009, 2008; Qin et al. 2007). In the case of flavonoids and compounds found in cigarette smoke, these inhibitory effects were attributed to formation of oligomeric complexes stabilized by specific interactions between the chemical agents and α-synuclein. Various factors and mechanisms can act independently, additively, or even synergistically.

Accumulation of protein deposits is commonly associated with severe cellular degeneration at protein-deposition sites; however, precise mechanisms remain elusive (Sacchettini and Kelly 2002). It is not clear now whether amyloid fibrils trigger cellular degeneration or simply represent highly visible side products of disrupted cellular processes. However, an established fact is that protein misfolding/aggregation and cellular degeneration are coupled. As it was nicely summarized in a recent review (Bhak et al. 2009), there are several potential mechanisms whereby protein aggregation and deposition lead to cytotoxicity. These include disruption of tissue architecture and functions prompted by invasion of extracellular spaces by amyloids (Tan and Pepys 1994; Merlini and Bellotti 2003); destabilization of intracellular and extracellular membranes by oligomer formation which may precede or coincide with appearance of amyloid fibrils (Lashuel et al. 2002a; Caughey and Lansbury 2003); apoptotic cell death and receptor-mediated toxicity triggered by oligomer interactions with various neuronal receptors (Ferreira et al. 2007); oligomer-mediated impairment of the presynaptic P-/Q-type calcium currents (Nimmrich et al. 2008); impaired maturation of autophagosomes to lysosomes mediated by oligomer accumulation (Nixon 2006); dysfunction of autophagy, a lysosomal pathway for degrading organelles and proteins (Powers et al. 2009); oxidative-damage-induced disruption of cell viability prompted by incorporation of redox metals into amyloid fibrils and subsequent generation of reactive oxygen species (Huang et al. 1999b, a; Cuajungco et al. 2000; Opazo et al. 2002; Barnham et al. 2003); general disorganization of cellular protein homeostasis associated with exhaustion of cellular defense mechanisms such as the chaperone system (Macario and Conway de Macario 2005; Muchowski and Wacker 2005); proteasome inhibition (Almeida et al. 2006); and loss of crucial protein function(s) and/or gain of toxic function(s).

Because of their complexity and devastating health implications, the problems of protein misfolding, aggregation, and amyloid fibril formation have attracted researchers' considerable attention. An intriguing recent development in this field is the emerging recognition that soluble amyloid oligomers (which are oligomeric but the soluble states of amyloidogenic proteins) play multiple, unique roles as both crucial precursors of amyloid fibrils and as independent toxic agents. Despite these facts, information on the structural properties of soluble oligomers and mechanisms of their formation and inter-conversion is sparse, and understanding of the molecular mechanisms of their toxicity remains mostly elusive. This review provides an overview of some topics related to these issues.

6.2 Mechanisms of Amyloid Fibril Formation

It is recognized now that amyloid fibrillation is a highly dynamic process that represents the most dramatic consequence of protein misfolding and takes place in parallel with or as an alternative to physiologic protein folding. Amyloid fibrillation seems to be a common state for polypeptide chains to adopt (Dobson 1999), as the number of proteins shown to form such structures *in vitro* is constantly increasing (Uversky and Fink 2004). Despite unique chemical features of the causative proteins, amyloid fibrils of different origins have rather similar [but not identical (!)] morphologies, consisting of 2–6 unbranched protofilaments 2–5 nm in diameter associated laterally or twisted together to form fibrils with 4–13 nm diameter (Dobson 1999). They also display many common properties, including a core cross-β-sheet structure in which continuous β-sheets are formed with β-strands running perpendicular to the long axis of the fibrils (Serpell et al. 1997). Besides their characteristic appearance in images captured by electron (EM) and atomic-force (AFM) microscopy (often observed as long, twisted rope-like structures), amyloids are easily recognizable by their apple-green birefringence by polarized light microscopy after staining with a specific fluorescent dye, Congo red. Although fibrillation of various proteins produces fibrils with generally similar morphology, the phenomenon of amyloid fibrillar polymorphism has been recently recognized: Fibrillation of a single amyloidogenic protein may result in the appearance of multiple forms of amyloid fibrils depending on fibrillation conditions (Kodali and Wetzel 2007; Petkova et al. 2005; Wetzel et al. 2007; Goldsbury et al. 2005; Gosal et al. 2005; Bhak et al. 2009). Such polymorphism is likely due to the existence of multiple, independent, and competing assembly pathways that can lead to amyloidogenesis (Goldsbury et al. 2005; Gosal et al. 2005; Bhak et al. 2009).

In addition to the amyloid fibrils discussed above, proteins can self-assemble to form several alternative types of aggregates, including soluble oligomers and amorphous aggregates. Amorphous aggregates typically form at a faster rate than fibrils as there is no special conformational prerequisite for amorphous aggregation to occur, and many destabilized and partially unfolded proteins precipitate out of solution in the form of amorphous aggregates. On the other hand, fibrillation requires special conditions promoting formation of specific amyloidogenic conformations (Uversky et al. 2006). The choice between the three aggregation pathways—fibrillation, amorphous aggregate formation, or oligomerization—is determined both by the amino-acid sequence and by the peculiarities of protein environment.

Fibrils are proposed to be formed through template-dependent and template-independent pathways [reviewed in (Bhak et al. 2009)]. In template-dependent fibrillation, interactions with a pre-existing template bring about conformational changes in an amyloidogenic protein that exposes interactive regions for consecutive self-assembly, thereby promoting its accommodation to the template (Griffith 1967). Here, the template is taken in a broad sense because almost any conformational species involved in amyloid fibrillation (i.e., altered monomeric conformation, oligomeric forms, immature fibrils, protofibrils, and fragments of fibrils) could facilitate this process. It is likely that very early stages of template-dependent fibril formation are in fact template-independent, since in the earliest stages of fibril formation one or more soluble subunits must undergo self-assembly to form the initial template.

In template-independent fibrillation, amyloidogenic conformations are formed spontaneously in the absence of a template. Template-independent fibrillation requires an amyloidogenic self-interactive conformer, which favors self-assembly processes eventually leading to amyloid fibril formation (Bhak et al. 2009). The mentioned conformational transition from soluble, biologically active forms to amyloidogenic species is a great illustration of the protein-misfolding concept. Obviously, such a misfolding process can be triggered by a multitude of extrinsic and intrinsic factors. For example, α-synuclein fibrillation was shown to be dramatically accelerated under conditions that favor transformation of this natively unfolded protein into amyloidogenic forms characterized as a partially folded monomeric conformation, resembling a pre-molten globule state (Uversky 2007; Uversky and Eliezer 2009). It is clear that only the very early stages of template-independent fibril formation are truly "template-independent". Following formation of monomeric amyloidogenic species and assembly into amyloidogenic oligomers, these initial amyloidogenic oligomers will immediately act as templates for sequential addition of monomers through induced conformational transitions.

6.3 Inevitable Oligomerization of Aggregating Proteins

Recently, in an excellent review, Morris, Watzky, and Finke provided an outstanding summary of the major models proposed for description and analysis of protein-aggregation mechanisms and kinetics (Morris et al. 2009). The authors distilled the enormous literature on protein aggregation (as of March 2010, there were >53,500 papers in PubMed discussing various aspects of this phenomenon) down to several major classes of kinetic mechanisms. In this review, interested readers can find thoughtful analyses and comparisons of various protein-aggregation models together with the formalisms proposed for the quantitative description of aggregation kinetics (Morris et al. 2009). Protein aggregation is an irreversible nucleated (or autocatalytic) process, which resembles a condensation reaction, since it typically occurs only above a critical concentration. Typically, protein fibrillation is described by a sigmoidal curve and is considered as a nucleation–polymerization reaction where the monomer-addition steps are assumed to be thermodynamically unfavorable until a critical nucleus is formed (i.e., during the nucleation stage). However, aggregation is a thermodynamically favorable process during the polymerization stage (i.e., after the critical nucleus formation). The critical nucleus is defined as the least thermodynamically stable species in solution, that is, the oligomer of minimal size capable of initiating further growth (Morris et al. 2009). The nucleus can also be defined as the aggregate size after which the association rate exceeds the dissociation rate for the first time (Ferrone 1999). In addition to homogeneous nucleation, heterogeneous nucleation (or seeding) also can take place on the surface of existing polymers (Ferrone et al. 1980). Furthermore, aggregation can be accelerated by fragmentation of existing aggregates (Wegner and Savko 1982; Baskakov 2007). For some proteins with specific distribution of polar and hydrophobic residues (e.g., for $A\beta_{1-40}$ peptide), fibrillation can start only above a certain critical micelle concentration at which the peptide micelles are formed. Formation of these micelles represents a crucial step since fibrils nucleate inside them and then grow by irreversible binding of additional monomers to fibril ends (Lomakin et al. 1996).

All the models mentioned above were developed to describe an aggregation process, which obviously starts with a monomeric protein and ends with aggregate formation. Figure 6.1 represents an idealized model of amyloid fibril formation and clearly shows that fibrillation is a directed process with a series of consecutive steps, including formation of several different oligomers. In this model, various oligomers comprise structurally identical monomers and formation of these oligomers constitutes productive steps of the fibrillation pathway. However, aggregation is known to induce dramatic structural changes in the aggregating protein. Therefore, monomers at different aggregation stages are not identical. In addition, recent studies clearly show that a given protein can self-assemble into various aggregated forms depending on the peculiarities of its environment. In fact, typical aggregation processes only rarely result in the appearance of a homogeneous product where at the end of the reaction only one aggregate species (amyloid fibrils, amorphous aggregates, or soluble oligomers) would be present. More often, heterogeneous mixtures of various aggregated forms are observed. Furthermore, each aggregated form can have multiple morphologies and monomers comprising morphologically distinct aggregated forms can be structurally distinct as well. All of this suggests that aggregation is not a simple reaction, but a very complex process with multiple related and unrelated pathways, which can be connected or disjoined. However, appearance of a large aggregate inevitably involves formation of some small oligomeric species regardless of the models or pathways considered.

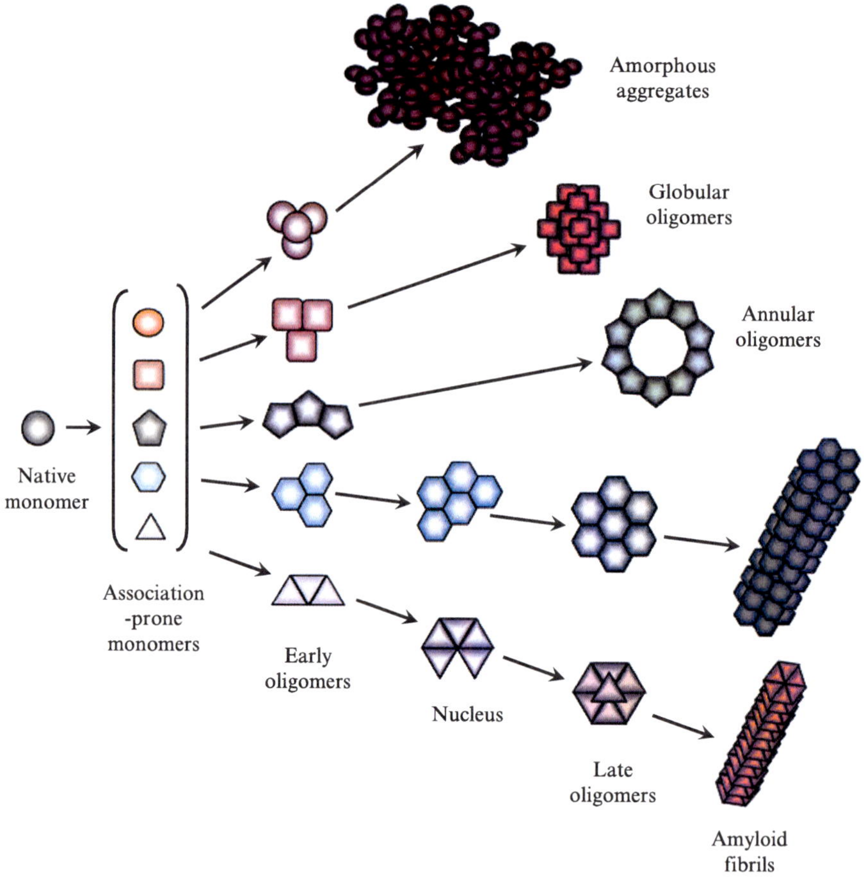

Fig. 6.1 An oversimplified schematic representation of protein self-association processes. Formation of multiple association-prone monomeric forms of proteins leads to multiple aggregation pathways. There are three major products of the aggregation reaction—amorphous aggregates (*top pathway*), morphologically different soluble oligomers (*second and third pathways from the top*) and morphologically different amyloid fibrils (*two bottom pathways*). Two types of soluble oligomers (spheroidal and annular) and two morphologically different amyloid fibrils are shown. Changes in color reflect potential structural changes within a monomer taking place at each elementary step. In reality, the picture is more complex and many species can be observed. Interconversions between various species in different pathways are also possible

6.4 A Brief Overview of Toxic, Non-toxic, Productive, and Off-Pathway Oligomers

Recent studies have indicated that small oligomeric species are potentially more cytotoxic than mature fibrils, which have been proposed to be products of detoxification (Lashuel et al. 2002a; Caughey and Lansbury 2003; Bucciantini et al. 2002).

In fact, very often, clinical manifestations of amyloidosis-related neurodegenerative diseases precede detectable accumulation of the fibrillar protein aggregates. In AD, the number of senile plaques in the affected brain regions was shown to poorly correlate with the local extent of neuron death or synaptic loss, or with cognitive impairment (Terry 1996). More generally, the amount of fibrillar deposits found at autopsy does not typically correlate with the clinical severity of AD or PD (Lashuel et al. 2002a). Furthermore, some patients with abundant amyloid deposits do not show neurodegenerative symptoms, suggesting that mature amyloid fibrils do not cause disease onset. In some transgenic animal and cell-culture models, pathological changes are frequently observed before accumulation of amyloid-plaques (Westermark et al. 2002; Billings et al. 2005). On the other hand, a robust correlation between levels of soluble Aβ oligomers and the extent of synaptic loss or severity of cognitive impairment has been established (Sakono and Zako 2010; Caughey and Lansbury 2003; Haass and Selkoe 2007; LaFerla et al. 2007; Klein et al. 2001; Chiti and Dobson 2006; Ferreira et al. 2007). Furthermore, the dopamine-dependent neurotoxicity of α-synuclein in PD was shown to be mediated by 54–83-kDa soluble protein complexes that contain α-synuclein and 14-3-3 protein, which are elevated selectively in the *substantia nigra* in PD (Dev et al. 2003). In general, some dimers or oligomers of α-synuclein or other amyloidogenic proteins are regarded now as real neurotoxins [reviewed in (Klein et al. 2001)]. For example, Hoshi et al. reported that spherical oligomers of Aβ peptide were the toxic moiety, whereas other Aβ aggregates, including fibrils, were not toxic (Hoshi et al. 2003). In agreement with these earlier studies, the late-stage, non-fibrillar oligomers of α-synuclein were shown to disrupt the integrity of biological membranes (Hong et al. 2010).

Therefore, soluble oligomers are important players in protein aggregation and in the associated cytotoxicity. The term "soluble oligomer" is used here to describe any non-monomeric form of an amyloidogenic protein that is soluble in aqueous solutions and remains in solution after high-speed centrifugation, indicating that it is not insoluble fibrillar or aggregated species. Several illustrative examples of these mysterious species are briefly described below.

6.4.1 *Aβ Oligomers*

Among various amyloidogenic proteins, oligomerization and its potential consequences are best documented for natural and synthetic Aβ peptides. Since various aspects of Aβ oligomer formation and toxicity were considered recently in an excellent review by Sakono and Zako (2010), only some key observations are presented below.

Several different oligomeric forms of natural and synthetic Aβ peptides ranging from dimers to 24-mers all the way up to soluble high-molecular-weight (MW) species have been reported (Sakono and Zako 2010; Haass and Selkoe 2007; Glabe 2008; Roychaudhuri et al. 2009). These oligomers have highly diverse structures, sizes, and shapes. For example, natural Aβ oligomers with a wide-ranging MW distribution (from <10 to >100 kDa) have been found in the AD brain (Kuo et al. 1996). This structural and morphological diversity is believed to be responsible for the diverse biological effects ascribed to these oligomers and for the related complexity of AD pathology (Sakono and Zako 2010; Caughey and Lansbury 2003; Haass and Selkoe 2007; Glabe 2008; Roychaudhuri et al. 2009; Ferreira et al. 2007). For example, analysis of soluble fractions of human brain and amyloid-plaque extracts revealed presence of SDS-stable dimers and trimers, suggesting that these oligomers could play crucial roles as the fundamental building blocks in the formation of larger oligomers or insoluble amyloid fibrils (Podlisny et al. 1995; Walsh et al. 2000; Bernstein et al. 2009). Cultured cells have also been shown to secrete the similarly sized Aβ oligomers, which have been shown to inhibit long-term potentiation (LTP) *in vitro* (Walsh et al. 2002a, b). *In vitro* studies revealed that Aβ dimers were three-fold more toxic than monomers, and that Aβ tetramers were 13-fold more toxic, clearly supporting the concept of the high toxicity of low-MW Aβ oligomers (Ono et al. 2009). The levels of the Aβ*56 oligomeric form (which corresponded to the SDS-stable Aβ nonamers and dodecamers) in the brain of the APP transgenic Tg2576 mice were shown to be correlated with memory deficits in this animal model (Lesné et al. 2006). Purified dodecamers were shown to induce a significant fall-off in the spatial memory performance of wild-type rats, suggesting that nonamers and dodecamers are potentially associated with deleterious effects on cognition. Morphological analyses revealed substantial shape variability among soluble Aβ oligomers. For example, small globular Aβ oligomers (5 nm in diameter), referred to as Aβ-derived diffusible ligands (ADDLs), were frequently found in Hams-F12 medium (Lambert et al. 1998). ADDLs were shown to interact strongly with the dendritic arbors of cultured neurons triggering neuronal death and LTP blocking. Furthermore, analysis of soluble brain extracts using an ADDL-specific antibody established presence of ADDLs in human AD brain, suggesting that formation and existence of ADDLs in the human AD brain can cause disease (Lambert et al. 1998).

The largest globular Aβ assemblies are amylospheroids. These are highly neurotoxic, off-pathway, spheroidal structures with diameters of 10–15 nm (Hoshi et al. 2003). Finally, various annular Aβ oligomers, with MW ranging from 150 to 250 kDa and with an outer diameter of 8–12 nm and an inner diameter of 2.0–2.5 nm, have also been described in the literature (Lashuel and Lansbury 2006; Bitan et al. 2003). Such doughnut-like oligomers are preferentially formed from mutant Aβ (such as those carrying the Arctic mutation). These annular Aβ oligomers can act as non-specific amyloid pores, which structurally resemble pores formed by the bacterial cytolytic β-barrel pore-forming toxins, and which could be responsible for the Aβ-associated cytotoxicity (Lashuel and Lansbury 2006).

6.4.2 Oligomers of Tau Protein

Appearance of neurofibrillary tangles (NFTs) or insoluble intracellular fibers of paired helical filaments (PHF) that arise from misfolding and aggregation of the neuron-specific, microtubule-associated protein tau is closely correlated with AD progression (Selkoe 1986; DeMager et al. 2002; LaFerla and Oddo 2005). As with Aβ and α-synuclein, formation of soluble tau oligomers rather than mature fibrils, is the causative agent of cell death associated with tau aggregation (Santacruz et al. 2005). Recently, structures of tau oligomers, which appear in response to heparin-induced aggregation were analyzed by multidimensional NMR (Peterson et al. 2008). This study revealed that the regions $VQIINK^{280}$ and $VQIVYK^{311}$ of the tau protein were the major sites of intermolecular interaction in the oligomer and that these intermolecular interactions were triggered in response to heparin addition (Peterson et al. 2008). It has also been shown that tau assembly involves two distinct dimers (cysteine-dependent and cysteine-independent) that differ in resistance to reduction (Sahara et al. 2007). Interestingly, the population of cysteine-dependent tau oligomers was shown to increase before detection of fibrils and was accompanied by an increase in the amount of cysteine-independent dimers (Sahara et al. 2007). In addition to the small oligomers discussed above, a granular tau oligomer having a pre-filamentous structure was also found (Maeda et al. 2006). Quantification of frontal-cortex samples displaying varying degrees of NFT pathology revealed significantly increased levels of granular tau oligomers even in brains at very early neuropathological stages when clinical symptoms of AD and NFTs in frontal cortex were absent (Maeda et al. 2006). Based on these observations, it was concluded that granular tau oligomer levels increase before detectable formation of NFTs and before manifestation of the clinical symptoms of AD, suggesting that granular tau oligomer levels may represent an early sign of NFT formation and AD (Maeda et al. 2006).

6.4.3 Prion Protein Oligomers

In the prion diseases, the autocatalytic conversion of the cellular form of the prion protein, PrP^C, to an alternative conformation, PrP^{Sc}, must take place (Prusiner 1998). Here, PrP^C is monomeric and sensitive to protease digestion, whereas PrP^{Sc} is rich in β-sheets, has a low solubility, and is resistant to protease digestion (Prusiner 1998). Furthermore, PrP^{Sc} can convert PrP^C into the pathogenic PrP^{Sc} form (Prusiner 1998). By analogy with other protein-conformational diseases, PrP oligomers and/or pre-fibrillar aggregates have been proposed to be cytotoxic (Kayed et al. 2003; Eghiaian 2005). In agreement with this hypothesis, the most infectious species were shown to be soluble oligomers of the prion protein, 17–27 nm in diameter and 300–600 kDa in mass, derived from the disaggregation of PrP^{Sc} (Silveira et al. 2005). Furthermore, these oligomers were able to convert efficiently the PrP^C into a protease-resistant form in an *in vitro* assay (Silveira et al. 2005). Incubation of the full-length mouse PrP at low pH, transforms the protein into an equilibrium mixture of soluble, β-sheet-rich oligomers and α-helix-rich monomers (Jain and Udgaonkar 2008). With time, these β-rich oligomers assemble into the worm-like amyloid fibrils in a stepwise manner and potentially *via* multiple routes (Jain and Udgaonkar 2008).

The fact that amyloids in general and PrP^{Sc} in particular can self-propagate suggests that the structure of the final fibrillar state might be determined by the structural information encoded in the initial nucleus formed in the very early phase of protein fibrillation. A systematic analysis of fibrillation of yeast prion protein, Sup35, supported this hypothesis and showed that the structural variability in the initial nucleus was a crucial determining factor for the diversity of prion strain conformations and resulting strain phenotypes (Ohhashi et al. 2010). In fact, an intriguing correlation was found between the reversible formation of soluble oligomers at low temperature and the ability of Sup35 to form the Sc4 prion conformation that leads to induction of strong [PSI^+] phenotypes. Oligomer formation was driven by the non-native aromatic interactions outside the amyloid core, which specifically led to formation of highly infectious strain conformations with more limited amyloid cores (Ohhashi et al. 2010). Based on this interesting study, the authors concluded that transient non-native interactions in the initial oligomers could play a crucial role for subsequent determination of the diversity of amyloid conformations and resulting prion strain phenotypes (Ohhashi et al. 2010).

6.4.4 Insulin Oligomers

The physiological form of insulin is a zinc-stabilized hexamer. However, in 20% acetic acid, insulin is monomeric (Nielsen et al. 2001). A systematic analysis of insulin fibrillation in 20% acetic acid revealed noticeable structural changes occurring before the onset of fibril formation (Ahmad et al. 2005). In this study, at least two different types of oligomeric intermediates between the native monomer and fibrils were detected. These intermediates had distinct underlying structures, being easily distinguishable by Fourier-transform infrared spectroscopy (FTIR), circular dichroism (CD), and 8-anilinonaphthalenesulfonic acid (ANS) binding, and corresponded to the significantly different association states as determined by the dynamic light scattering. Both oligomeric intermediates had non-native conformations, indicating that fibrillation occurred from a β-rich structure that is distinct from the native fold (Ahmad et al. 2005). Existence of significant amounts of oligomeric species of insulin before appearance of mature fibrils and throughout the fibril-elongation process was further confirmed by dynamic light scattering (Ahmad et al. 2005) and time-lapse AFM (Podesta et al. 2006). Small-angle X-ray scattering analysis of oligomeric species accumulated at the early fibrillation stages of insulin revealed the peculiar morphology where oligomers appeared as a bead-on-a-string assembly of six units, each with dimensions comparable to those of insulin monomers (Vestergaard et al. 2007).

6.4.5 Oligomers of the Immunoglobulin Light Chain

Light-chain amyloidosis (AL or primary amyloidosis) originates from the formation and systemic deposition (especially in the kidneys) of fibrils of monoclonal immunoglobulin light-chain variable domains in patients suffering from multiple myeloma (Solomon 1986). LEN is a κIV immunoglobulin light-chain variable domain found in patients suffering from multiple myeloma. The recombinant form of LEN is a dimer that represents an established system for the *in vitro* characterization of light-chain amyloid proteins (Stevens et al. 1995). A systematic analysis of the LEN fibrillation process revealed an inverse dependence on concentration due to formation of off-pathway soluble oligomers (probably octamers) at high protein concentrations (Souillac et al. 2003, 2002). In fact, these soluble, off-pathway oligomeric species formed at high protein concentrations before fibrils appeared, significantly slowing down the kinetics of fibril formation, as compared to the fibrillation rates measured at much lower protein concentrations. Interestingly, however, fibrils were still observed eventually at high protein concentrations, despite the initial trapping of most of the protein as soluble off-pathway oligomers. Because most of the protein was present in these off-pathway intermediates at relatively early aggregation times, and since eventually all the protein formed fibrils, a structural rearrangement from the non-fibril-prone, off-pathway oligomers to a more fibril-prone species must occur at later aggregation stages. The corresponding structural changes, monitored by a variety of techniques, revealed a significant increase in the disordered secondary structure content, an increase in the solvent accessibility, and a decrease in the intrinsic stability of the soluble oligomers (Souillac et al. 2003). More specifically, the fibrillation of the dimeric LEN can be described by the following model. First, the transition to a partially folded but relatively native-like conformation, *I*, takes place. At this stage, LEN preserves its dimeric state. Then, a soluble oligomer (an octamer), I_8, is formed from the partially folded dimer. This octamer is composed of partially folded LEN molecules, the conformations of which correspond to those of the initially formed partially folded intermediate *I*. With time, I_8 oligomers transform into a second class of soluble oligomers $(I^*_8)_n$ and the component molecules undergo a conformational change leading to a less ordered structure, *I**. At the final stage, an exponential growth of fibrils occurs. In contrast to *I*, the conformation of *I** is much more disordered as detected by probes of secondary structure, increased susceptibility to proteolysis, increased hydrogen–deuterium exchange, and decreased conformational stability (Souillac et al. 2003). This structural reorganization is likely to be driven by the self-association of the I_8 oligomers. Based on these observations it has been concluded that LEN represents a unique fibrillating system in which soluble off-pathway oligomeric intermediates are shown to be the major transient species and in which fibrillation occurs from a relatively unfolded conformation present in these intermediates (Souillac et al. 2003). LEN fibrillation at low protein concentrations (i.e., in the absence of dimers) likely follows the more "traditional" pathway, where partial unfolding of a monomeric protein precedes subsequent aggregation.

6.5 The Multitude of Oligomeric Forms of α-Synuclein

6.5.1 Dimers

α-Synuclein (which is believed to be a major player in the pathogenesis of PD and other synucleinopathies) fibrillization *in vitro* is not a simple, two-state transition from monomer to fibrils, but represents instead a rather complex process involving formation of oligomeric intermediates of various sizes and morphologies. Prolonged incubation of α-synuclein at different temperatures result in a temperature-dependent, progressive aggregation, with dimers forming first (Uversky et al. 2001a). This temperature-modulated oligomerization was shown to be accompanied by a small but reproducible increase in the ordered secondary structure content. Interestingly, the trapped oligomeric conformation was structurally similar to the pre-molten globule-like partially folded monomeric confomer induced by low pH or high temperature (Uversky et al. 2001a). Therefore, it has been concluded that the partially folded pre-molten globule-like conformation of α-synuclein can be stabilized as the protein undergoes a highly selective self-assembly process during prolonged incubation at elevated temperatures (Uversky et al. 2001a). Formation of oxidative dimers and high-order oligomers with dityrosine cross-links in α-synuclein under the conditions of oxidative stress has also been reported (Souza et al. 2000).

6.5.2 High-Order Spherical and Annular Oligomers

Various α-synuclein oligomers were separated from fibrillar and monomeric α-synuclein by sedimentation followed by gel-filtration chromatography (Conway et al. 2000a, b). AFM analysis revealed the great morphological diversity of these oligomers, including various spheres (with heights ranging from 2 to 6 nm), chains of spheres (protofibrils), and rings with height ranging from 3 to 7 nm (Conway et al. 2000a, b). In this process, spherical protofibrils formed rapidly, whereas annular species appeared during prolonged incubation of the spheres (Ding et al. 2002). In addition to the completed rings or doughnuts, partially formed rings (crescents) have also been observed (Ding et al. 2002). Formation of both doughnuts and fibrils was shown to require initial formation of spherical, β-structure-rich, α-synuclein oligomers. Morphology of oligomers was found to be affected by the solution conditions, including presence of lipids (Jo et al. 2000; Lee et al. 2002; Ding et al. 2002), organic solvents (Munishkina et al. 2003), or metal ions (Lowe et al. 2004). In fact, incubation of α-synuclein with different metals for 1 day at 4°C produced three different classes of oligomers. The first class of oligomers formed by incubation with Cu^{2+}, Fe^{3+}, and Ni^{2+} and yielded 0.8–4-nm spherical particles, similar to α-synuclein incubated without metal ions. Incubation with Mg^{2+}, Cd^{2+}, and Zn^{2+} gave larger, 5–8-nm spherical oligomers. Whereas Co^{2+} and Ca^{2+} frequently yielded a third class of annular (doughnut-like) oligomers, 70–90 nm in diameter with Ca^{2+}

and 22–30 nm in diameter with Co^{2+} (Lowe et al. 2004). Analysis of different α-synuclein oligomers by Raman microscopy revealed that spheroidal oligomers contained a significant amount of α-helical structure (~47%) and β-sheet structure (~29%). Formation of protofibrils was accompanied by a decrease in the α-helical content to ~37% and a concomitant increase in the β-sheet content to ~54% (Apetri et al. 2006). Both spheroidal and annular oligomers were proposed to be cytotoxic. In fact, the spherical protofibrils were shown to bind to brain-derived membrane fractions much more tightly than did monomeric or fibrillar α-synuclein (Ding et al. 2002). Annular oligomers, known as amyloid pores are also able to bind to membranes (Ding et al. 2002), significantly affecting membrane stability (Lashuel et al. 2002a; Caughey and Lansbury 2003).

6.5.3 On-Pathway and Off-Pathway Oligomers

Often, soluble oligomers are transient, because they are consumed as fibrillization proceeds (Conway et al. 2000b; Ding et al. 2002; Caughey and Lansbury 2003). *In situ* AFM analysis showed that formation of globular oligomers precedes appearance of amyloid fibrils and is systematically observed under conditions that accelerate fibrillation, suggesting that α-synuclein oligomers can act as on-pathway intermediates during amyloidogenesis (Hoyer et al. 2004). However, soluble oligomerization can also be an off-pathway reaction, and under some conditions the productive α-synuclein fibrillation is known to be inhibited in favor of formation of soluble oligomers. For example, nitrated α-synuclein remains assembled as oligomeric spheroids even after incubation for a prolonged time (Yamin et al. 2003). Furthermore, addition of nitrated α-synuclein substantially inhibited fibrillation of the non-modified protein in a concentration-dependent manner (Uversky et al. 2005; Yamin et al. 2003). Preferential oligomerization was also found when α-synuclein was co-incubated with components of cigarette smoke such as nicotine and hydroquinone (Hong et al. 2009), various flavonoids (Meng et al. 2009), or substoichiometric concentrations of 3,4-dihydroxyphenylacetic acid (Zhou et al. 2009), or as a result of the methionine oxidation (Zhou et al. 2010).

6.5.4 Soluble, Non-fibrillar, Late-Stage α-Synuclein Species in the Supernatant Fractions

In addition to transient oligomers, which are present during the lag-phase of α-synuclein amyloid fibril formation *in vitro* (Conway et al. 2000b; Ding et al. 2002; Caughey and Lansbury 2003), a substantial fraction of α-synuclein molecules (10–20%) fails to form fibrils and exists as soluble oligomers even in the absence of stabilizing interactions with small molecules. Structural peculiarities of these soluble oligomers

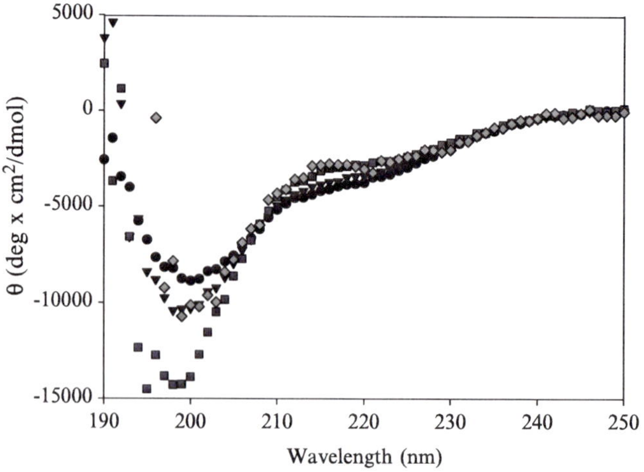

Fig. 6.2 Far-UV CD spectra of oligomers remaining in the supernatant after the completion of the fibrillation of 1.0 mg/mL α-synuclein in 20 mM (*black circles*), 50 mM (*gray triangles*), or 100 mM (*dark-gray squares*) NaCl. Corresponding spectra of the natively unfolded monomeric α-synuclein are shown for comparison (*light-gray diamonds*)

were characterized by a variety of biophysical techniques including CD, FTIR, AFM, and thioflavin-T (ThT) fluorescence (Hong et al. 2010).

Existence of a small but reproducible increase in the ThT fluorescence intensity during the fibrillation lag time in the presence of low salt concentrations has been reported (Hong et al. 2010). This pre-transition increase in the ThT fluorescence was attributed to formation of α-synuclein aggregates, which can affect fibrillation rates leading to longer lag times (Hong et al. 2010). This analysis revealed that while the fibrillation rate was dramatically accelerated by an increase in the NaCl concentration, the amount of non-fibrillar soluble α-synuclein in the supernatant fractions slightly decreased, suggesting that there may be more non-fibrillar interactions between proteins at low salt contents. These data also indicated that as fibrils formed more rapidly, the shorter period available for the "non-specific" intermolecular interactions resulted in less "late" oligomer build up. When the supernatant fractions containing the non-fibrillar protein were concentrated and incubated with or without the preformed α-synuclein fibrils, no noticeable increase in the ThT fluorescence was detected (even after the incubation for 4 days), suggesting that the non-fibrillar supernatant oligomers were stable and non-amyloidogenic. These stable oligomers were characterized by negligible dissociation rates (Hong et al. 2010).

Far-UV CD and FTIR were also used for structural characterization of the non-fibrillar supernatant oligomers. Figure 6.2 represents the far-UV CD spectra of α-synuclein measured under a variety of conditions. The monomeric protein possesses a spectrum typical of a highly disordered polypeptide chain, characterized by the presence of an intense minimum in the 198-nm region and the absence of noticeable bands in the vicinity of 210–230 nm. On the other hand, the supernatant oligomers possess a slightly more ordered secondary structure than the natively unfolded monomeric α-synuclein and the amount of residual structure decreases with the increase in NaCl concentration. The analysis of various α-synuclein species by FTIR further showed that the supernatant oligomers were mostly disordered, since oligomers formed in the presence of 20 mM NaCl have 30.1% of β-structures and 48.8% random coil, whereas the supernatant oligomers obtained from the 100-mM NaCl solutions contain 18.6% of β-structures and 54% random coil (see Fig. 6.3) (Hong et al. 2010).

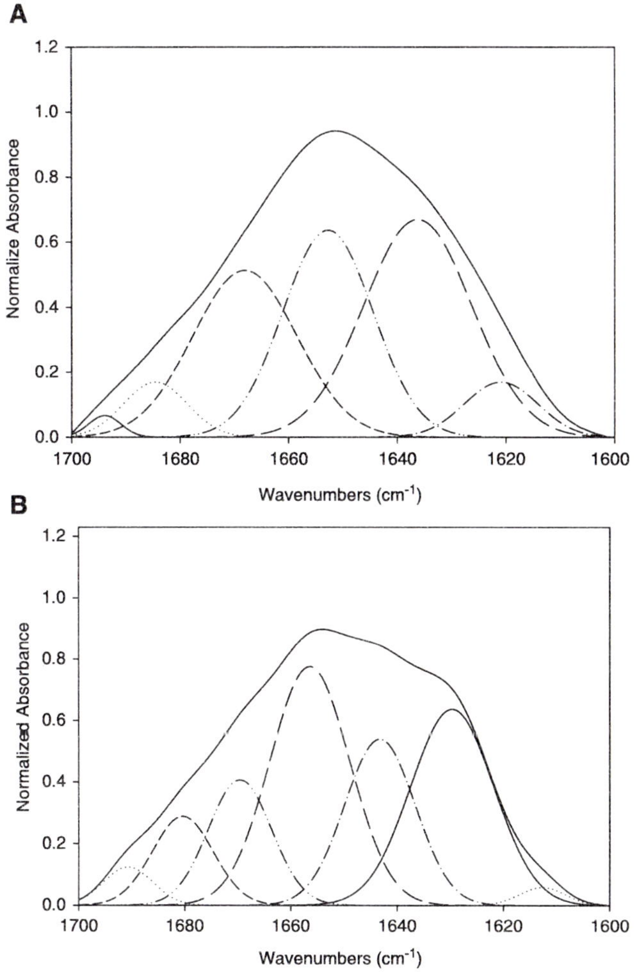

Fig. 6.3 FTIR spectra of monomeric α-synuclein (**A**) and soluble oligomers present in the supernatants at the completion of the fibrillation process in the presence of 20 mM (**B**), 50 mM (**C**), or 100 mM NaCl (**D**)

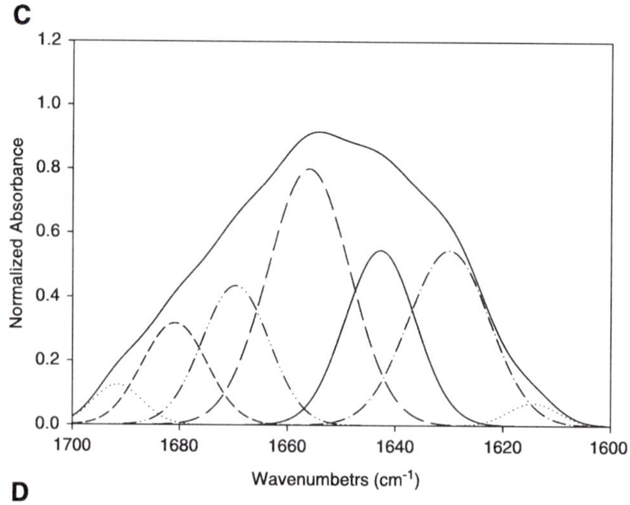

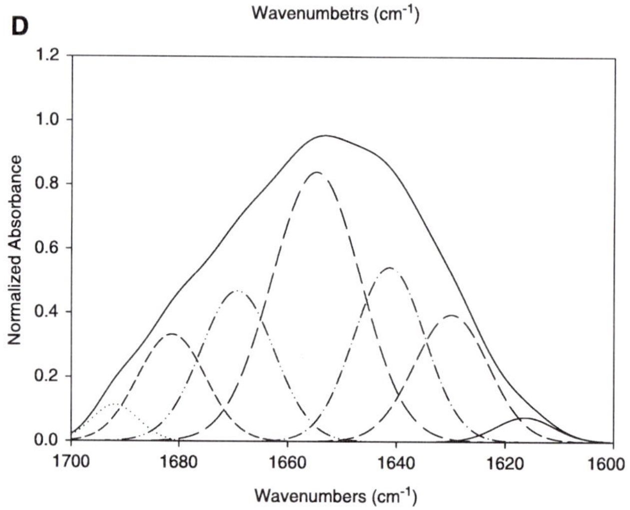

Fig. 6.3 (continued)

Figure 6.4 shows AFM images and height histograms of oligomeric species formed in the presence of 20, 50, and 100 mM NaCl. This analysis revealed that both the average size and the morphology of oligomers depended on salt concentrations. In fact, oligomers formed at low NaCl concentrations were noticeably smaller and did not show homology in size or shape in comparison with the more homogeneous oligomers formed at higher NaCl concentrations (cf. Fig. 6.4a–c) (Hong et al. 2010).

Finally, the effect of non-fibrillar supernatant oligomers on the integrity of a lipid membrane was analyzed. To this end, large unilamellar vesicles of 1,2-dipalmitoyl-*sn*-glycero-3-phosphate (PA)/1,2-dipalmitoyl-*sn*-glycero-3-phosphocholine (PC) encapsulating calcein were prepared (Hong et al. 2010). In these experiments, the

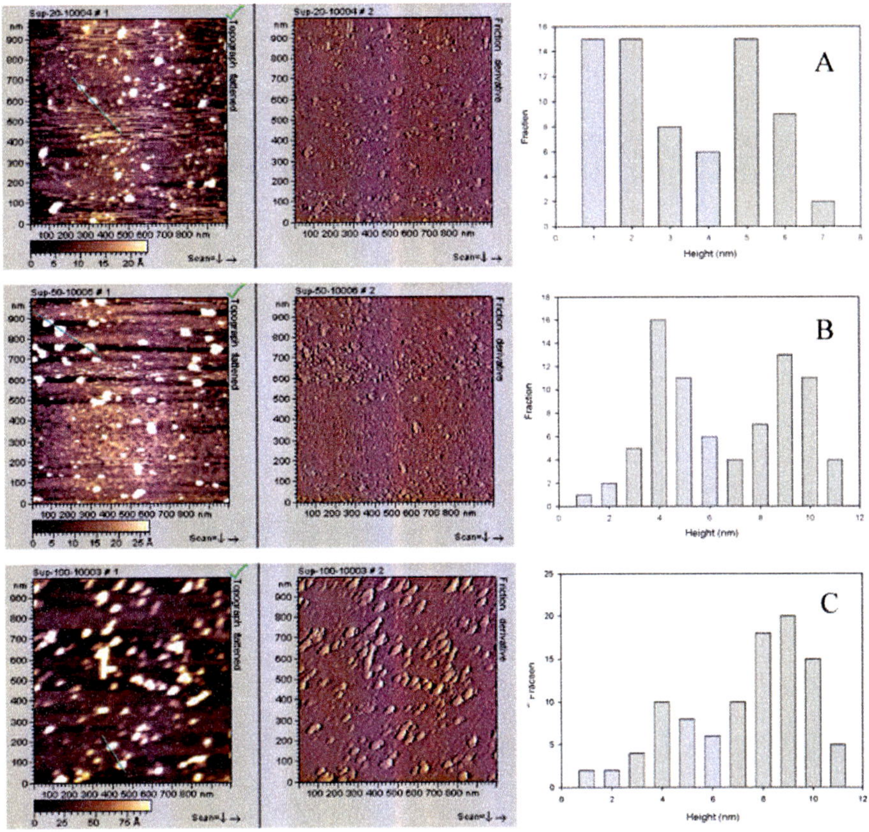

Fig. 6.4 AFM images and height distributions of oligomers present in the supernatants at the completion of the fibrillation of α-synuclein in the presence of 20 mM (**A**), 50 mM (**B**), or 100 mM NaCl (**C**)

encapsulated fluorophore is self-quenched due to its high concentration in the vesicle and therefore has low fluorescence intensity. Therefore, the release of the dye from the vesicle is accompanied by a noticeable increase in the fluorescence intensity. This analysis revealed that addition of monomeric, fibrillar, or oligomeric α-synuclein species resulted in noticeable release of the dye. The most effective disruptors of the membrane integrity were fibrils, followed by oligomers and monomers. The membrane permeability of supernatant oligomers depended on the concentration of salt at which they formed. The supernatant oligomers formed in low salt concentration (20 mM NaCl) showed stronger effects than those formed in high salt concentration (100 mM NaCl). These data suggested that the impact of soluble oligomers on membrane integrity might depend on their sizes or structures (Hong et al. 2010). Although the disruptive effect of these non-fibrillar oligomers on the membrane integrity was not as high as that of fibrils, it clearly suggested that the off-pathway stable oligomers might also be toxic. Note, since fibrils do not play a major role in neurodegeneration, the fact that they show increased membrane permeabilization might infer that the membrane-permeabilizing activity is not predictive of neurotoxicity.

Data presented above suggest that the non-fibrillar supernatant oligomers might be the off-pathway products of the α-synuclein fibrillation processes. The mechanism of late-stage oligomer formation can be understood considering the intrinsically disordered nature of α-synuclein. In fact, since α-synuclein is a natively unfolded protein which is able to interact in a condition-dependent manner with various partners, it is highly possible that under some conditions effective interactions between α-synuclein molecules could lead to formation of stable oligomers (Hong et al. 2010).

6.5.5 Non-toxic α-Synuclein Oligomers

Although in the majority of analyzed cases, soluble α-synuclein oligomers affect membrane permeability dramatically (Ding et al. 2002; Lashuel et al. 2002a; Caughey and Lansbury 2003), there are several cases where α-synuclein oligomerization results in formation of non-toxic species. In fact, this finding is rather expected due to the highly heterogeneous nature of protein aggregates (including oligomers) caused either by the heterogeneous starting materials or by multiple pathways of assembly, or both these factors. Therefore, it is difficult to believe that all soluble oligomers, with their astonishing morphological variability, will be cytotoxic similarly (Hong et al. 2009). In agreement with this reasoning, the flavonoid baicalein was shown to inhibit α-synuclein fibrillation by stabilizing soluble oligomers with specific structural features—being spherical, having relatively globular structure with packing density intermediate between that of pre-molten globules and typical globular proteins, having a relatively well-developed secondary structure, and being characterized by high thermodynamic stability (Hong et al. 2008). These oligomers were able to inhibit fibrillation of baicalein-untreated α-synuclein and, most importantly, did not disrupt the integrity of biological membranes (Hong et al. 2008). Similarly, oxidation of α-synuclein methionine residues by H_2O_2 greatly inhibit fibrillation of this protein *in vitro*, leading to formation of relatively stable oligomers, which are not toxic to dopaminergic or GABAergic neurons (Zhou et al. 2010). Lack of cytotoxicity in this case was confirmed using primary cultures of ventromedial cells from rat brains. In these mixed cultures, which contain dopaminergic and non-dopaminergic (e.g., GABAergic) neurons, loss of neuronal integrity was assessed as a decrease in neurotransmitter (dopamine or GABA) uptake. The methionine-oxidized α-synuclein showed no toxicity at 5 μM and a relatively little toxicity at 10 μM. Importantly, the mentioned toxicity of the methionine-oxidized α-synuclein at 10 μM was not specific for dopaminergic cells based on lack of difference between dopamine and GABA uptake. This suggested that methionine oxidation is able to shift α-synuclein away from fibrillation pathway and toward formation of relatively harmless oligomers (Zhou et al. 2010). Altogether, these observations clearly show that formation of soluble oligomers of certain amyloidogenic proteins does not always create harm and can be beneficial.

6.6 Concluding Remarks

Accumulating evidence indicates that uncontrolled (or unwanted) self-aggregation of amyloidogenic proteins represents a fundamental basis for the development of various types of amyloid-related degenerative diseases. Aggregation is a complex self-assembly process characterized by an astonishing polymorphism of final products. Proteins are known to self-assemble into various aggregated forms, e.g., soluble oligomers, amyloid fibrils, and amorphous aggregates. The choice between these three aggregation pathways—fibrillation, amorphous aggregate formation, and oligomerization—is determined by the amino-acid sequence and by the peculiarities of protein environment. However, often all these forms are found at the end of the aggregation reaction. In addition, each of these aggregated species can be present in several morphologically and structurally distinct forms. This polymorphism reflects a wide variation in association states of amyloidogenic proteins and diversity of intermolecular interactions stabilizing final self-aggregated forms. For example, Aβ oligomers are characterized by a wide range of associations (from <10 to >100 kDa), and biochemical properties of Aβ oligomers and their pathogenicity depend on their sizes and structures. Furthermore, even similarly sized Aβ oligomers can be characterized by a dramatic structural (and potentially pathogenic) polymorphism. Likely, this polymorphism reflects existence of multiple independent and competing assembly pathways that lead to aggregation.

For a long time it was believed that amyloid fibrils are harmful. However, a novel emerging paradigm favors the idea that the deposited proteinaceous inclusions (such as senile plaques in AD or Lewy bodies or Lewy neurites in PD, etc.) are not cytotoxic, and that formation of these large deposits represents a defensive mechanism for effective sequestration of misfolded species. Instead, amyloid oligomers (which are oligomeric but soluble states of amyloidogenic proteins), rather than insoluble amyloid fibrils, are believed now to be responsible for initiation of neurodegenerative cascades of corresponding diseases (Urbanc et al. 2002; Mucke et al. 2000, Lashuel et al. 2002b, a).

In some cases, formation of stable soluble oligomers occurs in parallel with amyloid fibril formation, and a significant portion of α-synuclein molecules (15–20%) remains non-fibrillar at the end of the fibrillation process (Hong et al. 2010). Therefore, at least two different types of oligomers can be formed during the fibrillation process: Some soluble oligomers are on the pathway for fibrillation, and some are on the pathway to non-fibrillar oligomers. These non-fibrillar late-stage species are soluble oligomers, the structural properties, morphology, and membrane-permeating capabilities of which are all dramatically affected by the conditions whereby fibrils form. Obviously, these soluble oligomers are important players in the multi-pathway aggregation of α-synuclein and they definitely should be taken into account in studies of the molecular mechanisms of α-synuclein fibrillation.

Importantly, oligomers inevitably form during the aggregation processes, both as crucial intermediates in the fibrillogenesis and as independent off-pathway entities. The fact that various amyloidogenic proteins can form toxic soluble oligomers clearly suggests that amyloid oligomers can be considered key factors in the pathogenesis of various degenerative diseases. Recent studies clearly demonstrate that formation and toxicity mechanisms of various amyloid oligomers can also differ from one another. Therefore, therapeutic strategies targeted at inhibition of fibrillation or at dissolution of preformed fibrils can be potentially harmful because prevention of aggregation may cause formation or stabilization of toxic oligomeric states. On the other hand, since oligomers form *via* various pathways, not all oligomers are equally cytotoxic and several cases have been reported that amyloidogenic proteins can form some non-toxic oligomers. Importantly, under some conditions, formation of such non-toxic oligomers can effectively compete with fibrillation processes and the preformed non-toxic oligomers can serve as potent fibrillation inhibitors.

Acknowledgements This work was supported in part by the Program of the Russian Academy of Sciences for the "Molecular and cellular biology" (V.N.U.), by grants R01 LM007688-01A1 (V.N.U.) and GM071714-01A2 (V.N.U.) from the National Institutes of Health and the grant EF 0849803 (V.N.U.) from the National Science Foundation. We gratefully acknowledge support of the IUPUI Signature Centers Initiative.

References

Ahmad A, Uversky VN, Hong D, Fink AL (2005) Early events in the fibrillation of monomeric insulin. J Biol Chem 280:42669–42675

Almeida CG, Takahashi RH, Gouras GK (2006) β-Amyloid accumulation impairs multivesicular body sorting by inhibiting the ubiquitin–proteasome system. J Neurosci 26:4277–4288

Apetri MM, Maiti NC, Zagorski MG, Carey PR, Anderson VE (2006) Secondary structure of α-synuclein oligomers: characterization by Raman and atomic force microscopy. J Mol Biol 355:63–71

Barnham KJ, Ciccotosto GD, Tickler AK, Ali FE, Smith DG, Williamson NA, Lam YH, Carrington D, Tew D, Kocak G, Volitakis I, Separovic F, Barrow CJ, Wade JD, Masters CL, Cherny RA, Curtain CC, Bush AI, Cappai R (2003) Neurotoxic, redox-competent Alzheimer's β-amyloid is released from lipid membrane by methionine oxidation. J Biol Chem 278:42959–42965

Baskakov IV (2007) Branched chain mechanism of polymerization and ultrastructure of prion protein amyloid fibrils. FEBS J 274:3756–3765

Bellotti V, Mangione P, Stoppini M (1999) Biological activity and pathological implications of misfolded proteins. Cell Mol Life Sci 55:977–991

Bernstein SL, Dupuis NF, Lazo ND, Wyttenbach T, Condron MM, Bitan G, Teplow DB, Shea J-E, Ruotolo BT, Robinson CV, Bowers MI (2009) Amyloid-β protein oligomerization and the importance of tetramers and dodecamers in the aetiology of Alzheimer's disease. Nat Chem 1:326–331

Bhak G, Choe YJ, Paik SR (2009) Mechanism of amyloidogenesis: nucleation-dependent fibrillation *versus* double-concerted fibrillation. BMB Rep 42:541–551

Billings LM, Oddo S, Green KN, Mcgaugh JL, Laferla FM (2005) Intraneuronal Aβ causes the onset of early Alzheimer's disease-related cognitive deficits in transgenic mice. Neuron 45:675–688
Bitan G, Kirkitadze MD, Lomakin A, Vollers SS, Benedek GB, Teplow DB (2003) Amyloid β-protein (Aβ) assembly: Aβ40 and Aβ42 oligomerize through distinct pathways. Proc Natl Acad Sci USA 100:330–335
Bucciantini M, Giannoni E, Chiti F, Baroni F, Formigli L, Zurdo J, Taddei N, Ramponi G, Dobson CM, Stefani M (2002) Inherent toxicity of aggregates implies a common mechanism for protein misfolding diseases. Nature 416:507–511
Caughey B, Lansbury PT (2003) Protofibrils, pores, fibrils, and neurodegeneration: separating the responsible protein aggregates from the innocent bystanders. Annu Rev Neurosci 26:267–298
Chen F, David D, Ferrari A, Gotz J (2004) Posttranslational modifications of tau—role in human tauopathies and modeling in transgenic animals. Curr Drug Targets 5:503–515
Chiti F, Dobson CM (2006) Protein misfolding, functional amyloid, and human disease. Annu Rev Biochem 75:333–366
Conway KA, Lee SJ, Rochet JC, Ding TT, Harper JD, Williamson RE, Lansbury PT Jr (2000a) Accelerated oligomerization by Parkinson's disease linked α-synuclein mutants. Ann NY Acad Sci 920:42–45
Conway KA, Lee SJ, Rochet JC, Ding TT, Williamson RE, Lansbury PT Jr (2000b) Acceleration of oligomerization, not fibrillization, is a shared property of both α-synuclein mutations linked to early-onset Parkinson's disease: implications for pathogenesis and therapy. Proc Natl Acad Sci USA 97:571–576
Cuajungco MP, Goldstein LE, Nunomura A, Smith MA, Lim JT, Atwood CS, Huang X, Farrag YW, Perry G, Bush AI (2000) Evidence that the β-amyloid plaques of Alzheimer's disease represent the redox-silencing and entombment of Aβ by zinc. J Biol Chem 275:19439–19442
Demager PP, Penke B, Walter R, Harkany T, Hartignny W (2002) Pathological peptide folding in Alzheimer's disease and other conformational disorders. Curr Med Chem 9:1763–1780
Dev KK, Hofele K, Barbieri S, Buchman VL, Van Der Putten H (2003) Part II: α-Synuclein and its molecular pathophysiological role in neurodegenerative disease. Neuropharmacology 45:14–44
Di Monte DA, Lavasani M, Manning-Bog AB (2002) Environmental factors in Parkinson's disease. Neurotoxicology 23:487–502
Ding TT, Lee SJ, Rochet JC, Lansbury PT Jr (2002) Annular α-synuclein protofibrils are produced when spherical protofibrils are incubated in solution or bound to brain-derived membranes. Biochemistry 41:10209–10217
Dobson CM (1999) Protein misfolding, evolution and disease. Trends Biochem Sci 24:329–332
Eghiaian F (2005) Structuring the puzzle of prion propagation. Curr Opin Struct Biol 15:724–730
Farrer M, Kachergus J, Forno L, Lincoln S, Wang DS, Hulihan M, Maraganore D, Gwinn-Hardy K, Wszolek Z, Dickson D, Langston JW (2004) Comparison of kindreds with parkinsonism and α-synuclein genomic multiplications. Ann Neurol 55:174–179
Ferreira ST, Vieira MN, De Felice FG (2007) Soluble protein oligomers as emerging toxins in Alzheimer's and other amyloid diseases. IUBMB Life 59:332–345
Ferrone F (1999) Analysis of protein aggregation kinetics. Methods Enzymol 309:256–274
Ferrone FA, Hofrichter J, Sunshine HR, Eaton WA (1980) Kinetic studies on photolysis-induced gelation of sickle cell hemoglobin suggest a new mechanism. Biophys J 32:361–380
Fink AL (2006) The aggregation and fibrillation of α-synuclein. Acc Chem Res 39:628–634
Glabe CG (2008) Structural classification of toxic amyloid oligomers. J Biol Chem 283:29639–29643
Goldsbury C, Frey P, Olivieri V, Aebi U, Muller SA (2005) Multiple assembly pathways underlie amyloid-β fibril polymorphisms. J Mol Biol 352:282–298
Gosal WS, Morten IJ, Hewitt EW, Smith DA, Thomson NH, Radford SE (2005) Competing pathways determine fibril morphology in the self-assembly of β2-microglobulin into amyloid. J Mol Biol 351:850–864

Griffith JS (1967) Self-replication and scrapie. Nature 215:1043–1044
Haass C, Selkoe DJ (2007) Soluble protein oligomers in neurodegeneration: lessons from the Alzheimer's amyloid β-peptide. Nat Rev Mol Cell Biol 8:101–112
Hong DP, Fink AL, Uversky VN (2008) Structural characteristics of α-synuclein oligomers stabilized by the flavonoid baicalein. J Mol Biol 383:214–223
Hong DP, Fink AL, Uversky VN (2009) Smoking and Parkinson's disease: does nicotine affect α-synuclein fibrillation? Biochim Biophys Acta 1794:282–290
Hong DP, Han S, Fink AL, Uversky VN (2010) Characterization of the non-fibrillar α-synuclein oligomers. Prot Pept Lett 18:230–240
Hoshi M, Sato M, Matsumoto S, Noguchi A, Yasutake K, Yoshida N, Sato K (2003) Spherical aggregates of β-amyloid (amylospheroid) show high neurotoxicity and activate tau protein kinase I/glycogen synthase kinase-3β. Proc Natl Acad Sci USA 100:6370–6375
Hoyer W, Cherny D, Subramaniam V, Jovin TM (2004) Rapid self-assembly of α-synuclein observed by *in situ* atomic force microscopy. J Mol Biol 340:127–139
Huang X, Atwood CS, Hartshorn MA, Multhaup G, Goldstein LE, Scarpa RC, Cuajungco MP, Gray DN, Lim J, Moir RD, Tanzi RE, Bush AI (1999a) The Aβ peptide of Alzheimer's disease directly produces hydrogen peroxide through metal ion reduction. Biochemistry 38:7609–7616
Huang X, Cuajungco MP, Atwood CS, Hartshorn MA, Tyndall JD, Hanson GR, Stokes KC, Leopold M, Multhaup G, Goldstein LE, Scarpa RC, Saunders AJ, Lim J, Moir RD, Glabe C, Bowden EF, Masters CL, Fairlie DP, Tanzi RE, Bush AI (1999b) Cu(II) potentiation of Alzheimer Aβ neurotoxicity. Correlation with cell-free hydrogen peroxide production and metal reduction. J Biol Chem 274:37111–37116
Jain S, Udgaonkar JB (2008) Evidence for stepwise formation of amyloid fibrils by the mouse prion protein. J Mol Biol 382:1228–1241
Jo E, Mclaurin J, Yip CM, St George-Hyslop P, Fraser PE (2000) α-Synuclein membrane interactions and lipid specificity. J Biol Chem 275:34328–34334
Kayed R, Head E, Thompson JL, Mcintire TM, Milton SC, Cotman CW, Glabe CG (2003) Common structure of soluble amyloid oligomers implies common mechanism of pathogenesis. Science 300:486–489
Kelly JW (1998) The alternative conformations of amyloidogenic proteins and their multi-step assembly pathways. Curr Opin Struct Biol 8:101–106
Klein WL, Krafft GA, Finch CE (2001) Targeting small Aβ oligomers: the solution to an Alzheimer's disease conundrum? Trends Neurosci 24:219–224
Kodali R, Wetzel R (2007) Polymorphism in the intermediates and products of amyloid assembly. Curr Opin Struct Biol 17:48–57
Kovacech B, Novak M (2010) Tau truncation is a productive posttranslational modification of neurofibrillary degeneration in Alzheimer's disease. Curr Alzheimer Res 7:708–716
Kuo YM, Emmerling MR, Vigo-Pelfrey C, Kasunic TC, Kirkpatrick JB, Murdoch GH, Ball MJ, Roher AE (1996) Water-soluble Aβ (N-40, N-42) oligomers in normal and Alzheimer disease brains. J Biol Chem 271:4077–4081
Laferla FM, Oddo S (2005) Alzheimer's disease: Aβ, tau and synaptic dysfunction. Trends Mol Med 11:170–176
Laferla FM, Green KN, Oddo S (2007) Intracellular amyloid-β in Alzheimer's disease. Nat Rev Neurosci 8:499–509
Lambert MP, Barlow AK, Chromy BA, Edwards C, Freed R, Liosatos M, Morgan TE, Rozovsky I, Trommer B, Viola KL, Wals P, Zhang C, Finch CE, Krafft GA, Klein WL (1998) Diffusible, nonfibrillar ligands derived from Aβ1–42 are potent central nervous system neurotoxins. Proc Natl Acad Sci USA 95:6448–6453
Lashuel HA, Lansbury PT Jr (2006) Are amyloid diseases caused by protein aggregates that mimic bacterial pore-forming toxins? Q Rev Biophys 39:167–201
Lashuel HA, Hartley D, Petre BM, Walz T, Lansbury PT Jr (2002a) Neurodegenerative disease: amyloid pores from pathogenic mutations. Nature 418:291

Lashuel HA, Petre BM, Wall J, Simon M, Nowak RJ, Walz T, Lansbury PT Jr (2002b) α-Synuclein, especially the Parkinson's disease-associated mutants, forms pore-like annular and tubular protofibrils. J Mol Biol 322:1089–1102

Lee HJ, Choi C, Lee SJ (2002) Membrane-bound α-synuclein has a high aggregation propensity and the ability to seed the aggregation of the cytosolic form. J Biol Chem 277:671–678

Lesné S, Koh MT, Kotilinek L, Kayed R, Glabe CG, Yang A, Gallagher M, Ashe KH (2006) A specific amyloid-β protein assembly in the brain impairs memory. Nature 440:352–357

Lichtenthaler SF (2010) α-Secretase in Alzheimer's disease: molecular identity, regulation and therapeutic potential. J Neurochem 116:10–21

Lomakin A, Chung DS, Benedek GB, Kirschner DA, Teplow DB (1996) On the nucleation and growth of amyloid β-protein fibrils: detection of nuclei and quantitation of rate constants. Proc Natl Acad Sci USA 93:1125–1129

Lowe R, Pountney DL, Jensen PH, Gai WP, Voelcker NH (2004) Calcium(II) selectively induces α-synuclein annular oligomers *via* interaction with the *C*-terminal domain. Protein Sci 13:3245–3252

Macario AJ, Conway De Macario E (2005) Sick chaperones, cellular stress, and disease. N Engl J Med 353:1489–1501

Maeda S, Sahara N, Saito Y, Murayama S, Ikai A, Takashima A (2006) Increased levels of granular tau oligomers: an early sign of brain aging and Alzheimer's disease. Neurosci Res 54:197–201

Martinez Z, Zhu M, Han S, Fink AL (2007) GM1 specifically interacts with α-synuclein and inhibits fibrillation. Biochemistry 46:1868–1877

Mccormack AL, Thiruchelvam M, Manning-Bog AB, Thiffault C, Langston JW, Cory-Slechta DA, Di Monte DA (2002) Environmental risk factors and Parkinson's disease: selective degeneration of nigral dopaminergic neurons caused by the herbicide paraquat. Neurobiol Dis 10:119–127

Medeiros R, Baglietto-Vargas D, Laferla FM (2010) The role of tau in Alzheimer's disease and related disorders. CNS Neurosci Ther. doi:10.1111/j.1755-5949.2010.00177.x

Meng X, Munishkina LA, Fink AL, Uversky VN (2009) Molecular mechanisms underlying the flavonoid-induced inhibition of α-synuclein fibrillation. Biochemistry 48:8206–8224

Merlini G, Bellotti V (2003) Molecular mechanisms of amyloidosis. N Engl J Med 349:583–596

Morris AM, Watzky MA, Finke RG (2009) Protein aggregation kinetics, mechanism, and curve-fitting: a review of the literature. Biochim Biophys Acta 1794:375–397

Muchowski PJ, Wacker JL (2005) Modulation of neurodegeneration by molecular chaperones. Nat Rev Neurosci 6:11–22

Mucke L, Masliah E, Yu GQ, Mallory M, Rockenstein EM, Tatsuno G, Hu K, Kholodenko D, Johnson-Wood K, Mcconlogue L (2000) High-level neuronal expression of Aβ1–42 in wild-type human amyloid protein precursor transgenic mice: synaptotoxicity without plaque formation. J Neurosci 20:4050–4058

Munishkina LA, Phelan C, Uversky VN, Fink AL (2003) Conformational behavior and aggregation of α-synuclein in organic solvents: modeling the effects of membranes. Biochemistry 42:2720–2730

Nielsen L, Khurana R, Coats A, Frokjaer S, Brange J, Vyas S, Uversky VN, Fink AL (2001) Effect of environmental factors on the kinetics of insulin fibril formation: elucidation of the molecular mechanism. Biochemistry 40:6036–6046

Nimmrich V, Grimm C, Draguhn A, Barghorn S, Lehmann A, Schoemaker H, Hillen H, Gross G, Ebert U, Bruehl C (2008) Amyloid β oligomers ($A\beta_{1-42}$ globulomer) suppress spontaneous synaptic activity by inhibition of P/Q-type calcium currents. J Neurosci 28:788–797

Nixon RA (2006) Autophagy in neurodegenerative disease: friend, foe or turncoat? Trends Neurosci 29:528–535

Ohhashi Y, Ito K, Toyama BH, Weissman JS, Tanaka M (2010) Differences in prion strain conformations result from non-native interactions in a nucleus. Nat Chem Biol 6:225–230

Ono K, Condron MM, Teplow DB (2009) Structure–neurotoxicity relationships of amyloid β-protein oligomers. Proc Natl Acad Sci USA 106:14745–14750

Opazo C, Huang X, Cherny RA, Moir RD, Roher AE, White AR, Cappai R, Masters CL, Tanzi RE, Inestrosa NC, Bush AI (2002) Metalloenzyme-like activity of Alzheimer's disease β-amyloid. Cu-dependent catalytic conversion of dopamine, cholesterol, and biological reducing agents to neurotoxic H_2O_2. J Biol Chem 277:40302–40308

Peterson DW, Zhou H, Dahlquist FW, Lew J (2008) A soluble oligomer of tau associated with fiber formation analyzed by NMR. Biochemistry 47:7393–7404

Petkova AT, Leapman RD, Guo Z, Yau WM, Mattson MP, Tycko R (2005) Self-propagating, molecular-level polymorphism in Alzheimer's β-amyloid fibrils. Science 307:262–265

Podesta A, Tiana G, Milani P, Manno M (2006) Early events in insulin fibrillization studied by time-lapse atomic force microscopy. Biophys J 90:589–597

Podlisny MB, Ostaszewski BL, Squazzo SL, Koo EH, Rydell RE, Teplow DB, Selkoe DJ (1995) Aggregation of secreted amyloid β-protein into sodium dodecyl sulfate-stable oligomers in cell culture. J Biol Chem 270:9564–9570

Powers ET, Morimoto RI, Dillin A, Kelly JW, Balch WE (2009) Biological and chemical approaches to diseases of proteostasis deficiency. Annu Rev Biochem 78:959–991

Prusiner SB (1998) Prions. Proc Natl Acad Sci USA 95:13363–13383

Qin Z, Hu D, Han S, Reaney SH, Di Monte DA, Fink AL (2007) Effect of 4-hydroxy-2-nonenal modification on α-synuclein aggregation. J Biol Chem 282:5862–5870

Rochet JC, Lansbury PT Jr (2000) Amyloid fibrillogenesis: themes and variations. Curr Opin Struct Biol 10:60–68

Roychaudhuri R, Yang M, Hoshi MM, Teplow DB (2009) Amyloid β-protein assembly and Alzheimer disease. J Biol Chem 284:4749–4753

Sacchettini JC, Kelly JW (2002) Therapeutic strategies for human amyloid diseases. Nat Rev Drug Discov 1:267–275

Sahara N, Maeda S, Murayama M, Suzuki T, Dohmae N, Yen SH, Takashima A (2007) Assembly of two distinct dimers and higher-order oligomers from full-length tau. Eur J Neurosci 25:3020–3029

Sakono M, Zako T (2010) Amyloid oligomers: formation and toxicity of Aβ oligomers. FEBS J 277:1348–1358

Santacruz K, Lewis J, Spires T, Paulson J, Kotilinek L, Ingelsson M, Guimaraes A, Deture M, Ramsden M, Mcgowan E, Forster C, Yue M, Orne J, Janus C, Mariash A, Kuskowski M, Hyman B, Hutton M, Ashe KH (2005) Tau suppression in a neurodegenerative mouse model improves memory function. Science 309:476–481

Saraiva MJ (2001) Transthyretin amyloidosis: a tale of weak interactions. FEBS Lett 498:201–203

Selkoe DJ (1986) Altered structural proteins in plaques and tangles: what do they tell us about the biology of Alzheimer's disease? Neurobiol Aging 7:425–432

Serpell LC, Sunde M, Blake CC (1997) The molecular basis of amyloidosis. Cell Mol Life Sci 53:871–887

Shikama Y, Kitazawa J, Yagihashi N, Uehara O, Murata Y, Yajima N, Wada R, Yagihashi S (2010) Localized amyloidosis at the site of repeated insulin injection in a diabetic patient. Intern Med 49:397–401

Silveira JR, Raymond GJ, Hughson AG, Race RE, Sim VL, Hayes SF, Caughey B (2005) The most infectious prion protein particles. Nature 437:257–261

Singleton AB, Farrer M, Johnson J, Singleton A, Hague S, Kachergus J, Hulihan M, Peuralinna T, Dutra A, Nussbaum R, Lincoln S, Crawley A, Hanson M, Maraganore D, Adler C, Cookson MR, Muenter M, Baptista M, Miller D, Blancato J, Hardy J, Gwinn-Hardy K (2003) α-Synuclein locus triplication causes Parkinson's disease. Science 302:841

Singleton A, Gwinn-Hardy K, Sharabi Y, Li ST, Holmes C, Dendi R, Hardy J, Crawley A, Goldstein DS (2004) Association between cardiac denervation and parkinsonism caused by α-synuclein gene triplication. Brain 127:768–772

Solomon A (1986) Clinical implications of monoclonal light chains. Semin Oncol 13:341–349

Souillac PO, Uversky VN, Millett IS, Khurana R, Doniach S, Fink AL (2002) Elucidation of the molecular mechanism during the early events in immunoglobulin light chain amyloid fibrillation. Evidence for an off-pathway oligomer at acidic pH. J Biol Chem 277:12666–12679

Souillac PO, Uversky VN, Fink AL (2003) Structural transformations of oligomeric intermediates in the fibrillation of the immunoglobulin light chain LEN. Biochemistry 42:8094–8104

Souza JM, Giasson BI, Chen Q, Lee VM, Ischiropoulos H (2000) Dityrosine cross-linking promotes formation of stable α-synuclein polymers. Implication of nitrative and oxidative stress in the pathogenesis of neurodegenerative synucleinopathies. J Biol Chem 275:18344–18349

Stevens PW, Raffen R, Hanson DK, Deng YL, Berrios-Hammond M, Westholm FA, Murphy C, Eulitz M, Wetzel R, Solomon A, Eulitz M, Wetzel R (1995) Recombinant immunoglobulin variable domains generated from synthetic genes provide a system for *in vitro* characterization of light-chain amyloid proteins. Protein Sci 4:421–432

Tan SY, Pepys MB (1994) Amyloidosis. Histopathology 25:403–414

Terry RD (1996) The pathogenesis of Alzheimer disease: an alternative to the amyloid hypothesis. J Neuropathol Exp Neurol 55:1023–1025

Urbanc B, Cruz L, Le R, Sanders J, Ashe KH, Duff K, Stanley HE, Irizarry MC, Hyman BT (2002) Neurotoxic effects of thioflavin S-positive amyloid deposits in transgenic mice and Alzheimer's disease. Proc Natl Acad Sci USA 99:13990–13995

Uversky VN (2003) A protein-chameleon: conformational plasticity of α-synuclein, a disordered protein involved in neurodegenerative disorders. J Biomol Struct Dyn 21:211–234

Uversky VN (2007) Neuropathology, biochemistry, and biophysics of α-synuclein aggregation. J Neurochem 103:17–37

Uversky VN (2008) α-Synuclein misfolding and neurodegenerative diseases. Curr Protein Pept Sci 9:507–540

Uversky VN, Eliezer D (2009) Biophysics of Parkinson's disease: structure and aggregation of α-synuclein. Curr Protein Pept Sci 10:483–499

Uversky VN, Fink AL (2004) Conformational constraints for amyloid fibrillation: the importance of being unfolded. Biochim Biophys Acta 1698:131–153

Uversky VN, Talapatra A, Gillespie JR, Fink AL (1999a) Protein deposits as the molecular basis of amyloidosis. I. Systemic amyloidoses. Med Sci Monitor 5:1001–1012

Uversky VN, Talapatra A, Gillespie JR, Fink AL (1999b) Protein deposits as the molecular basis of amyloidosis. II. Localized amyloidosis and neurodegenerative disorders. Med Sci Monitor 5:1001–1012

Uversky VN, Lee HJ, Li J, Fink AL, Lee SJ (2001a) Stabilization of partially folded conformation during α-synuclein oligomerization in both purified and cytosolic preparations. J Biol Chem 276:43495–43498

Uversky VN, Li J, Fink AL (2001b) Evidence for a partially folded intermediate in α-synuclein fibril formation. J Biol Chem 276:10737–10744

Uversky VN, Yamin G, Munishkina LA, Karymov MA, Millett IS, Doniach S, Lyubchenko YL, Fink AL (2005) Effects of nitration on the structure and aggregation of α-synuclein. Brain Res Mol Brain Res 134:84–102

Uversky VN, Fernandez A, Fink AL (2006) Structural and conformational prerequisites of amyloidogenesis. In: Uversky VN, Fink AL (eds) Protein misfolding, aggregation and conformational diseases. Springer Science+Business Media, LLC, New York

Verdone G, Corazza A, Viglino P, Pettirossi F, Giorgetti S, Mangione P, Andreola A, Stoppini M, Bellotti V, Esposito G (2002) The solution structure of human β2-microglobulin reveals the prodromes of its amyloid transition. Protein Sci 11:487–499

Vestergaard B, Groenning M, Roessle M, Kastrup JS, Van De Weert M, Flink JM, Frokjaer S, Gajhede M, Svergun DI (2007) A helical structural nucleus is the primary elongating unit of insulin amyloid fibrils. PLoS Biol 5:e134

Walsh DM, Tseng BP, Rydel RE, Podlisny MB, Selkoe DJ (2000) The oligomerization of amyloid β-protein begins intracellularly in cells derived from human brain. Biochemistry 39:10831–10839

Walsh DM, Klyubin I, Fadeeva JV, Cullen WK, Anwyl R, Wolfe MS, Rowan MJ, Selkoe DJ (2002a) Naturally secreted oligomers of amyloid β protein potently inhibit hippocampal long-term potentiation *in vivo*. Nature 416:535–539

Walsh DM, Klyubin I, Fadeeva JV, Rowan MJ, Selkoe DJ (2002b) Amyloid-β oligomers: their production, toxicity and therapeutic inhibition. Biochem Soc Trans 30:552–557

Wegner A, Savko P (1982) Fragmentation of actin filaments. Biochemistry 21:1909–1913
Westermark P, Benson MD, Buxbaum JN, Cohen AS, Frangione B, Ikeda S, Masters CL, Merlini G, Saraiva MJ, Sipe JD (2002) Amyloid fibril protein nomenclature—2002. Amyloid 9:197–200
Wetzel R, Shivaprasad S, Williams AD (2007) Plasticity of amyloid fibrils. Biochemistry 46:1–10
Yamin G, Uversky VN, Fink AL (2003) Nitration inhibits fibrillation of human α-synuclein *in vitro* by formation of soluble oligomers. FEBS Lett 542:147–152
Zhou W, Gallagher A, Hong DP, Long C, Fink AL, Uversky VN (2009) At low concentrations, 3,4-dihydroxyphenylacetic acid (DOPAC) binds non-covalently to α-synuclein and prevents its fibrillation. J Mol Biol 388:597–610
Zhou W, Long C, Reaney SH, Di Monte DA, Fink AL, Uversky VN (2010) Methionine oxidation stabilizes non-toxic oligomers of α-synuclein through strengthening the auto-inhibitory intra-molecular long-range interactions. Biochim Biophys Acta 1802:322–330
Zhu M, Li J, Fink AL (2003) The association of α-synuclein with membranes affects bilayer structure, stability, and fibril formation. J Biol Chem 278:40186–40197
Zhu M, Rajamani S, Kaylor J, Han S, Zhou F, Fink AL (2004) The flavonoid baicalein inhibits fibrillation of α-synuclein and disaggregates existing fibrils. J Biol Chem 279:26846–26857

Chapter 7
Cytotoxic Mechanisms of Islet Amyloid Polypeptide in the Pathogenesis of Type-2 Diabetes Mellitus (T2DM)

Theri Leica Degaki, Ph.D., Dahabada H.J. Lopes, Ph.D., and Mari Cleide Sogayar, Ph.D.

Abstract Diabetes mellitus (DM) is a chronic syndrome that occurs due to loss of the insulin-producing β-cells in the islets of Langerhans of the pancreas by a β-cell-specific autoimmune process, leading to insulin deficiency (type-1 DM, T1DM) or when insufficient amounts of insulin are produced by β-cells, or when resistance to the action of insulin occurs in different tissues (type-2 DM, T2DM). Several mechanisms may contribute to the progressive β-cell failure in T2DM, including loss of β-cell mass, β-cell exhaustion, and the cytotoxic effects of elevated glucose and lipid levels. Another hallmark of T2DM is the accumulation of β-cell-produced amylin, also called islet amyloid polypeptide (IAPP), which forms amyloid that is present in approximately 95% of T2DM patients.

IAPP is a 37-amino-acid peptide, which is co-synthesized, co-packaged within the Golgi apparatus, and co-expressed with insulin upon exposure to glucose stimulation (nutrient stimuli). IAPP aggregation induces β-cell death and decreases β-cell proliferation. Several mechanisms have been hypothesized to explain IAPP cytotoxicity. First, IAPP may induce apoptosis through an extrinsic pathway involving overexpression of Fas ligand and IL-1 in β-cells due to glucose toxicity; and an intrinsic pathway, causing endoplasmic reticulum stress due to accumulation of aggregated proteins. Second, IAPP interaction with the cell membrane may lead to membrane permeabilization by IAPP oligomers and cause formation of membrane pores, or the process of IAPP aggregation itself can cause membrane disruption.

T.L. Degaki, Ph.D. (✉) • M.C. Sogayar, Ph.D.
Chemistry Institute, Pos doc of NUCEL (Cell and Molecular Therapy Center), University of São Paulo, Av. Prof. Lineu Prestes, 748 Bld 9 room 964, 05508-000 São Paulo, SP, Brazil
e-mail: theride@iq.usp.br; mcsoga@iq.usp.br; http://www.usp.br/nucel/

D.H.J. Lopes, Ph.D.
Department of Neurology, David Geffen School of Medicine at UCLA, Neuroscience Research Building 1, Room 455, 635 Charles E. Young Drive South, Los Angeles, CA 90095-7334, USA
e-mail: dahabadalopes@ucla.edu

F. Rahimi and G. Bitan (eds.), *Non-fibrillar Amyloidogenic Protein Assemblies—Common Cytotoxins Underlying Degenerative Diseases*, DOI 10.1007/978-94-007-2774-8_7,

Development of new methods to decrease islet amyloid formation is important for prevention of β-cell mass loss and functional decline due to aggregated IAPP cytotoxicity. Reduction of amyloid formation may lead to effective treatment for T2DM and to a better understanding of the exact mechanism of IAPP cytotoxicity, which currently is debatable.

Keywords Amylin/IAPP • Amyloid • Type-2 diabetes mellitus • Cytotoxicity

Abbreviations

AP	Area Postrema
A1c	Glycated hemoglobin
C/EBP	CCAAT-Enhancer-Binding Protein
CHOP	C/EBP Homologous Protein
DISC	Death-Inducing Signaling Complex
DM	Diabetes Mellitus
T1DM	Type-1 Diabetes Mellitus
T2DM	Type-2 Diabetes Mellitus
ER	Endoplasmatic Reticulum
FDA	US Food and Drug Administration
GAGs	Glycosaminoglycans
IAPP	Islet Amyloid Polypeptide
hIAPP	Human IAPP
rIAPP	Rat IAPP
IDE	Insulin Degrading Enzyme
lPBN	Parabrachial Nucleus
JNK1	Jun *N*-terminal Kinase 1
MAPK	p38 Mitogen-Activated Protein Kinase
NTS	Nucleus of the Solitary Tract
TUNEL	Terminal deoxynucleotidyl transferase-mediated dUTP Nick-End Labeling
TXNIP	Thioredoxin-Interacting Protein
UPR	Unfolded-protein Response

7.1 Diabetes Mellitus

Diabetes mellitus (DM) is responsible for about 5% of all deaths occurring globally each year. Currently, more than 220 million people worldwide are victims of DM, and the prevalence of the disease is projected to increase by 2030 to ~366 million people according to the World Health Organization. Despite the very high numbers, many millions of people still remain undiagnosed and untreated (King et al. 1998; Wild et al. 2004).

Diabetes is a chronic syndrome which occurs as a result of loss of the insulin-producing β-cells in the islets of Langerhans in the pancreas by a β-cell-specific autoimmune process, leading to insulin deficiency (type-1 diabetes mellitus—T1DM); or when insufficient production of insulin occurs in β-cells, or resistance to the action of insulin occurs in different tissues of the body (type-2 diabetes mellitus—T2DM). Hyperglycemia is the main symptom present in both types of diabetic patients, which, together with other related disturbances in the body's metabolism, can lead to serious damage to many of the body's systems, especially the nerves and blood vessels, leading to a high risk of cardiovascular and metabolic diseases. T2DM is, by far, the most prevalent (~95%), while T1DM affects ~5% of the patients. T2DM is characterized as an acquired syndrome closely associated with obesity and elevated blood glucose levels due to progressive insulin resistance and β-cell exhaustion. Several mechanisms may contribute to the progressive β-cell failure in T2DM, including loss of β-cell mass, β-cell exhaustion, and the cytotoxic effects of elevated glucose and lipid levels (Marzban et al. 2003; Porte and Kahn 1991).

Genetic studies indicate that T2DM has a strong hereditary component. "Diabetes genes" may show only a subtle variation in the gene sequence, and these variations may be extremely common. The difficulty lies in linking such common gene variations, known as single nucleotide polymorphisms (SNPs), with an increased risk of developing diabetes (Dean and McEntyre 2004). With the ability to interrogate most alterations present in the genome, the number of genetic variants linked to T2DM has grown to 19 genes, many of which with multiple variants (Elbein 2009; Scott et al. 2007). These genes were associated with well-known biological processes of diabetes, namely β-cell function and insulin performance, as well as cell-cycle regulation. An unexpected finding was that one of the genes resided on the X chromosome, possibly indicating maternal inheritance (Voight et al. 2010).

Several risk factors for T2DM are related to peripheral insulin resistance, namely, obesity, acromegaly, Cushing's syndrome, and muscular dystrophies (Hansen et al. 1986; Boscaro et al. 2001; Chan et al. 1994; Perseghin et al. 2003), with diseases related to the pancreas (pancreatitis, cystic fibrosis, and pancreatic cancer), with hormonal syndromes, which interfere with insulin secretion (pheochromocytoma) (Resmini et al. 2009), and with long-term use of some drugs (glucocorticoids, estrogens, phenytoin) (DeFronzo 2004).

Development of T2DM only appears in those individuals who are genetically predisposed (Bonora et al. 1998). External factors, such as body weight, food intake, and physical activity seem to influence symptoms (Stridsberg and Wilander 1991), and together with the genetic component, lead T2DM to be considered a multifactorial disease. In addition to hyperglycemia, another hallmark of T2DM is accumulation of β-cells-produced human amylin/islet amyloid polypeptide (IAPP) in the amyloid form. It is important to note that amyloid plaques formed by IAPP are present in more than 95% of T2DM subjects (Lopes et al. 2004), while destruction of islet β-cells in T1DM removes the source of human IAPP (hIAPP). It has recently been found that inhibition of formation of aggregated and toxic forms of hIAPP by Zn^{2+}, which is found at millimolar concentrations in the secretory granules, provides a possible association between the deleterious mutation in the SLC30A8 Zn^{2+} transporter, which transports Zn^{2+} into the secretory granule, and T2DM (Brender et al. 2010).

Successful management of T2DM requires strict control of glycemia as well as other risk factors to prevent disease complications (UKPDS 1998). Despite the availability of insulin and of multiple classes of anti-diabetic drugs, the majority of the patients fail to attain or maintain tight glycemic control over time, raising their risk for serious complications (Fleming et al. 2001; Rossi et al. 2008; UKPDS 1998). New therapies are mandatory in order to improve diabetes outcome (Smith and Ashiya 2007). To face this challenge, new targets are currently recognized as playing a vital role in T2DM pathogenesis.

7.2 T2DM and Islet Amyloid

7.2.1 Historical Background

It has now been more than a century since Eugene L. Opie first described the presence of a hyaline-staining substance within the pancreatic islets in patients with hyperglycemia (Opie 1901), which later on, would be associated with the development of DM.

In 1869, Paul Langerhans was the first to describe the endocrine pancreas (pancreatic islets) and how these bundled cells appeared to be inserted into an ocean of acinar cells (exocrine pancreas). These findings led Laguesse, in 1893, to name these mysterious cells as the islands or islets of Langerhans (*iles de Langerhans*) in honor of his colleague. In 1889, Oskar Minkowski was the first to make the discovery that connected the pancreas to diabetes in his de-pancreatized dogs (Bliss 1982). In 1901, while at Johns Hopkins University, Eugene Opie supplied the missing link, by showing a pathological connection between diabetes and the hyaline degeneration within the islets of Langerhans. The amyloid nature of this hyaline material was established by Ahronheim in 1943 and later confirmed, using alkaline Congo Red staining, by Ehrlick and Ratner in 1961 (Ahronheim 1943; Ehrlich and Ratner 1961). In 1973, using electron microscopy, Westermark was able to demonstrate the fibrillar structure in the pancreatic amyloid (Westermark 1973).

Due to the extreme insolubility of the amyloid fibrils, it was not until 1987 that the peptide responsible for the deposits was identified. Two contemporary investigators (Westermark and Cooper), in separate laboratories, discovered that the hyaline-stained material consisted of the 37-amino-acid polypeptide referred to as islet amyloid polypeptide (IAPP) by Westermark and as amylin by Cooper (Westermark et al. 1987; Cooper et al. 1987). The two names have been used interchangeably in the literature since then. However, the pathophysiologic features that result in aberrant folding of the amyloid peptide to form fibrils in diabetes and other conditions remain remarkably elusive.

There is an urgent need to learn more about the relationship between IAPP-derived amyloid deposition, which results in alteration of islet structure and function, and the associated β-cell dysfunction that is consistently observed in T2DM.

7.2.2 Pancreatic Islet Amyloid and Diabetes

The term amyloidosis had been considered a differential diagnosis for those patients with unexplained organ failures, such as cardiac, hepatic, and renal failure. This dictum should be pondered since T2DM and Alzheimer's disease are two very common diseases incidence of which is exponentially growing as our society ages, and share common characteristics. Amyloid is the common thread interweaving the two diseases, being central to their origin and transforming histological changes (Gotz et al. 2009; Zhao and Townsend 2009; Gregg and Narayan 2000), rendering it, necessary to add, to the list of diabetic - opathies, the term "cognopathy".

Amyloidosis refers to a variety of conditions characterized by proteinaceous tissue deposits with common morphological, structural, and staining properties, but with variable protein composition. The amyloid deposits are literally defined as being a "starch-like" material, a definition that comes from the Greek root *amylo,* since these areas turned blue when iodine was applied to the tissue. However, this definition is a misnomer since amyloid is a proteinaceous extracellular deposit resulting from the polymerization of polypeptides, which undergo aggregation into cross-β-pleated sheets. Histologically, a characteristic feature of amyloid is its positive staining with Congo red and birefringence upon examination under polarized light. Transmission electron microscopy (TEM) reveals interlacing bundles of parallel arrays of fibrils with a diameter of 7–10 nm and X-ray diffraction shows that the adjacent amyloid fibrils are organized in a cross-β structure (Fig. 7.1).

Pancreatic islet amyloid deposits are restricted to the islets of Langerhans, which are the sites of insulin production in the body. Pancreatic islets (average diameter ranging from 150 to 300 μm) are distributed throughout the exocrine pancreas (weight 90–150 g in man) accounting for 3–5% of the total organ weight (Clark and Nilsson 2004).

Similar to cerebral amyloidosis in Alzheimer's disease, pancreatic amyloid deposits are small and difficult to identify *post mortem.* Unlike the invasive nature of some forms of systemic amyloidosis, pancreatic amyloid remains inside the confines of the islet and does not spread to the exocrine pancreatic tissue. These deposits cannot be identified easily *in vivo* and the islets do not increase in size as a result of the deposits. Rather, endocrine cells are replaced (Westermark and Wilander 1978; Clark et al. 1988; Rocken et al. 1992). Therefore, the toxic nature of islet amyloid deposits has important implications for maintenance of insulin reserves in diabetes. Indeed, islet amyloid deposition has been postulated to be an etiological

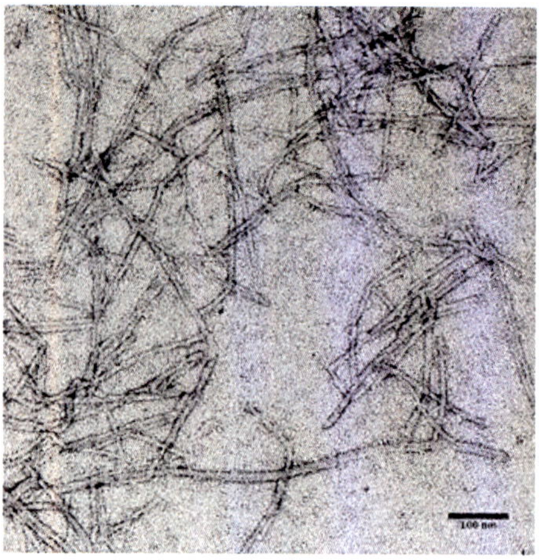

Fig. 7.1 Transmission electron micrograph (TEM) showing fibrils formed from synthetic hIAPP. Fibrils in the form of twisted protofilaments and ribbons formed when synthetic hIAPP was solubilized in 10 mM phosphate buffer, resulting in a final IAPP concentration of 10 μM (Lopes et al., unpublished data). Scale bar = 100 nm

factor for the loss of insulin-producing β-cells and inadequate insulin secretion which are characteristics of T2DM (Hoppener et al. 2002). However, the relationship of islet amyloid formation to the onset and progression of T2DM is largely unknown, and the formation of amyloid cannot be directly related to the pathophysiology of the disease.

Despite these difficulties, several lines of evidence suggest that islet amyloid deposition can be correlated with disease progression: Firstly, species barrier exists whereby in certain animals, such as rodents, IAPP does not form amyloid because of sequence alterations in critical residues (O'Brien et al. 1993). However, when a transgene for hIAPP is expressed in mice or rats, amyloid is deposited in pancreatic islets leading to loss of β-cell mass (Butler et al. 2004; Verchere et al. 1996; Howard 1986). Secondly, in experiments with non-human primates, the concentration of amyloid in the pancreas correlates with the severity of diabetes and β-cell function (Howard 1986). Thirdly, in humans, the rare S20G substitution in IAPP is associated with greater incidence or severity of T2DM in some populations (Seino 2001), which correlates with a greater propensity of this mutant form of IAPP to form amyloid (Ma et al. 2001).

Identification of the protein component of islet amyloid deposits as IAPP (Westermark et al. 1987; Cooper et al. 1987) was a breakthrough in islet amyloid research. This finding allowed measurement of the circulating peptide to be made in clinical observations (Sanke et al. 1991), production of synthetic analogues for commercial exploitation (Butler et al. 1990), and further investigations have been carried out using genetically modified animal models (de Koning et al. 1993a, b, 1994a, b; Janson et al. 1996; D'Alessio et al. 1994; Fox et al. 1993; Howard 1986; O'Brien et al. 1986) in order to determine the relationship of amyloidosis to the pathophysiology of diabetes. In addition, this small 37-amino-acid peptide has been extensively used in biophysical experiments *in vitro* to identify amyloidogenic folding processes and its potential action in different cellular systems.

7.3 IAPP

IAPP is the peptide component of the islet amyloid deposits found in the pancreas of cats and humans, and in insulinomas (Cooper et al. 1987; Westermark et al. 1987). IAPP is structurally related to calcitonin-gene-related peptide (CGRP) and shares more than 40% sequence homology with CGRP (Martinez-Alvarez et al. 2008; Cooper 1994). The gene encoding IAPP consists of three exons and is located on the short arm of chromosome 12. Chromosomes 11 (containing the CGRP family of genes) and 12 are thought to share some ancestral homologies. These similarities are also demonstrated in some of the proposed actions of IAPP, such as calcium metabolism (Gilbey et al. 1991), and in shared receptor binding with the CGRP family of peptides (Barth et al. 2004). IAPP has been identified in all the species in which it has been examined (Jaikaran and Clark 2001). The primary structure of IAPP is well-conserved among species (Christmanson et al. 1993) (Fig. 7.2) although some important species-specific substitutions have been implicated in the IAPP aggregation process, as discussed in the next topic.

IAPP is co-synthesized and co-packaged within the Golgi apparatus of β-cells (Clark 1989; Lukinius et al. 1989), and is co-expressed, along with insulin, by β-cells, upon their exposure to glucose stimulation (nutrient stimuli) (Butler et al. 1990; Sanke et al. 1991; Kahn et al. 1998). In normally functioning β-cells, the 37-residue, *C*-terminally amidated IAPP is derived from a 69-residue precursor polypeptide called proIAPP, which is endoproteolytically processed at the *C*-terminus of dibasic residue pairs K12 and R13 and K52 and R53, predominantly by the prohormone convertases PC 2 and PC 1/3, respectively (Fig. 7.3). Both enzymes are believed to be capable of cleaving other sites as well as a third site at the pair K60

IAPP	1							8												20									29								37
HUMAN	K	C	N	T	A	T	C	A	T	Q	R	L	A	N	F	L	V	H	S	S	N	N	F	G	A	I	L	S	S	T	N	V	G	S	N	T	Y
MONKEY	-	-	-	-	-	-	-	-	-	-	-	-	-	-	-	-	-	R	-	-	-	-	-	-	T	-	-	-	-	-	-	-	-	-	D	-	-
CAT	-	-	-	-	-	-	-	-	-	-	-	-	-	-	-	-	I	R	-	-	-	-	L	-	-	-	-	-	P	-	-	-	-	-	-	-	-
DOG	-	-	-	-	-	-	-	-	-	-	-	-	-	-	-	-	-	R	T	-	-	-	L	-	-	-	-	-	P	-	-	-	-	-	-	-	-
RAT	-	-	-	-	-	-	-	-	-	-	-	-	-	-	-	-	-	R	-	-	-	-	L	-	P	V	-	P	P	-	-	-	-	-	-	-	-
MOUSE	-	-	-	-	-	-	-	-	-	-	-	-	-	-	-	-	-	R	-	-	-	-	L	-	P	V	-	P	P	-	-	-	-	-	-	-	-
HAMSTER	-	-	-	-	-	-	-	-	-	-	-	-	-	-	-	-	-	-	-	N	-	L	-	P	V	-	P	-	-	-	-	-	-	-	-	-	-
GUINEA PIG	-	-	-	-	-	-	-	-	-	-	-	I	-	-	-	-	-	R	-	-	H		L		A		L	P		D					-	-	-
DEGU	-	-	-	-	-	-	-	-	-	-	-	T	-	-	-	-	-	R	-	-	H	-	L	-	A	-	P	P	-	K	-	-	-	-	-	-	-
RABBIT	-	-	-	-	V	-	-	-	-	-	-	-	-	-	-	-	I	-	-	-	-	-	-	-	-	-	F	S	P	P	S	-	-	-	-	S	-
HARE									-	-	-	-	-	-	-	-	I	-	-	-	-	-	-	-	-	-	F	-	P	P	-						
COUGAR	-	*	-	-	-	-	*	-	-	-	-	-	-	-	-	-	I	R	S	S	*	-	L	-	-	-	-	S									
PIG	-	-	-	M	-	-	-	-	-	-	H	-	-	-	-	-	D	R	-	R	-	N	L	-	T	-	F	-	P	-	K	-	-	-	-	-	-
CATTLE	-	-	G	-	-	-	-	E	-	-	-	-	-	-	-	-	A	P	-	-	-	K	L	-	-	-	F	-	P	-	K	M	-	-	-	-	-
BABOON	I	-	-	-	-	-	-	-	-	-	-	-	-	-	-	-	-	R	-	-	-	-	-	-	T	-	-	-	-	-	-	-	-	-	-	-	-
FISH	-	-	-	-	-	-	-	V	-	-	-	-	-	D	-	-	-	R	-	-	-	T	I	-	T	V	Y	A	P	-	-	-	-	-	S	-	-
CHICKEN	-	-	-	-	-	-	-	V	-	-	-	-	-	D	-	-	-	R	-	-	S	-	I	-	-	-	Y	-	P	-	-	-	-	-	-	-	-
KANGAROO									-	-	-	-	-	D	-	-	-	R	-	N	N	N	M	-	-	-	F	-	P	-	-	-	-				
HEDGEHOG	R	-	-	-	-	-	-	-	-	-	-	-	V	-	-	-	S	R	-	-	-	-	L	-	-	-	-	-	P	-	D	-	-				

Fig. 7.2 Amino-acid sequences of IAPP in different species given in single-letter code. Amino acids common to the human sequence are indicated by (–). Asterisks (*) represent yet-undetermined amino acids. Only partial sequences are available for hare, cougar, kangaroo, and hedgehog

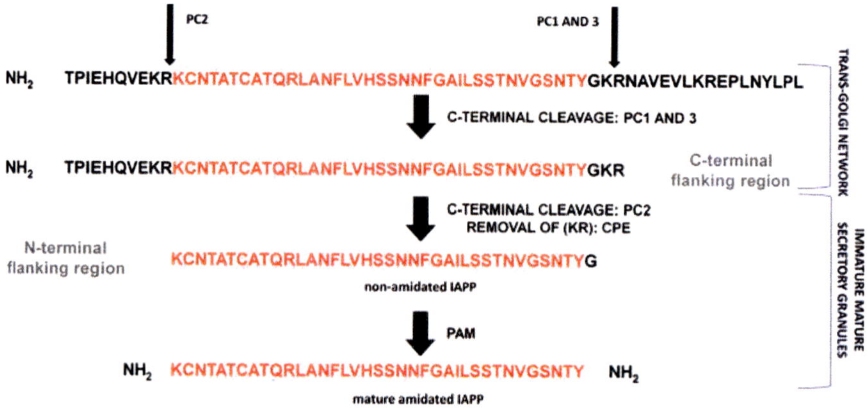

Fig. 7.3 Schematic representation of proIAPP processing in β-cells. ProIAPP processing is initiated by cleavage at its *C*-terminus preferentially by PC1 or PC3, followed by cleavage of the *N*-terminally unprocessed proIAPP intermediate by PC2 in immature and/or mature secretory granules. After cleavage by PC1/3, the *C*-terminal dibasic residues (KR) are removed by the action of CPE. This step is essential for removal of Gly49 and amidation of IAPP at the *C*-terminus by the PAM complex (Marzban et al. 2005a)

and R61, albeit at lower efficiencies (Marzban et al. 2004; Higham et al. 2000). Residue R53 is further processed by carboxypeptidase E (CPE), which removes the dibasic PC 1/3 recognition motif (KR) (Marzban et al. 2005a). The resulting peptide, containing an extra *C*-terminal glycine residue (G38) is referred to as IAPP free acid. The peptidyl amidating monooxygenase (PAM) complex ablates the CH_2COOH substructure of the *C*-terminal G38 as a glyoxylate in the last processing step, affording IAPP, which has an amide functional group at the new *C*-terminus (Y37). The proteolytic modifications are performed by the same enzymes responsible for proinsulin processing and are believed to occur in the acidified, immature secretory granules (Orci et al. 1994). In addition to these proteolytic modifications, native IAPP also has a disulfide bond between C2 and C7 residues, presumably formed under the oxidizing environment of the endoplasmic reticulum, possibly *via* an enzyme-mediated reaction carried through in the oxidized state throughout the aforementioned processing steps (Sanke et al. 1988; Badman et al. 1996; Higham et al. 2000; Marzban et al. 2005b). The soluble form of IAPP is normally found in the circulation at 3–5 pmol/L in humans, at a ratio of IAPP: insulin ≈1:100 (Butler et al. 1990; Ogawa et al. 1990). IAPP excretion is predominantly renal (Kautzky-Willer et al. 1994). Elevations of plasma glucose lead to an immediate rise in endogenous plasma IAPP levels to about 15–20 pmol/L (Lutz 2010) exhibiting a short half-life of about 15 min in blood circulation (Young et al. 1993).

Since the discovery of IAPP, a multitude of physiological functions have been associated with the soluble form of this hormone, but it is important to note that most of these studies are based on responses to exogenous, pharmacological doses of synthetic IAPP, rather than to endogenous levels (Gebre-Medhin et al. 2000). The

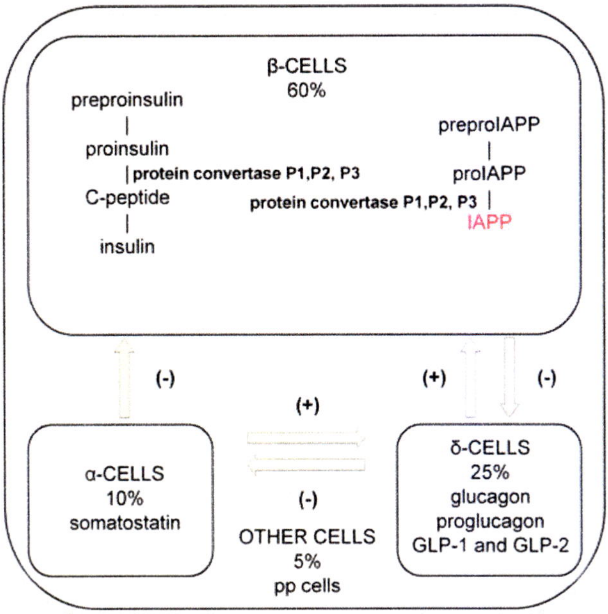

Fig. 7.4 Schematic representation of interrelationships of the different types of islet cells (Hayden and Tyagi 2001)

only notable feature in an IAPP-knockout mouse was moderately increased insulin secretion in response to glucose (Gebre-Medhin et al. 1998). In addition, islet amyloid deposits do not form in type-1 diabetic subjects, in whom β-cells (and, therefore, insulin and IAPP) are absent and these subjects do not have pathophysiologic features that could be attributed directly to IAPP deficiency. Unlike many other forms of amyloidosis, islet amyloid formation in T2DM cannot be directly related to post-translational modification of IAPP or gene mutations that would confer increased amyloidogenicity to the peptide.

Although IAPP remains a peptide without clearly identified functions *in vivo* (Gebre-Medhin et al. 2000), its putative physiological role is believed to be control of hyperglycemia. IAPP has been considered as the third hormone, together with insulin and glucagon, controlling glucose homeostasis (Fig. 7.4) (Cooper et al. 1987; Colin et al. 2008; Lorenzo et al. 1994). It also potentially inhibits gastric emptying, thus being important in controlling and delaying the rate of meal-derived glucose in blood circulation. Experimental evidence clearly indicates that this action is triggered by IAPP receptors in the *area postrema* (AP), with subsequent activation involving the solitary tract (NTS), parabrachial nucleus (lPBN), and other brain areas (Lutz et al. 1998c).

Chronic IAPP treatment was shown to decrease eating and body-weight gain in both experimental animals and humans. Administration of exogenous IAPP has an acute onset of action, decreasing eating in rats within a few minutes (Lutz et al. 1995), whereas administration of IAPP antagonists increases eating and causes weight gain in rats (Mollet et al. 2004; Reidelberger et al. 2004). These and other data suggest that IAPP also may function as an adiposity signal (Lutz 2006; Lutz et al. 1998c; Mollet et al. 2004; Riediger et al. 2001, 2004). In addition, IAPP has been shown to inhibit hepatic release and production of glucose in the postprandial period and inhibit glucagon and somatostatin secretion (Hayden and Tyagi 2001).

IAPP levels are elevated in T2DM patients, in insulin-resistant obese patients, and in patients with impaired glucose tolerance (Ludvik et al. 1997). In addition to producing satiety, IAPP also increases thirst, which indicates it has an action within the central nervous system (Lutz et al. 1998a, b; Arnelo et al. 1996; Riediger et al. 1999). IAPP has been shown to have binding sites within the renal cortex in the area of the juxtaglomerular apparatus and to activate the rennin–angiotensin–aldosterone system (Wookey et al. 1996; Cooper et al. 1995). Recent reports also indicate that IAPP exerts trophic effects in a variety of tissues and organs. It has been shown that IAPP influences the development of the kidneys, bones, and pancreas (Lutz et al. 1998c). Lutz also showed that IAPP might exert trophic effects supporting normal development of brainstem neuronal pathways (Lutz et al. 1998c), specifically for projections from the AP to the NTS. This phenomenon may be comparable to the effect of leptin in the hypothalamus, which has been extensively investigated by Bouret and colleagues (Bouret et al. 2004, 2008; Bouret 2010).

It is noteworthy that pancreatic β-cells are the major source of circulating IAPP. Changes in circulating IAPP levels are thought to reflect mainly changes in β-cell secretion, and it is generally believed that these fluctuations are the physiological basis for IAPP's effect on eating and energy homeostasis (Young 2005). Whether centrally synthesized IAPP also contributes to this control is still a matter of debate. In fact, it is not entirely clear whether IAPP synthesis occurs in the mammalian brain at all. It is important to note that most studies have been performed in male rats. A recent study in females suggested that central IAPP production in female rats may contribute to the control of maternal regulations because IAPP was specifically up-regulated in the preoptic area of the hypothalamus in the early postpartum period (Dobolyi 2009).

7.4 Therapeutic Use of IAPP

Despite the debate over hIAPP's role in the pathogenesis of T2DM, IAPP plays a clear function as a hormone in normal human physiology, being co-secreted with insulin by the pancreatic islets β-cells and stored, along with insulin, in the secretory granules (Lukinius et al. 1989). The release of both hormones occurs in response to nutrient stimuli and in the fasting state, with hIAPP levels being 10–15% those of insulin (Hanabusa et al. 1992), thus complementing postprandial glucose control. hIAPP levels are virtually zero in patients with type-1 diabetes, while T2DM patients display a relative hIAPP deficiency, when compared with normal subjects (Koda et al. 1992). It has been proposed that replacement of IAPP, together with insulin, in type-1 and in already insulin-dependent type-2 diabetes patients, could promote beneficial clinical effects (Scherbaum 1998). A synthetic analog of hIAPP, denominated pramlintide (Symilin®), was approved in April 2004 by the US Food and Drug Administration (FDA) to be administered together with mealtime insulin therapy to patients with type-1 and insulin-dependent type-2 diabetes. Pramlintide was generated by substitution of the hIAPP residues Ala-25, Ser-28, and Ser-29 for three proline residues, resulting in a soluble and stable synthetic non-amyloidogenic analog, which is suitable for pharmacological use (Kolterman et al. 1995). Studies have revealed that pramlintide decreased postprandial glucagon release and slowed gastric emptying in patients with T1DM and T2DM (Kong et al. 1997, 1998; Fineman et al. 2002a, b) and that upon administration with insulin, pramlintide reduced postprandial glucose levels without increasing insulin levels (Thompson et al. 1997a, b). Joint insulin and pramlintide therapy leads to reduction of average glycated hemoglobin (A1c) and of postprandial glucose levels, and potentially reduces the total daily administered insulin dose, with possible weight loss, increasing treatment safety with minimal contraindications (Ryan et al. 2009).

7.5 Structural Properties of IAPP

A major problem found in studies involving a range of amyloidogenic proteins, including IAPP, is the significant differences in assembly kinetics observed using synthetic peptides from different manufacturers or even using different lots from the same manufacturer (Padrick and Miranker 2002). This irreproducibility likely results from the presence of pre-existing aggregates in the peptide. Such aggregates serve as seeds for fibril formation making it very difficult to define a soluble 'native state' of these peptides. Furthermore, *in vitro* observations are confounded by the choice of methods for peptide preparation, the pH, the use of concentrated stock solutions, which may contain seeds, and the vast range of concentrations and solvents used in different laboratories. As an example, in the presence of helix-inducing organic solvents, such as 1,1,1,3,3,3-hexafluoroisopropanol (HFIP) or 2,2,2-trifluoroethanol (TFE), the circular dichroism (CD) spectra of hIAPP suggested an initial α-helical structure (Hubbard et al. 1991; McLean and Balasubramaniam 1992). These solvents, together with DMSO, have been used extensively to examine the conformation of IAPP but do not allow extrapolation to *in vivo* conditions, in which the three-dimensional structure will differ from those studied in non-aqueous solvents.

Even though the native structure of IAPP is unknown, more recent studies performed in aqueous buffers, pointed to an equilibrium between an unstructured form and a helical population (Mishra et al. 2009a; Wiltzius et al. 2009). Several factors have been speculated to alter this equilibrium, inducing IAPP conformational transitions and culminating in fibril formation, such as pH, concentration, and balance between synthesis, degradation, and interaction with membranes (Engel et al. 2008, 2006; Jeworrek et al. 2009; Khemtemourian et al. 2010; Radovan et al.

2009; Smith et al. 2009; Sparr et al. 2004). In spite of the biological importance of IAPP, details of the nature of the interactions involved in fibril formation, its crystal structure, and the state in which hIAPP is found inside cells and in the circulation are still unknown.

Based on current experimental data, a number of speculations have been made about the structure, spatial arrangement, and importance of motifs in the fibril formation process of IAPP. A number of studies (Jaikaran and Clark 2001; Azriel and Gazit 2001; Tenidis et al. 2000) support the idea of a helical intermediate spanning residue 8–18 and 22–27, just prior to IAPP conversion to a β-sheet conformation. This conversion is likely to be promoted by the condensation of the IAPP *N*-terminal helical domains, forming the contiguous β-sheet spanning residues 20–37 (Jarrett and Lansbury 1993), thereby bringing together a hydrophobic portion of IAPP monomers and forming the steric zipper spine of the fibril. The fibril core region was initially considered to comprise amino-acid residues 20–29 (Westermark et al. 1990; Ma et al. 2001). The importance of this sequence emerged from the observations of proline substitutions in this region, which were thought to be responsible for the lack of fibril formation in rats, mice, and other species (Betsholtz et al. 1990; Westermark et al. 1990; Moriarty and Raleigh 1999; Green et al. 2003). These findings were emphasized by the fact that peptide fragments including IAPP(23–27) (FGAIL) and IAPP(22–27) (NFGAIL) were found to be sufficient for the formation of amyloid-like structures *in vitro* (Glenner et al. 1998; Griffiths et al. 1995; Kayed et al. 1999; Goldsbury et al. 2000; Tenidis et al. 2000). However, with the use of peptide models that efficiently reproduced the conformational polymorphism displayed by the full-length IAPP, several sequences, corresponding to residues 8–20, 10–19, 13–18, 17–29, and 30–37 of hIAPP, have been identified as 'hot spots' for amyloid formation (Jaikaran et al. 2001; Scrocchi et al. 2003; Kajava et al. 2005; Mazzaglia et al. 2010; Gilead and Gazit 2008). These findings supported the ideas that the 20–29 region within IAPP could not be considered as the sole amyloidogenic core of the polypeptide and that IAPP had not one, but several amyloidogenic cores, which interact to form an organized fibrillar structure.

Nuclear magnetic resonance (NMR) analysis performed in phosphate buffer revealed that the *N*-terminus of IAPP preferentially populated backbone dihedral angles indicative of α-helical conformations, with no detectable β-strand structure (Yonemoto et al. 2008). Unambiguous long-range side-chain to side-chain nuclear Overhauser effect (NOE) resonances were not observed in soluble hIAPP, suggesting that the peptide does not adopt a unique 3D structure or folding. Amide proton chemical-shift deviations in hIAPP were similar to those found in the homologous rat IAPP (rIAPP), consistent with an *N*-terminus (residues 1–20) with a modest helical propensity and a *C*-terminus that is less structured (residues 21–37) (Williamson and Miranker 2007; Nanga et al. 2009). However, Cα proton chemical-shift deviations suggested that hIAPP had greater helical propensity than rIAPP across the entire peptide, markedly so in the *C*-terminal half of hIAPP. Hence, hIAPP seems to adopt a conformation similar to, but seemingly more structured than, that found in the rat homologue, or in IAPP's nearest paralogous relative, calcitonin (Breeze et al. 1991; Amodeo et al. 1994). This increased helical propensity of the *C*-terminus of hIAPP relative to rIAPP can be rationalized by the presence of three proline residues in the latter, which would act as helix breakers. The presence of these proline residues in the rat peptide is also thought to abrogate its amyloidogenicity (Betsholtz et al. 1998; Abedini and Raleigh 2006). It is possible that the higher amyloidogenicity of the human peptide necessitates the observed enhanced helicity in its *C*-terminus in order to protect it from aberrant aggregation along the secretory pathway.

NMR data also revealed that the disulfide bond at the *N*-terminus of hIAPP imparts an unusual structure to this region. Unlike almost all the other residues in IAPP, T4 and A5 show no amide–amide NOE. Instead, a T4–A8 NOE is observed. Strong sequential amide NOEs are found between residues T6–A8 and a strong NOE occurs between the amide protons of residues T4 and A5. Together, these data suggest that the disulfide bond orientates the amide bonds of residues 4 and 5 in a direction, which is opposite to the alignment established by the helix sampled by the remainder of the peptide and residues 6–8 are twisted to accommodate this change in direction. This unusual conformation likely results in a net destabilization of the helical conformation ensemble in this region.

Secondary structure predictions indicated that three potential β-strand regions are present in hIAPP (Jaikaran et al. 2001): A β-turn has been predicted at N31, which would result in two adjacent β-strands, comprising residues 32–37 and 24–29, creating an anti-parallel β-sheet, and a third β-strand was proposed to exist in the region 18–23. The turn of the predicted linker region (Wiltzius et al. 2009) centered at residue 20, would nucleate the second layer of β-sheet and initiate the elongation process, culminating in IAPP fibrillization. The residues D22, E24, I26, and L27 were identified to be important in this process (Westermark et al. 1990; Moriarty and Raleigh 1999). These residues would lie in the second strand region and localize to the outer edge of the β-sheet where they would be involved in interactions between protofilaments necessary for assembly of two or more protofilaments into fibrils. In rIAPP, the β-sheet would be disrupted by the proline substitutions in the region 24–29, which would prevent both intra- and intermolecular hydrogen bonding (Westermark et al. 1990; Moriarty and Raleigh 1999). The S20G substitution, found in a small number of T2DM Japanese, Chinese, and Mexican subjects (Garcia-Gonzalez et al. 2007; Seino 2001; Sakagashira et al. 1996; Lee et al. 2001), is thought to accelerate hIAPP fibril formation *in vitro* (Sakagashira et al. 2000; Ma et al. 2001). In the proposed model, the S20G substitution would result in a lower angle on the loop, due the smaller size of glycine, rendering this loop more flexible (Wiltzius et al. 2009). However, the mutation is relatively uncommon in T2DM and not all individuals with this mutation develop severe insulin-requiring diabetes, which would be associated with extensive amyloid deposition (Lee et al. 2001). Therefore, it is difficult to reconcile the biophysical data with diabetes pathophysiology.

The evolutionarily conserved disulfide loop between C2 and C7 has been shown to be essential for IAPP biological activity (Roberts et al. 1989) and is important for IAPP interaction with lipid membranes (Khemtemourian et al. 2010), but its role in the amyloid formation is still controversial. Although it has been shown that the disulfide bridge prevents the *N*-terminal region of IAPP from adopting a β-sheet-rich structure (characteristic of amyloid fibrils), the IAPP *N*-terminus is involved in amyloid formation in solution, albeit not as part of the core, altering two apparently distinct processes, namely the lag-phase activation and secondary nucleation (Koo and Miranker 2005).

7.6 Mechanisms of IAPP Fibril Formation

IAPP displays amyloidogenic properties in humans, primates, and cats (Nishi et al. 1989), while mice, rats, and dogs do not develop islet amyloid deposits (Gilead and Gazit 2008). The amyloidogenic potential of porcine IAPP is reduced (Potter et al. 2010). The most studied sequences are those of rodents and humans. These sequences differ in 6 out of 37 amino acids, 5 of which locate at a specific region between residues 20 and 29. The presence of three proline residues in the rodent sequence within this region does not favor β-sheet structures, suggesting that hIAPP(20–29) is responsible for amyloid aggregation (Betsholtz et al. 1989).

A rare IAPP mutant displays increased tendency to oligomerize, as is the case in the familial form of T2DM. The serine–glycine substitution at position 20 (S20G) in the IAPP molecule is associated with an earlier onset and a more severe form of the disease in a Japanese subpopulation (Sakagashira et al. 2000; Ma et al. 2001; Seino 2001) and Chinese people with T2DM (Lee et al. 2001). In the majority of the population with T2DM, no genetic determinant or aberrant IAPP sequence has been found (Nishi et al. 1990). In fact, it appears that rare mutations in the IAPP gene cannot explain amyloid formation in most T2DM patients, but, rather, may be more of a contributory factor for some patients (Marzban et al. 2003). The S20G mutation is associated with moderate alterations in insulin secretion in a small number of patients examined so far, therefore, its role in T2DM is unclear (Yamada et al. 1998). The Akita mouse model of diabetes has a mutation in the insulin gene leading to endoplasmatic reticulum (ER) aggregation of insulin, ER stress-induced β-cell apoptosis, and finally diabetes (Araki et al. 2003). Other mutant proteins designed to display an increased tendency to oligomerize are the Aβ gene product from familial Alzheimer's disease (Chartier-Harlin et al. 1991; Imai et al. 2001) and the gene products corresponding to the cystic-fibrosis gene (Knorre et al. 2002) and Huntington's disease (Kouroku et al. 2002). A mutation found in the IAPP promoter was associated with T2DM in a Spanish population, suggesting an important gene-expression regulation (Novials et al. 2001).

In acidic pH, hIAPP is largely in the soluble form. Under normal conditions, IAPP is secreted together with insulin from the acidic secretory granules (pH 5.5) to the neutral pH of the extracellular space. These changes in pH, together with increased accumulation of extracellular hIAPP in diabetes, may promote amyloid formation (Charge et al. 1995), though in some *in vivo* and *in vitro* studies no fibril formation was observed for IAPP (Jaikaran et al. 2001). Some studies pointed to the formation of heteromolecular complexes *in vitro* between IAPP and insulin, suggesting inhibition of IAPP fibril formation by insulin (Jaikaran et al. 2004). One widely accepted hypothesis is that the increased demand for insulin (insulin resistance) in T2DM increases production and secretion of IAPP, which might result in accumulation and aggregation of IAPP, exceeding the ER capacity to fold and traffic the protein. This notion is also consistent with the deficit in β-cell mass after partial pancreatectomy (Matveyenko et al. 2006).

Marchetti et al. evaluated some features of the ER function/dysfunction in human β-cells from T2DM organ donors (Marchetti et al. 2007) and confirmed that diabetic β-cells have functional and survival defects (decreased glucose-stimulated insulin release and increased apoptosis), as previously shown by the same group and others (Marchetti et al. 2004, 2007; Del Guerra et al. 2005; Deng et al. 2004). These defects may be associated with reduced insulin content, lower amounts of insulin granules, diminished islet insulin mRNA, and increased caspase activation (Pirot et al. 2007; Marchetti et al. 2004; Del Guerra et al. 2005). This study concluded that diabetic β-cells display a higher susceptibility to ER stress induced by metabolic perturbations, which would be consistent with a scenario in which T2DM β-cells face a condition of ER stress, which may be relatively compensated for, for a while, but is exacerbated in case of metabolic decompensation (i.e., high plasma glucose levels), thus contributing to β-cell damage, a predisposition that is absent in islets of non-diabetic individuals (Marchetti et al. 2007). ER stress is characteristic of β-cells in humans with T2DM, though interestingly, not in patients with T1DM (Huang et al. 2007b).

Another possible mechanism of islet amyloid formation in T2DM is impaired processing of proIAPP by islet β-cells, which may lead to hypersecretion of unprocessed proIAPP. The unprocessed polypeptide may have a higher propensity for aggregation compared to mature IAPP, since proIAPP is processed in parallel with proinsulin, by the same β-cell prohormone convertase enzymes (PC2 and PC3) and proinsulin production is impaired in T2DM (Porte and Kahn 1989; Kahn and Halban 1997). The *N*-terminal flanking region of proIAPP has been found in pancreatic islet amyloid deposits from T2DM patients, suggesting that proIAPP may be an important molecule in islet amyloid formation (Westermark et al. 1989). Some intracellular fibrils show immunoreactivity to the *N*-terminus of proIAPP and have been shown to be ubiquitinated, suggesting that proIAPP is located in a pre-Golgi part of the secretory pathway, possibly the ER (O'Brien et al. 1994). Other studies showed that prolonged exposure of human β-cells to high glucose concentrations increases the levels of proIAPP molecules containing the *N*-terminally extended form (Hou et al. 1999), and that prolonged exposure of β-cells to free fatty acids results in impaired processing of proinsulin due to decreased activation of PC2 and PC3 (Furukawa et al. 1999). The *N*-terminal flanking region of proIAPP contains a heparin-binding domain, which could bind to basement membrane heparan sulfate proteoglycans, another component of islet amyloid, forming a nidus or template for amyloid formation. Alternatively, glycosaminoglycans (GAGs) produced by β-cells could act as binding sites for IAPP (Park and Verchere 2001).

7.7 Roles of IAPP in Decreased β-Cell Mass and Function

Autopsy studies suggested that at death, ≤20–50% of β-cells have been lost after many years of T2DM (Westermark and Wilander 1978). A significant loss of these cells does not seem likely at early phases of clinical hyperglycemia (Porte and Kahn 2001). Decreased β-cell number occurs in individuals with amyloid deposition (Clark et al. 1988; Butler et al. 2003), suggesting that hIAPP deposition could be associated with decreased β-cell mass. This has been supported by *in vitro* studies showing that species which do not spontaneously develop T2DM, such as rats and mice, have an IAPP variant that is non-amyloidogenic (Westermark et al. 1990).

Genetic manipulations of mice and rats may induce diabetes. In addition, mice and rats, which are transgenic for hIAPP with sufficient expression levels, develop diabetes characterized by increased β-cell apoptosis and an islet phenotype, which is reminiscent of that in human T2DM patients (Butler et al. 2004). Moreover, hIAPP transgenic rats, similarly to T2DM patients, display increased rates of β-cell apoptosis and characteristics of the ER stress (Lin et al. 2007).

In addition to inducing β-cell death, IAPP aggregation may also reduce β-cell replication and function. β-cell deficiency in T2DM may result, in part, from failure to increase adaptively the β-cell mass due to increased vulnerability of replicating β-cells for apoptosis (Ritzel and Butler 2003). These data suggested that under conditions in which β-cell replication should be increased to compensate for hIAPP-induced β-cell loss, the newly replicating β-cells may be preferentially targeted by the same hIAPP-induced cytotoxic processes, leading to both increased β-cell death and decreased β-cell replication (Hull et al. 2004).

7.7.1 IAPP Cytotoxicity

Studies with hIAPP addressed its toxic effects in several primary cultures of human islet cells (Lorenzo et al. 1994; Janson et al. 1999), rat islet cells (Janson et al. 1999), and several different cell lines (Janciauskiene and Ahren 2000; Hiddinga and Eberhardt 1999; Kapurniotu et al. 1998; May et al. 1993; Saafi et al. 2001; Schubert et al. 1995; Zhang et al. 1999, Muthusamy et al. 2010a). Nonetheless, the exact mechanism of IAPP cytotoxicity and membrane disruption is far from understood. Various mechanisms have been hypothesized and are discussed below.

7.7.1.1 hIAPP Induces Apoptosis

Apoptosis is a morphologically stereotyped form of cell death with specific morphological characteristics, including cell shrinkage, nuclear condensation, chromatin margination, and clumping and blebbing of the cell surface, by which single cells are deleted by phagocytosis, without disturbance of tissue architecture or function and without initiating inflammation in living tissues (Bellamy et al. 1995). The extrinsic pathway of apoptosis is present as a mediator of β-cell glucose toxicity in both types 1 and 2 of diabetes, in which β-cells overexpress Fas ligand and IL1-β (Maedler et al. 2001, 2002) and in autoimmune-mediated β-cell death (Meier et al. 2005; Augstein et al. 1998). The intrinsic pathway of apoptosis is related to specific β-cell vulnerability as "professional" secretory cells known to be particularly sensitive to perturbations of the ER function and ER stress as a result of accumulation of aggregated proteins (Huang et al. 2007a, b). The first line of defense against ER-stress-induced apoptosis is the unfolded-protein response (UPR) (Harding and Ron 2002). UPR is an adaptive process, which takes place when there is increased burden of ER protein trafficking, including folding and post-translational modifications due to an increased rate of protein synthesis, which in diabetes constitutes a common mechanism for insulin production in response to insulin resistance (Huang et al. 2007a). When the UPR is unable to prevent and clear the ER of unfolded and aggregated proteins, ER stress may ensue (Haataja et al. 2008). ER stress includes the early and transient suppression of protein synthesis, activation of genes encoding components of the ER protein translocation, folding, secretion, and degradation machinery, and induction of programmed cell death. The failure of amyloidogenic protein trafficking through the ER induces formation of protein oligomers, which may cause membrane disruption. Moreover, addition of oligomers extracellularly to the culture medium may cause ER stress, possibly by causing Ca^{2+} influx (Demuro et al. 2005) and plasma-membrane disruption (Janson et al. 1999; Gurlo et al. 2010).

ER stress is a strong candidate for the mechanisms whereby IAPP induces apoptosis in T2DM. Both Alzheimer's (Costa et al. 2010; Haapasalo et al. 2010) and Parkinson's diseases (Shastry 2003) share the same characteristics of increased apoptosis with respect to protein aggregation and intracellular or extracellular accumulation of misfolded and aggregated amyloidogenic proteins. Toxic oligomers formed by different amyloidogenic proteins, including hIAPP, Aβ, α-synuclein, transthyretin, and prion protein, share a common epitope recognized by a polyclonal antibody, A11, which is also capable of blocking the cytotoxic effects of these oligomers (Kayed et al. 2003). This antibody was used to detect toxic oligomers in pancreas in hIAPP-transgenic mouse models of T2DM by immunohistochemistry. Oligomers were found intracellularly in 20–40% of hIAPP-transgenic mice but were absent in non-transgenic mice and in rIAPP-transgenic mice (Lin et al. 2007). In addition, vaccination of hIAPP-transgenic mice failed to prevent, and rather exacerbated hIAPP-induced β-cell apoptosis, leading to the conclusion that toxic oligomers may be located intracellularly, and thus are inaccessible to circulating antibodies (Lin et al. 2007). As suggested by previous studies, these findings agree with an intracellular location for the earliest stages of IAPP amyloidogenesis (O'Brien et al. 1995).

It is still unknown how oligomer formation induces ER stress. One proposed mechanism suggests that toxic oligomers interact with the ER membrane, leading to Ca^{2+} leakage, altering mitochondrial permeability and releasing cytochrome C, consequently inducing caspase-3 activation and amplification of the apoptotic signal (Orrenius et al. 2003; Demuro et al. 2005). Ca^{2+} leakage from the ER may also activate

ER-associated calpain, which mediates cytotoxic signals through mobilization of pro-apoptotic factors in a caspase-independent manner (Takano et al. 2005), or mediates activation of caspase-12 (Nakagawa et al. 2000) and caspase-4 (Hitomi et al. 2004). Caspase-12 expression was detected in hIAPP-, but not in rIAPP-transgenic mice, therefore, it could mediate β-cell apoptosis (Huang et al. 2007a).

β-Cell apoptosis and some characteristics of the ER-stress pathway have been observed in β-cell lines transduced with IAPP and in primary cultures of human pancreatic islets treated with exogenous hIAPP (Casas et al. 2007), as well as in hIAPP-transgenic mice and rats and in humans with T2DM (Laybutt et al. 2007; Huang et al. 2007b, a). ER-stress responses include the up-regulation of C/EBP homologous protein (CHOP) expression, followed by nuclear translocation and apoptosis. This was demonstrated in the INS-1 rat insulinoma cell line induced to express hIAPP and in pancreatic sections of a T2DM subject, but was not observed in INS-1 cells similarly transfected with rIAPP or green fluorescent protein (Huang et al. 2007a, b). In addition, apoptosis was decreased by knockdown of CHOP by small interfering RNA, which reduces the cleavage of caspase-3 in INS-1 cells overexpressing hIAPP.

It is important to note that levels of hIAPP in patients with T2DM are equivalent (or even decreased) relative to those of controls and that all of the abovementioned studies associating IAPP with β-cell ER stress were performed using supraphysiological levels of hIAPP. It has been postulated that ER stress was not an obligatory pathway mediating the toxic effects of amyloid formation at physiological levels of hIAPP, while hIAPP production and/or aggregation may be associated with ER stress under certain specific conditions (Hull et al. 2009).

In the cytoplasm, the ubiquitin–proteasome proteolytic pathway is responsible for the clearance of intracellular misfolded and aggregated proteins (Ciechanover 2005). Thioredoxin-interacting protein (TXNIP) is a ubiquitously expressed 50-kDa protein that binds to, and inhibits, thioredoxin, thereby modulating the cellular redox state and inducing oxidative stress (Chen et al. 2008). TXNIP undergoes proteasomal degradation in cells (Zhang et al. 2010). Extracellular addition of hIAPP oligomers has been shown to impair the ubiquitin–proteasomal pathway (Huang et al. 2007a), increasing levels of TXNIP. Recently, TXNIP was described as a potentially important link between glucose toxicity and apoptosis, based on the observation that high glucose concentrations induced apoptosis and increased the levels of this protein in INS-1 cells (Minn et al. 2005). Similarly, TXNIP-deficient islets of HcB-19 mice were protected against glucose-induced apoptosis (Chen et al. 2008). Therefore, TXNIP may represent a link between glucotoxicity and β-cell apoptosis, two critical mechanisms in the pathogenesis of T2DM (Minn et al. 2005).

7.7.1.2 IAPP–Membrane Interactions

The observation that IAPP fibrils are located at the cellular membrane in the islets of Langerhans, accompanied by alterations in membrane morphology (Khemtemourian et al. 2008) led researchers to hypothesize that the cell membrane may be the target of IAPP. These studies also revealed that IAPP-mediated degeneration of β-cells does not necessarily require amyloid fibril formation. In fact, small, pre-fibrillar oligomeric aggregates of hIAPP have been shown to be more cytotoxic than the large, fibrillar hIAPP deposits, by causing membrane disruption (Kapurniotu et al. 2002). Similar findings have been reported for other amyloidogenic proteins, such as the ABri peptide (involved in familial British dementia), Aβ (Alzheimer's disease), and prion protein (Creutzfeldt–Jakob disease).

The interaction between extracellular islet amyloid fibrils and β-cell membranes was first described using electron microscopy, as bundles of fibrils perpendicularly penetrating the cell membrane and causing distinct changes in cell morphology. Curiously, in the vicinity of other types of islet cells, such as α-cells, the fibrils were randomly oriented (Westermark 1973). Only after 14 years, the major component of islet amyloid was discovered (Westermark et al. 1987; Cooper et al. 1987) and hIAPP was generally considered to be toxic to β-cells, contributing to T2DM (Lorenzo et al. 1994; Clark et al. 1987). The toxic hIAPP peptide adopts a transmembrane orientation, which is absent in the nontoxic rIAPP (Nanga et al. 2008).

Results from several studies hypothesized that IAPP interacts with membranes (Mirzabekov et al. 1996; Janson et al. 1999; Kayed et al. 1999; Porat et al. 2003; Balali-Mood et al. 2005; Anguiano et al. 2002; Knight and Miranker 2004; Kayed et al. 2004) and that this interaction promotes IAPP aggregation and membrane leakage (Engel et al. 2008, 2006; Jeworrek et al. 2009; Khemtemourian et al. 2010; Radovan et al. 2009; Smith et al. 2009; Sparr et al. 2004). However, it is important to note that those are two separate processes: Engel et al. (2006) and Lopes et al. (2007) addressed this question by showing that the non-amyloidogenic and non-cytotoxic mouse IAPP (mIAPP) may insert into a phospholipid monolayer and increase the surface pressure at approximately the same rate as hIAPP, but fails to permeabilize lipid vesicles (Sparr et al. 2004; Anguiano et al. 2002; Green et al. 2004) or increase membrane conductance (Mirzabekov et al. 1996; Janson et al. 1999), causing only low membrane leakage (Engel et al. 2008).

Understanding the specific cause of IAPP membrane permeabilization is currently under debate. Hypotheses include intermediate species, the mature fibrils, and membrane-associated fibrillization as the cause of membrane permeabilization. These scenarios are discussed below.

Membrane Permeabilization by hIAPP Oligomers

Currently, a well-documented hypothesis is that soluble hIAPP oligomers are the toxic species. This hypothesis suggests that hIAPP fibrils are biologically inert (Haataja et al. 2008), a notion reinforced by the fact that the amount of hIAPP does not correlate with the decrease in the number of healthy β-cells (Ritzel et al. 2007). Transgenic rodent models of islet amyloidosis, which develop spontaneous diabetes, have been shown to exhibit β-cell loss without visible islet amyloid fibrils by light microscopy (Janson et al. 1996), suggesting that an intermediate in the fibrillization pathway could be the toxic species. Recent studies have shown cytotoxicity and membrane fragmentation in the absence of IAPP fibril formation (Brender et al. 2007, 2008). hIAPP preparations displaying low toxicity to rat RINm5F insulinoma β-cells were found to contain predominantly mature, β-sheet-rich fibrils, whereas preparations with high toxicity contained fewer, and contained IAPP predominantly in a random-coil state (Konarkowska et al. 2006). When freshly prepared aqueous solution of hIAPP was added to human or mouse islet cells in culture, apoptosis was induced by small, non-fibrillar hIAPP oligomers which formed non-selective ion channels and disrupted the β-cell membranes (Janson et al. 1999). Toxic oligomer formation would be increased if the capacity of β-cells to neutralize the toxic oligomers decreased. Consistent with this idea, insulin-degrading enzyme (IDE) shows a linkage to both Alzheimer's disease and T2DM (Qiu and Folstein 2006), inhibiting hIAPP and Aβ aggregation and cytotoxicity (Bennett et al. 2003; Leissring et al. 2003).

Interestingly, hIAPP and Aβ share similar structural properties, which may correlate with the observation that patients with Alzheimer's disease are more vulnerable to T2DM. In both diseases, a locally expressed protein (Aβ in AD and IAPP in T2DM) is deposited as amyloid with an accompanying gradual decline in the number of cells in the affected tissue. Amyloid deposits of both Aβ and IAPP (or more likely their oligomeric precursors) are cytotoxic by a mechanism that likely relate to membrane disruption (Janson et al. 2004). Mechanisms proposed for the oligomer-induced membrane-permeabilizing action of hIAPP include formation of cation-selective channels (Mirzabekov et al. 1996; Kagan et al. 2004; Hirakura et al. 2000; Quist et al. 2005), distortion of the phospholipid bilayer packing causing membrane instability (Janson et al. 1999; Kayed et al. 2004), and interaction of amyloid oligomers with specific membrane receptors (Engel 2009). Additional relevant observations are that pre-assembled hIAPP oligomers disrupt membranes (Kayed et al. 2004; Porat et al. 2003), pre-formed oligomers convert into a larger and more stable annular hIAPP oligomer (annular protofibrils) which insert into the membrane (Kayed et al. 2009), and interaction of hIAPP monomers with the membrane to form oligomeric forms in an α-helix-rich state with disruptive capacity (Knight et al. 2006; Knight and Miranker 2004). These models of membrane disruption by hIAPP oligomers are schematically depicted in Fig. 7.5a–c.

Taken together, multiple studies suggest that early hIAPP assemblies are involved in β-cell toxicity. However, more direct evidence is required and other possibilities have to be considered before concluding that hIAPP oligomers are the real culprit. For example, cytotoxicity assays involve incubation of "toxic" oligomers with cells for periods ranging from hours to days. Results of such experiments should be analyzed carefully to address the question: Do oligomers remain as oligomers throughout the incubation period or do they aggregate further into fibrils? And if they do, what species actually harms the cells? The answers for these questions have not been established yet. Similarly, in experiments using aggregation inhibitors, the exact form of hIAPP, which is inhibited by the inhibitor typically, is not known. An additional limitation of current knowledge stems from the difficulty to detect hIAPP oligomers *in vitro* despite availability of a number of methods. In fact, a large part of hIAPP data interpretation relies on comparison to studies of Aβ structure because Aβ aggregation is substantially slower *in vitro* compared with hIAPP (hour to days

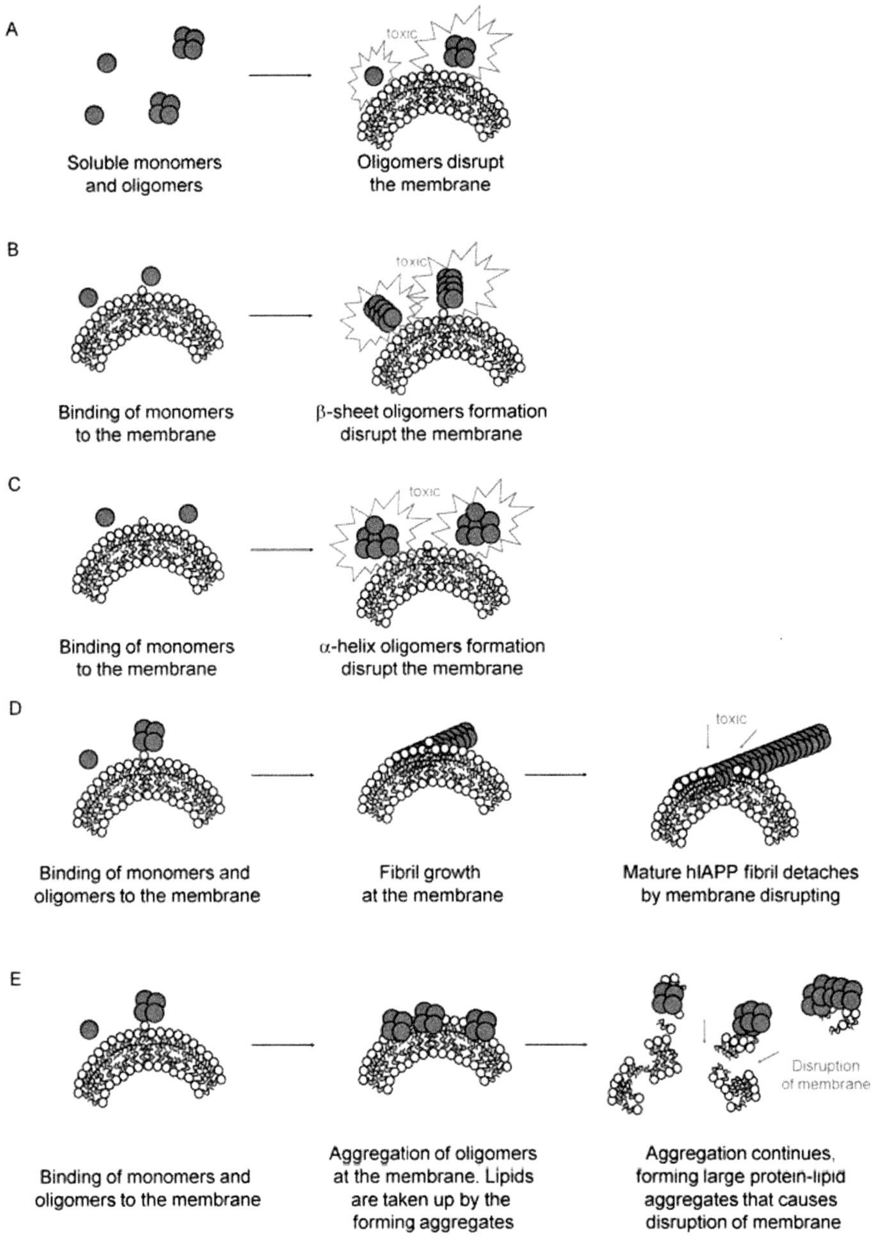

Fig. 7.5 Schematic representation of suggested mechanisms causing membrane damage by different toxic hIAPP species. (**A**) soluble, pre-assembled hIAPP oligomers and monomers adsorb onto, or insert into the membrane. (**B**) Preformed oligomers convert into large and more stable annular hIAPP oligomers or β-sheet oligomers, which insert into the membrane. (**C**) hIAPP monomers cooperatively interact with the membrane, folding into helical oligomeric forms (α-helical state) with membrane-disrupting capacity. (**D**) Interaction of monomeric and oligomeric hIAPP with the membrane initiates the process of fibril elongation at the membrane, resulting in a forced change in membrane curvature, concomitant membrane disruption, and detachment of mature fibrils from the distorted membrane. (**E**) Interaction of monomeric and oligomeric hIAPP with the membrane initiates the process of amyloid fibril formation using membrane lipids as a template

versus minutes, respectively) (Zraika et al. 2010). In addition, Fibrils could be a source of cytotoxic monomers or oligomers, which are in equilibrium with fibrils (Carulla et al. 2005), and amyloid fibrils could break down, as has been shown for insulin fibrils (Smith et al. 2006). Finally, there is no direct evidence for the existence of toxic hIAPP oligomers *in vivo*.

Membrane Permeabilization by the Process of hIAPP Aggregation

Recent reports suggest an alternative mechanism for hIAPP cytotoxicity, according to which membrane damage is not caused by specific hIAPP species, such as oligomers or fibrils, but rather, by the process of fibril growth at the cellular membrane (Fig. 7.5d). By comparing the kinetics of hIAPP fibril growth (using thioflavin T fluorescence assay) and membrane permeabilization, a similar sigmoidal profile was demonstrated, characterized by an initial lag phase followed by exponential acceleration and finally a plateau (Engel et al. 2008). In this model, the first steps of the hIAPP interaction with membranes are adsorption, followed by insertion of hIAPP into the membrane, either as monomers or as oligomers. Only monomeric hIAPP has a strong tendency to insert into phospholipid monolayers upon interacting with membranes (Lopes et al. 2007). In the following step, interaction of membrane-bound hIAPP monomers/oligomers with each other, or with solution hIAPP, leads to fibril growth at the membrane and lipid uptake into the growing fibrils. The mechanism of membrane damage may entail growth of a rigid hIAPP fibril on a flexible phospholipid bilayer, resulting in a forced change in membrane curvature. This change of the bilayer membrane curvature to unfavorable angles leads to deformation of the membrane shape (Engel et al. 2008). Uptake of lipids by hIAPP fibril elongation (Fig. 7.5e) has been observed *in vitro* (Domanov and Kinnunen 2008; Sparr et al. 2004; Zhao et al. 2004) and in some types of amyloids isolated from patients (Gellermann et al. 2005). Using molecular dynamics simulations, the tendency of amyloidogenic peptides to fibrillate on the surface of lipid vesicles, inducing damage to the lipid bilayer, was observed (Friedman et al. 2009).

These new ideas may direct current studies towards mechanisms suggesting that membrane permeabilization is governed by structural conversion along the fibril fibril-formation pathway. In such mechanisms, membranes could have an important function as mediators or accelerators of the conversion of one hIAPP species into another, possibly representing a cytotoxic event, because phospholipid bilayers have the ability to catalyze hIAPP fibril formation by reducing the lag phase, an effect which is most pronounced with negatively charged lipids (Knight et al. 2006; Knight and Miranker 2004). In addition, it has been suggested that other factors can significantly affect the interaction between hIAPP and membranes, such as Ca^{2+} ions (Sciacca et al. 2008) and crystalline insulin (Knight et al. 2008). Aggregation of peptides and proteins may be affected by interfaces serving as templates with preferential orientation, which promotes aggregation, as has been suggested for hIAPP (Knight and Miranker 2004; Linse et al. 2007). The effect of interface-mediated aggregation catalysis may be a very important factor in the mechanism of amyloid-induced cytotoxicity. With membranes acting as templates for amyloid aggregation, it is not surprising that conversion of amyloid species also affects the barrier properties of the bilayer (Engel 2009).

7.8 Inhibition of hIAPP Formation

Increasing evidence linking aggregation and cytotoxicity of IAPP and islet amyloid formation to a decline in β-cell mass and function underscores the importance of new methods to decrease or prevent islet amyloid formation. One approach is to decrease IAPP formation/release by using anti-diabetic agents, which act by reducing β-cell secretory demand. For example, treatment with metformin or the thiazolidinedione rosiglitazone for 1 year markedly reduced islet amyloid formation in hIAPP transgenic mice, decreasing both the number of amyloid-containing islets and the proportion of islet area replaced by amyloid (Hull et al. 2005). The capability of these interventions to reduce islet amyloid formation was greater than their ability to reduce the β-cell secretory demand alone, suggesting that they have additional, as yet unidentified, effects resulting in reduced islet amyloid deposition. However, these approaches did not prevent completely the development of amyloid deposition. Therefore, it appears that the development of approaches to prevent β-cell loss due to amyloid formation should also focus on the very early stages of amyloid fibrillogenesis, rather than solely aim to dissociate mature islet amyloid deposits.

7.8.1 IAPP and Drug Design: State of the Art

Development of small molecules to modulate the damaging effects of protein-aggregation processes is a high-priority goal of contemporary medicinal chemistry (Cohen and Kelly 2003). A large body of work on inhibitors of Aβ is available, but less attention has been focused on development of IAPP aggregation inhibitors. To date, two classes of inhibitors of hIAPP amyloidogenicity and/or cytotoxicity have been described. The first class includes aromatic organic compounds, such as Congo red, rifampicin, resveratrol, and acid fuchsin, which bind to amyloid fibrils and/or suppress amyloid fibril formation (Lorenzo and Yankner 1994; Tomiyama et al. 1997; Kudva et al. 1998; Aitken et al. 2003; Harroun et al. 2001; Mishra et al. 2008; Porat et al. 2004; Levy et al. 2008; Mishra et al. 2009b; Meng et al. 2010). The second class of inhibitors consists of synthetic peptides which have been derived from hIAPP itself and contain hIAPP self-recognition domains, such as those which incorporate proline residues or *N*-methylated amino acids into full-length IAPP or IAPP fragments (Scrocchi et al. 2002; Gilead and Gazit 2004; Tatarek-Nossol et al. 2005; Yan et al. 2006; Saraogi et al. 2009; Potter et al. 2009, Muthusamy et al. 2010b).

Most selected sequences have been based on the 20–29 region of hIAPP (Kapurniotu et al. 2002; Scrocchi et al. 2002). A recent report described the effect of peptides spanning the length of $hIAPP_{1-37}$ for their ability to inhibit toxicity of full-length hIAPP. $hIAPP_{3-6}$, $hIAPP_{9-13}$, $hIAPP_{22-27}$, $hIAPP_{29-34}$, and ST(*N*-Me)NV(*N*-Me) G(*N*-Me)S(*N*-Me) were identified as potential inhibitors of amylin-mediated cytotoxicity, with the latter, the *N*-methylated peptide corresponding to the 29–34 region, reported as most promising due to its biological stability (Muthusamy et al. 2010b). If the initial amyloid formation is intracellular, the use of peptide inhibitors may be inefficient because they are expected to be unable to enter the cell.

The importance of understanding the mechanism of inhibition has recently been highlighted (Rishton 2008; Feng et al. 2008) following findings that small-molecule inhibitors of fibrillogenesis may act non-specifically, likely rendering them unsuitable for treating amyloid-related disorders (Feng et al. 2008). If apoptosis and β-cell mass loss are important contributors to the loss of islet function observed in T2DM, then the use of agents that slow down or prevent these effects is an attractive possibility. The real challenge will be to determine which individuals would benefit from their use and whether these approaches should be initiated at an early stage to prevent the progressive decline of β-cell function that characterizes the development and progression of hyperglycemia (Hull et al. 2004).

7.9 Final Considerations

Although the discovery of new agents to prevent or slow down hIAPP aggregation is fundamental for future treatment of diabetic patients, it is important to understand the exact cytotoxic mechanism of hIAPP. Conflicting results obtained in studies of cytotoxicity and membrane interaction of hIAPP species most likely arise due to ill-defined or impure hIAPP samples, such as those containing traces of pre-existing aggregates (Konarkowska et al. 2006). Due to the often rapid and uncontrollable aggregation of amyloidogenic proteins and peptides, it is difficult to obtain pure, structurally uniform samples of IAPP monomers, oligomers, or fibrils. Development of molecules capable of diminishing the rapid aggregation of hIAPP in conjunction with available techniques, such as high-resolution microscopy or cross-linking methods, may prove useful in identifying the species responsible for toxicity. Novel purification strategies to obtain pure samples containing a single species and development of specific antibodies (Glabe 2004) may be very useful in identifying unique (oligomeric) hIAPP species. The ability to isolate hIAPP species is crucial for future studies attempting to link structure and cytotoxicity. Despite several studies discussed in this chapter, the exact cause of β-cell toxicity during hIAPP formation still remains controversial.

As previously mentioned, various studies have shown that it was not necessarily one particular species that might be cytotoxic, but, rather, the conversion from one species to another could also be cytotoxic. Future studies into these "amyloid conversions," in particular, membrane-mediated conversions, could promote new insights into the cytotoxicity mechanisms. Moreover, the importance of hIAPP–membrane interaction indicates that inhibition or alteration of this interaction may be an alternative strategy for reducing amyloid cytotoxicity and preventing β-cell death in T2DM (Khemtemourian et al. 2008).

Another important point to consider is the extrapolation of results of *in vitro* studies, particularly studies using artificial membranes, to *in vivo* hIAPP toxicity, because artificial membrane permeability is not identical to that of physiologic membranes. Therefore, the number of hypotheses based on model-membrane systems regarding physiological β-cell membranes is likely to increase in the future. Identification of the structure of hIAPP oligomers and fibrils produced *in vivo* would provide valuable information, contributing to understanding of the process, which currently is obtained mostly using synthetic peptides and *in vitro* conditions. In addition, physiologically relevant information from *in vivo* hIAPP structures may open new avenues in the development of inhibitors of hIAPP-induced cytotoxicity.

Several studies of Aβ oligomers related to Alzheimer's disease are available which allow better understanding of the amyloidogenesis process, but the aggregation pathways of other amyloidogenic systems are not necessarily the same as those of Aβ. There has been considerable progress in the field of hIAPP–membrane interaction during the past several years. However, it is still far from clear how these interactions relate to cytotoxicity in T2DM.

References

Abedini A, Raleigh DP (2006) Destabilization of human IAPP amyloid fibrils by proline mutations outside of the putative amyloidogenic domain: is there a critical amyloidogenic domain in human IAPP? J Mol Biol 355:274–281

Ahronheim JH (1943) The nature of the hyaline material in the pancreatic islands in diabetes mellitus. Am J Pathol 19:873–882

Aitken JF, Loomes KM, Konarkowska B, Cooper GJ (2003) Suppression by polycyclic compounds of the conversion of human amylin into insoluble amyloid. Biochem J 374:779–784

Amodeo P, Morelli MA, Castiglione Motta A (1994) Multiple conformations and proline *cis–trans* isomerization in salmon calcitonin: a combined nuclear magnetic resonance, distance geometry, and molecular mechanics study. Biochemistry 33:10754–10762

Anguiano M, Nowak RJ, Lansbury PT Jr (2002) Protofibrillar islet amyloid polypeptide permeabilizes synthetic vesicles by a pore-like mechanism that may be relevant to type II diabetes. Biochemistry 41:11338–11343

Araki E, Oyadomari S, Mori M (2003) Endoplasmic reticulum stress and diabetes mellitus. Intern Med 42:7–14

Arnelo U, Permert J, Adrian TE, Larsson J, Westermark P, Reidelberger RD (1996) Chronic infusion of islet amyloid polypeptide causes anorexia in rats. Am J Physiol 271:R1654–R1659

Augstein P, Stephens LA, Allison J, Elefanty AG, Ekberg M, Kay TW, Harrison LC (1998) β-Cell apoptosis in an accelerated model of autoimmune diabetes. Mol Med 4:495–501

Azriel R, Gazit E (2001) Analysis of the minimal amyloid-forming fragment of the islet amyloid polypeptide. An experimental support for the key role of the phenylalanine residue in amyloid formation. J Biol Chem 276:34156–34161

Badman MK, Shennan KI, Jermany JL, Docherty K, Clark A (1996) Processing of pro-islet amyloid polypeptide (proIAPP) by the prohormone convertase PC2. FEBS Lett 378:227–231

Balali-Mood K, Ashley RH, Hauss T, Bradshaw JP (2005) Neutron diffraction reveals sequence-specific membrane insertion of pre-fibrillar islet amyloid polypeptide and inhibition by rifampicin. FEBS Lett 579:1143–1148

Barth SW, Riediger T, Lutz TA, Rechkemmer G (2004) Peripheral amylin activates circumventricular organs expressing calcitonin receptor a/b subtypes and receptor-activity modifying proteins in the rat. Brain Res 997:97–102

Bellamy CO, Malcomson RD, Harrison DJ, Wyllie AH (1995) Cell death in health and disease: the biology and regulation of apoptosis. Semin Cancer Biol 6:3–16

Bennett RG, Hamel FG, Duckworth WC (2003) An insulin-degrading enzyme inhibitor decreases amylin degradation, increases amylin-induced cytotoxicity, and increases amyloid formation in insulinoma cell cultures. Diabetes 52:2315–2320

Betsholtz C, Christmansson L, Engstrom U, Rorsman F, Svensson V, Johnson KH, Westermark P (1989) Sequence divergence in a specific region of islet amyloid polypeptide (IAPP) explains differences in islet amyloid formation between species. FEBS Lett 251:261–264

Betsholtz C, Christmanson L, Engstrom U, Rorsman F, Jordan K, O'Brien TD, Murtaugh M, Johnson KH, Westermark P (1990) Structure of cat islet amyloid polypeptide and identification of amino acid residues of potential significance for islet amyloid formation. Diabetes 39:118–122

Betsholtz C, Christmansson L, Engström U, Rorsman F, Svensson V, Johnson KH, Westermark P (1998) Sequence divergence in a specific region of islet amyloid polypeptide (IAPP) explains differences in islet amyloid formation between species. FEBS Lett 17:261–264

Bliss M (1982) Banting's, Best's, and Collip's accounts of the discovery of insulin. Bull Hist Med 56:554–568

Bonora E, Kiechl S, Willeit J, Oberhollenzer F, Egger G, Targher G, Alberiche M, Bonadonna RC, Muggeo M (1998) Prevalence of insulin resistance in metabolic disorders: the Bruneck Study. Diabetes 47:1643–1649

Boscaro M, Barzon L, Fallo F, Sonino N (2001) Cushing's syndrome. Lancet 357:783–791

Bouret SG (2010) Development of hypothalamic neural networks controlling appetite. Forum Nutr 63:84–93

Bouret SG, Draper SJ, Simerly RB (2004) Trophic action of leptin on hypothalamic neurons that regulate feeding. Science 304:108–110

Bouret SG, Gorski JN, Patterson CM, Chen S, Levin BE, Simerly RB (2008) Hypothalamic neural projections are permanently disrupted in diet-induced obese rats. Cell Metab 7:179–185

Breeze AL, Harvey TS, Bazzo R, Campbell ID (1991) Solution structure of human calcitonin gene-related peptide by 1 H NMR and distance geometry with restrained molecular dynamics. Biochemistry 30:575–582

Brender JR, Durr UH, Heyl D, Budarapu MB, Ramamoorthy A (2007) Membrane fragmentation by an amyloidogenic fragment of human islet amyloid polypeptide detected by solid-state NMR spectroscopy of membrane nanotubes. Biochim Biophys Acta 1768:2026–2029

Brender JR, Hartman K, Reid KR, Kennedy RT, Ramamoorthy A (2008) A single mutation in the nonamyloidogenic region of islet amyloid polypeptide greatly reduces toxicity. Biochemistry 47:12680–12688

Brender JR, Hartman K, Nanga RP, Popovych N, de la Salud Bea R, Vivekanandan S, Marsh EN, Ramamoorthy A (2010) Role of zinc in human islet amyloid polypeptide aggregation. J Am Chem Soc 132:8973–8983

Butler PC, Chou J, Carter WB, Wang YN, Bu BH, Chang D, Chang JK, Rizza RA (1990) Effects of meal ingestion on plasma amylin concentration in NIDDM and nondiabetic humans. Diabetes 39:752–756

Butler AE, Janson J, Bonner-Weir S, Ritzel R, Rizza RA, Butler PC (2003) β-Cell deficit and increased β-cell apoptosis in humans with type 2 diabetes. Diabetes 52:102–110

Butler AE, Jang J, Gurlo T, Carty MD, Soeller WC, Butler PC (2004) Diabetes due to a progressive defect in β-cell mass in rats transgenic for human islet amyloid polypeptide (HIP Rat): a new model for type 2 diabetes. Diabetes 53:1509–1516

Carulla N, Caddy GL, Hall DR, Zurdo J, Gairi M, Feliz M, Giralt E, Robinson CV, Dobson CM (2005) Molecular recycling within amyloid fibrils. Nature 436:554–558

Casas S, Gomis R, Gribble FM, Altirriba J, Knuutila S, Novials A (2007) Impairment of the ubiquitin–proteasome pathway is a downstream endoplasmic reticulum stress response induced by extracellular human islet amyloid polypeptide and contributes to pancreatic β-cell apoptosis. Diabetes 56:2284–2294

Chan JM, Rimm EB, Colditz GA, Stampfer MJ, Willett WC (1994) Obesity, fat distribution, and weight gain as risk factors for clinical diabetes in men. Diabetes Care 17:961–969

Charge SB, de Koning EJ, Clark A (1995) Effect of pH and insulin on fibrillogenesis of islet amyloid polypeptide *in vitro*. Biochemistry 34:14588–14593

Chartier-Harlin MC, Crawford F, Hamandi K, Mullan M, Goate A, Hardy J, Backhovens H, Martin JJ, Broeckhoven CV (1991) Screening for the β-amyloid precursor protein mutation (APP717: Val→Ile) in extended pedigrees with early onset Alzheimer's disease. Neurosci Lett 129:134–135

Chen J, Hui ST, Couto FM, Mungrue IN, Davis DB, Attie AD, Lusis AJ, Davis RA, Shalev A (2008) Thioredoxin-interacting protein deficiency induces Akt/Bcl-xL signaling and pancreatic β-cell mass and protects against diabetes. FASEB J 22:3581–3594

Christmanson L, Betsholtz C, Leckstrom A, Engstrom U, Cortie C, Johnson KH, Adrian TE, Westermark P (1993) Islet amyloid polypeptide in the rabbit and European hare: studies on its relationship to amyloidogenesis. Diabetologia 36:183–188

Ciechanover A (2005) Intracellular protein degradation: from a vague idea thru the lysosome and the ubiquitin–proteasome system and onto human diseases and drug targeting. Cell Death Differ 12:1178–1190

Clark A (1989) Islet amyloid and type 2 diabetes. Diabet Med 6:561–567

Clark A, Nilsson MR (2004) Islet amyloid: a complication of islet dysfunction or an aetiological factor in type 2 diabetes? Diabetologia 47:157–169

Clark A, Cooper GJ, Lewis CE, Morris JF, Willis AC, Reid KB, Turner RC (1987) Islet amyloid formed from diabetes-associated peptide may be pathogenic in type-2 diabetes. Lancet 2:231–234

Clark A, Wells CA, Buley ID, Cruickshank JK, Vanhegan RI, Matthews DR, Cooper GJ, Holman RR, Turner RC (1988) Islet amyloid, increased α-cells, reduced β-cells and exocrine fibrosis: quantitative changes in the pancreas in type 2 diabetes. Diabetes Res 9:151–159

Cohen FE, Kelly JW (2003) Therapeutic approaches to protein-misfolding diseases. Nature 426:905–909

Colin C, Demasi MA, Degaki TL, Bustos-Valenzuela JC, Figueira RC, Montor WR, Cruz LO, Lojudice FH, Muras AG, Pereira TM, Winnischofer SM, Hasegawa AP, Carreira AC, Verbisck NV, Correa RG, Garay-Malpartida HM, Mares-Guia TR, Correa-Giannella ML, Granjeiro JM, Sogayar MC (2008) NUCEL (Cell and Molecular Therapy Center): a multidisciplinary center for translational research in Brazil. Mol Biotechnol 39:89–95

Cooper GJ (1994) Amylin compared with calcitonin gene-related peptide: structure, biology, and relevance to metabolic disease. Endocr Rev 15:163–201

Cooper GJ, Willis AC, Clark A, Turner RC, Sim RB, Reid KB (1987) Purification and characterization of a peptide from amyloid-rich pancreases of type 2 diabetic patients. Proc Natl Acad Sci USA 84:8628–8632

Cooper ME, McNally PG, Phillips PA, Johnston CI (1995) Amylin stimulates plasma renin concentration in humans. Hypertension 26:460–464

Costa RO, Ferreiro E, Cardoso SM, Oliveira CR, Pereira CMF (2010) ER stress-mediated apoptotic pathway induced by amyloid-β. J Alzheimers Dis 20:625–636

D'Alessio DA, Verchere CB, Kahn SE, Hoagland V, Baskin DG, Palmiter RD, Ensinck JW (1994) Pancreatic expression and secretion of human islet amyloid polypeptide in a transgenic mouse. Diabetes 43:1457–1461

de Koning EJ, Bodkin NL, Hansen BC, Clark A (1993a) Diabetes mellitus in *Macaca mulatta* monkeys is characterised by islet amyloidosis and reduction in β-cell population. Diabetologia 36:378–384

de Koning EJ, Hoppener JW, Oosterwijk C, Verbeek SJ, Visser HJ, Jansz HS, Lips CJ, Morris JF, Clark A (1993b) Localisation of islet amyloid polypeptide (IAPP) in pancreatic islets of transgenic mice expressing the human or rat IAPP gene. Biochem Soc Trans 21:26S de Koning EJ, Hoppener JW, Verbeek JS, Oosterwijk C, van Hulst KL, Baker CA, Lips CJ, Morris JF, Clark A (1994a) Human islet amyloid polypeptide accumulates at similar sites in islets of transgenic mice and humans. Diabetes 43:640–644

de Koning EJ, Morris ER, Hofhuis FM, Posthuma G, Hoppener JW, Morris JF, Capel PJ, Clark A, Verbeek JS (1994b) Intra- and extracellular amyloid fibrils are formed in cultured pancreatic islets of transgenic mice expressing human islet amyloid polypeptide. Proc Natl Acad Sci USA 91:8467–8471

Dean L, McEntyre J (2004) The genetic landscape of diabetes. National Center for Biotechnology Information (US), National Library of Medicine, National Institutes of Health, Bethesda

DeFronzo RA (2004) Pathogenesis of type 2 diabetes mellitus. Med Clin North Am 88:787–835, ix

Del Guerra S, Lupi R, Marselli L, Masini M, Bugliani M, Sbrana S, Torri S, Pollera M, Boggi U, Mosca F, Del Prato S, Marchetti P (2005) Functional and molecular defects of pancreatic islets in human type 2 diabetes. Diabetes 54:727–735

Demuro A, Mina E, Kayed R, Milton SC, Parker I, Glabe CG (2005) Calcium dysregulation and membrane disruption as a ubiquitous neurotoxic mechanism of soluble amyloid oligomers. J Biol Chem 280:17294–17300

Deng S, Vatamaniuk M, Huang X, Doliba N, Lian MM, Frank A, Velidedeoglu E, Desai NM, Koeberlein B, Wolf B, Barker CF, Naji A, Matschinsky FM, Markmann JF (2004) Structural and functional abnormalities in the islets isolated from type 2 diabetic subjects. Diabetes 53:624–632

Dobolyi A (2009) Central amylin expression and its induction in rat dams. J Neurochem 111:1490–1500

Domanov YA, Kinnunen PK (2008) Islet amyloid polypeptide forms rigid lipid-protein amyloid fibrils on supported phospholipid bilayers. J Mol Biol 376:42–54

Ehrlich JC, Ratner IM (1961) Amyloidosis of the islets of Langerhans. A restudy of islet hyalin in diabetic and non-diabetic individuals. Am J Pathol 38:49–59

Elbein SC (2009) Genetics factors contributing to type 2 diabetes across ethnicities. J Diabetes Sci Technol 3:685–689

Engel MF (2009) Membrane permeabilization by islet amyloid polypeptide. Chem Phys Lipids 160:1–10

Engel MF, Yigittop H, Elgersma RC, Rijkers DT, Liskamp RM, de Kruijff B, Hoppener JW, Antoinette Killian J (2006) Islet amyloid polypeptide inserts into phospholipid monolayers as monomer. J Mol Biol 356:783–789

Engel MF, Khemtemourian L, Kleijer CC, Meeldijk HJ, Jacobs J, Verkleij AJ, de Kruijff B, Killian JA, Hoppener JW (2008) Membrane damage by human islet amyloid polypeptide through fibril growth at the membrane. Proc Natl Acad Sci USA 105:6033–6038

Feng BY, Toyama BH, Wille H, Colby DW, Collins SR, May BC, Prusiner SB, Weissman J, Shoichet BK (2008) Small-molecule aggregates inhibit amyloid polymerization. Nat Chem Biol 4:197–199

Fineman M, Weyer C, Maggs DG, Strobel S, Kolterman OG (2002a) The human amylin analog, pramlintide, reduces postprandial hyperglucagonemia in patients with type 2 diabetes mellitus. Horm Metab Res 34:504–508

Fineman MS, Koda JE, Shen LZ, Strobel SA, Maggs DG, Weyer C, Kolterman OG (2002b) The human amylin analog, pramlintide, corrects postprandial hyperglucagonemia in patients with type 1 diabetes. Metabolism 51:636–641

Fleming BB, Greenfield S, Engelgau MM, Pogach LM, Clauser SB, Parrott MA (2001) The diabetes quality improvement project: moving science into health policy to gain an edge on the diabetes epidemic. Diabetes Care 24:1815–1820

Fox N, Schrementi J, Nishi M, Ohagi S, Chan SJ, Heisserman JA, Westermark GT, Leckstrom A, Westermark P, Steiner DF (1993) Human islet amyloid polypeptide transgenic mice as a model of non-insulin-dependent diabetes mellitus (NIDDM). FEBS Lett 323:40–44

Friedman R, Pellarin R, Caflisch A (2009) Amyloid aggregation on lipid bilayers and its impact on membrane permeability. J Mol Biol 387:407–415

Furukawa H, Carroll RJ, Swift HH, Steiner DF (1999) Long-term elevation of free fatty acids leads to delayed processing of proinsulin and prohormone convertases 2 and 3 in the pancreatic β-cell line MIN6. Diabetes 48:1395–1401
Garcia-Gonzalez CL, Montoya-Fuentes H, Padilla-Rosas M, Sanchez-Corona J (2007) Amylin S20G mutation in Mexican population. Diabetes Res Clin Pract 76:146–148
Gebre-Medhin S, Mulder H, Pekny M, Westermark G, Tornell J, Westermark P, Sundler F, Ahren B, Betsholtz C (1998) Increased insulin secretion and glucose tolerance in mice lacking islet amyloid polypeptide (amylin). Biochem Biophys Res Commun 250:271–277
Gebre-Medhin S, Olofsson C, Mulder H (2000) Islet amyloid polypeptide in the islets of Langerhans: Friend or foe? Diabetologia 43:687–695
Gellermann GP, Appel TR, Tannert A, Radestock A, Hortschansky P, Schroeckh V, Leisner C, Lutkepohl T, Shtrasburg S, Rocken C, Pras M, Linke RP, Diekmann S, Fandrich M (2005) Raft lipids as common components of human extracellular amyloid fibrils. Proc Natl Acad Sci USA 102:6297–6302
Gilbey SG, Ghatei MA, Bretherton-Watt D, Zaidi M, Jones PM, Perera T, Beacham J, Girgis S, Bloom SR (1991) Islet amyloid polypeptide: production by an osteoblast cell line and possible role as a paracrine regulator of osteoclast function in man. Clin Sci (Lond) 81:803–808
Gilead S, Gazit E (2004) Inhibition of amyloid fibril formation by peptide analogues modified with α-aminoisobutyric acid. Angew Chem Int Ed Engl 43:4041–4044
Gilead S, Gazit E (2008) The role of the 14–20 domain of the islet amyloid polypeptide in amyloid formation. Exp Diabetes Res 2008:256954
Glabe CG (2004) Conformation-dependent antibodies target diseases of protein misfolding. Trends Biochem Sci 29:542–547
Glenner GG, Eanes ED, Wiley CA (1998) Amyloid fibrils formed from a segment of the pancreatic islet amyloid protein. Biochem Biophys Res Commun 155:608–614
Goldsbury C, Goldie K, Pellaud J, Seelig J, Frey P, Muller SA, Kistler J, Cooper GJ, Aebi U (2000) Amyloid fibril formation from full-length and fragments of amylin. J Struct Biol 130:352–362
Gotz J, Ittner LM, Lim YA (2009) Common features between diabetes mellitus and Alzheimer's disease. Cell Mol Life Sci 66:1321–1325
Green J, Goldsbury C, Mini T, Sunderji S, Frey P, Kistler J, Cooper G, Aebi U (2003) Full-length rat amylin forms fibrils following substitution of single residues from human amylin. J Mol Biol 326:1147–1156
Green JD, Kreplak L, Goldsbury C, Li Blatter X, Stolz M, Cooper GS, Seelig A, Kistler J, Aebi U (2004) Atomic force microscopy reveals defects within mica supported lipid bilayers induced by the amyloidogenic human amylin peptide. J Mol Biol 342:877–887
Gregg EW, Narayan KMV (2000) Type 2 diabetes and cognitive function: are cognitive impairment and dementia complications of type 2 diabetes? Clin Geriatr 8:57–72
Griffiths JM, Ashburn TT, Auger M, Costa PR, Griffin RG, Lansbury PT (1995) Rotational resonance solidstate NMR elucidates a structural model of pancreatic amyloid. J Am Chem Soc 12:3539–3546
Gurlo T, Ryazantsev S, Huang CJ, Yeh MW, Reber HA, Hines OJ, O'Brien TD, Glabe CG, Butler PC (2010) Evidence for proteotoxicity in β cells in type 2 diabetes: toxic islet amyloid polypeptide oligomers form intracellularly in the secretory pathway. Am J Pathol 176:861–869
Haapasalo A, Viswanathan J, Bertram L, Soininen H, Tanzi RE, Hiltunen M (2010) Emerging role of Alzheimer's disease-associated ubiquilin-1 in protein aggregation. Biochem Soc Trans 38:150–155
Haataja L, Gurlo T, Huang CJ, Butler PC (2008) Islet amyloid in type 2 diabetes, and the toxic oligomer hypothesis. Endocr Rev 29:303–316
Hanabusa T, Kubo K, Oki C, Nakano Y, Okai K, Sanke T, Nanjo K (1992) Islet amyloid polypeptide (IAPP) secretion from islet cells and its plasma concentration in patients with non-insulin-dependent diabetes mellitus. Diabetes Res Clin Pract 15:89–96
Hansen I, Tsalikian E, Beaufrere B, Gerich J, Haymond M, Rizza R (1986) Insulin resistance in acromegaly: defects in both hepatic and extrahepatic insulin action. Am J Physiol 250:E269–E273

Harding HP, Ron D (2002) Endoplasmic reticulum stress and the development of diabetes: a review. Diabetes 51(Suppl 3):S455–S461
Harroun TA, Bradshaw JP, Ashley RH (2001) Inhibitors can arrest the membrane activity of human islet amyloid polypeptide independently of amyloid formation. FEBS Lett 507:200–204
Hayden MR, Tyagi SC (2001) “A” is for amylin and amyloid in type 2 diabetes mellitus. JOP 2:124–139
Hiddinga HJ, Eberhardt NL (1999) Intracellular amyloidogenesis by human islet amyloid polypeptide induces apoptosis in COS-1 cells. Am J Pathol 154:1077–1088
Higham CE, Hull RL, Lawrie L, Shennan KI, Morris JF, Birch NP, Docherty K, Clark A (2000) Processing of synthetic pro-islet amyloid polypeptide (proIAPP) ‘amylin’ by recombinant prohormone convertase enzymes, PC2 and PC3, *in vitro*. Eur J Biochem 267:4998–5004
Hirakura Y, Yiu WW, Yamamoto A, Kagan BL (2000) Amyloid peptide channels: blockade by zinc and inhibition by Congo red (amyloid channel block). Amyloid 7:194–199
Hitomi J, Katayama T, Eguchi Y, Kudo T, Taniguchi M, Koyama Y, Manabe T, Yamagishi S, Bando Y, Imaizumi K, Tsujimoto Y, Tohyama M (2004) Involvement of caspase-4 in endoplasmic reticulum stress-induced apoptosis and Aβ-induced cell death. J Cell Biol 165:347–356
Hoppener JW, Nieuwenhuis MG, Vroom TM, Ahren B, Lips CJ (2002) Role of islet amyloid in type 2 diabetes mellitus: consequence or cause? Mol Cell Endocrinol 197:205–212
Hou X, Ling Z, Quartier E, Foriers A, Schuit F, Pipeleers D, Van Schravendijk C (1999) Prolonged exposure of pancreatic β cells to raised glucose concentrations results in increased cellular content of islet amyloid polypeptide precursors. Diabetologia 42:188–194
Howard CF Jr (1986) Longitudinal studies on the development of diabetes in individual *Macaca nigra*. Diabetologia 29:301–306
Huang CJ, Haataja L, Gurlo T, Butler AE, Wu X, Soeller WC, Butler PC (2007a) Induction of endoplasmic reticulum stress-induced β-cell apoptosis and accumulation of polyubiquitinated proteins by human islet amyloid polypeptide. Am J Physiol Endocrinol Metab 293: E1656–E1662
Huang CJ, Lin CY, Haataja L, Gurlo T, Butler AE, Rizza RA, Butler PC (2007b) High expression rates of human islet amyloid polypeptide induce endoplasmic reticulum stress mediated β-cell apoptosis, a characteristic of humans with type 2 but not type 1 diabetes. Diabetes 56: 2016–2027
Hubbard JA, Martin SR, Chaplin LC, Bose C, Kelly SM, Price NC (1991) Solution structures of calcitonin-gene-related-peptide analogues of calcitonin-gene-related peptide and amylin. Biochem J 275(Pt 3):785–788
Hull RL, Westermark GT, Westermark P, Kahn SE (2004) Islet amyloid: a critical entity in the pathogenesis of type 2 diabetes. J Clin Endocrinol Metab 89:3629–3643
Hull RL, Shen ZP, Watts MR, Kodama K, Carr DB, Utzschneider KM, Zraika S, Wang F, Kahn SE (2005) Long-term treatment with rosiglitazone and metformin reduces the extent of, but does not prevent, islet amyloid deposition in mice expressing the gene for human islet amyloid polypeptide. Diabetes 54:2235–2244
Hull RL, Zraika S, Udayasankar J, Aston-Mourney K, Subramanian SL, Kahn SE (2009) Amyloid formation in human IAPP transgenic mouse islets and pancreas, and human pancreas, is not associated with endoplasmic reticulum stress. Diabetologia 52:1102–1111
Imai Y, Soda M, Inoue H, Hattori N, Mizuno Y, Takahashi R (2001) An unfolded putative transmembrane polypeptide, which can lead to endoplasmic reticulum stress, is a substrate of Parkin. Cell 105:891–902
Jaikaran ET, Clark A (2001) Islet amyloid and type 2 diabetes: from molecular misfolding to islet pathophysiology. Biochim Biophys Acta 1537:179–203
Jaikaran ET, Higham CE, Serpell LC, Zurdo J, Gross M, Clark A, Fraser PE (2001) Identification of a novel human islet amyloid polypeptide β-sheet domain and factors influencing fibrillogenesis. J Mol Biol 308:515–525
Jaikaran ET, Nilsson MR, Clark A (2004) Pancreatic β-cell granule peptides form heteromolecular complexes which inhibit islet amyloid polypeptide fibril formation. Biochem J 377:709–716

Janciauskiene S, Ahren B (2000) Fibrillar islet amyloid polypeptide differentially affects oxidative mechanisms and lipoprotein uptake in correlation with cytotoxicity in two insulin-producing cell lines. Biochem Biophys Res Commun 267:619–625

Janson J, Soeller WC, Roche PC, Nelson RT, Torchia AJ, Kreutter DK, Butler PC (1996) Spontaneous diabetes mellitus in transgenic mice expressing human islet amyloid polypeptide. Proc Natl Acad Sci USA 93:7283–7288

Janson J, Ashley RH, Harrison D, McIntyre S, Butler PC (1999) The mechanism of islet amyloid polypeptide toxicity is membrane disruption by intermediate-sized toxic amyloid particles. Diabetes 48:491–498

Janson J, Laedtke T, Parisi JE, O'Brien P, Petersen RC, Butler PC (2004) Increased risk of type 2 diabetes in Alzheimer disease. Diabetes 53:474–481

Jarrett JT, Lansbury PT Jr (1993) Seeding "one-dimensional crystallization" of amyloid: a pathogenic mechanism in Alzheimer's disease and scrapie? Cell 73:1055–1058

Jeworrek C, Hollmann O, Steitz R, Winter R, Czeslik C (2009) Interaction of IAPP and insulin with model interfaces studied using neutron reflectometry. Biophys J 96:1115–1123

Kagan BL, Azimov R, Azimova R (2004) Amyloid peptide channels. J Membr Biol 202:1–10

Kahn SE, Halban PA (1997) Release of incompletely processed proinsulin is the cause of the disproportionate proinsulinemia of NIDDM. Diabetes 46:1725–1732

Kahn SE, Verchere CB, Andrikopoulos S, Asberry PJ, Leonetti DL, Wahl PW, Boyko EJ, Schwartz RS, Newell-Morris L, Fujimoto WY (1998) Reduced amylin release is a characteristic of impaired glucose tolerance and type 2 diabetes in Japanese Americans. Diabetes 47:640–645

Kajava AV, Aebi U, Steven AC (2005) The parallel superpleated β-structure as a model for amyloid fibrils of human amylin. J Mol Biol 348:247–252

Kapurniotu A, Bernhagen J, Greenfield N, Al-Abed Y, Teichberg S, Frank RW, Voelter W, Bucala R (1998) Contribution of advanced glycosylation to the amyloidogenicity of islet amyloid polypeptide. Eur J Biochem 251:208–216

Kapurniotu A, Schmauder A, Tenidis K (2002) Structure-based design and study of non-amyloidogenic, double *N*-methylated IAPP amyloid core sequences as inhibitors of IAPP amyloid formation and cytotoxicity. J Mol Biol 315:339–350

Kautzky-Willer A, Thomaseth K, Pacini G, Clodi M, Ludvik B, Streli C, Waldhausl W, Prager R (1994) Role of islet amyloid polypeptide secretion in insulin-resistant humans. Diabetologia 37:188–194

Kayed R, Bernhagen J, Greenfield N, Sweimeh K, Brunner H, Voelter W, Kapurniotu A (1999) Conformational transitions of islet amyloid polypeptide (IAPP) in amyloid formation *in vitro*. J Mol Biol 287:781–796

Kayed R, Head E, Thompson JL, McIntire TM, Milton SC, Cotman CW, Glabe CG (2003) Common structure of soluble amyloid oligomers implies common mechanism of pathogenesis. Science 300:486–489

Kayed R, Sokolov Y, Edmonds B, McIntire TM, Milton SC, Hall JE, Glabe CG (2004) Permeabilization of lipid bilayers is a common conformation-dependent activity of soluble amyloid oligomers in protein misfolding diseases. J Biol Chem 279:46363–46366

Kayed R, Pensalfini A, Margol L, Sokolov Y, Sarsoza F, Head E, Hall J, Glabe C (2009) Annular protofibrils are a structurally and functionally distinct type of amyloid oligomer. J Biol Chem 284:4230–4237

Khemtemourian L, Killian JA, Hoppener JW, Engel MF (2008) Recent insights in islet amyloid polypeptide-induced membrane disruption and its role in β-cell death in type 2 diabetes mellitus. Exp Diabetes Res 2008:421287

Khemtemourian L, Engel MF, Kruijtzer JA, Hoppener JW, Liskamp RM, Killian JA (2010) The role of the disulfide bond in the interaction of islet amyloid polypeptide with membranes. Eur Biophys J 39:1359–1364

King H, Aubert RE, Herman WH (1998) Global burden of diabetes, 1995–2025: prevalence, numerical estimates, and projections. Diabetes Care 21:1414–1431

Knight JD, Miranker AD (2004) Phospholipid catalysis of diabetic amyloid assembly. J Mol Biol 341:1175–1187

Knight JD, Hebda JA, Miranker AD (2006) Conserved and cooperative assembly of membrane-bound α-helical states of islet amyloid polypeptide. Biochemistry 45:9496–9508

Knight JD, Williamson JA, Miranker AD (2008) Interaction of membrane-bound islet amyloid polypeptide with soluble and crystalline insulin. Protein Sci 17:1850–1856

Knorre A, Wagner M, Schaefer HE, Colledge WH, Pahl HL (2002) ΔF508-CFTR causes constitutive NF-κB activation through an ER-overload response in cystic fibrosis lungs. Biol Chem 383:271–282

Koda JE, Fineman M, Rink TJ, Dailey GE, Muchmore DB, Linarelli LG (1992) Amylin concentrations and glucose control. Lancet 339:1179–1180

Kolterman OG, Gottlieb A, Moyses C, Colburn W (1995) Reduction of postprandial hyperglycemia in subjects with IDDM by intravenous infusion of AC137, a human amylin analogue. Diabetes Care 18:1179–1182

Konarkowska B, Aitken JF, Kistler J, Zhang S, Cooper GJ (2006) The aggregation potential of human amylin determines its cytotoxicity towards islet β-cells. FEBS J 273:3614–3624

Kong MF, King P, Macdonald IA, Stubbs TA, Perkins AC, Blackshaw PE, Moyses C, Tattersall RB (1997) Infusion of pramlintide, a human amylin analogue, delays gastric emptying in men with IDDM. Diabetologia 40:82–88

Kong MF, Stubbs TA, King P, Macdonald IA, Lambourne JE, Blackshaw PE, Perkins AC, Tattersall RB (1998) The effect of single doses of pramlintide on gastric emptying of two meals in men with IDDM. Diabetologia 41:577–583

Koo BW, Miranker AD (2005) Contribution of the intrinsic disulfide to the assembly mechanism of islet amyloid. Protein Sci 14:231–239

Kouroku Y, Fujita E, Jimbo A, Kikuchi T, Yamagata T, Momoi MY, Kominami E, Kuida K, Sakamaki K, Yonehara S, Momoi T (2002) Polyglutamine aggregates stimulate ER stress signals and caspase-12 activation. Hum Mol Genet 11:1505–1515

Kudva YC, Mueske C, Butler PC, Eberhardt NL (1998) A novel assay *in vitro* of human islet amyloid polypeptide amyloidogenesis and effects of insulin secretory vesicle peptides on amyloid formation. Biochem J 331(Pt 3):809–813

Laybutt DR, Preston AM, Akerfeldt MC, Kench JG, Busch AK, Biankin AV, Biden TJ (2007) Endoplasmic reticulum stress contributes to β cell apoptosis in type 2 diabetes. Diabetologia 50:752–763

Lee SC, Hashim Y, Li JK, Ko GT, Critchley JA, Cockram CS, Chan JC (2001) The islet amyloid polypeptide (amylin) gene S20G mutation in Chinese subjects: evidence for associations with type 2 diabetes and cholesterol levels. Clin Endocrinol (Oxf) 54:541–546

Leissring MA, Farris W, Chang AY, Walsh DM, Wu X, Sun X, Frosch MP, Selkoe DJ (2003) Enhanced proteolysis of β-amyloid in APP transgenic mice prevents plaque formation, secondary pathology, and premature death. Neuron 40:1087–1093

Levy M, Porat Y, Bacharach E, Shalev DE, Gazit E (2008) Phenolsulfonphthalein, but not phenolphthalein, inhibits amyloid fibril formation: implications for the modulation of amyloid self-assembly. Biochemistry 47:5896–5904

Lin CY, Gurlo T, Kayed R, Butler AE, Haataja L, Glabe CG, Butler PC (2007) Toxic human islet amyloid polypeptide (h-IAPP) oligomers are intracellular, and vaccination to induce anti-toxic oligomer antibodies does not prevent h-IAPP-induced β-cell apoptosis in h-IAPP transgenic mice. Diabetes 56:1324–1332

Linse S, Cabaleiro-Lago C, Xue WF, Lynch I, Lindman S, Thulin E, Radford SE, Dawson KA (2007) Nucleation of protein fibrillation by nanoparticles. Proc Natl Acad Sci USA 104:8691–8696

Lopes DH, Colin C, Degaki TL, de Sousa AC, Vieira MN, Sebollela A, Martinez AM, Bloch C Jr, Ferreira ST, Sogayar MC (2004) Amyloidogenicity and cytotoxicity of recombinant mature human islet amyloid polypeptide (rhIAPP). J Biol Chem 279:42803–42810

Lopes DH, Meister A, Gohlke A, Hauser A, Blume A, Winter R (2007) Mechanism of islet amyloid polypeptide fibrillation at lipid interfaces studied by infrared reflection absorption spectroscopy. Biophys J 93:3132–3141

Lorenzo A, Yankner BA (1994) β-Amyloid neurotoxicity requires fibril formation and is inhibited by congo red. Proc Natl Acad Sci USA 91:12243–12247
Lorenzo A, Razzaboni B, Weir GC, Yankner BA (1994) Pancreatic islet cell toxicity of amylin associated with type-2 diabetes mellitus. Nature 368:756–760
Ludvik B, Kautzky-Willer A, Prager R, Thomaseth K, Pacini G (1997) Amylin: history and overview. Diabet Med 14(Suppl 2):S9–S13
Lukinius A, Wilander E, Westermark GT, Engstrom U, Westermark P (1989) Co-localization of islet amyloid polypeptide and insulin in the B cell secretory granules of the human pancreatic islets. Diabetologia 32:240–244
Lutz TA (2006) Amylinergic control of food intake. Physiol Behav 89:465–471
Lutz TA (2010) The role of amylin in the control of energy homeostasis. Am J Physiol Regul Integr Comp Physiol 298:R1475–R1484
Lutz TA, Geary N, Szabady MM, Del Prete E, Scharrer E (1995) Amylin decreases meal size in rats. Physiol Behav 58:1197–1202
Lutz TA, Althaus J, Rossi R, Scharrer E (1998a) Anorectic effect of amylin is not transmitted by capsaicin-sensitive nerve fibers. Am J Physiol 274:R1777–R1782
Lutz TA, Rossi R, Althaus J, Del Prete E, Scharrer E (1998b) Amylin reduces food intake more potently than calcitonin gene-related peptide (CGRP) when injected into the lateral brain ventricle in rats. Peptides 19:1533–1540
Lutz TA, Senn M, Althaus J, Del Prete E, Ehrensperger F, Scharrer E (1998c) Lesion of the area postrema/nucleus of the solitary tract (AP/NTS) attenuates the anorectic effects of amylin and calcitonin gene-related peptide (CGRP) in rats. Peptides 19:309–317
Ma Z, Westermark GT, Sakagashira S, Sanke T, Gustavsson A, Sakamoto H, Engstrom U, Nanjo K, Westermark P (2001) Enhanced *in vitro* production of amyloid-like fibrils from mutant (S20G) islet amyloid polypeptide. Amyloid 8:242–249
Maedler K, Spinas GA, Lehmann R, Sergeev P, Weber M, Fontana A, Kaiser N, Donath MY (2001) Glucose induces β-cell apoptosis via upregulation of the Fas receptor in human islets. Diabetes 50:1683–1690
Maedler K, Sergeev P, Ris F, Oberholzer J, Joller-Jemelka HI, Spinas GA, Kaiser N, Halban PA, Donath MY (2002) Glucose-induced β cell production of IL-1β contributes to glucotoxicity in human pancreatic islets. J Clin Invest 110:851–860
Marchetti P, Del Guerra S, Marselli L, Lupi R, Masini M, Pollera M, Bugliani M, Boggi U, Vistoli F, Mosca F, Del Prato S (2004) Pancreatic islets from type 2 diabetic patients have functional defects and increased apoptosis that are ameliorated by metformin. J Clin Endocrinol Metab 89:5535–5541
Marchetti P, Bugliani M, Lupi R, Marselli L, Masini M, Boggi U, Filipponi F, Weir GC, Eizirik DL, Cnop M (2007) The endoplasmic reticulum in pancreatic β cells of type 2 diabetes patients. Diabetologia 50:2486–2494
Martinez-Alvarez RM, Volkoff H, Cueto JA, Delgado MJ (2008) Molecular characterization of calcitonin gene-related peptide (CGRP) related peptides (CGRP, amylin, adrenomedullin and adrenomedullin-2/intermedin) in goldfish (Carassius auratus): cloning and distribution. Peptides 29:1534–1543
Marzban L, Park K, Verchere CB (2003) Islet amyloid polypeptide and type 2 diabetes. Exp Gerontol 38:347–351
Marzban L, Trigo-Gonzalez G, Zhu X, Rhodes CJ, Halban PA, Steiner DF, Verchere CB (2004) Role of β-cell prohormone convertase (PC)1/3 in processing of pro-islet amyloid polypeptide. Diabetes 53:141–148
Marzban L, Soukhatcheva G, Verchere CB (2005a) Role of carboxypeptidase E in processing of pro-islet amyloid polypeptide in β-cells. Endocrinology 146:1808–1817
Marzban L, Trigo-Gonzalez G, Verchere CB (2005b) Processing of pro-islet amyloid polypeptide in the constitutive and regulated secretory pathways of β cells. Mol Endocrinol 19:2154–2163
Matveyenko AV, Veldhuis JD, Butler PC (2006) Mechanisms of impaired fasting glucose and glucose intolerance induced by an approximate 50% pancreatectomy. Diabetes 55:2347–2356

May PC, Boggs LN, Fuson KS (1993) Neurotoxicity of human amylin in rat primary hippocampal cultures: similarity to Alzheimer's disease amyloid-β neurotoxicity. J Neurochem 61:2330–2333
Mazzaglia A, Micali N, Scolaro LM, Attanasio F, Magri A, Pappalardo G, Villari V (2010) Aggregation properties of the peptide fragments derived from the 17–29 region of the human and rat IAPP: a comparative study with two PEG-conjugated variants of the human sequence. J Phys Chem B 114:705–713
McLean LR, Balasubramaniam A (1992) Promotion of β-structure by interaction of diabetes associated polypeptide (amylin) with phosphatidylcholine. Biochim Biophys Acta 1122:317–320
Meier JJ, Bhushan A, Butler AE, Rizza RA, Butler PC (2005) Sustained β cell apoptosis in patients with long-standing type 1 diabetes: indirect evidence for islet regeneration? Diabetologia 48:2221–2228
Meng F, Abedini A, Plesner A, Middleton CT, Potter KJ, Zanni MT, Verchere CB, Raleigh DP (2010) The sulfated triphenyl methane derivative acid fuchsin is a potent inhibitor of amyloid formation by human islet amyloid polypeptide and protects against the toxic effects of amyloid formation. J Mol Biol 400:555–566
Minn AH, Hafele C, Shalev A (2005) Thioredoxin-interacting protein is stimulated by glucose through a carbohydrate response element and induces β-cell apoptosis. Endocrinology 146:2397–2405
Mirzabekov TA, Lin MC, Kagan BL (1996) Pore formation by the cytotoxic islet amyloid peptide amylin. J Biol Chem 271:1988–1992
Mishra R, Bulic B, Sellin D, Jha S, Waldmann H, Winter R (2008) Small-molecule inhibitors of islet amyloid polypeptide fibril formation. Angew Chem Int Ed Engl 47:4679–4682
Mishra R, Geyer M, Winter R (2009a) NMR spectroscopic investigation of early events in IAPP amyloid fibril formation. Chembiochem 10:1769–1772
Mishra R, Sellin D, Radovan D, Gohlke A, Winter R (2009b) Inhibiting islet amyloid polypeptide fibril formation by the red wine compound resveratrol. Chembiochem 10:445–449
Mollet A, Gilg S, Riediger T, Lutz TA (2004) Infusion of the amylin antagonist AC 187 into the area postrema increases food intake in rats. Physiol Behav 81:149–155
Moriarty DF, Raleigh DP (1999) Effects of sequential proline substitutions on amyloid formation by human amylin$_{20-29}$. Biochemistry 38:1811–1818
Muthusamy K, Albericio F, Arvidsson PI, Govender P, Kruger HG, Maguire GE, Govender T (2010a) Microwave assisted SPPS of amylin and its toxicity of the pure product to RIN-5 F cells. Biopolymers 94:323–330
Muthusamy K, Arvidsson PI, Govender P, Kruger HG, Maguire GE, Govender T (2010b) Design and study of peptide-based inhibitors of amylin cytotoxicity. Bioorg Med Chem Lett 20:1360–1362
Nakagawa T, Zhu H, Morishima N, Li E, Xu J, Yankner BA, Yuan J (2000) Caspase-12 mediates endoplasmic-reticulum-specific apoptosis and cytotoxicity by amyloid-β. Nature 403:98–103
Nanga RP, Brender JR, Xu J, Veglia G, Ramamoorthy A (2008) Structures of rat and human islet amyloid polypeptide IAPP(1–19) in micelles by NMR spectroscopy. Biochemistry 47:12689–12697
Nanga RP, Brender JR, Xu J, Hartman K, Subramanian V, Ramamoorthy A (2009) Three-dimensional structure and orientation of rat islet amyloid polypeptide protein in a membrane environment by solution NMR spectroscopy. J Am Chem Soc 131:8252–8261
Nishi M, Chan SJ, Nagamatsu S, Bell GI, Steiner DF (1989) Conservation of the sequence of islet amyloid polypeptide in five mammals is consistent with its putative role as an islet hormone. Proc Natl Acad Sci USA 86:5738–5742
Nishi M, Bell GI, Steiner DF (1990) Islet amyloid polypeptide (amylin): no evidence of an abnormal precursor sequence in 25 type 2 (non-insulin-dependent) diabetic patients. Diabetologia 33:628–630
Novials A, Rojas I, Casamitjana R, Usac EF, Gomis R (2001) A novel mutation in islet amyloid polypeptide (IAPP) gene promoter is associated with type II diabetes mellitus. Diabetologia 44:1064–1065
O'Brien TD, Hayden DW, Johnson KH, Fletcher TF (1986) Immunohistochemical morphometry of pancreatic endocrine cells in diabetic, normoglycaemic glucose-intolerant and normal cats. J Comp Pathol 96:357–369

O'Brien TD, Butler PC, Westermark P, Johnson KH (1993) Islet amyloid polypeptide: a review of its biology and potential roles in the pathogenesis of diabetes mellitus. Vet Pathol 30:317–332

O'Brien TD, Butler AE, Roche PC, Johnson KH, Butler PC (1994) Islet amyloid polypeptide in human insulinomas. Evidence for intracellular amyloidogenesis. Diabetes 43:329–336

O'Brien TD, Butler PC, Kreutter DK, Kane LA, Eberhardt NL (1995) Human islet amyloid polypeptide expression in COS-1 cells. A model of intracellular amyloidogenesis. Am J Pathol 147:609–616

Ogawa A, Harris V, McCorkle SK, Unger RH, Luskey KL (1990) Amylin secretion from the rat pancreas and its selective loss after streptozotocin treatment. J Clin Invest 85:973–976

Opie EL (1901) The relation of diabetes mellitus to lesions of the pancreas. Hyaline degeneration of the islands of langerhans. J Exp Med 5:527–540

Orci L, Halban P, Perrelet A, Amherdt M, Ravazzola M, Anderson RG (1994) pH-Independent and -dependent cleavage of proinsulin in the same secretory vesicle. J Cell Biol 126:1149–1156

Orrenius S, Zhivotovsky B, Nicotera P (2003) Regulation of cell death: the calcium-apoptosis link. Nat Rev Mol Cell Biol 4:552–565

Padrick SB, Miranker AD (2002) Islet amyloid: phase partitioning and secondary nucleation are central to the mechanism of fibrillogenesis. Biochemistry 41:4694–4703

Park K, Verchere CB (2001) Identification of a heparin binding domain in the *N*-terminal cleavage site of pro-islet amyloid polypeptide. Implications for islet amyloid formation. J Biol Chem 276:16611–16616

Perseghin G, Caumo A, Arcelloni C, Benedini S, Lanzi R, Pagliato E, Sereni LP, Testolin G, Battezzati A, Comi G, Comola M, Luzi L (2003) Contribution of abnormal insulin secretion and insulin resistance to the pathogenesis of type 2 diabetes in myotonic dystrophy. Diabetes Care 26:2112–2118

Pirot P, Naamane N, Libert F, Magnusson NE, Orntoft TF, Cardozo AK, Eizirik DL (2007) Global profiling of genes modified by endoplasmic reticulum stress in pancreatic β cells reveals the early degradation of insulin mRNAs. Diabetologia 50:1006–1014

Porat Y, Kolusheva S, Jelinek R, Gazit E (2003) The human islet amyloid polypeptide forms transient membrane-active prefibrillar assemblies. Biochemistry 42:10971–10977

Porat Y, Mazor Y, Efrat S, Gazit E (2004) Inhibition of islet amyloid polypeptide fibril formation: a potential role for heteroaromatic interactions. Biochemistry 43:14454–14462

Porte D Jr, Kahn SE (1989) Hyperproinsulinemia and amyloid in NIDDM. Clues to etiology of islet β-cell dysfunction? Diabetes 38:1333–1336

Porte D Jr, Kahn SE (1991) Mechanisms for hyperglycemia in type II diabetes mellitus: therapeutic implications for sulfonylurea treatment—an update. Am J Med 90:8S–14S

Porte D Jr, Kahn SE (2001) β-Cell dysfunction and failure in type 2 diabetes: potential mechanisms. Diabetes 50(Suppl 1):S160–S163

Potter KJ, Scrocchi LA, Warnock GL, Ao Z, Younker MA, Rosenberg L, Lipsett M, Verchere CB, Fraser PE (2009) Amyloid inhibitors enhance survival of cultured human islets. Biochim Biophys Acta 1790:566–574

Potter KJ, Abedini A, Marek P, Klimek AM, Butterworth S, Driscoll M, Baker R, Nilsson MR, Warnock GL, Oberholzer J, Bertera S, Trucco M, Korbutt GS, Fraser PE, Raleigh DP, Verchere CB (2010) Islet amyloid deposition limits the viability of human islet grafts but not porcine islet grafts. Proc Natl Acad Sci USA 107:4305–4310

Qiu WQ, Folstein MF (2006) Insulin, insulin-degrading enzyme and amyloid-β peptide in Alzheimer's disease: review and hypothesis. Neurobiol Aging 27:190–198

Quist A, Doudevski I, Lin H, Azimova R, Ng D, Frangione B, Kagan B, Ghiso J, Lal R (2005) Amyloid ion channels: a common structural link for protein-misfolding disease. Proc Natl Acad Sci USA 102:10427–10432

Radovan D, Opitz N, Winter R (2009) Fluorescence microscopy studies on islet amyloid polypeptide fibrillation at heterogeneous and cellular membrane interfaces and its inhibition by resveratrol. FEBS Lett 583:1439–1445

Reidelberger RD, Haver AC, Arnelo U, Smith DD, Schaffert CS, Permert J (2004) Amylin receptor blockade stimulates food intake in rats. Am J Physiol Regul Integr Comp Physiol 287:R568–R574

Resmini E, Minuto F, Colao A, Ferone D (2009) Secondary diabetes associated with principal endocrinopathies: the impact of new treatment modalities. Acta Diabetol 46:85–95
Riediger T, Schmid HA, Young AA, Simon E (1999) Pharmacological characterisation of amylin-related peptides activating subfornical organ neurones. Brain Res 837:161–168
Riediger T, Schmid HA, Lutz T, Simon E (2001) Amylin potently activates AP neurons possibly via formation of the excitatory second messenger cGMP. Am J Physiol Regul Integr Comp Physiol 281:R1833–R1843
Riediger T, Zuend D, Becskei C, Lutz TA (2004) The anorectic hormone amylin contributes to feeding-related changes of neuronal activity in key structures of the gut-brain axis. Am J Physiol Regul Integr Comp Physiol 286:R114–R122
Rishton GM (2008) Aggregator compounds confound amyloid fibrillization assay. Nat Chem Biol 4:159–160
Ritzel RA, Butler PC (2003) Replication increases β-cell vulnerability to human islet amyloid polypeptide-induced apoptosis. Diabetes 52:1701–1708
Ritzel RA, Meier JJ, Lin CY, Veldhuis JD, Butler PC (2007) Human islet amyloid polypeptide oligomers disrupt cell coupling, induce apoptosis, and impair insulin secretion in isolated human islets. Diabetes 56:65–71
Roberts AN, Leighton B, Todd JA, Cockburn D, Schofield PN, Sutton R, Holt S, Boyd Y, Day AJ, Foot EA et al (1989) Molecular and functional characterization of amylin, a peptide associated with type 2 diabetes mellitus. Proc Natl Acad Sci USA 86:9662–9666
Rocken C, Linke RP, Saeger W (1992) Immunohistology of islet amyloid polypeptide in diabetes mellitus: semi-quantitative studies in a post-mortem series. Virchows Arch A Pathol Anat Histopathol 421:339–344
Rossi MC, Nicolucci A, Arcangeli A, Cimino A, De Bigontina G, Giorda C, Meloncelli I, Pellegrini F, Valentini U, Vespasiani G (2008) Baseline quality-of-care data from a quality-improvement program implemented by a network of diabetes outpatient clinics. Diabetes Care 31:2166–2168
Ryan G, Briscoe TA, Jobe L (2009) Review of pramlintide as adjunctive therapy in treatment of type 1 and type 2 diabetes. Drug Des Devel Ther 2:203–214
Saafi EL, Konarkowska B, Zhang S, Kistler J, Cooper GJ (2001) Ultrastructural evidence that apoptosis is the mechanism by which human amylin evokes death in RINm5F pancreatic islet β-cells. Cell Biol Int 25:339–350
Sakagashira S, Sanke T, Hanabusa T, Shimomura H, Ohagi S, Kumagaye KY, Nakajima K, Nanjo K (1996) Missense mutation of amylin gene (S20G) in Japanese NIDDM patients. Diabetes 45:1279–1281
Sakagashira S, Hiddinga HJ, Tateishi K, Sanke T, Hanabusa T, Nanjo K, Eberhardt NL (2000) S20G mutant amylin exhibits increased *in vitro* amyloidogenicity and increased intracellular cytotoxicity compared to wild-type amylin. Am J Pathol 157:2101–2109
Sanke T, Bell GI, Sample C, Rubenstein AH, Steiner DF (1988) An islet amyloid peptide is derived from an 89-amino acid precursor by proteolytic processing. J Biol Chem 263:17243–17246
Sanke T, Hanabusa T, Nakano Y, Oki C, Okai K, Nishimura S, Kondo M, Nanjo K (1991) Plasma islet amyloid polypeptide (amylin) levels and their responses to oral glucose in type 2 (non-insulin-dependent) diabetic patients. Diabetologia 34:129–132
Saraogi I, Hebda JA, Becerril J, Estroff LA, Miranker AD, Hamilton AD (2009) Synthetic α-helix mimetics as agonists and antagonists of islet amyloid polypeptide aggregation. Angew Chem Int Ed Engl 49:736–739
Scherbaum WA (1998) The role of amylin in the physiology of glycemic control. Exp Clin Endocrinol Diabetes 106:97–102
Schubert D, Behl C, Lesley R, Brack A, Dargusch R, Sagara Y, Kimura H (1995) Amyloid peptides are toxic via a common oxidative mechanism. Proc Natl Acad Sci USA 92:1989–1993
Sciacca MF, Pappalardo M, Milardi D, Grasso DM, La Rosa C (2008) Calcium-activated membrane interaction of the islet amyloid polypeptide: implications in the pathogenesis of type II diabetes mellitus. Arch Biochem Biophys 477:291–298

Scott LJ, Mohlke KL, Bonnycastle LL, Willer CJ, Li Y, Duren WL, Erdos MR, Stringham HM, Chines PS, Jackson AU, Prokunina-Olsson L, Ding CJ, Swift AJ, Narisu N, Hu T, Pruim R, Xiao R, Li XY, Conneely KN, Riebow NL, Sprau AG, Tong M, White PP, Hetrick KN, Barnhart MW, Bark CW, Goldstein JL, Watkins L, Xiang F, Saramies J, Buchanan TA, Watanabe RM, Valle TT, Kinnunen L, Abecasis GR, Pugh EW, Doheny KF, Bergman RN, Tuomilehto J, Collins FS, Boehnke M (2007) A genome-wide association study of type 2 diabetes in Finns detects multiple susceptibility variants. Science 316:1341–1345

Scrocchi LA, Chen Y, Waschuk S, Wang F, Cheung S, Darabie AA, McLaurin J, Fraser PE (2002) Design of peptide-based inhibitors of human islet amyloid polypeptide fibrillogenesis. J Mol Biol 318:697–706

Scrocchi LA, Ha K, Chen Y, Wu L, Wang F, Fraser PE (2003) Identification of minimal peptide sequences in the (8–20) domain of human islet amyloid polypeptide involved in fibrillogenesis. J Struct Biol 141:218–227

Seino S (2001) S20G mutation of the amylin gene is associated with type II diabetes in Japanese. Study group of comprehensive analysis of genetic factors in diabetes mellitus. Diabetologia 44:906–909

Shastry BS (2003) Neurodegenerative disorders of protein aggregation. Neurochem Int 43:1–7

Smith RE, Ashiya M (2007) Antihypertensive therapies. Nat Rev Drug Discov 6:597–598

Smith JF, Knowles TP, Dobson CM, Macphee CE, Welland ME (2006) Characterization of the nanoscale properties of individual amyloid fibrils. Proc Natl Acad Sci USA 103:15806–15811

Smith PE, Brender JR, Ramamoorthy A (2009) Induction of negative curvature as a mechanism of cell toxicity by amyloidogenic peptides: the case of islet amyloid polypeptide. J Am Chem Soc 131:4470–4478

Sparr E, Engel MF, Sakharov DV, Sprong M, Jacobs J, de Kruijff B, Hoppener JW, Killian JA (2004) Islet amyloid polypeptide-induced membrane leakage involves uptake of lipids by forming amyloid fibers. FEBS Lett 577:117–120

Stridsberg M, Wilander E (1991) Islet amyloid polypeptide (IAPP). A short review. Acta Oncol 30:451–456

Takano J, Tomioka M, Tsubuki S, Higuchi M, Iwata N, Itohara S, Maki M, Saido TC (2005) Calpain mediates excitotoxic DNA fragmentation via mitochondrial pathways in adult brains: evidence from calpastatin mutant mice. J Biol Chem 280:16175–16184

Tatarek-Nossol M, Yan LM, Schmauder A, Tenidis K, Westermark G, Kapurniotu A (2005) Inhibition of hIAPP amyloid-fibril formation and apoptotic cell death by a designed hIAPP amyloid- core-containing hexapeptide. Chem Biol 12:797–809

Tenidis K, Waldner M, Bernhagen J, Fischle W, Bergmann M, Weber M, Merkle ML, Voelter W, Brunner H, Kapurniotu A (2000) Identification of a penta- and hexapeptide of islet amyloid polypeptide (IAPP) with amyloidogenic and cytotoxic properties. J Mol Biol 295:1055–1071

Thompson RG, Gottlieb A, Organ K, Koda J, Kisicki J, Kolterman OG (1997a) Pramlintide: a human amylin analogue reduced postprandial plasma glucose, insulin, and C-peptide concentrations in patients with type 2 diabetes. Diabet Med 14:547–555

Thompson RG, Pearson L, Kolterman OG (1997b) Effects of 4 weeks' administration of pramlintide, a human amylin analogue, on glycaemia control in patients with IDDM: effects on plasma glucose profiles and serum fructosamine concentrations. Diabetologia 40:1278–1285

Tomiyama T, Kaneko H, Kataoka K, Asano S, Endo N (1997) Rifampicin inhibits the toxicity of pre-aggregated amyloid peptides by binding to peptide fibrils and preventing amyloid-cell interaction. Biochem J 322(Pt 3):859–865

UKPDS (1998) Intensive blood-glucose control with sulphonylureas or insulin compared with conventional treatment and risk of complications in patients with type 2 diabetes (UKPDS 33). Lancet 352:837–853

Verchere CB, D'Alessio DA, Palmiter RD, Weir GC, Bonner-Weir S, Baskin DG, Kahn SE (1996) Islet amyloid formation associated with hyperglycemia in transgenic mice with pancreatic β cell expression of human islet amyloid polypeptide. Proc Natl Acad Sci USA 93:3492–3496

Voight BF, Scott LJ, Steinthorsdottir V, Morris AP, Dina C, Welch RP, Zeggini E, Huth C, Aulchenko YS, Thorleifsson G, McCulloch LJ, Ferreira T, Grallert H, Amin N, Wu G, Willer

CJ, Raychaudhuri S, McCarroll SA, Langenberg C, Hofmann OM, Dupuis J, Qi L, Segre AV, van Hoek M, Navarro P, Ardlie K, Balkau B, Benediktsson R, Bennett AJ, Blagieva R, Boerwinkle E, Bonnycastle LL, Bengtsson Bostrom K, Bravenboer B, Bumpstead S, Burtt NP, Charpentier G, Chines PS, Cornelis M, Couper DJ, Crawford G, Doney AS, Elliott KS, Elliott AL, Erdos MR, Fox CS, Franklin CS, Ganser M, Gieger C, Grarup N, Green T, Griffin S, Groves CJ, Guiducci C, Hadjadj S, Hassanali N, Herder C, Isomaa B, Jackson AU, Johnson PR, Jorgensen T, Kao WH, Klopp N, Kong A, Kraft P, Kuusisto J, Lauritzen T, Li M, Lieverse A, Lindgren CM, Lyssenko V, Marre M, Meitinger T, Midthjell K, Morken MA, Narisu N, Nilsson P, Owen KR, Payne F, Perry JR, Petersen AK, Platou C, Proenca C, Prokopenko I, Rathmann W, Rayner NW, Robertson NR, Rocheleau G, Roden M, Sampson MJ, Saxena R, Shields BM, Shrader P, Sigurdsson G, Sparso T, Strassburger K, Stringham HM, Sun Q, Swift AJ, Thorand B et al (2010) Twelve type 2 diabetes susceptibility loci identified through large-scale association analysis. Nat Genet 42:579–589

Westermark P (1973) Fine structure of islets of Langerhans in insular amyloidosis. Virchows Arch A Pathol Pathol Anat 359:1–18

Westermark P, Wilander E (1978) The influence of amyloid deposits on the islet volume in maturity onset diabetes mellitus. Diabetologia 15:417–421

Westermark P, Wernstedt C, O'Brien TD, Hayden DW, Johnson KH (1987) Islet amyloid in type 2 human diabetes mellitus and adult diabetic cats contains a novel putative polypeptide hormone. Am J Pathol 127:414–417

Westermark P, Engstrom U, Westermark GT, Johnson KH, Permerth J, Betsholtz C (1989) Islet amyloid polypeptide (IAPP) and pro-IAPP immunoreactivity in human islets of Langerhans. Diabetes Res Clin Pract 7:219–226

Westermark P, Engstrom U, Johnson KH, Westermark GT, Betsholtz C (1990) Islet amyloid polypeptide: pinpointing amino acid residues linked to amyloid fibril formation. Proc Natl Acad Sci USA 87:5036–5040

Wild S, Roglic G, Green A, Sicree R, King H (2004) Global prevalence of diabetes: estimates for the year 2000 and projections for 2030. Diabetes Care 27:1047–1053

Williamson JA, Miranker AD (2007) Direct detection of transient α-helical states in islet amyloid polypeptide. Protein Sci 16:110–117

Wiltzius JJ, Sievers SA, Sawaya MR, Eisenberg D (2009) Atomic structures of IAPP (amylin) fusions suggest a mechanism for fibrillation and the role of insulin in the process. Protein Sci 18:1521–1530

Wookey PJ, Tikellis C, Du HC, Qin HF, Sexton PM, Cooper ME (1996) Amylin binding in rat renal cortex, stimulation of adenylyl cyclase, and activation of plasma renin. Am J Physiol 270:F289–F294

Yamada K, Yuan X, Ishiyama S, Nonaka K (1998) Glucose tolerance in Japanese subjects with S20G mutation of the amylin gene. Diabetologia 41:125

Yan LM, Tatarek-Nossol M, Velkova A, Kazantzis A, Kapurniotu A (2006) Design of a mimic of nonamyloidogenic and bioactive human islet amyloid polypeptide (IAPP) as nanomolar affinity inhibitor of IAPP cytotoxic fibrillogenesis. Proc Natl Acad Sci USA 103:2046–2051

Yonemoto IT, Kroon GJ, Dyson HJ, Balch WE, Kelly JW (2008) Amylin proprotein processing generates progressively more amyloidogenic peptides that initially sample the helical state. Biochemistry 47:9900–9910

Young A (2005) Amylin and the integrated control of nutrient influx. Adv Pharmacol 52:67–77

Young AA, Rink TJ, Wang MW (1993) Dose response characteristics for the hyperglycemic, hyperlactemic, hypotensive and hypocalcemic actions of amylin and calcitonin gene-related peptide-I (CGRP α) in the fasted, anaesthetized rat. Life Sci 52:1717–1726

Zhang S, Liu J, Saafi EL, Cooper GJ (1999) Induction of apoptosis by human amylin in RINm5F islet β-cells is associated with enhanced expression of p53 and p21WAF1/CIP1. FEBS Lett 455:315–320

Zhang P, Wang C, Gao K, Wang D, Mao J, An J, Xu C, Wu D, Yu H, Liu JO, Yu L (2010) The ubiquitin ligase itch regulates apoptosis by targeting thioredoxin-interacting protein for ubiquitin-dependent degradation. J Biol Chem 285:8869–8879

Zhao WQ, Townsend M (2009) Insulin resistance and amyloidogenesis as common molecular foundation for type 2 diabetes and Alzheimer's disease. Biochim Biophys Acta 1792: 482–496
Zhao H, Tuominen EK, Kinnunen PK (2004) Formation of amyloid fibers triggered by phosphatidylserine-containing membranes. Biochemistry 43:10302–10307
Zraika S, Hull RL, Verchere CB, Clark A, Potter KJ, Fraser PE, Raleigh DP, Kahn SE (2010) Toxic oligomers and islet β cell death: guilty by association or convicted by circumstantial evidence? Diabetologia 53:1046–1056

Chapter 8
Protein Misfolding and Toxicity in Amyotrophic Lateral Sclerosis

Aaron Kerman and Avijit Chakrabartty

Abstract Amyotrophic lateral sclerosis (ALS) is a devastating adult-onset disease characterized by the degeneration of upper and lower motor neurons in the brain, brainstem, and spinal cord, resulting in paralysis and death. The misfolding and aggregation of proteins such as Cu-Zn superoxide dismutase (SOD1) and TAR-DNA-binding protein of 43 kDa (TDP-43) are believed to be key etiological features of familial and sporadic ALS (fALS and sALS), respectively. These proteins are deposited in intracellular inclusion bodies in the remaining motor neurons of affected patients. Recent evidence from cellular and rodent models of ALS resulting from mutant SOD1 suggests the existence of soluble and insoluble misfolded forms of SOD1. These SOD1 species, which accumulate specifically in spinal cord, are present long before microscopically visible inclusion bodies are seen, and their presence is better correlated with toxicity both *in vitro* and *in vivo*. Studies have elucidated the key structural features of these misfolded species: namely, they are non-amyloidogenic structures characterized by significant misfolding of SOD1, including monomerization, metal loss, reduction of the SOD1 intrasubunit disulfide bond, and exposure of hydrophobic surfaces. The potential toxic effects of these species, including effects on mitochondria and oxidative stress, are discussed.

Keywords Amyotrophic lateral sclerosis • Protein misfolding • Amyloid • Cu-Zn superoxide dismutase • TDP-43

A. Kerman • A. Chakrabartty (✉)
Department of Medical Biophysics, University of Toronto, TMDT-MaRS, 101 College St., Toronto, ON, Canada M5G 1L7

Department of Biochemistry, University of Toronto, TMDT-MaRS, 101 College St., Toronto, ON, Canada M5G 1L7
e-mail: chakrab@uhnres.utoronto.ca

F. Rahimi and G. Bitan (eds.), *Non-fibrillar Amyloidogenic Protein Assemblies—Common Cytotoxins Underlying Degenerative Diseases*, DOI 10.1007/978-94-007-2774-8_8,

8.1 Introduction

Amyotrophic lateral sclerosis (ALS) is the most common adult-onset motor-neuron disease. It is characterized by the degeneration and death of both upper motor neurons in the primary motor cortex, and lower motor neurons in the brainstem and spinal cord. Degeneration of upper motor neurons leads to spasticity and hyperreflexia, while lower motor-neuron degeneration leads to weakness and muscle atrophy, ultimately resulting in paralysis. Onset is generally in midlife, and death due to respiratory failure (or related complications) usually occurs 3–5 years after onset/diagnosis. The incidence of ALS is 1–2/100,000 population per year. The incidence is slightly higher for men than for women, with a ratio of approximately 1.5:1 (Mitchell and Borasio 2007).

The cause(s) of the majority (~90%) of ALS cases are not known; these cases are referred to as sporadic ALS (sALS). Generally, sALS is believed to be caused by a combination of genetic and environmental factors, and recent whole-genome-association studies of ALS patients have begun to identify genetic polymorphisms that have a modest impact on the risk of developing ALS (Dunckley et al. 2007; van Es et al. 2007). The environmental factors that may contribute to ALS development are not well-known; however, smoking has recently been identified as having a statistically significant impact on risk of ALS (Armon 2009). An exception is the ALS–parkinsonism–dementia complex (ALS-PDC) that occurred with unusually high incidence in the Chamorro Indian population on the island of Guam in the mid-twentieth century (the incidence has begun to decline in recent decades). This disease is believed to be caused by a dietary neurotoxin, although the identity of the toxin is still controversial (Steele and McGeer 2008).

The remaining 10% of ALS cases are familial (fALS), usually with an autosomal-dominant pattern of inheritance (Beleza-Meireles and Al-Chalabi 2009). Several genes have now been associated with fALS. Mutations in the antioxidant enzyme Cu-Zn superoxide dismutase (SOD1) are responsible for 20–25% of fALS cases, as well as 2–5% of apparently sporadic cases. Other genes that have been linked to fALS are *TDP-43* (the transactivation region DNA-binding protein of 43 kDa) (Sreedharan et al. 2008; Kabashi et al. 2008), *FUS/TLS* (Vance et al. 2009; Kwiatkowski et al. 2009), *ANG* (angiogenin), *ALS2* (alsin), *VAPB*, *SETX* (senataxin) (Beleza-Meireles and Al-Chalabi 2009), and *OPTN* (optineurin) (Maruyama et al. 2010). There are still several variants of fALS for which the causative gene has not yet been identified (Beleza-Meireles and Al-Chalabi 2009). The clinical features of many fALS variants differ significantly from the sporadic disease, in that they often are characterized by onset at younger ages and very long disease courses (i.e., slow progression) (Beleza-Meireles and Al-Chalabi 2009). However, the fALS variants caused by mutations in *SOD1* and *TDP-43* are remarkably similar, clinically and pathologically, to sALS. Therefore, the study of these two variants of fALS may shed light on the disease mechanisms involved in the vast majority of ALS cases.

There are several pathological mechanisms that are believed to be involved in the etiology of ALS. In addition to protein misfolding and aggregation, which are the focus of this chapter, these include oxidative stress, mitochondrial dysfunction, defects in axonal transport, and glutamate excitotoxicity (Rothstein 2009). A variety of pharmacological treatments based on one or more of these mechanisms have been investigated in a plethora of clinical trials. Unfortunately, there is currently no highly effective pharmacological treatment for ALS. The only approved drug for use in ALS is riluzole, which is believed to mitigate glutamate excitotoxicity by enhancing glutamate uptake by astrocytes. Riluzole modestly extends survival by a few months (Lacomblez et al. 1996).

This chapter will discuss the role of protein misfolding and aggregation in ALS, particularly familial ALS caused by *SOD1* mutations (SOD1-ALS). In particular, recent evidence that soluble species and small insoluble aggregates, rather than large inclusion bodies, are the key toxic species will be reviewed. The structure, possible mechanisms of formation, as well as the association of these SOD1 species with mitochondrial dysfunction and oxidative stress, will also be discussed.

8.2 Protein Misfolding and Aggregation in ALS

In addition to motor-neuron loss from the motor cortex, brainstem, and spinal cord, a near-universal pathological hallmark of ALS is the presence of inclusion bodies (i.e., abnormal intracellular deposits of aggregated proteins) in the remaining motor neurons of patients (Strong et al. 2005). Several types of inclusions have been observed in ALS. These include Lewy-body-like hyaline inclusions (LBHIs), hyaline conglomerate inclusions (HCIs), skein-like inclusions, Bunina bodies, and axonal spheroids. LBHIs, HCIs, and skeins are generally immunopositive for ubiquitin, but differ in size and shape. LBHIs are small round inclusions with a dense core surrounded by a more diffuse "halo". HCIs are much larger, neurofilament-rich multifocal inclusions that occur frequently in fALS caused by *SOD1* mutations, but are rarer in other forms of ALS. Skeins are small threadlike structures that are believed to be precursors of LBHIs and also primarily occur in sALS (Xiao et al. 2006; Strong et al. 2005). Bunina bodies are a highly ALS-specific phenomenon. They are small, round eosinophilic inclusions that are not ubiquitin-positive, but are positive for the lysosomal protease inhibitor cystatin C, suggesting a lysosomal origin for these inclusions (Wada et al. 1999). Axonal spheroids, as their name implies, are large round axonal swellings rich in neurofilaments. The presence of such inclusions suggests that protein misfolding and aggregation are key etiological features of ALS and places ALS in the category of diseases known as 'conformational diseases' (Carrell and Lomas 1997).

8.2.1 SOD1

Mutations in *SOD1* were discovered as a cause of fALS in 1993 (Rosen et al. 1993). SOD1 is a 32-kDa homodimeric enzyme that is expressed in most tissues and is predominantly cytosolic. Each monomer contains an eight-stranded β-barrel, as well as

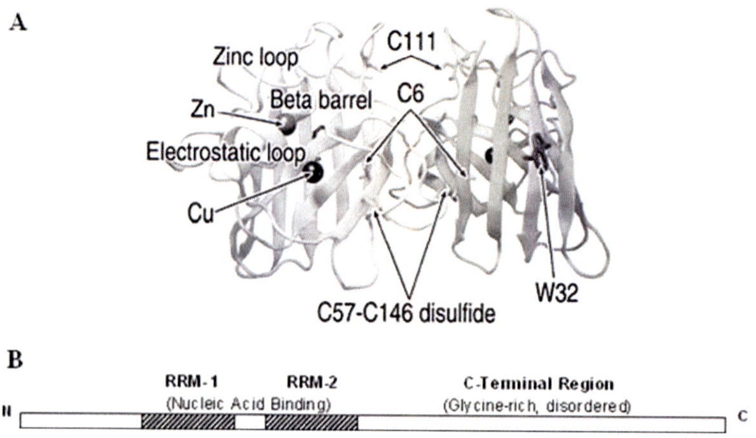

Fig. 8.1 (**A**) Structure of SOD1. The key features of the SOD1 structure that are highlighted in this figure include the eight-stranded β-barrel, two bound metal ions (Cu and Zn), the four cysteine residues, two of which form an intrasubunit disulfide bond (C57–C146), and the two long loops (zinc loop and electrostatic loop) that are involved in metal coordination and formation of the active site. The two monomer subunits are related by a rotational axis of symmetry, so that the structures highlighted here for the monomer on the *left*, face away from the reader in the monomer on the *right*. Also indicated is the position of the single tryptophan residue of SOD1 (W32), useful for monitoring conformational changes of the SOD1 β-barrel using fluorescence spectroscopy (Reproduced, with permission, from Mulligan et al. (2008), Copyright 2008 Elsevier B.V). (**B**) Domain architecture of TDP-43. The structure of TDP-43 can be divided into four main domains: The *N*-terminal domain, two RNA-recognition motifs (RRM) (RRM1 and RRM2) that are involved in the binding of nucleic acids, and the glycine-rich *C*-terminal domain. Many of the ALS-causing mutations that have been discovered in TDP-43 reside in the *C*-terminal domain

binding sites for one copper and one zinc ion per monomer. The copper ion is essential for the catalytic activity of SOD1, while both metal ions contribute to its structural stability and specificity. In addition to the β-barrel, each monomer contains two long loops that are critical to the structure and function of the enzyme. The electrostatic loop is involved in binding of the metal ions and also contributes to the electrostatic guidance of the superoxide anion to the active site, allowing SOD1 to achieve near-diffusion-limited reaction rates. The zinc-binding loop is primarily involved in the coordination of the zinc ion. Another key structural feature of the monomer is the presence of an intrasubunit disulfide bond, linking Cys57 to Cys146 (Fig. 8.1a) (Rakhit and Chakrabartty 2006). In mammalian cells, the delivery of copper into the active site of SOD1 is mediated by the copper chaperone for SOD1 (CCS) (Brown et al. 2004; Wong et al. 2000). CCS also plays a role in facilitating the formation of the intrasubunit disulfide bond. However, an alternate mechanism of SOD1 copper loading (and disulfide oxidation), which is CCS-independent and glutathione-dependent, is also believed to exist in mammalian cells (Carroll et al. 2004).

SOD1 plays a role in cellular antioxidant defense. The enzymatic function of SOD1 is the dismutation of superoxide radicals, which are toxic by-products of oxidative respiration. They are produced as a result of “leakage” of electrons from the electron-transport chain, leading to the one-electron reduction of molecular oxygen to superoxide. SOD1 detoxifies superoxide by first oxidizing one molecule of superoxide to molecular oxygen, resulting in the reduction of the catalytic copper ion from Cu(II) to Cu(I). The reduced enzyme then reduces a second molecule of superoxide to hydrogen peroxide, resulting in the conversion of the catalytic copper back to the Cu(II) state (Hart et al. 1999). The physiological importance of SOD1 function is suggested by the fact that SOD1-knockout mice develop a range of pathologies, including hepatocarcinogenesis (Elchuri et al. 2005) and increased neuronal death after axonal injury (Reaume et al. 1996).

Over 100 ALS-causing mutations in SOD1 have now been discovered (Wroe et al. 2008). These mutations do not preferentially occur near the active site—rather, they are spread throughout the sequence of the enzyme. Although some mutants have a significant impact on SOD1 enzyme activity, the activities of many of the mutant proteins are close to that of the wild-type enzyme (Borchelt et al. 1994; Marklund et al. 1997). Furthermore, SOD1-knockout mice do not develop motor-neuron disease (Reaume et al. 1996), suggesting that SOD1-ALS is not caused by a loss of SOD1 activity. Several lines of mice that overexpress human SOD1 with ALS-causing mutations have now been developed. Despite the presence of normal amounts of wild-type mouse SOD1, these mice develop features of human ALS, including motor-neuron degeneration, muscle atrophy, paralysis, and intraneuronal inclusion bodies (Gurney et al. 1994; Jonsson et al. 2004, 2006b; Bruijn et al. 1997b; Wang et al. 2002, 2005; Wong et al. 1995). All of these results suggest that SOD1-ALS is causes by a "toxic gain of function," rather than a loss of function.

The prevailing theory in the field of SOD1-ALS research is that the toxic gain of function is an increased propensity to misfold and/or aggregate. The initial evidence in favor of this theory was the discovery that in SOD1-ALS, LBHIs and HCIs are often immunopositive for the SOD1 protein. The presence of SOD1 protein in HCIs of human cases has been confirmed for several SOD1 mutations, including A4V (Shibata et al. 1996b), G127X (a truncation mutant resulting from a four base-pair insertion at codon 127) (Jonsson et al. 2004), H46R (Ohi et al. 2004), I113T (Kokubo et al. 1999), G72C (Stewart et al. 2006), L126Z (a truncation mutant resulting from a two base-pair deletion at codon 126) (Kato et al. 1996) and L126S (Takehisa et al. 2001).

8.2.2 *Mechanisms of SOD1 Misfolding and Aggregation*

In its properly folded form, SOD1 is a highly soluble protein. Therefore, the presence of aggregated mutant SOD1 in ALS suggests that at least a portion of mutant SOD1 is somehow misfolded *in vivo*. Many groups have studied the thermodynamics and kinetics of SOD1 folding and unfolding *in vitro* in order to identify SOD1 conformations that might be present *in vivo*, and that could be the precursors to the observed inclusion bodies. Our initial attempts at identifying such SOD1 conformations involved studying the denaturant-induced unfolding kinetics of purified, wild-type SOD1 (Mulligan et al. 2008). Several different

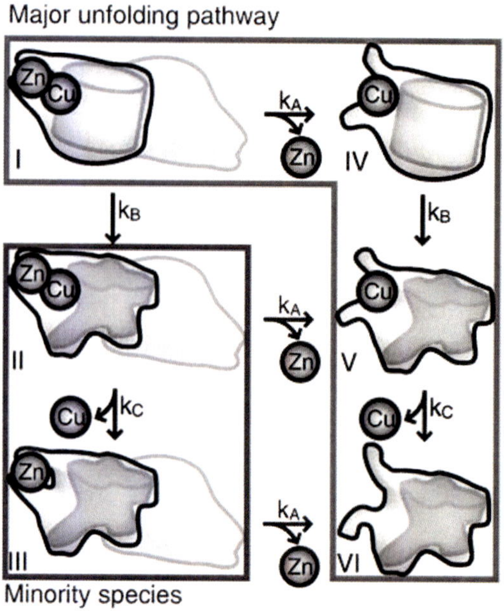

Fig. 8.2 Denaturant-induced unfolding mechanism of wild-type SOD1. The kinetics of copper and zinc release, as well as dimer dissociation and conformational changes in the β-barrel (monitored by tryptophan fluorescence), were determined in the presence of the denaturant guanidinium chloride. Determination of rate constants for each of these processes allowed a kinetic unfolding model to be established. In the above diagram, the three structures in the left column are dimers, and the three structures in the right column are monomers. The cylinders represent a folded β-barrel, while the other structures possess a misfolded β-barrel. The model proposes that zinc loss and dimer dissociation occur early and simultaneously, while copper loss and conformational changes in the β-barrel occur later (Reproduced, with permission, from Mulligan et al. (2008), Elsevier B.V.)

aspects of SOD1 structure were monitored upon unfolding, including metal release, dimer dissociation, and conformational changes of the β-barrel (monitored by observing changes in the fluorescence of the single tryptophan residue of SOD1: see Fig. 8.1a). Determination of rate constants for each of these processes suggested that the initial steps in SOD1 unfolding are release of the zinc ion and dissociation of the dimer, which occur nearly simultaneously. Conformational changes of the β-barrel and release of the copper ion occur subsequently (Fig. 8.2). These experiments allowed the proposal of a detailed unfolding mechanism where zinc-deficient monomers of SOD1 were the most highly populated, long-lived kinetic intermediates along the unfolding pathway (Mulligan et al. 2008). Preliminary data on the effects of ALS-causing mutations on this unfolding mechanism suggest that mutations increase many of the unfolding rates, leading to more rapid accumulation of the relevant intermediates.[1] As discussed throughout

[1]P. Ip, V.K. Mulligan and A. Chakrabartty, unpublished data.

the rest of this chapter, the presence of such metal-deficient monomers has been corroborated in mouse models of SOD1-ALS.

Other *in vitro* experiments on the folding and unfolding of wild-type and mutant SOD1 have also confirmed that the pathway(s) between fully folded and fully unfolded (or aggregated) SOD1 are populated by partially folded, metal-deficient monomeric intermediates (Rumfeldt et al. 2006; Khare et al. 2004; Rakhit et al. 2004). Such studies have also shown that in SOD1, the presence of bound metals, as well as the intrasubunit disulfide bond, has a strong influence on both dimer stability and the stability of the monomers (Arnesano et al. 2004; Doucette et al. 2004). Reduction of the disulfide bond was found to impair severely the ability of apoSOD1 to form dimers, and the resulting monomers were also highly destabilized (Lindberg et al. 2004). The presence of bound metals, however, does promote the formation of dimers from reduced SOD1 monomers, suggesting that metal binding and disulfide status can each affect monomer and dimer stability by independent mechanisms (Doucette et al. 2004).

Although these experiments allow the identification of folding/misfolding intermediates, how and why mis/unfolding of mutant SOD1 actually occurs *in vivo* is not immediately obvious. This is perplexing in the case of SOD1-ALS, because in its natively folded, metal-loaded form, SOD1 is an exceptionally stable protein. The melting temperature of fully metallated wild-type SOD1 is over 90°C, and although mutations generally lead to a lowering of this melting temperature, most ALS-related SOD1 mutants, in their metallated states, still have melting temperatures that would allow them to remain folded under physiological conditions (Rodriguez et al. 2002). Furthermore, that properly metallated species of mutant SOD1 adopt a native structure is also suggested by the demonstration that many such mutants display high levels of dismutase activity (Borchelt et al. 1994; Marklund et al. 1997).

One possible answer to the question of how mutant SOD1 misfolds *in vivo* is that misfolding and/or aggregation occurs during the folding process, before the stable, metal-loaded form of the protein is even formed. It has been shown that, in a cell-free reticulocyte extract, mutation slows the rate of post-translational folding of SOD1. In the presence of zinc, the translated SOD1 folds into a stable structure in several kinetic phases that could be distinguished based on protease sensitivity (Bruns and Kopito 2007). The presence of mutations significantly slows the rate of folding, suggesting that mutations in SOD1 may allow the accumulation of toxic, aggregation-prone folding intermediates.

Another explanation for how significant unfolding of folded SOD1 might occur *in vivo* is as a result of oxidative damage to the protein. SOD1's normal cellular role of converting superoxide anion into hydrogen peroxide and oxygen puts SOD1 at high risk for oxidative damage. It is known that exposure to hydrogen peroxide leads to the enzymatic inactivation of SOD1 (Kim and Han 2000) due to the generation of free radicals at the active site. Furthermore, SOD1 must have a rather long half-life in motor neurons with long axons, because SOD1 is transported down axons at the rate of only a few millimeters per day (Borchelt et al. 1998). Finally, SOD1 is present in high concentrations throughout motor neurons (Pardo et al. 1995). These properties of motor neurons and SOD1 suggest that significant amounts of oxidized SOD1 could accumulate over time in these cells (Rakhit et al. 2002), leading to misfolding and aggregation.

We have carried out extensive analyses of the misfolding and aggregation of SOD1 resulting from oxidation. In particular, we have looked at two different *in vitro* models of oxidation: metal-catalyzed oxidation (MCO) and hydrogen peroxide.

MCO of SOD1 is a relatively mild oxidation system that involves exposure of SOD1 to copper and ascorbate (Rakhit et al. 2002, 2004). Mass spectrometric analysis of SOD1 treated in this manner shows that the primary sites for oxidation are three histidine residues near the active site of the enzyme. Over the course of 48 h at 37°C, MCO of SOD1 leads to aggregation of the protein as measured by right-angle light scattering. Dynamic light scattering, as well as sedimentation-equilibrium analysis of this process shows that monomerization of the protein is an early event that precedes extensive aggregation (Rakhit et al. 2004). Zinc-deficient forms of SOD1 were much more susceptible to MCO-induced aggregation, suggesting that zinc-deficient monomers may be the most aggregation-prone SOD1 species in this system. Metal analysis of the resulting aggregates further indicates that some metal loss occurs, as expected, based on the fact that one or more of the metal-binding histidine residues are oxidized in the process (Rakhit and Chakrabartty 2006). In particular, the aggregates were zinc-deficient, consistent with the idea that zinc-deficient monomers are the precursors of the aggregates. Importantly, the resulting aggregates do not bind Congo red, and only bind thioflavin T rather weakly, suggesting that this method of inducing SOD1 misfolding/aggregation does not lead to amyloid formation, consistent with what is observed *in vivo* (see Sect. 8.2.4) (Kerman et al. 2010). The amorphous, non-fibrillar nature of these aggregates was further confirmed by atomic-force microscopy (Rakhit et al. 2002). Several mutations in SOD1 were found to increase the propensity of the protein to aggregate, suggesting either that mutations increase the susceptibility to oxidation, or that oxidized mutant SOD1 is more susceptible to misfolding/aggregation, or both.

Oxidation of SOD1 with hydrogen peroxide was also investigated. This method allowed us to investigate the processes of metal release and conformational change in a detailed manner, which was not possible in the case of the MCO system (because of the presence of large amounts of copper). In this case, lower temperatures and lower protein concentrations were employed, in the interest of identifying misfolding intermediates that are present before significant aggregation takes place. As was the case with MCO, mass spectrometric analysis suggested that oxidation is limited primarily to three histidine residues, suggesting that the oxidation mechanisms in MCO and peroxide oxidation are likely quite similar, and involve the catalytic copper ion. A detailed analysis of metal release induced by peroxide revealed that, even in the presence of high concentrations of peroxide, only partial metal release occurs, with the final product being a mixture of copper-deficient and zinc-deficient protein. This metal release is accompanied by partial monomerization, consistent with the results of MCO treatment. Surprisingly, however, monitoring of the conformational state of the β-barrel using tryptophan fluorescence indicated that no significant conformational change takes place.[2] This contrasts with the results of the MCO experiment, where

[2] V.K. Mulligan, P. Sharda, K. Hadley and A. Chakrabartty, manuscript in preparation.

CD spectroscopy showed unfolding of the β-barrel (Rakhit et al. 2002). The difference in these results may arise from differences in temperature, protein concentration and time, although both experiments confirm that relatively mild oxidative treatment of SOD1 leads to the formation of metal-deficient monomers.

The hypothesis that mutant SOD1 misfolding is caused by oxidative modification *in vivo* is supported by the findings of two groups who investigated protein carbonyl content in the spinal cords of mice overexpressing SOD1-G93A. Spinal cord protein carbonyl levels were found to be increased, relative to non-transgenic mice or mice overexpressing wild-type SOD1, as early as 30 days of age. Proteomic techniques were used to identify the proteins with increased carbonyl levels and it was found that SOD1 itself was one of the major targets of oxidation (Poon et al. 2005; Andrus et al. 1998). In order to determine if the mechanism of oxidation is the same as our MCO and peroxide experiments, it will be necessary to identify the exact sites at which SOD1 is modified *in vivo*.

8.2.3 In Vivo Conformation of Misfolded SOD1

To determine whether the conformational changes observed *in vitro*, such as dimer dissociation and β-barrel unfolding, are responsible for SOD1 misfolding and aggregation *in vivo*, we developed two conformation-specific antibodies that report on different aspects of SOD1 structure. The occurrence of dimer dissociation as an early step in the mechanism of SOD1 *in vitro* unfolding and aggregation led us to develop the first antibody, SEDI (for SOD1 Exposed Dimer Interface). This antibody was raised against a peptide comprising residues 143–151 of SOD1 (Rakhit et al. 2007). This sequence is normally buried in the SOD1 dimer interface, but becomes exposed for antibody interactions upon dimer dissociation. Immunohistochemistry using this antibody allowed us to detect SOD1 with an exposed dimer interface both in mouse models of ALS and in human patient tissue. In the G93A, G85R and G37R mouse models of SOD1-ALS, dimer-interface-exposed SOD1 was specifically detected in the ventral horn and ventral roots of mice that were slightly younger than the age at which overt symptoms of motor-neuron degeneration first appear. Misfolded SOD1 was not, however, detected in young asymptomatic mice, indicating that the appearance of symptoms is strongly correlated with the appearance of misfolded SOD1. Misfolded SOD1 could also be immunoprecipitated from the soluble fraction of spinal-cord extracts of these mice, suggesting that soluble species of dimer-interface-exposed SOD1 are also present (Rakhit et al. 2007). Such species are presumably the precursors of the aggregates seen upon immunohistochemical staining of the tissue.

The SEDI antibody was also used to detect dimer-interface-exposed SOD1 in several different human cases of SOD1-ALS (Liu et al. 2009; Kerman et al. 2010). In each of the cases, HCIs in the remaining motor neurons were stained strongly by SEDI. As mentioned above, other investigators have detected the presence of SOD1 in HCIs in SOD1-ALS cases resulting from other mutations, although in those cases, the antibodies used do not allow us to infer the conformation of the aggregated SOD1. It is therefore

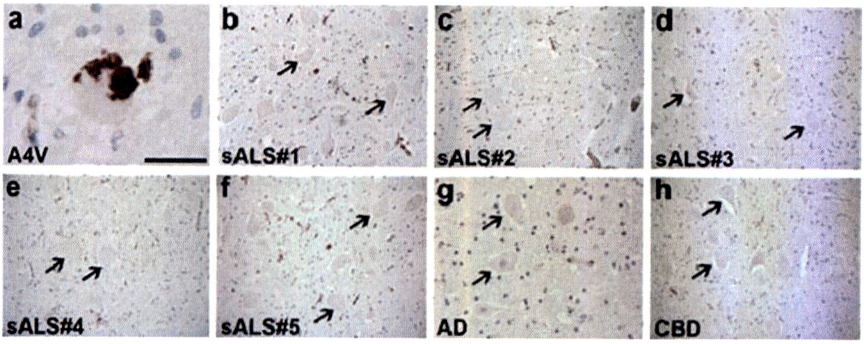

Fig. 8.3 Misfolded SOD1 is present in SOD1-ALS but not sporadic ALS. Immunohistochemistry using the USOD antibody was performed on spinal-cord sections of SOD1-ALS and sporadic ALS cases. As shown in *panel a*, USOD intensely stains large hyaline inclusions in motor neurons. Other panels show that motor neurons (in some cases indicated by *arrows*) in sporadic ALS cases are not stained by USOD. Scale bar: 35 μm (*panel a*); 70 μm (all other panels) (Reproduced with permission from Kerman et al. (2010), Copyright 2010 Springer-Verlag)

tempting to conclude that misfolding of SOD1 that includes exposure of the dimer interface is a common event in most cases of ALS caused by SOD1 mutations.

While the SEDI antibody reports on the exposure of the dimer interface, it does not report on the conformational integrity of the monomers (i.e., the β-barrel). To this end, we designed a second antibody, named USOD (for Unfolded SOD1) that was raised against SOD1 residues 42–48 (Kerman et al. 2010). This sequence, which forms part of the fourth β-strand and contains two of the copper-chelating histidine residues, is normally deeply buried within the SOD1 monomer structure. Exposure of this sequence for antibody interactions would require extensive unfolding of the β-barrel, including loss of the copper ion. In the G93A mouse model, USOD stained ventral horns and ventral roots in a pattern very similar to that obtained with SEDI. In two human cases of SOD1-ALS [A4V and ΔG27/P28 (Zinman et al. 2009)], the use of serial spinal cord sections allowed HCIs to be stained by both SEDI and USOD. Strong immunopositivity was observed with both SEDI and USOD (Fig. 8.3, panel A; Fig. 8.4, panels E and F), indicating that SOD1 deposited in those inclusions was extensively misfolded, with exposure of the dimer interface as well as unfolding of the β-barrel (Kerman et al. 2010). Use of the USOD antibody in other cases of SOD1-ALS will reveal whether unfolding of the β-barrel is a general phenomenon in SOD1-ALS.

8.2.4 The Non-amyloid Nature of Aggregated SOD1 in Human ALS

Ultrastructurally, SOD1-immunopositive LBHIs and HCIs comprise 15–25-nm-thick, granule-coated fibers, as well as other granular material. The presence of SOD1 in the fibers has been confirmed by immunoelectron microscopy (Kato et al. 2000).

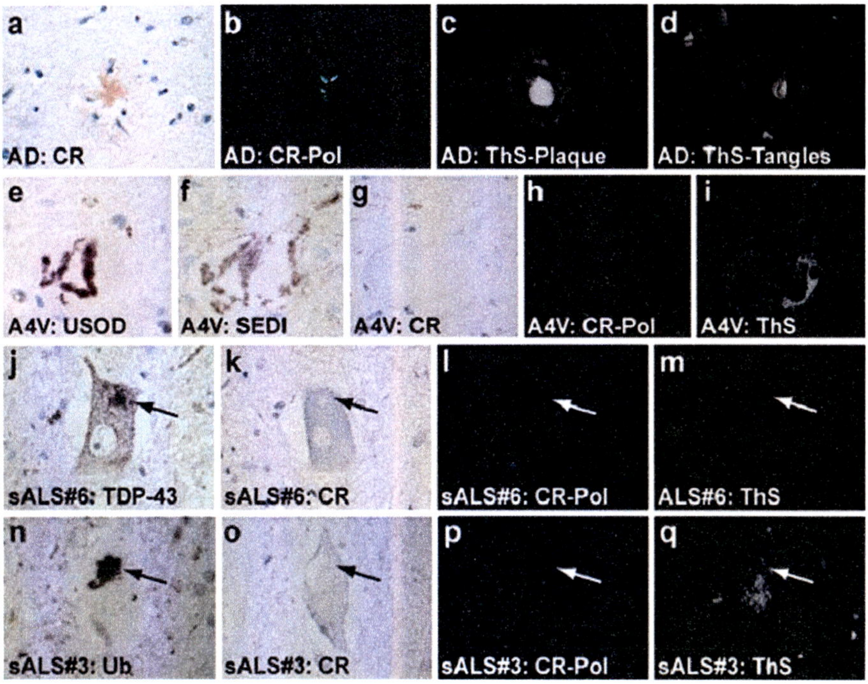

Fig. 8.4 Absence of amyloid in ALS. The *top row* shows the staining of amyloid plaques and neurofibrillary tangles from Alzheimer's-disease brain tissue. *Panel a* shows Congo red (CR) staining of a plaque, while *panel b* shows the same plaque viewed through crossed polarizers ('Pol'), resulting in characteristic apple-green birefringence. *Panels c and d* show thioflavin S (ThS) staining of plaques and tangles, respectively. The *second row* shows staining of serial sections (i.e., staining of the same motor neuron) from a SOD1-A4V ALS case with USOD (*panel e*), SEDI (*panel f*), Congo red (*panel g*) viewed through crossed polarizers (*panel h*), and with thioflavin S (*panel l*). Similarly, the *bottom two rows* show staining of serial spinal-cord sections from two different sALS cases. *Panel j* shows the presence of a ubiquitin-positive inclusion, while *panel n* shows a TDP-43-positive inclusion. The absence of Congo red and thioflavin S staining in each case indicates that amyloid structure is not present in ALS motor neurons. The diffuse, granular fluorescence seen in *panels i and q* is not thioflavin S staining—it arises from lipofuscin, an autofluorescent pigment often present in aged cells (Reproduced with permission from Kerman et al. (2010), Copyright 2010 Springer-Verlag)

The fibrous nature of some of the aggregated proteins in these inclusions has raised the question of whether or not aggregated SOD1 in ALS assumes an amyloid structure. If true, this would make SOD1-ALS similar to other neurodegenerative diseases such as Alzheimer's, Parkinson's, and Huntington's diseases, where aggregates of the disease-causing protein(s) invariably contain amyloid structure. The key characteristics of amyloid structure are (1) the presence of long, unbranched 5–10-nm fibrils, confirmed by electron microscopy or atomic-force microscopy; (2) β-sheet secondary structure, confirmed by circular-dichroism or infra-red spectroscopy; and (3) binding and apple-green birefringence of the amyloid-selective dye, Congo red (Sipe 1992). The presence of amyloid in Alzheimer's,

Parkinson's, and Huntington's diseases has been confirmed by staining postmortem tissue samples with Congo red. In each case, Congo-red binding, as well as apple-green birefringence when viewing through crossed polarizers, has confirmed the presence of amyloid fibrils *in vivo* in these diseases (Sipe 1992; Hashimoto et al. 1998; McGowan et al. 2000).

In the case of ALS, several investigators had suggested that amyloid was not a key feature of either SOD1-ALS or sporadic ALS. This was based on anecdotal reports of the absence of Congo red staining in ALS postmortem tissue (Kato et al. 2000), as well as the solubility properties of aggregated protein in ALS (Kwong et al. 2008). The amyloid plaques that are present in Alzheimer's disease are highly insoluble—they cannot be dissolved in 8 M urea, a very strong chaotrope/denaturant that can generally solubilize other aggregated proteins. In their efforts to identify biochemically the proteins present LBHIs of sporadic ALS, Lee, Trojanowski, and colleagues attempted to use a differential solubilization approach that had been successful in Alzheimer's disease, owing to the unique (in)solubility properties of Aβ plaques. This approach was unsuccessful in the case of sALS, leading the authors to conclude that amyloid is not present in sALS (Kwong et al. 2008).

In recent years, however, the issue of amyloid in ALS has been revisited. In some mouse models of SOD1-ALS, some of the inclusions that are present in spinal motor neurons have been shown to bind thioflavin S (Wang et al. 2002, 2003), suggesting that amyloid structure may be present. Also, several different groups have discovered that SOD1 can be driven to form amyloid fibrils *in vitro* (Oztug Durer et al. 2009; Furukawa et al. 2008; Chattopadhyay et al. 2008). In these experiments, fully metallated SOD1 with an intact intrasubunit disulfide bond could not be induced to form fibrils under physiological solution conditions. Amyloid formation generally required varying degrees of metal depletion and/or disulfide-bond reduction, with fully reduced apoSOD1 being most susceptible to formation of fibrils (Furukawa et al. 2008). Agitation of the protein(s) in the presence of Teflon beads also was required in most cases. The ability to form amyloid structures *in vitro*, however, does not necessarily imply the presence of amyloid in patients. The formation of amyloid *in vitro* has been reported for several proteins that are not known to be involved in any disease, and the ability to form amyloid has been suggested to be an inherent property of the polypeptide backbone (Stefani and Dobson 2003). Whether amyloid is actually formed by a protein *in vivo*, however, would presumably depend on the precise mechanisms responsible for the misfolding and aggregation of the protein in the disease state.

In light of these considerations, we systematically investigated whether amyloid fibrils were present in postmortem spinal-cord tissue of both SOD1-ALS and sALS (Kerman et al. 2010). Our approach was to stain SOD1-immunopositive inclusions (in SOD1-ALS) and ubiquitin-positive/TDP-43-positive inclusions (in sALS) with the amyloid-selective dyes Congo red and thioflavin S (Elghetany and Saleem 1988). In the case of SOD1-ALS, we investigated two cases arising from different SOD1 mutations—the A4V mutation, and a deletion of G27/P28 (Zinman et al. 2009). We identified SOD1-immunopositive inclusions using the SEDI or USOD antibodies described above (Fig. 8.4, panels E and F). In the case of sALS, we investigated several LBHIs in 2 different cases. The LBHIs were identified immunohistochemically using anti-ubiquitin and/or anti-TDP-43 antibodies [LBHI in sALS invariably contain TDP-43 protein (see Sect. 8.6)] (Fig. 8.4, panels J and N).

For all of the inclusions that we investigated in both SOD1-ALS and sALS, we did not observe any Congo-red staining or birefringence (Fig. 8.4). This was corroborated by the absence of thioflavin S staining of the same inclusions (Fig. 8.4, panels I, M, and Q). By contrast, we were readily able to detect the presence of amyloid structures in Alzheimer's brain using the same staining techniques. This was true of both amyloid plaques and neurofibrillary tangles (Fig. 8.4, top row). We concluded that inclusion-forming proteins such as SOD1 and TDP-43 do not form amyloid fibrils in human cases of ALS (Kerman et al. 2010).

8.3 Soluble and Insoluble Forms of Misfolded SOD1

There is mounting evidence that large deposits of insoluble protein (such as intracytoplasmic inclusions), when they are present at all, are a late-stage phenomenon, and that their presence is not required for SOD1-mediated motor-neuron degeneration. The most straightforward evidence for this idea is that in the SOD1-G93A mouse model of ALS, significant pathological abnormalities can be detected long before large inclusion bodies are present. Such abnormalities include vacuolar degeneration of motor-neuron mitochondria (Dal Canto and Gurney 1995) as well as muscle atrophy and motor deficits (Hayworth and Gonzalez-Lima 2009). Similar findings have been made in other mouse models, such as the SOD1-G37R model, where motor-neuron loss occurs at a time (~133 days of age) where SOD1-containing inclusion are extremely rare or undetectable (Watanabe et al. 2001), thus suggesting that other forms of insoluble and/or soluble mutant protein are present that are better correlated with toxicity.

There is also evidence from human cases of SOD1-ALS that the formation of SOD1-containing intracytoplasmic inclusions may not be a necessary step in SOD1-mediated motor-neuron degeneration. In many cases of SOD1-ALS examined to date, LBHIs and/or HCIs are prominent features. Furthermore, these inclusions are frequently immunopositive for SOD1 protein. However, cases of SOD1-ALS have been reported where LBHI and/or HCI were either absent, or did not stain with antibodies against SOD1 (Ohi et al. 2002; Shaw et al. 1997). This lends further support to the idea that the formation of large, microscopically visible aggregates may not be the common denominator underlying motor-neuron degeneration in SOD1-ALS.

Johnston et al. used a cellular model of mutant SOD1 expression to demonstrate that the formation of large inclusion bodies and accumulation of insoluble protein are separable events (Johnston et al. 2000). When SOD1 mutants G93A and G85R were expressed in HEK293 cells, a fraction of the mutant protein was insoluble in mild, non-ionic detergents, and the insoluble protein also formed characteristic SDS-resistant, high-molecular-weight bands upon electrophoresis. Cells expressing the mutant proteins also frequently formed large, juxtanuclear inclusion bodies.

Treatment of the cells with nocodazole, a drug that induces microtubule depolymerization, completely abrogated the formation of inclusion bodies, suggesting that the formation of the large inclusions is an active process dependent on microtubule-based transport. However, the treatment with nocodazole did not affect the detergent solubility of the mutant protein, indicating that insolubility and inclusion-body formation are mechanistically separate events. In G93A transgenic mice, the SDS-resistant high-molecular-weight bands could be detected in spinal-cord extracts at postnatal day 30, long before the presence of microscopically visible inclusion bodies (Johnston et al. 2000).

Both Karch et al. (2009) and Jonsson et al. (2006a) also demonstrated the presence of detergent insoluble, high-molecular-weight forms of mutant SOD1 in spinal-cord extracts of mouse models of SOD1-ALS. As reported by Johnston et al., these insoluble species were present as early as 50 days of age in all of the mouse models tested. Significant increases the abundance of such species were noted near the end-stage of disease, suggesting that they may play an important role during later, rapidly progressing phases of disease. However, Jonsson et al. showed that, compared to mice expressing mutant SOD1, similar amounts of insoluble SOD1 were also present in mice overexpressing wild-type SOD1 at earlier time points (i.e., 50–200 days of age) (Jonsson et al. 2006a). Mice overexpressing wild-type SOD1 do experience subtle motor-neuron pathology, but this generally occurs much later than in mutant-overexpressing mice (Jaarsma et al. 2000). Therefore, the presence of similar amounts of insoluble SOD1 suggests that other forms of SOD1 may be responsible for motor-neuron pathology.

Further evidence that insoluble forms of SOD1 may not be the key toxic species comes from the work of Witan et al., who studied the effects of wild-type SOD1 on the toxicity and solubility of mutant SOD1. In one study (Witan et al. 2008), obligate homodimers and heterodimers of SOD1 containing several different mutant SOD1 species were expressed in HEK293 cells. Cells expressing heterodimers comprising mutant and wild-type SOD1 accumulated significantly less insoluble protein than those expressing homodimers of mutant SOD1. Furthermore, in N2A cells, wild-type-mutant heterodimers conferred a greater susceptibility to oxidative stress than mutant homodimers, thus indicating that cellular toxicity and aggregation propensity are not correlated. Similar results with respect to protein solubility and cellular toxicity were obtained when these homo- and heterodimers were expressed in the neurons of *Caenorhabditis elegans*: wild-type–mutant heterodimers were less prone to form intracellular aggregates, but conferred greater susceptibility to paraquat-mediated toxicity. A subsequent study (Witan et al. 2009) confirmed that wild-type SOD1 can form heterodimers with mutant SOD1 expressed in HEK293 cells, and that this heterodimer formation is correlated with reduced accumulation of insoluble SOD1. Furthermore, little or no accumulation of wild-type SOD1 in the insoluble fraction was observed in these experiments, further suggesting that the effects of wild-type SOD1 on the solubility and toxicity of mutant SOD1 is related to the formation of soluble heterodimers. Similar results were obtained by Prudencio et al. (2009), who also showed that the co-expression of wild-type SOD1 and mutant SOD1 in cell culture led to a significant slowing of the aggregation rate of the mutant SOD1. Therefore, soluble, misfolded forms of SOD1 are likely candidates for the toxic species.

A striking example of a lack of correlation between the accumulation of insoluble SOD1 aggregates and toxicity is the work of Son et al., who attempted to delineate the role of CCS in SOD1-ALS pathogenesis (Son et al. 2007). Because CCS is involved in the maturation of SOD1, it was hypothesized that simultaneous overexpression of CCS along with mutant SOD1 might impact the disease course in transgenic mice by influencing the proportion of properly metallated SOD1. Surprisingly, however, crossing CCS-overexpressing mice with SOD1-G93A mice led to a much more rapid disease course, despite the fact that the CCS-overexpressing mice themselves had a normal lifespan and no evidence of neurological deficits. The age-at-disease-onset of the SOD1-G93A mouse used in the study dropped from 180 days to 11 days, and the mean survival time dropped from 242 days to 36 days. Rather than being associated with an increased amount of protein aggregation, ubiquitin-positive inclusion bodies were essentially undetectable, as was the presence of SDS-resistant, high-molecular-weight SOD1 species in spinal-cord extracts, suggesting that a detergent-soluble species of SOD1-G93A was responsible for the rapid disease course. Proescher et al. (2008) confirmed the near absence of detergent-insoluble SOD1 in these mice, and also determined that, rather than increasing the proportion of SOD1 with the proper intrasubunit disulfide bond, the presence of CCS led to a small but significant increase in the amount of disulfide-reduced SOD1. These effects of CCS on mutant SOD1 were confirmed in cell culture, where the co-expression of CCS led to a reduction in the amount of detergent-insoluble mutant SOD1. The formation of a CCS-SOD1 heterodimer is believed to be required for CCS-mediated copper loading and disulfide bond formation of SOD1 (Lamb et al. 2000; Casareno et al. 1998). Therefore, a possible explanation for disease acceleration by CCS is the stabilization of soluble, toxic SOD1 species that might otherwise be degraded by the ubiquitin-proteasome system, or be sequestered into large aggregates. The CCS–SOD1 double transgenic mice have mitochondrial pathology that is much more severe than that of other mutant SOD1 transgenic mice (Son et al. 2007), indicating that the toxicity of such soluble SOD1 species may arise from their interaction with mitochondria. It is unlikely, however, that a SOD1–CCS interaction is a general requirement for SOD1-mediated toxicity. SOD1-G93A mice that have reduced CCS expression still develop motor-neuron disease (Subramaniam et al. 2002), and although CCS is immunochemically detected in inclusions in mouse models of ALS (Watanabe et al. 2001), its presence in the inclusions of human patients is uncertain (Watanabe et al. 2001; Kato et al. 2001).

The studies described here also provide a possible explanation for the effect of wild-type SOD1 overexpression on disease severity in mutant SOD1-ALS mouse models. The co-expression of high levels of wild-type SOD1 has been found to accelerate the disease course in several mouse models of SOD1-ALS, including lines expressing G93A (Jaarsma et al. 2000), G85R (Wang et al. 2009), A4V, and L126Z (Deng et al. 2006). Part of this effect may be due to wild-type SOD1 itself being toxic when expressed at high levels—wild-type SOD1 was, in fact, found to be recruited into the detergent-insoluble fraction in some of these cases (Deng et al. 2006).

However, it was also found that levels of SOD1-G93A protein were increased in the presence of wild-type SOD1, suggesting a stabilizing effect of wild-type SOD1 on the mutant SOD1 (Fukada et al. 2001). Combined with the results of the cell-culture experiments described above, such data suggests that soluble, misfolded SOD1 species are better correlated with toxicity than insoluble aggregates or inclusions. In particular, because heterodimer formation between mutant SOD1 and wild-type SOD1 or CCS may be involved, it is possible that the toxic species is in a near-native conformation that would be capable of forming such heterodimers.

8.3.1 Isolation and Characterization of Soluble, Misfolded SOD1

Isolation and characterization of soluble, misfolded SOD1 species have been attempted in both mouse models and human cases of ALS. Marklund and colleagues have used hydrophobic-interaction chromatography (HIC) employing octyl-sepharose matrices to isolate soluble, misfolded species of SOD1 from the spinal cords of SOD1-ALS mouse models (Zetterstrom et al. 2007). The species that were isolated by HIC were in the monomer–trimer range (determined by size-exclusion chromatography), metal-depleted and disulfide-reduced. Not all of the disulfide-reduced SOD1 in the extract was recovered by HIC (Jonsson et al. 2006a; Zetterstrom et al. 2007), suggesting that disulfide reduction alone was not sufficient to expose enough hydrophobic surface sufficient for binding to the HIC matrix. They found that in the case of relatively stable SOD1 mutants that are expressed at high levels (such as G93A), these species represented only a small fraction of the total soluble SOD1, whereas in the case of unstable mutants (such as G85R and G127X) that are expressed at low levels, the HIC-isolated species represented the majority of these proteins. The species isolated by HIC were present throughout the life of the mice, and their amount increased over time specifically in spinal-cord tissue (Zetterstrom et al. 2007).

We have also used our conformation-specific antibodies to detect misfolded SOD1 species in detergent-soluble fractions of spinal-cord homogenates from both ALS mice and human patients (Liu et al. 2009; Kerman et al. 2010; Rakhit et al. 2007). Both SEDI and USOD could detect misfolded SOD1 in these homogenates. Consistent with the findings of the Marklund group, the immunoprecipitated species only represented a small fraction of the total soluble SOD1. An interesting aspect of the immunoprecipitation experiments and the HIC experiments is that in each case, only full-length, apparently unmodified SOD1 was detected. The USOD antibody recognizes an epitope in the *N*-terminal half of the protein, while the SEDI antibody recognizes the extreme *C*-terminus, suggesting that our immunoprecipitation experiments should have been able to detect *N*- or *C*-terminal fragments of the protein if they were present. By contrast, when Borchelt and colleagues investigated the detergent-insoluble (i.e., presumably inclusion-containing) fraction, some proteolytic fragments and oligo-ubiquitinated species of SOD1 were also detected (Wang et al. 2003). This suggests that much of the ubiquitination and proteolytic processing takes place after the SOD1 molecules have already become part of larger, detergent-insoluble aggregates.

As discussed above, *in vitro* studies on the misfolding of SOD1 suggest that the bound metals, as well as the presence of the intrasubunit disulfide bond, have a major impact on the overall stability of the SOD1 molecule. Therefore, one possible reason for the accumulation of disulfide-reduced, metal-deficient misfolded SOD1 may be an increased propensity of mutant SOD1 for metal release and/or disulfide bond reduction. Mutations have been found to reduce significantly the affinity of SOD1 for zinc (Crow et al. 1997), and there is evidence that mutations differentially affect the ability of SOD1 to either obtain copper initially, or to maintain it in a bound state *in vivo* (Hayward et al. 2002). While certain mutant SOD1 species appear to be efficiently metallated *in vivo*, such as G93A, other mutations such as G85R are significantly metal-depleted (Hayward et al. 2002; Borchelt et al. 1994). Tiwari and Hayward showed that mutations had a significant effect on the ability of reducing agents such as TCEP to reduce the SOD1 disulfide bond *in vitro* (Tiwari and Hayward 2003). Marklund and colleagues confirmed the presence of significant amounts of disulfide reduced SOD1 in the spinal cords of several lines of mutant SOD1 transgenic mice (Jonsson et al. 2006a). These studies support the idea that the ability of mutations in SOD1 to cause ALS is related to increased propensity to form metal-deficient, disulfide-reduced species of the protein. In addition to increasing the propensity of SOD1 to form such species, mutation also decreases their stability, leading to increased populations of monomeric, misfolded SOD1 (Kayatekin et al. 2010; Svensson et al. 2010).

8.3.2 The Role of Disulfide Cross-Linking in SOD1 Misfolding and Aggregation

The presence of disulfide-reduced SOD1 in affected tissues suggests that aberrant disulfide cross-linking could be involved in the misfolding and aggregation of mutant SOD1. Such cross-linking could involve any of the four cysteine residues of SOD1. In addition to the two cysteine residues involved in the intrasubunit disulfide bond (C57 and C146), there are two free cysteine residues—C6, which is buried in the protein interior, and C111, which is near the dimer interface and is somewhat exposed to solvent. The two free cysteine residues could presumably be involved in intermolecular cross-linking even if the intrasubunit disulfide is still intact. Some investigators have found that replacement of C6 and/or C111 has a significant impact on the aggregation and toxicity of SOD1 expressed in cell culture (Niwa et al. 2007), while other investigators have shown that replacement of the SOD1 cysteine residues can modulate the aggregation of the protein, but does not prevent aggregation from occurring (Karch and Borchelt 2008). In transgenic mice, detergent-insoluble SOD1 contains both disulfide cross-linked (Furukawa et al. 2006) and disulfide-reduced (Karch et al. 2009) species of SOD1, although detergent-soluble SOD1 did not contain any significant amounts of intermolecular cross-linked species (Karch et al. 2009). Furthermore, the detergent insolubility of disulfide-cross-linked SOD1 was maintained upon reduction of the cross-links with a reducing agent, suggesting that aggregation is primarily mediated by protein–protein interactions and not covalent cross-linking. It can be concluded that the toxicity of disulfide-reduced SOD1 does not result from intermolecular disulfide cross-linking *per se*; rather, reduction of the disulfide destabilizes the protein structure, facilitating toxicity and aggregation.

8.4 Association of (Misfolded) SOD1 with Spinal-Cord Mitochondria

Mitochondrial dysfunction is believed to be one of the mechanisms resulting in motor-neuron degeneration in ALS (Rothstein 2009; Hervias et al. 2006). Defects in mitochondrial respiration, increased production of reactive oxygen species, and loss of mitochondrial membrane potential have been demonstrated in both cellular and mouse models of SOD1-ALS (Hervias et al. 2006). In addition to functional defects, mitochondria undergo morphological alterations in ALS. In mouse models of SOD1-ALS caused by the G93A and G37R mutations, mitochondria undergo significant vacuolization (Wong et al. 1995; Dal Canto and Gurney 1995), although other mouse models (as well as human patients) do not exhibit this phenomenon, suggesting that it is a result of artificially high levels of protein overexpression in these mice, and not a general mechanism of SOD1 toxicity.

Several lines of evidence from cellular and rodent models of SOD1-ALS suggest that biochemically and/or conformationally altered SOD1 is responsible for the toxic effects of SOD1 on mitochondria. Several different groups have shown that in spinal cord homogenates from SOD1-ALS mice, as well as in cellular models of mutant SOD1 expression, there is a preferential association of mutant SOD1 with mitochondria, and that this association is specific to the spinal cord (Liu et al. 2004; Ferri et al. 2006; Jung et al. 2002; Higgins et al. 2002; Deng et al. 2006). The mitochondrion-associated SOD1 is often present as disulfide-linked oligomers, suggesting that misfolding of the protein is involved. Vande Velde et al. confirmed an association of mutant SOD1 with mitochondria by demonstrating that when mitochondria are floated on a sucrose gradient, mutant SOD1 remains associated with the mitochondria (Vande Velde et al. 2008). This indicates that previous findings of the association of SOD1 with mitochondria did not result from co-sedimentation of SOD1 aggregates with mitochondria during centrifugation. Furthermore, Vande Velde et al. investigated mitochondria-associated SOD1 using a monoclonal antibody, raised against the electrostatic loop of SOD1, that specifically recognizes misfolded SOD1 and does not recognize natively folded holoSOD1. They found that SOD1 associated with spinal-cord mitochondria was indeed misfolded, and that this misfolded SOD1 was protease-sensitive and associated with the cytoplasmic face of the outer mitochondrial membrane (Vande Velde et al. 2008). This association was very strong, in that it could not be disrupted by high salt or alkali treatments that are

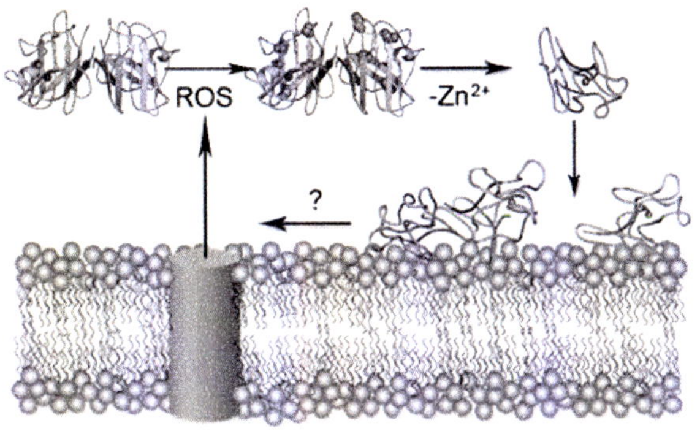

Fig. 8.5 A model for SOD1-ALS pathogenesis, highlighting the interplay between oxidative stress, protein misfolding, and mitochondrial dysfunction. Reactive oxygen species (ROS) produced by mitochondria react with folded SOD1, leading to metal release (particularly zinc release) and misfolding. Misfolded SOD1 then interacts with spinal-cord mitochondria, leading to mitochondrial dysfunction, including increased production of ROS. This establishes a positive feedback cycle where modest initial amounts of ROS and/or misfolded protein ultimately result in pathological amounts of oxidative stress, protein misfolding, and mitochondrial dysfunction

typically used to remove peripheral proteins from membrane. Rakhit et al. also found that there is a preferential association of misfolded SOD1 with spinal-cord mitochondria. Using the SEDI antibody described earlier, they found that dimer-interface-exposed SOD1 that could be immunoprecipitated from spinal-cord extracts of SOD1-G93A rats was predominantly present in the mitochondrial fraction (Rakhit et al. 2007).

The significant association between misfolded SOD1 species and mitochondria suggests a mechanism of SOD1 toxicity that links together the phenomena of protein misfolding, mitochondrial dysfunction, and oxidative stress (Fig. 8.5). The initial misfolding of SOD1, which may itself be initiated by low levels of oxidative stress, leads to the association of misfolded SOD1 with mitochondria. This aberrant interaction may lead to various forms of mitochondrial dysfunction, including increased production of superoxide anion by the electron-transport chain. These free radicals may then cause further SOD1 misfolding, thus leading to a positive feedback cycle involving significant mitochondrial dysfunction, oxidative stress, and protein misfolding.

8.5 The "Pro-oxidant" Hypothesis for Mutant SOD1 Toxicity

An alternative proposal for the toxic function that is acquired by SOD1 as a result of mutation is an alteration of its enzymatic activity, resulting in its conversion to a "pro-oxidant" enzyme. Beckman and colleagues have demonstrated that zinc-deficient SOD1 has an increased propensity for tyrosine nitration of proteins (Beckman and Koppenol 1996). This stems from a greater accessibility of the active-site copper to cellular reductants such as ascorbate, which is in turn caused by greater disorganization of the zinc-binding and electrostatic loops. This disorganization is caused by zinc loss (which is more probable in the case of mutant SOD1 species that have significantly lower affinities for zinc than wild-type SOD1), or in some cases, by mutation *per se*. The high levels of ascorbate in neurons would then maintain the active-site copper in the reduced [Cu(I)] state. In this state, the enzyme can become a net producer of superoxide (analogous to the metal-catalyzed oxidation system). Peroxynitrite, which is capable of nitrating tyrosine, would then form from the reaction of superoxide and nitric oxide (Trumbull and Beckman 2009). There is some evidence for this mechanism *in vivo*, as it was shown that the levels of free nitrotyrosine are increased in mutant SOD1 mouse spinal cords (Bruijn et al. 1997a). Also, the delivery of zinc-deficient SOD1 into cells in culture leads to nitric-oxide synthase-dependent apoptosis, consistent with the above mechanism (Estevez et al. 1999). It has also been shown that mutant SOD1s have a lowered Km for hydrogen peroxide. It has been suggested that as a result of this lowered Km, the peroxidase activity that is normally exhibited by SOD1 could become enhanced, leading to increased production of free radicals and oxidative stress (Yim et al. 1996).

This "pro-oxidant" hypothesis for the toxicity of mutant SOD1 is consistent with the presence of metal-deficient (i.e., zinc-deficient) destabilized SOD1 species in ALS spinal-cord tissue. It is also consistent with our proposed model of mutant SOD1 toxicity (Rakhit et al. 2002) in which oxidative stress, possibly caused by mutant SOD1 itself, leads to accumulation of oxidized/misfolded SOD1. However, the hypothesis presumably requires the presence of bound, active copper ions in the protein. This is at odds, however, with several results obtained by other groups. Borchelt et al. showed that, when expressed in cell culture, certain mutants (such as G85R) possess almost no superoxide scavenging activity (Borchelt et al. 1994), suggesting a complete lack of bound copper. Also, Wang et al. have shown that engineered SOD1s that lack 2 or 4 of the copper-chelating histidine residues are capable of causing motor-neuron disease when overexpressed in mice (Wang et al. 2002, 2003). This is despite the fact that these mutant enzymes have severely reduced copper affinity and are essentially dismutase-inactive (Wang et al. 2007). It is possible that the pro-oxidant hypothesis may be more relevant for wild-type-like SOD1 mutants that retain copper-binding ability and dismutase activity, whereas protein misfolding and aggregation may play a more predominant role in severely destabilized mutants that may not bind copper.

8.6 SOD1 and TDP-43 in Sporadic ALS

The discovery that mutations in SOD1 cause 2–5% of all ALS cases was a major breakthrough in ALS research. Importantly, the similarity of the clinical features of SOD1-ALS and sporadic ALS suggested that studying cellular and rodent models of SOD1-ALS would also shed light on the mechanisms involved in sporadic ALS.

Many of the *in vitro* studies on the misfolding and aggregation of SOD1 employed wild-type SOD1. Although mutant SOD1 is generally more prone to misfolding and aggregation than wild-type SOD1, such studies nevertheless indicated that wild-type SOD1 could also be induced to misfold and/or aggregate, and that the mechanisms involved were similar, if not identical (Stathopulos et al. 2003; Khare et al. 2004; Rakhit et al. 2004). For example, the work of Rakhit et al. showed that high concentrations of wild-type SOD1 could undergo misfolding and aggregation induced by MCO (Rakhit et al. 2004). As discussed above, wild-type SOD1 also undergoes metal release and partial monomerization upon treatment with hydrogen peroxide. In a similar vein, Julien and colleagues demonstrated that peroxide-mediated oxidation of wild-type SOD1 imparted several mutant SOD1-like properties on the wild-type protein, including the ability to interact with chromogranins, and the ability to induce cellular toxicity in culture (Ezzi et al. 2007). Studies in mice showed that overexpression of wild-type SOD1 could induce motor-neuron pathology in the absence of mutant SOD1, as well as hasten the disease course in mice co-expressing mutant SOD1 (Jaarsma et al. 2000). Such data has led to the suggestion that misfolding and aggregation of wild-type SOD1 could be involved in the pathogenesis of sporadic ALS (Rakhit et al. 2004; Kabashi et al. 2007).

If this were the case, one might expect to find intracytoplasmic inclusions containing misfolded SOD1 in cases of sporadic ALS. Shibata and colleagues did in fact report the occurrence of SOD1-immunoreactive inclusions in sporadic ALS cases (Shibata et al. 1994, 1996a). However, such results have not been reproduced by other investigators. Rothstein and colleagues failed to detect SOD1 immunoreactivity in motor neurons in a total of 17 sporadic ALS cases (Watanabe et al. 2001). The antibody used could, however, detect SOD1 in HCIs in a case of A4V SOD1-ALS. We employed our conformation-specific antibodies, SEDI and USOD, in an attempt to detect misfolded SOD1 in a series of 10 sporadic ALS cases. Although these antibodies could detect misfolded SOD1 in the motor neurons of 5 SOD1-ALS cases, no misfolded SOD1 could be detected in any of the sporadic cases studied (Fig. 8.3, panels B–H) (Liu et al. 2009; Kerman et al. 2010). These results suggest that the extensive misfolding of SOD1, ultimately resulting in deposition in HCIs, which is involved in the etiology of most SOD1-ALS cases, does not occur in sporadic ALS. Although cases of SOD1-ALS have been described in which HCIs are absent or do not stain with antibodies against SOD1 (Ohi et al. 2002; Shaw et al. 1997), these cases tend to be rare exceptions rather than the rule. Furthermore, in the sporadic ALS cases investigated using our antibodies, immunoprecipitation experiments failed to detect any misfolded SOD1 in detergent-soluble extracts of spinal cord tissue (Kerman et al. 2010; Liu et al. 2009). This further suggests that, in addition to being absent from inclusions, extensively misfolded soluble SOD1 species are also absent in these cases. Despite this, we cannot rule out the possibility that, in both SOD1-ALS and sporadic ALS, there are soluble toxic species of SOD1 that are only subtly different from native holoSOD1. Such species may include those that are properly folded but lack one or both metal ions, or lack the intrasubunit disulfide bond. A particular example could be zinc-deficient SOD1, as discussed above. Detecting such species may require antibodies directed against other epitopes of SOD1 that only require subtle misfolding of the protein in order to be exposed for antibody binding. It is also possible that soluble, misfolded SOD1 could be present at low levels that cannot be detected by our antibodies in immunoprecipitation experiments. This issue could be addressed in the future by developing high-affinity monoclonal antibodies against the SEDI and USOD epitopes.

Ubiquitin-positive inclusions, including skein-like inclusions and LBHIs, are a key pathological feature of sporadic ALS. However, the identification of one of the main protein components of these inclusions is a relatively recent discovery. In 2006, it was discovered that ubiquitin-positive inclusions in sporadic ALS and in frontotemporal lobar degeneration (FTLD) were immunopositive for TDP-43 (Neumann et al. 2006; Arai et al. 2006). It was subsequently discovered that, in addition to the full-length 43-kDa protein, detergent-insoluble extracts from ALS and FTLD brains also included 25-kDa and 35-kDa fragments, as well as TDP-43 that is hyperphosphorylated at several residues near the *C*-terminus (Arai et al. 2010; Nishimoto et al. 2010; Inukai et al. 2008). In ALS spinal cord, however, the insoluble TDP-43 is primarily the full-length protein (Igaz et al. 2008). Normally, TDP-43 is involved in RNA processing and has a predominantly nuclear location. The domain structure of TDP-43 (Fig. 8.1b) comprises an *N*-terminal region, two RNA-recognition motifs (RRM1 and RRM2) and a long *C*-terminal domain which is rich in glycine residues and is intrinsically disordered. In FTLD and ALS tissue, there is a marked redistribution of TDP-43 from the nucleus to the cytosol (Sasaki et al. 2010; Igaz et al. 2008). This redistribution precedes the formation of visible inclusions (Giordana et al. 2009) and is better correlated with toxicity (Barmada et al. 2010), suggesting that, as is the case with SOD1, the inclusions themselves are likely not the key toxic species. While ubiquitinated TDP-43 has been detected (Neumann et al. 2006), it is unclear whether it is the main ubiquitinated target in the inclusions. Sanelli et al. have shown, using high-resolution fluorescence microscopy, that ubiquitin and TDP-43 immunoreactivity do not overlap within the inclusions (Sanelli et al. 2007), suggesting that there are yet-to-be identified ubiquitinated proteins present.

That mislocalization and/or aggregation of TDP-43 is actually involved in the etiology of ALS, and is not simply a downstream consequence of neuronal degeneration, has been suggested by the discovery that mutations in TDP-43 are a cause of familial ALS, as well as some apparently "sporadic" cases (Kabashi et al. 2008; Sreedharan et al. 2008). The majority of the mutations are located in the disordered *C*-terminal domain. In cell culture, TDP-43 with ALS-associated mutations has a higher propensity to mislocalize and aggregate than wild-type TDP-43 (Barmada et al. 2010; Nonaka et al. 2009), further suggesting a direct relationship between toxicity and mislocalization/aggregation of TDP-43.

The presence of insoluble TDP-43 in ALS/FTLD neural tissue suggests the possibility that, like SOD1, TDP-43 causes ALS by a toxic gain of function—that is, an increased propensity to misfold and aggregate, either as a result of mutation, or other disease-related factors. Although this is not as well-established for TDP-43 as it is for SOD1, there are several lines of evidence that support it. Expression of human mutant or wild-type TDP-43 in mice leads to motor-neuron degeneration even in the presence of endogenous mouse TDP-43 (Wegorzewska et al. 2009; Wils et al. 2010). An extensive study of TDP-43 expression in cell culture suggested that cellular toxicity correlated with the presence of TDP-43 in the cytosol, and not with depletion of TDP-43 from the nucleus (Barmada et al. 2010), although it remains to be determined whether the TDP-43 that remains in the nucleus is still functional.

It was recently discovered that the protein ataxin-2, which causes the polyglutamine-expansion disease spinocerebellar ataxia type-2, interacts with TDP-43 in an RNA-dependent manner, and that this interaction enhances TDP-43 toxicity (Elden et al. 2010). Although the interaction with ataxin-2 increased accumulation of TDP-43 protein, enhanced toxicity was shown to be critically dependent on the RNA-binding function of TDP-43, suggesting that cytoplasmic accumulation *per se* may not be enough for TDP-43 toxicity. Rather, the formation of specific complexes containing RNA, TDP-43, and ataxin-2 may be a necessary step in the disease mechanism. It was further shown that ataxin-2 with intermediate-length polyglutamine expansions significantly increased the risk for ALS in human patients, further attesting to the potential importance of this TDP-43–ataxin-2 interaction in the disease (Elden et al. 2010).

In the case of SOD1, accumulation of insoluble SOD1 is preceded by misfolding of the enzyme. In the case of TDP-43, the recombinant-expressed protein has been shown to possess a significant propensity to aggregate. This, in turn, is dependent on the glycine-rich (i.e., disordered) *C*-terminus of the protein, and is enhanced by ALS-associated mutations in this region (Johnson et al. 2009). It is therefore not currently known whether mislocalization and aggregation of TDP-43 is related to misfolding of the other stably folded RRM domains.

In addition to TDP-43 and SOD1, mutations in two other proteins have recently been discovered that cause ALS, involving deposition of the disease-causing protein in neuronal inclusions. The first is FUS/TLS, which is an RNA-/DNA-binding protein with several similarities to TDP-43 (Vance et al. 2009; Kwiatkowski et al. 2009). The second is optineurin, a ubiquitin-binding protein (Maruyama et al. 2010). The discovery of ALS caused by mutations in these proteins is quite recent and, as is the case with TDP-43, it has not yet been proven that these mutations cause ALS by a "gain-of-toxic-function" mechanism, or whether deposition of these proteins in inclusions is a result of their misfolding. The FUS protein has been shown to be associated in a complex with TDP-43 (Ling et al. 2010), and this complex was shown to be involved in the regulation of histone deacetylase-6 mRNA (Kim et al. 2010), suggesting that TDP-43 and FUS may cause ALS by a related mechanism involving their RNA-binding function.

A remarkable finding is that SOD1-ALS generally does not involve TDP-43 mislocalization or inclusion-body formation. TDP-43-immunopositive inclusions are found in sporadic ALS cases and many non-SOD1 familial ALS cases, but are only rarely associated with SOD1-ALS (Mackenzie et al. 2007). This finding was corroborated in SOD1 mouse models of ALS, where mislocalization and/or aggregation of TDP-43 is generally absent (Robertson et al. 2007)—it was only observed sparsely in mice at the end-stage of the disease (Shan et al. 2009). The same is true of the FUS protein, which is commonly found in the inclusions of sporadic ALS and non-SOD1 familial ALS, but not in the inclusions of SOD1-ALS (Deng et al. 2010). Such findings suggest that the aberrant behavior of these proteins cannot be explained by "general" ALS-related defects in protein homeostasis—rather, there must be specific cellular mechanisms that are unique to each ALS subtype.

8.7 Summary and Conclusions

Since the discovery that mutations in SOD1 cause a subset of familial ALS, 17 years of research have begun to shed light on the identity of the toxic SOD1 species that are responsible for motor-neuron degeneration. As is the case with other neurodegenerative diseases, the large, microscopically visible accumulations of misfolded protein(s) (inclusion bodies) are no longer believed to be the primary toxic species. Extensive studies on rodent and cellular models of mutant SOD1-ALS now point to the existence of soluble, misfolded SOD1 species as the least common denominator underlying mutant SOD1-mediated motor-neuron degeneration. These species are conformationally altered, probably as a result of loss of metals and reduction of the intrasubunit disulfide bond. The misfolding of SOD1 may be triggered by oxidative damage, possibly initiated by mutant SOD1 itself, and the motor-neuron-specific accumulation of oxidized SOD1 could be due to the high concentration and long half-life of SOD1 in these cells. The key toxic species may be as simple as misfolded SOD1 monomers, although larger (insoluble) complexes of SOD1 may also be involved, particularly in the later stages of disease. Importantly, the toxic protein species in ALS are fundamentally different from those in other neurodegenerative diseases in that they are not amyloidogenic. This could mean that the mechanisms involved in ALS pathogenesis are fundamentally different than those in other diseases (such as Parkinson's and Huntington's diseases). Alternatively, it could also mean that many of the mechanisms involved in all of these diseases do not depend on amyloidogenic conformations *per se*, but on more general properties of misfolded proteins.

Despite the characterization of such species, the precise mechanisms whereby they exert their toxic effects still remain largely unknown. Interaction of misfolded SOD1 with mitochondria is now well-established, as is the occurrence of mitochondrial dysfunction in the disease. However, how this dysfunction is caused by misfolded SOD1, as well as the relative importance of mitochondrial dysfunction (as compared to axonal transport defects, oxidative stress, excitotoxicity, etc.) remains to be determined.

References

Andrus PK, Fleck TJ, Gurney ME, Hall ED (1998) Protein oxidative damage in a transgenic mouse model of familial amyotrophic lateral sclerosis. J Neurochem 71:2041–2048

Arai T, Hasegawa M, Akiyama H, Ikeda K, Nonaka T, Mori H, Mann D, Tsuchiya K, Yoshida M, Hashizume Y, Oda T (2006) TDP-43 is a component of ubiquitin-positive tau-negative inclusions in frontotemporal lobar degeneration and amyotrophic lateral sclerosis. Biochem Biophys Res Commun 351:602–611

Arai T, Hasegawa M, Nonoka T, Kametani F, Yamashita M, Hosokawa M, Niizato K, Tsuchiya K, Kobayashi Z, Ikeda K, Yoshida M, Onaya M, Fujishiro H, Akiyama H (2010) Phosphorylated and cleaved TDP-43 in ALS, FTLD and other neurodegenerative disorders and in cellular models of TDP-43 proteinopathy. Neuropathology 30:170–181

Armon C (2009) Smoking may be considered an established risk factor for sporadic ALS. Neurology 73:1693–1698

Arnesano F, Banci L, Bertini I, Martinelli M, Furukawa Y, O'halloran TV (2004) The unusually stable quaternary structure of human Cu, Zn-superoxide dismutase 1 is controlled by both metal occupancy and disulfide status. J Biol Chem 279:47998–48003

Barmada SJ, Skibinski G, Korb E, Rao EJ, Wu JY, Finkbeiner S (2010) Cytoplasmic mislocalization of TDP-43 is toxic to neurons and enhanced by a mutation associated with familial amyotrophic lateral sclerosis. J Neurosci 30:639–649

Beckman JS, Koppenol WH (1996) Nitric oxide, superoxide, and peroxynitrite: the good, the bad, and ugly. Am J Physiol 271:C1424–C1437

Beleza-Meireles A, Al-Chalabi A (2009) Genetic studies of amyotrophic lateral sclerosis: controversies and perspectives. Amyotroph Lateral Scler 10:1–14

Borchelt DR, Lee MK, Slunt HS, Guarnieri M, Xu ZS, Wong PC, Brown RH Jr, Price DL, Sisodia SS, Cleveland DW (1994) Superoxide dismutase 1 with mutations linked to familial amyotrophic lateral sclerosis possesses significant activity. Proc Natl Acad Sci USA 91:8292–8296

Borchelt DR, Wong PC, Becher MW, Pardo CA, Lee MK, Xu ZS, Thinakaran G, Jenkins NA, Copeland NG, Sisodia SS, Cleveland DW, Price DL, Hoffman PN (1998) Axonal transport of mutant superoxide dismutase 1 and focal axonal abnormalities in the proximal axons of transgenic mice. Neurobiol Dis 5:27–35

Brown NM, Torres AS, Doan PE, O'halloran TV (2004) Oxygen and the copper chaperone CCS regulate posttranslational activation of Cu, Zn superoxide dismutase. Proc Natl Acad Sci USA 101:5518–5523

Bruijn LI, Beal MF, Becher MW, Schulz JB, Wong PC, Price DL, Cleveland DW (1997a) Elevated free nitrotyrosine levels, but not protein-bound nitrotyrosine or hydroxyl radicals, throughout amyotrophic lateral sclerosis (ALS)-like disease implicate tyrosine nitration as an aberrant *in vivo* property of one familial ALS-linked superoxide dismutase 1 mutant. Proc Natl Acad Sci USA 94:7606–7611

Bruijn LI, Becher MW, Lee MK, Anderson KL, Jenkins NA, Copeland NG, Sisodia SS, Rothstein JD, Borchelt DR, Price DL, Cleveland DW (1997b) ALS-linked SOD1 mutant G85R mediates damage to astrocytes and promotes rapidly progressive disease with SOD1-containing inclusions. Neuron 18:327–338

Bruns CK, Kopito RR (2007) Impaired post-translational folding of familial ALS-linked Cu, Zn superoxide dismutase mutants. EMBO J 26:855–866

Carrell RW, Lomas DA (1997) Conformational disease. Lancet 350:134–138

Carroll MC, Girouard JB, Ulloa JL, Subramaniam JR, Wong PC, Valentine JS, Culotta VC (2004) Mechanisms for activating Cu- and Zn-containing superoxide dismutase in the absence of the CCS Cu chaperone. Proc Natl Acad Sci USA 101:5964–5969

Casareno RL, Waggoner D, Gitlin JD (1998) The copper chaperone CCS directly interacts with copper/zinc superoxide dismutase. J Biol Chem 273:23625–23628

Chattopadhyay M, Durazo A, Sohn SH, Strong CD, Gralla EB, Whitelegge JP, Valentine JS (2008) Initiation and elongation in fibrillation of ALS-linked superoxide dismutase. Proc Natl Acad Sci USA 105:18663–18668

Crow JP, Sampson JB, Zhuang Y, Thompson JA, Beckman JS (1997) Decreased zinc affinity of amyotrophic lateral sclerosis-associated superoxide dismutase mutants leads to enhanced catalysis of tyrosine nitration by peroxynitrite. J Neurochem 69:1936–1944

Dal Canto MC, Gurney ME (1995) Neuropathological changes in two lines of mice carrying a transgene for mutant human Cu, Zn SOD, and in mice overexpressing wild type human SOD: a model of familial amyotrophic lateral sclerosis (FALS). Brain Res 676:25–40

Deng HX, Shi Y, Furukawa Y, Zhai H, Fu R, Liu E, Gorrie GH, Khan MS, Hung WY, Bigio EH, Lukas T, Dal Canto MC, O'halloran TV, Siddique T (2006) Conversion to the amyotrophic lateral sclerosis phenotype is associated with intermolecular linked insoluble aggregates of SOD1 in mitochondria. Proc Natl Acad Sci USA 103:7142–7147

Deng HX, Zhai H, Bigio EH, Yan J, Fecto F, Ajroud K, Mishra M, Ajroud-Driss S, Heller S, Sufit R, Siddique N, Mugnaini E, Siddique T (2010) FUS-immunoreactive inclusions are a common feature in sporadic and non-SOD1 familial amyotrophic lateral sclerosis. Ann Neurol 67:739–748

Doucette PA, Whitson LJ, Cao X, Schirf V, Demeler B, Valentine JS, Hansen JC, Hart PJ (2004) Dissociation of human copper-zinc superoxide dismutase dimers using chaotrope and reductant. Insights into the molecular basis for dimer stability. J Biol Chem 279:54558–54566

Dunckley T, Huentelman MJ, Craig DW, Pearson JV, Szelinger S, Joshipura K, Halperin RF, Stamper C, Jensen KR, Letizia D, Hesterlee SE, Pestronk A, Levine T, Bertorini T, Graves MC, Mozaffar T, Jackson CE, Bosch P, Mcvey A, Dick A, Barohn R, Lomen-Hoerth C, Rosenfeld J, O'connor DT, Zhang K, Crook R, Ryberg H, Hutton M, Katz J, Simpson EP, Mitsumoto H, Bowser R, Miller RG, Appel SH, Stephan DA (2007) Whole-genome analysis of sporadic amyotrophic lateral sclerosis. N Engl J Med 357:775–788

Elchuri S, Oberley TD, Qi W, Eisenstein RS, Jackson Roberts L, Van Remmen H, Epstein CJ, Huang TT (2005) CuZnSOD deficiency leads to persistent and widespread oxidative damage and hepatocarcinogenesis later in life. Oncogene 24:367–380

Elden AC, Kim HJ, Hart MP, Chen-Plotkin AS, Johnson BS, Fang X, Armakola M, Geser F, Greene R, Lu MM, Padmanabhan A, Clay-Falcone D, Mccluskey L, Elman L, Juhr D, Gruber PJ, Rub U, Auburger G, Trojanowski JQ, Lee VM, Van Deerlin VM, Bonini NM, Gitler AD (2010) Ataxin-2 intermediate-length polyglutamine expansions are associated with increased risk for ALS. Nature 466:1069–1075

Elghetany MT, Saleem A (1988) Methods for staining amyloid in tissues: a review. Stain Technol 63:201–212

Estevez AG, Crow JP, Sampson JB, Reiter C, Zhuang Y, Richardson GJ, Tarpey MM, Barbeito L, Beckman JS (1999) Induction of nitric oxide-dependent apoptosis in motor neurons by zinc-deficient superoxide dismutase. Science 286:2498–2500

Ezzi SA, Urushitani M, Julien JP (2007) Wild-type superoxide dismutase acquires binding and toxic properties of ALS-linked mutant forms through oxidation. J Neurochem 102:170–178

Ferri A, Cozzolino M, Crosio C, Nencini M, Casciati A, Gralla EB, Rotilio G, Valentine JS, Carri MT (2006) Familial ALS-superoxide dismutases associate with mitochondria and shift their redox potentials. Proc Natl Acad Sci USA 103:13860–13865

Fukada K, Nagano S, Satoh M, Tohyama C, Nakanishi T, Shimizu A, Yanagihara T, Sakoda S (2001) Stabilization of mutant Cu/Zn superoxide dismutase (SOD1) protein by coexpressed wild SOD1 protein accelerates the disease progression in familial amyotrophic lateral sclerosis mice. Eur J Neurosci 14:2032–2036

Furukawa Y, Fu R, Deng HX, Siddique T, O'halloran TV (2006) Disulfide cross-linked protein represents a significant fraction of ALS-associated Cu, Zn-superoxide dismutase aggregates in spinal cords of model mice. Proc Natl Acad Sci USA 103:7148–7153

Furukawa Y, Kaneko K, Yamanaka K, O'halloran TV, Nukina N (2008) Complete loss of post-translational modifications triggers fibrillar aggregation of SOD1 in the familial form of amyotrophic lateral sclerosis. J Biol Chem 283:24167–24176

Giordana MT, Piccinini M, Grifoni S, De Marco G, Vercellino M, Magistrello M, Pellerino A, Buccinna B, Lupino E, Rinaudo MT (2009) TDP-43 redistribution is an early event in sporadic amyotrophic lateral sclerosis. Brain Pathol 20:351–360

Gurney ME, Pu H, Chiu AY, Dal Canto MC, Polchow CY, Alexander DD, Caliendo J, Hentati A, Kwon YW, Deng H-X, Chen W, Zhai P, Sufit RL, Siddique T (1994) Motor neuron degeneration in mice that express a human Cu, Zn superoxide dismutase mutation. Science 264:1772–1775

Hart PJ, Balbirnie MM, Ogihara NL, Nersissian AM, Weiss MS, Valentine JS, Eisenberg D (1999) A structure-based mechanism for copper-zinc superoxide dismutase. Biochemistry 38:2167–2178

Hashimoto M, Hsu LJ, Sisk A, Xia Y, Takeda A, Sundsmo M, Masliah E (1998) Human recombinant NACP/α-synuclein is aggregated and fibrillated *in vitro*: relevance for Lewy body disease. Brain Res 799:301–306

Hayward LJ, Rodriguez JA, Kim JW, Tiwari A, Goto JJ, Cabelli DE, Valentine JS, Brown RH Jr (2002) Decreased metallation and activity in subsets of mutant superoxide dismutases associated with familial amyotrophic lateral sclerosis. J Biol Chem 277:15923–15931

Hayworth CR, Gonzalez-Lima F (2009) Pre-symptomatic detection of chronic motor deficits and genotype prediction in congenic B6.SOD1(G93A) ALS mouse model. Neuroscience 164:975–985

Hervias I, Beal MF, Manfredi G (2006) Mitochondrial dysfunction and amyotrophic lateral sclerosis. Muscle Nerve 33:598–608

Higgins CM, Jung C, Ding H, Xu Z (2002) Mutant Cu, Zn superoxide dismutase that causes motoneuron degeneration is present in mitochondria in the CNS. J Neurosci 22:RC215

Igaz LM, Kwong LK, Xu Y, Truax AC, Uryu K, Neumann M, Clark CM, Elman LB, Miller BL, Grossman M, Mccluskey LF, Trojanowski JQ, Lee VM (2008) Enrichment of *C*-terminal fragments in TAR DNA-binding protein-43 cytoplasmic inclusions in brain but not in spinal cord of frontotemporal lobar degeneration and amyotrophic lateral sclerosis. Am J Pathol 173:182–194

Inukai Y, Nonaka T, Arai T, Yoshida M, Hashizume Y, Beach TG, Buratti E, Baralle FE, Akiyama H, Hisanaga S, Hasegawa M (2008) Abnormal phosphorylation of Ser409/410 of TDP-43 in FTLD-U and ALS. FEBS Lett 582:2899–2904

Jaarsma D, Haasdijk ED, Grashorn JA, Hawkins R, Van Duijn W, Verspaget HW, London J, Holstege JC (2000) Human Cu/Zn superoxide dismutase (SOD1) overexpression in mice causes mitochondrial vacuolization, axonal degeneration, and premature motoneuron death and accelerates motoneuron disease in mice expressing a familial amyotrophic lateral sclerosis mutant SOD1. Neurobiol Dis 7:623–643

Johnson BS, Snead D, Lee JJ, Mccaffery JM, Shorter J, Gitler AD (2009) TDP-43 is intrinsically aggregation-prone, and amyotrophic lateral sclerosis-linked mutations accelerate aggregation and increase toxicity. J Biol Chem 284:20329–20339

Johnston JA, Dalton MJ, Gurney ME, Kopito RR (2000) Formation of high molecular weight complexes of mutant Cu, Zn-superoxide dismutase in a mouse model for familial amyotrophic lateral sclerosis. Proc Natl Acad Sci USA 97:12571–12576

Jonsson PA, Ernhill K, Andersen PM, Bergemalm D, Brannstrom T, Gredal O, Nilsson P, Marklund SL (2004) Minute quantities of misfolded mutant superoxide dismutase-1 cause amyotrophic lateral sclerosis. Brain 127:73–88

Jonsson PA, Graffmo KS, Andersen PM, Brannstrom T, Lindberg M, Oliveberg M, Marklund SL (2006a) Disulphide-reduced superoxide dismutase-1 in CNS of transgenic amyotrophic lateral sclerosis models. Brain 129:451–464

Jonsson PA, Graffmo KS, Brannstrom T, Nilsson P, Andersen PM, Marklund SL (2006b) Motor neuron disease in mice expressing the wild type-like D90A mutant superoxide dismutase-1. J Neuropathol Exp Neurol 65:1126–1136

Jung C, Higgins CM, Xu Z (2002) Mitochondrial electron transport chain complex dysfunction in a transgenic mouse model for amyotrophic lateral sclerosis. J Neurochem 83:535–545

Kabashi E, Valdmanis PN, Dion P, Rouleau GA (2007) Oxidized/misfolded superoxide dismutase-1: the cause of all amyotrophic lateral sclerosis? Ann Neurol 62:553–559

Kabashi E, Valdmanis PN, Dion P, Spiegelman D, Mcconkey BJ, Vande Velde C, Bouchard JP, Lacomblez L, Pochigaeva K, Salachas F, Pradat PF, Camu W, Meininger V, Dupre N, Rouleau GA (2008) TARDBP mutations in individuals with sporadic and familial amyotrophic lateral sclerosis. Nat Genet 40:572–574

Karch CM, Borchelt DR (2008) A limited role for disulfide cross-linking in the aggregation of mutant SOD1 linked to familial amyotrophic lateral sclerosis. J Biol Chem 283:13528–13537

Karch CM, Prudencio M, Winkler DD, Hart PJ, Borchelt DR (2009) Role of mutant SOD1 disulfide oxidation and aggregation in the pathogenesis of familial ALS. Proc Natl Acad Sci USA 106:7774–7779

Kato S, Shimoda M, Watanabe Y, Nakashima K, Takahashi K, Ohama E (1996) Familial amyotrophic lateral sclerosis with a two base pair deletion in superoxide dismutase 1 gene: multisystem degeneration with intracytoplasmic hyaline inclusions in astrocytes. J Neuropathol Exp Neurol 55:1089–1101

Kato S, Takikawa M, Nakashima K, Hirano A, Cleveland DW, Kusaka H, Shibata N, Kato M, Nakano I, Ohama E (2000) New consensus research on neuropathological aspects of familial amyotrophic lateral sclerosis with superoxide dismutase 1 (SOD1) gene mutations: inclusions containing SOD1 in neurons and astrocytes. Amyotroph Lateral Scler Other Motor Neuron Disord 1:163–184

Kato S, Sumi-Akamaru H, Fujimura H, Sakoda S, Kato M, Hirano A, Takikawa M, Ohama E (2001) Copper chaperone for superoxide dismutase co-aggregates with superoxide dismutase 1 (SOD1) in neuronal Lewy body-like hyaline inclusions: an immunohistochemical study on familial amyotrophic lateral sclerosis with SOD1 gene mutation. Acta Neuropathol 102:233–238

Kayatekin C, Zitzewitz JA, Matthews CR (2010) Disulfide-reduced ALS variants of Cu, Zn superoxide dismutase exhibit increased populations of unfolded species. J Mol Biol 398:320–331

Kerman A, Liu HN, Croul S, Bilbao J, Rogaeva E, Zinman L, Robertson J, Chakrabartty A (2010) Amyotrophic lateral sclerosis is a non-amyloid disease in which extensive misfolding of SOD1 is unique to the familial form. Acta Neuropathol 119:335–344

Khare SD, Caplow M, Dokholyan NV (2004) The rate and equilibrium constants for a multistep reaction sequence for the aggregation of superoxide dismutase in amyotrophic lateral sclerosis. Proc Natl Acad Sci USA 101:15094–15099

Kim YS, Han S (2000) Nitric oxide protects Cu, Zn-superoxide dismutase from hydrogen peroxide-induced inactivation. FEBS Lett 479:25–28

Kim SH, Shanware N, Bowler MJ, Tibbetts RS (2010) Amyotrophic lateral sclerosis-associated proteins TDP-43 and FUS/TLS function in a common biochemical complex to co-regulate HDAC6 mRNA. J Biol Chem 285:34097–34105

Kokubo Y, Kuzuhara S, Narita Y, Kikugawa K, Nakano R, Inuzuka T, Tsuji S, Watanabe M, Miyazaki T, Murayama S, Ihara Y (1999) Accumulation of neurofilaments and SOD1-immunoreactive products in a patient with familial amyotrophic lateral sclerosis with I113T SOD1 mutation. Arch Neurol 56:1506–1508

Kwiatkowski TJ Jr, Bosco DA, Leclerc AL, Tamrazian E, Vanderburg CR, Russ C, Davis A, Gilchrist J, Kasarskis EJ, Munsat T, Valdmanis P, Rouleau GA, Hosler BA, Cortelli P, De Jong PJ, Yoshinaga Y, Haines JL, Pericak-Vance MA, Yan J, Ticozzi N, Siddique T, Mckenna-Yasek D, Sapp PC, Horvitz HR, Landers JE, Brown RH Jr (2009) Mutations in the FUS/TLS gene on chromosome 16 cause familial amyotrophic lateral sclerosis. Science 323:1205–1208

Kwong LK, Uryu K, Trojanowski JQ, Lee VM (2008) TDP-43 proteinopathies: neurodegenerative protein misfolding diseases without amyloidosis. Neurosignals 16:41–51

Lacomblez L, Bensimon G, Leigh PN, Guillet P, Meininger V (1996) Dose-ranging study of riluzole in amyotrophic lateral sclerosis. Amyotrophic lateral sclerosis/riluzole study group II. Lancet 347:1425–1431

Lamb AL, Torres AS, O'halloran TV, Rosenzweig AC (2000) Heterodimer formation between superoxide dismutase and its copper chaperone. Biochemistry 39:14720–14727

Lindberg MJ, Normark J, Holmgren A, Oliveberg M (2004) Folding of human superoxide dismutase: disulfide reduction prevents dimerization and produces marginally stable monomers. Proc Natl Acad Sci USA 101:15893–15898

Ling SC, Albuquerque CP, Han JS, Lagier-Tourenne C, Tokunaga S, Zhou H, Cleveland DW (2010) ALS-associated mutations in TDP-43 increase its stability and promote TDP-43 complexes with FUS/TLS. Proc Natl Acad Sci USA 107:13318–13323

Liu J, Lillo C, Jonsson PA, Vande Velde C, Ward CM, Miller TM, Subramaniam JR, Rothstein JD, Marklund S, Andersen PM, Brannstrom T, Gredal O, Wong PC, Williams DS, Cleveland DW (2004) Toxicity of familial ALS-linked SOD1 mutants from selective recruitment to spinal mitochondria. Neuron 43:5–17

Liu HN, Sanelli T, Horne P, Pioro EP, Strong MJ, Rogaeva E, Bilbao J, Zinman L, Robertson J (2009) Lack of evidence of monomer/misfolded superoxide dismutase-1 in sporadic amyotrophic lateral sclerosis. Ann Neurol 66:75–80

Mackenzie IR, Bigio EH, Ince PG, Geser F, Neumann M, Cairns NJ, Kwong LK, Forman MS, Ravits J, Stewart H, Eisen A, Mcclusky L, Kretzschmar HA, Monoranu CM, Highley JR, Kirby J, Siddique T, Shaw PJ, Lee VM, Trojanowski JQ (2007) Pathological TDP-43 distinguishes sporadic amyotrophic lateral sclerosis from amyotrophic lateral sclerosis with SOD1 mutations. Ann Neurol 61:427–434

Marklund SL, Andersen PM, Forsgren L, Nilsson P, Ohlsson PI, Wikander G, Oberg A (1997) Normal binding and reactivity of copper in mutant superoxide dismutase isolated from amyotrophic lateral sclerosis patients. J Neurochem 69:675–681

Maruyama H, Morino H, Ito H, Izumi Y, Kato H, Watanabe Y, Kinoshita Y, Kamada M, Nodera H, Suzuki H, Komure O, Matsuura S, Kobatake K, Morimoto N, Abe K, Suzuki N, Aoki M, Kawata A, Hirai T, Kato T, Ogasawara K, Hirano A, Takumi T, Kusaka H, Hagiwara K, Kaji R, Kawakami H (2010) Mutations of optineurin in amyotrophic latcral sclerosis. Nature 465:223–226

Mcgowan DP, Van Roon-Mom W, Holloway H, Bates GP, Mangiarini L, Cooper GJ, Faull RL, Snell RG (2000) Amyloid-like inclusions in Huntington's disease. Neuroscience 100:677–680

Mitchell JD, Borasio GD (2007) Amyotrophic lateral sclerosis. Lancet 369:2031–2041

Mulligan VK, Kerman A, Ho S, Chakrabartty A (2008) Denaturational stress induces formation of zinc-deficient monomers of Cu, Zn superoxide dismutase: implications for pathogenesis in amyotrophic lateral sclerosis. J Mol Biol 383:424–436

Neumann M, Sampathu DM, Kwong LK, Truax AC, Micsenyi MC, Chou TT, Bruce J, Schuck T, Grossman M, Clark CM, Mccluskey LF, Miller BL, Masliah E, Mackenzie IR, Feldman H, Feiden W, Kretzschmar HA, Trojanowski JQ, Lee VM (2006) Ubiquitinated TDP-43 in frontotemporal lobar degeneration and amyotrophic lateral sclerosis. Science 314:130–133

Nishimoto Y, Ito D, Yagi T, Nihei Y, Tsunoda Y, Suzuki N (2010) Characterization of alternative isoforms and inclusion body of the TAR DNA-binding protein-43. J Biol Chem 285:608–619

Niwa J, Yamada S, Ishigaki S, Sone J, Takahashi M, Katsuno M, Tanaka F, Doyu M, Sobue G (2007) Disulfide bond mediates aggregation, toxicity, and ubiquitylation of familial amyotrophic lateral sclerosis-linked mutant SOD1. J Biol Chem 282:28087–28095

Nonaka T, Kametani F, Arai T, Akiyama H, Hasegawa M (2009) Truncation and pathogenic mutations facilitate the formation of intracellular aggregates of TDP-43. Hum Mol Genet 18:3353–3364

Ohi T, Saita K, Takechi S, Nabesima K, Tashiro H, Shiomi K, Sugimoto S, Akematsu T, Nakayama T, Iwaki T, Matsukura S (2002) Clinical features and neuropathological findings of familial amyotrophic lateral sclerosis with a His46Arg mutation in Cu/Zn superoxide dismutase. J Neurol Sci 197:73–78

Ohi T, Nabeshima K, Kato S, Yazawa S, Takechi S (2004) Familial amyotrophic lateral sclerosis with His46Arg mutation in Cu/Zn superoxide dismutase presenting characteristic clinical features and Lewy body-like hyaline inclusions. J Neurol Sci 225:19–25

Oztug Durer ZA, Cohlberg JA, Dinh P, Padua S, Ehrenclou K, Downes S, Tan JK, Nakano Y, Bowman CJ, Hoskins JL, Kwon C, Mason AZ, Rodriguez JA, Doucette PA, Shaw BF, Selverstone Valentine J (2009) Loss of metal ions, disulfide reduction and mutations related to familial ALS promote formation of amyloid-like aggregates from superoxide dismutase. PLoS One 4:e5004

Pardo CA, Xu Z, Borchelt DR, Price DL, Sisodia SS, Cleveland DW (1995) Superoxide dismutase is an abundant component in cell bodies, dendrites, and axons of motor neurons and in a subset of other neurons. Proc Natl Acad Sci USA 92:954–958

Poon HF, Hensley K, Thongboonkerd V, Merchant ML, Lynn BC, Pierce WM, Klein JB, Calabrese V, Butterfield DA (2005) Redox proteomics analysis of oxidatively modified proteins in G93A-SOD1 transgenic mice—a model of familial amyotrophic lateral sclerosis. Free Radic Biol Med 39:453–462

Proescher JB, Son M, Elliott JL, Culotta VC (2008) Biological effects of CCS in the absence of SOD1 enzyme activation: implications for disease in a mouse model for ALS. Hum Mol Genet 17:1728–1737

Prudencio M, Durazo A, Whitelegge JP, Borchelt DR (2009) Modulation of mutant superoxide dismutase 1 aggregation by co-expression of wild-type enzyme. J Neurochem 108:1009–1018

Rakhit R, Chakrabartty A (2006) Structure, folding, and misfolding of Cu, Zn superoxide dismutase in amyotrophic lateral sclerosis. Biochim Biophys Acta 1762:1025–1037

Rakhit R, Cunningham P, Furtos-Matei A, Dahan S, Qi XF, Crow JP, Cashman NR, Kondejewski LH, Chakrabartty A (2002) Oxidation-induced misfolding and aggregation of superoxide dismutase and its implications for amyotrophic lateral sclerosis. J Biol Chem 277:47551–47556

Rakhit R, Crow JP, Lepock JR, Kondejewski LH, Cashman NR, Chakrabartty A (2004) Monomeric Cu, Zn-superoxide dismutase is a common misfolding intermediate in the oxidation models of sporadic and familial amyotrophic lateral sclerosis. J Biol Chem 279:15499–15504

Rakhit R, Robertson J, Vande Velde C, Horne P, Ruth DM, Griffin J, Cleveland DW, Cashman NR, Chakrabartty A (2007) An immunological epitope selective for pathological monomer-misfolded SOD1 in ALS. Nat Med 13:754–759

Reaume AG, Elliott JL, Hoffman EK, Kowall NW, Ferrante RJ, Siwek DF, Wilcox HM, Flood DG, Beal MF, Brown RH Jr, Scott RW, Snider WD (1996) Motor neurons in Cu/Zn superoxide dismutase-deficient mice develop normally but exhibit enhanced cell death after axonal injury. Nat Genet 13:43–47

Robertson J, Sanelli T, Xiao S, Yang W, Horne P, Hammond R, Pioro EP, Strong MJ (2007) Lack of TDP-43 abnormalities in mutant SOD1 transgenic mice shows disparity with ALS. Neurosci Lett 420:128–132

Rodriguez JA, Valentine JS, Eggers DK, Roe JA, Tiwari A, Brown RH Jr, Hayward LJ (2002) Familial amyotrophic lateral sclerosis-associated mutations decrease the thermal stability of distinctly metallated species of human copper/zinc superoxide dismutase. J Biol Chem 277:15932–15937

Rosen DR, Siddique T, Patterson D, Figlewicz DA, Sapp P, Hentati A, Donaldson D, Goto J, O'regan JP, Deng HX, Rahmani Z, Krizus A, McKenna-Yasek D, Cayabyab A, Gaston SM, Berger R, Tanzi RE, Halperin JJ, Herzfeldt B, Van den Bergh R, Hung W-Y, Bird T, Deng G, Mulder DW, Smyth C, Laing NG, Soriano E, Pericak–Vanceparallelparallel MA, Haines J, Rouleau GA, Gusella JS, Horvitzparallel HR, Brown RH Jr (1993) Mutations in Cu/Zn superoxide dismutase gene are associated with familial amyotrophic lateral sclerosis. Nature 362:59–62

Rothstein JD (2009) Current hypotheses for the underlying biology of amyotrophic lateral sclerosis. Ann Neurol 65(Suppl 1):S3–S9

Rumfeldt JA, Stathopulos PB, Chakrabarrty A, Lepock JR, Meiering EM (2006) Mechanism and thermodynamics of guanidinium chloride-induced denaturation of ALS-associated mutant Cu, Zn superoxide dismutases. J Mol Biol 355:106–123

Sanelli T, Xiao S, Horne P, Bilbao J, Zinman L, Robertson J (2007) Evidence that TDP-43 is not the major ubiquitinated target within the pathological inclusions of amyotrophic lateral sclerosis. J Neuropathol Exp Neurol 66:1147–1153

Sasaki S, Takeda T, Shibata N, Kobayashi M (2010) Alterations in subcellular localization of TDP-43 immunoreactivity in the anterior horns in sporadic amyotrophic lateral sclerosis. Neurosci Lett 478(2):72–76

Shan X, Vocadlo D, Krieger C (2009) Mislocalization of TDP-43 in the G93A mutant SOD1 transgenic mouse model of ALS. Neurosci Lett 458:70–74

Shaw CE, Enayat ZE, Powell JF, Anderson VE, Radunovic A, Al-Sarraj S, Leigh PN (1997) Familial amyotrophic lateral sclerosis. Molecular pathology of a patient with a SOD1 mutation. Neurology 49:1612–1616

Shibata N, Hirano A, Kobayashi M, Sasaki S, Kato T, Matsumoto S, Shiozawa Z, Komori T, Ikemoto A, Umahara T, Asayama K (1994) Cu/Zn superoxide dismutase-like immunoreactivity in Lewy body-like inclusions of sporadic amyotrophic lateral sclerosis. Neurosci Lett 179:149–152

Shibata N, Asayama K, Hirano A, Kobayashi M (1996a) Immunohistochemical study on superoxide dismutases in spinal cords from autopsied patients with amyotrophic lateral sclerosis. Dev Neurosci 18:492–498

Shibata N, Hirano A, Kobayashi M, Siddique T, Deng HX, Hung WY, Kato T, Asayama K (1996b) Intense superoxide dismutase-1 immunoreactivity in intracytoplasmic hyaline inclusions of familial amyotrophic lateral sclerosis with posterior column involvement. J Neuropathol Exp Neurol 55:481–490

Sipe JD (1992) Amyloidosis. Annu Rev Biochem 61:947–975

Son M, Puttaparthi K, Kawamata H, Rajendran B, Boyer PJ, Manfredi G, Elliott JL (2007) Overexpression of CCS in G93A-SOD1 mice leads to accelerated neurological deficits with severe mitochondrial pathology. Proc Natl Acad Sci USA 104:6072–6077

Sreedharan J, Blair IP, Tripathi VB, Hu X, Vance C, Rogelj B, Ackerley S, Durnall JC, Williams KL, Buratti E, Baralle F, De Belleroche J, Mitchell JD, Leigh PN, Al-Chalabi A, Miller CC, Nicholson G, Shaw CE (2008) TDP-43 mutations in familial and sporadic amyotrophic lateral sclerosis. Science 319:1668–1672

Stathopulos PB, Rumfeldt JA, Scholz GA, Irani RA, Frey HE, Hallewell RA, Lepock JR, Meiering EM (2003) Cu/Zn superoxide dismutase mutants associated with amyotrophic lateral sclerosis show enhanced formation of aggregates *in vitro*. Proc Natl Acad Sci USA 100:7021–7026

Steele JC, Mcgeer PL (2008) The ALS/PDC syndrome of Guam and the cycad hypothesis. Neurology 70:1984–1990

Stefani M, Dobson CM (2003) Protein aggregation and aggregate toxicity: new insights into protein folding, misfolding diseases and biological evolution. J Mol Med 81:678–699

Stewart HG, Mackenzie IR, Eisen A, Brannstrom T, Marklund SL, Andersen PM (2006) Clinicopathological phenotype of ALS with a novel G72C SOD1 gene mutation mimicking a myopathy. Muscle Nerve 33:701–706

Strong MJ, Kesavapany S, Pant HC (2005) The pathobiology of amyotrophic lateral sclerosis: a proteinopathy? J Neuropathol Exp Neurol 64:649–664

Subramaniam JR, Lyons WE, Liu J, Bartnikas TB, Rothstein J, Price DL, Cleveland DW, Gitlin JD, Wong PC (2002) Mutant SOD1 causes motor neuron disease independent of copper chaperone-mediated copper loading. Nat Neurosci 5:301–307

Svensson AK, Bilsel O, Kayatekin C, Adefusika JA, Zitzewitz JA, Matthews CR (2010) Metal-free ALS variants of dimeric human Cu, Zn-superoxide dismutase have enhanced populations of monomeric species. PLoS One 5:e10064

Takehisa Y, Ujike H, Ishizu H, Terada S, Haraguchi T, Tanaka Y, Nishinaka T, Nobukuni K, Ihara Y, Namba R, Yasuda T, Nishibori M, Hayabara T, Kuroda S (2001) Familial amyotrophic lateral sclerosis with a novel Leu126Ser mutation in the copper/zinc superoxide dismutase gene showing mild clinical features and lewy body-like hyaline inclusions. Arch Neurol 58:736–740

Tiwari A, Hayward LJ (2003) Familial amyotrophic lateral sclerosis mutants of copper/zinc superoxide dismutase are susceptible to disulfide reduction. J Biol Chem 278:5984–5992

Trumbull KA, Beckman JS (2009) A role for copper in the toxicity of zinc-deficient superoxide dismutase to motor neurons in amyotrophic lateral sclerosis. Antioxid Redox Signal 11:1627–1639

Van Es MA, Van Vught PW, Blauw HM, Franke L, Saris CG, Andersen PM, Van Den Bosch L, De Jong SW, Van'T Slot R, Birve A, Lemmens R, De Jong V, Baas F, Schelhaas HJ, Sleegers K, Van Broeckhoven C, Wokke JH, Wijmenga C, Robberecht W, Veldink JH, Ophoff RA, Van Den Berg LH (2007) ITPR2 as a susceptibility gene in sporadic amyotrophic lateral sclerosis: a genome-wide association study. Lancet Neurol 6:869–877

Vance C, Rogelj B, Hortobagyi T, De Vos KJ, Nishimura AL, Sreedharan J, Hu X, Smith B, Ruddy D, Wright P, Ganesalingam J, Williams KL, Tripathi V, Al-Saraj S, Al-Chalabi A, Leigh PN, Blair IP, Nicholson G, De Belleroche J, Gallo JM, Miller CC, Shaw CE (2009) Mutations in FUS, an RNA processing protein, cause familial amyotrophic lateral sclerosis type 6. Science 323:1208–1211

Vande Velde C, Miller TM, Cashman NR, Cleveland DW (2008) Selective association of misfolded ALS-linked mutant SOD1 with the cytoplasmic face of mitochondria. Proc Natl Acad Sci USA 105:4022–4027

Wada M, Uchihara T, Nakamura A, Oyanagi K (1999) Bunina bodies in amyotrophic lateral sclerosis on Guam: a histochemical, immunohistochemical and ultrastructural investigation. Acta Neuropathol 98:150–156

Wang J, Xu G, Gonzales V, Coonfield M, Fromholt D, Copeland NG, Jenkins NA, Borchelt DR (2002) Fibrillar inclusions and motor neuron degeneration in transgenic mice expressing superoxide dismutase 1 with a disrupted copper-binding site. Neurobiol Dis 10:128–138

Wang J, Slunt H, Gonzales V, Fromholt D, Coonfield M, Copeland NG, Jenkins NA, Borchelt DR (2003) Copper-binding-site-null SOD1 causes ALS in transgenic mice: aggregates of non-native SOD1 delineate a common feature. Hum Mol Genet 12:2753–2764

Wang J, Xu G, Li H, Gonzales V, Fromholt D, Karch C, Copeland NG, Jenkins NA, Borchelt DR (2005) Somatodendritic accumulation of misfolded SOD1-L126Z in motor neurons mediates degeneration: αB-crystallin modulates aggregation. Hum Mol Genet 14:2335–2347

Wang J, Caruano-Yzermans A, Rodriguez A, Scheurmann JP, Slunt HH, Cao X, Gitlin J, Hart PJ, Borchelt DR (2007) Disease-associated mutations at copper ligand histidine residues of superoxide dismutase 1 diminish the binding of copper and compromise dimer stability. J Biol Chem 282:345–352

Wang L, Deng HX, Grisotti G, Zhai H, Siddique T, Roos RP (2009) Wild-type SOD1 overexpression accelerates disease onset of a G85R SOD1 mouse. Hum Mol Genet 18:1642–1651

Watanabe M, Dykes-Hoberg M, Culotta VC, Price DL, Wong PC, Rothstein JD (2001) Histological evidence of protein aggregation in mutant SOD1 transgenic mice and in amyotrophic lateral sclerosis neural tissues. Neurobiol Dis 8:933–941

Wegorzewska I, Bell S, Cairns NJ, Miller TM, Baloh RH (2009) TDP-43 mutant transgenic mice develop features of ALS and frontotemporal lobar degeneration. Proc Natl Acad Sci USA 106:18809–18814

Wils H, Kleinberger G, Janssens J, Pereson S, Joris G, Cuijt I, Smits V, Ceuterick-De Groote C, Van Broeckhoven C, Kumar-Singh S (2010) TDP-43 transgenic mice develop spastic paralysis and neuronal inclusions characteristic of ALS and frontotemporal lobar degeneration. Proc Natl Acad Sci USA 107:3858–3863

Witan H, Kern A, Koziollek-Drechsler I, Wade R, Behl C, Clement AM (2008) Heterodimer formation of wild-type and amyotrophic lateral sclerosis-causing mutant Cu/Zn-superoxide dismutase induces toxicity independent of protein aggregation. Hum Mol Genet 17:1373–1385

Witan H, Gorlovoy P, Kaya AM, Koziollek-Drechsler I, Neumann H, Behl C, Clement AM (2009) Wild-type Cu/Zn superoxide dismutase (SOD1) does not facilitate, but impedes the formation of protein aggregates of amyotrophic lateral sclerosis causing mutant SOD1. Neurobiol Dis 36:331–342

Wong PC, Pardo CA, Borchelt DR, Lee MK, Copeland NG, Jenkins NA, Sisodia SS, Cleveland DW, Price DL (1995) An adverse property of a familial ALS-linked SOD1 mutation causes motor neuron disease characterized by vacuolar degeneration of mitochondria. Neuron 14:1105–1116

Wong PC, Waggoner D, Subramaniam JR, Tessarollo L, Bartnikas TB, Culotta VC, Price DL, Rothstein J, Gitlin JD (2000) Copper chaperone for superoxide dismutase is essential to activate mammalian Cu/Zn superoxide dismutase. Proc Natl Acad Sci USA 97:2886–2891

Wroe R, Wai-Ling Butler A, Andersen PM, Powell JF, Al-Chalabi A (2008) ALSOD: the amyotrophic lateral sclerosis online database. Amyotroph Lateral Scler 9:249–250

Xiao S, Mclean J, Robertson J (2006) Neuronal intermediate filaments and ALS: a new look at an old question. Biochim Biophys Acta 1762:1001–1012

Yim MB, Kang JH, Yim HS, Kwak HS, Chock PB, Stadtman ER (1996) A gain-of-function of an amyotrophic lateral sclerosis-associated Cu, Zn-superoxide dismutase mutant: an enhancement of free radical formation due to a decrease in Km for hydrogen peroxide. Proc Natl Acad Sci USA 93:5709–5714

Zetterstrom P, Stewart HG, Bergemalm D, Jonsson PA, Graffmo KS, Andersen PM, Brannstrom T, Oliveberg M, Marklund SL (2007) Soluble misfolded subfractions of mutant superoxide dismutase-1s are enriched in spinal cords throughout life in murine ALS models. Proc Natl Acad Sci USA 104:14157–14162

Zinman L, Liu HN, Sato C, Wakutani Y, Marvelle AF, Moreno D, Morrison KE, Mohlke KL, Bilbao J, Robertson J, Rogaeva E (2009) A mechanism for low penetrance in an ALS family with a novel SOD1 deletion. Neurology 72:1153–1159

Chapter 9
Structural Studies of Prion Proteins and Prions

Giuseppe Legname, Gabriele Giachin, and Federico Benetti

Abstract Prion diseases are a group of fatal and incurable neurodegenerative disorders of mammals. They uniquely manifest as sporadic, genetic, and infectious maladies. The agent responsible for prion diseases is the prion. A prion is defined as a proteinaceous infectious particle, which is solely constituted by an alternately folded form of the prion protein (PrP) (Prusiner 1982).

In diseased animals and humans, PrP exists in two forms, the physiological, cellular form of PrP, PrP^{C}, and the pathological prion form denoted as PrP^{Sc}. The mechanism whereby nascent PrP^{Sc} is generated is currently unknown. Structural studies of either isoform are of great importance for understanding the biology of prion diseases since they may clarify the molecular mechanisms responsible for these pathologies. In this chapter, we present an overview of the studies into PrP^{C} as well as structures of prions.

Keywords Prion protein • Prion • Prion diseases • Disease-linked mutations • NMR • Limited proteolysis • Fourier-transform infrared spectroscopy • Antibody labeling • Electron microscopy • Atomic-force microscopy • Small-angle X-ray scattering • X-ray fiber diffraction

9.1 Prion Biology and Diseases

Prion diseases or transmissible spongiform encephalopathies (TSEs) are a group of rare neuropathies characterized by a spongiform neurodegeneration of the central nervous system (CNS) caused by prions. These disorders include Creutzfeldt–Jakob disease (CJD), Gerstmann–Sträussler–Scheinker (GSS) syndrome, fatal familial

G. Legname (✉) • G. Giachin • F. Benetti
Scuola Internazionale Superiore di Studi Avanzati (SISSA), via Bonomea 265, I-34136 Trieste, Italy
e-mail: legname@sissa.it

F. Rahimi and G. Bitan (eds.), *Non-fibrillar Amyloidogenic Protein Assemblies—Common Cytotoxins Underlying Degenerative Diseases*, DOI 10.1007/978-94-007-2774-8_9,

insomnia (FFI), and kuru in humans, bovine spongiform encephalopathy in cattle, scrapie in sheep and goats, and chronic wasting disease in elk, deer, and moose (Prusiner 1998).

Prions are thought to consist solely of a misfolded isoform (PrP^{Sc}) of the normal, host-encoded cellular protein (PrP^{C}), function of which is still unknown. Despite a wealth of supporting data, some investigators still resist the postulate that prions are infectious proteins (Manuelidis and Fritch 1996; Kimberlin 1990; Chesebro 1992). Nonetheless, since the introduction of the 'protein-only hypothesis' formulated by Stanley B. Prusiner in 1982 as mechanistic explanation for prion diseases, its formal experimental demonstration remained elusive for quite some time (Prusiner 1982).

A major feature that distinguishes infectious prions from viruses is that both PrP^{C} and PrP^{Sc} are encoded by a gene (Prusiner et al. 1998). In humans, the PrP gene, designated *PRNP*, is located on the short arm of chromosome 20.

Human (Hu) PrP^{C} is a 209-residues sialoglycoprotein, tethered to the outer leaflet of the plasma membrane by a glycosylphosphatidylinositol (GPI) anchoring and its primary structure is highly conserved among mammals (Wopfner et al. 1999; Premzl et al. 2005). In the CNS, several relevant processes are influenced by PrP^{C}, although its precise physiological functions still remain elusive (Aguzzi et al. 2008). According to the 'protein-only hypothesis' during the course of prion diseases PrP^{C} is converted into the abnormal form by a process whereby most α-helical motifs are replaced by β-sheet secondary structures (Kuwata et al. 2002). The PrP^{C}-to-PrP^{Sc} conversion leads to altered biochemical properties, such as resistance to limited proteolysis and insolubility in non-denaturing detergents (Caughey et al. 1991; Pan et al. 1993).

In disease-affected brain homogenates, limited proteolysis completely hydrolyzes PrP^{C} and produces a smaller, protease-resistant PrP^{Sc} molecule of ~142 amino acids, designated PrP27–30. In the presence of detergent, PrP27–30 polymerizes into amyloid (McKinley et al. 1991a). Prion amyloid, or rods, formed by limited proteolysis and detergent extraction are indistinguishable from the filaments that aggregate to form PrP amyloid plaques in the CNS. Both rods and filaments found in brain tissue exhibit similar ultrastructural morphology and green–gold birefringence after staining with Congo red dye.

Although the molecular mechanisms leading to the disease are still controversial, many lines of evidence suggest that generation of prion diseases is dependent only on PrP^{C}. Mice devoid of PrP^{C} are resistant to scrapie, and reintroduction of the PrP gene (*Prnp* in mice) restores TSEs susceptibility (Bueler et al. 1993; Fischer et al. 1996). Moreover, amyloid deposits may parallel the pathology and are mainly composed of the abnormal PrP^{Sc}. In fact, amyloid PrP fibrils generated *in vitro* induced prion diseases in transgenic (Tg) mice overexpressing PrP, which was subsequently transmissible to wild-type mice (Legname et al. 2004, 2005, 2006).

Production of synthetic prions has brought a wealth of support for the contention that prions are infectious proteins composed exclusively of PrP^{Sc}. Comparison of secondary structures of PrP^{C} and PrP^{Sc} were performed on proteins purified from Syrian hamster (SHa) brains (Pan et al. 1993). Based on these data, structural models for PrP^{C} and PrP^{Sc} were proposed (Huang et al. 1995). Subsequently, solution NMR structures of recombinant SHa and mouse PrPs produced in bacteria showed that it is likely that PrP^{C} has three α-helices and not four as predicted by molecular modeling

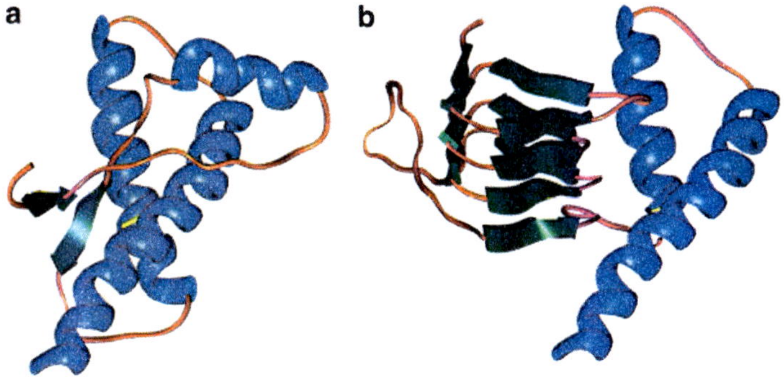

Fig. 9.1 (**a**) Recombinant prion protein, PrP (James et al. 1997). (**b**) A proposed model for monomeric PrP^{Sc} (Govaerts et al. 2004)

(Fig. 9.1a) (Riek et al. 1996; Liu et al. 1999). The computational model of PrP^{Sc} is supported by studies with recombinant antibody fragments, which have been used to map the surfaces of PrP^{C} and PrP^{Sc} (Peretz et al. 1997), and by electron crystallography (Wille et al. 2002). This α-to-β transition in PrP structure is the fundamental event underlying prion diseases.

In contrast to pathogens with a nucleic acid genome that encodes strain-specific properties, prions encipher these properties in their tertiary structure (Fig. 9.1b) (Bessen and Marsh 1992; Telling et al. 1996; Prusiner 1997). The amino acid sequence of PrP^{Sc} corresponds to that encoded by the PrP gene of the mammalian host in which it last replicated. Several studies argue that PrP^{Sc} acts as a template upon which PrP^{C} is refolded into a nascent PrP^{Sc} molecule through a process facilitated by an as-yet unidentified protein.

In addition to impressive advances in our understanding of the molecular structure and biology of prions, important progress has been made in studies of the pathogenesis of prion diseases (Telling 2000). At least three different pathogenic mechanisms have been found in natural and experimentally induced prion diseases. In the first form of prion diseases, PrP^{C} is thought to be converted into PrP^{Sc} in caveolae-like domains (CLDs) (Gorodinsky and Harris 1995; Taraboulos et al. 1992; Vey et al. 1996; Kaneko et al. 1997a) and subsequently trafficked to endosomes and lysosomes (Caughey 1991, McKinley et al. 1991b). Accumulation of PrP^{Sc} in CLDs and lysosomes is thought to cause CNS dysfunction, but the mechanism remains unclear (Campana et al. 2005).

A second form of prion disease is a PrP-storage-like disorder caused by the deletion of either helix B or C of PrP^{C} (Muramoto et al. 1997). As yet, this disease has been seen only in mice expressing PrP transgenes with a deletion mutation. In these mice, a massive proliferation of the endoplasmic reticulum (ER) was found.

The third form of disease involves adoption of a transmembrane topology by PrP (Hegde et al. 1998, 1999). A transmembrane form of PrP accumulated in mice expressing several different mutant PrP transgenes, and increased levels of a transmembrane form of PrP were found in humans with GSS(A117V), and were suggested to be pathogenic.

9.2 Structural Studies of Recombinant Prion Proteins

A wealth of various structural data exists from studies carried out on wild-type and mutant recombinant PrP molecules from different species. As revealed by its atomic structure, the mature PrP^C expressed by different mammals are very similar to each other (Table 9.1). The full-length PrP^C has a unique structure: while the *C*-terminus possesses well-defined secondary and tertiary structures, the *N*-terminal half is unstructured (Donne et al. 1997). The latter contains a highly conserved octapeptide-repeat region (PHGGGWGQ in human sequence), which enables the protein to coordinate Cu(II) ions (Viles et al. 1999, 2001). Several studies have clarified the mode of binding of Cu(II) ions, indicating well-defined structural features (Burns et al. 2003). NMR studies of the globular (residue 124–231, in human numbering) domain revealed a structure containing three α-helices, comprising the residues 144–154, 173–194, and 200–228, and very short anti-parallel β-strands, comprising residues 128–131 and 161–164. A disulfide bond, C179–C214, connects helices 2 and 3 (Fig. 9.1a) (Zahn et al. 1997, 2000).

One of the strongest arguments supporting the 'protein-only hypothesis' is the link between inherited prion diseases and mutations in the *PRNP* gene. Currently, 55 pathogenic mutations and 16 polymorphisms have been identified in the *PRNP* gene (Kovacs et al. 2002). They include missense point-mutations, most of them located in the globular part, insertion or deletion mutations involving the *N*-terminal domain, and non-sense mutations resulting in the premature termination of PrP synthesis. Twelve polymorphisms are silent, while four of them alter the amino-acid sequence: the M/V polymorphism at codon 129, the N/S at position 171, the E/K at position 219, and the deletion of one octarepeat (Collinge 2001). The M/V polymorphism at codon 129 is common; the homozygous M/M and V/V and the heterozygous M/V subjects account for 43%, 8%, and 49%, respectively, in the Caucasian population (Zimmermann et al. 1999). This polymorphism is a key determinant of genetic susceptibility to acquired and sporadic prion diseases, the large majority of which occur in homozygous individuals (Collinge et al. 1991; Palmer et al. 1991; Windl et al. 1996). The *PRNP* heterozygotes appear to be protected from sporadic CJD compared to the *PRNP* homozygotes (Kobayashi et al. 2009; Baker et al. 1991; Hsiao et al. 1992). The M/V polymorphism at position 129, as well as the E/K at position 219, affects the disease phenotype when it is located on the mutant allele: D178N–129V causes fCJD, while D179N–129M is responsible for FFI. The M/V polymorphism located on the normal allele affects the onset age and duration of the disease. Patients carrying either M or V129 codon have been observed in all inherited prion diseases. Tg mice carrying pathological PrP mutations develop a spectrum of neurological diseases sharing some features with TSEs (Chiesa et al. 1998; Dossena et al. 2008; Hsiao et al. 1990, 1994). Our understanding of the mechanisms whereby mutations induce the disease still remains limited. Mutations may increase the likelihood of misfolding by the thermodynamic destabilization of PrP^C (Apetri et al. 2004; Liemann and Glockshuber 1999; Swietnicki et al. 1998; Vanik and Surewicz 2002). PrP mutants may escape cellular

Table 9.1 List of prion protein molecules resolved by NMR or X-ray crystallography

Prion protein molecule	Species	Method	pdb	References
Ovine prion protein (residue 167–234)	Sheep	NMR	2KTM	Adrover et al. (2010)
Solution structure of synthetic 26-mer peptide containing 142–166 sheep prion protein segment and *C*-terminal cysteine with Y155A mutation	Sheep	NMR	2RMV	Bertho et al. (2008)
Solution structure of synthetic 26-mer peptide containing 142–166 sheep prion protein segment and *C*-terminal cysteine with R156A mutation	Sheep	NMR	2RMW	Bertho et al. (2008)
Solution structure of the sheep prion protein with polymorphism H168	Sheep	NMR	1XYU	Lysek et al. (2005)
Ovine prion protein variant R168	Sheep	NMR	1Y2S	Lysek et al. (2005)
Ovine recombinant prion protein (114–234), ARQ variant in complex with the Fab of the VRQ14 antibody	Sheep	X-ray	1TPX	Eghiaian et al. (2004)
Ovine recombinant prion protein (114–234), VRQ variant in complex with the Fab of the VRQ14 antibody	Sheep	X-ray	1TQB	Eghiaian et al. (2004)
Ovine recombinant prion protein (114–234), ARR variant in complex with the VRQ14 Fab fragment (IGg2a)	Sheep	X-ray	1TQC	Eghiaian et al. (2004)
The crystal structure of the globular domain of sheep prion protein	Sheep	X-ray	1UW3	Haire et al. (2004)
Solution structure of synthetic 21-mer peptide spanning region 135–155 (in human numbering) of sheep prion protein	Sheep	NMR	1S4T	Kozin et al. (2004)
Structure of synthetic 26-mer peptide containing 145–169 sheep prion protein segment and *C*-terminal cysteine in TFE solution	Sheep	NMR	1M25	Megy et al. (2004)
Solution structure of synthetic 26-mer peptide containing 145–169 sheep prion protein segment and *C*-terminal cysteine	Sheep	NMR	1G04	Kozin et al. (2001)
Human prion protein fragment 121–230	Human	NMR	1QM2	Zahn et al. (2000)
Human prion protein residue 90–230	Human	NMR	1QLZ	Zahn et al. (2000)
Human prion protein crystal dimer	Human	X-ray	1I4M	Knaus et al. (2001)
Human prion protein 121–230 M166C/E221C	Human	NMR	1H0L	Zahn et al. (2003)
Human prion protein at pH 7.0	Human	NMR	1HJN	Calzolai and Zahn (2003)
Human prion protein 61–68	Human	NMR	1OEH	Zahn (2003)
Human prion protein 61–84	Human	NMR	1OEI	Zahn (2003)
Peptide Corresponding to Residues 170–175 of human prion protein	Human	X-ray	2OL9	Sawaya et al. (2007)

(continued)

Table 9.1 (continued)

Prion protein molecule	Species	Method	pdb	References
Human prion protein 180–195 structure	Human	NMR	2IV4	Ronga et al. (2008)
Human prion protein 173–195 structure	Human	NMR	2IV5	Ronga et al. (2008)
Human prion protein 173–195, D178N structure	Human	NMR	2IV6	Ronga et al. (2008)
Human prion protein variant V129 domain-swapped dimer (residues 90–231)	Human	X-ray	3HAF	Lee et al. (2010)
Human prion protein variant V129 (residue 125–227)	Human	X-ray	3HAK	Lee et al. (2010)
Human prion protein variant D178N with M129 (residues 90–231)	Human	X-ray	3HEQ	Lee et al. (2010)
Human prion protein variant F198S with V129 (residues 90–231)	Human	X-ray	3HER	Lee et al. (2010)
Human prion protein variant F198S with M129 (residues 90–231)	Human	X-ray	3HES	Lee et al. (2010)
Human prion protein variant V129 domain-swapped dimer (residues 90–231)	Human	X-ray	3HJ5	Lee et al. (2010)
Human prion protein variant D178N with V129 (residues 90–231)	Human	X-ray	3HJX	Lee et al. (2010)
Fab fragment complexed with human prion protein fragment 119–231	Human	X-ray	2W9E	Antonyuk et al. (2009)
Human prion protein (mutant E200K) fragment 90–231	Human	NMR	1FO7	Zhang et al. (2000)
Human prion protein variant S170N	Human	NMR	1E1S	Calzolai et al. (2000)
Human prion protein variant M166V	Human	NMR	1E1G	Calzolai et al. (2000)
Human prion protein variant R220K	Human	NMR	1E1U	Calzolai et al. (2000)
NMR studies of a pathogenic mutant (D178N) of the human prion protein	Human	NMR	2K1D	Mills et al. (2009)
Horse prion protein	Horse	NMR	2KU4	Perez et al. (2010)
Bank vole prion protein (121–231)	Bank vole	NMR	2 K56	Christen et al. (2008)
NMR structure of rabbit prion protein mutation S173N	Rabbit	NMR	2JOH	Wen et al. (2010)
NMR structure of rabbit prion protein mutation I214V	Rabbit	NMR	2JOM	Li et al. (2007)
NMR solution of rabbit prion protein (91–228)	Rabbit	NMR	2FJ3	Wen et al. (2010)
Bovine prion protein fragment 121–230	Cow	NMR	1DWY	Lopez Garcia et al. (2000)
N-terminal fragment (1–30) of bovine prion protein	Cow	NMR	1SKH	Biverstahl et al. (2004)
Quadruplex structure of an RNA aptamer against bovine prion protein	Cow	NMR	2RQJ	Mashima et al. (2009)
Elk prion protein	Elk	NMR	1XYW	Gossert et al. (2005)
NNQNTF segment from Elk prion	Elk	X-ray	3FVA	Wiltzius et al. (2009)

Tammar wallaby prion protein (121–230)	Tammar wallaby	NMR	2KFL	Christen et al. (2009)
Mouse prion protein fragment 121–231	Mouse	NMR	1XYX	Gossert et al. (2005)
Mouse prion protein with mutation N174T	Mouse	NMR	1Y15	Gossert et al. (2005)
Mouse prion protein (121–231) with mutations Y225A and Y226A	Mouse	NMR	2KFM	Christen et al. (2009)
Mouse prion protein (121–231) with mutation V166A	Mouse	NMR	2KFO	Christen et al. (2009)
Mouse prion protein with mutations S170N and N174T	Mouse	NMR	1Y16	Gossert et al. (2005)
Mouse prion protein (121–231) with mutation S170N	Mouse	NMR	2K5O	Christen et al. (2008)
Mouse prion protein (121–231) with mutations D167S and N173K	Mouse	NMR	2KU6	Perez et al. (2010)
Mouse prion protein (121–231) with mutation D167S	Mouse	NMR	2KU5	Perez et al. (2010)
NMR structure of the chicken prion protein fragment 128–242	Chicken	NMR	1U3M	Calzolai et al. (2005)
Solution structure of the turtle prion protein fragment (121–226)	Turtle	NMR	1U5L	Calzolai et al. (2005)
Solution structure of *Xenopus laevis* prion protein	Frog	NMR	1XU0	Calzolai et al. (2005)
NMR structure of the cat prion protein	Cat	NMR	1XYJ	Lysek et al. (2005)
NMR structure of the canine prion protein	Dog	NMR	1XYK	Lysek et al. (2005)
NMR structure of the swine prion protein	Pig	NMR	1XYQ	Lysek et al. (2005)
Syrian hamster prion protein (90–231)	Syrian hamster	NMR	1B10	James et al. (1997)
Syriam hamster prion protein 23–106 bound to pentosan polysulfate	Syrian hamster	NMR	2KKG	Taubner et al. (2010)

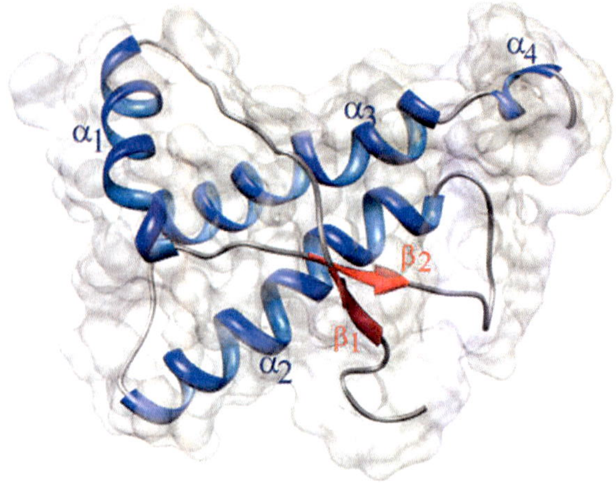

Fig. 9.2 High-resolution structure of human PrP (90–231, M129, Q212P). Cartoon representation of the lowest-energy structure on the van der Waals surface (Adapted from Ilc et al. 2010)

quality-control pathways and accumulate intracellularly (Ashok and Hegde 2009; Hegde et al. 1998; Heske et al. 2004; Mishra et al. 2003). In addition, mutations may change surface properties promoting an abnormal interaction between PrP^C and other not-yet identified factors (Kaneko et al. 1997b; Telling et al. 1995; Zhang et al. 2000).

Therefore, further studies on PrP variants containing familial mutations should provide important clues regarding the molecular basis of the disease. Up to date in fact there is no evidence showing a pathological point-mutation causing substantial structural differences in PrP folding. Though indeed, the solution structures of some pathogenic human PrP mutants exhibit conformations (Zhang et al. 2000) and dynamics (Bae et al. 2009) similar to the wild-type protein.

To shed new light on the role of pathological point-mutations on PrP structure, we recently determined and examined a high-resolution three-dimensional structure of the truncated recombinant human PrP(90–231) conformation containing the pathological Q212P mutation (Ilc et al. 2010). This mutation is responsible for a GSS syndrome characterized by mild PrP amyloid deposition in patients (Piccardo et al. 1998; Young et al. 1998). The high-resolution NMR structure of Q212P mutant revealed unique conformational features compared to the known structures of either human or other mammalian PrP^C (Christen et al. 2008, 2009; Gossert et al. 2005; Lopez Garcia et al. 2000; Riek et al. 1996).

The most remarkable differences involve the *C*-terminal end of the protein and the β_2–α_2 loop region. The Q212P mutant is the first known example of PrP structure where the α_3 helix between E200 and Y226 is broken into two helices (Fig. 9.2). This breakage brings about dramatic changes in the hydrophobic interactions between α_3 helix and β_2–α_2-loop region. In the wild-type protein, long-range interactions between Y225 and M166 define the position of β_2–α_2 loop and thus the tertiary structure of the protein. The distance between C_α atoms of Y225 and M166 in the structures of prion proteins of different mammals is typically 8.4 Å, whereas it is 17.6 Å in the

Q212P mutant. The longer distance is correlated with a marked twist of Y225 away from the β_2–α_2 loop. In the Q212P mutant, Y225 forms hydrophobic interactions with residues in the α_3 helix (e.g., I215). These interactions define the mutual orientation of α_3 and α_4 helices. Since Y225 is unable to form contact with M166, the hydrophobic cluster is opened and accessible to solvent. Exposure of the hydrophobic surface is tightly correlated with the orientation of aromatic residues Y163, Y169, and F175 in the Q212P mutant. This opened cleft has been proposed as the binding site for a hypothetical facilitator of prion conversion playing a role in pathogenic PrP^{Sc} formation (Kaneko et al. 1997b; Telling et al. 1995). In the wild-type protein, the solvent-exposed surface of the β_2–α_2 loop and the α_3 helical region is smaller, and Y169 is buried inside the hydrophobic cluster.

We then compared our structural findings with the already-resolved NMR structures of human PrP carrying respectively the CJD-related E200K (Zhang et al. 2000) and the artificial R220K mutations (Calzolai et al. 2000). In the HuPrP(90–231, M129, R220K) artificial mutant, for example, the α_3 helix is well-ordered up to the point-mutation (Fig. 9.3b) (Calzolai et al. 2000). After this mutation, the α_3 helix shows increased flexibility and it is much less ordered. At the same time, the R220K mutation does not alter the hydrophobic interactions between the aromatic residues of the β_2–α_2 loop and α_3 helix. On the other hand, the structure of CJD-related mutant HuPrP(90–231, M129, E200K) has revealed features in the β_2–α_2 loop region similar to those of the Q212P mutant (Zhang et al. 2000). In the E200K variant, Y169 is exposed to solvent and shows increased flexibility (Fig. 9.3c). The remaining aromatic residues in the β_2–α_2 loop (Y163 and F175) form a hydrophobic cluster by interacting with Y218 and Y225 in the α_3 helix. The effects of both Q212P and R220K mutations in terms of unstructured *C*-terminal parts of proteins are comparable (Fig. 9.3a, b). In support of this, recently published crystal structures of the pathological HuPrP (90–231, D178N, M/V129) and HuPrP(90–231, F198S, M/V129) mutants, demonstrated similar orientations of Y169 outside the globular part (Lee et al. 2010).

Special interest in prion biology is focused on the epitope formed by the β_2–α_2 loop and the α_3 helix, because this surface has been implicated in interactions with a hypothetical facilitator of prion conversion involved in the development of TSEs (Kaneko et al. 1997b; Telling et al. 1995). Thus, the plasticity of the loop may modulate the susceptibility of a given species to prion disease. Within the highly conserved PrP scaffold, the β_2–α_2 loop residues (165–173, in human numbering) show local structural variations among species. While in PrP^C from most mammalian species this loop is flexible, it is well-defined in PrP^C of elk (Gossert et al. 2005), bank vole (Christen et al. 2008), Tammar Wallaby (Christen et al. 2009), and, very recently, horse (Perez et al. 2010), and rabbit (Wen et al. 2010). Interestingly, elk and bank vole are highly susceptible to TSEs, but in contrast there have been no cases of prion diseases neither in marsupials nor in horse or rabbit. It has been proposed that TSEs susceptibility may be correlated both with the amino-acid composition of the β_2–α_2 loop and with long-range interactions between residues in the β_2–α_2 loop and residues at the *C*-terminus of the α_3 helix. Indeed, horse and wallaby PrP present unique amino-acid substitutions in the β_2–α_2 loop and long-range interactions between this loop and residues at the *C*-terminal end of the α_3 helix. In contrast to

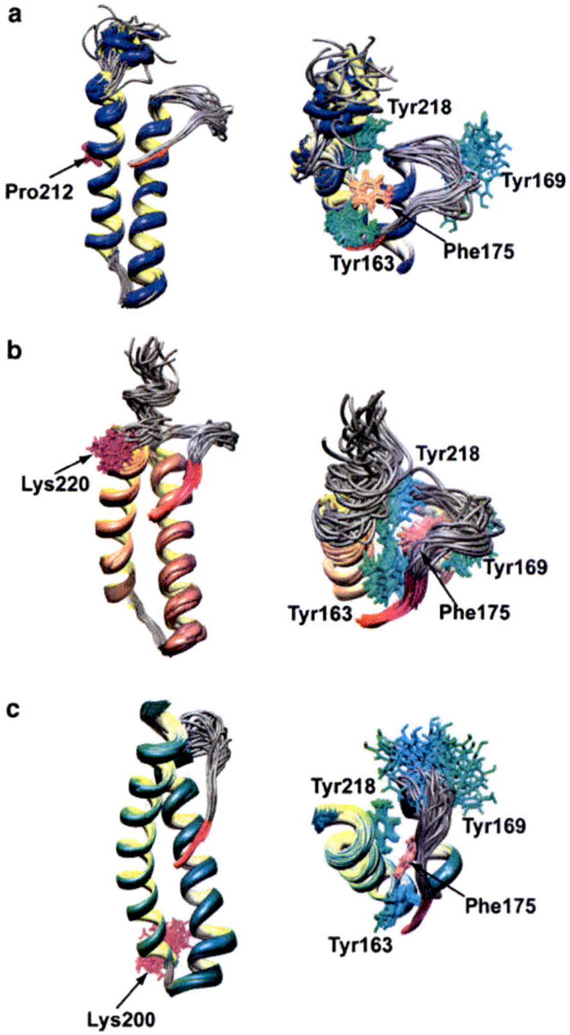

Fig. 9.3 Structural comparison of human PrP mutants. (**a**) Structural details of β2α2α3α4 region (161–228) of 20 lowest-energy structures for human PrP(90–231, M129, Q212P) mutant. (**b**) Structural details of β2α2α3 region (161–228) of 20 lowest-energy structures for human PrP(90–231, M129, R220K) (pdb ID: 1E1U) mutant (Zhang et al. 2000). (**c**) Structural details of β2α2α3 region (161–228) of 20 lowest-energy structures for human PrP(90–231, M129, E200K) (pdb ID: 1FO7) mutant (Calzolai et al. 2000). In all three panels the point-mutation is indicated in magenta (*left*). Top view of the hydrophobic core composed of aromatic amino-acid residues (*center*) (Adapted from Ilc et al. 2010)

horse and wallaby, other mammalian PrPs show a β_2–α_2 loop spatially separated from α_3 helix and readily accessible on the protein surface.

In the case of inherited prion diseases, it is possible to argue that mutations may exacerbate the flexibility of the β_2–α_2 loop and enhance the distance between this loop and the α_3 helix. In the wild-type PrP, the side-chains of residues 214 and 218, belonging to α_3 helix, form a discontinuous epitope with residues 167 and 171, localized in the β_2–α_2 loop, for binding with a protein X. The latter has been hypothesized to be involved in prion formation (Kaneko et al. 1997b; Telling et al. 1995). Experimental evidence showed that Tg mice expressing HuPrP, Tg(HuPrP), were not susceptible to human prions, while Tg mice expressing a chimeric Hu/MoPrP transgene (MHu2M) became susceptible to human prions.

Moreover, Tg(HuPrP) crossed with PrP-null mice became more susceptible to human prions than Tg(MHu2M)/*Prnp*$^{0/0}$ mice. These findings suggested that mouse PrP^{C} binds to murine protein X. This binding has higher affinity than does human PrP^{C}, inhibiting the transmission of human prions in Tg(HuPrP)/*Prnp*$^{+/+}$ mice. Therefore, the larger solvent exposure of this epitope observed in mutant PrPs might cause an altered interaction with other yet-unknown cellular factors, whether chaperones or PrP^{C} ligands. Recent NMR studies provided evidence that some GSS mutations increase the binding affinity of PrP to lipid structures (Hornemann et al. 2009). Moreover, the altered conformation observed in the case of the Q212P mutant may cause a different affinity for cellular membranes and, consequently, an aberrant localization of PrP in cellular compartments, favoring formation of altered ER topologies (Hegde et al. 1998; Heske et al. 2004). Independent evidence derived from cultured cells expressing the mutant Q212P showed that this point-mutation affects folding and maturation of PrP^{C} in the secretory pathway of neurons (Ashok and Hegde 2009; Mishra et al. 2003). These authors investigated the generation and the turnover of Q212P and other mutants in mouse neuroblastoma N2a cells, discovering an intracellular post-ER control pathway that selectively routes aberrant PrP species to lysosomes (Borchelt et al. 1992).

The structure–function relationship suggested by these studies may provide a molecular basis for understanding the generation of PrP^{Sc} in inherited prion diseases. Indeed, characterization of high-resolution structures of pathological mutants of PrP and their comparison with the overall wild-type folding highlight *hot-spot* regions in these proteins that could be involved in early events of PrP misfolding. It may also provide a molecular explanation for prion formation in the sporadic forms of prion disease.

9.3 Biophysical and Structural Characteristics of Toxic/Infective Prions

As mentioned before, PrP^{Sc} and its physiological counterpart, PrP^{C}, share a common sequence and pattern of post-translational modifications, but have different secondary and tertiary structures. These two conformational isoforms can be distinguished in view of their different physicochemical properties (Prusiner 1998; Hill et al. 1999; Cohen and Prusiner 1998). The conformer PrP^{Sc} is stable at high temperatures for extended periods of time, and resistant to reagents that disrupt nucleic-acid polymers such as nucleases, psoralens, and UV irradiation (Alper et al. 1967; Latarjet et al. 1970; Prusiner 1982; Diener et al. 1982; Bellinger-Kawahara et al. 1987).

In conclusion, the molecular mechanisms of PrP conversion and propagation still remain elusive. They may be understood once the whole structure of PrP^{Sc} is gathered. The insoluble nature of PrP^{Sc} hampers the use of classic techniques such as nuclear magnetic resonance or X-ray crystallography. Therefore, alternative approaches such as limited proteolysis, Fourier-transform infrared spectroscopy, antibody labeling, electron microscopy, fiber X-ray diffraction, and small-angle X-ray scattering have been used.

9.3.1 Limited Proteolysis: Prions Contain a Proteinase-K-Resistant Core

The infectious PrP^{Sc} is distinguished from PrP^{C} by its resistance to protease digestion in non-denaturing conditions (McKinley et al. 1983). Limited proteolysis permitted to identify PrP^{Sc} as the etiological agent of prion diseases. This technique is a useful tool for obtaining insights on the structural features of proteins. In fact, proteolytic enzymes cleave proteins more readily at exposed sites, preferentially within loops, and rarely in β-strands. Full-length PrP^{Sc} is partially resistant to proteases, particularly to proteinase K (PK). After hydrolysis, PK-resistant PrP^{Sc} appears as a truncated form, lacking its *N*-terminal region (Requena 2009). This PK-resistant core, called PrP27–30 because of its apparent size in SDS-PAGE, retains infectivity. Limited-proteolysis studies on prions revealed the presence of two domains: one labile, spanning from the *N*-terminus to residues 78–103 (McKinley et al. 1983; Zanusso et al. 2004; Bessen and Marsh 1994; Zou et al. 2003; Parchi et al. 2000), and one protease-resistant. Analyses using NMR reveal that the *N*-terminal region of PrP^{C} is unstructured. Its susceptibility to PK digestion suggests that part of it remains unfolded also in PrP^{Sc}. Identification of PK-cleavage sites offer a valuable approach to the study of PrP^{Sc} structural features.

In human prion diseases, two PK-resistant fragments with distinct relative molecular masses have been detected: type 1, which has a molecular mass of 21 kDa, and type 2, characterized by a proteolytic fragment of 19 kDa (Worrall et al. 1999). In PrP27–30 type 1, the *N*-terminus commonly starts at residue G82, but several secondary species with the *N*-terminus starting between residues 78 and 97 do exist. PrP27–30 type 2 commonly starts at residue S97 with secondary cleavable sites between residues 92 and 103 (Zou et al. 2003; Parchi et al. 2000). The differences in size reflect distinct conformations or different ligand interactions.

Zou and coworkers identified two novel PK-resistant *C*-terminal fragments of PrP (PrP-CTF 12/13) in brains of subjects with sporadic CJD (sCJD) (Zou et al. 2003). The non-glycosylated PrP-CTF 12/13 migrates with a molecular mass of 12 and 13 kDa, and have an *N*-terminus at residues 162/167 and 154/156, respectively. PrP-CTF12/13 may represent the *C*-terminal region of a subpopulation of PrP^{Sc} in which the PK-resistant core is displaced by 64–76 residues toward the *C*-terminus. Therefore, the tertiary structure of these fragments must be different from that of PrP^{Sc} generating PrP27–30. PrP-CTF12/13 fragments were localized prevalently in neocortex (Zou et al. 2003).

In order to relate the variety of neurological signs caused by sCJD and the molecular properties of PrP^{Sc}, Zanusso and colleagues highlighted the presence of distinct *N*-terminally truncated prion forms in different CJD subtypes (Zanusso et al. 2004). All sCJD type 1 PrP27–30 cases, and MM homozygote subjects belonging to type-2

PrP27–30, were characterized by the presence of non-glycosylated fragments of 16–17 kDa. Patients VV and MV with type-2 PrP27–30 were characterized by the presence of non-glycosylated fragments of 17.5–18 kDa. These last ones were associated with plaque-like aggregates or amyloid plaques. Distinct from CJD subjects, patients with GSS also showed the presence of a PK-resistant fragment spanning from 74/80 to 146/153 (Zou et al. 2003).

Limited-proteolysis experiments facilitated identification of different PrP^{Sc} strains. Prion strains are distinct varieties of prions with the same primary structure, exhibiting characteristic biological properties, such as tropism for specific brain areas and incubation period, peculiar neurological signs, as well as distinct physicochemical features. Considerable evidence argues that the biological information of a prion strain is encoded in its conformation.

To decipher the molecular basis of strain variations in prion disorders, Bessen and Marsh analyzed the biochemical differences between two Syrian hamster (SHa) strains, called hyper (HY) and drowsy (DY) (Bessen and Marsh 1994). HY causes hyper-excitability, head and body tremors, incoordination, and ataxia. It has an incubation period of 65 days. DY is characterized by a progressive lethargy and an incubation period of 165 days. HY-infected brains contain a 100-fold higher concentration of infectivity than DY-infected brains at the onset of clinical disease. Limited proteolysis coupled to Edman protein sequencing revealed that the major *N*-terminal end of the HY strain starts at least 10 residues prior to that of DY. In particular, the HY *N*-termini are located inside the octarepeat region that extends from P52 to Q91. The PK-cleavable sites of DY are located at G92, Q98, and K101 of hamster PrP. Bessen and Marsh also reported differences in sedimentation properties between HY and DY (Bessen and Marsh 1992). Since the ability to aggregate is correlated to the β-sheet content, as well as the size of PK-resistant core, HY has higher β-sheet content than DY.

Recently, Sajnani and colleagues performed limited-proteolysis and mass-spectrometry experiments on PrP^{Sc} isolated from brains of hamsters infected with 263 K and DY strains (Sajnani et al. 2008). To avoid the heterogeneity induced by carbohydrate chains and GPI anchor, PK-resistant fragments were denatured, reduced and cleaved at position 178 using 2-nitro-5-thiocyanatobenzoic acid. Besides the known cleavage sites at positions 86, 90, and 92 for 263 K and at positions 86, 90, 92, 98, and 101 for DY, authors identified other internal positions in both strains: 117, 119, 135, 139, 142, and 154. These cleavage sites in the resistant core correspond to flexible sub-domains, like loops and turns, interspersed with short β-sheet stretches. Therefore, these structural elements indicate that PrP^{Sc} conforms to a succession of β-strand–turn/loop–β-strand assemblies forming more than one layer, like the solenoid structure of amyloid fibrils of the HET-s (218–289) prion (Sajnani et al. 2008).

9.3.2 *Chemical Cross-Linking*

Chemical cross-linking, in combination with proteolytic digestion and mass spectrometry, is a useful technique to identify pairs of specific amino-acid residues that are close enough to each other to react with a bifunctional reagent of a given chain length. Onisko et al. performed cross-linking on PrP27–30 purified from scrapie-infected SHa, using the Bis(sulfosucinimidyl)suberate (BS3) as the bifunctional group (Onisko et al. 2005). Mass spectral analysis of proteolytic fragments showed that BS3 reacted preferentially with G90. This cross-link was found only in PrP27–30, but not in intramolecularly cross-linked monomers or control samples. This finding provides a distance constraint of ~1.14 nm between consecutive G90 in the PrP27–30 stack. Considering the left-handed β-helical model described by Govaerts (Govaerts et al. 2004) and for HET-s prions (Wasmer et al. 2008), the distance of 1.14 nm may correspond to two rungs of β-stand (~0.94 nm). Moreover, the cross-linking involving G90 indicates that this amino acid is exposed to solvent in PrP27–30 fibers.

9.3.3 Fourier-Transform Infrared Spectroscopy: Prions Are β-Sheet Enriched

Infrared spectroscopy (IR) is a useful technique in providing structural information on proteins, especially on secondary structure. Conversely to X-ray crystallography and NMR, where well-ordered crystals and soluble proteins are respectively required, IR can be used on insoluble molecules without problems of background fluorescence or light scattering, and issues related to the size of the proteins.

The two major bands of protein infrared spectra are amide I, and amide II. The amide-I band (between 1,600 and 1,700 cm^{-1}) is mainly associated with the C=O stretching vibration and is directly related to the backbone conformation as well as to the hydrogen-bonding pattern. Amide-I band is a sharp sensor of secondary structure. Amide II, located in the 1,510- and 1,580-cm^{-1} regions, results from the N–H-bending vibration and from the C–N stretching vibration.

Numerous attempts to identify post-translational chemical modifications causing the conversion of PrP^{C} into PrP^{Sc} have been unsuccessful. Thus, the physicochemical properties of PrP^{Sc} must only be due to conformational changes. Caughey and colleagues described the secondary structure of the PK-resistant core, PrP27–30, of SHa 263 K strain (Caughey et al. 1991). The IR primary spectrum of PrP27–30 contained major amide-I and -II bands, with maxima at 1,636 and 1,549 cm^{-1}. This spectrum was corrected subtracting the *N*-acetylglucosamide contribution (~6%), as well as that of side-chains of asparagine, glutamine, arginine, and lysine (less than 3%) residues. Both of them absorb at specific frequencies in the amide-I region. The maximum amide-I absorbance at 1,636 cm^{-1} and the shoulder at 1,627 cm^{-1} indicated a predominance of β-sheet structures. The weaker peak at 1,657 cm^{-1} and the shoulder on the high-wave-number side suggested the presence of α-helix and turns. Authors quantified the amide-I infrared bands corresponding to secondary structures performing the second-derivative spectrum. This analysis showed that SHa PrP27–30 is composed of β-sheet (47%), α-helix (17%), and turns (31%). Analysis of the second-derivative spectrum obtained in deuterium oxide (D_2O) did not reveal gross changes, thus highlighting a well-packed structure.

Gasset and coworkers generated PrP27–30 from scrapie-infected hamster brains (Gasset et al. 1993) and analyzed the secondary structure using attenuated total-reflection (ATR)–Fourier-transform infrared spectroscopy (FTIR). This technique has a major advantage over the conventional transmission mode for insoluble samples, because the analyte is manipulated as a thin hydrated film (Goormaghtigh et al. 1990). Secondary structure of PrP27–30 obtained from this analysis confirmed the higher β-content (54%) than that of PrP^{C} (3%), with 25% α-helix, 11% random coil, and 10% turn structures. Almost 60% of the β-sheet (~33%) was low-frequency, infrared-absorbing, reflecting intermolecular associations. In the presence of denaturing conditions such as SDS, there was a decrease in the low-frequency β-sheet with a subsequent increase of turns. In these conditions, PrP27–30 was more accessible to hydrogen exchange, less compact, and more soluble, but with a diminished infectivity. Same effects were obtained in the presence of alkaline conditions (pH > 10), where PrP27–30 lost the major part of β-sheet content (39%) favoring α-helix (30%) and turn (21%) formation.

A comparison between PrP^{C} and PrP^{Sc}/PrP27–30 purified from scrapie-infected Syrian-hamster brains was carried out by Pan and colleagues (Pan et al. 1993). While PrP^{C} has a high α-helix (42%) and low β-sheet (3%) content, as also confirmed by circular-dichroism measurements, PrP^{Sc} has a higher β-sheet (43%) and lower α-helix (30%) content than PrP^{C}. Moreover, PrP27–30 has even higher β-sheet content (54%) and lower α-helix content (21%).

As mentioned above, strain diversity in TSEs might be due to variations in the conformation of the abnormal, PK-resistant PrP^{Sc}. Evidence for the existence of prion strains derived from multiple observations, indicate that different PrP^{Sc} strains present different *N*-terminal PK-cleavable sites (*see limited proteolysis*).

Caughey and colleagues investigated the secondary structural differences existing between three different strains isolated from infected brains of hamsters: HY, DY, and 263 K (Caughey et al. 1998). While HY and 263 K cause similar symptoms such as hyper-excitability, ataxia, and a widespread distribution of PrP^{Sc} in the brain gray matter, DY causes progressive lethargy, and prominent PrP-res deposits along white-matter tracts in the brain. The FTIR spectra of the PK-treated 263 K and HY were similar; major bands were found at 1,626, 1,636, and 1,657 cm^{-1}. The bands at 1,626 and 1,636 cm^{-1} are indicative of β-sheet, and the 1,657 cm^{-1} band is generally assigned to α-helix. Another β-sheet band, at 1,694–1,695 cm^{-1}, was more intense in the HY spectra than in the 263 K spectra. In the DY spectra, the peaks at 1,626 and 1,636 cm^{-1} were absent. Prominent bands were observed at 1,616, 1,629–1,630, and 1,694–1,695 cm^{-1}. The 1,629 cm^{-1} band was restricted to the DY spectrum, but the 1,616 and 1,695 cm^{-1} bands were also present, at lower intensity, in the HY spectra. Absorbance values in the 1,616–1,636 cm^{-1} spectral region are attributed to β-sheets, suggesting that the strain-dependent conformational differences are associated with β-sheet secondary structures. The secondary-structure content of both untreated and PK-treated HY, 263 K, and DY strains is very similar. All strains revealed a β-sheet content of 50%, α-helix 14%, and turns or undefined structures around 36%. After PK digestion, the β-sheet content is 60%, α-helix 18%, and turns or undefined structures 22%. The authors of these studies hypothesized that differences between HY, DY, and 263 K were due to the type of interactions rather than their β-sheet content.

More recently, Spassov and coworkers extended the FTIR study on prion strains, exploring the secondary structures of four Syrian-hamster-adapted TSE agents: 263 K, ME7-H, 22A-H, and BSE-H (Spassov et al. 2006). The second-derivative FTIR spectra at different experimental conditions—samples hydrated in H_2O or D_2O, and different temperatures—exhibited strain-specific infrared characteristics, in both the secondary-structure-sensitive amide-I region, and in the amide-II and amide-A absorption regions.

9.3.4 Antibody Labeling: The Region 90–120 Is Involved in the Conversion

Monoclonal antibodies are sensitive probes of protein conformation. Therefore, another approach to investigate the differences between PrP^{C} and PrP^{Sc} conformations is to generate antibodies to diverse epitopes. Williamson, Peretz, and their colleagues demonstrated that monoclonal antibodies recognizing the PrP region 96–104 (R10, D4, and D13) were unable to interact with native PrP27–30 (Peretz et al. 1997; Williamson et al. 1998). The same pattern was observed for the 3 F4 antibody that recognizes the region 109–112. The monoclonal antibodies recognized these regions only after guanidine denaturation. On the other hand, monoclonal antibodies recognizing the *C*-terminal region 225–231 (R1, R2, and D2) were able to bind both PrP^{C} and PrP27–30. The region of PrP27–30 encompassing residues 152–163 is partially accessible to monoclonal antibody R72, although the binding is weaker than with PrP^{C}. These findings suggest that the *C*-terminal portion of PrP^{C} remains unaltered during the conversion to PrP^{Sc}, while the conformational rearrangements toward the pathological form involve the region encompassing residues 90–120.

9.3.5 Electron Microscopy, Atomic-Force Microscopy, and Small-Angle X-Ray Scattering

Many investigators have used electron microscopy to search for a scrapie-specific particle. The first scrapie-specific structures to be identified were spherical particles within post-synaptic evaginations of scrapie-infected mouse brains, sheep brains, and brain tissues of patients affected by CJD (David-Ferreira et al. 1968; Bignami and Parry 1971; Bots et al. 1971).

Godsave et al. performed cryoimmunogold electron microscopy on hippocampal slices from RML-infected mice (Godsave et al. 2008). They used two antibodies: R2 that recognizes both PrP forms and F4-31, which detects only PrP^{C} in non-denatured sections. At a late subclinical stage of infection, they found PrP^{C} and

PrP^{Sc} on neuronal plasma membranes and on early endocytic or recycling vesicles in the neuropil. After trypsin digestion of infected hippocampal slices, authors found a reduction of >85% in R2 labeling and hypothesized that a high proportion of PrP^{Sc} may be oligomeric, protease-sensitive PrP^{Sc} ($sPrP^{Sc}$).

In crude extracts of scrapie-infected rodent brains, fibrillar structures were observed. These structures differ from amyloids and cytoskeletal elements by their well-defined morphology. In purified fractions prepared from brains of scrapie-infected SHa, rod-shaped particles with the tinctorial properties of amyloid were found. McKinley et al., using a hamster-adapted isolate of scrapie prions, found that the formation of prion rods *in vitro* requires both detergent extraction and limited proteolysis (McKinley et al. 1991a). Same results are described in Meyer's work (Meyer et al. 1986), where prion rods were obtained only after microsome solubilization by using either anionic detergents such as sarkosyl, or nonionic detergents such as octylglucoside and limited proteolysis.

In humans, PrP is characterized by heavy surface glycosylation because of two large *N*-linked sugar moieties at positions 181 and 197. Therefore, glycosylation may impede the study of the underlying protein core of fibrils. Although recombinant PrP fibrils may facilitate this study for lack of these glycan moieties, their infectivity may be lower than native PrP^{Sc}. This makes it difficult to determine how recombinant PrP fibril structures relate to the infectious forms (Anderson et al. 2006; Novitskaya et al. 2006). On the other hand, Chesebro et al. developed anchorless PrP transgenic mice expressing the protein without GPI anchor (Chesebro et al. 2005), mostly in an unglycosylated form. Once inoculated with scrapie, these mice were able to propagate infectivity and develop large perivascular amyloid plaques of mostly unglycosylated PrP^{Sc}. Anchorless transgenic mice are susceptible to propagate different prion strains, such as the mouse-adapted scrapie strains ME7, 22L, and RML. Both 22L and RML strains have the same incubation period (~150 days post inoculation in wild-type C57Bl/6), similar clinical signs and glycosylation patterns, but different regions of accumulation. These variations may arise from differences in PrP^{Sc} conformation. Sim and coworkers investigated the ultrastructure of ME7, 22L, and RML strains (Sim and Caughey 2009) by using transmission-electron microscopy (TEM) and atomic-force microscopy (AFM). They isolated the prion strains from both wild-type and anchorless transgenic mice. TEM images carried out on 22L and RML, coming from infected wild-type mice, revealed fibrils of 100–150 nm in length with fibrils occasionally approaching 300 nm in the 22L preparations. TEM images of anchorless 22L and RML showed long fibrils of several hundred nanometers. All fibrils appeared to be composed of thinner strands (protofilaments) combined in either twisting or straight associations. The average widths of anchorless protofilaments were: 3.0 ± 0.5 nm for ME7, 3.1 ± 0.7 nm for 22L, and 3.5 ± 0.6 nm in the case of RML strain. In all prion strains, wild-type fibrils were larger than anchorless fibrils: 3.4 ± 0.6 nm for wild-type 22L and 3.7 ± 0.6 nm for wild-type RML. Both wild-type and anchorless RML fibrils are larger than their 22L counterparts. The majority of wild-type and anchorless fibrils were of the twisting variety and were found spiraling in either right- or left-hand directions. The anchorless 22L fibrils had a lower percentage of right-handed twist,

16%, compared with all other fibrils. Moreover, both wild-type and anchorless 22L fibrils revealed 11% more of straight protofilaments than their RML counterparts. Fibrils of 22L were also characterized by either a partial twist in one direction, and then a fold back toward other directions, or a half-twist in one direction occurring in the middle of a fibril that was primarily twisting in the opposite direction. RML and 22L fibrils presented different periodicities. Anchorless 22L had a periodicity of 106±23 nm, RML a periodicity of 64±18 nm, and ME7 had a periodicity of 66±11 nm. Authors investigated the heights of such fibrils using AFM. This technique provides the resolution of TEM, but does not require staining. From this analysis, Sim and colleagues found the same height for both anchorless fibrils of 22L and RML: 5.5±0.6 nm and 5.6±0.7 nm, respectively.

Recently, Requena et al. investigated the architecture of PrP27–30 fibrils extracted from 263 K-infected SHa. The PrP27–30 fibrils were generated after PrP^{Sc} extraction in the presence of 10% sarkosyl and PK digestion. PrP27–30 appeared, under negative-stain TEM and cryo-EM quasi-native conditions, as twisted 12–15 nm wide fibers composed of two ~5 nm wide individual fibrils. To investigate better the structure of SHa PrP27–30, authors performed small-angle X-ray scattering (SAXS). This technique is a fundamental tool in studies of the structure of biological macromolecules with sizes ranging from a few kilodaltons to several megadaltons. SAXS data were best fitted to a three-term model consisting of interacting polydisperse cylinders, with a radius of 5 nm and a cylinder-to-cylinder distance of 10.6±2 nm. SAXS analysis revealed that PrP27–30, in aqueous suspension, consists of cylindrical fibers with a radius of 5 nm, in excellent agreement with cryo-EM images. These fibers were composed of two twisted or intertwined parallel protofibrils, each ~5 nm wide (Benetti et al. 2010).

During examination of negatively stained TEM images, Wille et al. discovered two-dimensional (2D) crystals of the *N*-terminally truncated PrP^{Sc}, PrP27–30 and of the miniprion PrP106 (Δ23–88, Δ141–176) with an apparent hexagonal lattice (α and β=69 Å and γ=120 Å) (Wille et al. 2002). Immunogold labeling with several antibodies such as R1, R2, 3F4, and 28D established that PrP is an integral part of these crystals. Moreover, immunolabeling with 3F4 was possible only after urea denaturation, arguing that PrP27–30 was present in these crystals. Two-dimensional crystals were found in the direct proximity of prion rods, suggesting a transition between them. These transitional aggregates suggest a stacking of the disk-like oligomers into protofilaments. Crystals were visualized with negative stains such as uranyl acetate. The dark area in the center of 2D crystals was due to negative charges within the crystal lattice. The PrP27–30 sequence contains several negatively charged residues lying between positions 143 and 177. *N*-linked glycans were located outside the oligomers, as detected by labeling them with 1.4-nm gold particles (monoamino nanogold). Sugars are linked at positions N181 and N197 in hamster PrP27–30. These residues lie in helix 2 (residues 179–193), and helix 3 (residues 200–217). Therefore, these helices seem to be preserved in PrP27–30 and localized outside the crystal. Modeling these data, authors suggested a parallel β-helical fold for PrP27–30 with the *C*-terminal helices and glycans on the periphery of oligomers. Parallel β-helices have little twist or bend, with subsequent planar faces that permit stacking along the fiber axis.

Furthermore, Wille et al. investigated the structures of PrP27–30 and PrP106 2D crystals using various heavy metals: Uranyl acetate, uranyl oxalate, uranyl phthalate, and uranyl citrate (Wille et al. 2007). These analyses allowed localization of the internal deletion of PrP106 at the center of the trimeric oligomers. The negative stain, ammonium molybdate, confirmed the threefold symmetry of the unit cell. To conclude, PrP27–30 would be a solenoid formed by several rungs composed of a certain number of β-strands, an architecture that is similar to that of HET-s prions. The 141–176 deletion in PrP106 corresponds to one rung or layer of β-strands that does not affect the overall structure of the molecule.

9.3.6 Fiber X-Ray Diffraction

As mentioned before, limited proteolysis and detergent extraction of PrP^{Sc} generate prion rods of molecular mass of 27–30 kDa. Although it has been impossible to obtain crystals from both PrP^{Sc} and PrP27–30, it has been possible to obtain X-ray diffraction patterns from infectious fibers (Nguyen et al. 1995; Wille et al. 2009).

In 1995, Nguyen and coworkers performed X-ray diffraction and EM on rods purified from scrapie-infected SHa brains and from synthetic SHa PrP peptides (Nguyen et al. 1995). They synthesized three SHa peptides corresponding to the 90–145 region: 113–120, representing an octamer composed of glycine and alanine residues; 109–122, the first predicted α-helical region of PrP^{C}; and 90–145, a 56-residue peptide containing both the first and the second α-helical regions. EM measurements revealed that all peptides and PrP27–30 formed linear polymers, which were ~6–20 nm wide with fibrillar or ribbon-like morphology. X-ray diffraction patterns indicated a β-sheet conformation with a hydrogen-bond distance of 4.72 Å, and with an inter-sheet distance of 8.82 Å. The three peptides showed a wide range of inter-sheet distance, 5.13–9.15 Å, owing to different side-chains affecting β-sheet interactions.

More recently, Wille et al. obtained X-ray diffraction patterns from infectious prions extracted from Sc237-infected SHa, recombinant (rec) mouse (Mo)-PrP (89–230) and recSHa-PrP (90–231) as well as from synthetic prions recovered from recMoPrP89–230 amyloid-fiber-infected brains of Tg9949 mice overexpressing *N*-terminally truncated MoPrP (Wille et al. 2009). Fiber-diffraction patterns of PrP27–30 exhibited a marked intensity maximum at 4.8 Å resolution, indicating presence of β-strands running at right angles to the filament axis and characteristic of amyloid structures. Diffraction patterns exhibited a series of equatorial maxima diminishing in intensity with increasing resolution. These were measured at 30.9, 20.3, 15.5, 11.9, 9.3, and 7.9 Å. Equatorial diffraction from many fibers also included an intense, moderately sharp, low-angle reflection (63.3 Å). The presence of reflection at low angle is typical of fibers with poorly ordered paracrystalline packing.

Diffraction patterns from recSHaPrP (90–231) amyloid differed from those of brain-derived prions. These fibers also showed a well-defined 4.8 Å meridional layer line, but with a broad equatorial maximum at 10.5 Å comparable to those seen in diffraction from short-peptide amyloids. These differences in the equatorial diffraction among PrP27–30 and recSHaPrP (90–231) imply that the majority of recombinant fibrils bear different structures compared to brain-derived prions. The strong intensity at 10.5 Å, together with the 4.8 Å meridional diffraction, is characteristic of a stacked-sheet amyloid structure: β-sheets packed together with inter-sheet spacing close to 10 Å. On the contrary, PrP27–30 equatorial diffraction becomes progressively weaker as the distance from the origin increases, with no evidence for any internal, regularly spaced structure at right angles to the fiber axis. The β-helical model is consistent with the diffraction data reported above.

9.3.7 *Solid-State NMR: Structure of Amyloid Fibrils of the HET-s (218–289) Prion*

Although PrP^{C} and its pathological form, PrP^{Sc}, differ solely in their three-dimensional structure, no atomically resolved structure of infectious fibrillar state has been described to date. However, Wasmer et al. reported a structural model based on solid-state NMR restraints of the HET-s prion-forming domain (residues 218–289) (Wasmer et al. 2008). HET-s is a protein of the filamentous fungus *Podospora anserine* that, in its prion form, plays a role in heterokaryon incompatibility. In particular, the prion form of HET-s is involved in the fungal self-/non-self-recognition phenomena, which prevent different forms of parasitism. The PK-resistant core of HET-s is formed by the 72 *C*-terminal amino acids (218–289) (Balguerie et al. 2003). The PK-digested fragment displayed infectivity in the biolistic assay (Maddelein et al. 2002) and caused aggregation of GFP, *in vivo*. The HET-s fibril organization is a left-handed β-solenoid with two windings per molecule. Each winding is composed of three β-strands (β1a–β1b–β2a and β3a–β3b–β4a) that form continuous, in-register, parallel β-sheets. The segments β1a–β1b, and then β3a–β3b are connected by a two-residue β arch, forming an approximately rectangular kink in the strand. The connection between β1b–β2a (β3b–β4a) is provided by a three-residue arch, allowing an orientation change of the polypeptide backbone by around 150°. An additional β-sheet is located outside the solenoid and is formed by two β-strands, β2b and β4b. A disruption of the β-sheet pattern is observed between β2a–β4a and β2b–β4b, leading to a 90° β arch. This arrangement is stabilized by side-chain contacts. Each winding forms a triangular hydrophobic core, which is tightly packed and contains almost exclusively hydrophobic residues (alanine, leucine, isoleucine, and valine) with numerous external restraints between hydrophobic side-chains. All charged residues are located in β arches, where the solvent accessibility is high. Several experimental restraints support the existence of three salt bridges: K229–E265, E234–K270, and R236–E272. Two asparagine residues close to the hydrophobic core are stacked and can form a ladder (N226–N262), contributing to the fibril stability through side-chain hydrogen bonds. Another asparagine ladder can be formed outside the hydrophobic core (N243–N279). Moreover, at least 23 hydrogen bonds are formed along the β-solenoid.

A left-handed β-solenoid, on the basis of modeling and electron micrographs, has been also proposed for human PrP^{Sc} (Govaerts et al. 2004).

Unfortunately, the approach used for HET-s prion cannot be applied to mammalian prions. While recombinant HET-s (218–289) fibrils *in vitro* have the same properties of naturally occurring fibers, recombinant mammalian PrP amyloid differs substantially from highly infectious brain-derived prions, both in structure as demonstrated by the diffraction data, and in heterogeneity as shown by electron microscopy (Wille et al. 2009; Smirnovas et al. 2009).

9.4 Prion Models

Up to date, there are two models describing the mammalian prion structures: the spiral model and the β-helix model (DeMarco et al. 2006; Govaerts et al. 2004).

DeMarco et al. developed the molecular model called spiral model (DeMarco et al. 2006). This model has a spiraling core of extended sheets formed by parallel and anti-parallel extended strands. The monomer is derived from an all-atom, explicit-solvent molecular dynamics simulation (DeMarco and Daggett 2004). It has a height of 5.8 nm and a width of 3 nm. Simulations were also used to model a protofibril, docking hydrophobic patches of the template structure to form hydrogen-bonded sheets spanning adjacent subunits. The resulting model provided a non-branching aggregate with a 3_1 axis of symmetry.

Govaerts et al., took advantage of structural studies at low resolution and delineated the molecular model of *N*-terminally truncated PrP27–30 (Govaerts et al. 2004). Their model, called left-handed β-helix, is similar to the HET-s structure. The left-handed β-helix model originated from a study of 119 all-β folds observed in globular proteins. In this model, PrP residues from 90 to 170 are converted into β-strands and subsequently in β-helices, while the *C*-terminal region maintains its α-helical structure with the disulfide bond and the glycan moieties located outside the oligomeric core. Each monomer is 4.2 nm wide with a height of 6.8 nm. Left-handed β-helices readily form trimers, providing a template for a trimeric model of PrP^{Sc}.

9.5 Conclusions

Prion diseases, like many other neurodegenerative disorders such as Parkinson's and Alzheimer's diseases, are classified as protein-misfolding diseases. These pathologies are characterized by accumulation of abnormal conformers of cellular proteins. In the case of prion diseases, the cellular form of PrP is converted into its pathological form, PrP^{Sc}. Currently little information is available regarding the conversion process and the high-resolution structures. Here, we reported the most important advances in the field, obtained using low-resolution techniques. These data have so far allowed outlining the major structural characteristics of prions. Further studies are required to solve the remaining questions. Important contributions may come from recent advances in cryo-electron microscopy, tomography, small-angle X-ray scattering, and scanning electron microscopy as well as from refinement of X-ray fiber-diffraction data. These techniques, coupled with more traditional approaches such as hydrogen–deuterium exchange, limited proteolysis, and cross-linking, may help defining PrP^{Sc} structures. Although PrP^{Sc} extracted from brains of infected animals and synthetic prions have different infectivity and structure, progress could be achieved by employing recombinant amyloid. The latter in fact may be useful in performing solid-state NMR and describing prion fibers at high resolution. Moreover, the use of pathological PrP mutants may clarify the rearrangements occurring during the early stages of conversion.

Since PrP^{Sc} acts as template to convert into prions either PrP^{C} or recombinant PrP while maintaining its own structural features, ^{13}C- and ^{15}N-labeled recombinant PrP may be used in seeding assays to form fibrils closely related to the template. The resulting labeled fibrils could then be exploited to unveil their structure by solid-state NMR.

All these pieces of information taken together may provide indications useful to elucidate the PrP^{Sc} structure and its molecular mechanisms of conversion, hopefully paving the way to better rational design of more efficient anti-prion drugs.

Acknowledgments The authors wish to thank Cedric Govaerts for kindly providing Fig. 9.1, and Gabriella Furlan for editing and proofreading the manuscript.

References

Adrover M, Pauwels K, Pringent S, De Chiara C, Xu Z, Chapuis C, Pastore A, Rezaei H (2010) Prion fibrillization is mediated by a native structural element which comprises the helices H2 and H3. J Biol Chem 285:21004–21012

Aguzzi A, Sigurdson C, Heikenwaelder M (2008) Molecular mechanisms of prion pathogenesis. Annu Rev Pathol 3:11–40

Alper T, Cramp WA, Haig DA, Clarke MC (1967) Does the agent of scrapie replicate without nucleic acid? Nature 214:764–766

Anderson M, Bocharova OV, Makarava N, Breydo L, Salnikov VV, Baskakov IV (2006) Polymorphism and ultrastructural organization of prion protein amyloid fibrils: An insight from high resolution atomic force microscopy. J Mol Biol 358:580–596

Antonyuk SV, Trevitt CR, Strange RW, Jackson GS, Sangar D, Batchelor M, Cooper S, Fraser C, Jones S, Georgiou T, Khalili-Shirazi A, Clarke AR, Hasnain SS, Collinge J (2009) Crystal structure of human prion protein bound to a therapeutic antibody. Proc Natl Acad Sci USA 106:2554–2558

Apetri AC, Surewicz K, Surewicz WK (2004) The effect of disease-associated mutations on the folding pathway of human prion protein. J Biol Chem 279:18008–18014

Ashok A, Hegde RS (2009) Selective processing and metabolism of disease-causing mutant prion proteins. PLoS Pathog 5:e1000479

Bae SH, Legname G, Serban A, Prusiner SB, Wright PE, Dyson HJ (2009) Prion proteins with pathogenic and protective mutations show similar structure and dynamics. Biochemistry 48:8120–8128
Baker HE, Poulter M, Crow TJ, Frith CD, Lofthouse R, Ridley RM (1991) Aminoacid polymorphism in human prion protein and age at death in inherited prion disease. Lancet 337:1286
Balguerie A, Dos Reis S, Ritter C, Chaignepain S, Coulary-Salin B, Forge V, Bathany K, Lascu I, Schmitter JM, Riek R, Saupe SJ (2003) Domain organization and structure–function relationship of the HET-s prion protein of Podospora anserina. EMBO J 22:2071–2081
Bellinger-Kawahara C, Diener TO, Mckinley MP, Groth DF, Smith DR, Prusiner SB (1987) Purified scrapie prions resist inactivation by procedures that hydrolyze, modify, or shear nucleic acids. Virology 160:271–274
Benetti F, Amenitsch H, Vos M, Peters P, Legname G, Requena JR (2010) SAXS study of Syrian hamster prion fibrils and recombinant truncated prion protein in the presence of transition metals (in press). In: Sartori BRM, Amenitsch H, Bernstorff S (ed) Annual report of the Austrian SAXS beamline 2009. Institute of Biophysics and Nanosystems Research, Graz
Bertho G, Bouvier G, Hui Bon Hoa G, Girault JP (2008) The key-role of tyrosine 155 in the mechanism of prion transconformation as highlighted by a study of sheep mutant peptides. Peptides 29:1073–1084
Bessen RA, Marsh RF (1992) Biochemical and physical properties of the prion protein from two strains of the transmissible mink encephalopathy agent. J Virol 66:2096–2101
Bessen RA, Marsh RF (1994) Distinct PrP properties suggest the molecular basis of strain variation in transmissible mink encephalopathy. J Virol 68:7859–7868
Bignami A, Parry HB (1971) Aggregations of 35-nanometer particles associated with neuronal cytopathic changes in natural scrapie. Science 171:389–399
Biverstahl H, Andersson A, Graslund A, Maler L (2004) NMR solution structure and membrane interaction of the *N*-terminal sequence (1–30) of the bovine prion protein. Biochemistry 43:14940–14947
Borchelt DR, Taraboulos A, Prusiner SB (1992) Evidence for synthesis of scrapie prion proteins in the endocytic pathway. J Biol Chem 267:16188–16199
Bots GT, De Man JC, Verjaal A (1971) Virus-like particles in brain tissue from two patients with Creutzfeldt–Jakob disease. Acta Neuropathol 18:267–270
Bueler H, Aguzzi A, Sailer A, Greiner RA, Autenried P, Aguet M, Weissmann C (1993) Mice devoid of PrP are resistant to scrapie. Cell 73:1339–1347
Burns CS, Aronoff-Spencer E, Legname G, Prusiner SB, Antholine WE, Gerfen GJ, Peisach J, Millhauser GL (2003) Copper coordination in the full-length, recombinant prion protein. Biochemistry 42:6794–6803
Calzolai L, Zahn R (2003) Influence of pH on NMR structure and stability of the human prion protein globular domain. J Biol Chem 278:35592–35596
Calzolai L, Lysek DA, Guntert P, Von Schroetter C, Riek R, Zahn R, Wuthrich K (2000) NMR structures of three single-residue variants of the human prion protein. Proc Natl Acad Sci USA 97:8340–8345
Calzolai L, Lysek DA, Perez DR, Guntert P, Wuthrich K (2005) Prion protein NMR structures of chickens, turtles, and frogs. Proc Natl Acad Sci USA 102:651–655
Campana V, Sarnataro D, Zurzolo C (2005) The highways and byways of prion protein trafficking. Trends Cell Biol 15:102–111
Caughey B (1991) Cellular metabolism of PrP. Prion Diseases in Humans and Animals Conference, London
Caughey BW, Dong A, Bhat KS, Ernst D, Hayes SF, Caughey WS (1991) Secondary structure analysis of the scrapie-associated protein PrP 27–30 in water by infrared spectroscopy. Biochemistry 30:7672–7680
Caughey B, Raymond GJ, Bessen RA (1998) Strain-dependent differences in β-sheet conformations of abnormal prion protein. J Biol Chem 273:32230–32235
Chesebro B (1992) PrP and the scrapie agent. Nature 356:560
Chesebro B, Trifilo M, Race R, Meade White K, Teng C, Lacasse R, Raymond L, Favara C, Baron G, Priola S, Caughey B, Masliah E, Oldstone M (2005) Anchorless prion protein results in infectious amyloid disease without clinical scrapie. Science 308:1435–1439

Chiesa R, Piccardo P, Ghetti B, Harris DA (1998) Neurological illness in transgenic mice expressing a prion protein with an insertional mutation. Neuron 21:1339–1351

Christen B, Perez DR, Hornemann S, Wuthrich K (2008) NMR structure of the bank vole prion protein at 20 °C contains a structured loop of residues 165–171. J Mol Biol 383:306–312

Christen B, Hornemann S, Damberger FF, Wuthrich K (2009) Prion protein NMR structure from Tammar Wallaby (*Macropus eugenii*) shows that the β2–α2 loop is modulated by long-range sequence effects. J Mol Biol 389:833–845

Cohen FE, Prusiner SB (1998) Pathologic conformations of prion proteins. Annu Rev Biochem 67:793–819

Collinge J (2001) Prion diseases of humans and animals: their causes and molecular basis. Annu Rev Neurosci 24:519–550

Collinge J, Palmer MS, Dryden AJ (1991) Genetic predisposition to iatrogenic Creutzfeldt–Jakob disease. Lancet 337:1441–1442

David-Ferreira JF, David-Ferreira KL, Gibbs CJ Jr, Morris JA (1968) Scrapie in mice: Ultrastructural observations in the cerebral cortex. Proc Soc Exp Biol Med 127:313–320

Demarco ML, Daggett V (2004) From conversion to aggregation: protofibril formation of the prion protein. Proc Natl Acad Sci USA 101:2293–2298

Demarco ML, Silveira J, Caughey B, Daggett V (2006) Structural properties of prion protein protofibrils and fibrils: an experimental assessment of atomic models. Biochemistry 45:15573–15582

Diener TO, Mckinley MP, Prusiner SB (1982) Viroids and prions. Proc Natl Acad Sci USA 79:5220–5224

Donne DG, Viles JH, Groth D, Mehlhorn I, James TL, Cohen FE, Prusiner SB, Wright PE, Dyson HJ (1997) Structure of the recombinant full-length hamster prion protein PrP(29–231): the *N*-terminus is highly flexible. Proc Natl Acad Sci USA 94:13452–13457

Dossena S, Imeri L, Mangieri M, Garofoli A, Ferrari L, Senatore A, Restelli E, Balducci C, Fiordaliso F, Salio M, Bianchi S, Fioriti L, Morbin M, Pincherle A, Marcon G, Villani F, Carli M, Tagliavini F, Forloni G, Chiesa R (2008) Mutant prion protein expression causes motor and memory deficits and abnormal sleep patterns in a transgenic mouse model. Neuron 60:598–609

Eghiaian F, Grosclaude J, Lesceu S, Debey P, Doublet B, Treguer E, Rezaei H, Knossow M (2004) Insight into the $PrP^{C} \rightarrow PrP^{Sc}$ conversion from the structures of antibody-bound ovine prion scrapie-susceptibility variants. Proc Natl Acad Sci USA 101:10254–10259

Fischer M, Rulicke T, Raeber A, Sailer A, Moser M, Oesch B, Brandner S, Aguzzi A, Weissmann C (1996) Prion protein (PrP) with amino-proximal deletions restoring susceptibility of PrP knockout mice to scrapie. EMBO J 15:1255–1264

Gasset M, Baldwin MA, Fletterick RJ, Prusiner SB (1993) Perturbation of the secondary structure of the scrapie prion protein under conditions that alter infectivity. Proc Natl Acad Sci USA 90:1–5

Godsave SF, Wille H, Kujala P, Latawiec D, Dearmond SJ, Serban A, Prusiner SB, Peters PJ (2008) Cryo-immunogold electron microscopy for prions: toward identification of a conversion site. J Neurosci 28:12489–12499

Goormaghtigh E, Cabiaux V, Ruysschaert J-M (1990) Secondary structure and dosage of soluble membrane proteins by attenuated total reflection Fourier-transform infrared spectroscopy on hydrated films. Eur J Biochem 193:409–420

Gorodinsky A, Harris DA (1995) Glycolipid-anchored proteins in neuroblastoma cells form detergent-resistant complexes without caveolin. J Cell Biol 129:619–627

Gossert AD, Bonjour S, Lysek DA, Fiorito F, Wuthrich K (2005) Prion protein NMR structures of elk and of mouse/elk hybrids. Proc Natl Acad Sci USA 102:646–650

Govaerts C, Wille H, Prusiner SB, Cohen FE (2004) Evidence for assembly of prions with left-handed β-helices into trimers. Proc Natl Acad Sci USA 101:8342–8347

Haire LF, Whyte SM, Vasisht N, Gill AC, Verma C, Dodson EJ, Dodson GG, Bayley PM (2004) The crystal structure of the globular domain of sheep prion protein. J Mol Biol 336:1175–1183

Hegde RS, Mastrianni JA, Scott MR, Defea KA, Tremblay P, Torchia M, Dearmond SJ, Prusiner SB, Lingappa VR (1998) A transmembrane form of the prion protein in neurodegenerative disease. Science 279:827–834

Hegde RS, Tremblay P, Groth D, Dearmond SJ, Prusiner SB, Lingappa VR (1999) Transmissible and genetic prion diseases share a common pathway of neurodegeneration. Nature 402:822–826

Heske J, Heller U, Winklhofer KF, Tatzelt J (2004) The *C*-terminal globular domain of the prion protein is necessary and sufficient for import into the endoplasmic reticulum. J Biol Chem 279:5435–5443

Hill AF, Antoniou M, Collinge J (1999) Protease-resistant prion protein produced in vitro lacks detectable infectivity. J Gen Virol 80(Pt 1):11–14

Hornemann S, Von Schroetter C, Damberger FF, Wuthrich K (2009) Prion protein-detergent micelle interactions studied by NMR in solution. J Biol Chem 284:22713–22721

Hsiao KK, Scott M, Foster D, Groth DF, Dearmond SJ, Prusiner SB (1990) Spontaneous neurodegeneration in transgenic mice with mutant prion protein. Science 250:1587–1590

Hsiao K, Dlouhy SR, Farlow MR, Cass C, Da Costa M, Conneally PM, Hodes ME, Ghetti B, Prusiner SB (1992) Mutant prion proteins in Gerstmann–Straüssler–Scheinker disease with neurofibrillary tangles. Nat Genet 1:68–71

Hsiao KK, Groth D, Scott M, Yang SL, Serban H, Rapp D, Foster D, Torchia M, Dearmond SJ, Prusiner SB (1994) Serial transmission in rodents of neurodegeneration from transgenic mice expressing mutant prion protein. Proc Natl Acad Sci USA 91:9126–9130

Huang Z, Prusiner SB, Cohen FE (1995) Scrapie prions: a three-dimensional model of an infectious fragment. Fold Des 1:13–19

Ilc G, Giachin G, Jaremko M, Jaremko L, Benetti F, Plavec J, Zhukov I, Legname G (2010) NMR structure of the human prion protein with the pathological Q212P mutation reveals unique structural features. PLoS One 5:e11715

James TL, Liu H, Ulyanov NB, Farr-Jones S, Zhang H, Donne DG, Kaneko K, Groth D, Mehlhorn I, Prusiner SB, Cohen FE (1997) Solution structure of a 142-residue recombinant prion protein corresponding to the infectious fragment of the scrapie isoform. Proc Natl Acad Sci USA 94:10086–10091

Kaneko K, Vey M, Scott M, Pilkuhn S, Cohen FE, Prusiner SB (1997a) COOH-terminal sequence of the cellular prion protein directs subcellular trafficking and controls conversion into the scrapie isoform. Proc Natl Acad Sci USA 94:2333–2338

Kaneko K, Zulianello L, Scott M, Cooper CM, Wallace AC, James TL, Cohen FE, Prusiner SB (1997b) Evidence for protein X binding to a discontinuous epitope on the cellular prion protein during scrapie prion propagation. Proc Natl Acad Sci USA 94:10069–10074

Kimberlin RH (1990) Scrapie and possible relationships with viroids. Semin Virol 1:153–162

Knaus KJ, Morillas M, Swietnicki W, Malone M, Surewicz WK, Yee VC (2001) Crystal structure of the human prion protein reveals a mechanism for oligomerization. Nat Struct Biol 8:770–774

Kobayashi A, Hizume M, Teruya K, Mohri S, Kitamoto T (2009) Heterozygous inhibition in prion infection: the stone fence model. Prion 3:27–30

Kovacs GG, Trabattoni G, Hainfellner JA, Ironside JW, Knight RS, Budka H (2002) Mutations of the prion protein gene phenotypic spectrum. J Neurol 249:1567–1582

Kozin SA, Bertho G, Mazur AK, Rabesona H, Girault JP, Haertle T, Takahashi M, Debey P, Hoa GH (2001) Sheep prion protein synthetic peptide spanning helix 1 and β-strand 2 (residues 142–166) shows β-hairpin structure in solution. J Biol Chem 276:46364–46370

Kozin SA, Lepage C, Hui Bon Hoa G, Rabesona H, Mazur AK, Blond A, Cheminant M, Haertle T, Debey P, Rebuffat S (2004) Solution structure of synthetic 21mer peptide spanning region 135–155 (in human numbering) of sheep prion protein http://www.pdb.org/pdb/explore/explore.do?structureId=1S4T

Kuwata K, Li H, Yamada H, Legname G, Prusiner SB, Akasaka K, James TL (2002) Locally disordered conformer of the hamster prion protein: a crucial intermediate to PrP^{Sc}? Biochemistry 41:12277–12283

Latarjet R, Muel B, Haig DA, Clarke MC, Alper T (1970) Inactivation of the scrapie agent by near monochromatic ultraviolet light. Nature 227:1341–1343

Lee S, Antony L, Hartmann R, Knaus KJ, Surewicz K, Surewicz WK, Yee VC (2010) Conformational diversity in prion protein variants influences intermolecular β-sheet formation. EMBO J 29:251–262

Legname G, Baskakov IV, Nguyen HO, Riesner D, Cohen FE, Dearmond SJ, Prusiner SB (2004) Synthetic mammalian prions. Science 305:673–676

Legname G, Nguyen HO, Baskakov IV, Cohen FE, Dearmond SJ, Prusiner SB (2005) Strain-specified characteristics of mouse synthetic prions. Proc Natl Acad Sci USA 102:2168–2173

Legname G, Nguyen HO, Peretz D, Cohen FE, Dearmond SJ, Prusiner SB (2006) Continuum of prion protein structures enciphers a multitude of prion isolate-specified phenotypes. Proc Natl Acad Sci USA 103:19105–19110

Li J, Mei FH, Xiao GF, Guo CY, Lin DH (2007) ^{1}H, ^{13}C and ^{15}N resonance assignments of rabbit prion protein (91–228). J Biomol NMR 38:181

Liemann S, Glockshuber R (1999) Influence of amino acid substitutions related to inherited human prion diseases on the thermodynamic stability of the cellular prion protein. Biochemistry 38:3258–3267

Liu H, Farr-Jones S, Ulyanov NB, Llinas M, Marqusee S, Groth D, Cohen FE, Prusiner SB, James TL (1999) Solution structure of Syrian hamster prion protein rPrP(90–231). Biochemistry 38:5362–5377

Lopez Garcia F, Zahn R, Riek R, Wuthrich K (2000) NMR structure of the bovine prion protein. Proc Natl Acad Sci USA 97:8334–8339

Lysek DA, Schorn C, Nivon LG, Esteve-Moya V, Christen B, Calzolai L, Von Schroetter C, Fiorito F, Herrmann T, Guntert P, Wuthrich K (2005) Prion protein NMR structures of cats, dogs, pigs, and sheep. Proc Natl Acad Sci USA 102:640–645

Maddelein ML, Dos Reis S, Duvezin-Caubet S, Coulary-Salin B, Saupe SJ (2002) Amyloid aggregates of the HET-s prion protein are infectious. Proc Natl Acad Sci USA 99:7402–7407

Manuelidis L, Fritch W (1996) Infectivity and host responses in Creutzfeldt–Jakob disease. Virology 216:46–59

Mashima T, Matsugami A, Nishikawa F, Nishikawa S, Katahira M (2009) Unique quadruplex structure and interaction of an RNA aptamer against bovine prion protein. Nucleic Acids Res 37:6249–6258

Mckinley MP, Bolton DC, Prusiner SB (1983) A protease-resistant protein is a structural component of the scrapie prion. Cell 35:57–62

Mckinley MP, Meyer RK, Kenaga L, Rahbar F, Cotter R, Serban A, Prusiner SB (1991a) Scrapie prion rod formation in vitro requires both detergent extraction and limited proteolysis. J Virol 65:1340–1351

Mckinley MP, Taraboulos A, Kenaga L, Serban D, Stieber A, Dearmond SJ, Prusiner SB, Gonatas N (1991b) Ultrastructural localization of scrapie prion proteins in cytoplasmic vesicles of infected cultured cells. Lab Invest 65:622–630

Megy S, Bertho G, Kozin SA, Debey P, Hoa GH, Girault JP (2004) Possible role of region 152–156 in the structural duality of a peptide fragment from sheep prion protein. Protein Sci 13:3151–3160

Meyer RK, Mckinley MP, Bowman KA, Braunfeld MB, Barry RA, Prusiner SB (1986) Separation and properties of cellular and scrapie prion proteins. Proc Natl Acad Sci USA 83:2310–2314

Mills NL, Surewicz K, Surewicz WK, Sonnichsen FD (2009) Residue 129 polymorphism and conformational dynamics of familial prion diseases associated with the human prion protein variant D178N. doi:10.2210/pdb2k1d/pdb http://www.pdb.org/pdb/explore/explore.do?structureId=2K1D

Mishra RS, Bose S, Gu Y, Li R, Singh N (2003) Aggresome formation by mutant prion proteins: the unfolding role of proteasomes in familial prion disorders. J Alzheimers Dis 5:15–23

Muramoto T, Dearmond SJ, Scott M, Telling GC, Cohen FE, Prusiner SB (1997) Heritable disorder resembling neuronal storage disease in mice expressing prion protein with deletion of an α-helix. Nat Med 3:750–755

Nguyen JT, Inouye H, Baldwin MA, Fletterick RJ, Cohen FE, Prusiner SB, Kirschner DA (1995) X-ray diffraction of scrapie prion rods and PrP peptides. J Mol Biol 252:412–422

Novitskaya V, Makarava N, Bellon A, Bocharova OV, Bronstein IB, Williamson RA, Baskakov IV (2006) Probing the conformation of the prion protein within a single amyloid fibril using a novel immunoconformational assay. J Biol Chem 281:15536–15545

Onisko B, Fernandez EG, Freire ML, Schwarz A, Baier M, Camina F, Garcia JR, Rodriguez-Segade Villamarin S, Requena JR (2005) Probing PrP^{Sc} structure using chemical cross-linking and mass spectrometry: Evidence of the proximity of Gly90 amino termini in the PrP 27–30 aggregate. Biochemistry 44:10100–10109

Palmer MS, Dryden AJ, Hughes JT, Collinge J (1991) Homozygous prion protein genotype predisposes to sporadic Creutzfeldt–Jakob disease. Nature 352:340–342

Pan K-M, Baldwin M, Nguyen J, Gasset M, Serban A, Groth D, Mehlhorn I, Huang Z, Fletterick RJ, Cohen FE, Prusiner SB (1993) Conversion of α-helices into β-sheets features in the formation of the scrapie prion proteins. Proc Natl Acad Sci USA 90:10962–10966

Parchi P, Zou W, Wang W, Brown P, Capellari S, Ghetti B, Kopp N, Schulz-Schaeffer WJ, Kretzschmar HA, Head MW, Ironside JW, Gambetti P, Chen SG (2000) Genetic influence on the structural variations of the abnormal prion protein. Proc Natl Acad Sci USA 97:10168–10172

Peretz D, Williamson RA, Matsunaga Y, Serban H, Pinilla C, Bastidas RB, Rozenshteyn R, James TL, Houghten RA, Cohen FE, Prusiner SB, Burton DR (1997) A conformational transition at the *N*-terminus of the prion protein features in formation of the scrapie isoform. J Mol Biol 273:614–622

Perez DR, Damberger FF, Wuthrich K (2010) Horse prion protein NMR structure and comparisons with related variants of the mouse prion protein. J Mol Biol 400:121–128

Piccardo P, Dlouhy SR, Lievens PM, Young K, Bird TD, Nochlin D, Dickson DW, Vinters HV, Zimmerman TR, Mackenzie IR, Kish SJ, Ang LC, De Carli C, Pocchiari M, Brown P, Gibbs CJ Jr, Gajdusek DC, Bugiani O, Ironside J, Tagliavini F, Ghetti B (1998) Phenotypic variability of Gerstmann–Straüssler–Scheinker disease is associated with prion protein heterogeneity. J Neuropathol Exp Neurol 57:979–988

Premzl M, Delbridge M, Gready JE, Wilson P, Johnson M, Davis J, Kuczek E, Marshall Graves JA (2005) The prion protein gene: identifying regulatory signals using marsupial sequence. Gene 349:121–134

Prusiner SB (1982) Novel proteinaceous infectious particles cause scrapie. Science 216:136–144

Prusiner SB (1997) Biology of prions. In: Rosenberg RN, Prusiner SB, Dimauro S, Barchi RL (eds) The molecular and genetic basis of neurological disease, 2nd edn. Butterworth Heinemann, Stoneham

Prusiner SB (1998) Prions. Proc Natl Acad Sci USA 95:13363–13383

Prusiner SB, Scott MR, Dearmond SJ, Cohen FE (1998) Prion protein biology. Cell 93:337–348

Requena JR (2009) Structure of mammalian prions. Future Virol 4:295–307

Riek R, Hornemann S, Wider G, Billeter M, Glockshuber R, Wüthrich K (1996) NMR structure of the mouse prion protein domain PrP(121–231). Nature 382:180–182

Ronga L, Palladino P, Saviano G, Tancredi T, Benedetti E, Ragone R, Rossi F (2008) Structural characterization of a neurotoxic threonine-rich peptide corresponding to the human prion protein α-2 helical 180–195 segment, and comparisonwith full length α2-helix-derived peptides. J. Pept. Sci. 14:1096–1102

Sajnani G, Pastrana MA, Dynin I, Onisko B, Requena JR (2008) Scrapie prion protein structural constraints obtained by limited proteolysis and mass spectrometry. J Mol Biol 382:88–98

Sawaya MR, Sambashivan S, Nelson R, Ivanova MI, Sievers SA, Apostol MI, Thompson MJ, Balbirnie M, Wiltzius JJ, Mcfarlane HT, Madsen AO, Riekel C, Eisenberg D (2007) Atomic structures of amyloid cross-β spines reveal varied steric zippers. Nature 447:453–457

Sim VL, Caughey B (2009) Ultrastructures and strain comparison of under-glycosylated scrapie prion fibrils. Neurobiol Aging 30:2031–2042

Smirnovas V, Kim JI, Lu X, Atarashi R, Caughey B, Surewicz WK (2009) Distinct structures of scrapie prion protein (PrP^{Sc})-seeded versus spontaneous recombinant prion protein fibrils revealed by hydrogen/deuterium exchange. J Biol Chem 284:24233–24241

Spassov S, Beekes M, Naumann D (2006) Structural differences between TSEs strains investigated by FT-IR spectroscopy. Biochim Biophys Acta 1760:1138–1149

Swietnicki W, Petersen RB, Gambetti P, Surewicz WK (1998) Familial mutations and the thermodynamic stability of the recombinant human prion protein. J Biol Chem 273:31048–31052

Taraboulos A, Raeber AJ, Borchelt DR, Serban D, Prusiner SB (1992) Synthesis and trafficking of prion proteins in cultured cells. Mol Biol Cell 3:851–863

Taubner LM, Bienkiewicz EA, Copie V, Caughey B (2010) Structure of the flexible amino-terminal domain of prion protein bound to a sulfated glycan. J Mol Biol 395:475–490

Telling GC (2000) Prion protein genes and prion diseases: studies in transgenic mice. Neuropathol Appl Neurobiol 26:209–220

Telling GC, Scott M, Mastrianni J, Gabizon R, Torchia M, Cohen FE, Dearmond SJ, Prusiner SB (1995) Prion propagation in mice expressing human and chimeric PrP transgenes implicates the interaction of cellular PrP with another protein. Cell 83:79–90

Telling GC, Scott M, Prusiner SB (1996) Deciphering prion diseases with transgenic mice. In: Gibbs CJ Jr (ed) Bovine spongiform encephalopathy: the BSE dilemma. Springer Verlag, New York

Vanik DL, Surewicz WK (2002) Disease-associated F198S mutation increases the propensity of the recombinant prion protein for conformational conversion to scrapie-like form. J Biol Chem 277:49065–49070

Vey M, Pilkuhn S, Wille H, Nixon R, Dearmond SJ, Smart EJ, Anderson RG, Taraboulos A, Prusiner SB (1996) Subcellular colocalization of the cellular and scrapie prion proteins in caveolae-like membranous domains. Proc Natl Acad Sci USA 93:14945–14949

Viles JH, Cohen FE, Prusiner SB, Goodin DB, Wright PE, Dyson HJ (1999) Copper binding to the prion protein: structural implications of four identical cooperative binding sites. Proc Natl Acad Sci USA 96:2042–2047

Viles JH, Donne D, Kroon G, Prusiner SB, Cohen FE, Dyson HJ, Wright PE (2001) Local structural plasticity of the prion protein. Analysis of NMR relaxation dynamics. Biochemistry 40:2743–2753

Wasmer C, Lange A, Van Melckebeke H, Siemer AB, Riek R, Meier BH (2008) Amyloid fibrils of the HET-s(218–289) prion form a β solenoid with a triangular hydrophobic core. Science 319:1523–1526

Wen Y, Li J, Yao W, Xiong M, Hong J, Peng Y, Xiao G, Lin D (2010) Unique structural characteristics of the rabbit prion protein. J Biol Chem 285(41):31682–31693

Wille H, Michelitsch MD, Guenebaut V, Supattapone S, Serban A, Cohen FE, Agard DA, Prusiner SB (2002) Structural studies of the scrapie prion protein by electron crystallography. Proc Natl Acad Sci USA 99:3563–3568

Wille H, Govaerts C, Borovinskiy A, Latawiec D, Downing KH, Cohen FE, Prusiner SB (2007) Electron crystallography of the scrapie prion protein complexed with heavy metals. Arch Biochem Biophys 467:239–248

Wille H, Bian W, Mcdonald M, Kendall A, Colby DW, Bloch L, Ollesch J, Borovinskiy AL, Cohen FE, Prusiner SB, Stubbs G (2009) Natural and synthetic prion structure from X-ray fiber diffraction. Proc Natl Acad Sci USA 106:16990–16995

Williamson RA, Peretz D, Pinilla C, Ball H, Bastidas RB, Rozenshteyn R, Houghten RA, Prusiner SB, Burton DR (1998) Mapping the prion protein using recombinant antibodies. J Virol 72:9413–9418

Wiltzius JJ, Landau M, Nelson R, Sawaya MR, Apostol MI, Goldschmidt L, Soriaga AB, Cascio D, Rajashankar K, Eisenberg D (2009) Molecular mechanisms for protein-encoded inheritance. Nat Struct Mol Biol 16:973–978

Windl O, Dempster M, Estibeiro JP, Lathe R, De Silva R, Esmonde T, Will R, Springbett A, Campbell TA, Sidle KC, Palmer MS, Collinge J (1996) Genetic basis of Creutzfeldt–Jakob disease in the United Kingdom: a systematic analysis of predisposing mutations and allelic variation in the PRNP gene. Hum Genet 98:259–264

Wopfner F, Weidenhofer G, Schneider R, Von Brunn A, Gilch S, Schwarz TF, Werner T, Schatzl HM (1999) Analysis of 27 mammalian and 9 avian PrPs reveals high conservation of flexible regions of the prion protein. J Mol Biol 289:1163–1178

Worrall BB, Herman ST, Capellari S, Lynch T, Chin S, Gambetti P, Parchi P (1999) Type 1 protease resistant prion protein and valine homozygosity at codon 129 of PRNP identify a subtype of sporadic Creutzfeldt–Jakob disease. J Neurol Neurosurg Psychiatry 67:671–674

Young K, Piccardo P, Kish SJ, Ang LC, Dlouhy S, Ghetti B (1998) Gerstmann–Sträussler–Scheinker disease (GSS) with a mutation at prion protein (PrP) residue 212. J Neuropathol Exp Neurol 57:518

Zahn R (2003) The octapeptide repeats in mammalian prion protein constitute a pH-dependent folding and aggregation site. J Mol Biol 334:477–488

Zahn R, Von Schroetter C, Wuthrich K (1997) Human prion proteins expressed in *Escherichia coli* and purified by high-affinity column refolding. FEBS Lett 417:400–404

Zahn R, Liu A, Luhrs T, Riek R, Von Schroetter C, Lopez Garcia F, Billeter M, Calzolai L, Wider G, Wuthrich K (2000) NMR solution structure of the human prion protein. Proc Natl Acad Sci USA 97:145–150

Zahn R, Guntert P, Von Schroetter C, Wuthrich K (2003) NMR structure of a variant human prion protein with two disulfide bridges. J Mol Biol 326:225–234

Zanusso G, Farinazzo A, Prelli F, Fiorini M, Gelati M, Ferrari S, Righetti PG, Rizzuto N, Frangione B, Monaco S (2004) Identification of distinct *N*-terminal truncated forms of prion protein in different Creutzfeldt–Jakob disease subtypes. J Biol Chem 279:38936–38942

Zhang Y, Swietnicki W, Zagorski MG, Surewicz WK, Sonnichsen FD (2000) Solution structure of the E200K variant of human prion protein. Implications for the mechanism of pathogenesis in familial prion diseases. J Biol Chem 275:33650–33654

Zimmermann K, Turecek PL, Schwarz HP (1999) Genotyping of the prion protein gene at codon 129. Acta Neuropathol 97:355–358

Zou WQ, Capellari S, Parchi P, Sy MS, Gambetti P, Chen SG (2003) Identification of novel proteinase K-resistant *C*-terminal fragments of PrP in Creutzfeldt–Jakob disease. J Biol Chem 278:40429–40436

Chapter 10
Role of Prion Protein Oligomers in the Pathogenesis of Transmissible Spongiform Encephalopathies

Rodrigo Morales, Claudia A. Duran-Aniotz, and Claudio Soto

Abstract Prion diseases, also known as transmissible spongiform encephalopathies, are a group of neurodegenerative disorders associated with misfolding and aggregation of prion proteins. Although it is not completely known how structural changes in the prion protein induce neurodegeneration, it is widely accepted that formation of the misfolded prion protein (termed PrP^{Sc}) is both the triggering event in the disease and the main component of the infectious agent responsible for disease transmission. A long-debated issue in prion diseases has been the exact composition and size of the PrP^{Sc} particle required for initiating brain degeneration and propagating disease. Old and recent evidence show that PrP^{Sc} is an oligomer composed of several units of the prion protein monomer, folded into a β-sheet-rich conformation. In this article we discuss the potential roles of prion oligomers in both neurotoxicity and infectivity and the similarities of prion diseases to other neurodegenerative diseases associated with protein misfolding and aggregation.

Keywords Prions • Misfolded oligomers • Neurodegeneration • Seeding • Protein misfolding

R. Morales • C.A. Duran-Aniotz • C. Soto (✉)
Mitchell Center for Alzheimer's Disease and Related Brain Disorders,
Department of Neurology, University of Texas Medical School at Houston,
6431 Fannin Street, Houston, TX 77030, USA

Facultad de Medicina, Universidad de los Andes, Santiago, Chile
e-mail: Claudio.Soto@uth.tmc.edu.

F. Rahimi and G. Bitan (eds.), *Non-fibrillar Amyloidogenic Protein Assemblies—Common Cytotoxins Underlying Degenerative Diseases*, DOI 10.1007/978-94-007-2774-8_10,

10.1 Transmissible Spongiform Encephalopathies: Infectious Protein Aggregates

Correct folding into a native three-dimensional structure is a requirement for a protein to exert its biological function. Incorrectly folded proteins may present unusual properties, sometimes leading to disease. Protein-misfolding disorders (PMDs) are a group of diseases triggered by accumulation of misfolded proteins (Luheshi et al. 2008; Soto 2003). Examples of PMDs are Alzheimer's disease (AD), Parkinson's disease (PD), Huntington's disease, type-2 diabetes, amyotrophic lateral sclerosis, and other rarer disorders. Among them, transmissible spongiform encephalopathies (TSEs), also known as prion diseases, are a group of infectious and fatal transmissible neurodegenerative disorders that affect both humans and various species of animals (Aguzzi et al. 2008; Prusiner 1998). The human TSEs include Kuru, Creutzfeldt–Jakob disease (CJD), Gerstmann–Sträussler–Scheinker syndrome, and fatal familial insomnia (Collinge 2001). In other mammals, scrapie affects sheep and goats, bovine spongiform encephalopathy (BSE) or "mad-cow disease" in cattle, and chronic wasting disease in elk and deer (Collinge 2001). Clinical signs of prion diseases principally comprise loss of brain function, resulting in dementia and/or ataxia, deterioration of physical and mental abilities and finally, death of the individual (Collinge 2001). Current evidence suggests that the infectious agent in prion diseases is composed predominantly or exclusively by an abnormal form of the prion protein, called PrP^{Sc} (Diaz-Espinoza and Soto 2010). It is proposed that PrP^{Sc} acts as a template to promote conversion of PrP^{C}, a normal host-encoded glycoprotein, to PrP^{Sc}. No differences in chemical post-translational modifications have been found between PrP^{C} and PrP^{Sc} and, apparently, conformational changes are the exclusive characteristic discriminating both PrP isoforms (Stahl et al. 1993).

A common feature observed in TSEs and other PMDs is the presence of misfolded protein deposits in affected tissues (Soto 2001). The classical organization of these abnormal protein aggregates consists of stacks of misfolded units organized in a polymeric arrangement known as a "cross-β" structure (Soto 2001). These structures are able to stabilize intermolecular interactions leading to formation of aggregates commonly referred to as "amyloid". Between the native monomeric protein and the large amyloid fibrils there are several intermediates, including misfolded monomers, soluble oligomers, protofibrils, and fibrillar polymers (Fig. 10.1) (Caughey and Lansbury 2003; Glabe 2005; Haass and Selkoe 2007; Soto and Estrada 2008). Many of these intermediates are in a dynamic equilibrium among each other and recent evidence indicate that oligomeric units, rather than the large fibrillar structures, are the main toxic species responsible for cell damage and disease (Glabe 2006; Haass and Selkoe 2007; Walsh and Selkoe 2007).

The main goal of this chapter is to discuss the contribution of oligomeric PrP^{Sc} in TSEs. We will compare the cases of prion oligomers with experimental data found in other PMDs, principally with AD's amyloid β-protein (Aβ), which have been studied extensively. The relationship between toxicity and propagation abilities will also be discussed, taking into account the latest experimental evidence and comparing the roles of fibrillar *versus* small oligomeric structures.

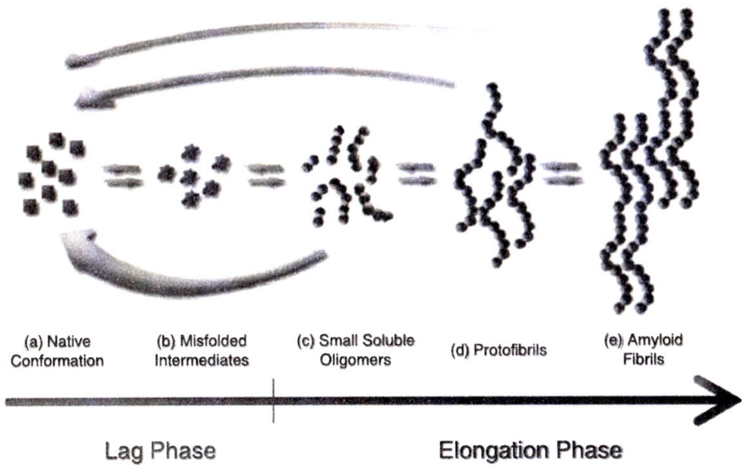

Fig. 10.1 Model of amyloid formation. Among the many conformations that a protein can adopt, only a limited number can be considered as physiological and usually a single structure, called the native fold, is the biologically active one (**a**). However, the normal folding of a protein is in equilibrium with several other possible protein conformations. Unfolded or partially unfolded structures (**b**) are often produced in the pathway to correct folding. However, these structures may also be important intermediates in the process of protein misfolding and aggregation. Misfolded structures can be formed by intermolecular interactions among unfolded or abnormally folded monomers, leading to formation of small oligomeric units (**c**), which are the minimally stable misfolded structures. By further growth, oligomers produce higher and complex structures such as protofibrils (**d**) and mature fibrils (**e**). The nucleation-polymerization model is the leading hypothesis explaining how amyloids form. Two kinetically different phases can be identified: a lag phase and an exponential phase. The lag phase is often a slow process requiring unfavorable interactions among misfolded monomers to form the stable oligomeric seeds. Preformed aggregates are able to seed the oligomerization of soluble monomeric units, bypassing the lag phase. In the figure, seeding capabilities of the different protein aggregates are represented by *arrows*. Extensive experimental evidence suggests that misfolded oligomers are better seeds than protofibrils and fibrils

10.2 Mechanisms of, and Intermediates in, Protein Misfolding and Aggregation

Protein misfolding and aggregation in PMDs follow a kinetic pathway known as seeding nucleation (Jarrett and Lansbury 1993; Soto et al. 2006), whereby two clear stages can be identified (Fig. 10.1). The first one, termed the lag phase, corresponds to the process in which the initial misfolding and formation of minimally stable, misfolded structures take place. Once misfolded seeds are established at a suitable concentration, a phase of exponential recruitment of normally folded protein into the growing polymers starts to occur, resulting in a large burden of protein aggregates. Misfolded protein aggregates grow to different degrees producing a heterogeneous mixture of polymeric structures (Caughey and Lansbury 2003). As a result, an extensive diversity of structures can be found, including misfolded monomers, soluble small oligomeric units, protofibrils, fibrils, and finally an intertwined mesh of fibrils which accumulate into the tissue as amyloid plaques (Fig. 10.1). Protein misfolding is induced and stabilized by protein aggregation and the existence of misfolded monomers is questionable. However, there have been reports of some proteins needing a partial conformational change before aggregates appear (Fink 1998; Jackson et al. 1999; Jiang et al. 2001). In these cases, a prior conformational change to generate the pathological protein is required, but in order to complete the misfolding process a polymerization step is needed. Soluble oligomers are small assemblies of misfolded proteins that are present in the buffer or in soluble fractions of brain extracts and include structures of various sizes. A well-characterized example can be found for Aβ where misfolded oligomers have been described to exist from dimers to 24-mers (Chromy et al. 2003). Recent compelling evidence collected in several fields indicate that oligomers might be the most neurotoxic species in the misfolding and aggregation pathways (Glabe 2006; Haass and Selkoe 2007; Walsh and Selkoe 2007). Indeed, low concentrations of both synthetic and natural oligomers have been shown to induce apoptosis in cell cultures (Bucciantini et al. 2004; Demuro et al. 2005), block long-term potentiation in brain-slice cultures (Wang et al. 2002), and impair synaptic plasticity and memory in animals (Cleary et al. 2004; Shankar et al. 2008). Soluble oligomers then aggregate into protofibrils, which have been seen using electron microscopy (EM) as curvilinear structures of 4–11 nm diameter and <200 nm long (Walsh et al. 1999). Protofibrils grow in size by increasing time and protein concentration and are elongated by recruitment of new misfolded units at their ends (Harper et al. 1999). Annular protofibrils are pore-like assemblies that form in the cell membrane and may contribute to cell death (Lashuel et al. 2002; Srinivasan et al. 2004). Protofibrils then aggregate into fibrils, which have been shown to be the most stable species (Modler et al. 2003). Protofibrils and annular protofibrils have also been shown to be highly neurotoxic *in vitro* (Hartley et al. 1999; Lashuel et al. 2002).

In the biological context, it is likely that normally folded and misfolded protein species are in a dynamic equilibrium. This equilibrium is the result of a struggle between different driving forces within the cell and the involvement of various quality-control cellular processes. On one hand, normal protein structures are maintained among other mechanisms by the proper action of chaperone proteins, which participate in their biosynthesis (Morales et al. 2005). On the other hand, misfolding events are favored by mutations (Gotz et al. 2004; Hsiao and Prusiner 1990), abnormalities of the protein biosynthetic machinery (Lee et al. 2006), and impairments of the cellular clearance mechanisms (Bukau et al. 2006). When equilibrium is displaced towards the misfolded form of a protein, toxic events are manifested triggering disease. Currently, there is no clear consensus about how misfolded aggregates exert their toxic properties. Nevertheless, latest experimental evidence place small soluble oligomers as key pieces in this puzzle (Glabe 2006; Haass and Selkoe 2007; Walsh and Selkoe 2007). Some evidence suggest that protein oligomers can form annular pores in cell membranes (Caughey et al. 2009; Lashuel and Lansbury 2006), whereas others support the hypothesis that membrane receptors generate signaling cascades leading to cell death (Lauren et al. 2009). It is also known that misfolded-protein accumulation can activate the "unfolded-protein response"

(UPR), which initially tries to turn cells back into a normal situation (Lindholm et al. 2006; Morales et al. 2005; Salminen et al. 2009). As a last call and when damage is extensive, cells activate autophagic and death-signaling pathways leading to apoptosis (Hetz and Soto 2006; Morales et al. 2005). As the systems that maintain protein homeostasis become less functional with advanced age, the chances for misfolding events to succeed increase.

Oligomers may not only play crucial roles in neurodegeneration, but may also be important for self-replication of misfolded aggregates, which is the basis for prion infectivity. *In vitro* experiments have shown that the lag phase observed in amyloid formation can be decreased by addition of pre-formed, misfolded seeds (Jarrett and Lansbury 1993). This process provides a plausible explanation for the infectious events observed in TSEs (Soto et al. 2006). Interestingly, the seeding effects of smaller aggregates is far greater than the ones observed for fibrils (Cheon et al. 2008; Silveira et al. 2005; Tixador et al. 2010).

Understanding how oligomers form and their actual role in disease will not only help in defining their contribution to pathological decline, but also in finding cures for these devastating diseases. In fact, some of the current therapeutic strategies under development for AD are focused towards eliminating protein oligomers (Walsh et al. 2005). The presence of misfolded oligomers and their critical roles in pathological progression appear to be a common factor among PMDs. Prion diseases are not an exception. In the following section we will discuss the presence, characterization, and contribution of PrP^{Sc} oligomers to disease, focusing on their infective, replicative, and toxic properties.

10.3 Prion Oligomers

Experimental evidence suggests that PrP^{Sc} oligomers play fundamental roles in the progression of TSEs. Many reports indirectly mention these structures by their biochemical properties, such as their sensitivity to proteinase K (PK) or their small molecular weights. The recent discoveries in AD and other PMDs prompted prion researchers to investigate the oligomeric path. As a result, the existence and characterization of prion oligomers is now well-documented. Considering the particular infectious feature of prions, the subject is even more interesting, since oligomers might not only be the intermediates in the aggregation pathway or the most neurotoxic species, but may also be the carriers of biological information in the prion infectious units. Nevertheless, the relationship among the size of the oligomeric misfolded prion particles, infectivity, propagation, and toxicity of these structures remain to be elucidated.

10.3.1 Roles of Prion Oligomers in Infectivity

Existence of soluble PK-sensitive prion oligomers was a matter of controversy some years ago. Traditionally, prions have been characterized as "protease-resistant and detergent-insoluble infectious proteins" (Prusiner 2004). The absence of PK-resistant PrP^{Sc} in brains from some rare cases was often used as an argument against the protein-only hypothesis of prion transmission (Soto and Castilla 2004). However, the recent characterization of infectious PK-sensitive PrP^{Sc} units has partially settled this controversy (Cronier et al. 2008; Pastrana et al. 2006; Tzaban et al. 2002).

In order to analyze the presence of PK-sensitive infectious material, size fractionation of prion aggregates (by sucrose gradients, size exclusion, flow–field–flow fractionation, etc.) and conformation-dependent immunoassay (CDI—a technique used to differentiate PrP^{Sc} species by their conformational stability) have been used to detect this form of the protein. Studies performed by the group of Stanley Prusiner suggested for the first time the presence of low-molecular–weight, PK-sensitive species when PrP^{Sc} strains were differentiated using CDI (Safar et al. 1998). In 2002, PK-sensitive PrP^{Sc} forms were identified in brains of experimental rodents as well as in prion-infected cells (Tzaban et al. 2002). In this report, PK-sensitive PrP^{Sc} molecules with molecular sizes as small as 600 kDa were identified. However, the authors were unable to address whether these aggregates were a dynamic part of the misfolded-protein assembly or intermediates in the formation or degradation of larger structures. Later, we confirmed the presence of PK-sensitive PrP^{Sc} by fractionating non-PK–treated, purified prion preparations (Pastrana et al. 2006). In these experiments, oligomers were separated from fibrillar structures by ultracentrifugation, clearly identifying two PrP^{Sc} populations in regards to their resistance to PK activity. Further characterization, including *in vitro* replication by protein-misfolding cyclic-amplification assay, allowed us to differentiate between PK-sensitive prions and PrP^C (Pastrana et al. 2006). Existence of PK-sensitive prions has been recently confirmed by thermolysin treatment, an enzyme able to digest PrP^C, allowing isolation of full-length PK-sensitive PrP^{Sc} species (Cronier et al. 2008).

PK-sensitive prions have been identified in patients and animal models (Collinge et al. 1990; Lasmezas et al. 1997; Safar et al. 2005). Human cases of dementia with rapid clinical decline and neuropathological features such as spongiform degeneration and inflammation, all characteristics of TSEs, failed to show PrP deposits or PK-resistant prions. However, recent studies involving use of thermolysin and CDI suggest that up to 80–90% of CJD-related prions are destroyed by PK (Cronier et al. 2008; Safar et al. 2005). The presence of PK-sensitive prions has been further suggested in experimental rodent models (Hill and Collinge 2003; Makarava et al. 2010). An interesting study by Lasmezas et al. showed that BSE transmission to wild-type mice generated two different phenotypes associated with disease (Lasmezas et al. 1997): PK-sensitive and PK-resistant PrP^{Sc}. In a second infectivity passage of the PK-sensitive material both phenotypes were observed. Third infectivity passages generated only the PK-resistant phenotype. Despite the prevalence of the PK-resistant phenotype in this case, the existence of PK-sensitive prions transmissible to other individuals was evident since inoculation of this material produced a typical form of prion disease.

In an elegant study, Silveira et al. identified what they called "the most infectious prion protein particles" (Silveira et al. 2005). By using sophisticated fractionation studies, the authors found that oligomeric structures ranging from 500 to 600 kDa, corresponding to particles composed of approximately 20–30 PrP monomers (if the particles comprised PrP only), presented the highest infectivity and the best *in vitro* converting activity (Silveira et al. 2005). Transmission-electron-microscopy images of these molecules showed small spherical or ellipsoid structures compared to fibrillar prion aggregates obtained from subsequent fractions. One concern in this study was the use of PK–treated, purified samples. PrP^{Sc} processing in this way results in a truncated form of the protein that is known to have increased aggregation (McKinley et al. 1991), which could affect the distribution of the infectious aggregates found in brain. In addition, this procedure discards the possibility to find protease-sensitive infectious units. Despite these issues, these findings were important to set prion oligomers as important, if not the most pathogenic, structures found in TSEs. Similar results were recently published using a sedimentation-velocity technique to study the oligomerization state of PrP^{Sc} using a panel of natural and biologically cloned strains (Tixador et al. 2010). For this study, detergent-solubilized, infected brain homogenates were used as starting material and the distribution of PrP^{Sc} and infectivity in the gradient were determined by immunoblotting and mouse bioassays. A major PK-resistant PrP^{Sc} peak was observed in the middle part of the gradient, which corresponds to polymers of 12–30 PrP molecules (Tixador et al. 2010). The most infectious form of PrP^{Sc} in strains able to propagate rapidly and efficiently corresponded to small-size oligomers, but substantial differences were observed in the distribution and size of oligomers in other prion strains. These findings suggest that prion infectious particles are subject to marked strain-dependent variations, which could determine the biological phenotype of the strains, in particular their replication dynamics (Tixador et al. 2010). The critical roles of PrP oligomers *versus* fibrils are also supported in GPI-anchorless PrP transgenic mice (Chesebro et al. 2005). These animals do not develop disease (or develop disease after very long incubation times), despite the fact that they have high quantities of PrP^{Sc} in their brains. Although the low brain damage in these transgenic animals have been mostly attributed to the lack of PrP^{C} attached to cell membranes, the absence of neurodegeneration in this model could also be ascribed to the formation of very large PrP^{Sc} deposits, mostly composed of compacted amyloid plaques. This hypothesis is further supported by recent data from our laboratory suggesting that there is a direct correlation between the size of PrP aggregates and the incubation period necessary to produce the disease (RM and CS, unpublished results).

10.3.2 Roles of Prion Oligomers in Neurotoxicity

Despite many studies designed to evaluate the toxicity of prion aggregates, little information is available about the toxic properties of brain-derived oligomeric PrP^{Sc}.

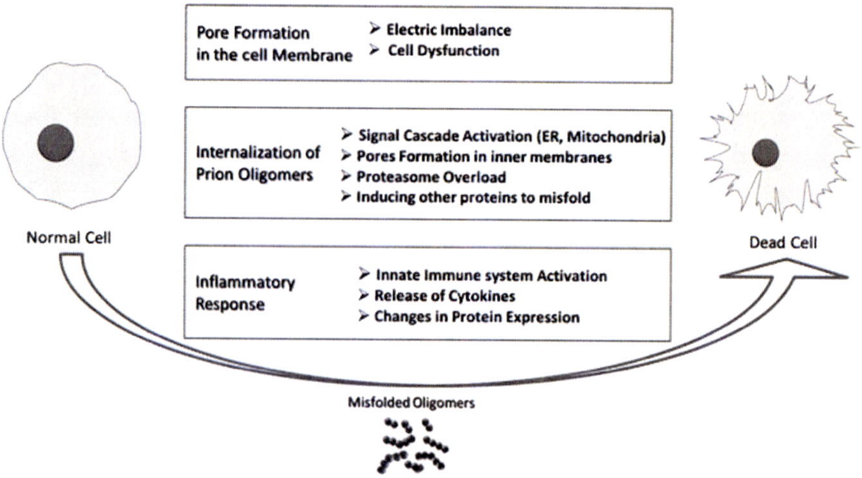

Fig. 10.2 Putative mechanisms whereby misfolded protein oligomers may be toxic to cells. Although compelling evidence suggests that misfolded oligomers are intrinsically associated with toxicity in PMDs, the mechanisms whereby oligomers induce cellular dysfunction are largely unknown. Despite the fact that misfolded oligomers derived from different proteins have striking similarities, it is still unknown whether all oligomeric structures promote the same biological responses. One of the hypotheses proposed to explain the cellular dysfunction produced by misfolded oligomers involves the formation of pores in the cell membranes, leading to an imbalance between the intra- and extracellular environments. Another model suggests that oligomers could lead to apoptosis by induction of ER or mitochondrial stress-signaling pathways. Misfolded structures produced or internalized by cells may also overload the proteasome, impairing the biological clearance mechanisms. The latter as well as direct cross-seeding interactions may induce the misfolding of other proteins that can subsequently contribute to cellular dysfunction. Finally, misfolded protein oligomers could activate immune responses leading to tissue dysfunction by indirect routes, such as the chronic release of pro-inflammatory cytokines

In addition, although several models have been hypothesized to explain the mechanisms whereby misfolded and oligomeric PrP^{Sc} may induce neurodegeneration, a consensus on the actual mechanisms operating *in vivo* is far from clear. Many hypotheses have been formulated in order to understand how PrP oligomers would exert their deleterious effects in cells (Fig. 10.2). Among them, the activation of signaling cascades leading to apoptosis, especially through induction of stress responses in the endoplasmic reticulum or mitochondria, have been described (Hetz et al. 2003; O'Donovan et al. 2001). In addition, the putative generation of pores in cell membranes (Solomon et al. 2010) and activation of inflammatory responses (Soto and Satani 2011) by misfolded PrP^{Sc} are commonly associated with the intrinsic toxicity of these structures.

Toxicity of prion aggregates was first characterized using preparations obtained from infected brains (Hetz et al. 2003) or using synthetic peptides (Forloni et al. 1993). The synthetic PrP106–126 peptide, related to the possible amyloidogenic core of PrP^{Sc}, was described to induce apoptosis in cell cultures (Forloni et al. 1993; O'Donovan et al. 2001). However, the toxic effects of oligomeric preparations of this peptide have not yet been reported. In addition, the lack of infectivity of PrP106–126 amyloidogenic preparations decreases the impact of these results. The first studies describing the toxicity of prion oligomeric species was performed using pre-fibrillar, recombinant hamster prions (Kazlauskaite et al. 2005). β-Sheet-rich pre-fibrillar aggregates were isolated by ultracentrifugation and tested in PC12 neuroblastoma cells. As expected, the highest toxicity was found for this subset of aggregates. Later, a controversy arose when a second report suggested that oligomeric and fibrillar synthetic prion preparations had similar toxicity indexes in cell lines and primary neuron cultures (Novitskaya et al. 2006). These authors reported that toxicity of prion aggregates was mediated by PrP^{C} since administration of siRNA against PrP considerably decreased cell death. A more recent report described the toxicity of PrP oligomers in animal models (Simoneau et al. 2007). Stereotaxically injected recombinant ovine and murine prion oligomers were toxic in brain, whereas other species showed little or no effects. These results were confirmed using primary mouse cortical cultures. The same report showed that the toxic effects of PrP aggregates did not depend on PrP^{C} since wild-type and PrP-knockout mice responded in the same way to prion-mediated insults (Simoneau et al. 2007). Surprisingly, antibodies against the 106–126-prion domain rescued cells from toxicity, indicating a structural mechanism whereby PrP oligomers could act. However, we need to be cautious about extrapolating these findings to bona fide PrP^{Sc} aggregates because most of "artificial" PrP preparations (from either recombinant or synthetic origins) are not infectious.

The concept that neurodegeneration is caused exclusively by the formation and accumulation of PrP^{Sc} oligomers is probably an oversimplification. Despite good correlations between PrP^{Sc} accumulation, neurodegeneration, and disease in the large majority of cases, there are rare cases in which disease appears without detectable PrP^{Sc} (Lasmezas et al. 1997) or also cases in which abundant PrP^{Sc} deposition is observed with no neurodegeneration or disease (Piccardo et al. 2007). These results suggest that PrP^{Sc} is neither necessary nor sufficient for the disease. However, an alternative explanation is that the replicative and infectious forms of PrP^{Sc} are not the same as the neurotoxic forms (Chiesa et al. 2003, 2008; Harris and True 2006). The most toxic forms could be composed of small, misfolded oligomers, which might not be detectable as PrP^{Sc} by classical biochemical assays. Alternatively, minor PrP^{Sc} isoforms, including transmembrane or cytosolic versions of the misfolded protein, could be the neurotoxic species (Hegde et al. 1998; Ma et al. 2002). Another possibility is that PrP^{Sc} subverts or modifies the normal functions of PrP^{C} (Chesebro et al. 2005; Westergard et al. 2007). In this scenario, the biological functions of PrP^{C} might be altered by binding to PrP^{Sc}, and for example, this may activate alternative signal-transduction pathways that lead to neurotoxicity rather than neuroprotection. PrP^{Sc} might produce this effect by cross-linking cell-surface PrP^{C}, which has been shown to induce neuronal apoptosis *in vivo* (Solforosi et al. 2004) or by binding to and blocking specific functional PrP^{C} domains.

10.4 Interactions Between Different Misfolded Proteins at the Oligomeric Level

As stated above, all PMDs involve formation of misfolded oligomeric and larger polymers through a seeding–nucleation mechanism of polymerization (Soto and Estrada 2008). The striking similarities in the mechanisms and the characteristics of the aggregates suggest that PMDs may interact at the level of protein misfolding and be a risk factor for each other (Morales et al. 2009b; Tsigelny et al. 2008). Multiple *in vitro*, *in vivo*, and clinical evidence suggest an interaction at the level of protein misfolding between different PMDs, through a phenomenon called "cross-seeding" (Morales et al. 2009b). This process occurs when oligomeric seeds composed by one protein nucleate the misfolding and aggregation of a second protein.

While it is well-established that fibril formation is enhanced by adding preformed homologous or heterologous seeds, it has also been shown *in vitro* that sequence similarities could affect seeding efficiency. By cross-seeding hen lysozyme with a series of proteins, it was demonstrated that differing sequences had lower efficiencies of seeding (Krebs et al. 2004). *In vivo* studies revealed that non-mammalian protein fibrils can cross-seed amyloid A protein (AA) in a murine experimental AA amyloidosis model (Johan et al. 1998). In addition, several reports showing the co-existence of different amyloid pathologies in the same tissue, further support the existence of cross-seeding as a disease mechanism (Morales et al. 2009b). In some cases, cross-seeding occurs reciprocally, i.e., both proteins can cross-seed each other (Morales et al. 2009b). Conversely, in other cases cross-seeding is unidirectional, i.e., aggregates from one protein seed the misfolding of the second protein, but aggregates composed of the second protein cannot seed misfolding and aggregation of the first protein (O'Nuallain et al. 2004; Yan et al. 2007). Moreover in some cases, these interactions inhibit misfolding of one or both proteins (Yan et al. 2007).

As the seeding phenomenon was discussed above, it is likely that the key elements responsible for cross-seeding are oligomeric species with the capacity to nucleate polymerization. A study by Tsigelny et al. convincingly showed *in vitro* and *in vivo* cross-seeding interactions between α-synuclein and Aβ (Tsigelny et al. 2008), implicated in the pathogenesis of PD and AD, respectively. The phenomenon of co-existence of amyloids has been described for several other amyloidogenic proteins. Among them, the simultaneous presence of Aβ and prions has been extensively documented in patients affected with CJD, Gerstmann-Sträussler-Scheinker syndrome, and AD (Debatin et al. 2008; Ferrer et al. 2001; Hainfellner et al. 1998; Leuba et al. 2000; Miyazono et al. 1992; Paquet et al. 2008; Tsuchiya et al. 2004). To study the possibility of a molecular interaction between Alzheimer's and prion-protein misfolding processes we inoculated prions in an AD transgenic mouse model that developed typical amyloid plaques (Morales et al. 2010). This experiment showed a dramatic acceleration and exacerbation of both pathologies. The onset of prion-disease symptoms in transgenic mice appeared significantly faster with a concomitant increase in the levels of misfolded prion protein in the brain (Morales et al. 2010). A striking increase in amyloid-plaque deposition was observed in prion-infected mice compared with their non-inoculated counterparts. Histological and biochemical studies showed association of the two misfolded proteins in the brain and *in vitro* experiments showed that protein misfolding could be enhanced by cross-seeding mechanisms (Morales et al. 2010).

It seems likely that several other PMDs could be interacting at the protein-oligomeric level, indicating that one protein-misfolding process may be an important risk factor for the development of a second one. These findings may have important implications to understanding of origins and progression of PMDs.

10.5 Concluding Remarks

Misfolded proteins accumulate in a heterogeneous population of aggregates of different sizes. Due to their striking biological effects compared to larger polymers, small soluble oligomers are the ones taking most of the attention of the scientific community (Glabe 2006; Haass and Selkoe 2007; Walsh and Selkoe 2007). Recent findings by studies of proteins associated with different PMDs indicate that oligomers are the key structures responsible for the toxic activities of misfolded proteins and the critical intermediates for propagation of the misfolding processes (Glabe 2006; Haass and Selkoe 2007; Walsh and Selkoe 2007). However, there is still a long way to go in order to understand fully how these aggregates exert their biological functions. Despite the importance of oligomers, it is imperative to pay attention to all the dynamic populations of protein aggregates generated during the protein-misfolding processes. Several interesting issues involving these still-enigmatic particles such as their structural properties, stability, biological functions, putative transmissibility, and interactions with other structures are yet to be resolved. Over the years, the importance of large fibrillar structures has changed from being thought to be the culprits of tissue degeneration and disease, to biologically inert structures to acting as protective entities that encapsulate toxic oligomers. However, much more research is needed to elucidate the exact contributions of each of the distinct misfolded aggregates to the disease. In the case of large fibrils deposited in amyloid plaques, it is difficult to imagine that these structures are indeed protective, since at the very least they occupy substantial tissue space and disrupt neuronal connections. Large amyloid aggregates may also act as a reservoir for toxic oligomers.

Recent exciting findings have shown that misfolded proteins associated with various PMDs may have an element of transmissibility similar to prions in TSEs (Aguzzi 2009; Brundin et al. 2010; Frost and Diamond 2010; Soto et al. 2006). The case of Aβ aggregation in AD is perhaps the most emblematic due to the high prevalence of this disease in the population. Interestingly, a considerable number of similarities between prion and AD in terms of propagation and pathological outcomes have been listed (Morales et al. 2009a; Soto et al. 2006). Nevertheless, up to now experiments showing a prion-like propagation of Aβ aggregates have only been performed in animal models using biologically irrelevant routes of exposure to the infectious material (Eisele et al. 2010; Kane et al. 2000; Meyer-Luehmann et al. 2006) and thus it is difficult to assess whether AD transmissibility may occur in real life. In this sense, an additional and important evidence related to the possible transmissibility of misfolded protein aggregates has been described for α-synuclein in PD, where grafted tissue in patients who received cells devoid of any type of aggregates were later shown to develop misfolded structures (Kordower et al. 2008). Other natural amyloidoses such as AA secondary systemic amyloidosis strongly support the idea that PMDs would have an infectious component, which is composed principally of a misfolded protein aggregate (Ganowiak et al. 1994; Kisilevsky and Boudreau 1983). Although at this time, it is debatable whether other diseases may indeed propagate by an infectious process, the prion phenomenon is at least acting at the molecular and cellular level to contribute to the spread of the disease within the organism. In this scenario, oligomers may be the key structures responsible for transmission through their ability to nucleate protein misfolding and aggregation. It seems that smaller and more soluble oligomers may be the best seeds to propagate misfolding processes; however, it is also likely that larger and more stable structures may be more relevant to propagate these processes *in vivo* due to their higher resistance to biological clearance.

Considering the similarities between misfolded protein oligomers among PMDs, a better understanding of their biology could lead to novel and perhaps common therapeutic approaches for all diseases in this group. The recent discovery of molecules able to specifically bind oligomeric species (Kayed et al. 2003) open new and promising strategies to attack many of the most devastating diseases manifested at later stages of life.

Acknowledgments We thank Natalia Salvadores for suggestions and Maria-Jose Liberona for critical review of the manuscript. This work was supported in part by the NIH grants R01NS049173 and R01NS050349.

References

Aguzzi A (2009) Cell biology: beyond the prion principle. Nature 459:924–925

Aguzzi A, Sigurdson C, Heikenwaelder M (2008) Molecular mechanisms of prion pathogenesis. Annu Rev Pathol 3:11–40

Brundin P, Melki R, Kopito R (2010) Prion-like transmission of protein aggregates in neurodegenerative diseases. Nat Rev Mol Cell Biol 11:301–307

Bucciantini M, Calloni G, Chiti F, Formigli L, Nosi D, Dobson CM, Stefani M (2004) Prefibrillar amyloid protein aggregates share common features of cytotoxicity. J Biol Chem 279: 31374–31382

Bukau B, Weissman J, Horwich A (2006) Molecular chaperones and protein quality control. Cell 125:443–451

Caughey B, Lansbury PT (2003) Protofibrils, pores, fibrils, and neurodegeneration: separating the responsible protein aggregates from the innocent bystanders. Annu Rev Neurosci 26:267–298

Caughey B, Baron GS, Chesebro B, Jeffrey M (2009) Getting a grip on prions: oligomers, amyloids, and pathological membrane interactions. Annu Rev Biochem 78:177–204
Cheon M, Favrin G, Chang I, Dobson CM, Vendruscolo M (2008) Calculation of the free energy barriers in the oligomerization of Aβ peptide fragments. Front Biosci 13:5614–5622
Chesebro B, Trifilo M, Race R, Meade-White K, Teng C, LaCasse R, Raymond L, Favara C, Baron G, Priola S, Caughey B, Masliah E, Oldstone M (2005) Anchorless prion protein results in infectious amyloid disease without clinical scrapie. Science 308:1435–1439
Chiesa R, Piccardo P, Quaglio E, Drisaldi B, Si-Hoe SL, Takao M, Ghetti B, Harris DA (2003) Molecular distinction between pathogenic and infectious properties of the prion protein. J Virol 77:7611–7622
Chiesa R, Piccardo P, Biasini E, Ghetti B, Harris DA (2008) Aggregated, wild-type prion protein causes neurological dysfunction and synaptic abnormalities. J Neurosci 28:13258–13267
Chromy BA, Nowak RJ, Lambert MP, Viola KL, Chang L, Velasco PT, Jones BW, Fernandez SJ, Lacor PN, Horowitz P, Finch CE, Krafft GA, Klein WL (2003) Self-assembly of $A\beta_{1-42}$ into globular neurotoxins. Biochemistry 42:12749–12760
Cleary JP, Walsh DM, Hofmeister JJ, Shankar GM, Kuskowski MA, Selkoe DJ, Ashe KH (2004) Natural oligomers of the amyloid-β protein specifically disrupt cognitive function. Nat Neurosci 8:79–84
Collinge J (2001) Prion diseases of humans and animals: their causes and molecular basis. Annu Rev Neurosci 24:519–550
Collinge J, Owen F, Poulter M, Leach M, Crow TJ, Rossor MN, Hardy J, Mullan MJ, Janota I, Lantos PL (1990) Prion dementia without characteristic pathology. Lancet 336:7–9
Cronier S, Gros N, Tattum MH, Jackson GS, Clarke AR, Collinge J, Wadsworth JD (2008) Detection and characterization of proteinase K-sensitive disease-related prion protein with thermolysin. Biochem J 416:297–305
Debatin L, Streffer J, Geissen M, Matschke J, Aguzzi A, Glatzel M (2008) Association between deposition of β-amyloid and pathological prion protein in sporadic Creutzfeldt–Jakob disease. Neurodegener Dis 5:347–354
Demuro A, Mina E, Kayed R, Milton SC, Parker I, Glabe CG (2005) Calcium dysregulation and membrane disruption as a ubiquitous neurotoxic mechanism of soluble amyloid oligomers. J Biol Chem 280:17294–17300
Diaz-Espinoza R, Soto C (2010) Generation of prions *in vitro* and the protein-only hypothesis. Prion 4:53–59
Eisele YS, Obermuller U, Heilbronner G, Baumann F, Kaeser SA, Wolburg H, Walker LC, Staufenbiel M, Heikenwalder M, Jucker M (2010) Peripherally applied Aβ-containing inoculates induce cerebral β-amyloidosis. Science 330:980–982
Ferrer I, Blanco R, Carmona M, Puig B, Ribera R, Rey MJ, Ribalta T (2001) Prion protein expression in senile plaques in Alzheimer's disease. Acta Neuropathol (Berl) 101:49–56
Fink AL (1998) Protein aggregation: folding aggregates, inclusion bodies and amyloid. Fold Des 3:R9–R23
Forloni G, Angeretti N, Chiesa R, Monzani E, Salmona M, Bugiani O, Tagliavini F (1993) Neurotoxicity of a prion protein fragment. Nature 362:543–546
Frost B, Diamond MI (2010) Prion-like mechanisms in neurodegenerative diseases. Nat Rev Neurosci 11:155–159
Ganowiak K, Hultman P, Engstrom U, Gustavsson A, Westermark P (1994) Fibrils from synthetic amyloid-related peptides enhance development of experimental AA-amyloidosis in mice. Biochem Biophys Res Commun 199:306–312
Glabe CC (2005) Amyloid accumulation and pathogenesis of Alzheimer's disease: significance of monomeric, oligomeric and fibrillar Aβ. Subcell Biochem 38:167–177
Glabe CG (2006) Common mechanisms of amyloid oligomer pathogenesis in degenerative disease. Neurobiol Aging 27:570–575
Gotz J, Streffer JR, David D, Schild A, Hoerndli F, Pennanen L, Kurosinski P, Chen F (2004) Transgenic animal models of Alzheimer's disease and related disorders: histopathology, behavior and therapy. Mol Psychiatry 9:664–683

Haass C, Selkoe DJ (2007) Soluble protein oligomers in neurodegeneration: lessons from the Alzheimer's amyloid β-peptide. Nat Rev Mol Cell Biol 8:101–112

Hainfellner JA, Wanschitz J, Jellinger K, Liberski PP, Gullotta F, Budka H (1998) Coexistence of Alzheimer-type neuropathology in Creutzfeldt-Jakob disease. Acta Neuropathol (Berl) 96: 116–122

Harper JD, Wong SS, Lieber CM, Lansbury PT Jr (1999) Assembly of Aβ amyloid protofibrils: an *in vitro* model for a possible early event in Alzheimer's disease. Biochemistry 38:8972–8980

Harris DA, True HL (2006) New insights into prion structure and toxicity. Neuron 50:353–357

Hartley DM, Walsh DM, Ye CP, Diehl T, Vasquez S, Vassilev PM, Teplow DB, Selkoe DJ (1999) Protofibrillar intermediates of amyloid β-protein induce acute electrophysiological changes and progressive neurotoxicity in cortical neurons. J Neurosci 19:8876–8884

Hegde RS, Mastrianni JA, Scott MR, DeFea KA, Tremblay P, Torchia M, DeArmond SJ, Prusiner SB, Lingappa VR (1998) A transmembrane form of the prion protein in neurodegenerative disease. Science 279:827–834

Hetz CA, Soto C (2006) Stressing out the ER: a role of the unfolded protein response in prion-related disorders. Curr Mol Med 6:37–43

Hetz C, Russelakis-Carneiro M, Maundrell K, Castilla J, Soto C (2003) Caspase-12 and endoplasmic reticulum stress mediate neurotoxicity of pathological prion protein. EMBO J 22:5435–5445

Hill AF, Collinge J (2003) Subclinical prion infection in humans and animals. Br Med Bull 66:161–170

Hsiao K, Prusiner SB (1990) Inherited human prion diseases. Neurology 40:1820–1827

Jackson GS, Hosszu LL, Power A, Hill AF, Kenney J, Saibil H, Craven CJ, Waltho JP, Clarke AR, Collinge J (1999) Reversible conversion of monomeric human prion protein between native and fibrilogenic conformations. Science 283:1935–1937

Jarrett JT, Lansbury PT Jr (1993) Seeding "one-dimensional crystallization" of amyloid: a pathogenic mechanism in Alzheimer's disease and scrapie? Cell 73:1055–1058

Jiang X, Smith CS, Petrassi HM, Hammarstrom P, White JT, Sacchettini JC, Kelly JW (2001) An engineered transthyretin monomer that is nonamyloidogenic, unless it is partially denatured. Biochemistry 40:11442–11452

Johan K, Westermark G, Engstrom U, Gustavsson A, Hultman P, Westermark P (1998) Acceleration of amyloid protein A amyloidosis by amyloid-like synthetic fibrils. Proc Natl Acad Sci USA 95:2558–2563

Kane MD, Lipinski WJ, Callahan MJ, Bian F, Durham RA, Schwarz RD, Roher AE, Walker LC (2000) Evidence for seeding of β-amyloid by intracerebral infusion of Alzheimer brain extracts in β-amyloid precursor protein-transgenic mice. J Neurosci 20:3606–3611

Kayed R, Head E, Thompson JL, McIntire TM, Milton SC, Cotman CW, Glabe CG (2003) Common structure of soluble amyloid oligomers implies common mechanism of pathogenesis. Science 300:486–489

Kazlauskaite J, Young A, Gardner CE, Macpherson JV, Venien-Bryan C, Pinheiro TJ (2005) An unusual soluble β-turn-rich conformation of prion is involved in fibril formation and toxic to neuronal cells. Biochem Biophys Res Commun 328:292–305

Kisilevsky R, Boudreau L (1983) Kinetics of amyloid deposition. I. The effects of amyloid-enhancing factor and splenectomy. Lab Invest 48:53–59

Kordower JH, Chu Y, Hauser RA, Freeman TB, Olanow CW (2008) Lewy body-like pathology in long-term embryonic nigral transplants in Parkinson's disease. Nat Med 14:504–506

Krebs MR, Morozova-Roche LA, Daniel K, Robinson CV, Dobson CM (2004) Observation of sequence specificity in the seeding of protein amyloid fibrils. Protein Sci 13:1933–1938

Lashuel HA, Lansbury PT Jr (2006) Are amyloid diseases caused by protein aggregates that mimic bacterial pore-forming toxins? Q Rev Biophys 39:167–201

Lashuel HA, Hartley D, Petre BM, Walz T, Lansbury PT Jr (2002) Neurodegenerative disease: amyloid pores from pathogenic mutations. Nature 418:291

Lasmezas CI, Deslys JP, Robain O, Jaegly A, Beringue V, Peyrin JM, Fournier JG, Hauw JJ, Rossier J, Dormont D (1997) Transmission of the BSE agent to mice in the absence of detectable abnormal prion protein. Science 275:402–405

Lauren J, Gimbel DA, Nygaard HB, Gilbert JW, Strittmatter SM (2009) Cellular prion protein mediates impairment of synaptic plasticity by amyloid-β oligomers. Nature 457:1128–1132

Lee JW, Beebe K, Nangle LA, Jang J, Longo-Guess CM, Cook SA, Davisson MT, Sundberg JP, Schimmel P, Ackerman SL (2006) Editing-defective tRNA synthetase causes protein misfolding and neurodegeneration. Nature 443:50–55

Leuba G, Saini K, Savioz A, Charnay Y (2000) Early-onset familial Alzheimer disease with coexisting β-amyloid and prion pathology. JAMA 283:1689–1691

Lindholm D, Wootz H, Korhonen L (2006) ER stress and neurodegenerative diseases. Cell Death Differ 13:385–392

Luheshi LM, Crowther DC, Dobson CM (2008) Protein misfolding and disease: from the test tube to the organism. Curr Opin Chem Biol 12:25–31

Ma J, Wollmann R, Lindquist S (2002) Neurotoxicity and neurodegeneration when PrP accumulates in the cytosol. Science 298:1781–1785

Makarava N, Kovacs GG, Bocharova O, Savtchenko R, Alexeeva I, Budka H, Rohwer RG, Baskakov IV (2010) Recombinant prion protein induces a new transmissible prion disease in wild-type animals. Acta Neuropathol 119:177–187

McKinley MP, Meyer RK, Kenaga L, Rahbar F, Cotter R, Serban A, Prusiner SB (1991) Scrapie prion rod formation *in vitro* requires both detergent extraction and limited proteolysis. J Virol 65:1340–1351

Meyer-Luehmann M, Coomaraswamy J, Bolmont T, Kaeser S, Schaefer C, Kilger E, Neuenschwander A, Abramowski D, Frey P, Jaton AL et al (2006) Exogenous induction of cerebral β-amyloidogenesis is governed by agent and host. Science 313:1781–1784

Miyazono M, Kitamoto T, Iwaki T, Tateishi J (1992) Colocalization of prion protein and β protein in the same amyloid plaques in patients with Gerstmann-Straüssler syndrome. Acta Neuropathol (Berl) 83:333–339

Modler AJ, Gast K, Lutsch G, Damaschun G (2003) Assembly of amyloid protofibrils via critical oligomers—a novel pathway of amyloid formation. J Mol Biol 325:135–148

Morales R, Hetz C, Soto C (2005) Signaling pathways controlling prion neurotoxicity: role of endoplasmic reticulum stress-mediated apoptosis. In: Brown DR (ed) Neurodegeneration and prion disease. Springer, New York, pp 319–344

Morales R, Chen B, Soto C (2009a) Are amyloids infectious? In: Maccioni RB, Perry G (eds) Current hypotheses and research milestones in Alzheimer's disease. Springer, New York, pp 171–180

Morales R, Green KM, Soto C (2009b) Cross currents in protein misfolding disorders: interactions and therapy. CNS Neurol Disord Drug Targets 8:363–371

Morales R, Estrada LD, Diaz-Espinoza R, Morales-Scheihing D, Jara MC, Castilla J, Soto C (2010) Molecular cross talk between misfolded proteins in animal models of Alzheimer's and prion diseases. J Neurosci 30:4528–4535

Novitskaya V, Bocharova OV, Bronstein I, Baskakov IV (2006) Amyloid fibrils of mammalian prion protein are highly toxic to cultured cells and primary neurons. J Biol Chem 281:13828–13836

O'Donovan CN, Tobin D, Cotter TG (2001) Prion protein fragment PrP-(106–126) induces apoptosis via mitochondrial disruption in human neuronal SH-SY5Y cells. J Biol Chem 276:43516–43523

O'Nuallain B, Williams AD, Westermark P, Wetzel R (2004) Seeding specificity in amyloid growth induced by heterologous fibrils. J Biol Chem 279:17490–17499

Paquet C, Privat N, Kaci R, Polivka M, Dupont O, Haik S, Laplanche JL, Hauw JJ, Gray F (2008) Cerebral amyloid angiopathy with co-localization of prion protein and β-amyloid in an 85-year-old patient with sporadic Creutzfeldt-Jakob disease. Acta Neuropathol 116:567–573

Pastrana MA, Sajnani G, Onisko B, Castilla J, Morales R, Soto C, Requena JR (2006) Isolation and Characterization of a Proteinase K-Sensitive PrP^{Sc} Fraction. Biochemistry 45:15710–15717

Piccardo P, Manson JC, King D, Ghetti B, Barron RM (2007) Accumulation of prion protein in the brain that is not associated with transmissible disease. Proc Natl Acad Sci USA 104:4712–4717

Prusiner SB (1998) Prions. Proc Natl Acad Sci USA 95:13363–13383
Prusiner SB (2004) Early evidence that a protease-resistant protein is an active component of the infectious prion. Cell 116:S109–S111
Safar J, Wille H, Itri V, Groth D, Serban H, Torchia M, Cohen FE, Prusiner SB (1998) Eight prion strains have PrP^{Sc} molecules with different conformations. Nat Med 4:1157–1165
Safar JG, Geschwind MD, Deering C, Didorenko S, Sattavat M, Sanchez H, Serban A, Vey M, Baron H, Giles K et al (2005) Diagnosis of human prion disease. Proc Natl Acad Sci USA 102:3501–3506
Salminen A, Kauppinen A, Suuronen T, Kaarniranta K, Ojala J (2009) ER stress in Alzheimer's disease: a novel neuronal trigger for inflammation and Alzheimer's pathology. J Neuroinflammation 6:41
Shankar GM, Li S, Mehta TH, Garcia-Munoz A, Shepardson NE, Smith I, Brett FM, Farrell MA, Rowan MJ, Lemere CA, Regan CM, Walsh DM, Sabatini BL, Selkoe DJ (2008) Amyloid-β protein dimers isolated directly from Alzheimer's brains impair synaptic plasticity and memory. Nat Med 14:837–842
Silveira JR, Raymond GJ, Hughson AG, Race RE, Sim VL, Hayes SF, Caughey B (2005) The most infectious prion protein particles. Nature 437:257–261
Simoneau S, Rezaei H, Sales N, Kaiser-Schulz G, Lefebvre-Roque M, Vidal C, Fournier JG, Comte J, Wopfner F, Grosclaude J, Schätzl H, Lasmézas CI (2007) *In vitro* and *in vivo* neurotoxicity of prion protein oligomers. PLoS Pathog 3:e125
Solforosi L, Criado JR, McGavern DB, Wirz S, Sanchez-Alavez M, Sugama S, DeGiorgio LA, Volpe BT, Wiseman E, Abalos G, Masliah E, Gilden D, Oldstone MB, Conti B, Williamson RA (2004) Cross-linking cellular prion protein triggers neuronal apoptosis *in vivo*. Science 303:1514–1516
Solomon IH, Huettner JE, Harris DA (2010) Neurotoxic mutants of the prion protein induce spontaneous ionic currents in cultured cells. J Biol Chem 285:26719–26726
Soto C (2001) Protein misfolding and disease; protein refolding and therapy. FEBS Lett 498:204–207
Soto C (2003) Unfolding the role of protein misfolding in neurodegenerative diseases. Nat Rev Neurosci 4:49–60
Soto C, Castilla J (2004) The controversial protein-only hypothesis of prion propagation. Nat Med 10:S63–S67
Soto C, Estrada LD (2008) Protein misfolding and neurodegeneration. Arch Neurol 65:184–189
Soto C, Satani N (2011) The intricate mechanisms of neurodegeneration in prion diseases. Trends Mol Med 17:14–24
Soto C, Estrada L, Castilla J (2006) Amyloids, prions and the inherent infectious nature of misfolded protein aggregates. Trends Biochem Sci 31:150–155
Srinivasan R, Marchant RE, Zagorski MG (2004) ABri peptide associated with familial British dementia forms annular and ring-like protofibrillar structures. Amyloid 11:10–13
Stahl N, Baldwin MA, Teplow DB, Hood L, Gibson BW, Burlingame AL, Prusiner SB (1993) Structural studies of the scrapie prion protein using mass spectrometry and amino acid sequencing. Biochemistry 32:1991–2002
Tixador P, Herzog L, Reine F, Jaumain E, Chapuis J, Le DA, Laude H, Beringue V (2010) The physical relationship between infectivity and prion protein aggregates is strain-dependent. PLoS Pathog 6:e1000859
Tsigelny IF, Crews L, Desplats P, Shaked GM, Sharikov Y, Mizuno H, Spencer B, Rockenstein E, Trejo M, Platoshyn O, Yuan JX, Masliah E (2008) Mechanisms of hybrid oligomer formation in the pathogenesis of combined Alzheimer's and Parkinson's diseases. PLoS One 3:e3135
Tsuchiya K, Yagishita S, Ikeda K, Sano M, Taki K, Hashimoto K, Watabiki S, Mizusawa H (2004) Coexistence of CJD and Alzheimer's disease: an autopsy case showing typical clinical features of CJD. Neuropathology 24:46–55
Tzaban S, Friedlander G, Schonberger O, Horonchik L, Yedidia Y, Shaked G, Gabizon R, Taraboulos A (2002) Protease-sensitive scrapie prion protein in aggregates of heterogeneous sizes. Biochemistry 41:12868–12875
Walsh DM, Selkoe DJ (2007) Aβ oligomers—a decade of discovery. J Neurochem 101:1172–1184

Walsh DM, Hartley DM, Kusumoto Y, Fezoui Y, Condron MM, Lomakin A, Benedek GB, Selkoe DJ, Teplow DB (1999) Amyloid β-protein fibrillogenesis. Structure and biological activity of protofibrillar intermediates. J Biol Chem 274:25945–25952

Walsh DM, Klyubin I, Shankar GM, Townsend M, Fadeeva JV, Betts V, Podlisny MB, Cleary JP, Ashe KH, Rowan MJ, Selkoe DJ (2005) The role of cell-derived oligomers of Aβ in Alzheimer's disease and avenues for therapeutic intervention. Biochem Soc Trans 33:1087–1090

Wang HW, Pasternak JF, Kuo H, Ristic H, Lambert MP, Chromy B, Viola KL, Klein WL, Stine WB, Krafft GA, Trommer BL (2002) Soluble oligomers of β amyloid (1–42) inhibit long-term potentiation but not long-term depression in rat dentate gyrus. Brain Res 924:133–140

Westergard L, Christensen HM, Harris DA (2007) The cellular prion protein (PrP^C): its physiological function and role in disease. Biochim Biophys Acta 1772:629–644

Yan J, Fu X, Ge F, Zhang B, Yao J, Zhang H, Qian J, Tomozawa H, Naiki H, Sawashita J, Mori M, Higuchi K (2007) Cross-seeding and cross-competition in mouse apolipoprotein A-II amyloid fibrils and protein A amyloid fibrils. Am J Pathol 171:172–180

Chapter 11
When More Is Not Better: Expanded Polyglutamine Domains in Neurodegenerative Disease

Regina M. Murphy, Robert H. Walters, Matthew D. Tobelmann, and Joseph P. Bernacki

Abstract Expanded polyglutamine domains are implicated in nine progressive neurodegenerative disorders, of which Huntington's disease is the best known. Nine proteins are associated with these disorders; the proteins share no sequence or structural homology except for a polyglutamine tract that is unusually lengthened in those affected by the diseases. A common feature of these disorders is the presence of inclusions containing the associated protein in the nucleus or cytoplasm of affected neurons. Why do expanded polyglutamine proteins cause neurodegenerative disease? This remains an unsettled question, but most researchers now suspect that long polyglutamine domains cause protein misfolding and aggregation, and that there is a gain-of-toxic-function upon aggregation. Two specific hypotheses attribute toxicity to sequestration of transcription factors and subsequent transcriptome dysregulation, or to overloading and subsequent impairment of the ubiquitin–proteasome system, leading to loss of protein homeostasis. Most but not all evidence points to aggregation as an essential component of toxicity; many researchers suggest that less-ordered, smaller, and/or soluble intermediates are the toxic species. Detailed biophysical studies with synthetic peptides and proteins have uncovered many details about the conformation of polyglutamine domains, and the structure of aggregates, but much remains to be done. Significant evidence points to the important role of protein context on modulating the effect of polyglutamine. Although early studies with model peptides suggested that aggregation occurred by a simple nucleation–elongation mechanism, more recent studies clearly indicate that a more complex pathway is involved. There are a few other expanded poly-amino-acid domains associated with disease, particularly polyalanine, although these are much less common than polyglutamine-related disorders, and even less is known about

R.M. Murphy (✉) • R.H. Walters • M.D. Tobelmann • J.P. Bernacki
University of Wisconsin, 1415 Engineering Drive, Madison, WI 53706, USA
e-mail: regina@engr.wisc.edu

F. Rahimi and G. Bitan (eds.), *Non-fibrillar Amyloidogenic Protein Assemblies—Common Cytotoxins Underlying Degenerative Diseases*, DOI 10.1007/978-94-007-2774-8_11,

them. A comparison of expanded polyglutamine and expanded polyalanine may illuminate their common features as well as those features unique to each.

Keywords Homo-amino-acid repeat tracts • Polyalanine • Polyglutamine • Trinucleotide repeats

11.1 Trinucleotide Repeats

Trinucleotide repeat elements are commonplace in eukaryotic genomes (Richard et al. 2008). When in coding regions, the trinucleotide repeats are translated into proteins with homo-amino-acid repeat tracts, typically defined as at least 5–7 identical amino acids in a row. These repeat tracts are not unusual. A search of over 2.5 million protein sequences deposited in GENPEPT revealed 1.4% of proteins are repeat-containing proteins (RCPs) (Faux et al. 2005), and for human proteins the frequency is estimated at 3% (Faux et al. 2007). This number may underestimate the true frequency because of bias in selection of proteins in the database: one recent survey of the draft human genome identified a remarkable 20% of genes as containing one or more homo-amino-acid repeat tract (Karlin et al. 2002).

The amino-acid composition of repeat tracts is skewed relative to the typical amino-acid composition of all proteins (Table 11.1). Aromatic (F, W, Y) and hydrophobic (I, V, M, L) residues are sparse; cationic residues (K, R) are also underrepresented. This skewness becomes more apparent as the length of the repeat tract increases. The frequency of short (5–9 residues) repeat tracts generally reflects the overall amino-acid composition; but among longer (10 or more residues) repeat tracts, glutamine is by far the most common despite its relative overall rarity in proteins (Green and Wang 1994). Interestingly, polyasparagine is common in eukaryotes but rare in humans; this could be related to the fact that asparagine is a glycosylation site whereas glutamine is not. Within eukaryotes, a striking 24% of all RCPs have multiple repeats (Faux et al. 2005). Of those human proteins with multiple repeat tracts, proline is the dominant residue (Karlin et al. 2002); it has been hypothesized that polyproline (polyP) tracts may counteract the conformational effects of other poly-amino-acid domains (Siwach et al. 2009). Multiple repeat tracts are much less common in prokaryotes than in eukaryotes (Faux et al. 2005), leading to the suggestion that multiple repeat tracts may be associated with complex brain development or could be important in embryogenesis or neurogenesis (Karlin et al. 2002).

The range of lengths of poly-amino-acid tracts in proteins varies considerably. Hydrophobic or other rare repeats are generally short (<15 residues), whereas N, Q, T, S, and E exhibit a broad range of lengths, even above 100 residues (Faux et al. 2005). Most repeat tracts in a given protein are identical in length, but a few are polymorphic (size-variant), particularly polyQ (Siwach et al. 2009). Whether or not repeat tracts serve a unique structural or functional role is not known. In proteins, repeat sequences are more likely to be disordered rather than folded (Huntley and

Table 11.1 Distribution of amino acids in repeat tracts compared to overall protein composition

Amino acid	Q	N	A	S	G	E	P	T	K	D	L	H	R	F	V	Y	C	I	M	W
Occurrence in repeat tracts (%)[a]	16.5	14.2	11.2	11.1	10.3	8.9	8.5	5.1	3.9	3.2	3.0	2.2	0.9	0.4	0.2	0.08	0.08	0.07	0.04	0.006
Typical protein composition (%)[b]	3.9	4.4	7.8	6.8	6.7	6.0	4.9	5.7	5.9	5.1	9.3	2.2	5.8	3.9	6.7	3.2	1.8	5.7	2.6	1.4
Ratio	4.2	3.2	1.4	1.6	1.5	1.5	1.7	0.9	0.7	0.6	0.3	1.0	0.2	0.1	0.03	0.03	0.06	0.01	0.02	0.004

[a]Determined from analysis of eukaryotic proteins listed in GENPEPT. % is the percent of all identified repeats that are composed of the indicated amino acids. Runs are 7 or more amino acids (From Faux et al. 2007)

[b]The frequency of appearance of each amino acid in all proteins. Determined from analysis of 2,912 protein sequences (From White 1994)

Golding 2002), leading to the speculation that repeat tracts serve as flexible linkers between two folded domains (Karlin et al. 2002; Faux et al. 2005). Many RCPs are involved in transcription, translation, and/or assembly of large macromolecular complexes (Faux et al. 2005).

11.1.1 Expanded Trinucleotide Repeats and Disease

More than half of all homopeptide sequences are translated from homocodons, making for relatively facile expansion (and contraction) of trinucleotide repeats during DNA replication and repair. Because not all homopeptides are entirely homogeneous in codon makeup, both strand slippage and point-mutation are likely contributors to expansion (Faux et al. 2007). Repeat length instability can be at both meiotic and somatic levels, and repeat length may vary between organs in a single affected individual (Kahlem and Djian 2000). Dynamic mutation can occur with any repeat sequence, wherein the rate of increase in codon number increases as the length increases (Richards and Sutherland 1997); in other words, an expanded homopeptide will tend to expand further and faster than a "normal–length" homopeptide. One intriguing hypothesis is that DNA-repair proteins may cause expansion when they are active in cells that are not dividing (Sinden 2001). This hypothesis could explain the observation that expansion occurs more frequently in cells, like neurons, that do not normally divide.

Abnormal expansion or contraction of homo-amino-acid tracts in RCPs has been linked to a variety of diseases, including cancers and neurodegenerative disorders (Karlin et al. 2002). Interestingly, there is evidence that older people have decreased repeat-tract lengths genome-wide, suggesting that less is better, and that people with, on average, relatively shorter repeat tracts will live longer (Richards and Sutherland 1997).

11.1.1.1 Expanded Polyglutamine and Disease

Polyglutamine (polyQ) is both meiotically and somatically unstable (Brown and Brown 2004) but only the homogenous CAG repeats have somatic instability (Gray et al. 2008). Expanded polyglutamine domains are now implicated in nine progressive neurodegenerative diseases (Table 11.2), of which the best known is Huntington's disease (HD). Each disease is associated with a unique protein and unique symptoms, but in all cases expansion of the polyglutamine tract beyond the normal length is considered to be the proximal cause of disease. For example, the HD protein, huntingtin (htt), commonly contains a stretch of 17–30 glutamine residues; 27–35 is rare and not disease-associated but is meiotically unstable (Imarisio et al. 2008). Although it is common to see reference in the literature to a "critical" length of 36–40 residues in polyglutamine stretches, perusal of Table 11.2 shows that this is

Table 11.2 Expanded-polyglutamine diseases [From Bauer and Nukina (2009) and Gatchel and Zoghbi (2005)]

Disease	Affected protein	Function	No. of residues in protein	Normal polyQ length	Pathological polyQ length
Huntington's disease (HD)	Huntingtin (htt)	Signaling, transcription	3,144 (Q23)	6–35	36–121
Kennedy's disease, or SBMA (spinal bulbar muscular atrophy)	Androgen receptor (AR)	Steroid hormone receptor	917 (Q20)	6–36	38–62
DRPLA, dentatorubro-pallidoluysian atrophy	Atrophin-1	Transcription	1,191 (Q20)	3–38	49–88
SCA1 (spinocerebellar ataxia)	Ataxin-1	Transcription	815 (Q27)	6–39	41–83
SCA2	Ataxin-2	RNA metabolism	1,313 (Q23)	14–32	34–77
Machado-Joseph Disease (MJD, or SCA3)	Ataxin-3	De-ubiquitylating activity	361 (Q13)	12–40	62–86
SCA6	CACNA1A	Calcium channel subunit	2,506 (Q13)	4–18	21–30
SCA7	Ataxin-7	Transcription	892 (Q10)	7–18	38–200
SCA17	TATA-binding protein	Transcription	339 (Q40)	25–43	45–63

an oversimplification. Pathological polyQ lengths range from 21 (in SCA6) to over 60 (in Machado–Joseph disease).

Whether there is an evolutionary advantage for longer polyQ domains is an open question. PolyQ domains in proteins appear to serve as regulators of transcriptional activity, as seen most clearly with the androgen receptor (AR). AR contains multiple polyQ tracts, which all suppress transcriptional activity (Harada et al. 2010); deletion of any one of these tracts leads to higher transcriptional activity. Shorter-than-normal polyglutamine domains have been associated with prostate cancer (Harada et al. 2010). Longer domains, by suppressing transcription, could serve as an anti-cancer mechanism.

There is strong epidemiological evidence correlating the length of the polyQ domain with the age of onset of the disease (Gusella and MacDonald 2000). For example, symptoms of Huntington's disease are typically observed around midlife but can occur as early as adolescence at the high end of the polyQ length range; it is estimated that about 70% of the variance in age of onset is attributable to the length of the domain (Imarisio et al. 2008). In addition, the length of the normal allele appears to affect the disease severity and age of onset in a complex pattern (Aziz et al. 2009). For HD patients with relatively short mutant alleles (e.g., 40), the age of onset was later, and the severity of disease was less, if the normal allele was also short (e.g., 10). However, if the mutant allele was longer (e.g., 60), then a longer normal allele (e.g., 30) delayed age of onset and reduced severity relative to a shorter normal allele. The reason for this transition is unknown but suggests that interactions between mutant and normal htt are length-dependent and can be both beneficial (if the normal htt reduces toxicity of mutant) or detrimental (if the mutant htt inhibits normal htt function) (Aziz et al. 2009).

Studies in transgenic animals also demonstrate the length- and age-dependence of disease. As just one of many examples, in rats infected with a lentiviral vector with the *N*-terminal 171 residues of htt (exon 1 of htt, or httex1) containing either Q18 or Q82, there was greater loss of striatal neurons with the longer construct, and greater neuronal death in older compared to younger rats (Diguet et al. 2009). A curious feature of many of the animal studies is that toxicity requires polyQ lengths much greater than 35–40, the putative "critical length" for human disease. For example, in a mouse model, a Q76 expansion in atrophin-1 led to some neuronal inclusions that increased with age but no neurodegeneration, while a Q129 expansion produced massive intraneuronal accumulation and brain atrophy (Sato et al. 2009). Androgen receptor with Q113 was needed to observe symptoms similar to Kennedy disease in mice; no disease was observed if the expansion was limited to Q48 (Lieberman and Robins 2008) even though in humans the normal length is Q20 and disease symptoms will be observed with 40 glutamines (Harada et al. 2010). The reason for this difference between animal and human is not known. It could be simply due to the inverse correlation between age-of-onset and Q length: the animal experiments are not long enough. Or, human proteins could be more sensitive to changes in structure or stability due to expansion of the polyQ region, or human neurons could be more susceptible to insults.

Table 11.3 Polyalanine-associated diseases (From Brown and Brown 2004)

Disease	Affected protein	Normal polyA length	Pathological polyA length
Synpolydactyly type II	Transcription factor	15	22–29
Cleidocranial dysplasia	Transcription factor	17	27
Oculopharyngeal muscular dystrophy (OPMD)	Polyadenylate-binding protein	10	11–17
Holoprosencephaly (HPE5)	Transcription factor	15	25
Hand-foot-genital syndrome	Transcription factor	18	24–26
Blepharophimosis, ptosis and epicanthus inversus	Transcription factor	14	22–24
X-linked mental retardation	Transcription factor	15	26
X-linked infantile spasm syndrome, Partington syndrome, lissencephaly	Transcription factor	12–16	18–23
Congenital central hypoventilation syndrome/Ondine curse	Transcription factor	20	25–29

11.1.1.2 Expanded Polyalanine and Disease

As a useful point of comparison to polyQ, we will very briefly discuss polyalanine (polyA) tracts, which have been identified in almost 500 human proteins (Amiel et al. 2004). Besides the difference in hydrophobicity, polyA differs from polyQ in several important ways. First, polyA is generally not homogeneous in codon usage, so replication slippage may not explain expansion in all cases; unequal crossover between mispaired alleles has been proposed as a mechanism (Brown and Brown 2004; Messaed and Rouleau 2009). Second, polyA tracts are generally not polymorphic, and expanded polyA is stable whereas polyQ is meiotically and somatically unstable (Brown and Brown 2004). Third, polyA tracts are shorter than polyQ: the longest normal polyA domain contains 20 residues (Albrecht and Mundlos 2005; Amiel et al. 2004).

Nine rare human diseases have been linked to expanded polyA (Table 11.3). Only oculopharyngeal muscular dystrophy (OPMD) is a progressive late-onset disease (Brown and Brown 2004), whereas all the polyQ diseases are progressive degenerative disorders. However, in OPMD it is muscles rather than neurons that die (Brown and Brown 2004). As with polyQ, the severity of the disease tends to increase as the length of the polyA tract increases. The other eight polyA expansions occur in transcription factors and cause congenital malformations and birth defects (Brown and Brown 2004; Albrecht and Mundlos 2005; Amiel et al. 2004). This suggests perhaps that proteins can more readily accommodate expanded polyQ than polyA, and/or the consequences of expanded polyA are more severe.

It has been suggested that polyA and polyQ diseases may be linked. A frameshift from CAG to GCA would result in polyA expression. There is indirect evidence that frameshifts do occur, with the frequency of frameshifts increasing with increasing CAG-repeat length (Toulouse et al. 2005; Gaspar et al. 2000; Davies and Rubinsztein 2006). CAA-repeat-based polyQ proteins were less susceptible to frameshifts and less toxic to cells, suggesting that frameshift products such as polyA might have a significant adverse effect on polyQ disorders (Toulouse et al. 2005).

11.1.2 Mechanisms of Toxicity

Why do proteins with expanded polyglutamine domains cause neurodegenerative disease, and why is age of onset and severity of symptoms related to the length of the domain? The argument surrounding this question has not yet been settled definitively: both gain-of-toxic-function and loss-of-normal-function hypotheses have been put forward, although most researchers in the field now seem to support the former idea as the predominant effect (Gatchel and Zoghbi 2005). There are several excellent and detailed review articles discussing the various hypotheses regarding mechanisms of toxicity as well as possible therapeutic approaches (e.g., Gatchel and Zoghbi 2005; Gusella and MacDonald 2000; Imarisio et al. 2008; Lima and Pimentel 2004; Bauer and Nukina 2009; Okazawa 2003; Orr and Zoghbi 2007; Rosas et al. 2008; Ross 2002; Usdin 2008; Fecke et al. 2009), and the reader desiring a more complete discussion is referred to these sources.

One compelling piece of evidence for the gain-of-toxic-function hypothesis was an experiment in which Q146 was introduced into a non-disease-related protein (hypoxanthine phosphoribosyltransferase); transgenic mice carrying this protein developed a typical disease phenotype of nuclear inclusions and late-onset neurodegeneration (Ordway et al. 1997). Other investigators have shown that knockout mice lacking the disease-related protein do not develop the neurodegeneration typical of the polyQ phenotype (Matilla et al. 1998). Further complicating the issue is that there is generally no correlation between protein expression and the site of damage: polyQ proteins are expressed broadly, but toxicity is restricted to specific neuronal subtypes, unique to each disease.

Nuclear and/or cytoplasmic inclusions of polyQ-containing proteins are commonplace features of the disease. The aggregates are usually detected in the affected neurons (Bauer and Nukina 2009), although there are a few exceptions, and the inclusions typically also contain chaperones and components of the ubiquitin–proteasome system. Thus, a common hypothesis is that these are protein-conformation disorders: the abnormally long polyQ domain causes conformational changes in the host protein, or prevents folding to the native structure (Orr and Zoghbi 2007). Misfolded proteins could lose normal functions, or could form toxic aggregates. Other proteins could be recruited into these aggregates (Raspe et al. 2009), and thus be prevented from normal functioning. Supporting the protein-misfolding hypothesis are several studies showing that chaperone proteins reduce aggregation, accumulation, and toxicity (Bauer and Nukina 2009; Muchowski et al. 2000).

One hypothesis is that the misfolding and aggregation of mutant proteins lead to overloading and subsequent impairment of the ubiquitin–proteasome system, leading to loss of protein homeostasis. For example, Bonini and coworkers have shown a strong link between ubiquitin activity and polyQ-dependent neurodegeneration (Warrick et al. 2005). There is evidence that polyQ protein aggregates become ensnared in proteasomes (Holmberg et al. 2004), inhibiting their normal proteodegrading function. Mutant htt expressed in both PC-12 and yeast cells was observed to cause strong impairment of ER-associated protein degradation that was Q-length dependent (Duennwald and Lindquist 2008). Others have suggested that expanded polyQ proteins disrupt calcium homeostasis; for example, Zhang et al. observed disrupted Ca^{2+} signaling and increased toxic sensitivity to glutamate in mice expressing htt with Q120 (Zhang et al. 2008). Mitochondrial dysfunction and oxidative injury are widespread, but are most likely downstream effects subsequent to the primary cause of disease (Imarisio et al. 2008; Rosas et al. 2008).

Other researchers believe that different mechanisms of toxicity are in play. Several of the proteins listed in Tables 11.2 and 11.3 are transcription factors, and many transcriptional regulators such as CREB-binding protein and Sp-1 contain glutamine-rich domains (Imarisio et al. 2008). Thus, one reasonable hypothesis is that mutant proteins with expanded polyQ tracts interact with these regulators, altering transcription in adverse ways. There is good experimental evidence indicating down-regulation of gene expression in animal models of Huntington's disease (Diguet et al. 2009) and DRPLA (Sato et al. 2009). Runne et al. expressed htt171 mutants with variable-length polyQ domains in primary striatal neurons and observed widespread changes in the transcriptome that were time- and length-dependent (Runne et al. 2008). The activity of androgen receptor, a ligand-activated nuclear receptor transcription factor, is inversely related to the Q length even within the normal range; shortened polyQ domains are associated with prostate neoplasia (Lieberman and Robins 2008). Expanded polyQ domains could lead to transcriptome dysregulation through changes in activities of DNA-binding transcription factors, or sequestration of transcription factors in aggregates.

Currently available therapies for the expanded polyglutamine diseases treat only symptoms. Proposed treatment approaches are diverse, and include RNAi to prevent production of mutant protein; compounds that interfere with aggregation, remove aggregates faster, or prevent interaction with transcription factors; or drugs to prevent the associated mitochondrial dysfunction (Fecke et al. 2009).

Even less is known about the cause of pathology in the expanded polyalanine disorders. Like polyglutamine, aggregation of the affected proteins is believed to contribute to the disease. At one point the favored hypothesis invoked loss of normal function of the affected transcription factors due to aggregation (Amiel et al. 2004; Albrecht and Mundlos 2005; Brown and Brown 2004). More recent evidence suggests instead that the affected proteins retain normal function but gain toxic function (Messaed and Rouleau 2009), similar to the expanded polyQ diseases and other aggregation-related disorders. Still, there are likely to be some fundamental differences in disease mechanisms with expanded polyalanine compared to polyglutamine, because all of the polyQ disorders are progressive and late-onset, whereas only one of the polyA diseases falls into this category.

11.1.3 Is Aggregation Necessary for Toxicity?

Nuclear and cytoplasmic inclusions of expanded polyQ proteins in neurons are commonplace features of these diseases. Their presence raises the obvious question: are aggregates necessary—and sufficient—for disease? Numerous experiments have attempted to answer this question, and the evidence correlating the presence of inclusions with toxicity is substantial, but not universal.

Some convincing data that aggregates are early disease markers come from studies of presymptomatic patients with the mutant huntingtin gene who die from other causes. In these patients, aggregates that contain chaperones, htt, transcription factors, and proteasomal proteins were present, and there was evidence of early abnormal morphological changes in striatal and cortical pyramidal neurons before degeneration (Rosas et al. 2008).

Taking a very different approach, Siwach et al. constructed GFP mutants fused to *C*-terminal poly-amino-acids A20, D20, L20, P20, Q20, or S20. Cells transfected with A20, D20, or L20 mutants died, whereas those transfected with P20, Q20, or S20 mutants did not (Siwach et al. 2009). Toxicity correlated strongly with the presence of cytoplasmic aggregates. Interestingly, homo-amino-acid-repeat tracts placed in truncated GFP were more toxic than those placed in full-length protein, possibly because they comprise a larger percent of the polypeptide and therefore have a larger effect on conformation (Siwach et al. 2009). These experiments argue in favor of a hypothesis that aggregates *per se* are toxic, independent of the amino-acid composition of the expanded domain.

It may be that inclusions are necessary but must be present in large quantities to cause symptoms. Zhang et al. (2008) saw nuclear inclusions in mice expressing either full-length htt or htt fragments (both with Q120), but cellular dysfunction was observed only in mice with full-length htt. Sato et al. (2009) worked with transgenic mice with Q76 and Q129 expansions in atrophin-1. Neuronal intranuclear inclusions were present in both the Q76 and Q129 mice that increased with age, but the mass of inclusions was much greater with the longer construct. Only the Q129 mice suffered the progressive neurological phenotype reminiscent of DRPLA. Transcriptional dysregulation and brain atrophy, but no obvious neuronal loss, were observed. This group concluded that polyQ diseases are neuronal dysfunction without neuronal death.

On the other hand, several reports show lack of correlation between aggregation and toxicity. In one of the earliest studies demonstrating this idea, Saudou et al. showed that blocking nuclear localization of mutant htt suppressed nuclear inclusion formation but did not prevent toxicity (Saudou et al. 1998); in fact, death was accelerated in this cellular model. Takahashi et al. demonstrated that a soluble form of expanded polyQ repressed transcription before formation of macroscopic aggregates, and that formation of large aggregates was not required for transcriptional repression (Takahashi et al. 2005). In experiments with polyA fused to fluorescent proteins, aggregates were observed with longer (A70) but not shorter (A29) sequences, but toxicity was seen with both lengths, thus decoupling aggregation from cellular toxicity (Toriumi et al. 2008, 2009). Transgenic HD mice had some pathology without evidence of aggregated htt, as detected by antibody staining (Gray et al. 2008). In these mice, large inclusions were detected months later, after the onset of symptoms. The researchers concluded that neuronal dysfunction precedes degeneration. These data might lead to the hypothesis that inclusions are associated with degeneration and are a product, not a cause, of pathology.

The specific nature of the aggregates likely influences their toxicity, and several lines of evidence point to the role of smaller, less-ordered, and/or soluble oligomers as the toxic species. In fact, nuclear inclusions may be protective (Saudou et al. 1998). For example, working in yeast cells, Behrends and colleagues showed that overexpression of a chaperone protein altered the morphology of htt-exon1-Q53 oligomers and prevented toxicity (Behrends et al. 2006). Similarly, Wacker et al. demonstrated that heat-shock proteins reduce formation of soluble oligomers, and reduce toxicity, without affecting formation of mature fibrils (Wacker et al. 2004). As an interesting aside, low-level expression of a polyalanine protein triggered a heat-shock response in *Drosophila* and protected against polyglutamine toxicity (Berger et al. 2006). In further support of the need for specific morphologies, Nekooki-Machida et al. produced alternate fibrillar morphologies of htt-exon1-Q62 fibrils by aggregating at different temperatures: fibrils grown at 4°C were less ordered and more toxic than those grown at 37°C (Nekooki-Machida et al. 2009). In a study using androgen-receptor fragments with expanded polyQ domains, the greatest toxicity was observed in those proteins that formed large soluble oligomers (100–600 kDa) (Schiffer et al. 2008). Diguet et al. created a rat model of HD by injecting a lentiviral vector encoding htt-171-82Q into young and old rats (Diguet et al. 2009). Older rats had more, but smaller, inclusions and suffered greater neuron death. The authors speculated that older rats were less able to sequester mutant htt in large, non-toxic inclusions.

In a clever study, Takahashi et al. tested expanded polyQ-containing mutants of both htt and truncated atrophin-1 in SH-SY5Y cells, and used FRET confocal microscopy to distinguish oligomers from monomers and inclusion bodies (Takahashi et al. 2008). More oligomers and more inclusion bodies were observed with the longer Q stretches. Importantly, they observed more cell death in cells with oligomers, than with monomers or inclusion bodies. Arrasate et al. transiently infected striatal neurons with exon 1 htt with expanded Q linked to GFP (Arrasate et al. 2004). Inclusion bodies increased in both number and size over time in a Q-length-dependent manner. There were twice as many inclusion bodies in the Q103 mutant than in the Q47 mutant, but the death risk was not much different. These researchers suggest that inclusion bodies were not required for cell death. Rather, diffuse aggregates correlated strongly with cell death while inclusion-body formation reduced diffuse aggregates and extended cell survival (Arrasate et al. 2004). Olshina et al. also used GFP-htt exon 1 constructs and were able to track aggregate size over time in mammalian cells using fluorescent-adapted sedimentation velocity (Olshina et al. 2010). They detected three populations: monomer, soluble oligomers, and insoluble inclusion bodies, and they demonstrated that the cells were exposed to a constant amount of soluble oligomers. They further showed that a chaperone protein increased the rate of conversion of oligomers to inclusion bodies (Olshina et al. 2010), a result suggesting that chaperone-mediated protection is afforded through a decrease in the toxic oligomer population and a concomitant rise in inclusion-body formation.

Some studies argue that a specific conformation is required for cell death. Poirier et al. constructed htt exon1 mutants with a modified polyQ domain engineered to fold into a β-turn/β-sheet conformation; these constructs were highly toxic (Poirier et al. 2005). One study went further and proposed that there is indeed a toxic monomeric conformer. Nagai et al. designed thioredoxin–polyQ fusion proteins with variable number of Q residues (Nagai et al. 2007). The freshly prepared material, with a strong α-helical content, was not toxic, whereas soluble material containing β-sheet monomers and oligomers, as well as the fibrils, was toxic.

Taken in its entirety, this body of work points to the role of immature aggregates in triggering neuronal dysfunction. Such aggregates may be difficult to detect, and likely co-exist with monomers, fibrils, and/or inclusion bodies. The existence of these aggregates has only been recognized fairly recently, and they may explain the discrepancies in studies, which argue for or against the necessity of inclusion bodies for toxicity. The tentative answer to the question "Is aggregation necessary for toxicity?" is "Yes, but it depends on the nature of the aggregates." Assessing the conformation, size, and morphology of aggregates in living cells or in animals is a daunting technical challenge. Yet, such an assessment is what is needed to definitively answer the question as to whether, and which kind, of aggregates are essential for toxicity. There are several very recent contributions that demonstrate technical advances in tracking aggregates in cells (Ramdzan et al. 2010; Ossato et al. 2010; Sathasivam et al. 2010; Olshina et al. 2010). Data using these and similar methods should lead to a definitive answer to this important question.

11.2 Conformation and Conversion

Given the clear roles of expanded polyglutamine and polyalanine in multiple diseases, and the likely role of aggregation, many research groups have begun a detailed inquiry into the conformation of the expanded poly-amino-acid domain, the effect of that expansion on the structure and stability of the remainder of the protein, the kinetics of conversion to aggregated species, and the structure and morphology of the resultant aggregates. For biophysical studies, researchers have chosen to work with either synthetic peptides or host proteins. We will review in some detail the history and the current understanding (and controversies).

11.2.1 Synthetic Peptides for Biophysical Studies

Synthetic polyQ and polyA peptides have been utilized as model systems in multiple investigations. The advantages are several. The researcher has full control over the design of the peptide, its length, and any flanking residues. Unique residues including non-native amino acids or fluorescent tags can be engineered into the peptide to probe specific questions. Considerable insight into the role of homo-amino-acid length has been gleaned from these studies. Another advantage with synthetic peptides is that the homo-amino-acid domain is isolated from any contributions imposed by the protein. But this is also the disadvantage: the material is taken out of the context of a real protein. Recent work with synthetic peptides highlights the critical role of context in modulating conformation and aggregation.

11.2.1.1 Polyglutamine Monomer Conformation

Characterization of polyQ peptide monomer has not been without its challenges. Noting the lack of regular structural elements in CD spectra, Altschuler et al. concluded that monomeric polyglutamine is a random coil (Altschuler et al. 1997). Results from NMR spectra led to a similar conclusion: polyQ is disordered (Klein et al. 2007). This group further made the important observation that there was no length-dependent change in solution structure conformation. A closer examination of CD spectra led Chelgren et al. to conclude that, although mostly disordered, there was some evidence for extended polyP II helical content in polyQ peptides that decreased with increasing length (Chellgren et al. 2006).

Although lacking regular secondary structural features, polyQ monomers are likely not true random coils, as a polymer physicist would define the term. Rather the monomers could be either extended or collapsed coils. To probe this question, Singh and Lapidus used polyQ peptides of the type $K_2CQ_xWK_2$ where X equals 4–16, and a tryptophan quenching technique. They concluded that polyQ monomers adopt unusually extended, rigid conformations in aqueous solutions (Singh and Lapidus 2008). This result suggests that the amide side-chain facilitates strong solvation of the peptide in water. However, Crick et al. came to the opposite conclusion—polyQ peptides are very collapsed (Crick et al. 2006). This conclusion was reached based on rotational diffusion measurements using fluorescence correlation spectroscopy, with peptides of the type GQ_xCK_2 where X equals 15–53. The researchers concluded that water is a poor solvent for polyQ peptides, resulting in the peptides adopting a heterogeneous collection of collapsed structures. Further evidence that polyQ peptides adopt collapsed structures was provided in a study by Dougan et al. in which single-molecule force-clamp spectroscopy was used to study a polyQ region containing 15–75 glutamine residues attached to the protein I27 titin (Dougan et al. 2009). Although the folded structure of the protein was found to unfold and expand when pulled at 180 pN, the polyQ regions resisted expansion above 900 pN, suggesting the preference for collapsed conformations is extremely strong. Work in our laboratory may help resolve some of the apparent conflicts regarding whether polyQ peptides adopt extended or collapsed conformations (Walters and Murphy 2009). Using peptides of the type $K_2WQ_xAK_2$ with X equaling 8–24, we measured intramolecular end-to-end distances using FRET. We found that peptides with fewer than 16 residues were relatively extended and well solvated, peptides with more than 16 residues preferred collapsed structures, and that water acted as a theta solvent with Q16. Furthermore, our data demonstrated that the flanking lysine residues strongly affected the conformation: neutralizing the charge led to reduction in the end-to-end distance, with the greatest reduction observed with the longer peptides (Walters and Murphy 2009).

Theoreticians have explored the question of polyQ conformation. Using realistic force fields and explicit solvents in molecular dynamics simulations, two research groups concluded that long polyQ peptides collapse into a flattened β-helix, with the stability increasing with increased length (Merlino et al. 2006; Zanuy et al. 2006). However, a cautionary note regarding these studies is needed: to be computationally tractable the groups needed to start with *a priori* assumption of a β-helical structure (Perutz et al. 2002); thus raising the concern that the initialization dictated the outcome. Other groups have sacrificed resolution to gain computational tractability, using simplified atom representations and potentials, and implicit solvent models. Marchut and Hall used these techniques to conclude that isolated polyQ peptides exist as random coils or as collapsed coils with partial intramolecular hydrogen bonding but no regular secondary structure, depending on temperature, solvent, and Q length (Marchut and Hall 2006a, b, 2007). Pappu and coworkers have completed extensive molecular dynamic simulations to obtain conformational ensembles of polyQ peptides; these researchers concluded that water is a poor solvent for polyQ, and that polyQ attains a conformational ensemble of collapsed coils that resemble "loosely packed globules" lacking regular secondary structure (Vitalis et al. 2007, 2008; Tran et al. 2008). Recent theoretical studies by the same group demonstrated that the total number of intramolecular hydrogen bonds increased with increasing chain length, but these peptides retained a disordered rather than a regular β-sheet structure (Vitalis et al. 2009).

Together, experimental data support a picture of polyQ monomers as structurally disordered, populating a distribution of conformations, and becoming increasingly collapsed as the length increases or as repulsive interactions in the flanking residues decrease. The theoretical results are consistent with the picture emerging from experimental data, and explain collapse with increasing Q length as facilitated by formation of multiple intramolecular hydrogen bonds. Because the side-chain of glutamine is an amide, the hydrogen bonding possibilities are many—side-chain–side-chain, main-chain–main-chain, and side-chain–main-chain.

11.2.1.2 Structure and Morphology of Polyglutamine Peptide Aggregates

PolyQ peptides will spontaneously assemble into fibrillar aggregates, as first demonstrated by Perutz (1994). These aggregates bind thioflavin T and will precipitate out of solution, but do not show the Congo-red birefringence typical of conventional amyloid fibrils (Chen et al. 2002a). Mature polyQ peptide aggregates are known to contain a large amount of β-sheet content, but there is still little data as to the detailed structure of the aggregates. Perutz originally suggested a polar zipper model, based largely on X-ray diffraction data (Perutz et al. 1994), but later reanalyzed the data and concluded that the aggregates were water-filled nanotubes, with polyQ folding into a single-sheet β-helix of 3-nm diameter with 20 residues per turn and side-chains pointing alternately into and out of the cylinder (Perutz et al. 2002). Using the same X-ray diffraction data, Sikorski and Atkins concluded that the most likely structure of the aggregates was antiparallel β-sheets, with side-chain–side-chain hydrogen bonding causing tighter-than-normal inter-sheet compaction (Sikorski and Atkins 2005). The fact that one set of data can result in three distinctly different interpretations nicely illustrates the difficulty in achieving a consensus structure of the aggregates! Independent X-ray diffraction examination of peptides containing 8–45 glutamine residues concurred with the Sikorski and Atkins analysis, that the aggregates were β-sheet slabs, and further established that the longer peptides fold *via* multiple reverse turns (Sharma et al. 2005).

Depending on the length of the polyQ domain as well as solvent conditions, thin linear aggregates, ribbon-like assemblies of these aggregates, and bundles of fibrils have been observed in TEM images (Lee et al. 2007). In one study, after a few hours of incubation, solutions of Q20 and Q24 contained loose clusters of linear aggregates; the Q20 clusters contained short, poorly defined chains whereas Q24 were more clearly defined and longer. Over several days these matured into more clearly fibrillar aggregates with well-defined laterally aligned rods (Walters and Murphy 2009). Consistent with these reports, in a recent AFM study, Legleiter et al. showed that a peptide as small as Q7 could form small oligomers and short, metastable fibrils, while Q23 assembled into a mix of small oligomers, fibrils, and fibril bundles (Legleiter et al. 2010).

Synthetic polyQ peptides will also assemble into soluble aggregates. Laser light scattering demonstrated the presence of soluble aggregates in solutions of polyQ peptides containing 20 or more glutamine residues before onset of sedimentation (Lee et al. 2007; Walters and Murphy 2009). These soluble aggregates did not have any regular secondary structure, distinguishing them from mature fibrils. From detailed analysis of data from multiangle laser light scattering, Lee et al. concluded that these soluble aggregates formed rapidly and then increased in size slowly over time; the aggregates were semiflexible chains, with characteristic lengths on the order of 1 μm and diameters ranging from 4 to 10 nm (Lee et al. 2007).

If computational studies of polyQ monomers is challenging, then simulation of aggregation is an even more daunting task. Still, some progress has been made. An intermediate-resolution algorithm was developed specifically to simulate polypeptide aggregation; the method relies on a simplified pseudo-atom representation and empirical estimates of hydrophobic interaction and hydrogen-bond interaction energies (Marchut and Hall 2006b, 2007). Using this algorithm, a "phase diagram" was generated describing regions of amorphous aggregates, β-sheet aggregates, and disordered chains, as a function of chain length and temperature. Very recently, Pappu and coworkers used atomistic description and an implicit solvent model to simulate dimer formation from polyQ peptides (Williamson et al. 2010). They saw strong association with Q15 or longer that depended on temperature. Most experimentalists have out of necessity added flanking lysine residues when synthesizing polyQ peptides; this theoretical study illustrated that aggregation is strongly inhibited by the presence of these lysine residues (Williamson et al. 2010).

11.2.1.3 Kinetics and Pathways of Aggregation

The Wetzel group pioneered studying the kinetics of polyQ peptide aggregation. Using techniques such as thioflavin-T binding to measure fibril growth, and sedimentation to measure monomer loss, they proposed a nucleation–elongation model for polyQ peptide aggregation (Chen et al. 2002b). In this model, the nucleus is a thermodynamically unfavorable conformation of the monomer. Elongation is achieved as monomer binds to the nucleus and undergoes consolidation, providing a new binding site for more monomers. The peptides used in this study had glutamine lengths of 28–46 residues, and were flanked by 2 lysine residues on each side to confer solubility. These peptides were found by CD to lack regular secondary structure in the monomeric state, regardless of the length of the glutamine region. Aggregation was associated with an increase in β-sheet content, and a related study revealed fibrillar aggregate structures by TEM (Chen et al. 2002a). Formation of aggregates was preceded by a characteristic lag phase during which aggregates were not detected. The kinetic data at early times (10–20% aggregated) were analyzed using a simple linearized model. From these data, Chen concluded that the nucleus is a "thermodynamic monomer" in an energetically unfavorable conformation. This description of polyglutamine-peptide aggregation kinetics has been widely cited.

This analysis rests on two fundamental assumptions: first, that the model is reliable, and second, that there are only monomers present during the putative lag phase. Both of these assumptions have been challenged. First, the kinetic data were re-analyzed using a statistical model discrimination algorithm (Bernacki and Murphy 2009). This quantitative analysis demonstrated that several models other than the nucleated growth mechanism provided equally good statistical fits to the experimental data. In other words, the experimental data were insufficient for determination of a unique mechanism, and the conclusion that polyQ peptides aggregate *via* a nucleation–elongation model with a monomeric nucleus is not justifiable from a statistical point of view (Bernacki and Murphy 2009). Second, new experimental data challenged the idea that the putative lag phase of aggregation was truly devoid of aggregates (Walters and Murphy 2009; Lee et al. 2007). As described above, soluble oligomers were detected by laser light scattering in solutions containing peptides with 16 or more glutamine residues; these aggregates were present well before sedimentation onset. Soluble oligomers were detected by TEM, consisting of linear structures, which become better defined and formed laterally aligned structures over time. The presence of a mix of oligomers, fibrils, and fibril bundles in solutions of short polyQ peptides has been confirmed by AFM (Legleiter et al. 2010).

From these combined experimental and theoretical studies, a clearer picture is emerging, although some details remain obscure. Although glutamine is a polar side-chain, monomeric polyQ peptides tend to prefer collapsed conformations. This is likely driven by intramolecular hydrogen bonds, which are likely a mixture of backbone–backbone, backbone–side-chain, or side-chain–side-chain bonds. The preference for collapsed coil becomes stronger as length increases, and is weaker if there are charged flanking residues. Even well below the critical length for human disease, polyQ peptides will self-associate *via* hydrophobic interactions into soluble oligomers. For sufficiently long peptides (probably somewhere between 12 and 20 glutamine residues), a structural re-arrangement to increase β-sheet content will lead to formation of insoluble fibrillar aggregates. The structural re-arrangement could happen within the soluble oligomers, as intramolecular hydrogen bonds slowly re-form to a more thermodynamically favorable β-sheet structure and dehydration occurs. Alternatively, conformational fluctuations within the monomeric population could lead to formation of a rare thermodynamically unfavorable β-sheet monomer. In the former scenario, the soluble oligomers are "on-pathway" while in the latter they are "off-pathway". Arguments can be made in favor of either pathway, and it is possible that both pathways are operable; this remains one of the key open questions.

11.2.1.4 PolyQ Peptides in Cellular Environment

Raspe et al. utilized a clever system to study the release of pure polyQ peptides in a cellular system (Raspe et al. 2009). They produced GFP–ubiquitin–polyQ constructs, in which *C*-terminal hydrolases were used to remove the polyQ section from the rest of the protein, resulting in the expression of pure polyQ peptides with lengths from 16 to 112 glutamine residues. They found that polyQ peptides formed a core of aggregates that then recruited other glutamine-containing proteins and the ubiquitin–GFP construct. This technique provides for the study of polyQ peptides in cells without the context of an attached protein or even surrounding residues.

11.2.1.5 Making PolyQ Peptides More Like Proteins: Flanking and Interrupting Residues

All proteins associated with expanded polyglutamine diseases have multiple amino-acid repeat tracts (Karlin et al. 2002); polyP is particularly frequent in its pairing with polyglutamine. The reason for this is unknown, but there is ample evidence that flanking residues such as polyP modulate the physical properties of polyglutamine. Particular interest has been drawn towards the proline repeat on huntingtin that is *C*-terminal to the polyQ domain. The effect of polyP on polyQ has been addressed by several groups working with synthetic peptides.

Bhattacharyya et al. examined polyQ peptides with 40 glutamine residues, flanked by 10, 6, or 3 *C*-terminal proline residues, as well as peptides with 10 proline residues attached *N*-terminally or to a flexible linker (Bhattacharyya et al. 2006). Peptides containing 6 or more *C*-terminal proline residues aggregated more slowly and did not incorporate all available monomers, unlike peptides without the flanking prolines. Three proline residues, the *N*-terminal proline residues, and residues attached by a flexible linker had no effect on the aggregation kinetics, indicating the importance of not only the additional residues but also their position relative to the glutamine stretch. Mixtures of the slower aggregating, proline-containing peptides and peptides lacking proline aggregated at the same rate as if all of the peptides lacked proline. This suggests that *C*-terminal proline does not prevent the peptide from being incorporated into an already growing aggregate.

Darnell et al. examined polyQ peptides of the type $R_3GQ_xP_{11}GY$ with X equaling 3–15 glutamine residues (Darnell et al. 2007). Analysis of these peptides by CD and X-ray diffraction detected a decrease in the amount of β-sheet content of these peptides and an increase in polyP II structures in the peptides containing the *C*-terminal proline residues. This perturbation of the secondary structure reduces the aggregation potential of the peptide, as the proline-containing peptides failed to show aggregates for less than 15 glutamine residues, while peptides lacking proline aggregated with only 6 glutamine residues. The proline stretch also impacted on the morphology of the aggregates, as distinct soluble oligomers were detected in the peptides without proline, but a single, broad oligomeric peak eluted by SEC for the peptides containing 9 and 15 glutamine residues and the proline stretch. Additionally, TEM images revealed clearly fibrillar aggregates for peptides without proline, while those with proline formed amorphous or altered fibrillar aggregates with less well-defined edges. Chellgren et al. argue that this propensity to form polyP II structure is inducible in polyQ peptides (Chellgren et al. 2006). Using peptides of the type $P_3Q_xP_3GY$ where X is 1–15 glutamine residues, they detected polyP II structure using CD for each peptide up to 15 glutamine residues. The 3 *C*-terminal proline residues were not found to impact aggregation kinetics in the Bhattacharyya study (Bhattacharyya et al. 2006), but did impact the structure of these shorter peptides. Darnell et al. concluded that polyQ is conformationally adaptable, and that polyP pushes polyQ towards a PPII helix conformation and away from β-sheet (Darnell et al. 2009). This group also observed small oligomers with peptides containing as few as three glutamine residues; oligomerization was driven *via* both hydrophobic and steric effects (Darnell et al. 2009).

The threshold repeat length for polyQ diseases varies among the affected proteins, leading to the hypothesis that different flanking residues alter the stability of longer polyQ domains. To examine this question, Nozaki et al. created a polypeptide construct of 31 to 79 glutamine residues, flanked by the 8 nearest *N*-terminal residues and 9 nearest *C*-terminal residues for 4 polyQ proteins (ataxin-2, huntingtin, DRPLAP, and ataxin-3) (Nozaki et al. 2001). At the same glutamine length, huntingtin and ataxin-2 constructs aggregated more aggressively than the DRPLAP and ataxin-3 constructs, consistent with the threshold lengths of the associated diseases.

Some researchers have examined whether interrupting the glutamine stretch with other residues would impact the aggregation kinetics and morphology of the aggregates. Using Q45 peptides, Thakur and Wetzel inserted pairs of proline-glycine (PG) residues at various positions (Thakur and Wetzel 2002). All of the inserts slowed aggregation, but the decreases were less pronounced when the inserts were positioned in such a way that they could be incorporated into β-turns. Addition of individual proline residues into regions that were expected to be part of a β-sheet greatly reduced or stopped aggregation. They interpreted this result to indicate that the PG inserts will not prevent aggregation if the insert can be incorporated into a β-turn, but will inhibit aggregation when the insert interferes with the main portion of a β-sheet. Additionally, they studied a peptide containing D-proline, a template for β-turn formation. They found this peptide to aggregate more quickly, suggesting that forcing β-turn formation at the right position accelerates polyQ aggregation.

Jayaraman et al. examined the impact of histidine interruption on polyQ peptide aggregation (Jayaraman et al. 2009). This study is relevant for ataxin-1 aggregation, as histidine sometimes interrupts the glutamine region of ataxin-1. Using a Q30 peptide with a HQH insertion in the center, they found a decrease in the aggregation rate, but no major morphological changes at pH 7.5–8.5. At pH 6, near the pKa of histidine, the peptide aggregated *via* an intermediate that did not bind thioflavin T, similar to polyQ peptides without the insert. Additionally, the His-containing aggregates featured straight, thick fibrils that were unlike normal polyQ aggregates. Furthermore, histidine-interrupted aggregates were less stable.

In our lab, we studied aggregation kinetics of several interrupted polyQ peptides with Pro-Pro, Ala-Ala, or *D*-Pro-Gly inserts (Walters and Murphy 2011). With 2 proline residues, globular aggregates were detected by light scattering and TEM, but these did not develop insoluble aggregates. This result indicates that the polyQ regions of the peptides still interact despite the β-sheet-breaking proline residues, but that β-sheet formation is needed to progress to mature fibrils. Alanine interruptions had only a small effect on slowing down aggregation. The D-Pro-Gly insert, a β-turn template, greatly accelerated precipitation of the peptide. Additionally, the morphology of the aggregates changed slightly. While uninterrupted polyQ peptides mature into aggregates that show a high degree of lateral alignment between the fibrils, the D-Pro-Gly peptide failed to develop alignment between individual fibrils. As the peptides already have β-sheet structure when they associate, they do not interact as much to produce laterally aligned structures. These results suggest an aggregation mechanism in which hydrophobic collapse drives initial associations, but some β-sheet content is necessary to provide stability and provide for on-pathway aggregation. Uninterrupted polyQ peptides associate in a disordered manner, and then undergo a slow disorder-to-order transition resulting in laterally aligned fibrils with a large amount of β-sheet structure.

11.2.1.6 Polyalanine Peptides

Polyalanine-based peptides have long been of interest in fundamental studies of protein folding, since alanine is the simplest α-helix-forming residue. However, experimental studies on aggregation of polyalanine have been hampered by practical problems with difficult peptide synthesis and low solubility (Warrass et al. 2000; Heitmann et al. 2005). Blondelle and coworkers conducted a series of investigations using polyalanine peptides with flanking lysine residues (Blondelle et al. 1997). With fewer than 8 alanine residues, the peptides were monomeric and adopted an α-helical structure (Blondelle et al. 1997; Shinchuk et al. 2005). Longer peptides (A16 or greater) immediately assembled into large β-sheet-rich aggregates that were impervious to dissolution or degradation; their formation could be avoided only by interrupting the alanine repeat domain with a residue such as proline (Forood et al. 1995; Blondelle et al. 1997; Shinchuk et al. 2005). Intermediate-length peptides were a mix of monomers and aggregates, with the β-sheet-rich fraction increasing over time. Similarly, Giri et al. reported that polyalanine peptides ($AcKMA_nGY$) where $n=7$ were α-helical and monomeric, while expansion to $n=11$ led to increased β-sheet character and aggregation (Giri et al. 2003, 2007). In contrast, other investigators presented CD and NMR data on a peptide containing 7 alanine residues and argued for an extended polyP-II-like structure, whereas a slightly longer chain of 10 alanine residues was α-helical (Shi et al. 2002). The different conclusions in these studies may have arisen due to differences in flanking residues, temperature, or other experimental conditions. This is a potentially rich area for further exploration, because of the importance of expanded polyalanine in several diseases, and as a point of comparison to polyQ.

11.3 PolyQ Proteins

Synthetic peptides have provided considerable insight into the pathways of polyglutamine-driven aggregation, and the factors that modulate them. However, as is apparent from the studies on peptides with various flanking residues, the context in which the polyQ domain is placed has a major influence on the outcome. Furthermore, a central issue arises: does the expanded polyQ domain in a disease-associated protein directly trigger aggregation through specific polyQ-driven interactions, or does expansion of the domain cause changes in protein structure or stability that then trigger aggregation and changes in protein function? Work with synthetic polyglutamine implicitly postulates the former, while work with polyQ-containing proteins allows for either possibility—or a combination thereof.

The most obvious choices for proteins to study are those directly associated with diseases (Table 11.2). Working with the disease-implicated proteins offers the advantage of retaining native flanking sequences, which may be critical to aggregation characteristics, as determined from studies with synthetic peptides. Relevant structural changes caused by expansion of the polyQ domain are more readily assessed in the context of the actual protein. Finally, non-polyQ domains of the protein, even those non-contiguous with the polyQ tract, may greatly influence the overall folding and aggregation pathways. Thus, ideally one would choose a disease-related protein for study. Unfortunately, this presents particular challenges for those interested in detailed biophysical studies where one also ideally has (a) a robust expression system with reasonably good yields, (b) a method for generating mutants where the polyQ length can be easily and systematically manipulated, (c) a purification process that produces pure, monomeric, and natively folded protein, and (d) well-known structural characteristics and folding pathways to facilitate interpretation of biophysical data. With some exceptions, the disease-associated proteins do not fit the bill: structural data is scarce and the full-length proteins are generally difficult to express and purify.

11.3.1 Disease-Related PolyQ Proteins

Some success has been achieved by researchers working with disease-associated proteins. Generally speaking, one can divide up the published research on polyQ-containing proteins into (a) huntingtin fragment, (b) ataxin-3, (c) a smattering of other disease-associated proteins, and (d) model proteins that are unrelated to disease. We will review each category in turn.

11.3.1.1 Huntingtin (htt)

Huntingtin (htt) is a large (350 kDa) multi-domain protein made up of multiple HEAT domains (2 α-helices forming a helical hairpin) stacked into a rod, possessing a significant degree of conformational flexibility (Li and Li 2004; Li et al. 2006). Recombinant expression in *Escherichia coli* and purification of full-length htt has not, to our knowledge, been achieved; successful expression in insect cells has been reported, but yields were low and much protein was lost to proteolysis and aggregation (Li et al. 2006). While full-length htt has been used for *in vivo* studies, virtually all biophysical studies have employed the *N*-terminal htt exon 1 (httex1) or related fragments, which can be produced recombinantly and which contain the polyQ domain. This choice is biologically relevant as proteolytic fragments are commonly found in HD inclusions, and it has been hypothesized that proteolysis is needed for access to the nucleus and toxicity (see Bauer and Nukina 2009 for review).

Thakur et al. synthesized peptides that contained the 17-amino-acid-flanking region (Nt17) from httex1 along with the polyQ domain. Interestingly, they observed that the polyQ domain altered the conformation of the flanking region, causing it to convert from a collapsed to a more extended coil. It was the flanking regions that facilitated association into globular oligomers, which eventually matured into amyloid-like fibrils through polyQ contacts (Thakur et al. 2009). The *C*-terminal polyP sequence was found to reduce aggregation, consistent with work described earlier with synthetic peptides containing only polyQ and polyP domains. Aggregation rates increased with Q length; it was hypothesized that this was due both to a length-dependent disturbance of the collapsed coil of the flanking region and a length-dependent acceleration of conversion from a more amorphous to a more fibrillar structure. Poirier and coworkers transfected mammalian cells with httex1 constructs containing polyQ and/or proline segments (Poirier et al. 2005). Their work supported a model in which the longer polyQ domains fold into an anti-parallel β-sheet, which facilitates aggregation. However, this work was challenged by recent theoretical and experimental studies (Williamson et al. 2010), which reported that Nt17 and polyQ become increasingly disordered as Q length increases, and that the Nt17 segment suppresses polyQ aggregation. They explained the discrepancy between their results and that of Thakur et al. by pointing out that in their simulations the comparison was between Nt17-Q_n and pure Q_n, while in the experimental work of Thakur et al. the polyQ peptide contained flanking lysine residues. In other words, adding lysine residues inhibited polyQ aggregation more than Nt17.

Very recently, X-ray crystallography was used to determine the secondary structure of httex1 with a Q17 domain. The httex1 fragment contained an *N*-terminal α-helix and a polyP helix (Kim et al. 2009). The polyQ region was found to populate multiple conformations, including α-helix, disordered, and extended loop (Kim et al. 2009). This notion, that polyQ is conformationally adaptable and heterogeneous, may explain to a great extent why flanking residues and context are so important in modulating polyQ aggregation.

Several researchers have worked with larger constructs containing the full httex1 fragment. Because of its strong tendency to aggregate, fusion proteins with cleavable linkers have been developed. This approach solves the problem of generating soluble and monomeric proteins, but introduces added complexity through enzymatic kinetics and efficiencies. In one of the earliest studies, GST-httex1 fusion proteins were developed; proteolytic cleavage released httex1, which then aggregated into fibrillar materials in a Q-length-dependent manner (Scherzinger et al. 1997). A different MBP–httex1 fusion protein produced a similar result: proteolytic cleavage initiated assembly into high-molecular-weight aggregates (Poirier et al. 2002). By TEM and AFM, httex1 with expanded polyQ was shown to form spherical oligomers that eventually matured into larger ordered fibrils (Poirier et al. 2002). Oligomers and fibrils of similar morphology were observed in another study (Wacker et al. 2004), where it was also demonstrated that heat-shock proteins reduced formation of soluble oligomers by interaction with monomers.

More detailed AFM studies describing the morphology of httex1 aggregates as a function of Q-length have been published very recently (Legleiter et al. 2010). Small globular oligomers were observed at all Q lengths, whereas fibrils (defined here as aggregates of 4-nm diameter or larger, with aspect ratio greater than 2.5) were observed only with Q35 or larger. Larger amorphous aggregates were also observed. In most cases, a population of small oligomers built up early and then decayed concurrent with appearance of fibrils. Oligomers were highly heterogeneous, and the distribution of morphologies depended on both Q length and concentration. This group suggested that fibrils could arise by parallel pathways: from monomers or small multimers, or from larger spherical oligomers (Legleiter et al. 2010). Using a FRET technique, Takahashi et al. demonstrated that oligomers formed from httex1 even well below critical Q length (Q17), although the quantity of oligomers increased with increasing Q length (Takahashi et al. 2008). Their data suggested that soluble oligomers are eventually incorporated into larger insoluble inclusion bodies. Of interest, aggregates became more tightly packed and stable as Q length increased (Takahashi et al. 2008). Another important study demonstrated that all fibrils are not alike: fibrils of httex1 grown at different temperatures have different morphologies (Nekooki-Machida et al. 2009).

A question that arises is whether the morphology and structure of aggregates is the same *in vitro* as *in vivo*. Ossato et al. tagged httex1 with green fluorescent protein (GFP) and followed aggregation in cells using confocal microscopy (Ossato et al. 2010). They observed monomers, small oligomers, and inclusion bodies, in a concentration-, polyQ-length-, and cell-dependent manner, but never detected fibrils. In an important study, Sathasivam and coworkers demonstrated the presence of oligomers and aggregates in HD-knock-in mice that were morphologically similar to those prepared from recombinant htt protein (Sathasivam et al. 2010).

As described briefly in an earlier section, there are several studies that link httex1 aggregate morphology and toxicity (Nekooki-Machida et al. 2009; Takahashi et al. 2008), that demonstrate the influence of Q length and the effect of flanking residues (especially polyP) on both aggregate morphology and toxicity (Duennwald et al. 2006), and that show chaperones alter the balance between oligomers and fibrils and concurrently inhibit toxicity (Behrends et al. 2006; Wacker et al. 2004). Olshina et al. also used GFP–httex1 constructs and were able to track aggregate size over time in mammalian cells using fluorescent-adapted sedimentation velocity (Olshina et al. 2010). They detected three populations: monomer, soluble oligomers, and insoluble inclusion bodies, and demonstrated that the cells were exposed to a constant amount of soluble oligomers. They further showed that a chaperone protein increased the rate of conversion of oligomers to inclusion bodies (Olshina et al. 2010), a result suggesting that chaperone-mediated protection is afforded through a decrease in the toxic oligomer population and a concomitant rise in inclusion-body formation. In general, then, the aggregates that were more amorphous, or less ordered, were found to be more toxic.

11.3.1.2 Ataxin-3

Ataxin-3 is a relatively small (42 kDa) protein with two domains: a globular, predominantly α-helical *N*-terminal Josephin domain and a flexible *C*-terminal tail where the polyQ region is located (Albrecht et al. 2004; Nicastro et al. 2005, 2006). Full-length ataxin-3 has been successfully expressed recombinantly and purified, as has the Josephin domain (Gales et al. 2005; Nicastro et al. 2005), and a solution structure of the Josephin domain is available (Nicastro et al. 2005). Thus ataxin-3, like httex1, has been a popular target of study.

The Bottomley group was one of the first to employ ataxin-3 for aggregation studies *in vitro*, including the development of rigorous purification methods giving high protein yields and purities (Chow et al. 2006). There were no structural or stability differences between several expanded polyQ mutants above and below the critical length (Chow et al. 2004a). However, only expanded ataxin-3 (Q50) formed thioflavin-T-positive fibrils. Non-expanded ataxin-3 (Q15 and Q28) was able to form large fibrillar aggregates under modest chemical-denaturant concentrations, suggesting that expanded polyQ lowers the energy barrier for partial unfolding requisite for entry into the aggregation pathway (Chow et al. 2004a). Bevivino and Loll reported destabilization of the native helical structure of ataxin-3 with an expanded polyQ domain, and suggested that both destabilization and intermolecular glutamine contacts partner together to cause aggregation (Bevivino and Loll 2001). They observed soluble aggregates by TEM for both expanded and non-expanded ataxin-3, but only expanded (Q78) ataxin-3 aggregates were positive for Congo-red staining

(Bevivino and Loll 2001). Others, however, have stated that expansion of polyQ does not destabilize ataxin-3 (Chow et al. 2004a).

Aggregation of ataxin-3 occurs even in the absence of expanded polyQ through the Josephin domain, complicating any analysis of the role of polyQ length on aggregation (Masino et al. 2004; Chow et al. 2004b). Gales and colleagues were able to form thioflavin-T-positive aggregates with non-expanded (Q14) ataxin-3 (Gales et al. 2005). These aggregates were imaged by TEM and appeared to have similar morphologies and sizes to those captured by Bevivino and Loll, suggesting that the apparent differences in results could be due to inadequacies in the thioflavin-T assay. They concluded that oligomerization of ataxin-3 occurred independent of the polyQ domain. Ellisdon and colleagues used thioflavin-T fluorescence to monitor aggregation of ataxin-3 with both non-expanded and expanded polyQ domains (Ellisdon et al. 2006, 2007). However, only expanded ataxin-3 (Q64) formed insoluble aggregates whereas all other Q lengths formed soluble aggregates. By TEM, all proteins studied formed smaller intermediates, but only expanded ataxin-3 could progress to form ordered fibrils after longer aggregation times. Interestingly, QBP, a peptide shown to prevent polyQ-protein toxicity *in vivo* (Nagai et al. 2000), could not prevent oligomerization, but did prevent insoluble-aggregate formation in the expanded polyQ ataxin-3 (Ellisdon et al. 2007). The ability of *any* polyQ length of ataxin-3 to aggregate while only expanded lengths are pathogenic certainly suggests a specific role of length-dependent aggregate morphologies or aggregation kinetics in toxicity. Taken together, these data suggest that insoluble aggregates are the most toxic, which is different than the assessment from studies of httex1. Haacke and colleagues performed an interesting study comparing the aggregation of variously sized ataxin-3 truncation mutants with both expanded and non-expanded polyQ tracts (Haacke et al. 2006). Only the shortest truncation mutant containing expanded polyQ (Q71) was able to initiate insoluble-aggregate formation. However, full-length ataxin-3 was subsequently recruited into these aggregates. This study demonstrates that proteolysis may be an important step in the pathology of SCA3, and differentiates between initiation of aggregation and further growth.

11.3.1.3 Other Disease-Associated Proteins

Beyond ataxin-3 and huntingtin exon 1, the literature on aggregation of other disease-associated polyQ proteins is limited. This is at least in part due to the sizes and complexities of the proteins, and a lack of structural information. Expression and purification of the disease-associated proteins often necessitates construction of fusion proteins or affinity tags, or work with just a protein fragment (e.g., Bevivino and Loll 2001; Masino et al. 2002; Takahashi et al. 2008; Husain-Ponnampalam et al. 2010). We will briefly review the limited results on each of the remaining disease-associated proteins.

Ataxin-1 has a globular AXH domain near its *C*-terminus; this domain has been expressed and its crystal structure is available but little is known about the remainder of the protein (de Chiara et al. 2005). Like the Josephin domain in ataxin-3, the

AXH domain is aggregation-prone in the absence of the polyQ domain (de Chiara et al. 2005). Very recently, a method for expressing and purifying full-length ataxin-1 has been reported (Husain-Ponnampalam et al. 2010). Rich and colleagues studied ataxin-1–fluorescent-protein fusions and imaged oligomeric aggregates for both expanded and non-expanded polyQ tracts (Rich and Varadaraj 2007). These researchers also demonstrated that the identity of the fusion protein can affect aggregation characteristics, illustrating one of the difficulties of using fusion proteins for these studies. Krol and colleagues studied the aggregation characteristics of ataxin-1 as compared to several other expanded polyQ proteins (Krol et al. 2008). They argued ataxin-1 aggregates differently than other polyQ-disease proteins due to its inability to form insoluble aggregates.

Ataxin-2 (140 kDa) has 2 globular domains that are predicted to be α-helical but for which no high-resolution data are available; most of the protein is believed to be disordered including the *N*-terminal tail that contains the polyQ tract (Albrecht et al. 2004). Atrophin-1 is a large protein (1,100 residues) whose structure and function are poorly understood (Shen et al. 2007). In early work, Igarashi and colleagues showed that expanded polyQ atrophin-1 could form filamentous aggregates and induce cell death in COS cells (Igarashi et al. 1998). More recently, Takahashi et al. used a FRET confocal-microscopy technique with truncated atrophin-1; like their parallel work with httex1, oligomers were observed even with the shortest (Q12) polyQ domain (Takahashi et al. 2008). Nozaki and colleagues studied truncation mutants of ataxin-2 and atrophin-1 (along with ataxin-3 and htt) tethered to a fluorescent protein (Nozaki et al. 2001). The truncation mutants contained the polyQ region and the given protein's flanking sequences. Their results showed expanded polyQ constructs (Q56) expressed in COS cells yielded more aggregates than non-expanded constructs (Q41). The results are indicative of a length-dependent aggregation mechanism. Additionally, different quantities of aggregate formation between the different disease-associated proteins were recorded suggesting flanking residues may explain why the age-of-onset is different for each disease.

Androgen receptor (AR) is a large multi-domain, membrane-bound protein; its ligand-binding and DNA-binding domains are both predominantly α-helical; its *N*-terminal activation-function domain, where the polyQ region is located, appears to be unstructured or weakly folded (Bain et al. 2007). Studying AR in mice, Li and colleagues found that expanded androgen receptor (Q112) led to oligomer formation while non-expanded androgen receptor did not (Li et al. 2007). The appearance of oligomers preceded onset of disease symptoms by ~2 weeks. Davies and colleagues found few structural changes in the AR upon polyQ expansion; however, pathogenic-length (Q45) protein was more susceptible to proteolysis (Davies et al. 2008). Schiffer et al. expressed polyQ-expanded *N*-terminal fragments of AR that lacked its hormone-binding domain (Schiffer et al. 2008). AR fragments with Q67 formed large SDS-resistant inclusions. The most pronounced toxicity was seen with Q102 fragments, which formed exclusively soluble oligomers of 100–600 kDa. Full-length polyQ-expanded AR was fully functional in transactivation but became inactivated if the corresponding *N*-terminal fragment was present, suggesting that soluble fragments bind to full-length AR and inactivate it.

TBP, ataxin-7, and α_1A calcium-channel subunit are all components of larger protein complexes rather than isolated soluble proteins (Nikolov et al. 1992; Kubodera et al. 2003; Helmlinger et al. 2004). Very little data has been collected on these systems. Lundes et al. made a fragment of TBP (TATA-binding protein) with *N*-terminal Q38 and saw long filaments that were about 10–12 nm diameter that tended to organize in clusters (Lunkes et al. 1999). The authors suggested that the filaments grow by forming discrete branching points in a network, which eventually gather into larger filamentous networks. An antibody that recognizes polyQ, 1C2, prevented aggregation, suggesting that the polyQ domain was important in mediating aggregation.

11.3.1.4 PolyA Proteins

There are few studies with polyalanine proteins. Recombinant PABN1, the protein associated with OPMD, formed amorphous aggregates regardless of the presence or length of the polyA segment (Scheuermann et al. 2003). Using a stable *N*-terminal fragment, Scheuermann et al. demonstrated that expansion of the polyA domain to pathological lengths increased the α-helical content of the fragment and led to polyA-length-dependent aggregation (Scheuermann et al. 2003). Fibril formation was followed by NMR and fluorescence methods; no soluble pre-fibrillar oligomers were detected and the authors concluded that aggregation was a simple two-state conversion from unfolded monomers to folded insoluble fibrils (Rohrberg et al. 2008). This pattern distinctly differs from polyQ proteins, where soluble oligomers are repeatedly observed. In another study, it was reported that proteins fused to this fragment retained activity even after aggregation (Sackewitz et al. 2008). Cellular expression of PABPN1 with either normal or expanded polyA tracts produced intranuclear inclusions and cell toxicity in one study (Sasseville et al. 2006). Contradictory results were obtained in another study, in which expanded but not normal polyA lengths in PABPN1 caused significant toxicity (Klein et al. 2008). Only very long polyA segments, far longer than those required for human pathology, resulted in nuclear inclusion formation, and deletion of the entire polyA segment led to both nuclear and cytoplasmic inclusions (Klein et al. 2008). The inherent expression and stability issues with full-length PABPN1 add to the difficulties of these experiments. One brief study using the protein involved with BPES demonstrated that expression of the associated transcription factor with expanded but not normal polyalanine results in aggregation and lowered activity (Moumne et al. 2008).

11.3.2 Model PolyQ Proteins

A landmark study demonstrated that transgenic mice can develop classic symptoms of progressive neurodegeneration when an expanded polyQ domain is placed in an innocuous protein (Ordway et al. 1997). This result leads to the hypothesis that any protein could be toxic if it were mutated to contain an expanded polyQ tract. Given this hypothesis, and the practical difficulties associated with disease-associated proteins described above, several researchers have chosen to study model proteins to play "host" to non-native polyQ tracts. There are several advantages to this approach if the host protein is chosen carefully. A protein that is easily expressed and manipulated enables systematic studies of both polyQ length and context within a single protein. If available, high-resolution structural data and well-characterized folding pathways facilitate detailed interpretation of data on the effects of polyQ on folding, secondary structures, and tertiary structures. Finally, for many of the disease-associated proteins, a fusion protein construct and/or affinity tag was required, or a protein fragment rather than the full-length protein was studied. Indeed, some studies have demonstrated that the fusion partner strongly affects aggregation characteristics (e.g., Rich and Varadaraj 2007). In these cases, the fusion protein or the protein fragment might be considered to be a model protein, one that carries the disease-associated sequence but is not a completely accurate rendering of the disease-associated protein.

The choice of a host protein and the positioning of the polyQ domain are significant considerations. For the most part, researchers have chosen to place the polyQ domain at the *N*- or *C*-terminus of the host protein. This positioning is more likely to isolate the polyQ effects from the remainder of the protein.

GST–polyQ fusions have been popular. With this model system, Masino and colleagues showed the polyQ region to be flexible, solvent exposed, and in an unordered conformation by NMR and CD (Masino et al. 2002). Greater quantities of aggregates, as observed by TEM, were obtained with expanded polyQ lengths of Q41, after aggregation at 52°C. Klein and colleagues studied polyQ aggregation after cleavage from the GST tag (Klein et al. 2007). By NMR, the structure of polyQ did not change significantly after cleavage, suggesting that polyQ stays in an unordered conformation. Importantly, this group saw no difference in conformation with Q length, indicating that there is not a conformational "critical length". Both expanded and non-expanded (Q41 and Q22) were found to aggregate, although Q41 did so at greater rates. The aggregates were shown to be non-dynamic, insoluble, and SDS-resistant. Bulone and colleagues completed a rather detailed study of the aggregation kinetics of GST fused to either Q22 or Q41 (Bulone et al. 2006). By dynamic light scattering, a dimer and a large, soluble oligomer (~90–95 nm) were identified. The pool of soluble oligomers grew until a critical concentration was reached, allowing ordered fibrils to form. The Q41–GST fusion formed a larger pool of soluble oligomers than did Q22–GST; however, the shorter mutant had a surprisingly greater propensity to form ordered fibrils at lower temperatures. The authors hypothesized that the longer polyQ sequence traps aggregates at the soluble-oligomer stage. This hypothesis coupled with the protective effects of inclusion bodies *in vivo* suggests a toxic intermediate in the polyQ aggregation pathway.

Thioredoxin–polyQ fusions have been employed in several studies. These fusions were first used as a screening tool, in which phage display was used to search for a peptide capable of inhibiting aggregation of expanded polyQ (Q62 and Q81) (Nagai et al. 2000). This study led to the discovery of a Q-binding peptide (QBP).

Co-expression of QBP with expanded polyQ–thioredoxin fusions in cells resulted in fewer aggregates and lowered toxicity (Nagai et al. 2000). The effects of proline interruptions were studied in this same model system. Popiel et al. expressed polyQ tracts containing 19–81 glutamine residues in thioredoxin attached to GFP, and then inserted 1–7 proline residues evenly dispersed amongst the glutamine stretches (Popiel et al. 2004). By CD, proline interruptions caused the polyQ tracts to be more unordered. As the number of proline inserts increased, aggregation slowed or stopped for each of the peptides tested, and cytotoxicity decreased. Interestingly, when proteins with proline inserts were mixed with proteins without proline, aggregation was suppressed for the entire mixture. They found a short polyQ stretch containing 19 glutamine residues and 1 proline residue to be more effective as an aggregation inhibitor than longer polyQ stretches containing more proline residues. The researchers hypothesized molecules which force certain non-toxic polyQ conformations, replicating the effects of a proline interruption, may be potential therapeutics.

In a more recent study, also using thioredoxin–polyQ proteins, these researchers proposed that there is a toxic-monomer conformer that precedes aggregation (Nagai et al. 2007). Thioredoxin alone is predominantly β-sheet; interestingly, the addition of polyQ led to an increase in α-helical content in a Q-length-dependent manner. Incubation of the thioredoxin–Q62 fusion protein led to an increase in turbidity and eventual precipitation of a fraction of the material. The recovered soluble material contained a significant amount of β-sheet character. The authors concluded that the monomer underwent an α-helix to β-sheet transition preceding oligomerization. This interpretation of the data is somewhat tenuous, as the molecular sizes of the material determined by SEC and by SDS–PAGE do not correspond to the monomer size. The authors further observed that the aggregate material contained 10–40-nm diameter fibrils that were 100–3,000 nm long with some helical periodicity; fibrillar aggregates were not seen in the shorter (Q19) construct. QBP was shown to inhibit the helix-to-sheet conformational transition and to inhibit aggregation. Finally, the freshly prepared materials were not toxic, but both "aged" soluble and fibrillar materials were toxic.

In a recent study, polyQ of variable lengths was fused to the Sup 35 prion domain (Goehler et al. 2010). Expanded polyQ lengths (Q55 and Q85) sedimented with length-dependent kinetics and formed fibrils. Non-expanded lengths (Q19) did not sediment, though some non-fibrillar oligomers were observed by EM. Only expanded polyQ mutants were able to seed aggregation of full-length Sup35 protein. The researchers hypothesized that long polyQ lengths initiate aggregation by folding into β-sheet structures.

Taking a hybrid approach, httex1 was fused either *N*- or *C*-terminally to the host protein, cellular retinoic-acid-binding protein I (CRAPBI), a β-barrel protein (Ignatova and Gierasch 2006). Expanded polyQ mutants were found to be less stable and more structurally perturbed than non-expanded mutants. Only extended (Q53 and Q64) mutants formed inclusion bodies upon cellular expression. In a subsequent study in which aggregate morphology was monitored, a shift from spherical to curvilinear to fibrillar structures was observed (Ignatova et al. 2007). In seeding studies,

CRAPBI–Q53 aggregates taken at early times accelerated aggregation of a CRAPBI mutant lacking a polyQ domain better than CRAPBI–Q29, while older CRAPBI–Q53 aggregates were more effective at recruiting CRAPBI–Q29. The authors proposed that early-stage assembly was driven by the CRAPBI domain but in the later stages, the polyQ-domain interactions dominated (Ignatova et al. 2007).

A second set of studies utilized model proteins with non-terminal polyQ tracts. In an early experiment, short (Q4 or Q10) polyQ tracts were inserted into the flexible loop of chymotrypsin inhibitor 2 (Stott et al. 1995). The mutants associated into dimers and trimers with a small increase in β-sheet character upon oligomerization. Crystals of dimers were later analyzed with X-ray diffraction (Chen et al. 1999). The self-association was shown to be due to domain-swapping rather than inter-glutamine hydrogen bonding. Interestingly, computational studies have supported the domain-swapping behavior in small, but not expanded, polyQ tracts (Barton et al. 2007).

Tanaka and colleagues studied the aggregation characteristics of myoglobin with various lengths of polyQ inserted into its flexible loop between the C and D helices. Their initial study focused on the structural effects. From circular-dichroism spectra they inferred increased helical character in the monomer with greater polyQ length, whereas FTIR analysis of the aggregates revealed increased β-sheet character (Tanaka et al. 2001). Protein stability decreased with increasing polyQ length, and the tertiary structure was altered in a Q-length-dependent manner. In a subsequent study, they compared Q50 to other aggregation-inducing motifs inserted in the same loop location (Tanaka et al. 2002) and found that the Q50 conferred the fasted aggregation rate and highest Congo-red fluorescence. Employing small-angle X-ray scattering (SAXS) to examine Q35 and Q50 myoglobin mutants, this group observed rapid assembly into small oligomers followed by further association into "quasi-aggregates" containing ~90 monomers (Tanaka et al. 2003). Quasi-aggregates appeared to be requisite for fibril formation.

In perhaps the most systematic biophysical study to date that explains the relative importance of context and length, Robertson and colleagues created mutants based on the SpA B domain in which polyQ of variable length was either fused to the *C*-terminus or inserted into a flexible loop between helices (Robertson et al. 2008). In all cases, the polyQ domain was found to be in an unordered conformation by CD and NMR. There was little change in protein stability for the *C*-terminal mutants but greatly reduced chemical and thermal stability of the loop-insert mutants, suggesting that polyQ may perturb tertiary structure depending on its placement within the protein. *C*-terminal mutants with more than 35 glutamine residues aggregated at 37°C into ThT-positive fibrils. The aggregation properties of the loop-insert mutants were not studied, as the group was unsuccessful in their attempts to generate loop-insert mutants containing more than 25 Q residues.

In our lab, we studied the impact of polyQ length and position on the structure, stability, and aggregation of apomyoglobin (Tobelmann and Murphy 2011). A library of mutants were generated with varying polyQ lengths inserted either into the disordered loop between two helices, or at the *N*-terminus. With the loop-insert mutants, loss of native α-helix and gain of both disordered and β-sheet content was detected in a Q-length-dependent manner. All of these mutants were less stable than wild-type protein in urea, and showed evidence of tertiary structure perturbation. These mutants rapidly assembled into soluble chain-like oligomers that slowly clustered into insoluble aggregates. Interestingly, a glycine–serine mutant showed similar structural perturbations and aggregation characteristics, suggesting that for these mutants the primary aggregation mechanism is due to nonspecific partial loss of native fold. In sharp contrast, the *N*-terminal mutants gained α-helix relative to wild-type protein, and retained tertiary fold. However, unlike wild-type apomyoglobin, a fraction of these proteins assembled into long aggregates with classic fibrillar morphology. This study suggests that structural perturbations alone produce soluble chain-like oligomers while specific polyQ intermolecular interactions are likely more significant in formation of fibrillar aggregates.

PolyA expansion has been fruitfully studied using model proteins. When attached to green or yellow fluorescent protein, only those proteins with longer polyA tracts aggregated (Rankin et al. 2000; Toriumi et al. 2008, 2009). GST– or YFP–polyA fusion protein associated into oligomers with 23 or more alanine residues; the oligomers were resistant to trypsin proteolysis (Nojima et al. 2009). A similar threshold of 23 alanine residues was observed when intermolecular association was tested in a yeast two-hybrid system (Oma et al. 2007).

11.4 Concluding Comments

The connection between expanded polyQ (and polyA) and disease has been firmly established. This connection has motivated research into the underlying biophysical and biological mechanisms. Much progress has been achieved through the combined efforts of several research groups, and much remains to be learned. We conclude this chapter with final thoughts and our personal assessment of the current state of knowledge.

1. Length matters. Most evidence points to the absence of a "critical" or threshold length. Rather, there is a gradual transition in physicochemical properties. PolyQ is conformationally flexible and adaptable, which may explain its frequent appearance in natively disordered regions, or flexible linkers between domains of multi-domain proteins. As an isolated side-chain, glutamine is considered to be polar; its side-chain amide can hydrogen-bond with water as both donor and acceptor. However, in a homo-glutamine-repeat tract, the opportunities for intramolecular hydrogen bonding multiply rapidly as the length of the repeat increases. These hydrogen bonds can be heterogeneous: side-chain–side-chain, backbone–backbone, or side-chain–backbone. Accompanying the formation of intramolecular hydrogen bonds is the conversion to a nonpolar and collapsed conformation.
2. One can imagine two alternative pathways open to the collapsed polyQ domain with intramolecular hydrogen bonding. First is re-arrangement within the

monomer to an intramolecular β-sheet conformation. The β-sheet monomer could then serve as a template for growth of organized fibrils. Second is a rapid, hydrophobically driven association of the collapsed polyQ monomers into amorphous soluble oligomers, followed by slow re-arrangement within the oligomer to an intermolecular β-sheet, then dehydration and further growth to an insoluble fibril. There is indirect evidence to support either pathway, and both could be operative simultaneously. Sorting out this question has biological relevance, as it has been postulated that the soluble oligomers are toxic.

3. Much has been written about soluble oligomers, protofibrils, fibrils, and filaments, yet there is no firm and consistent consensus on how each is defined. Furthermore, there is emerging evidence that fibrils are morphologically and structurally heterogeneous, and it is even more likely that soluble oligomers are heterogeneous in size and structure. More specific definitions and multiple measures (AFM, CD, antibody-epitope mapping?) are needed.
4. Expanded polyQ may mediate aggregation either directly or indirectly. Indirectly, expansion of polyQ can cause structural perturbations and native-fold instability in the remainder of the protein, accelerating aggregation. Intriguingly, several of the disease-related proteins have non-polyQ domains that are aggregation-prone as is. Limited data suggests that this indirect mode of aggregation leads to amorphous (or non-fibrillar) aggregates. Directly, polyQ domains in different proteins can associate *via* intermolecular hydrogen bonds, forming extended β-sheets. Again, limited data suggests that direct polyQ-driven association produces more ordered, fibrillar aggregates. Some studies have indicated that both indirect and direct mechanisms are in operation: initial association is driven through hydrophobic interaction of the non-polyQ segments, and this is followed by structural re-arrangement and polyQ–polyQ association.
5. Context matters. The flanking residues and the position of the expanded domain in the protein have a large influence on the conformation of the polyQ region and its propensity to aggregate. The rules have not been fully worked out. It is clear that polyP inhibits polyQ-driven aggregation, and it is interesting to note that many proteins with polyQ domains also possess a polyP domain.
6. Lastly, polyglutamine domains *per se* are not sufficient to cause disease. The human genome encodes ~400 proteins with polyglutamine repeats of 7 or more (Faux et al. 2005). Analysis of data from another source reveals 33 proteins with polyQ tracts of 15 or longer, and two with 34 or more glutamine residues (Siwach et al. 2006). Fully half of these are polymorphic, yet none has been associated with any disease. Why is polyQ so prevalent? Why is it so likely to be polymorphic, unlike most homo-amino-acid repeat tracts? Why are there proteins with long polyQ tracts that are not associated with disease? Is there an evolutionary advantage to expansion of polyQ, and is expansion relatively benign compared to other amino acids? Any rigorous understanding, from a genetic or biochemical viewpoint, of the role of expanded polyglutamine in neurodegenerative disorders must incorporate these observations.

References

Albrecht A, Mundlos S (2005) The other trinucleotide repeat: polyalanine expansion disorders. Curr Opin Genet Dev 15:285–293

Albrecht M, Golatta M, Wullner U, Lengauer T (2004) Structural and functional analysis of ataxin-2 and ataxin-3. Eur J Biochem 271:3155–3170

Altschuler EL, Hud NV, Mazrimas JA, Rupp B (1997) Random coil conformation for extended polyglutamine stretches in aqueous soluble monomeric peptides. J Pept Res 50:73–75

Amiel J, Trochet D, Clement-Ziza M, Munnich A, Lyonnet S (2004) Polyalanine expansions in human. Hum Mol Genet 13:R235–R243

Arrasate M, Mitra S, Schweitzer ES, Segal MR, Finkbeiner S (2004) Inclusion body formation reduces levels of mutant huntingtin and the risk of neuronal death. Nature 431:805–810

Aziz NA, Jurgens CK, Landwehrmeyer GB, Van Roon-Mom WMC, Van Ommen GJB, Stijnen T, Roos RAC (2009) Normal and mutant htt interact to affect clinical severity and progression in huntington disease. Neurology 73:1280–1285

Bain DL, Heneghan AF, Connaghan-Jones KD, Miura MT (2007) Nuclear receptor structure: implications for function. Annu Rev Physiol 69:201–220

Barton S, Jacak R, Khare SD, Ding F, Dokholyan NV (2007) The length dependence of the polyQ-mediated protein aggregation. J Biol Chem 282:25487–25492

Bauer PO, Nukina N (2009) The pathogenic mechanisms of polyglutamine diseases and current therapeutic strategies. J Neurochem 110:1737–1765

Behrends C, Langer CA, Boteva R, Bottcher UM, Stemp MJ, Schaffar G, Rao BV, Giese A, Kretzschmar H, Siegers K, Hartl FU (2006) Chaperonin TRiC promotes the assembly of polyQ expansion proteins into nontoxic oligomers. Mol Cell 23:887–897

Berger Z, Davies JE, Luo SQ, Pasco MY, Majoul I, O'kane CJ, Rubinsztein DC (2006) Deleterious and protective properties of an aggregate-prone protein with a polyalanine expansion. Hum Mol Genet 15:453–465

Bernacki JP, Murphy RM (2009) Model discrimination and mechanistic interpretation of kinetic data in protein aggregation studies. Biophys J 96:2871–2887

Bevivino AE, Loll PJ (2001) An expanded glutamine repeat destabilizes native ataxin-3 structure and mediates parallel β-fibrils. Proc Natl Acad Sci USA 98:11955–11960

Bhattacharyya A, Thakur AK, Chellgren VM, Thiagarajan G, Williams AD, Chellgren BW, Creamer TP, Wetzel R (2006) Oligoproline effects on polyglutamine conformation and aggregation. J Mol Biol 355:524–535

Blondelle SE, Forood B, Houghten RA, Perezpaya E (1997) Polyalanine-based peptides as models for self-associated β-pleated-sheet complexes. Biochemistry 36:8393–8400

Brown LY, Brown SA (2004) Alanine tracts: the expanding story of human illness and trinucleotide repeats. Trends Genet 20:51–58

Bulone D, Masino L, Thomas DJ, Biagio PLS, Pastore A (2006) The interplay between polyQ and protein context delays aggregation by forming a reservoir of protofibrils. PLoS One 1:e111

Chellgren BW, Miller AF, Creamer TP (2006) Evidence for polyproline II helical structure in short polyglutamine tracts. J Mol Biol 361:362–371

Chen YW, Stott K, Perutz MF (1999) Crystal structure of a dimeric chymotrypsin inhibitor 2 mutant containing an inserted glutamine repeat. Proc Natl Acad Sci USA 96:1257–1261

Chen SM, Berthelier V, Hamilton JB, O'nuallain B, Wetzel R (2002a) Amyloid-like features of polyglutamine aggregates and their assembly kinetics. Biochemistry 41:7391–7399

Chen SM, Ferrone FA, Wetzel R (2002b) Huntington's disease age-of-onset linked to polyglutamine aggregation nucleation. Proc Natl Acad Sci USA 99:11884–11889

Chow MKM, Ellisdon AM, Cabrita LD, Bottomley SP (2004a) Polyglutamine expansion in ataxin-3 does not affect protein stability—implications for misfolding and disease. J Biol Chem 279:47643–47651

Chow MKM, Paulson HL, Bottomley SP (2004b) Destabilization of a non-pathological variant of ataxin-3 results in fibrillogenesis via a partially folded intermediate: a model for misfolding in polyglutamine disease. J Mol Biol 335:333–341

Chow MK, Ellisdon AM, Cabrita LD, Bottomley SP (2006) Purification of polyglutamine proteins. Methods Enzymol 413:1–19

Crick SL, Jayaraman M, Frieden C, Wetzel R, Pappu RV (2006) Fluorescence correlation spectroscopy shows that monomeric polyglutamine molecules form collapsed structures in aqueous solutions. Proc Natl Acad Sci USA 103:16764–16769

Darnell G, Orgel J, Pahl R, Meredith SC (2007) Flanking polyproline sequences inhibit β-sheet structure in polyglutamine segments by inducing PPII-like helix structure. J Mol Biol 374:688–704

Darnell GD, Derryberry J, Kurutz JW, Meredith SC (2009) Mechanism of *Cis*-inhibition of polyQ fibrillation by polyP: PPII oligomers and the hydrophobic effect. Biophys J 97:2295–2305

Davies JE, Rubinsztein DC (2006) Polyalanine and polyserine frameshift products in Huntington's disease. J Med Genet 43:893–896

Davies P, Watt K, Kelly SM, Clark C, Price NC, Mcewan IJ (2008) Consequences of poly-glutamine repeat length for the conformation and folding of the androgen receptor amino-terminal domain. J Mol Endocrinol 41:301–314

De Chiara C, Menon RP, Dal Piaz F, Calder L, Pastore A (2005) Polyglutamine is not all: the functional role of the AXH domain in the ataxin-1 protein. J Mol Biol 354:883–893

Diguet E, Petit F, Escartin C, Cambon K, Bizat N, Dufour N, Hantraye P, Deglon N, Brouillet E (2009) Normal aging modulates the neurotoxicity of mutant huntingtin. PLoS One 4(2):e4637

Dougan L, Li JY, Badilla CL, Berne BJ, Fernandez JM (2009) Single homopolypeptide chains collapse into mechanically rigid conformations. Proc Natl Acad Sci USA 106:12605–12610

Duennwald ML, Lindquist S (2008) Impaired ERAD and ER stress are early and specific events in polyglutamine toxicity. Genes Dev 22:3308–3319

Duennwald ML, Jagadish S, Muchowski PJ, Lindquist S (2006) Flanking sequences profoundly alter polyglutamine toxicity in yeast. Proc Natl Acad Sci USA 103:11045–11050

Ellisdon AM, Thomas B, Bottomley SP (2006) The two-stage pathway of ataxin-3 fibrillogenesis involves a polyglutamine-independent step. J Biol Chem 281:16888–16896

Ellisdon AM, Pearce MC, Bottomley SP (2007) Mechanisms of ataxin-3 misfolding and fibril formation: kinetic analysis of a disease-associated polyglutamine protein. J Mol Biol 368:595–605

Faux NG, Bottomley SP, Lesk AM, Irving JA, Morrison JR, De La Banda MC, Whisstock JC (2005) Functional insights from the distribution and role of homopeptide repeat-containing proteins. Genome Res 15:537–551

Faux NG, Huttley GA, Mahmood K, Webb GI, De La Banda MG, Whisstock JC (2007) RCPdb: an evolutionary classification and codon usage database for repeat-containing proteins. Genome Res 17:1118–1127

Fecke W, Gianfriddo M, Gaviraghi G, Terstappen GC, Heitz F (2009) Small molecule drug discovery for Huntington's disease. Drug Discov Today 14:453–464

Forood B, Perezpaya E, Houghten RA, Blondelle SE (1995) Formation of an extremely stable polyalanine β-sheet macromolecule. Biochem Biophys Res Commun 211:7–13

Gales L, Cortes L, Almeida C, Melo CV, Costa MD, Maciel P, Clarke DT, Damas AM, Macedo-Ribeiro S (2005) Towards a structural understanding of the fibrillization pathway in Machado-Joseph's disease: trapping early oligomers of non-expanded ataxin-3. J Mol Biol 353:642–654

Gaspar C, Jannatipour M, Dion P, Laganiere J, Sequeiros J, Brais B, Rouleau GA (2000) CAG tract of MJD-1 may be prone to frameshifts causing polyalanine accumulation. Hum Mol Genet 9:1957–1966

Gatchel JR, Zoghbi HY (2005) Diseases of unstable repeat expansion: mechanisms and common principles. Nat Rev Genet 6:743–755

Giri K, Ghosh U, Bhattacharyya NP, Basak S (2003) Caspase 8 mediated apoptotic cell death induced by β-sheet forming polyalanine peptides. FEBS Lett 555:380–384

Giri K, Bhattacharyya NP, Basak S (2007) pH-dependent self-assembly of polyalanine peptides. Biophys J 92:293–302

Goehler H, Droge A, Lurz R, Schnoegl S, Chernoff YO, Wanker EE (2010) Pathogenic polyglutamine tracts are potent inducers of spontaneous Sup35 and Rnq1 amyloidogenesis. PLoS One 5(3):e9642

Gray M, Shirasaki DI, Cepeda C, Andre VM, Wilburn B, Lu XH, Tao JF, Yamazaki I, Li SH, Sun YE, Li XJ, Levine MS, Yang XW (2008) Full-length human mutant huntingtin with a stable polyglutamine repeat can elicit progressive and selective neuropathogenesis in BACHD mice. J Neurosci 28:6182–6195

Green H, Wang N (1994) Codon reiteration and the evolution of proteins. Proc Natl Acad Sci USA 91:4298–4302

Gusella JF, Macdonald ME (2000) Molecular genetics: unmasking polyglutamine triggers in neurodegenerative disease. Nat Rev Neurosci 1:109–115

Haacke A, Broadley SA, Boteva R, Tzvetkov N, Hartl FU, Breuer P (2006) Proteolytic cleavage of polyglutamine-expanded ataxin-3 is critical for aggregation and sequestration of non-expanded ataxin-3. Hum Mol Genet 15:555–568

Harada N, Mitani T, Higashimura Y, Yamaji R, Okamoto K, Nakano Y, Inui H (2010) Involvement of three glutamine tracts in human androgen receptor transactivation. J Steroid Biochem Mol Biol 118:77–84

Heitmann B, Job GE, Kennedy RJ, Walker SM, Kemp DS (2005) Water-solubilized, cap-stabilized, helical polyalanines: calibration standards for NMR and CD analyses. J Am Chem Soc 127:1690–1704

Helmlinger D, Hardy S, Sasorith S, Klein F, Robert F, Weber C, Miguet L, Potier N, Van-Dorsselaer A, Wurtz JM, Mandel JL, Tora L, Devys D (2004) Ataxin-7 is a subunit of GCN5 histone acetyltransferase-containing complexes. Hum Mol Genet 13:1257–1265

Holmberg CI, Staniszewski KE, Mensah KN, Matouschek A, Morimoto RI (2004) Inefficient degradation of truncated polyglutamine proteins by the proteasome. EMBO J 23:4307–4318

Huntley MA, Golding GB (2002) Simple sequences are rare in the protein data bank. Proteins Struct Funct Genet 48:134–140

Husain-Ponnampalam R, Turnbull V, Tarlac V, Storey E (2010) Expression and purification of ataxin-1 protein. J Neurosci Methods 189:30–35

Igarashi S, Koide R, Shimohata T, Yamada M, Hayashi Y, Takano H, Date H, Oyake M, Sato T, Sato A, Egawa S, Ikeuchi T, Tanaka H, Nakano R, Tanaka K, Hozumi I, Inuzuka T, Takahashi H, Tsuji S (1998) Suppression of aggregate formation and apoptosis by transglutaminase inhibitors in cells expressing truncated DRPLA protein with an expanded polyglutamine stretch. Nat Genet 18:111–117

Ignatova Z, Gierasch LM (2006) Extended polyglutamine tracts cause aggregation and structural perturbation of an adjacent β barrel protein. J Biol Chem 281:12959–12967

Ignatova Z, Thakur AK, Wetzel R, Gierasch LM (2007) In-cell aggregation of a polyglutamine-containing chimera is a multistep process initiated by the flanking sequence. J Biol Chem 282:36736–36743

Imarisio S, Carmichael J, Korolchuk V, Chen CW, Saiki S, Rose C, Krishna G, Davies JE, Ttofi E, Underwood BR, Rubinsztein DC (2008) Huntington's disease: from pathology and genetics to potential therapies. Biochem J 412:191–209

Jayaraman M, Kodali R, Wetzel R (2009) The impact of ataxin-1-like histidine insertions on polyglutamine aggregation. Protein Eng Des Sel 22:469–478

Kahlem P, Djian P (2000) The expanded CAG repeat associated with juvenile Huntington disease shows a common origin of most or all neurons and glia in human cerebrum. Neurosci Lett 286:203–207

Karlin S, Brocchieri L, Bergman A, Mrazek J, Gentles AJ (2002) Amino acid runs in eukaryotic proteomes and disease associations. Proc Natl Acad Sci USA 99:333–338

Kim MW, Chelliah Y, Kim SW, Otwinowski Z, Bezprozvanny I (2009) Secondary structure of huntingtin amino-terminal region. Structure 17:1205–1212

Klein FAC, Pastore A, Masino L, Zeder-Lutz G, Nierengarten H, Oulad-Abdeighani M, Altschuh D, Mandel JL, Trottier Y (2007) Pathogenic and non-pathogenic polyglutamine tracts have similar structural properties: towards a length-dependent toxicity gradient. J Mol Biol 371:235–244

Klein AF, Ebihara M, Alexander C, Dicaire MJ, Sasseville AMJ, Langelier Y, Rouleau GA, Brais B (2008) PABPN1 polyalanine tract deletion and long expansions modify its aggregation pattern and expression. Exp Cell Res 314:1652–1666

Krol HA, Krawczyk PM, Bosch KS, Aten JA, Hol EM, Reits EA (2008) Polyglutamine expansion accelerates the dynamics of ataxin-1 and does not result in aggregate formation. PLoS One 3(1):e1503

Kubodera T, Yokota T, Ohwada K, Ishikawa K, Miura H, Matsuoka T, Mizusawa H (2003) Proteolytic cleavage and cellular toxicity of the human α1A calcium channel in spinocerebellar ataxia type 6. Neurosci Lett 341:74–78

Lee CC, Walters RH, Murphy RM (2007) Reconsidering the mechanism of polyglutamine peptide aggregation. Biochemistry 46:12810–12820

Legleiter J, Mitchell E, Lotz GP, Sapp E, Ng C, Difiglia M, Thompson LM, Muchowski PJ (2010) Mutant huntingtin fragments form oligomers in a polyglutamine length-dependent manner *in vitro* and *in vivo*. J Biol Chem 285:14777–14790

Li SH, Li XJ (2004) Huntingtin-protein interactions and the pathogenesis of Huntington's disease. Trends Genet 20:146–154

Li W, Serpell LC, Carter WJ, Rubinsztein DC, Huntington JA (2006) Expression and characterization of full-length human huntingtin, an elongated heat repeat protein. J Biol Chem 281:15916–15922

Li M, Chevalier-Larsen ES, Merry DE, Diamond MI (2007) Soluble androgen receptor oligomers underlie pathology in a mouse model of spinobulbar muscular atrophy. J Biol Chem 282:3157–3164

Lieberman AP, Robins DM (2008) The androgen receptor's CAG/glutamine tract in mouse models of neurological disease and cancer. J Alzheimers Dis 14:247–255

Lima MAC, Pimentel MMG (2004) Dynamic mutation and human disorders: the spinocerebellar ataxias (review). Int J Mol Med 13:299–302

Lunkes A, Trottier Y, Fagart J, Schultz P, Zeder-Lutz G, Moras D, Mandel JL (1999) Properties of polyglutamine expansion *in vitro* and in a cellular model for Huntington's disease. Philos Trans R Soc Lond B Biol Sci 354:1013–1019

Marchut AJ, Hall CK (2006a) Side-chain interactions determine amyloid formation by model polyglutamine peptides in molecular dynamics simulations. Biophys J 90:4574–4584

Marchut AJ, Hall CK (2006b) Spontaneous formation of annular structures observed in molecular dynamics simulations of polyglutamine peptides. Comput Biol Chem 30:215–218

Marchut AJ, Hall CK (2007) Effects of chain length on the aggregation of model polyglutamine peptides: molecular dynamics simulations. Proteins Struct Funct Bioinform 66:96–109

Masino L, Kelly G, Leonard K, Trottier Y, Pastore A (2002) Solution structure of polyglutamine tracts in GST-polyglutamine fusion proteins. FEBS Lett 513:267–272

Masino L, Nicastro G, Menon RP, Dal Piaz F, Calder L, Pastore A (2004) Characterization of the structure and the amyloidogenic properties of the Josephin domain of the polyglutamine-containing protein ataxin-3. J Mol Biol 344:1021–1035

Matilla A, Roberson ED, Banfi S, Morales J, Armstrong DL, Burright EN, Orr HT, Sweatt JD, Zoghbi HY, Matzuk MM (1998) Mice lacking ataxin-1 display learning deficits and decreased hippocampal paired-pulse facilitation. J Neurosci 18:5508–5516

Merlino A, Esposito L, Vitagliano L (2006) Polyglutamine repeats and β-helix structure: molecular dynamics study. Proteins Struct Funct Bioinform 63:918–927

Messaed C, Rouleau GA (2009) Molecular mechanisms underlying polyalanine diseases. Neurobiol Dis 34:397–405

Moumne L, Dipietromaria A, Batista F, Kocer A, Fellous M, Pailhoux E, Veitia RA (2008) Differential aggregation and functional impairment induced by polyalanine expansions in FOXL2, a transcription factor involved in cranio-facial and ovarian development. Hum Mol Genet 17:1010–1019

Muchowski PJ, Schaffar G, Sittler A, Wanker EE, Hayer-Hartl MK, Hartl FU (2000) Hsp70 and Hsp40 chaperones can inhibit self-assembly of polyglutamine proteins into amyloid-like fibrils. Proc Natl Acad Sci USA 97:7841–7846

Nagai Y, Tucker T, Ren HZ, Kenan DJ, Henderson BS, Keene JD, Strittmatter WJ, Burke JR (2000) Inhibition of polyglutamine protein aggregation and cell death by novel peptides identified by phage display screening. J Biol Chem 275:10437–10442

Nagai Y, Inui T, Popiel HA, Fujikake N, Hasegawa K, Urade Y, Goto Y, Naiki H, Toda T (2007) A toxic monomeric conformer of the polyglutamine protein. Nat Struct Mol Biol 14:332–340

Nekooki-Machida Y, Kurosawab M, Nukina N, Ito K, Oda T, Tanaka M (2009) Distinct conformations of *in vitro* and *in vivo* amyloids of huntingtin-exon1 show different cytotoxicity. Proc Natl Acad Sci USA 106:9679–9684

Nicastro G, Menon RP, Masino L, Knowles PP, Mcdonald NQ, Pastore A (2005) The solution structure of the Josephin domain of ataxin-3: structural determinants for molecular recognition. Proc Natl Acad Sci USA 102:10493–10498

Nicastro G, Habeck M, Masino L, Svergun DI, Pastore A (2006) Structure validation of the Josephin domain of ataxin-3: conclusive evidence for an open conformation. J Biomol NMR 36:267–277

Nikolov DB, Hu SH, Lin J, Gasch A, Hoffmann A, Horikoshi M, Chua NH, Roeder RG, Burley SK (1992) Crystal-structure of TFIID TATA-box binding-protein. Nature 360:40–46

Nojima J, Oma Y, Futai E, Sasagawa N, Kuroda R, Turk B, Ishiura S (2009) Biochemical analysis of oligomerization of expanded polyalanine repeat proteins. J Neurosci Res 87:2290–2296

Nozaki K, Onodera O, Takano H, Tsuji S (2001) Amino acid sequences flanking polyglutamine stretches influence their potential for aggregate formation. Neuroreport 12:3357–3364

Okazawa H (2003) Polyglutamine diseases: a transcription disorder? Cell Mol Life Sci 60:1427–1439

Olshina MA, Angley LM, Ramdzan YM, Tang JW, Bailey MF, Hill AF, Hatters DM (2010) Tracking mutant huntingtin aggregation kinetics in cells reveals three major populations that include an invariant oligomer pool. J Biol Chem 285:21807–21816

Oma Y, Kino Y, Toriumi K, Sasagawa N, Ishiura S (2007) Interactions between homopolymeric amino acids (HPAAS). Protein Sci 16:2195–2204

Ordway JM, Tallaksengreene S, Gutekunst CA, Bernstein EM, Cearley JA, Wiener HW, Dure LS, Lindsey R, Hersch SM, Jope RS, Albin RL, Detloff PJ (1997) Ectopically expressed cag repeats cause intranuclear inclusions and a progressive late onset neurological phenotype in the mouse. Cell 91:753–763

Orr HT, Zoghbi HY (2007) Trinucleotide repeat disorders. Annu Rev Neurosci 30:575–621

Ossato G, Digman MA, Aiken C, Lukacsovich T, Marsh JL, Gratton E (2010) A two-step path to inclusion formation of huntingtin peptides revealed by number and brightness analysis. Biophys J 98:3078–3085

Perutz MF, Johnson T, Suzuki M, Finch JT (1994) Glutamine repeats as polar zippers: their possible role in inherited neurodegenerative diseases. Proc Natl Acad Sci USA 91:5355–5358

Perutz MF, Finch JT, Berriman J, Lesk A (2002) Amyloid fibers are water-filled nanotubes. Proc Natl Acad Sci USA 99:5591–5595

Poirier MA, Li HL, Macosko J, Cai SW, Amzel M, Ross CA (2002) Huntingtin spheroids and protofibrils as precursors in polyglutamine fibrilization. J Biol Chem 277:41032–41037

Poirier MA, Jiang H, Ross CA (2005) A structure-based analysis of huntingtin mutant polyglutamine aggregation and toxicity: evidence for a compact β-sheet structure. Hum Mol Genet 14:765–774

Popiel HA, Nagai Y, Onodera O, Inui T, Fujikake N, Urade Y, Strittmatter WJ, Burke JR, Ichikawa A, Toda T (2004) Disruption of the toxic conformation of the expanded polyglutamine stretch leads to suppression of aggregate formation and cytotoxicity. Biochem Biophys Res Commun 317:1200–1206

Ramdzan YM, Nisbet RM, Miller J, Finkbeiner S, Hill AF, Hatters DM (2010) Conformation sensors that distinguish monomeric proteins from oligomers in live cells. Chem Biol 17:371–379

Rankin J, Wyttenbach A, Rubinsztein DC (2000) Intracellular green fluorescent protein–polyalanine aggregates are associated with cell death. Biochem J 348:15–19

Raspe M, Gillis J, Krol H, Krom S, Bosch K, Van Veen H, Reits E (2009) Mimicking proteasomal release of polyglutamine peptides initiates aggregation and toxicity. J Cell Sci 122: 3262–3271

Rich T, Varadaraj A (2007) Ataxin-1 fusion partners alter polyQ lethality and aggregation. PLoS One 2(10):e1014

Richard GF, Kerrest A, Dujon B (2008) Comparative genomics and molecular dynamics of DNA repeats in eukaryotes. Microbiol Mol Biol Rev 72:686–727

Richards RI, Sutherland GR (1997) Dynamic mutation: possible mechanisms and significance in human disease. Trends Biochem Sci 22:432–436

Robertson AL, Horne J, Ellisdon AM, Thomas B, Scanlon MJ, Bottomley SP (2008) The structural impact of a polyglutamine tract is location-dependent. Biophys J 95:5922–5930

Rohrberg J, Sachs R, Lodderstedt G, Sackewitz M, Balbach J, Schwarz E (2008) Monitoring fibril formation of the *N*-terminal domain of PABPN1 carrying an alanine repeat by tryptophan fluorescence and real-time NMR. FEBS Lett 582:1587–1592

Rosas HD, Salat DH, Lee SY, Zaleta AK, Hevelone N, Hersch SM (2008) Complexity and heterogeneity: what drives the ever-changing brain in Huntington's disease? In: Gibson GE, Ratan RR, Beal MF (eds) Mitochondria and oxidative stress in neurodegenerative disorders. Blackwell Publishing, Boston

Ross CA (2002) Polyglutamine pathogenesis: emergence of unifying mechanisms for Huntington's disease and related disorders. Neuron 35:819–822

Runne H, Regulier E, Kuhn A, Zala D, Gokce O, Perrin V, Sick B, Aebischer P, Deglon N, Luthi-Carter R (2008) Dysregulation of gene expression in primary neuron models of Huntington's disease shows that polyglutamine-related effects on the striatal transcriptome may not be dependent on brain circuitry. J Neurosci 28:9723–9731

Sackewitz M, Von Einem S, Hause G, Wunderlich M, Schmid FX, Schwarz E (2008) A folded and functional protein domain in an amyloid-like fibril. Protein Sci 17:1044–1054

Sasseville AMJ, Caron AW, Bourget L, Klein AF, Dicaire MJ, Rouleau GA, Massie B, Langelier Y, Brais B (2006) The dynamism of PABPN1 nuclear inclusions during the cell cycle. Neurobiol Dis 23:621–629

Sathasivam K, Lane A, Legleiter J, Warley A, Woodman B, Finkbeiner S, Paganetti P, Muchowski PJ, Wilson S, Bates GP (2010) Identical oligomeric and fibrillar structures captured from the brains of R6/2 and knock-in mouse models of Huntington's disease. Hum Mol Genet 19:65–78

Sato T, Miura M, Yamada M, Yoshida T, Wood JD, Yazawa I, Masuda M, Suzuki T, Shin RM, Yau HJ, Liu FC, Shimohata T, Onodera O, Ross CA, Katsuki M, Takahashi H, Kano M, Aosaki T, Tsuji S (2009) Severe neurological phenotypes of Q129 DRPLA transgenic mice serendipitously created by en masse expansion of CAG repeats in Q76 DRPLA mice. Hum Mol Genet 18:723–736

Saudou F, Finkbeiner S, Devys D, Greenberg ME (1998) Huntingtin acts in the nucleus to induce apoptosis but death does not correlate with the formation of intranuclear inclusions. Cell 95:55–66

Scherzinger E, Lurz R, Turmaine M, Mangiarini L, Hollenbach B, Hasenbank R, Bates GP, Davies SW, Lehrach H, Wanker EE (1997) Huntingtin-encoded polyglutamine expansions form amyloid-like protein aggregates *in vitro* and *in vivo*. Cell 90:549–558

Scheuermann T, Schulz B, Blume A, Wahle E, Rudolph R, Schwarz E (2003) Trinucleotide expansions leading to an extended poly-l-alanine segment in the poly (A) binding protein PABPN1 cause fibril formation. Protein Sci 12:2685–2692

Schiffer NW, Ceraline J, Hartl FU, Broadley SA (2008) *N*-terminal polyglutamine-containing fragments inhibit androgen receptor transactivation function. Biol Chem 389:1455–1466

Sharma D, Shinchuk LM, Inouye H, Wetzel R, Kirschner DA (2005) Polyglutamine homopolymers having 8–45 residues form slablike β-crystallite assemblies. Proteins Struct Funct Bioinform 61:398–411

Shen YG, Lee G, Choe Y, Zoltewicz JS, Peterson AS (2007) Functional architecture of atrophins. J Biol Chem 282:5037–5044

Shi ZS, Olson CA, Rose GD, Baldwin RL, Kallenbach NR (2002) Polyproline II structure in a sequence of seven alanine residues. Proc Natl Acad Sci USA 99:9190–9195

Shinchuk LM, Sharma D, Blondelle SE, Reixach N, Inouye H, Kirschner DA (2005) Poly-(l-alanine) expansions form core β-sheets that nucleate amyloid assembly. Proteins Struct Funct Bioinform 61:579–589

Sikorski P, Atkins E (2005) New model for crystalline polyglutamine assemblies and their connection with amyloid fibrils. Biomacromolecules 6:425–432

Sinden RR (2001) Neurodegenerative diseases—origins of instability. Nature 411:757–758

Singh VR, Lapidus LJ (2008) The intrinsic stiffness of polyglutamine peptides. J Phys Chem B 112:13172–13176

Siwach P, Pophaly SD, Ganesh S (2006) Genomic and evolutionary insights into genes encoding proteins with single amino acid repeats. Mol Biol Evol 23:1357–1369

Siwach P, Sengupta S, Parihar R, Ganesh S (2009) Spatial positions of homopolymeric repeats in the human proteome and their effect on cellular toxicity. Biochem Biophys Res Commun 380:382–386

Stott K, Blackburn JM, Butler PJG, Perutz M (1995) Incorporation of glutamine repeats makes protein oligomerize: implications for neurodegenerative diseases. Proc Natl Acad Sci USA 92:6509–6513

Takahashi T, Nozaki K, Tsuji S, Nishizawa M, Onodera O (2005) Polyglutamine represses cAMP-responsive-element-mediated transcription without aggregate formation. Neuroreport 16:295–299

Takahashi T, Kikuchi S, Katada S, Nagai Y, Nishizawa M, Onodera O (2008) Soluble polyglutamine oligomers formed prior to inclusion body formation are cytotoxic. Hum Mol Genet 17:345–356

Tanaka M, Morishima I, Akagi T, Hashikawa T, Nukina N (2001) Intra- and intermolecular β-pleated sheet formation in glutamine-repeat inserted myoglobin as a model for polyglutamine diseases. J Biol Chem 276:45470–45475

Tanaka M, Machida Y, Nishikawa Y, Akagi T, Morishima I, Hashikawa T, Fujisawa T, Nukina N (2002) The effects of aggregation-inducing motifs on amyloid formation of model proteins related to neurodegenerative diseases. Biochemistry 41:10277–10286

Tanaka M, Machida Y, Nishikawa Y, Akagi T, Hashikawa T, Fujisawa T, Nukina N (2003) Expansion of polyglutamine induces the formation of quasi-aggregate in the early stage of protein fibrillization. J Biol Chem 278:34717–34724

Thakur AK, Wetzel R (2002) Mutational analysis of the structural organization of polyglutamine aggregates. Proc Natl Acad Sci USA 99:17014–17019

Thakur AK, Jayaraman M, Mishra R, Thakur M, Chellgren VM, Byeon IJL, Anjum DH, Kodali R, Creamer TP, Conway JF, Gronenborn AM, Wetzel R (2009) Polyglutamine disruption of the huntingtin exon 1 *N*-terminus triggers a complex aggregation mechanism. Nat Struct Mol Biol 16:380–389

Tobelmann MD, Murphy RM (2011) Location trumps length: Polyglutamine-mediated changes in folding and aggregation of a host protein. Biophys J 100:2773–2782

Toriumi K, Oma Y, Kino Y, Futai E, Sasagawa N, Ishiura S (2008) Expression of polyalanine stretches induces mitochondrial dysfunction. J Neurosci Res 86:1529–1537

Toriumi K, Oma Y, Mimoto A, Futai E, Sasagawa N, Turk B, Ishiura S (2009) Polyalanine tracts directly induce the release of cytochrome C, independently of the mitochondrial permeability transition pore, leading to apoptosis. Genes Cells 14:751–757

Toulouse A, Au-Yeung F, Gaspar C, Roussel J, Dion P, Rouleau GA (2005) Ribosomal frameshifting on MJD-1 transcripts with long cag tracts. Hum Mol Genet 14:2649–2660

Tran HT, Mao A, Pappu RV (2008) Role of backbone–solvent interactions in determining conformational equilibria of intrinsically disordered proteins. J Am Chem Soc 130:7380–7392

Usdin K (2008) The biological effects of simple tandem repeats: lessons from the repeat expansion diseases. Genome Res 18:1011–1019

Vitalis A, Wang XL, Pappu RV (2007) Quantitative characterization of intrinsic disorder in polyglutamine: insights from analysis based on polymer theories. Biophys J 93:1923–1937

Vitalis A, Wang XL, Pappu RV (2008) Atomistic simulations of the effects of polyglutamine chain length and solvent quality on conformational equilibria and spontaneous homodimerization. J Mol Biol 384:279–297

Vitalis A, Lyle N, Pappu RV (2009) Thermodynamics of β-sheet formation in polyglutamine. Biophys J 97:303–311

Wacker JL, Zareie MH, Fong H, Sarikaya M, Muchowski PJ (2004) Hsp70 and Hsp40 attenuate formation of spherical and annular polyglutamine oligomers by partitioning monomer. Nat Struct Mol Biol 11:1215–1222

Walters RH, Murphy RM (2009) Examining polyglutamine peptide length: a connection between collapsed conformations and increased aggregation. J Mol Biol 393:978–992

Walters RH, Murphy RM (2011) Aggregation kinetics of interrupted polyglutamine peptides. J Mol Biol 412:505–519

Warrass R, Wieruszeski JM, Boutillon C, Lippens G (2000) High-resolution magic angle spinning NMR study of resin-bound polyalanine peptides. J Am Chem Soc 122:1789–1795

Warrick JM, Morabito LM, Bilen J, Gordesky-Gold B, Faust LZ, Paulson HL, Bonini NM (2005) Ataxin-3 suppresses polyglutamine neurodegeneration in *Drosophila* by a ubiquitin-associated mechanism. Mol Cell 18:37–48

White SH (1994) Global statistics of protein sequences: implications for the origin, evolution, and prediction of structure. Annu Rev Biophys Biomol Struct 23:407–439

Williamson TE, Vitalis A, Crick SL, Pappu RV (2010) Modulation of polyglutamine conformations and dimer formation by the *N*-terminus of huntingtin. J Mol Biol 396:1295–1309

Zanuy D, Gunasekaran K, Lesk AM, Nussinov R (2006) Computational study of the fibril organization of polyglutamine repeats reveals a common motif identified in β-helices. J Mol Biol 358:330–345

Zhang H, Li Q, Graham RK, Slow E, Hayden MR, Bezprozvannya I (2008) Full length mutant huntingtin is required for altered Ca^{2+} signaling and apoptosis of striatal neurons in the YAC mouse model of huntington's disease. Neurobiol Dis 31:80–88

Chapter 12
Protein Misfolding and Toxicity in Dialysis-Related Amyloidosis

John P. Hodkinson, Alison E. Ashcroft, and Sheena E. Radford

Abstract Amyloid formation from the protein precursor β_2-microglobulin (β_2m) is implicated in the human pathology dialysis-related amyloidosis (DRA). The first report of the clinical symptoms of the pathology was noted in 1975, but the amyloid basis for the pathology was not realized until 1980, and the precursor protein was identified as β_2m in 1985 (Warren and Otieno, Postgrad Med J 51:450–452, 1975; Assenat et al., Nouv Presse Med 9:1715, 1980; Gejyo et al., Biochem Biophys Res Commun 129:701–706, 1985). Here we discuss the physiological role of β_2m as a component of the major histocompatibility complex class I (MHC I); the clinical implications of DRA; the current knowledge of β_2m aggregation resulting from *in vitro* models; and how this has informed our understanding of the molecular basis of the disease. In particular, the role of toxic oligomers in the pathology of DRA is considered.

Keywords β_2-Microglobulin • Dialysis-related amyloidosis • Fibril structure Oligomer • Major histocompatibility complex I

12.1 Structure and Function of β_2m

12.1.1 MHC I Structure and Function

The MHC I is an essential molecular complex for the adaptive immune system, which allows differentiation between self and non-self. It is a membrane glycoprotein, which is made up of a transmembrane heavy chain (α-chain) and β_2m.

J.P. Hodkinson • A.E. Ashcroft • S.E. Radford (✉)
Astbury Centre for Structural Molecular Biology, Institute of Molecular and Cellular Biology, University of Leeds, Leeds LS2 9JT, UK
e-mail: s.e.radford@leeds.ac.uk

F. Rahimi and G. Bitan (eds.), *Non-fibrillar Amyloidogenic Protein Assemblies—Common Cytotoxins Underlying Degenerative Diseases*, DOI 10.1007/978-94-007-2774-8_12,

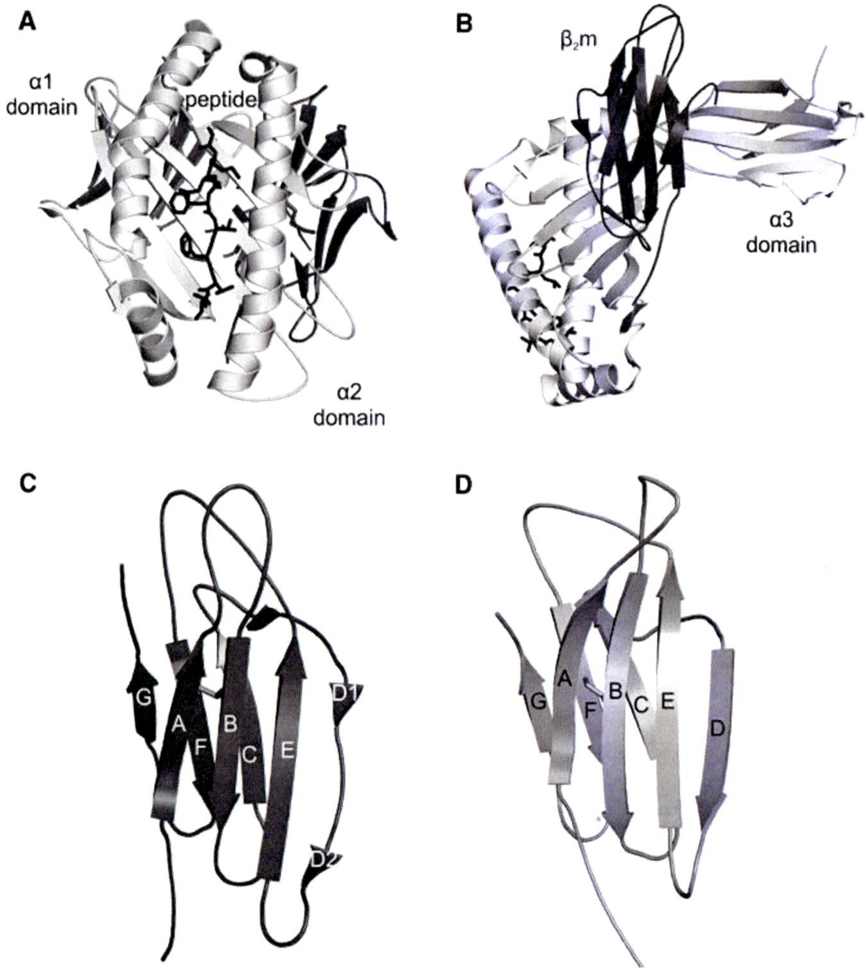

Fig. 12.1 (**A**) and (**B**) The crystal structure of the extracellular portion of the MHC-1 complex. The bound peptide from the human lymphotrophic virus is shown in *black wire*. β_2m is shown in *dark grey ribbon* and the heavy chain is shown in *white* (1IM3) (Gewurz et al. 2001). (**C**) The NMR solution structure of monomeric β_2m at pH 7.0 (1JNJ) (Verdone et al. 2002). (**D**) The X-ray crystal structure of monomeric β_2m at pH 5.6 (1LDS) (Trinh et al. 2002)

The extracellular portion of the complex has been crystallized, and the structure with a bound peptide from the human T lymphotrophic virus type 1 is shown in Fig. 12.1a, b (Gewurz et al. 2001; Sundaram et al. 2004). The heavy chain has three extracellular domains (α1, α2, and α3), a hydrophobic section, which anchors the molecule to the membrane, and a short hydrophilic sequence that protrudes into the cytoplasm. Of these, the α1 and α2 domains form a groove that can bind a range of peptides (Garcia and Adams 2005). Antigenic peptides, which are bound in this groove, are formed by proteasomal degradation of proteins from both self, and intracellular pathogens (Cascio et al. 2001). The role of the MHC complex is to present these peptides at the cell surface allowing their interrogation and, if necessary, the subsequent destruction of the cell by $CD8^+$ T-lymphocytes (Raghavan et al. 2008).

12.1.2 Role of β_2m Within the MHC I Complex

The pathway which brings the three components of the MHC complex together (the α-chain (heavy chain), β_2m (light chain) and an endogenous peptide) is complex, and only a brief overview is presented here. A detailed account of the process and an illustration of the vital role that β_2m plays in antigen presentation to cytotoxic $CD8^+$ T cells can be found in published reviews (Cresswell et al. 2005).

Both β_2m and the α-chain of the MHC complex are synthesized on the rough endoplasmic reticulum (ER) where folding of the α-chain is facilitated by the binding of calnexin. Upon correct folding of the α-chain, β_2m associates and calnexin dissociates (Peaper and Cresswell 2008). The α-chain/β_2m complex binds to the peptide-loading complex (PLC) the peptide is loaded into the groove formed by the a1/a2 domains (Dong et al. 2009). Peptide binding to the heavy chain triggers dissociation of the PLC and transit of the intact complex *via* the Golgi apparatus to the surface of all nucleated cells. Mouse models which have the β_2m gene knocked out show that no MHC class I complex is presented on the cell surface, indicating the essential nature of β_2m for transport of the α-chain to the cell surface (Zijlstra et al. 1990). Work on Daudi cells (in which β_2m is not expressed) showed that dimerization of the α-chain with β_2m is essential for association of the α-chain with the PLC and also that in the absence of β_2m, misfolding and degradation of the α-chain occurs rapidly (Hughes et al. 1997; Paulsson et al. 2001). It has also been shown that peptide binding is essential for cell surface expression of the MHC I at physiological temperature and it is thought that peptide binding is essential for complex stability (Ljunggren et al. 1990; Wearsch and Cresswell 2008). Moreover, MHC heavy chain binding has been shown to reduce the dynamic properties of β_2m leading to the view that dissociation of the light chain with concomitant increased dynamics is key to initiating β_2m aggregation (Hodkinson et al. 2009).

12.1.3 Structure of β_2m

β_2m is a 99 amino acid, non-polymorphic protein that exhibits a typical immunoglobulin fold comprising seven β-strands organized into a β-sandwich structure. This structure is stabilized by an inter-strand disulfide bridge between residues 25 and 80 (Fig. 12.1c, d). In both the MHC I bound form and in β_2m monomers in solution the β-strands are arranged in an all antiparallel arrangement and the edge strand 'D' exhibits a β-bulge (Khan et al. 2000; Verdone et al. 2002) (Fig. 12.1a–c). However, the structure of monomeric β_2m first determined by X-ray crystallography revealed a conformation in which the β-bulge was absent in this strand (Trinh et al. 2002) (Fig. 12.1d). Studies using NMR revealed that the conformation of the monomer with a straight D-strand was only rarely populated (Trinh et al. 2002); residues that form this strand appearing instead to be highly dynamic in this state (Verdone et al. 2002). As the β-bulge has been suggested as a feature, which protects against edge strand aggregation, it has been proposed that this continuous D-strand could facilitate self-association (Richardson and Richardson 2002; Trinh et al. 2002). The β_2m sequence contains five proline residues. While four of these are in the *trans* conformation, proline 32 (in the B–C loop) is in the *cis* conformation in native β_2m. As discussed in more detail below, the isomerization of this residue is very important in controlling the aggregation of β_2m into amyloid fibrils (Jahn et al. 2006).

12.1.4 Production/Life Time of β_2m in Healthy Individuals

In vivo, 2.4 ± 0.7 mg kg^{-1} day^{-1} of β_2m is synthesized and, upon dissociation from the MHC complex, free monomeric β_2m is distributed throughout the extracellular space and passes into the vasculature (Karlsson et al. 1980; Floege et al. 1991). The concentration of β_2m in the plasma is 1.9 ± 0.3 mg L^{-1}. In healthy individuals, β_2m is eliminated from the blood *via* the kidney by glomerular filtration and proteolytic catabolism following proximal tubular resorption (Floege et al. 1991; Kay 1997; Floege and Ketteler 2001). Kinetic studies indicate that an alternative pathway exists which accounts for a small fraction of β_2m catabolism and excretion; however, thus far this pathway has not been identified.

12.2 Dialysis-Related Amyloidosis

12.2.1 Production/Life Time of β_2m in Pathology

The first report of the clinical symptoms of the pathology of dialysis related amyloidosis was noted in 1975, but the amyloid basis for the pathology was not realized until 1980, and the precursor protein was first identified as β_2m in 1985 (Warren and Otieno 1975; Assenat et al. 1980; Gejyo et al. 1985). As noted previously, β_2m is eliminated from the body *via* the kidney. In disease states where kidney function is reduced, clearance of β_2m from the blood is, therefore, reduced and dialysis therapy designed to function in place of the kidney does not effectively remove the protein (Gejyo et al. 1986a; Ohashi 2001). This results in an increase in the plasma concentration of β_2m to 25–60-fold above normal levels, depending on the residual kidney function (Gejyo et al. 1986a, reviewed also in Jahn and Radford 2005). Depending on the type of dialysis procedure employed, the concentration of β_2m in serum is substantially reduced; however, serum β_2m concentrations remain at least tenfold greater than in patients with full renal function. The most effective form of dialysis is daily high flux hemodialysis, which results in an average β_2m serum level of <20 mg L^{-1}, while a number of methods (hemodiafiltration, high flux hemodialysis and peritoneal dialysis) result in average serum levels of 20–30 mg L^{-1}, and cuprophan hemodialysis results in >30 mg L^{-1} average β_2m serum concentration (Sethi et al. 1989;

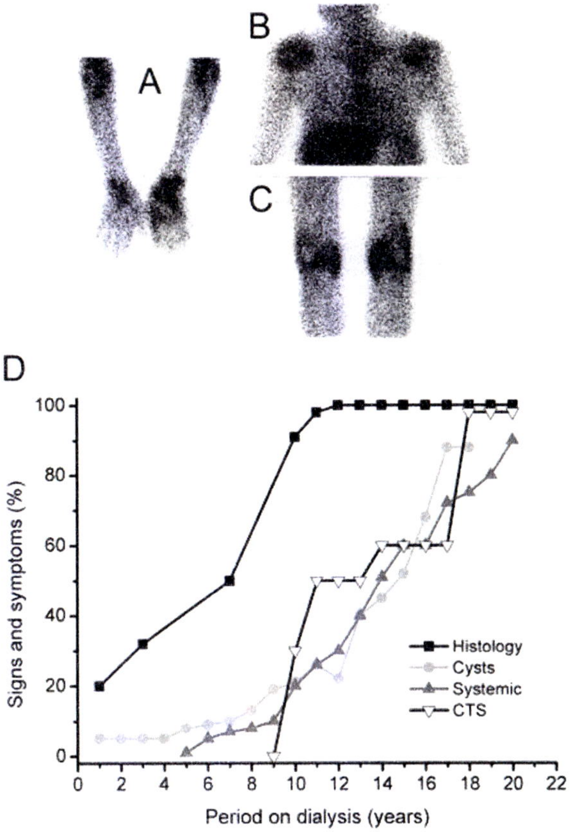

Fig. 12.2 (**A–C**) Images of a DRA patient taken using ^{111}In β_2m scintigraphy. Accumulation of β_2m is demonstrated in (**a**) the wrists and elbows, (**B**) the shoulders and (**C**) the knees. Accumulation in the liver results from uptake of the tracer by the reticuloendothelial system and does not indicate fibril deposition (Linke et al. 2000). (**D**) Graph sho wing the progression of DRA as monitored by histological detection of articular β_2m fibrils (■), radiological appearance of bone cysts (●), histological detection of systemic β_2m fibrils (▲), and carpal tunnel syndrome (CTS) (▽). [Figure adopted from Horl (2004) using data from (Charra et al. 1988; Destrihou et al. 1991; Jadoul et al. 1997b; Miyata et al. 1998)]

Tielemans et al. 1989; Koda et al. 1997; Locatelli et al. 1999; Raj et al. 2000; Lin et al. 2001). It must be pointed out that a general paucity of good quality studies in this area has been reported by a Cochrane review (Rabindranath et al. 2006).

12.2.2 Pathological Fibril Deposition

Over the course of many years of renal insufficiency the resulting increase in β_2m blood concentration coincides with the deposition of amyloid fibrils, composed primarily of full-length β_2m in the osteoarticular tissues, in a condition known as DRA (Otsubo et al. 2009) (Fig. 12.2). The concentration of β_2m in the blood does not correlate well with fibril deposition in the joints of DRA sufferers and, therefore, it is suspected that β_2m concentration is not the only factor, which influences the rate and extent of fibril formation and deposition (Gejyo et al. 1986b). Other factors include: truncation of the *N*-terminal six residues to produce the highly amyloidogenic variant ΔN6 β_2m which has been found in *ex vivo* amyloid fibrils (Esposito et al. 2000; Stoppini et al. 2005); modification of β_2m with advanced-glycation end-products (AGE), which has been observed in DRA patients and has been proposed as key to β_2m pathogenesis (Miyata et al. 1996; Niwa et al. 1997; Niwa 2001); the age of the patient (Destrihou et al. 1991); the duration of kidney failure (Davison 1995); and the interaction of β_2m monomer or aggregates with physiological factors, e.g., collagen (Relini et al. 2006, 2008), glycosaminoglycans (Athanasou et al. 1995; Yamamoto et al. 2005), apolipoprotein E (Stoppini et al. 2005), serum amyloid protein (SAP) (Pepys 2006), Cu^{2+} ions (Eakin and Miranker 2005), lysophospholipids (Ookoshi et al. 2008; Pal-Gabor et al. 2009), and non-esterified fatty acids (Hasegawa et al. 2008). Conflicting evidence exists as to whether dialysis itself is responsible for the disease. Deposition of β_2m amyloid fibrils has been reported in uremic patients who have not undergone dialysis, which indicates that dialysis itself is not the sole cause of the disorder (Zingraff et al. 1990; Moriniere et al. 1991). Conversely, high β_2m levels have been reported in other patients for years without DRA-like symptoms (Fuchs et al. 1992).

12.2.3 Clinical Manifestations of the Pathology of DRA

The affinity of β_2m for collagen results in the predominantly osteoarticular deposition of β_2m fibrils (Linke et al. 2000) and determines the clinical manifestations of DRA (Fig. 12.2). β_2m amyloid fibril deposition on the surface of cartilage in peripheral joints can be detected histologically following a few months of dialysis and later, deposits can be detected in the synovium and ultimately in adjacent bones (Jadoul et al. 1997a; Garbar et al. 2000). Joint and bone invasion gradually increase in prevalence with period on dialysis until up to 100% of the dialysis population is affected (Destrihou et al. 1991; Mccarthy et al. 1994). However, clinical symptoms are rarely experienced before 5 years of renal replacement therapy, with most patients presenting with carpal tunnel syndrome (CTS) which later progresses to arthropathy of the peripheral joints and spine (Floege and Ehlerding 1996; Floege and Ketteler 2001). Bilateral chronic arthalgias and arthropathy can result in joint destruction and severe morbidity, including pathological fractures due to bone cysts, particularly in the femoral neck (Bardin and Kuntz 1987; Onishi et al. 1991; Ferreira et al. 1995; Saito and Gejyo 2006). Deposition of β_2m amyloid has also been noted in visceral locations in long-term (>10 years) dialysis patients; however, these are largely asymptomatic (Bandini et al. 2001; Yusa et al. 2001; Mount et al. 2002; Shimizu et al. 2003).

Joint destruction is thought to result from inflammation caused by recruitment of macrophages and monocytes and the subsequent release of pro-inflammatory cytokines (Hou and Owen 2002; Kazama et al. 2006; Morten et al. 2007). AGE modification has been implicated in this process as it has been shown that AGE modified β_2m is a component of *ex vivo* amyloid, and can stimulate monocyte activation, as well as interleukin-6, tumor necrosis factor-α and interferon-γ release (Imani et al. 1993; Miyata et al. 1993, 1998). The sheer bulk of β_2m fibrillar material may cause some of the symptoms of DRA, including CTS due to median nerve compression, and fractures due to reduction in bone integrity (Onishi et al. 1991; Chary-Valckenaere et al. 1998). Almost 100% of patients exhibit some form of debilitating clinical symptoms by 15 years of treatment and an example of the bilateral β_2m amyloid deposition, which can cause considerable morbidity, is shown in Fig. 12.2 (Miyata et al. 1998).

Dialysis-related amyloidosis has a higher than average prevalence in Japan where a shortage of kidneys available for transplant and long life expectancy result in a large (17,527 in 2006) and increasing number of patients surviving on dialysis for more than 20 years (Yamamoto et al. 2009). Conversely, the incidence of DRA has been reported to be decreasing due to the improvement in dialysis technology (Yamamoto et al. 2009).

12.2.4 Current Treatment Options

Prevention of DRA has been based on a reduction in β_2m serum concentration by using improved high flux, high efficiency membranes which increase the rate of water and solute removal and, crucially for DRA, exhibit greater permeability to 'middle molecules' e.g., β_2m (Cheung and Leypoldt 1997; Leypoldt et al. 1999). However, this can only delay symptoms as the β_2m serum concentration remains tenfold higher than in patients with full renal function (Destrihou et al. 1991, 1994; Koda et al. 1997). Improved biocompatibility of membranes has also paid dividends in reducing disease progression (Hakim et al. 1996; Canaud 2009). These improvements have been combined with increased purity of dialysate leading to reduced endotoxin and AGE contamination (Baz et al. 1991; Schwalbe et al. 1997). β_2m adsorption columns such as Lixelle™ can be connected in series with a conventional dialysis set up (at a considerable expense), and reduce the circulating β_2m concentration (Abe et al. 2003). The manufacturers also report reduced circulating inflammatory markers, reduced morbidity and improved quality of life (Furuyoshi et al. 1998; Kutsuki 2005).

Currently treatment of DRA is largely symptomatic with non-steroidal anti-inflammatory drugs used to relieve chronic pain and inflammation, while intra-articular corticosteroids are used to reduce single joint inflammation (Horl 2004). Surgery can also be used to reduce fibril bulk and alleviate nerve compression. However, the only curative procedure currently is kidney transplantation (Jadoul et al. 1997a; Campistol 2001). Renal transplantation prevents further fibril deposition and improves osteoarticular symptoms (initially due to steroid therapy, but improvement persists following cessation of drug treatment); however, regression of fibril deposits is not expected (Bardin et al. 1995; Jadoul et al. 1997a). Demand for renal replacement therapy has been projected to increase by up to 50% within 10 years in developed countries. This will require improved therapies as the elderly will form the bulk of this increase and are less suitable for kidney transplant (Roderick et al. 2004).

12.3 β_2m Amyloid Formation *In Vitro*

12.3.1 Protein Misfolding

As discussed previously, the precise mechanism of amyloid formation in DRA is unknown, although numerous factors have been suggested to influence the advent and impact of β_2m amyloid in the disease. Biochemical and biophysical investigations *in vitro* have been used to shed light on the mechanism of β_2m fibril formation *in vivo*, and also as a model system, which may yield general insights into amyloid fibril formation, which could be applied to other pathologies (Platt and Radford 2009).

In vitro investigation has been undertaken since the discovery of β_2m as an *in vivo* amyloid-forming protein in 1985 (Gejyo et al. 1985; Gorevic et al. 1985), and has resulted in the surprising discovery that in the absence of denaturant, and at physiological temperature and pH, β_2m does not spontaneously form amyloid fibrils even when incubated for long periods of time at high protein concentration (McParland et al. 2000; Chiti et al. 2001). This realization has led to the investigation of alternative conditions under which β_2m can form fibrils *in vitro* and may indicate the important role of biological factors in the pathogenesis of DRA. Numerous sets of conditions have been discovered which result in the generation of β_2m amyloid fibrils *in vitro* and some have been suggested to relate to *in vivo* scenarios. The most important of these findings are discussed below.

The discovery that agitation at low pH results in the rapid and extensive formation of β_2m into amyloid fibrils has stimulated much research to shed light on the fibril-forming properties of β_2m and amyloid formation in general (Kad et al. 2001; Smith et al. 2003; Gosal et al. 2005; Radford et al. 2005). Up to 60% of the sequence of β_2m is predicted to be highly amyloidogenic judged by algorithms and by studies of synthetic peptides (Platt and Radford 2009; Routledge et al. 2009) and, therefore, it may not be surprising that unfolding the protein in acidic pH results in rapid amyloid formation. These studies also indicate that structure in the native state protects against fibril formation, while residual structure in acid-denatured β_2m also modulates the amyloid propensity of the unfolded state (Kad et al. 2001; Smith et al. 2007; Platt et al. 2008; Routledge et al. 2009).

Perturbation of the sequence of β_2m modulates the amyloid fibril-forming propensity of the acid-unfolded state and several regions have been purported to be important in the aggregation of the protein. Studies based on peptide fragments have suggested that the region 21–40 (that spans strands B and C in the intact protein) is able to aggregate into amyloid fibrils in isolation (Kozhukh et al. 2002), whilst similar studies of shorter peptides spanning the entire sequence of β_2m showed that residues 59–71, which encompass the native E strand, are highly aggregation prone (Jones et al. 2003b). This region is highly enriched in aromatic residues and an extensive mutagenesis study showed that it is uniquely important in facilitating both nucleation and elongation of fibrils in the intact protein, presumably by forming a key interacting surface for self-assembly (Platt et al. 2008; Routledge et al. 2009). Finally, residues in the DE loop have been shown to be strained in human β_2m, promoting its aggregation (Esposito et al. 2008), whilst a seven-residue stretch in the *C*-terminal region of human β_2m (residues 83–89) was shown to favor fibril formation (Ivanova et al. 2004). By contrast, with the above regions that appear to promote aggregation, the A-strand of β_2m has been proposed to have a protective role, stabilizing the native state and disfavoring aggregation (Esposito et al. 2000).

These findings indicate that exposure of key residues or stretches of amino acids are important for fibril formation from β_2m and that subtle interactions among those residues are influential in determining the amyloid-forming potential of the protein. The highly amyloidogenic nature of the acid-unfolded state of β_2m and the non-fibrillogenic nature of the native protein at pH 7 could indicate that unfolding of the native structure is a prerequisite for fibril formation *in vivo*. This mechanism has been demonstrated in other globular proteins (Kelly 1998; Uversky and Fink 2004; Chiti and Dobson 2009). It is important to note, however, that simply denaturing the folded state using urea does not result in extensive fibril formation (McParland 2001).

Several groups have investigated fibril formation from β_2m at pH 7 using a number of different approaches, the most significant of which will be considered here. One approach has been to fragment fibrils grown *de novo* at pH 2.5 and use these fibril fragments to seed fibril elongation at pH 7 in the presence of natively folded monomer (Myers et al. 2006a). The seeds were stabilized with heparin, and showed that although wild-type β_2m is not capable of nucleating fibril growth at neutral pH, elongation is possible. Seeded growth at pH 7 has been shown to be enhanced by trifluoroethanol (TFE) at concentrations less than 20% (v/v) and maximally at 10% (v/v) (Yamamoto et al. 2004b; Yamaguchi et al. 2006). A similar effect was observed upon addition of sodium dodecyl sulfate (SDS) below the critical micelle concentration (Yamamoto et al. 2004a). The investigators hypothesized that *in vivo*, phospholipids may have an effect analogous to SDS and have extended their research to include lysophospholipids, which facilitate nucleation of β_2m under physiological conditions (Ookoshi et al. 2008).

Numerous studies from the Miranker laboratory have implicated Cu^{2+} in β_2m fibril formation, including determination of the 3D structure of β_2m oligomers formed in the presence of Cu^{2+} (Morgan et al. 2001; Eakin et al. 2002, 2004, 2006; Eakin and Miranker 2005; Calabrese and Miranker 2007; Calabrese et al. 2008). These studies indicate that β_2m populates an alternative conformation or 'activated' state, which is native-like and, based on these findings the authors proposed a model involving the formation of fibrils composed of native-like monomers. The authors argue that Cu^{2+} ions released from dialysis membranes (e.g., Cuprophan) cause β_2m fibrillogenesis and thus propose that the effect of Cu^{2+} ions on β_2m fibrillogenesis *in vitro* is relevant to fibril formation *in vivo*. Studies evaluating the effect of cuprophan membranes and, therefore, Cu^{2+} on DRA progression are confounded by the relatively poor clearance of β_2m by Cuprophan membranes.

An extensively used approach to investigate fibril formation from β_2m at neutral pH is protein engineering. An alanine scan at 12 positions throughout the sequence of β_2m showed that all variants had decreased stability relative to wild-type β_2m; however, only the proteins containing mutations in the *N*- and *C*-terminal regions (I7A, V9A, V93A, R97A) led to *de novo* fibril growth at pH 7 (Jones et al. 2003a;

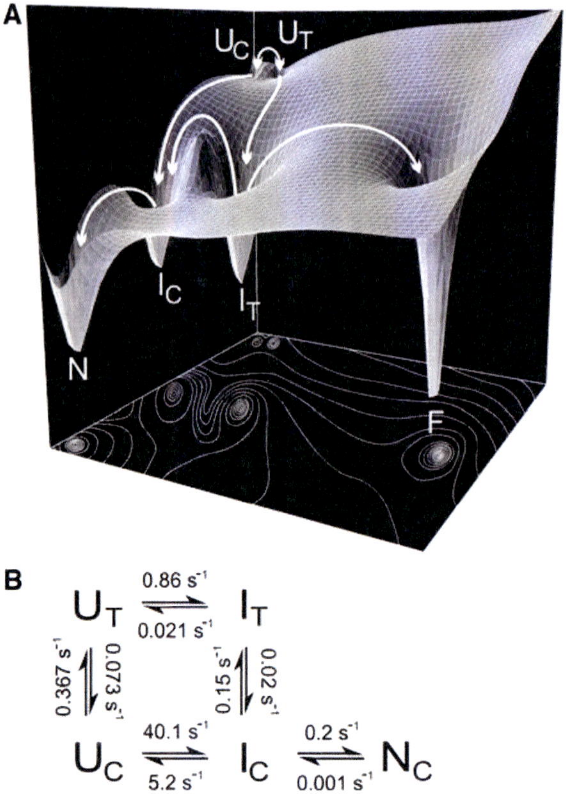

Fig. 12.3 Folding landscape of β_2m. (**A**) Proposed folding energy landscape for β_2m *in vitro* at pH 7, 37°C showing the pathway from the unfolded (U) to intermediate (I_T and I_C), native (N) and fibrillar (F) states. (**B**) Folding pathway with microscopic rate constants (Reproduced from Jahn et al. 2006)

Smith et al. 2003). A number of variants, which have been found *in vivo*, have also been investigated for their effect on amyloidogenicity *in vitro*. ΔK58 β_2m (removal of lysine at position 58) has been shown to be conformationally heterogeneous at pH 6.8 (Mimmi et al. 2006); however, the effect of this modification on fibril formation is not convincing (Corlin et al. 2009). N17D β_2m, which mimics a deamidation product of β_2m found in *ex vivo* fibrils, does not enhance fibril formation relative to the wild-type sequence (Kad et al. 2001). The most interesting variant, which has been identified in *ex vivo* fibrils and subsequently investigated, is the six-residue truncated variant ΔN6 β_2m. ΔN6 β_2m is the major modification found in *ex vivo* fibrils and has been shown to be highly amyloidogenic (Bellotti et al. 1998; Esposito et al. 2000; Eichner and Radford 2009). ΔN6 β_2m also has another intriguing property in that the native *cis* conformation of proline 32 which is seen in wild-type β_2m, is thought to be in the *trans* conformation in this variant (Eichner and Radford 2009). The significance of a *trans* proline at position 32 is realized when the folding pathway of wild-type β_2m is considered (Jahn et al. 2006) (Fig. 12.3). Refolding of β_2m from the unfolded state includes a slow proline isomerization step, which results in the accumulation of an intermediate in which proline 32 is in the *trans* conformation and, therefore, has been termed I_T. I_T has been suggested to be the link between the folding and aggregation landscapes for β_2m at neutral pH and its population has been correlated with the ability to form amyloid fibrils at neutral pH (Jahn et al. 2006). ΔN6 β_2m, the wild-type β_2m refolding intermediate and the wild-type Cu^{2+} activated state are analogous by analytical gel filtration, suggesting that the structural changes which result from proline 32 isomerization could be common in these proteins and, therefore, generally important in β_2m fibrillation at neutral pH.

12.3.2 Oligomerization

Oligomerization is an essential part of the transition from soluble monomer to insoluble fibril in any amyloid pathology, and much time has been invested in unraveling the nature of these sometimes ephemeral complexes. The simplest nucleation–polymerization model of amyloidogenesis requires the formation of a series of larger and thermodynamically favorable oligomers until the attainment of the nucleus, which is followed by rapid, thermodynamically favorable polymerization (Xue et al. 2008). Much evidence of this general form exists for β_2m fibril formation *in vitro*; however, nucleation-independent polymerization has also been observed under specific conditions (McParland et al. 2000; Gosal et al. 2005; Radford et al. 2005). Oligomers are fascinating to biophysicists as they pose interesting structural, thermodynamic, and kinetic challenges, and they are also very important in understanding and treating disease (Tsigelny et al. 2008; Haydar et al. 2009). If the formation of specific oligomers is necessary for amyloidogenesis then these entities may represent targets for pharmacological intervention. It is, of course, clear that many branched or alternative pathways may exist in any one pathology under a particular set of conditions, meaning that one target may be easily circumvented or superseded *in vivo*, or even be irrelevant to disease progression from the outset (Kelly 1998; Souillac et al. 2002; He et al. 2010). It has become increasingly evident that oligomers may play a direct role in toxicity with small soluble oligomers shown to be toxic, initially in Alzheimer's disease, and now in other degenerative amyloid diseases as well (Haass and Selkoe 2007). In DRA, the formation and role of oligomers in the disease is still under investigation and whether the amyloid cascade hypothesis (in which the insoluble fibrils are the primary cause of pathology), or the toxic-oligomer hypothesis (in which soluble oligomers are implicated as the primary pathological structure) will prevail remains unclear.

At acidic pH and with agitation, oligomers of β_2m have been identified by mass spectrometry and analytical ultra-centrifugation (AUC), showing that under conditions that favor non-nucleation-dependent amyloid fibril formation (pH 3.6, high salt concentration), species ranging from monomer through 13-mer could be detected. By contrast, under conditions which favor nucleation-dependent kinetics (pH 2.5, low salt concentration), only monomers, dimers, trimers and tetramers could be detected (Smith et al. 2006a). This study allows us to suppose that under conditions that result in the generation of long straight amyloid fibrils the nucleus, which precedes rapid elongation, is approximately composed of ≤ 8 β_2m monomers. This is consistent with a later study, which estimated the structural nucleus size under similar conditions to be a hexamer (Xue et al. 2008). Recent research carried out using mass spectrometry coupled with ion-mobility spectrometry (ESI–MS–IMS) concurs with the findings discussed above regarding the size and distribution of the oligomers formed during the lag phase at pH 2.5. However, this study goes much further and indicates that the oligomers formed prior to fibril formation are elongated as opposed to globular structures. The study also indicates that the oligomers formed increase in stability concomitant with their increase in size and that the monomeric components, which make up the oligomers are structurally collapsed compared with the free monomer in solution (Smith et al. 2010).

Much research has been published regarding the role of Cu^{2+} in β_2m oligomerization and a number of attractive structures have been proposed from chemical modification studies monitored by mass spectrometry, modeling, and X-ray crystallography (Fig. 12.4a, b). The formation of soluble β_2m oligomers at neutral pH in the presence of Cu^{2+} was first demonstrated using AUC which showed that dimeric and tetrameric oligomers are populated rapidly (<20 h) in the presence of 200 μM Cu^{2+} and 0.5 M urea at pH 7.4, 37°C (Eakin et al. 2004). These oligomers were shown to be unstable in the absence of Cu^{2+} as the addition of the chelating agent EDTA resulted in the disappearance of the oligomers. However, after 1 week the oligomers were shown to be stable in the absence of Cu^{2+} ions. The study suggested that these oligomers are obligate on-pathway intermediates to amyloid fibril formation. Further work has confirmed the rapid formation (<20 h) of β_2m oligomers in the presence of Cu^{2+} and their continued stability even after removal of the divalent metal (following 5 days incubation with Cu^{2+}) using analytical size-exclusion chromatography (Calabrese and Miranker 2007). Other similar studies have been carried out using mass spectrometry, dynamic light scattering (DLS) and analytical size-exclusion chromatography. They concur that rapid and extensive oligomerization of β_2m occurs in the presence of Cu^{2+} and indicate that species up to hexamer can be observed (Antwi et al. 2008; Srikanth et al. 2009). The structure of a Cu^{2+}-induced dimer of β_2m has been inferred by a covalent-labeling and mass-spectrometry study which suggested that the wild-type dimer formed in the presence Cu^{2+} consisted of a face-to-face ABED interface (Mendoza et al. 2010) (Fig. 12.4a). The structure of a hexameric species of the β_2m variant H31F formed in the presence of Cu^{2+} has also been solved by X-ray crystallography and has been shown to be a hollow ring with a central pore 14 Å wide and an outer diameter of 55 Å (Calabrese et al. 2008) (Fig. 12.4b). In this structure, the D-strand β-bulge present in the wild-type monomer is present and forms one of two interfaces between monomers. The second interface is a face-to-face interaction between the ABED β-sheets of two adjacent monomers. The oligomers formed in the presence of Cu^{2+}, however, do not appear to proceed to long, straight amyloid fibrils and could indeed reduce the ability of β_2m to form amyloid fibrils by sequestering the available pool of β_2m monomers in dead-end, off-pathway oligomers (Hodkinson 2009).

Other oligomers formed from β_2m variants have been reported including a crystallographic dimer of P32A β_2m which the authors suggest is prone to self-assembly due to the ablation of the β-bulge in strand-D and structural rearrangement

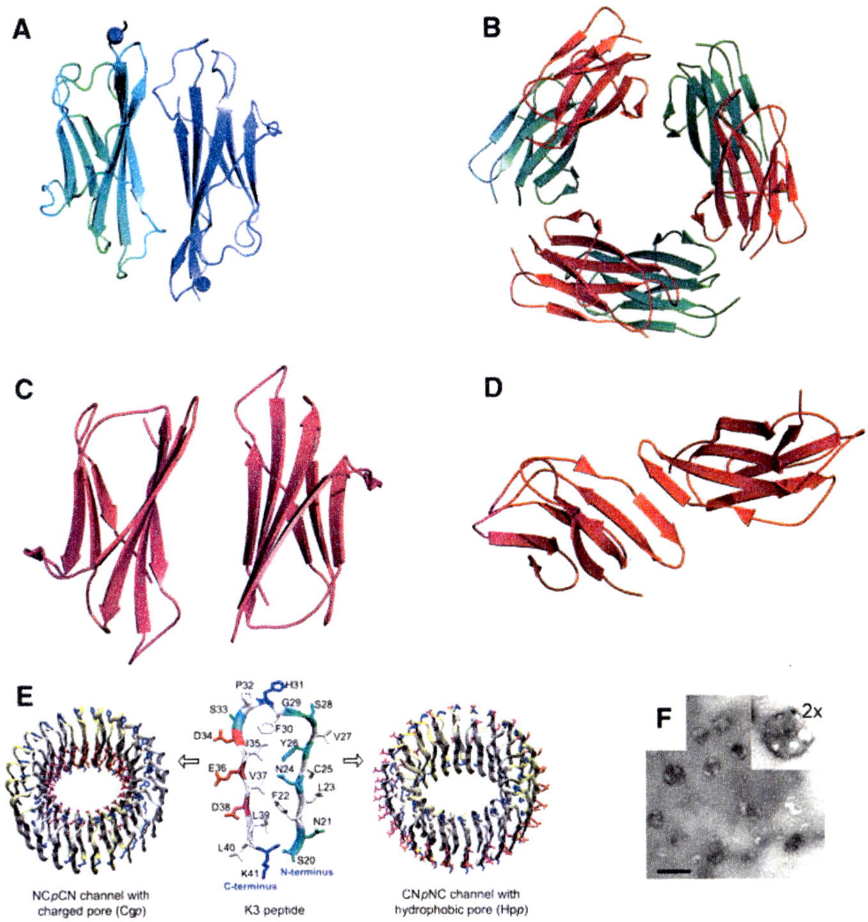

Fig. 12.4 Various oligomeric states of β_2m. (**A**) Model for a dimer of wild-type β_2m proposed based on MS methods and energy minimization. The model has the four stands, ABED acting as a face to face interacting surface (Mendoza et al. 2010). (**B**) Hexameric oligomer of H31F β_2m formed in the presence of Cu^{2+} ions solved using X-ray crystallography (3CIQ) (Calabrese et al. 2008). (**C**) Dimer of P32A β_2m formed as a result of crystallographic packing in the presence of Cu^{2+} ions (2F8O) (Eakin et al. 2006). (**D**) Dimer of H31Y β_2m in which the D-strand acts as the interaction surface during crystal packing (1PY4) (Rosano et al. 2004). (**E**) Proposed oligomers of the K3 peptide of β_2m which form into ion channels (Mustata et al. 2009). (**F**) Electron micrograph of large oligomers of P5G β_2m (Eichner and Radford 2009)

throughout the protein, notably the bond between residues 31–32 which is in the *trans* conformation (Eakin et al. 2006) (Fig. 12.4c). The authors have used this structure to speculate that the cross-β backbone of the fibril is composed only of the ABED strands in a head to head, tail to tail continuous sheet; however, this structure has been deemed improbable by the above mentioned MS and covalent labeling study (Mendoza et al. 2010). Earlier research on the H31Y β_2m variant also proposes a continuous β-sheet mediated by D-strand interaction; however, in this structure the β-bulge is retained and the interaction surface is solely the D2-strand (Rosano et al. 2004) (Fig. 12.4d).

Oligomers of ΔN6 and P5G β_2m have been detected at pH 7, 37°C in the absence of organic or inorganic additives using size-exclusion chromatography and DLS (Eichner and Radford 2009). These oligomers were shown to consist of up to 30 monomers and these large species have been captured using electron microscopy (EM) (Fig. 12.4f). The authors suggest that these oligomers are formed *via* a generic mechanism applicable to the formation of β_2m amyloid under the variety of conditions employed by different researchers, which have been suggested to involve the rearrangement of the hydrophobic packing in the monomer and isomerization of the bond between residues 31–32 to the *trans* conformation. This mechanism is consistent with previously published data from a number of laboratories (Kameda et al. 2005; Eakin et al. 2006; Jahn et al. 2006). β_2m oligomers have also been observed *in vitro* under physiological conditions when β_2m was incubated with type I collagen and heparin (Relini et al. 2008) or SDS (Hodkinson 2009). β_2m oligomers formed in the presence of SDS under physiological conditions caused a dose-dependent reduction in the lag time of *de novo* amyloid fibril formation and are, therefore, likely to be on-pathway intermediates.

12.3.3 β_2m Oligomer Toxicity

It has been reported that full-length β_2m could form non-selective ion channels in phospholipid bilayer membranes (Hirakura and Kagan 2001). More recently, a fragment of β_2m has been suggested to form pores in kidney cell membranes and facilitate calcium uptake (Mustata et al. 2009). The fragment used in these studies is termed K3 and represents the Ser20–Lys41 portion of β_2m. Using the structure of the fragment previously determined by solid-state NMR and molecular modeling, a structure for the pores has been suggested (Fig. 12.4e) (Iwata et al. 2006; Mustata et al. 2009). Such an organization of the peptides into a pore, however, has yet to be experimentally verified.

β_2m has been shown to induce apoptosis in human leukemia cell lines CCRF-HSB-2, K562, and HL-60; however, the assembly state of β_2m was not defined in these studies (Wu et al. 2001, 2002; Gordon et al. 2002, 2003). The effect of β_2m oligomers has been addressed in particular by two studies, which reached opposing conclusions. Using the MTT reduction assay (Mosmann 1983), multimeric species of β_2m were shown to be responsible for the reduction in viability of SH-SY5Y neuroblastoma cells by up to 25% in a dose-dependent manner. Although the authors concluded that this was due to toxic oligomers, the possibility of the toxicity resulting from fibrils was not excluded. No information was presented on the size, structure or role in fibril formation of the toxic species discussed in this study (Giorgetti et al. 2009). By contrast, a further study addressed these questions by using the MTT assay to test three separate cell lines: SH-SY5Y neuroblastoma cells, RAW

264.7 mouse macrophage cells, and HeLa human cervical carcinoma cells for the toxicity of β_2m monomers, oligomers, and fibrils of varying morphology (Xue et al. 2009). This study showed that cellular toxicity could only be attributed to long, straight β_2m fibrils and not to fibrils of other morphologies or to monomers or pre-fibrillar oligomers. Furthermore, this study concluded that fragmentation of long straight fibrils enhanced their ability to cause toxic effects through a number of mechanisms (Xue et al. 2009). Although oligomers of β_2m have been identified in numerous studies, the identity of those key to the ultimate formation of clinically relevant fibrils is unclear. Little has been reported which differentiates between on- and off-pathway species and the evidence for toxic oligomers of β_2m *in vitro* and *in vivo* is limited. There remains much to learn about species *en route* towards fibril formation and the role of these species in the development of DRA.

12.4 β_2m Fibril Structure

12.4.1 *Ex Vivo β_2m Fibrils*

In order to be designated amyloid, β_2m protein aggregates must be deposited *in vivo*, have a characteristic long, straight and unbranched fibrillar appearance by EM, and give rise to a typical cross-β X-ray fiber-diffraction pattern (Westermark et al. 2007). *Ex vivo* β_2m amyloid fibrils fulfill those criteria. The characteristic X-ray fiber-diffraction pattern which is known as the cross-β pattern shows a meridional reflection at 4.7 Å (which represents the inter-strand distance within the β-sheets) and an equatorial reflection around 10 Å perpendicular to the fibril axis (which represents the inter-sheet distance within protofilaments) (Eanes and Glenner 1968; Smith et al. 2003) (Fig. 12.5d). These reflections indicate that β_2m amyloid fibers have a cross-β structure, which means that the β-strands run perpendicular to the fibril axis (Geddes et al. 1968). *Ex vivo* β_2m amyloid fibers bind the amyloid indicative dyes Congo red and thioflavin T (ThT) showing birefringence upon Congo red binding when visualized using plane polarized light and a fluorescence shift upon ThT binding (Gejyo et al. 1985; Gorevic et al. 1985). Additionally, *ex vivo* β_2m amyloid fibrils are generally found decorated by numerous physiological ligands e.g., SAP, apolipoprotein E (Apo E) and glycosaminoglycans (GAGs) (Athanasou et al. 1995; Alexandrescu 2005; Pepys 2006). *Ex vivo* amyloid fibrils are generally composed of full-length, intact β_2m. However, a number of studies have identified truncations and other modifications of β_2m in *ex vivo* fibrils. The truncations make up 20–30% of recoverable material and are mainly composed of the *N*-terminal truncation ΔN6, though ΔN10, ΔN17 and ΔN19 are also found to significant extent (Linke et al. 1987, 1989a, b; Bellotti et al. 1998; Stoppini et al. 2000, 2005). To a lesser extent, the *C*-terminal truncations ΔC99 and ΔC87 have been reported, as have oxidation and deamidation products (Stoppini et al. 2005). AGE-modified β_2m has also been extracted from *ex vivo* fibrils, e.g., modification has occurred by imidazolone and N(ε)-(carboxymethyl)lysine (Niwa et al. 1997; Mironova and Niwa 2001). It has been suggested that AGE-modified β_2m enhances the inflammatory response to β_2m fibrils by inducing chemotaxis of monocytes and increasing tumor necrosis factor-α and interleukin-1β release from macrophages (Miyata et al. 1994). *Ex vivo* β_2m amyloid can be difficult to characterize structurally and morphologically due to the difficulty of removing amyloid from collagenous deposits. However, studies have shown that fibrils *ex vivo* are highly protease resistant and retain an intact disulfide bridge, and that the monomeric components can be refolded to their native structure following dissolution of the deposit (Bellotti et al. 1998).

12.4.2 In Vitro β_2m Fibrils

Due to the difficulties mentioned about obtaining and analyzing β_2m fibrils *ex vivo*, much scientific endeavor has been directed towards growth and analysis of β_2m amyloid fibrils *in vitro*. This initially led to the finding that intact β_2m does not readily form fibrils at pH 7 even when incubated at high concentration for extended times (McParland et al. 2000). Consequently, fibrils formed at low pH have been extensively researched. Two main morphologies of β_2m fibrils formed at low pH have been discovered: long straight fibrils, which are formed spontaneously in a nucleation-dependent manner at low pH (e.g., pH 2.5) and low ionic strength (Kad et al. 2001; Gosal et al. 2005), and curvilinear or worm-like fibrils which form spontaneously *via* a non-nucleation-dependent pathway at low pH (pH 3.6) and high salt concentrations (Smith et al. 2003; Gosal et al. 2005). These fibril types assemble *via* divergent mechanisms, exhibit distinct morphologies, and have different levels of protection from proteolysis (Myers et al. 2006b). The appeal of analyzing fibril assembly under conditions that more closely resemble those *in vivo* has led to numerous methods of forming β_2m fibrils at pH 7 and subsequent analysis of their relevance to *ex vivo* structures. These studies have shown that β_2m fibrils formed at pH 7 following seeding or protein engineering resemble the morphology of *ex vivo* fibrils and also that seen in fibrils generated *in vitro* at pH 2.5 (Fig. 12.5a–c). These similarities have been shown to extend further than gross morphological characteristics in a study which used EM and Fourier-transform infrared (FT-IR) spectroscopy and demonstrated that long, straight β_2m fibrils formed *in vitro* at pH 2.5, pH 7 or *in vivo* shared a common β-sheet architecture including the length and twist of the constituent β-strands and the secondary structure of the constituent proteins, despite their disparate assembly conditions. The stacking of the β-strands in pH 2.5 and pH 7 fibrils has been shown to be in a parallel, in-register arrangement, ruling out a structure for these fibrils based on the stacking of native-like monomers (Kardos et al. 2005; Fabian et al. 2008; Ladner et al. 2010).

The similarities in the conformational properties of fibrils formed from β_2m *in vitro* at pH 2.5 and pH 7 have led to the suggestion that the pathways of amyloid assembly from the unfolded precursor (pH 2.5) and the native precursor (pH 7) converge to form similar end products (Platt and Radford 2009). However, the end

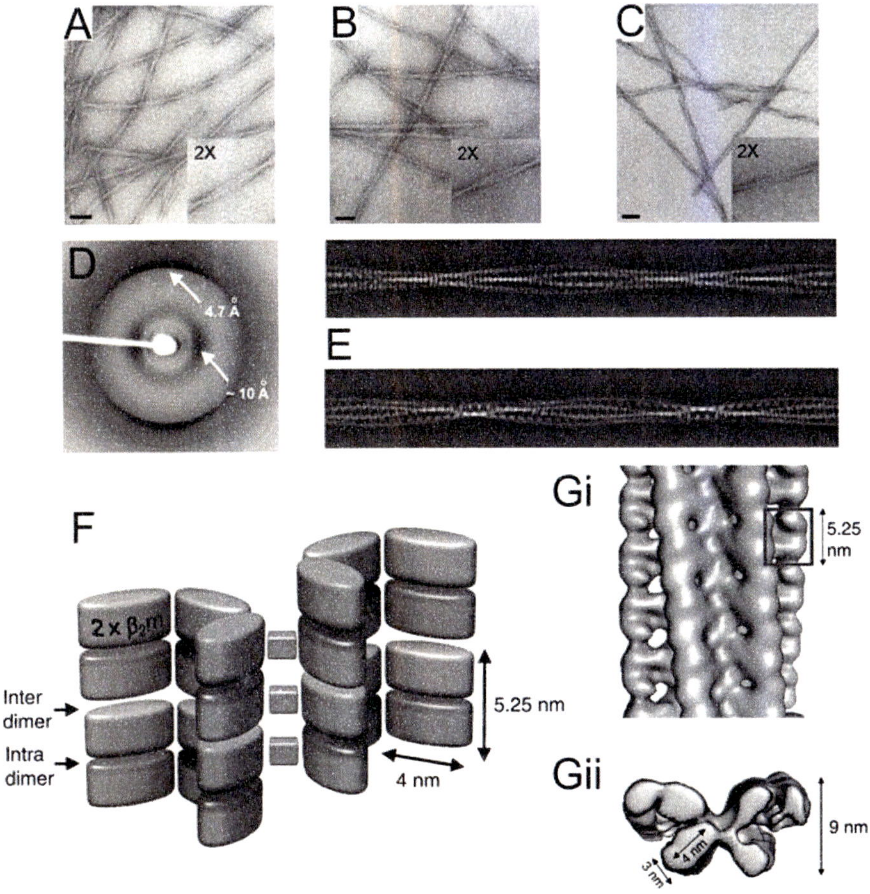

Fig. 12.5 Morphology and structural features of β_2m fibrils. (**A–C**) electron micrographs of fibrils formed from (**A**) wild-type β_2m at pH 2.5, (**B**) wild-type β_2m seeded at pH 7 with fibrils formed at pH 2.5, and (**C**) fibrils formed from a double mutant I7A/P32G unseeded at pH 7. *Scale bar* represents 100 nm (Taken from Jahn et al. 2008). (**D**) X-ray diffraction pattern of β_2m fibrils formed at pH 2.5 (Taken from Platt and Radford 2009). (**E**) Reprojections of type A (*top image*) and type B (*lower image*) β_2m fibrils formed at pH 2.5 (White et al. 2009). (**F**) Schematic showing the globular dimer-of-dimers repeat in β_2m fibrils formed at pH 2.5 (White et al. 2009). (**G**) Cryo-EM structure of β_2m fibrils formed at pH 2.5 showing the *side view* (*i*) and *top view* (*ii*) (Figure adapted from White et al. 2009)

products have been shown to be sufficiently different in that fibrils formed at pH 2.5 rapidly depolymerize at pH 7 and the environment of introduced tryptophan residues is significantly different in fibrils formed under the two conditions, leaving the idea open to debate (Eakin et al. 2006; Kihara et al. 2006; Myers et al. 2006a; Platt et al. 2008; Routledge et al. 2009). The long, straight β_2m fibrils formed at pH 2.5 and pH 7 *in vitro* have been shown to bind ThT and Congo red (and exhibit characteristic red/green birefringence), to exhibit an X-ray diffraction pattern indicative of cross-β structure, to bind the accessory molecules typically found associated with amyloid fibrils (SAP, ApoE and GAGs), and to be composed of two or more protofilaments possessing a left-handed twist (Smith et al. 2003; Gosal et al. 2006; Myers et al. 2006a; Jahn et al. 2008). This indicates that the β_2m fibrils formed *in vitro* are relevant to both DRA and amyloid pathologies in general. Furthermore, it has been demonstrated that antibodies that recognize a generic amyloid epitope, also bind β_2m fibrils formed *in vitro*. These amyloid fibril antibodies (named WO1) were originally raised against fibrils of Aβ1–40 but recognize a generic epitope in a range of amyloid fibrils (O'Nuallain and Wetzel 2002). The finding that these antibodies also bind to β_2m fibrils suggests that these fibrils formed *in vitro* share common structural features with many other fibril types. Importantly, the antibody WO1 does not recognize unfolded β_2m or the native β_2m monomer, suggesting a different structure in the fibrils compared with their precursors (Gosal et al. 2005). Other probes that recognize a fibril-specific epitope include RNA aptamers (Bunka and Stockley 2006). Such aptamers have been developed against worm-like and long, straight β_2m fibrils and shown to recognize some, but not all, amyloid fibrils, suggesting that either the structural feature recognized by the aptamer is not shared throughout all amyloid fibrils or is not solvent-accessible in some (Bunka et al. 2007).

Due to the insoluble and non-crystalline nature of amyloid fibrils, the traditional structural techniques of solution NMR and X-ray crystallography have proven impotent, at least for detailed structural analysis, but many alternative methods to gain structural information have been developed (Makin et al. 2005; Margittai and Langen 2006; Tycko 2006). An interesting method has been used to gain insight into β_2m amyloid structure and involves comparison of the hydrogen/deuterium exchange (HDX) protection of β_2m monomers when free in solution and when a component of the intact fibril (Hoshino et al. 2002). The HDX reaction is monitored using NMR and is made possible by dissolution of the fibrils after different periods of HDX using anhydrous DMSO, which quenches the HDX reaction and dissolves the fibrils into their constituent monomers. This method has been used to reveal which amino acids are integral to the core of β_2m fibrils. The results revealed that residues found in loops of the free monomer are located in the hydrogen-bonded structure within the fibril (Hoshino et al. 2002; Yamaguchi et al. 2004). These studies suggest that the β-sheet structure of the free monomer is remodeled prior to inclusion in the fibril and that the extensive hydrogen bonding network forms the basis of the intriguing mechanical properties of amyloid fibrils (Smith et al. 2006b). A drawback of this method is that NMR HDX measures the average protection at each residue and, therefore, cannot take detailed account of the inherent heterogeneity within amyloid fibrils (Kelly 2002; Goldsbury et al. 2005; Meinhardt et al. 2009).

The inherent heterogeneity of amyloid fibril formation even under controlled conditions poses a significant challenge to the determination of fibril structure. For β_2m fibrils this heterogeneity was revealed by detailed analysis of fibrils formed at pH 2.5 using cryo-EM (White et al. 2009). These studies revealed that β_2m fibrils have a diameter of approximately 20 nm and a typical pattern of cross-over repeats that varied in length between 120 and 185 nm, even within fibrils in the same preparation. In order to achieve a high quality cryo-EM density map, similar particles must be grouped and averaged (Egelman 2000). For β_2m, this type of analysis revealed two types of fibrils, classified as type A and type B, based upon the polarity of the constituent protofilaments. In addition, based on the precise orientation of the protofilaments, fibrils could be classified into 31 groups indicative of immense heterogeneity within these fibril samples (Fig. 12.5e). The class averages indicate that the fibrils are formed from globular units organized as dimer-of-dimers repeats, rather than the more common continuous cross-β core suggested as the underlying structure of other amyloid fibrils (Jimenez et al. 2002; Petkova et al. 2006; Zhang et al. 2009) (Fig. 12.5e–g). Modeling data indicate that β_2m cannot be accommodated into this 3D structure in a natively folded or globally unfolded conformation, nor do previously proposed domain swapped or crystallographic dimers or hexamers suffice (Eakin et al. 2006; Calabrese et al. 2008; White et al. 2009). This work represents an important step forward towards the atomic resolution structure of β_2m fibrils; however, much work remains to be done before a structure in atomic resolution can be achieved. Towards this goal, recent studies using solid-state NMR have yielded spectra of fibrils of β_2m formed at pH 2.5 of excellent quality, despite the relatively large size of the proteins subunits (100 residues), boding well for a full atomic structure of these fibrils in the future (Debelouchina et al. 2010).

12.5 The Impact of Research into Fibril Formation *In Vitro* on the Understanding of the Pathology of DRA

Great strides have been made in developing our understanding of amyloid formation and amyloidosis, in general (Pepys 2006; Chiti and Dobson 2009). However, a complete understanding of the mechanism and structures *en route* from the β_2m monomer to fully assembled amyloid fibrils remains elusive when considering fibril formation both *in vitro* and *in vivo* (Drueke and Massy 2009; Platt and Radford 2009) (Fig. 12.6). In the case of β_2m amyloidosis, the temporal disconnection between the histological appearance of amyloid deposits in the synovium and the onset of clinically relevant symptoms indicates that oligomers and fibrils of β_2m do not cause clinical pathology for many years (Fig. 12.2d). This can be rationalized by considering that oligomers and fibrils of β_2m may be largely biologically inert when stored in the synovium, whereas the infiltration of macrophages and the associated inflammatory response causes the clinically relevant pathology that occurs in later stages of DRA. It is of course possible that oligomers of β_2m are toxic and sequestration of these entities as fibrils allows the concentration of the relevant oligomeric species to remain low until the sheer bulk of fibrils and the resulting inflammatory response elucidates a long-evaded pathology. From the evidence currently available it seems that the biological response to the presence of β_2m fibrils is responsible for the symptoms of DRA, and the amyloid hypothesis remains to be disproved for this particular disorder. Methods of reducing the formation of β_2m fibrils by improving the dialysis procedure or removing monomers, as well as reducing uremia and the

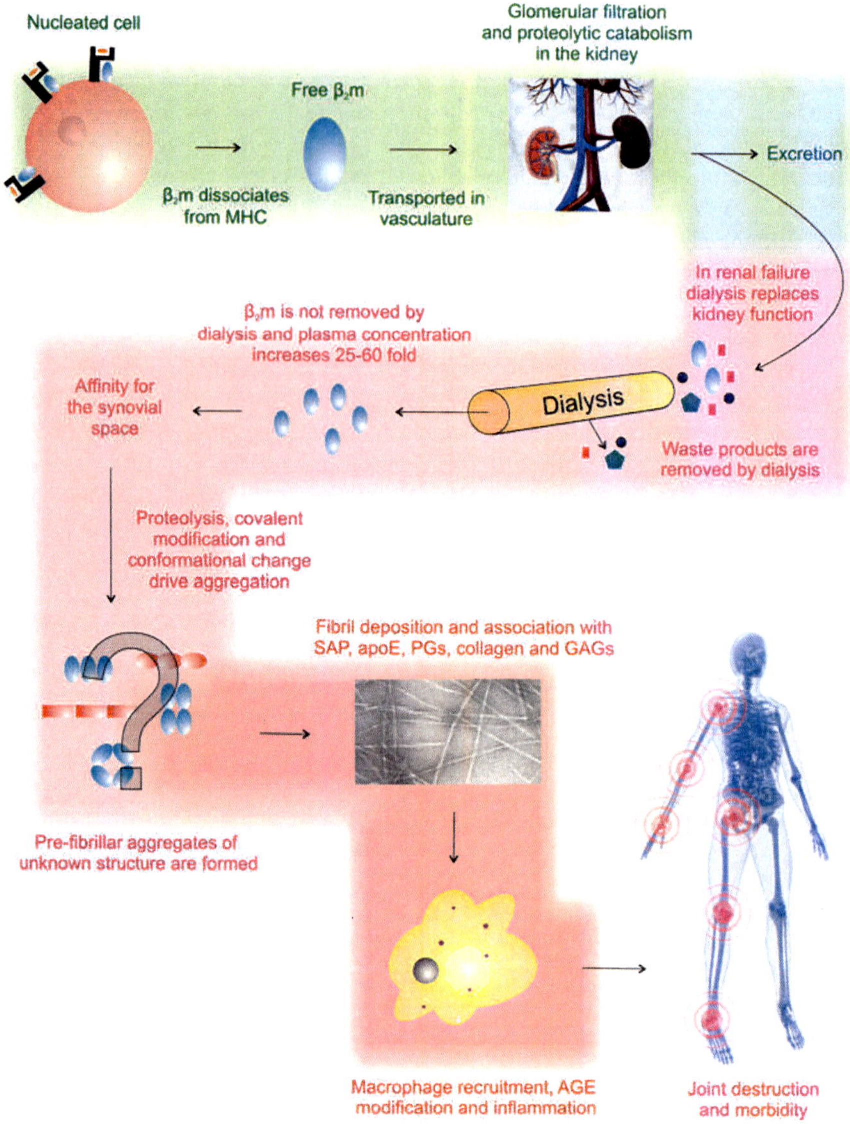

Fig. 12.6 Schematic of the key processes, which result in the pathological symptoms, experienced in DRA

associated causes of dialysis-induced inflammation remain the best routes for reduction of the incidence of DRA in the future. However, more innovative therapies based on RNA aptamers or antibodies, which could be targeted against the monomer, may provide possibilities for evading DRA in the future. Given the generic epitopes identified across amyloid fibrils, the emerging understanding of the mechanisms of amyloid formation in atomic detail, and the improvements in understanding of the factors that protect amyloid fibrils from degradation *in vivo*, the future looks bright for improved therapies against DRA and other amyloid disorders in the years ahead.

Acknowledgements We are very grateful to members of our research groups for many discussions during the preparation of this review. Our research is funded by the Biotechnological and Biological Sciences Research Council and the Wellcome Trust. Their support is gratefully acknowledged.

References

Abe T, Uchita K, Orita H, Kamimura M, Oda M, Hasegawa H, Kobata H, Fukunishi M, Shimazaki M, Abe T, Akizawa T, Ahmad S (2003) Effect of β_2-microglobulin adsorption column on dialysis-related amyloidosis. Kidney Int 64:1522–1528

Alexandrescu AT (2005) Amyloid accomplices and enforcers. Protein Sci 14:1–12

Antwi K, Mahar M, Srikanth R, Olbris MR, Tyson JF, Vachet RW (2008) Cu(II) organizes β_2-microglobulin oligomers but is released upon amyloid formation. Protein Sci 17:748–759

Assenat H, Calemard E, Charra B, Laurent G, Terrat JC, Vanel T (1980) Hemodialysis: carpal tunnel syndrome and amyloid substance. Nouv Presse Med 9:1715

Athanasou NA, Puddle B, Sallie B (1995) Highly sulfated glycosaminoglycans in articular cartilage and other tissues containing β_2-microglobulin dialysis amyloid deposits. Nephrol Dial Transplant 10:1672–1678

Bandini S, Bergesio F, Conti P, Mancini G, Cerretini C, Cirami C, Rosati A, Caselli GM, Arbustini E, Merlini G, Ficarra G, Salvadori M (2001) Nodular macroglossia with combined light chain and β_2-microglobulin deposition in a long-term dialysis patient. J Nephrol 14:128–131

Bardin T, Kuntz D (1987) The arthropathy of chronic haemodialysis. Clin Exp Rheumatol 5:379–386

Bardin T, Lebaildarne JL, Zingraff J, Laredo JD, Voisin MC, Kreis H, Kuntz D (1995) Dialysis arthropathy—outcome after renal-transplantation. Am J Med 99:243–248

Baz M, Durand C, Ragon A, Jaber K, Andrieu D, Merzouk T, Purgus R, Olmer M, Reynier JP, Berland Y (1991) Using ultrapure water in haemodialysis delays Carpal-Tunnel Syndrome. Int J Artif Organs 14:681–685

Bellotti V, Stoppini M, Mangione P, Sunde M, Robinson C, Asti L, Brancaccio D, Ferri G (1998) β_2-microglobulin can be refolded into a native state from *ex vivo* amyloid fibrils. Eur J Biochem 258:61–67

Bunka DHJ, Stockley PG (2006) Aptamers come of age—at last. Nat Rev Microbiol 4:588–596

Bunka DHJ, Mantle BJ, Morten IJ, Tennent GA, Radford SE, Stockley PG (2007) Production and characterization of RNA aptamers specific for amyloid fibril epitopes. J Biol Chem 282: 34500–34509

Calabrese MF, Miranker AD (2007) Formation of a stable oligomer of β_2-microglobulin requires only transient encounter with Cu(II). J Mol Biol 367:1–7

Calabrese MF, Eakin CM, Wang JM, Miranker AD (2008) A regulatable switch mediates self-association in an immunoglobulin fold. Nat Struct Mol Biol 15:965–971

Campistol JM (2001) Dialysis-related amyloidosis after renal transplantation. Semin Dial 14:99–102

Canaud BJM (2009) Changing paradigms of renal replacement therapy in chronic kidney disease patients: ultrapure dialysis fluid and high-efficiency hemodiafiltration for all? Kidney Int 76:591–593

Cascio P, Hilton C, Kisselev AF, Rock KL, Goldberg AL (2001) 26 S proteasomes and immunoproteasomes produce mainly *N*-extended versions of an antigenic peptide. EMBO J 20:2357–2366

Charra B, Calemard E, Laurent G (1988) Chronic renal-failure treatment duration and mode—their relevance to the late dialysis periarticular syndrome. Blood Purif 6:117–124
Chary-Valckenaere I, Kessler M, Mainard D, Schertz L, Chanliau J, Champigneulle J, Pourel J, Gaucher A, Netter P (1998) Amyloid and non-amyloid carpal tunnel syndrome in patients receiving chronic renal dialysis. J Rheumatol 25:1164–1170
Cheung AK, Leypoldt JK (1997) The hemodialysis membranes: a historical perspective, current state and future prospect. Semin Nephrol 17:196–213
Chiti F, Dobson CM (2009) Amyloid formation by globular proteins under native conditions. Nat Chem Biol 5:15–22
Chiti F, De Lorenzi E, Grossi S, Mangione P, Giorgetti S, Caccialanza G, Dobson CM, Merlini G, Ramponi G, Bellotti V (2001) A partially structured species of β_2-microglobulin is significantly populated under physiological conditions and involved in fibrillogenesis. J Biol Chem 276:46714–46721
Corlin DB, Johnsen CK, Nissen MH, Heegaard NHH (2009) A β_2-microglobulin cleavage variant fibrillates at near-physiological pH. Biochem Biophys Res Commun 381:187–191
Cresswell P, Ackerman AL, Giodini A, Peaper DR, Wearsch PA (2005) Mechanisms of MHC class I-restricted antigen processing and cross-presentation. Immunol Rev 207:145–157
Davison AM (1995) β_2-microglobulin and amyloidosis: who is at risk? Nephrol Dial Transplant 10:48–51
Debelouchina GT, Platt GW, Bayro MJ, Radford SE, Griffin RG (2010) Magic angle spinning NMR analysis of β_2-microglobulin amyloid fibrils in two distinct morphologies. J Am Chem Soc 132:10414–10423
Destrihou CV, Jadoul M, Malghem J, Maldague B, Jamart J (1991) Effect of dialysis membrane and patients age on signs of dialysis-related amyloidosis. Kidney Int 39:1012–1019
Destrihou CV, Floege J, Jadoul M, Koch KM (1994) Amyloidosis and its relationship to different dialyzers. Nephrol Dial Transplant 9:156–161
Dong G, Wearsch PA, Peaper DR, Cresswell P, Reinisch KM (2009) Insights into MHC Class I peptide loading from the structure of the tapasin-ERp57 thiol oxidoreductase heterodimer. Immunity 30:21–32
Drueke TB, Massy ZA (2009) β_2-microglobulin. Semin Dial 22:378–380
Eakin CM, Miranker AD (2005) From chance to frequent encounters: origins of β_2-microglobulin fibrillogenesis. Biochim Biophys Acta 1753:92–99
Eakin CM, Knight JD, Morgan CJ, Gelfand MA, Miranker AD (2002) Formation of a copper specific binding site in non-native states of β_2-microglobulin. Biochemistry 41:10646–10656
Eakin CM, Attenello FJ, Morgan CJ, Miranker AD (2004) Oligomeric assembly of native-like precursors precedes amyloid formation by β_2-microglobulin. Biochemistry 43:7808–7815
Eakin CM, Berman AJ, Miranker AD (2006) A native to amyloidogenic transition regulated by a backbone trigger. Nat Struct Mol Biol 13:202–208
Eanes ED, Glenner GG (1968) X-ray diffraction studies on amyloid filaments. J Histochem Cytochem 16:673–677
Egelman EH (2000) A robust algorithm for the reconstruction of helical filaments using single-particle methods. Ultramicroscopy 85:225–234
Eichner T, Radford SE (2009) A generic mechanism of β_2-microglobulin amyloid assembly at neutral pH involving a specific proline switch. J Mol Biol 386:1312–1326
Esposito G, Michelutti R, Verdone G, Viglino P, Hernandez H, Robinson CV, Amoresano A, Dal Piaz F, Monti M, Pucci P, Mangione P, Stoppini M, Merlini G, Ferri G, Bellotti V (2000) Removal of the *N*-terminal hexapeptide from human β_2-microglobulin facilitates protein aggregation and fibril formation. Protein Sci 9:831–845
Esposito G, Ricagno S, Corazza A, Rennella E, Gümral D, Mimmi MC, Betto E, Pucillo CE, Fogolari F, Viglino P, Raimondi S, Giorgetti S, Bolognesi B, Merlini G, Stoppini M, Bolognesi M, Bellotti V (2008) The controlling roles of Trp60 and Trp95 in β_2-microglobulin function, folding and amyloid aggregation properties. J Mol Biol 378:887–897
Fabian H, Gast K, Laue M, Misselwitz R, Uchanska-Ziegler B, Ziegler A, Naumann D (2008) Early stages of misfolding and association of β_2-microglobulin: insights from infrared spectroscopy and dynamic light scattering. Biochemistry 47:6895–6906

Ferreira A, Urena P, Ang KS, Simon P, Morieux C, Souberbielle JC, Devernejoul MC, Drueke TB (1995) Relationship between serum β_2-microglobulin, bone histology, and dialysis membranes in uremic patients. Nephrol Dial Transplant 10:1701–1707

Floege J, Ehlerding G (1996) β_2-Microglobulin-associated amyloidosis—discussion. Nephron 72:9–26

Floege J, Ketteler M (2001) β_2-Microglobulin-derived amyloidosis: an update. Kidney Int Suppl 59:S164–S171

Floege J, Bartsch A, Schulze M, Shaldon S, Koch KM, Smeby LC (1991) Clearance and synthesis rates of β_2-microglobulin in patients undergoing hemodialysis and in normal subjects. J Lab Clin Med 118:153–165

Fuchs D, Norkrans G, Wejstal R, Reibnegger G, Weiss G, Weiland O, Schvarcz R, Fryden A, Wachter H (1992) Changes of serum neopterin, β_2-microglobulin and interferon-γ in patients with chronic hepatitis C treated with interferon-α2b. Eur J Med 1:196–200

Furuyoshi S, Nakatani M, Taman J, Kutsuki H, Takata S, Tani N (1998) New adsorption column (Lixelle) to eliminate β_2-microglobulin for direct hemoperfusion. Ther Apher 2:13–17

Garbar C, Jadoul M, Noel H, De Strihou CV (2000) Histologic characteristics of sternoclavicular β_2-microglobulin amyloidosis—reply. Kidney Int 57:346–347

Garcia KC, Adams EJ (2005) How the T cell receptor sees antigen—a structural view. Cell 122:333–336

Geddes AJ, Parker KD, Atkins EDT, Beighton E (1968) Cross-β conformation in proteins. J Mol Biol 32:343–358

Gejyo F, Yamada T, Odani S, Nakagawa Y, Arakawa M, Kunitomo T, Kataoka H, Suzuki M, Hirasawa Y, Shirahama T, Cohen AS, Schmid K (1985) A new form of amyloid protein associated with chronic hemodialysis was identified as β_2-microglobulin. Biochem Biophys Res Commun 129:701–706

Gejyo F, Homma N, Suzuki Y, Arakawa M (1986a) Serum levels of β_2-microglobulin as a new form of amyloid protein in patients undergoing long-term hemodialysis. N Engl J Med 314:585–586

Gejyo F, Odani S, Yamada T, Honma N, Saito H, Suzuki Y, Nakagawa Y, Kobayashi H, Maruyama Y, Hirasawa Y, Suzuki M, Arakawa M (1986b) β_2-Microglobulin—a new form of amyloid protein associated with chronic hemodialysis. Kidney Int 30:385–390

Gewurz BE, Gaudet R, Tortorella D, Wang EW, Ploegh HL, Wiley DC (2001) Antigen presentation subverted: structure of the human cytomegalovirus protein US2 bound to the class I molecule HLA-A2. Proc Natl Acad Sci USA 98:6794–6799

Giorgetti S, Raimondi S, Cassinelli S, Bucciantini M, Stefani M, Gregorini G, Albonico G, Moratti R, Montagna G, Stoppini M, Bellotti V (2009) β_2-Microglobulin is potentially neurotoxic, but the blood brain barrier is likely to protect the brain from its toxicity. Nephrol Dial Transplant 24:1176–1181

Goldsbury C, Frey P, Olivieri V, Aebi U, Muller SA (2005) Multiple assembly pathways underlie amyloid-β fibril polymorphisms. J Mol Biol 352:282–298

Gordon J, Wu C, Rastegar M, Safa A (2002) β_2-Microglobulin induces apoptosis in CCRF-HSB-2 leukemia cells by molecular mechanisms different than doxorubicin- and taxol-induced apoptosis mechanisms. Eur J Cancer 38:S167

Gordon J, Wu CH, Rastegar M, Safa AR (2003) β_2-Microglobulin induces caspase-dependent apoptosis in the CCRF-HSB-2 human leukemia cell line independently of the caspase-3, -8 and -9 pathways but through increased reactive oxygen species. Int J Cancer 103:316–327

Gorevic PD, Casey TT, Stone WJ, Diraimondo CR, Prelli FC, Frangione B (1985) β_2-Microglobulin is an amyloidogenic protein in man. J Clin Invest 76:2425–2429

Gosal WS, Morten IJ, Hewitt EW, Smith DA, Thomson NH, Radford SE (2005) Competing pathways determine fibril morphology in the self-assembly of β_2-microglobulin into amyloid. J Mol Biol 351:850–864

Gosal WS, Myers SL, Radford SE, Thomson NH (2006) Amyloid under the atomic force microscope. Protein Pept Lett 13:261–270

Haass C, Selkoe DJ (2007) Soluble protein oligomers in neurodegeneration: lessons from the Alzheimer's amyloid β-peptide. Nat Rev Mol Cell Biol 8:101–112

Hakim RM, Wingard RL, Husni L, Parker RA, Parker TF (1996) The effect of membrane biocompatibility on plasma β_2-microglobulin levels in chronic hemodialysis patients. J Am Soc Nephrol 7:472–478

Hasegawa K, Tsutsumi-Yasuhara S, Ookoshi T, Ohhashi Y, Kimura H, Takahashi N, Yoshida H, Miyazaki R, Goto Y, Naiki H (2008) Growth of β_2-microglobulin-related amyloid fibrils by non-esterified fatty acids at a neutral pH. Biochem J 416:307–315

Haydar SN, Yun HED, Staal RGW, Hirst WD (2009) Small-molecule protein–protein interaction inhibitors as therapeutic agents for neurodegenerative diseases: recent progress and future directions. Annu Rep Med Chem 44:51–69

He X, Giurleo JT, Talaga DS (2010) Role of small oligomers on the amyloidogenic aggregation free-energy landscape. J Mol Biol 395:134–154

Hirakura Y, Kagan BL (2001) Pore formation by β_2-microglobulin: a mechanism for the pathogenesis of dialysis associated amyloidosis. Amyloid 8:94–100

Hodkinson JP (2009) β_2-Microglobulin from function to fibril: an investigation using hydrogen/deuterium exchange mass spectrometry. PhD thesis, University of Leeds, Leeds

Hodkinson JP, Jahn TR, Radford SE, Ashcroft AE (2009) HDX-ESI-MS reveals enhanced conformational dynamics of the amyloidogenic protein β_2-microglobulin upon release from the MHC-1. J Am Soc Mass Spectrom 20:278–286

Horl WH (2004) β_2-Microglobulin amyloidosis. In: Replacement of renal function by dialysis. Kluwer Academic Publishers, Dordrecht/London

Hoshino M, Katou H, Hagihara Y, Hasegawa K, Naiki H, Goto Y (2002) Mapping the core of the β_2-microglobulin amyloid fibril by H/D exchange. Nat Struct Biol 9:332–336

Hou FF, Owen WF (2002) β_2-Microglobulin amyloidosis: role of monocytes/macrophages. Curr Opin Nephrol Hypertens 11:417–421

Hughes EA, Hammond C, Cresswell P (1997) Misfolded major histocompatibility complex class I heavy chains are translocated into the cytoplasm and degraded by the proteasome. Proc Natl Acad Sci USA 94:1896–1901

Imani F, Horii Y, Suthanthiran M, Skolnik EY, Makita Z, Sharma V, Sehajpal P, Vlassara H (1993) Advanced glycosylation endproduct-specific receptors on human and rat T-lymphocytes mediate synthesis of interferon-γ—role in tissue remodeling. J Exp Med 178:2165–2172

Ivanova MI, Sawaya MR, Gingery M, Attinger A, Eisenberg D (2004) An amyloid-forming segment of β_2-microglobulin suggests a molecular model for the fibril. Proc Natl Acad Sci USA 101:10584–10589

Iwata K, Fujiwara T, Matsuki Y, Akutsu H, Takahashi S, Naiki H, Goto Y (2006) 3D structure of amyloid protofilaments of β_2-microglobulin fragment probed by solid-state NMR. Proc Natl Acad Sci USA 103:18119–18124

Jadoul M, Drueke T, Zingraff J, Destrihou CV (1997a) Does dialysis-related amyloidosis regress after transplantation? Nephrol Dial Transplant 12:655–657

Jadoul M, Garbar C, Noel H, Sennesael J, Vanholder R, Bernaert P, Rorive G, Hanique G, Destrihou CV (1997b) Histological prevalence of β_2-microglobulin amyloidosis in hemodialysis: a prospective post-mortem study. Kidney Int 51:1928–1932

Jahn TR, Radford SE (2005) The Yin and Yang of protein folding. FEBS J 272:5962–5970

Jahn TR, Parker MJ, Homans SW, Radford SE (2006) Amyloid formation under physiological conditions proceeds *via* a native-like folding intermediate. Nat Struct Mol Biol 13:195–201

Jahn TR, Tennent GA, Radford SE (2008) A common β-sheet architecture underlies *in vitro* and *in vivo* β_2-microglobulin amyloid fibrils. J Biol Chem 283:17279–17286

Jimenez JL, Nettleton EJ, Bouchard M, Robinson CV, Dobson CM, Saibil HR (2002) The protofilament structure of insulin amyloid fibrils. Proc Natl Acad Sci USA 99:9196–9201

Jones S, Smith DP, Radford SE (2003a) Role of the *N*- and *C*-terminal strands of β_2-microglobulin in amyloid formation at neutral pH. J Mol Biol 330:935–941

Jones S, Manning J, Kad NM, Radford SE (2003b) Amyloid-forming peptides from β_2-microglobulin—insights into the mechanism of fibril formation *in vitro*. J Mol Biol 325:249–257

Kad NM, Thomson NH, Smith DP, Smith DA, Radford SE (2001) β_2-Microglobulin and its deamidated variant, N17D form amyloid fibrils with a range of morphologies *in vitro*. J Mol Biol 313:559–571

Kameda A, Hoshino M, Higurashi T, Takahashi S, Naiki H, Goto Y (2005) Nuclear magnetic resonance characterization of the refolding intermediate of β_2-microglobulin trapped by non-native prolyl peptide bond. J Mol Biol 348:383–397
Kardos J, Okuno D, Kawai T, Hagihara Y, Yumoto N, Kitagawa T, Závodszky P, Naiki H, Goto Y (2005) Structural studies reveal that the diverse morphology of β_2-microglobulin aggregates is a reflection of different molecular architectures. Biochim Biophys Acta 1753:108–120
Karlsson FA, Groth T, Sege K, Wibell L, Peterson PA (1980) Turnover in humans of β_2-microglobulin—the constant chain of HLA-antigens. Eur J Clin Invest 10:293–300
Kay J (1997) β_2-Microglobulin amyloidosis. Amyloid 4:187–211
Kazama JJ, Yamamoto S, Takahashi N, Ito Y, Maruyama H, Narita I, Gejyo F (2006) Aβ-2 M-amyloidosis and related bone diseases. J Bone Miner Metab 24:182–184
Kelly JW (1998) The alternative conformations of amyloidogenic proteins and their multi-step assembly pathways. Curr Opin Struct Biol 8:101–106
Kelly JW (2002) Towards an understanding of amyloidogenesis. Nat Struct Biol 9:323–325
Khan AR, Baker BM, Ghos P, Biddison WE, Wiley DC (2000) The structure and stability of an HLA-A*0201/octameric tax peptide complex with an empty conserved peptide-*N*-terminal binding site. J Immunol 164:6398–6405
Kihara M, Chatani E, Iwata K, Yamamoto K, Matsuura T, Nakagawa A, Naiki H, Goto Y (2006) Conformation of amyloid fibrils of β_2-microglobulin probed by tryptophan mutagenesis. J Biol Chem 281:31061–31069
Koda Y, Nishi S, Miyazaki S, Haginoshita S, Sakurabayashi T, Suzuki M, Sakai S, Yuasa Y, Hirasawa Y, Nishi T (1997) Switch from conventional to high-flux membrane reduces the risk of carpal tunnel syndrome and mortality of hemodialysis patients. Kidney Int 52:1096–1101
Kozhukh GV, Hagihara Y, Kawakami T, Hasegawa K, Naiki H, Goto Y (2002) Investigation of a peptide responsible for amyloid fibril formation of β_2-microglobulin by *Achromobacter* protease I. J Biol Chem 277:1310–1315
Kutsuki H (2005) β_2-Microglobulin-selective direct hemoperfusion column for the treatment of dialysis-related amyloidosis. Biochim Biophys Acta 1753:141–145
Ladner CL, Chen M, Smith DP, Platt GW, Radford SE, Langen R (2010) Stacked sets of parallel, in register β-strands in β_2-microglobulin amyloid fibrils revealed by site-directed spin labelling and chemical labelling. J Biol Chem 285:17137–17147
Leypoldt JK, Cheung AK, Carroll CE, Stannard DC, Pereira BJG, Agodoa LY, Port FK (1999) Effect of dialysis membranes and middle molecule removal on chronic hemodialysis patient survival. Am J Kidney Dis 33:349–355
Lin CL, Yang CW, Chiang CC, Chang CT, Huang CC (2001) Long-term on-line hemodiafiltration reduces predialysis β_2-microglobulin levels in chronic hemodialysis patients. Blood Purif 19:301–307
Linke RP, Hampl H, Bartelschwarze S, Eulitz M (1987) β_2-Microglobulin: different fragments and polymers thereof in synovial amyloid in long-term hemodialysis. Biol Chem 368:137–144
Linke RP, Eulitz M, Hampl H, Lobeck H (1989a) Altered β_2-microglobulin in amyloid deposits of patients on long-term hemodialysis. Kidney Int 35:254
Linke RP, Hampl H, Lobeck H, Ritz E, Bommer J, Waldherr R, Eulitz M (1989b) Lysine-specific cleavage of β_2-microglobulin in amyloid deposits associated with Haemodialysis. Kidney Int 36:675–681
Linke RP, Schaeffer J, Gielow P, Lindner P, Lottspeich F, Pluckthun A, Weiss EH (2000) Production of recombinant human β_2-microglobulin for scintigraphic diagnosis of amyloidosis in uremia and hemodialysis. Eur J Biochem 267:627–633
Ljunggren HG, Stam NJ, Ohlen C, Neefjes JJ, Hoglund P, Heemels MT, Bastin J, Schumacher TNM, Townsend A, Karre K, Ploegh HL (1990) Empty MHC class-I molecules come out in the cold. Nature 346:476–480
Locatelli F, Marcelli D, Conte F, Limido A, Malberti F, Spotti D, Trapianto RLD (1999) Comparison of mortality in ESRD patients on convective and diffusive extracorporeal treatments. Kidney Int 55:286–293
Makin OS, Atkins E, Sikorski P, Johansson J, Serpell LC (2005) Molecular basis for amyloid fibril formation and stability. Proc Natl Acad Sci USA 102:315–320

Margittai M, Langen R (2006) Spin labeling analysis of amyloids and other protein aggregates. Methods Enzymol 413:122–139

Mccarthy JT, Williams AW, Johnson WJ (1994) Serum β_2-microglobulin concentration in dialysis patients—importance of intrinsic renal function. J Lab Clin Med 123:495–505

McParland VJ (2001) Conformational changes of β_2-microglobulin and implications for amyloid fibril formation. PhD thesis, University of Leeds, Leeds

McParland VJ, Kad NM, Kalverda AP, Brown A, Kirwin-Jones P, Hunter MG, Sunde M, Radford SE (2000) Partially unfolded states of β_2-microglobulin and amyloid formation *in vitro*. Biochemistry 39:8735–8746

Meinhardt J, Sachse C, Hortschansky P, Grigorieff N, Fandrich M (2009) Aβ1–40 fibril polymorphism implies diverse interaction patterns in amyloid fibrils. J Mol Biol 386:869–877

Mendoza VL, Antwi K, Baron-Rodriguez MA, Blanco C, Vachet RW (2010) Structure of the preamyloid dimer of β_2-microglobulin from covalent labeling and mass spectrometry. Biochemistry 49:1522–1532

Mimmi MC, Jorgensen TJD, Pettirossi F, Corazza A, Viglino P, Esposito G, De Lorenzi E, Giorgetti S, Pries M, Corlin DB, Nissen MH, Heegaard NHH (2006) Variants of β_2-microglobulin cleaved at lysine-58 retain the main conformational features of the native protein but are more conformationally heterogeneous and unstable at physiological temperature. FEBS J 273:2461–2474

Mironova R, Niwa T (2001) Molecular heterogeneity of amyloid β_2-microglobulin and modification with advanced glycation end products. J Chromatogr B 758:109–115

Miyata T, Oda O, Inagi R, Iida Y, Araki N, Yamada N, Horiuchi S, Taniguchi N, Maeda K, Kinoshita T (1993) β_2-Microglobulin modified with advanced glycation end-products is a major component of hemodialysis-associated amyloidosis. J Clin Invest 92:1243–1252

Miyata T, Inagi R, Iida Y, Sato M, Yamada N, Oda O, Maeda K, Seo H (1994) Involvement of β_2-microglobulin modified with advanced glycation end-products in the pathogenesis of hemodialysis-associated amyloidosis—induction of human monocyte chemotaxis and macrophage secretion of tumor necrosis factor-α and interleukin-1. J Clin Invest 93:521–528

Miyata T, Taneda S, Kawai R, Ueda Y, Horiuchi S, Hara M, Maeda K, Monnier VM (1996) Identification of pentosidine as a native structure for advanced glycation end products in β_2-microglobulin-containing amyloid fibrils in patients with dialysis-related amyloidosis. Proc Natl Acad Sci USA 93:2353–2358

Miyata T, Jadoul M, Kurokawa K, De Strihou CV (1998) β_2-Microglobulin in renal disease. J Am Soc Nephrol 9:1723–1735

Morgan CJ, Gelfand M, Atreya C, Miranker AD (2001) Kidney dialysis-associated amyloidosis: a molecular role for copper in fiber formation. J Mol Biol 309:339–345

Moriniere P, Marie A, Elesper N, Fardellone P, Deramond H, Remond A, Sebert JL, Fournier A (1991) Destructive spondyloarthropathy with β_2-microglobulin amyloid deposits in a uremic patient before chronic-hemodialysis. Nephron 59:654–657

Morten IJ, Gosal WS, Radford SE, Hewitt EW (2007) Investigation into the role of macrophages in the formation and degradation of β_2-microglobulin amyloid fibrils. J Biol Chem 282: 29691–29700

Mosmann T (1983) Rapid colorimetric assay for cellular growth and survival: application to proliferation and cytotoxicity assays. J Immunol Methods 65:55–63

Mount SL, Eltabbakh GH, Hardin NJ (2002) β_2-Microglobulin amyloidosis presenting as bilateral ovarian masses—a case report and review of the literature. Am J Surg Pathol 26:130–133

Mustata M, Capone R, Jang H, Arce FT, Ramachandran S, Lal R, Nussinov R (2009) K3 Fragment of Amyloidogenic β_2-microglobulin forms ion channels: implication for dialysis-related amyloidosis. J Am Chem Soc 131:14938–14945

Myers SL, Jones S, Jahn TR, Morten IJ, Tennent GA, Hewitt EW, Radford SE (2006a) A systematic study of the effect of physiological factors on β_2-microglobulin amyloid formation at neutral pH. Biochemistry 45:2311–2321

Myers SL, Thomson NH, Radford SE, Ashcroft AE (2006b) Investigating the structural properties of amyloid-like fibrils formed *in vitro* from β_2-microglobulin using limited proteolysis and electrospray ionisation mass spectrometry. Rapid Commun Mass Spectrom 20:1628–1636

Niwa T (2001) Dialysis-related amyloidosis: pathogenesis focusing on AGE modification. Semin Dial 14:123–126

Niwa T, Katsuzaki T, Miyazaki S, Momoi T, Akiba T, Miyazaki T, Nokura K, Hayase F, Tatemichi N, Takei Y (1997) Amyloid β_2-microglobulin is modified with imidazolone, a novel advanced glycation end product, in dialysis-related amyloidosis. Kidney Int 51:187–194

O'nuallain B, Wetzel R (2002) Conformational Abs recognizing a generic amyloid fibril epitope. Proc Natl Acad Sci USA 99:1485–1490

Ohashi K (2001) Pathogenesis of β_2-microglobulin amyloidosis. Pathol Int 51:1–10

Onishi S, Andress DL, Maloney NA, Coburn JW, Sherrard DJ (1991) β_2-Microglobulin deposition in bone in chronic renal failure. Kidney Int 39:990–995

Ookoshi T, Hasegawa K, Ohhashi Y, Kimura H, Takahashi N, Yoshida H, Miyazaki R, Goto Y, Naiki H (2008) Lysophospholipids induce the nucleation and extension of β_2-microglobulin-related amyloid fibrils at a neutral pH. Nephrol Dial Transplant 23:3247–3255

Otsubo S, Kimata N, Okutsu I, Oshikawa K, Ueda S, Sugimoto H, Mitobe M, Uchida K, Otsubo K, Nitta K, Akiba T (2009) Characteristics of dialysis-related amyloidosis in patients on haemodialysis therapy for more than 30 years. Nephrol Dial Transplant 24:1593–1598

Pal-Gabor H, Gombos L, Micsonai A, Kovacs E, Petrik E, Kovacs J, Graf L, Fidy J, Naiki H, Goto Y, Liliom K, Kardos J (2009) Mechanism of lysophosphatidic acid-induced amyloid fibril formation of β_2-microglobulin *in vitro* under physiological conditions. Biochemistry 48:5689–5699

Paulsson KM, Wang P, Anderson PO, Chen SW, Pettersson RF, Li SL (2001) Distinct differences in association of MHC class I with endoplasmic reticulum proteins in wild-type, and β_2-microglobulin- and TAP-deficient cell lines. Int Immunol 13:1063–1073

Peaper DR, Cresswell P (2008) Regulation of MHC class I assembly and peptide binding. Annu Rev Cell Dev Biol 24:343–368

Pepys MB (2006) Amyloidosis. Annu Rev Med 57:223–241

Petkova AT, Yau WM, Tycko R (2006) Experimental constraints on quaternary structure in Alzheimer's β-amyloid fibrils. Biochemistry 45:498–512

Platt GW, Radford SE (2009) Glimpses of the molecular mechanisms of β_2-microglobulin fibril formation *in vitro*: aggregation on a complex energy landscape. FEBS Lett 583:2623–2629

Platt GW, Routledge KE, Homans SW, Radford SE (2008) Fibril growth kinetics reveal a region of β_2-microglobulin important for nucleation and elongation of aggregation. J Mol Biol 378:251–263

Rabindranath KS, Strippoli GF, Daly C, Roderick PJ, Wallace S, Macleod AM (2006) Haemodiafiltration, haemofiltration and haemodialysis for end-stage kidney disease. Cochrane Database Syst Rev 4:CD006258

Radford SE, Gosal WS, Platt GW (2005) Towards an understanding of the structural molecular mechanism of β_2-microglobulin amyloid formation *in vitro*. Biochim Biophys Acta 1753:51–63

Raghavan M, Del Cid N, Rizvi SM, Peters LR (2008) MHC class I assembly: out and about. Trends Immunol 29:436–443

Raj DSC, Ouwendyk M, Francoeur R, Pierratos A (2000) β_2-Microglobulin kinetics in nocturnal haemodialysis. Nephrol Dial Transplant 15:58–64

Relini A, Canale C, De Stefano S, Rolandi R, Giorgetti S, Stoppini M, Rossi A, Fogolari F, Corazza A, Esposito G, Gliozzi A, Bellotti V (2006) Collagen plays an active role in the aggregation of β_2-microglobulin under physiopathological conditions of dialysis-related amyloidosis. J Biol Chem 281:16521–16529

Relini A, De Stefano S, Torrassa S, Cavalleri O, Rolandi R, Gliozzi A, Giorgetti S, Raimondi S, Marchese L, Verga L, Rossi A, Stoppini M, Bellotti V (2008) Heparin strongly enhances the formation of β_2-microglobulin amyloid fibrils in the presence of type I collagen. J Biol Chem 283:4912–4920

Richardson JS, Richardson DC (2002) Natural β-sheet proteins use negative design to avoid edge to edge aggregation. Proc Natl Acad Sci USA 99:2754–2759

Roderick P, Davies R, Jones C, Feest T, Smith S, Farrington K (2004) Simulation model of renal replacement therapy: predicting future demand in England. Nephrol Dial Transplant 19:692–701

Rosano C, Zuccotti S, Mangione P, Giorgetti S, Bellotti V, Pettirossi F, Corazza A, Viglino P, Esposito G, Bolognesi M (2004) β_2-Microglobulin H31Y variant 3D structure highlights the protein natural propensity towards intermolecular aggregation. J Mol Biol 335:1051–1064
Routledge KE, Tartaglia GG, Platt GW, Vendrusco M, Radford SE (2009) Competition between intramolecular and intermolecular interactions in an amyloid-forming protein. J Mol Biol 389:776–786
Saito A, Gejyo F (2006) Current clinical aspects of dialysis-related amyloidosis in chronic dialysis patients. Ther Apher Dial 10:316–320
Schwalbe S, Holzhauer M, Schaeffer J, Galanski M, Koch KM, Floege J (1997) β_2-Microglobulin associated amyloidosis: a vanishing complication of long-term hemodialysis? Kidney Int 52:1077–1083
Sethi D, Murphy CMB, Brown EA, Muller BR, Gower PE (1989) Clearance of β_2-microglobulin using continuous ambulatory peritoneal-dialysis. Nephron 52:352–355
Shimizu S, Yasui C, Yasukawa K, Nakamura H, Shimizu H, Tsuchiya K (2003) Subcutaneous nodules on the buttocks as a manifestation of dialysis-related amyloidosis: a clinicopathological entity? Br J Dermatol 149:400–404
Smith DP, Jones S, Serpell LC, Sunde M, Radford SE (2003) A systematic investigation into the effect of protein destabilisation on β_2-microglobulin amyloid formation. J Mol Biol 330:943–954
Smith AM, Jahn TR, Ashcroft AE, Radford SE (2006a) Direct observation of oligomeric species formed in the early stages of amyloid fibril formation using electrospray ionisation mass spectrometry. J Mol Biol 364:9–19
Smith JF, Knowles TPJ, Dobson CM, Macphee CE, Welland ME (2006b) Characterization of the nanoscale properties of individual amyloid fibrils. Proc Natl Acad Sci USA 103:15806–15811
Smith DP, Giles K, Bateman RH, Radford SE, Ashcroft AE (2007) Monitoring copopulated conformational states during protein folding events using electrospray ionization-ion mobility spectrometry-mass spectrometry. J Am Soc Mass Spectrom 18:2180–2190
Smith DP, Radford SE, Ashcroft AE (2010) Elongated oligomers in β_2-microglobulin amyloid assembly revealed by ion mobility spectrometry-mass spectrometry. Proc Natl Acad Sci USA 107:6794–6798
Souillac PO, Uversky VN, Millett IS, Khurana R, Doniach S, Fink AL (2002) Elucidation of the molecular mechanism during the early events in immunoglobulin light chain amyloid fibrillation—evidence for an off-pathway oligomer at acidic pH. J Biol Chem 277:12666–12679
Srikanth R, Mendoza VL, Bridgewater JD, Zhang GS, Vachet RW (2009) Copper binding to β_2-microglobulin and its pre-amyloid oligomer. Biochemistry 48:9871–9881
Stoppini M, Arcidiaco P, Mangione P, Giorgetti S, Brancaccio D, Bellotti V (2000) Detection of fragments of β_2-microglobulin in amyloid fibrils. Kidney Int 57:349–350
Stoppini M, Mangione P, Monti M, Giorgetti S, Marchese L, Arcidiaco P, Verga L, Segagni S, Pucci P, Merlini G, Bellotti V (2005) Proteomics of β_2-microglobulin amyloid fibrils. Biochim Biophys Acta 1753:23–33
Sundaram R, Lynch MP, Rawale S, Dakappagari N, Young D, Walker CM, Lemonnier F, Jacobson S, Kaumaya PTP (2004) Protective efficacy of multiepitope human leukocyte antigen-A*0201 restricted cytotoxic T-lymphocyte peptide construct against challenge with human T-cell lymphotropic virus type 1 Tax recombinant vaccinia virus. J Acquir Immune Defic Syndr 37:1329–1339
Tielemans C, Dratwa M, Bergmann P, Goldman M, Flamion B, Collart F, Wens R (1989) Continuous ambulatory peritoneal-dialysis, protective against developing dialysis-associated amyloid. Nephron 53:174–175
Trinh CH, Smith DP, Kalverda AP, Phillips SEV, Radford SE (2002) Crystal structure of monomeric human β_2-microglobulin reveals clues to its amyloidogenic properties. Proc Natl Acad Sci USA 99:9771–9776
Tsigelny IF, Crews L, Desplats P, Shaked GM, Sharikov Y, Mizuno H, Spencer B, Rockenstein E, Trejo M, Platoshyn O, Yuan JXJ, Masliah E (2008) Mechanisms of hybrid oligomer formation in the pathogenesis of combined Alzheimer's and Parkinson's diseases. PLoS One 3:e3135
Tycko R (2006) Solid-state NMR as a probe of amyloid structure. Protein Pept Lett 13:229–234

Uversky VN, Fink AL (2004) Conformational constraints for amyloid fibrillation: the importance of being unfolded. Biochim Biophys Acta 1698:131–153
Verdone G, Corazza A, Viglino P, Pettirossi F, Giorgetti S, Mangione P, Andreola A, Stoppini M, Bellotti V, Esposito G (2002) The solution structure of human β_2-microglobulin reveals the prodromes of its amyloid transition. Protein Sci 11:487–499
Warren DJ, Otieno LS (1975) Carpal-tunnel syndrome in patients on intermittent hemodialysis. Postgrad Med J 51:450–452
Wearsch PA, Cresswell P (2008) The quality control of MHC class I peptide loading. Curr Opin Cell Biol 20:624–631
Westermark P, Benson MD, Buxbaum JN, Cohen AS, Frangione B, Ikeda SI, Masters CL, Merlini G, Saraiva MJ, Sipeo JD (2007) A primer of amyloid nomenclature. Amyloid 14:179–183
White HE, Hodgkinson JL, Jahn TR, Cohen-Krausz S, Gosal WS, Muller S, Orlova EV, Radford SE, Saibil HR (2009) Globular tetramers of β_2-microglobulin assemble into elaborate amyloid fibrils. J Mol Biol 389:48–57
Wu CH, Rastegar M, Gordon J, Safa AR (2001) β_2-Microglobulin induces apoptosis in HL-60 human leukemia cell line and its multidrug resistant variants overexpressing MRP1 but lacking Bax or overexpressing P-glycoprotein. Oncogene 20:7006–7020
Wu CH, Gordon JD, Zhong XL, Safa AR (2002) Mechanism of β_2-microglobulin-induced apoptosis in the K562 leukemia cell line, defective in major histocompatibility class 1. Anticancer Res 22:2613–2621
Xue WF, Homans SW, Radford SE (2008) Systematic analysis of nucleation-dependent polymerization reveals new insights into the mechanism of amyloid self-assembly. Proc Natl Acad Sci USA 105:8926–8931
Xue WF, Hellewell AL, Gosal WS, Homans SW, Hewitt EW, Radford SE (2009) Fibril fragmentation enhances amyloid cytotoxicity. J Biol Chem 284:34272–34282
Yamaguchi KI, Katou H, Hoshino M, Hasegawa K, Naiki H, Goto Y (2004) Core and heterogeneity of β_2-microglobulin amyloid fibrils as revealed by H/D exchange. J Mol Biol 338:559–571
Yamaguchi K, Naiki H, Goto Y (2006) Mechanism by which the amyloid-like fibrils of a β_2-microglobulin fragment are induced by fluorine-substituted alcohols. J Mol Biol 363:279–288
Yamamoto S, Hasegawa K, Yamaguchi I, Tsutsumi S, Kardos J, Goto Y, Gejyo F, Naiki H (2004a) Low concentrations of sodium dodecyl sulfate induce the extension of β_2-microglobulin-related amyloid fibrils at a neutral pH. Biochemistry 43:11075–11082
Yamamoto S, Yamaguchi I, Hasegawa K, Tsutsumi S, Goto Y, Gejyo F, Naiki H (2004b) Glycosaminoglycans enhance the trifluoroethanol-induced extension of β_2-microglobulin-related amyloid fibrils at a neutral pH. J Am Soc Nephrol 15:126–133
Yamamoto S, Hasegawa K, Yamaguchi I, Goto Y, Gejyo F, Naiki H (2005) Kinetic analysis of the polymerization and depolymerization of β_2-microglobulin-related amyloid fibrils *in vitro*. Biochim Biophys Acta 1753:34–43
Yamamoto S, Kazama JJ, Narita I, Naiki H, Gejyo F (2009) Recent progress in understanding dialysis-related amyloidosis. Bone 45:S39–S42
Yusa H, Yoshida H, Kikuchi H, Onizawa K (2001) Dialysis-related amyloidosis of the tongue. J Oral Maxillocfac Surg 59:947–950
Zhang R, Hu XY, Khant H, Ludtke SJ, Chiu W, Schmid MF, Frieden C, Lee JM (2009) Interprotofilament interactions between Alzheimer's Aβ(1–42) peptides in amyloid fibrils revealed by cryoEM. Proc Natl Acad Sci USA 106:4653–4658
Zijlstra M, Bix M, Simister NE, Loring JM, Raulet DH, Jaenisch R (1990) β_2-Microglobulin deficient mice lack Cd4-8$^+$ cytolytic T-cells. Nature 344:742–746
Zingraff JJ, Noel LH, Bardin T, Atienza C, Zins B, Drueke TB, Kuntz D (1990) β_2-Microglobulin amyloidosis in chronic-renal-failure. N Engl J Med 323:1070–1071

Chapter 13
Transthyretin Aggregation and Toxicity

Maria João Saraiva and Isabel Santos Cardoso

Abstract In 1952, Corino de Andrade identified the first form of hereditary amyloidosis—familial amyloid polyneuropathy (FAP)—in the northern Portugal near Porto. The age-of-onset of clinical symptoms was described to be the third or fourth decade of life in the affected Portuguese kindreds. Typical clinical features of FAP included early impairment of thermoception or nociception in the feet, and autonomic dysfunctions, all leading to paresis, malabsorption, emaciation, and death.

The genetic defect in the Portuguese FAP kindreds was found to be heterozygosity for a single point-mutation in the transthyretin (TTR) gene, giving rise to a variant TTR (TTR-Val30Met). Although originally regarded as a rare disease, it has now become clear that many other affected kindreds exist worldwide with over 100 amyloidogenic TTR mutations have been described. The mechanisms whereby various TTR mutations lead to amyloid aggregation have been the focus of research for the last three decades. Development of therapeutic strategies for FAP entails, among others, not only the elucidation of molecular mechanisms leading to TTR fibril formation, but also understanding the cellular/tissular effects produced by TTR deposition. Whilst researchers have gained some insight into the former aspect due to intense research *in vitro*, the latter issue is now emerging.

M.J. Saraiva (✉)
Molecular Neurobiology, IBMC, Instituto de Biologia Molecular e Celular,
R. Campo Alegre 823, Porto, Portugal

ICBAS, Instituto de Ciências Biomédicas Abel Salazar, Universidade do Porto,
Porto, Portugal
e-mail: mjsaraiv@ibmc.up.pt

I.S. Cardoso
Molecular Neurobiology, IBMC, Instituto de Biologia Molecular e Celular,
R. Campo Alegre 823, Porto, Portugal

Escola Superior de Tecnologia da Saúde do Porto, Instituto Politecnico do Porto,
Porto, Portugal

F. Rahimi and G. Bitan (eds.), *Non-fibrillar Amyloidogenic Protein Assemblies—Common Cytotoxins Underlying Degenerative Diseases*, DOI 10.1007/978-94-007-2774-8_13,

The work presented here addresses pathophysiological mechanisms associated with cellular dysfunction in FAP and integrates the molecular and cellular aspects underlying FAP.

Keywords Familial amyloidotic polyneuropathy • Heat-shock response • Nerve Neuropathies • Transthyretin

13.1 Introduction

Familial amyloidotic polyneuropathy (FAP) is a late-onset, genetic neurodegenerative disorder associated with systemic deposition of mutated transthyretin (TTR) aggregates and amyloid fibrils, particularly in the peripheral nervous system (PNS). In FAP, axonal degeneration begins with unmyelinated and low-diameter myelinated fibers, leading to neuronal loss. Neuropathological studies of FAP, understanding of mechanisms of TTR aggregation, and delineation of molecular signaling mechanisms responsible for neurodegeneration guide novel approaches towards rational therapeutics, ranging from compounds directed to TTR aggregates and fibrils, to compounds that counteract cytotoxic cascades. These integrated studies are described in this chapter. Such studies are also highly relevant for non-hereditary TTR amyloidoses, which encompass deposition of non-mutated TTR in ~25% of aging individuals, affecting particularly the heart—a condition named senile systemic amyloidosis (SSA) (Westermark et al. 1990). Generally, such studies also contribute to research on aging and other neurodegenerative diseases.

TTR is a protein circulated in plasma and in cerebrospinal fluid where it transports thyroxine and retinol—in the latter case, by complexing with the retinol-binding protein (Raz and Goodman 1969). TTR also has a proteolytic activity (Liz et al. 2004) as it binds and cleaves the amyloid β-protein (Aβ) of Alzheimer's disease (Costa et al. 2008). TTR is synthesized in the liver, in the cerebral choroid plexus, in the retina, and in the pancreas. Liver is the main organ for TTR degradation, but up to now, no specific receptor for hepatic uptake of TTR has been identified (Sousa and Saraiva 2001). Uptake of TTR in both kidney and sensory neurons has been assigned to megalin, a member of the low-density-lipoprotein (LDL) family of receptors (Sousa et al. 2000; Fleming et al. 2007). Work in TTR-null mice has shown that lack of TTR does not affect fetal development, thyroid function, or vitamin-A levels (Episkopou et al. 1993), but suggests a role for TTR in the nervous system, particularly affecting behavior. TTR-null mice were shown to display: (i) reduced signs of depressive behavior as compared to control mice (Sousa et al. 2004); (ii) maintenance of cognitive functions in aging (Sousa et al. 2007); (iii) and increased processing of amydated peptides (Nunes et al. 2006). Neuroprotective roles have also been ascribed to TTR as it enhances peripheral nerve regeneration (Fleming et al. 2007), reduces plaque deposition in an Alzheimer's disease model (Choi et al. 2007; Buxbaum et al. 2008), and protects neurons in an ischemia animal model with compromised stress responses (Santos et al. 2010b).

13.2 Clinical, Genetic, and Pathological Features of FAP

The first form of hereditary amyloidosis, described by Corino de Andrade in Portuguese individuals, was a peculiar form of peripheral polyneuropathy, termed familial amyloidotic polyneuropathy (FAP). The onset of clinical symptoms generally occurs before the age of 40. In ~50% of the Portuguese FAP patients, the first symptoms relate to sensory neuropathy beginning at the extremities especially lower limbs, with paresthesia, dysesthesia, and losses of thermoception and nociception. These symptoms progress proximally in a "dying-back" fashion while the upper extremities become affected later after loss of sensation reaches the genicular level. Usually 2 or 3 years after initial sensory manifestations, motor disturbances are noticeable, with atrophy and muscular weakness, beginning also at the lower extremities. In the more advanced stages, patients become confined to wheelchairs (Andrade 1952).

Autonomic neuropathy, which can be one the first disease manifestations in many patients, is particularly severe in Portuguese FAP patients. Autonomic involvement manifests as gastrointestinal disturbances, with alternating periods of constipation and diarrhea, gastric stasis, nausea, and vomiting. Loss of weight and asthenia frequently precede gastrointestinal disturbances. Sphincter dysfunctions are responsible for urinary or fecal incontinence as well as impotence. Electrocardiographic abnormalities are common and manifest as cardiac conduction disturbances, leading to branch blocks, which demands pacemaker implantation in some cases.

Renal involvement varies from mild to moderate proteinuria. In some cases, as the disease progresses, massive proteinuria and nephrotic syndrome are observed. Ocular abnormalities may develop due to vitreous opacities and bilateral scalloping of the pupils. Visual acuity can be restored by vitrectomy. Trophic lesions are frequent in more advanced stages of the disease and vary from dermal atrophy to feet ulcers and even bone necrosis in the extremities. The disease progression is slow but relentless, leading to cachexia and death 10–15 years after the disease onset.

The genetic defect in Portuguese FAP kindreds is a heterozygous single point-mutation in *TTR*, giving rise to variant TTR Val30Met (Saraiva et al. 1984). In Portugal, where FAP is common, the gene-carrier frequency has been estimated to be 1 in 625 (Alves et al. 1997). FAP kindreds with TTR Val30Met have also been identified worldwide, with particular foci in northern Sweden, Japan, and Maiorca (Andersson 1970; Araki 1984; MunarQués et al. 1997). The TTR gene has four exons, each ~200 bp long (Sasaki et al. 1985), which allow easy diagnosis by DNA techniques such as PCR, restriction fragment-length polymorphism (RFLP), and sequencing. Over 100 TTR mutations have been identified throughout the TTR gene [http://www.ibmc.up.pt/mjsaraiva/ttrmut.html, (Benson and Kincaid 2007)]. Some of these mutations cause FAP that is indistinguishable clinically from the original description of this disease; others give rise to phenotypes that variously include neuropathy, cardiomyopathy, carpal-tunnel syndrome (entrapment of the median nerve at the carpal bones of the wrist), vitreous TTR deposition, or leptomeningeal involvement.

Few TTR mutations are associated predominantly with cardiomyopathy. The most common TTR mutation associated with cardiac amyloidosis is Val122Ile, described in the African-American population. After the age of 60, cases of cardiac amyloidoses are four times more common among African-Americans than Caucasian Americans in the USA, and 3.9% of African-Americans are heterozygous for Val122Ile (Jacobson et al. 1997).

FAP is associated with systemic extracellular deposition of TTR aggregates and amyloid fibrils throughout connective tissue, with the exception of cerebral and hepatic parenchyma. TTR amyloid deposits are distributed diffusely in the PNS, involving nerve trunks, plexuses, and sensory or autonomic ganglia (Said et al. 1984; Sobue et al. 1990). In the nerves, TTR is deposited in the epineurium and perineurium, but more prominently in the endoneurium. Endoneurial globoid TTR deposits occur perivascularly, and in close contact with Schwann cells and collagen fibers (Coimbra and Andrade 1971b, a). In severely affected nerves, endoneurial contents are replaced by amyloid, collagen bundles, Schwann cells without axons, and fibroblasts while few nerve fibers retain viability. No intracellular TTR deposits in axons, neuronal cell bodies, or Schwann cells were ever found. At the ultrastructural level, amyloid deposits present as typical non-branching fibrils of varied lengths with diameters ranging from 7 to 10 nm.

Axonal fiber degeneration begins in unmyelinated and low-diameter myelinated fibers and only in advanced cases, the large-diameter myelinated fibers are affected (Dyck and Lambert 1969; Said et al. 1984). The reason for the preferential loss of unmyelinated fibers remains unclear mainly due to the fact that most of the pathological analyses in FAP have been performed using sural-nerve biopsies and that comparatively little information is available on more proximal nerves or the nerve ganglia themselves. Availability of recent animal models with PNS involvement (described below) will allow studies addressing this and other questions—in particular whether nerve-fiber degeneration in FAP results from multifocal compression by amyloid deposits, or is unrelated to amyloid fibrils. The latter hypothesis is raised by the finding that in some asymptomatic cases, fiber degeneration can be found without endoneurial deposits of amyloid (Guimarães et al. 1988), whereas in other cases without signs of degeneration, myelinated fibers are displaced by amyloid deposits. These observations render highly unlikely the hypothesis that neurodegeneration in FAP arises solely as a result of compression by amyloid infiltrates.

In conclusion, in FAP a cause–effect relationship between amyloid deposition, structural nerve changes, and degeneration is difficult to establish. Extensive research on cytotoxicity of TTR aggregates and assemblies is pivotal to shedding light into the molecular mechanisms underlying neurodegeneration in the peripheral and autonomic nervous systems.

13.3 Presence of Toxic Non-fibrillar TTR Aggregates: Early Deposition of TTR in Asymptomatic Carriers and Animal Models as Non-fibrillar Material

TTR deposition was assessed in the nerves of asymptomatic TTR Val30Met carriers and FAP patients in different stages of disease progression. A scoring system of patients' material was established by morphometric measurements of myelinated

(MF) and unmyelinated (UF) nerve fibers. Accordingly, FAP 0 designates the least-affected nerves (no nerve degeneration, no amyloid), while stages one to three classify symptomatic individuals presenting: FAP 1—showing a discrete reduction in the number of nerve fibers and modest fibrillar deposits; FAP 2—evident reduction of nerve fibers and amyloid throughout the nerve; and FAP 3—severe reduction of nerve fibers and extensive amyloid deposition.

It was observed that in the disease stages before loss of UF or MF and major nerve-fiber degeneration (FAP 0), despite the absence of Congo-red birefringence (the characteristic amyloid staining), TTR was present as revealed by immunohistochemistry using an anti-TTR antibody. Therefore, TTR does deposit in a non-fibrillar, or pre-fibrillar form in early stages of FAP before assembling into mature amyloid fibrils; positive TTR immunocytolabeling was observed extracellularly in the proximity of Schwann cells, in a non-fibrillar form. Some small contiguous, fibril-like assemblies were noticeable, but these are most likely too small to emanate birefringence upon Congo-red staining. It was concluded that TTR deposits in a non-fibrillar form early in the asymptomatic phases (Sousa et al. 2001a).

In order to understand the mechanisms underlying fibril formation, deposition, and cytotoxic effects caused by TTR aggregates, many questions remain to be investigated. In an attempt to gain insights into FAP pathogenesis, several groups have generated different *in vivo* models. For example, models of *Drosophila melanogaster* carrying mutant TTR variants have been produced (Berg et al. 2009; Pokrzywa et al. 2007); however, these models do not present clear extracellular aggregates with amyloid-like properties. Furthermore, flies have a nervous system with very different properties as compared to other species. For example, they have a primitive form of myelin, which limits their usefulness in toxicity/neurodegeneration studies in TTR amyloidoses, especially regarding the PNS. In mice, the human homologous TTR-promoter sequences were used to generate TTR Val30Met transgenic animals. In these mice, amyloid was observed starting at 6 months of age. At 24 months, the pattern of amyloid deposition was similar to that observed in human FAP autopsy cases, except for amyloid absence in the choroid plexus, and in the peripheral and autonomic nervous systems (Yi et al. 1991). The same pattern of amyloid deposition was reported when these transgenic mice were backcrossed with TTR-null background mice (Kohno et al. 1997). These animals presented widespread TTR staining by immunohistochemistry, representing non-fibrillar TTR aggregates with particular involvement of the gastrointestinal tract, starting as early as 3 months. At this stage, no fibrillar material was detected by Congo-red staining or by immunocytolabelling. Congo-red-positive deposits were only observed at older ages, affecting some mice over 9 months and a majority of animals after 21 months (Sousa et al. 2002).

Presence of such non-fibrillar TTR aggregates was documented in other animal models of TTR amyloidosis, namely in a transgenic strain producing high levels of wild-type TTR (Teng et al. 2001). Given the increasing body of evidence supporting that the non-fibrillar aggregates may be the primary toxic species in other amyloid-associated disorders, possible pathological roles of non-fibrillar TTR aggregates were assessed *in vivo* and *in vitro*.

The toxicity of non-fibrillar TTR was first suggested by the observation of signs of oxidative and inflammatory stresses in FAP 0 nerves (presenting non-fibrillar TTR aggregates but lacking amyloid) (Sousa et al. 2001b). *In vitro* work with cell lines and primary neurons proved that mature TTR fibrils were essentially unable to damage cells whereas TTR aggregates were toxic (Sousa et al. 2001a; Andersson et al. 2002). Later Reixach et al. reported that TTR amyloid fibrils and soluble aggregates (>100 kDa) were not toxic, whereas transient low-molecular-mass assemblies (<100 kDa) were highly cytotoxic in tissue cultures (Reixach et al. 2004). Cytotoxicity may provide an explanation for what causes neurodegeneration, whereas physical effects of amyloid deposition fail to explain neurodegeneration in FAP.

13.4 Mechanisms of TTR Aggregation

TTR is a tetrameric protein with a molecular mass of 54,980 Da (Kanda et al. 1974); its 4 identical subunits comprise 127 amino-acid residues. The three-dimensional structure of TTR was made available at 1.8-Å resolution by X-ray-diffraction studies of the crystallized protein (Blake et al. 1978). A considerable number of the amino-acid residues in each TTR monomer (55%) are implicated in the formation of β-sheet structures. Each monomer contains two β-sheets formed by strands DAGH and CBEF. All, except strands A and G, display an antiparallel orientation, and are arranged in a topology similar to the classic Greek-key barrel. These strands are 7–8 residues long, except strand D, which, is only 3 residues in length. Only ~5% of the amino-acid residues are located in one segment of an α-helix comprising residues 75–83, at the end of strand E. Two monomers associate forming a dimer by interactions between chains F and H of each monomer; the arrangement within a dimer is DAGHH′G′A′D′ and CBEFF′E′B′C′ as presented in Fig. 13.1.

The tetramer consists of two dimers with connecting edges occurring between the AB loop of one dimer with the H strand of the other dimer. The quaternary structure of TTR has the shape of a globular protein with an overall size of $70 \times 55 \times 50$ Å. The two dimers are slightly rotated relatively to each other along the y-axis. Thyroxine binds to a hydrophobic channel that runs through the molecule.

13.4.1 Factors Influencing TTR Aggregation

Mechanisms leading to amyloid formation have been extensively studied over the later few years although they are not yet completely understood. It is believed that several factors contribute to amyloid formation with the amyloidogenic potential of the precursor protein being highly important. This potential can be enhanced or diminished by the presence of mutations or post-translational modifications. A large

Fig. 13.1 The TTR dimer; for explanation see text

number of proteins associated with amyloidotic diseases together with their unrelated primary structures and lack of a consensus sequence, suggest the presence of other particularities that may contribute to their tendencies to self-aggregate. The well-known high content of β-sheet structures is the most cited factor as deduced from the cross-β pattern present in all amyloid fibrils. Thus, proteins rich in β-sheet would be more prone to fibril formation. However, some proteins associated with amyloid disorders present predominantly α-helices in their secondary structures. Lysozyme, in its native form, contains only a small amount of β-structure (Artymiuk and Blake 1981). Modifications, particularly changes in the β-sheet content as occurs in the prion protein, trigger protein aggregation. In TTR, amyloidogenic mutations favor formation and/or exposure of new structural motifs that culminate to tetramer dissociation and formation of partially unfolded intermediates, as discussed below. The crystal structure of TTR Val30Met shows a large spacing between DAGH and CBEF sheets in the monomer together with a movement solvent-exposing residue 10 of strand A (Terry et al. 1993). In the other variant, TTR Val122Ile, an increase in the length of the hydrogen bonds between dimers is observed (Damas et al. 1996). Furthermore, the FG loop also differs in all amyloidogenic TTR variants while it is conserved in the wild-type or non-amyloidogenic TTR.

In general, specific regions of a protein act as "hot spots," which promote aggregation (Sánchez de Groot et al. 2005). This issue is most important for natively unfolded proteins or unfolded states of globular proteins because there is no tertiary structure to mask the referred regions involved in aggregation. Therefore, identification of such sequences is very important for developing therapeutic approaches. TTR reveals 3 regions relevant to protein aggregation—residues 10–20, 23–33, and 105–118 (Sánchez de Groot et al. 2005).

Other factors, such as the presence or absence of circulating factors or cellular receptors can contribute to TTR deposition and may determine organ specificity.

Additional factors, like environmental or genetic background, have been proposed and could account for the differences in disease onset between individuals carrying the same mutation.

Besides the intact protein, TTR-derived peptides have also been found in amyloid fibrils from SSA and some FAP patients, raising the hypothesis that proteolysis may trigger fibril formation by releasing amyloidogenic fragments (Saraiva and Costa 1991). Diffraction analyses of *ex vivo* fibrils (Inouye et al. 1998) showed 29 Å periodic lengths along the meridian axis, which are greater than the size measured for H-bonded β-chains, suggesting that shorter TTR fragments may fill in the spaces in between.

The released fragments result from cleavages that vary according to the TTR variants involved: in SSA fibrils, the fragments result from cleavages at positions 46, 49, and 52 (Cornwell et al. 1988); in the cases of TTR Val30Met (Wahlquist et al. 1991) TTR Val122Ile (Gorevic et al. 1989), or TTR Phe33Ile (Nakazato et al. 1984), fragments result from a cleavage at position 49; TTR Leu111Met carriers show amyloid fibrils containing TTR molecules cleaved at positions 46, 49, and 59 (Hermansen et al. 1995). SDS–PAGE analyses of vitreous FAP fibrils show peptides smaller than the monomer and dimer (Thylén et al. 1993). One of these corresponds to a peptide starting at Thr49 and others are a mixture of peptides starting at positions 1 and 3. However, others have reported uncleaved TTR *N*-terminus (Saraiva et al. 1984; Tawara et al. 1983). TTR cleavage may be organ-specific or relate to particular organs such as the heart and the vitreous humor. In summary, the role of proteolysis in TTR amyloidosis is still unclear and evidence suggests that release of TTR peptides is probably not essential for fibril formation.

More recently, several studies have addressed post-translational TTR modifications and their importance in amyloid formation. *S*-sulfonation of TTR through oxidative modifications of the thiol residue of the cysteine-10 of TTR may be an important triggering step in the formation of transthyretin-related amyloid fibrils (Nakanishi et al. 2010).

13.4.2 Stabilization of the Tetrameric Fold

It is commonly believed that the first step in TTR fibrillization is the dissociation of the TTR tetramer into non-native monomers, which then associate to form oligomers. Oligomerization then leads to formation of larger aggregates. With time, these aggregates organize into short fibrils, which grow in length and associate with other short fibrils to form, together with other components, the mature amyloid. The initial dissociation of the tetramer is directly related to the protein stability, as the amyloidogenic TTR variants have been shown to have a decreased stability.

For instance, Thr119Met TTR has been described as a non-amyloidogenic TTR variant. In Portugal, this variant has been found in compound heterozygous individuals carrying Val30Met TTR, the most prevalent variant associated with FAP. In these individuals, disease evolution seems to be more benign than in typical Val30Met

TTR patients, suggesting a protective effect of TTR Thr119Met against the pathogenic effects of TTR Val30Met. Furthermore, a higher resistance to dissociation in Thr119Met carriers, in contrast to a lower resistance in Val30Met TTR carriers, was found (Longo Alves et al. 1997). It was suggested that the protective clinical effects of the Thr119Met mutation possibly involve stabilization of the tetrameric structure of TTR. Subsequent *in vitro* studies have confirmed this finding (Hammarström et al. 2001, 2003b). Crystallographic studies of TTR-Thr119Met structure revealed new hydrogen bonds within each monomer and monomer–monomer inter-subunit contacts (Ser117–Ser117 and Met119–Tyr114), resulting in increased protein stability, possibly underlying the protective effect of TTR Val30Met/Thr119Met (Sebastião et al. 2001). The Thr119Met mutation is located in the H strand and participates in the binding to thyroxine, resulting in increased affinity for the hormone; therapeutic strategies aimed at stabilizing the TTR tetrameric fold screen small molecules that compete with thyroxine and present a higher affinity to the protein than the hormone (Almeida et al. 2004).

13.4.3 Structural Amyloidogenic Determinants in TTR: The Leu55Pro and the Tyr78Phe TTR Variants

TTR Leu55Pro is one of the most clinically aggressive TTR mutations characterized by an early age-of-onset between the ages of 15–20 years. The disease progresses very rapidly to death in 5–10 years (Jacobson et al. 1992; Yamamoto et al. 1994). Besides neuropathy, this variant is also associated with cardiomyopathy and vitreous opacities.

It was suggested that mutated Leu55Pro TTR significantly alters tetramer stability and increases amyloidogenicity. The Leu55Pro TTR tetramer was more unstable and prone to denaturation at different pH values (McCutchen et al. 1993). X-ray-diffraction studies have compared the TTR Leu55Pro with wild-type TTR structures. Crystallographic work on other amyloidogenic variants, such as TTR Val30Met (Terry et al. 1993), TTR Val122Ile (Damas et al. 1996), and TTR Ile84Ser (Hamilton et al. 1996), described a dimer in the asymmetric unit. An overview of a large number of crystal structures of mutant TTRs reported structural differences to be insignificant (Hörnberg et al. 2000). In contrast, the crystal structure of TTR Leu55Pro showed presence of eight monomers in the unit cell (Sebastião et al. 1998). In the mutant monomer, due to disruption of the hydrogen bonds between strands D and A, residues 54–56 are part of a long-surface loop that connects strands C and E. Thus, the TTR Leu55Pro monomer is organized in seven strands and one α-helix, instead of the normal 8 strands that constitute the wild-type monomer. This loop is involved in the crystallographic packing observed in the crystal structure, and contacts between CE loops are believed to be determinants of amyloid formation. Another element involved in the overall force in the assembly of units is the hydrogen bonding between the AB loop and the α-helix of the nearest neighbor. The topology presented by this mutant is similar to the classic β-barrel. Furthermore, the crystal packing shows several channels running parallel to each other, suggesting a tubular structure. The authors proposed that this variant can constitute an intermediate structure for TTR amyloid formation. In fact, the TTR Leu55Pro preparations, used for crystallization, bound thioflavin T, indicating presence of an amyloid-like oligomeric structure (Sebastião et al. 2000). Thus, changes in the D strand together with the contacts of the AB loop and the α-helix are pivotal in amyloid formation. This conclusion is further corroborated by subsequent data both by: (i) probing solvent accessibility of TTR amyloid by solution NMR spectroscopy, showing that strands C and D are dislocated from their native-edge region and become solvent-exposed, leaving a new interface involving strands A and B accessible for intermolecular interactions (Olofsson et al. 2004); and (ii) recent data using multiple molecular dynamics simulations of wild-type and Leu55Pro TTR, indicating that the unfolding behaviors of these two proteins are different. Unfolding of Leu55Pro TTR involves a premature disruption and displacement of β-strand D and separation of β-strand C from the core of the monomer β-sandwich. This exposes key residues and a hydrophobic interface, embodied by β-strands A and B, which can promote monomer–monomer interactions; unfolding of the α-helical motif into coils and turns is also observed. In contrast, in the wild-type TTR, only extensive unfolding of the entire monomer allows displacement of strands C and D from the structure; the α-helix is maintained in the folded state in most cases (Rodrigues et al. 2010).

TTR Tyr78Phe was designed in silico to destabilize the AB loop and thus the tetrameric TTR fold (Redondo et al. 2000a). The resulting protein presents a soluble tetrameric form as shown by resistance to dissociation by native and SDS–PAGE studies besides elution patterns on gel filtration. This protein also retains the ability to bind thyroxine, indicating a functional tetrameric folding. Interestingly, this variant is prone to amyloid formation as observed by thioflavin-T binding. Furthermore, the mutated protein is recognized by a monoclonal antibody known to react only with highly amyloidogenic, mutant TTR proteins and with TTR amyloid fibrils (Goldsteins et al. 1997). Tyr78Phe was later observed in an Italian patient with TTR-associated amyloidosis (Anesi et al. 2001).

13.4.4 Proposed Models for TTR Amyloid

Ultrastructural analyses by transmission-electron microscopy (TEM) of *ex vivo* FAP TTR Val30Met fibrils from sural nerve biopsies indicated that the amyloid fibrils are formed by a core composed of a tight helix of 3-nm-wide double tracks containing pentosomes (subunits of amyloid P component) at the center. Other types of double tracks, 4.5–5 nm wide, are observed made up of heparan sulfate proteoglycans (HSPG). Based on these observations, a model was proposed (Inoue et al. 1998), where TTR appears as filaments 0.5–1 nm wide at the periphery of the core. The dimensions of the TTR filaments made the authors suggest that the TTR basic unit in the FAP fibrils is a modified monomer.

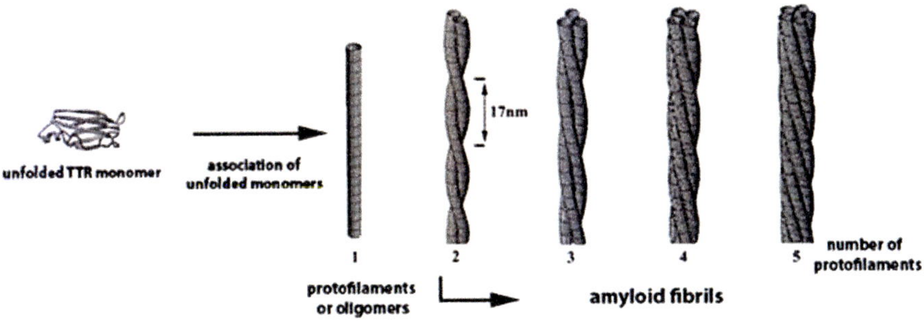

Fig. 13.2 Proposed model for TTR amyloid fibril formation based on STEM and mass-per-length analyses

In vitro studies using TEM, freeze-drying followed by unidirectional metal shadowing, and scanning transmission-electron microscopy (STEM) indicated that the prominent fibrils were 8 nm wide, left-hand-twisted with an axial substructure of 17 nm (Cardoso et al. 2002), as represented in Fig. 13.2.

13.4.5 Intermediate Species in TTR Amyloidogenesis

13.4.5.1 Monomeric Intermediates

In vitro dissociation of the TTR tetramers into monomers depends on pH and concentration and involves formation of several different intermediates having variable quaternary, tertiary, and secondary structures. Even though the use of low pH has been useful to study amyloid formation, physiological conditions are required to understand better the processes that takes place *in vivo*. Quintas et al., reported that TTR dissociates to a monomeric species at pH 7.0 and nearly physiological ionic strengths; tetramer dissociation is apparently irreversible and the generated monomer is non-native (Quintas et al. 1997). This monomeric species does not behave like a molten globule and leads to formation of partially unfolded monomeric species and high-molecular-mass soluble aggregates (Quintas et al. 1999). Aging experiments of tetrameric TTR under physiological conditions, and protein-unfolding experiments of the non-native monomeric forms, have shown that tetramer dissociation and partial unfolding of the monomer precede amyloid fibril formation (Quintas et al. 2001).

A different approach using STEM further demonstrated that the TTR monomer is the unit being added to the growing fibrils under physiological pH (Cardoso et al. 2002) because analysis of a heterogeneous population of amyloidogenic TTR species estimated a 14-Da difference in mass-per-length between the different entities, thus corresponding to the TTR monomer mass. Furthermore, the amyloidogenic Asp18Gly TTR mutation has been observed in Hungarian individuals and shown to be monomeric *in vitro* (Hammarström et al. 2003a). This TTR variant causes CNS amyloidosis and is present in low levels in serum and CSF, which may explain the absence of an early-onset systemic disease and may imply rapid degradation either after secretion or within the cell before secretion (Hammarström et al. 2003b; Sekijima et al. 2005).

13.4.5.2 Dimeric Intermediates

Several authors have questioned the monomer as the building block of fibrils. In this respect, the fact that the structure of TTR Val30Met reveals movement of strand A exposing Cys10 to the solvent led Terry et al. to consider that amyloidogenic fibrils could form after association of TTR molecules through disulfide bridges (Terry et al. 1993). However, this theory does not fit with the formation of fibrils by a TTR Cys10Arg variant. Redondo et al. showed that mutant dimers covalently linked through disulfide bonds were unable to polymerize into amyloid while disruption of the S–S bridges by a reducing agent lead to amyloid formation (Redondo et al. 2000b). These data implicate the need of a monomeric entity prior to TTR fibril assembly.

Olofsson et al. showed that substitution of two amino-acid residues in the hydrophobic core of TTR to generate a cross-linked engineered TTR mutant with extra cysteine residues, lead to a mutant very prone to form amyloid (Olofsson et al. 2001). In this mutant, before fibril formation, a stable dimeric species was detected at low temperatures. This data concord with the findings by Serag et al., suggesting that the native dimeric interactions are preserved within the amyloid fibrils (Serag et al. 2001). However, in the mutant described by Olofsson et al., a small population of monomers was also observed (Olofsson et al. 2001).

Recent studies using mathematic and simulation tools suggest that the TTR dimer can undergo initial fibril formation and that although the monomer may also form fibrils, the dissociation into monomers might not be a requirement for fibril formation (Sørensen et al. 2007).

13.4.5.3 Tetrameric Intermediates

A tetrameric TTR structure as the fibril building blocks has also been proposed by Ferrão-Gonçalves et al., based on *in vitro* experiments under high pressure in which an altered tetramer could be responsible for TTR aggregation (Ferrão-Gonçalves et al. 2000). However, this work did not present morphological evidence for fibril formation. A small population of monomers was also obtained and, therefore, this hypothesis may apply to very early events during fibril formation. Moreover, this work was performed using wild-type TTR and did not include an equal treatment of amyloidogenic TTR variants.

Other studies presenting the structure of a synthetic TTR variant, Gly53Ser-Glu54AspLeu55Ser, suggested an altered tetramer as the building block for TTR fibrils (Eneqvist et al. 2000). This mutant, generated according to a predicted mutational hot-spot region in the short strand D of the molecule, polymerizes at physiological conditions, yielding high-molecular-mass aggregates with amyloid characteristics (Goldsteins et al. 1997).

13.4.5.4 High-Molecular-Mass Intermediates

TTR oligomers and other high-molecular-mass structures of amyloidogenic proteins (including TTR) present in tissues together with mature fibrils have been intensively studied not only for their toxicity, as described above, but also for their mechanisms of formation. Epitopes displayed by oligomers are conformational and have been reported for several amyloidogenic systems, including Aβ, TTR, α-synuclein, and others. An antibody was reported to recognize a common generic structure of low-molecular-mass soluble amyloidogenic oligomers of different proteins (including TTR), thus suggesting a similar mechanism of pathogenesis for amyloid-related disorders (Kayed et al. 2003).

Large insoluble aggregates are also part of the dynamics of TTR fibril formation. Their place in the amyloidogenic cascade is, however, not clear. Most of the time, small oligomers and large aggregates co-exist in *in vitro* preparations, as observed by TEM, probably reflecting rapid interconversion of different species. Studies on different amyloidogenic proteins (Sorgjerd et al. 2008; Zako et al. 2009) have shown that oligomers are transiently present during fibrillization and, under the correct circumstances, can be exclusively populated without further conversion into amyloid fibrils (Lindgren and Hammarström 2010). *Ex vivo* analyses of tissues from FAP patients using specific detection by dyes or generic antibodies against different oligomeric TTR species remain to be performed.

13.4.6 Interference with the Dynamics of Fibril Formation

The knowledge acquired in the last few years on structural and biochemical features of amyloidogenic intermediates, has paved the way for investigation of drugs capable of interfering with amyloidogenic pathways. For TTR-associated amyloidoses, the first step of intervention is to obstruct tetramer dissociation and thus, prevent potential species from being generated. Several authors have identified compounds capable of inhibiting TTR fibril formation (Klabunde et al. 2000; Oza et al. 1999; Baures et al. 1999, 1998) based on stabilization of the tetrameric structure. The above compounds include non-steroidal anti-inflammatory drugs (NSAIDs) or derivatives thereof, and natural compounds such as plant-derived flavonoids and xanthones [for a review see Almeida et al. 2005; Connelly et al. 2010]. However, most of these studies were performed under non-physiological acidic conditions, which are believed to induce rapid conformational changes while no specific distinction was made regarding the species affected by those drugs. Alternative studies have been performed using *ex vivo* samples (sera) from normal and TTR-Val30Met carriers, and compounds such as iododiflunisal and epigallocatechin-3-gallate (EGCG) have shown effectiveness in TTR tetramer stabilization (Almeida et al. 2004; Ferreira et al. 2009). In contrast, compounds such flufenamic acid, diclofenac, resveratrol, ginestein, and DES (diethylstilbestrol) previously published as TTR stabilizers by *in vitro* assays using TTR acidification, did not present any anti-amyloidogenic activity under *ex vivo* analyses in the presence of serum. These results stress severe limitations of *in vitro* testing of compounds on isolated TTR preparations (Cardoso et al. 2007).

A cellular system was created using transfected cells expressing Leu55Pro TTR, precluding isolation of the target protein and representing a more physiological *in vitro* approach to study inhibitors of amyloidogenesis (Cardoso et al. 2007). Aggregated Leu55Pro accumulates in conditioned media and is retained in a filter; quantification of the retained material is indicative of the anti-amyloidogenic properties of the drugs under study. In this study, compounds such as triiodophenol (TIP), [2-(3,5-dichlorophenyl) amino] benzoic acid (DCPA), [4-(3,5-difluorophenyl)] benzoic acid (DFPB), and benzoxazole were identified as anti-amyloidogenic drugs corroborating data obtained by other *in vitro* methodologies.

13.5 Cytotoxicity in FAP

13.5.1 TTR Binding to Membrane Lipids

Generally, cytotoxicity of extracellular protein aggregates might be a direct consequence of binding to plasma membranes (Yip et al. 2002; Kakio et al. 2001). To further elucidate TTR interactions with plasma membranes, surface-plasmon resonance on neuroblastoma cells after incubation with wild-type TTR or aggregated TTR mutants, was performed (Hou et al. 2005). All TTR forms bound to the plasma membrane through electrostatic interactions as membrane binding was inversely correlated with ionic strength. After protein removal from plasma membrane preparations, the binding of all forms of TTR remained relatively unchanged when compared to binding to intact plasma membranes. As binding to the membrane-derived lipid extracts was similar to that obtained with intact membranes, it seemed likely that TTR mainly bound to lipid components in plasma membranes. The lipid species involved in TTR binding was not identified; however, it was proposed that TTR binds to the polar region of phospholipid head groups on the membrane surface. Interestingly, the extent of binding directly correlated with the extent of aggregation and with the cytotoxicity of each TTR form as assessed by cell-viability or cell-death assays (Hou et al. 2005). As membrane fluidity is known to influence cell viability, the effect of TTR on membrane fluidity was examined using fluorescence anisotropy. It was determined that binding of amyloidogenic TTR mutants increased membrane fluidity. As such, it was suggested that TTR interaction with a lipid fraction within the plasma membrane may alter its dynamic properties, thereby causing a decrease in cell viability (Hou et al. 2005). In another study, the same authors assessed the effect of TTR oligomers on Ca^{2+} homeostasis (Hou et al. 2007). TTR rapidly induced an increase in the intracellular Ca^{2+} concentration ($[Ca^{2+}]_i$) when applied to SH-SY5Y human neuroblastoma cells. The greatest effect on $[Ca^{2+}]_i$ was induced by a preparation that contained the highest concentration of TTR oligomers. The TTR-induced increase in $[Ca^{2+}]_i$ was due to an influx of extracellular Ca^{2+}, mainly *via* L- and N-type voltage-gated calcium channels (VGCCs). These results suggest that increasing $[Ca^{2+}]_i$ *via* VGCCs may be an important early event contributing to TTR-induced cytotoxicity, and that TTR oligomers, rather than mature fibrils, may be the major cytotoxic forms of TTR. Disruption of Ca^{2+} homeostasis induced by TTR oligomers was also attributable to mobilization from ER stores (Teixeira et al. 2006). In these studies, the authors showed that inhibitors of Ca^{2+} mobilization from ER to the cytosol could inhibit caspase-3 activation induced by TTR aggregates; both pathways could cause dysregulation of Ca^{2+} homeostasis in a chronologically distinct manner.

13.5.2 Contribution of Cellular Receptors: The Receptor for Advanced-Glycation End-Products (RAGE) in Inflammation and Oxidative Stress

Treatment of a rat Schwannoma cell line with TTR aggregates stimulated ERK1/2 activation, which was partially mediated by the receptor for advanced-glycation end-products (RAGE). This suggests that abnormally sustained activation of ERK in FAP may represent an early signaling cascade leading to neurodegeneration (Monteiro et al. 2006), in part mediated by the RAGE receptor. Increased ERK1/2 activation, and RAGE expression has been observed in human TTR amyloid-laden tissues, in particular in nerve biopsies (Sousa et al. 2001b; Monteiro et al. 2006). RAGE is a member of the immunoglobulin superfamily with a broad repertoire of ligands in addition to amyloid-associated macromolecules. These include products of non-enzymatic glycoxidation (advanced-glycation end-products, AGEs), pro-inflammatory mediators (S100/calgranulins), and amphotericin (Yan et al. 2009). In each case, the receptor recruits signal-transduction mechanisms, often resulting in sustained and pathogenic inflammatory and stress responses. RAGE binding by aggregated TTR triggers activation of the transcription factor NF-κB. This response might underlie peripheral nerve dysfunction. Analyses of nerve biopsy samples from Portuguese patients at different stages of FAP (including the pre-symptomatic stage), were compared with age-matched controls by semi-quantitative immunohistology and *in situ* hybridization. These experiments showed up-regulation in axons of pro-inflammatory cytokines (tumor necrosis factor and interleukin-1) and the inducible form of nitric oxide synthase (iNOS), as well as increased tyrosine nitration and activated caspase-3 (Sousa et al. 2001b).

13.5.3 Activation of Extracellular Matrix Remodeling Genes in FAP

Gene arrays using clinical material (salivary glands) and control tissues allow analyses of differential gene expression between FAP and healthy tissues. Among the differentially expressed genes, up-regulation of genes related to the extracellular-matrix (ECM) remodeling was evident in patients' tissues. Matrix metalloproteinase-9 (MMP-9), neutrophil gelatinase-associated lipocalin (NGAL), and biglycan were over-expressed in amyloid-laden tissues (Sousa et al. 2005), suggesting that fibrils trigger specific signaling pathways, leading to ECM remodeling and that this process causes tissue damage with pathological consequences. The same markers of ECM remodeling together with tissue inhibitor of metalloproteinase-1 (TIMP-1) and chondroitin-sulfate proteoglycan synthase (CSPG) were investigated in Val30Met TTR transgenic mouse model at different stages of TTR deposition, i.e., in tissues with non-fibrillar or fibrillar deposits. Immunohistochemistry or RT-PCR analyses showed that biglycan was already increased in animals presenting TTR deposited in a non-fibrillar state, whereas MMP-9, TIMP-1, NGAL, and CSPG were elevated only in mice with TTR amyloid deposits (Cardoso et al. 2008). The mechanism(s) involved in triggering ECM-related genes by TTR-aggregate deposition and the consequences in the degenerative process are presently unknown.

13.5.4 The Ubiquitin–Proteasome System in FAP

The ubiquitin–proteasome system (UPS) has been studied in FAP in TTR-synthesizing and TTR-non-synthesizing tissues from affected individuals and in the transgenic Val30Met mice model. Ubiquitin protein conjugates were up-regulated, the proteasome levels were decreased, and parkin and α-synuclein expression were both decreased (Santos et al. 2007). On the other hand, the liver, that normally synthesizes variant TTR Val30Met, did not show this response. When neuronal or Schwannoma cell lines were treated with TTR aggregates, an increase in ubiquitin and a decrease in parkin levels were seen. The overall results suggest extracellular TTR deposition as an external stimulus to an intracellular UPS response in FAP.

13.5.5 Endoplasmic Reticulum (ER) Stress Response in FAP

In the process of TTR translation and secretion, a competition between ER-assisted folding and ER-associated degradation were shown to occur intracellularly (Sekijima et al. 2005). To further assess the involvement of ER-stress-response molecules in

TTR secretion, co-expression of wild-type TTR, Leu55Pro, Ala25Thr, or Asp18Gly with the chaperone BiP was compared in prokaryotes (Sorgjerd et al. 2006). Only Asp18Gly TTR was significantly captured by BiP forming stable complexes followed by a sixfold smaller amount of captured Ala25Thr TTR, and negligible amounts of Leu55Pro TTR and wild-type TTR. This result showed that BiP recognition inversely correlates with the stability of the TTR variants, which in turn is clearly correlated with the secretion efficiency. The stoichiometry of captured Asp18Gly TTR *versus* BiP increased with increasing size of the oligomers, suggesting that following co-expression in prokaryotes, BiP can capture Asp18Gly TTR, and keep it sequestered. It was suggested that *in vivo*, this mechanism likely facilitates Asp18Gly TTR degradation and that this protective mechanism prevents an otherwise devastating form of early-onset FAP.

The involvement of the ER-stress response in FAP was studied in FAP samples and in cellular assays. Teixeira et al. demonstrated activation of the classical unfolded-protein-response pathway in tissues without TTR synthesis but presenting extracellular TTR deposition (Teixeira et al. 2006). Increased levels of the ER-resident chaperone BiP, ATF-6, and activation of eIF2α were found consistently in biopsies from FAP patients and asymptomatic carriers (FAP 0) as documented in Fig. 13.3. In the cellular models, it was observed that extracellular TTR oligomers induce BiP expression and activation of eIF2α. The increase in BiP levels involved Ca^{2+} mobilization from ER as it could be blocked by Ca^{2+}-channel inhibitors. This study clearly demonstrated the connection between the ER-stress response and FAP. In a broader perspective, the finding of induction of ER stress by an extracellular protein certainly has implications for understanding signal transduction in general.

13.5.6 The Unfolded-Protein Response in FAP: Activation of Heat-Shock Factor-1 (HSF-1) and Heat-Shock Proteins

The heat-shock response was investigated in FAP. Up-regulation of Hsp27 and Hsp70 expression related to the presence of extracellular TTR aggregates in human

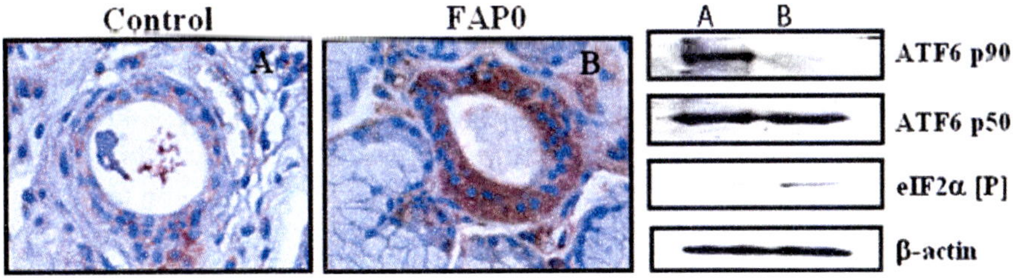

Fig. 13.3 ER stress observed by analyzing salivary gland from controls (**A**) and FAP 0 individuals (**B**) by immunohistochemistry for ATF6 and western blotting (ATF 6 and phosphorylated eIF2α) (This research was originally published by Teixeira et al. 2006)

FAP biopsies as compared to normal controls was documented. TTR aggregates did not co-localize with Hsps suggesting that extracellular TTR tissue deposits are able to induce an intracellular stress response. Moreover, the heat-shock transcription factor-1 (HSF-1) was up-regulated and localized to the nucleus. Transgenic mouse models expressing the Val30Met mutant TTR show a similar response: the presence of TTR deposits induces a stress response with activation of HSF-1 and increased synthesis of Hsps. In cell culture, up-regulation of Hsp70 and Hsp27 in the presence of toxic TTR aggregates was found (Santos et al. 2008).

It was hypothesized that HSF-1 could be involved in FAP pathogenesis as a cell-defense mechanism against the presence of extracellular TTR deposits and that disruption of the heat-shock response would aggravate TTR deposition. A mouse model expressing the human TTR-Val30Met in an HSF-1-null background was characterized. The lack of HSF-1 expression lead to extensive and earlier non-fibrillar TTR, evolving into fibrillar material in distinct organs, including the peripheral and autonomic nervous systems, which was not found in Val30Met animals in the wild-type background. As in the human disease, liver, brain, and spinal cord did not present deposition in this animal model. Furthermore, inflammatory stress and a reduction in unmyelinated nerve fibers were observed, as in human patients, indicating that HSF-1-regulated genes are involved in FAP, modulating TTR tissue deposition (Santos et al. 2010a).

The TTR/HSF-1 model allows different studies not possible with previous models. Sural-nerve biopsies, on which most of the pathological analyses in FAP have been performed, represent a restricted portion of the PNS, and it is clearly possible that TTR deposits in ganglia, or more proximally in nerve trunks, which could be responsible for distal nerve fiber loss. The possibility to analyze dorsal root ganglia (DRG) with deposition in the TTR/HSF-1 mouse model is invaluable, and permits a close evaluation of events occurring in sensory and autonomic neurons responsible for neurodegeneration and in the assessment of therapies.

13.6 Counteracting TTR Aggregation and Cytotoxicity: Lessons from Animal Models

In vitro studies have shown that doxycycline disrupts TTR amyloid fibrils. Moreover, the generated assemblies did not display cytotoxic properties in caspase-3-activation assays (Cardoso et al. 2003). *In vivo* studies corroborated the *in vitro* findings when doxycycline was administrated to 23–28-month-old Val30Met-TTR transgenic animals in the drinking water over a period of 3 months. Immunohistochemistry revealed no differences in non-fibrillar TTR deposition between treated and non-treated mice, but Congo-red-positive material was only observed in the control, non-treated group. Immunohistochemistry for several markers associated with amyloid, such as matrix MMP-9 and serum amyloid P component (SAP) was performed. Significantly lower levels of MMP-9 were found in the treated animals when compared with the control group. Mouse SAP was absent in treated animals, being only observed in non-treated animals with congophilic deposits. These results indicate that doxycycline is capable of disrupting Congo-red-positive TTR amyloid deposits and decreases standard markers associated with fibrillar deposition, being a potential drug in the treatment of amyloidosis (Cardoso and Saraiva 2006).

Another drug, which was tested in TTR transgenic animals and was effective to counteract TTR aggregation, was tauroursodeoxycholic acid (TUDCA). TUDCA is a unique natural compound that acts as a potent anti-apoptotic and anti-oxidant agent, reducing cytotoxicity in several neurodegenerative diseases (Keene et al. 2002). Since oxidative stress, apoptosis, and inflammation are associated with TTR deposition in FAP, the possible therapeutic application of TUDCA in this disease has been investigated. It was shown by semi-quantitative immunohistochemistry and western blotting that administration of TUDCA to a transgenic mouse model of FAP decreased apoptotic and oxidative biomarkers usually associated with TTR deposition, namely the ER chaperone BiP, the Fas death receptor, and oxidation products such as 3-nitrotyrosine. Most importantly, TUDCA treatment significantly reduced TTR toxic aggregates as much as 75% (Macedo et al. 2008). Since TUDCA has no effect on TTR aggregation *in vitro*, this finding pointed at the *in vivo* modulation of TTR aggregation by oxidative stress and apoptosis, and prompts for the use of this safe drug in prophylactic and therapeutic measures in the FAP population.

13.7 Clinical Trials

Based on the knowledge gathered from the various *in vitro* and *in vivo* approaches to TTR aggregation and toxicity, different multicenter clinical trials are underway. We might need more than a single approach to treat FAP effectively.

1. Approaches based on TTR tetramer stabilization:
   Diflunisal (http://clinicaltrials.gov) and Tafamidis (www.foldrx.com): in the latter case, the initial results for an 18-month phase-II/-III randomized double-blind trial comparing Tafamidis with placebo in 128 individuals in the early phase of the disease showed that 60% had no disease progression but improvement of nutritional status as compared to placebo (Coelho et al. 2010). As referred to previously, once the clinical symptoms are triggered in FAP, disease evolution takes several years; thus, the initial studies have been extended to assess the usefulness of this drug in FAP treatment and also to investigate the mechanisms underlying the results found in non-responders.

2. Approaches based on siRNA to silence liver TTR expression:
   ALN-TTR01 (www.Alnylam.com) is a systemically delivered RNAi therapeutic that targets the human TTR gene. As mentioned above, TTR is mainly synthesized in the liver. Studies with ALN-TTR01 in mice transgenic for the human Val30Met have shown durable suppression, as much as 90%, for both TTR mRNA and protein levels in plasma and reduction of deposition in extra-hepatic

tissues (Alvarez et al. 2010). Phase-I human clinical studies are underway to evaluate the safety and tolerability of this drug in FAP patients.

3. Approaches based on removal/reduction of TTR deposits:
   It was recently shown (Cardoso et al. 2010) that combined cyclic drug administration of doxycycline and TUDCA, resulted in significant reduction in TTR deposition and associated tissue markers in the TTR Val30Met transgenic mouse model. The observed synergistic effect of doxycycline/TUDCA within the range of human-tolerable quantities prompted their application in FAP, particularly in the early stages of disease. For this purpose, a phase-II clinical study is underway (www.clinicaltrials.gov; transthyretin amyloidosis).

13.8 Concluding Remark

Increasing evidence indicates that accumulation of misfolded proteins, protofibril formation, their interaction with membranes, and the consequential intracellular signaling cascades represent unifying events in many of slowly progressive neurodegenerative disorders. Extracellular protein misfolding and aggregation occurring in PNS triggers inflammation, oxidative stress, matrix remodeling, and stresses of unfolded-protein-response and ER pathways that resemble in many aspects, including common molecular players and scenarios, to those described in the CNS for protein-misfolding disorders, such as Alzheimer Disease. Thus, similarities and dissimilarities in toxicity found between the CNS and PNS are very useful to pinpoint and guide us to the treatment of aging-associated neurodegenerative disorders. At this point, the various studies on the molecular pathology associated with FAP and data from the use of anti-apoptotic agents in transgenic animals suggest that signaling pathways in FAP can contribute to extracellular polymerization of TTR in tissues, as depicted in Fig. 13.4.

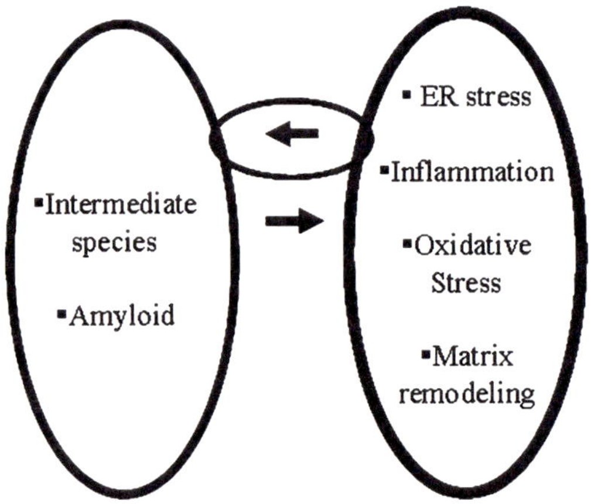

Fig. 13.4 Proposed working hypothesis for a two-way crosstalk between the extracellular milieu and the cell in the TTR amyloidoses

Understanding the two-way crosstalk between the extracellular milieu and the cell is a major trend in diseases related to protein aggregation.

References

Almeida MR, Macedo B, Cardoso I, Alves I, Valencia G, Arsequell G, Planas A, Saraiva MJ (2004) Selective binding to transthyretin and tetramer stabilization in serum from patients with familial amyloidotic polyneuropathy by an iodinated diflunisal derivative. Biochem J 381:351–356

Almeida MR, Gales L, Damas AM, Cardoso I, Saraiva MJ (2005) Small transthyretin (TTR) ligands as possible therapeutic agents in TTR amyloidoses. Curr Drug Targets CNS Neurol Disord 4:587–596

Alvarez R, Borland T, Chen Q, Milstein S, Nguyen T, Hinkle G, Kuchimanchi S, Costigan J, Ristoiu V, Wang G, Cole G, Dorkin R, Akinc A, Nechev L, Kosovrasti V, Tchangov L, Tracy M, Jeffs L, Maclachlan I, Lutwyche P, Martins D, Costelha S, Saraiva MJ, Sah DWY (2010) ALN-TTR, an RNAi therapeutic for the treatment of transthyretin-mediated amyloidosis. Amyloid J Protein Fold Disord 17:51–52

Alves IL, Altland K, Almeida MR, Winter P, Saraiva MJ (1997) Screening and biochemical characterization of transthyretin variants in the Portuguese population. Hum Mutat 9:226–233

Andersson R (1970) Hereditary amyloidosis with polyneuropathy. Acta Med Scand 1–2:85–94

Andersson K, Olofsson A, Nielsen EH, Svehag SE, Lundgren E (2002) Only amyloidogenic intermediates of transthyretin induce apoptosis. Biochem Biophys Res Commun 294:309–314

Andrade C (1952) A peculiar form of peripheral neuropathy; familiar atypical generalized amyloidosis with special involvement of the peripheral nerves. Brain 75:408–427

Anesi E, Palladini G, Perfetti V, Arbustini E, Obici L, Merlini G (2001) Therapeutic advances demand accurate typing of amyloid deposits. Am J Med 111:243–244

Araki S (1984) Type I familial amyloidotic polyneuropathy (Japanese type). Brain Dev 6:128–133

Artymiuk PJ, Blake CC (1981) Refinement of human lysozyme at 1.5 Å resolution analysis of non-bonded and hydrogen-bond interactions. J Mol Biol 152:737–762

Baures PW, Peterson SA, Kelly JW (1998) Discovering transthyretin amyloid fibril inhibitors by limited screening. Bioorg Med Chem 6:1389–1401

Baures PW, Oza VB, Peterson SA, Kelly JW (1999) Synthesis and evaluation of inhibitors of transthyretin amyloid formation based on the non-steroidal anti-inflammatory drug, flufenamic acid. Bioorg Med Chem 7:1339–1347

Benson MD, Kincaid JC (2007) The molecular biology and clinical features of amyloid neuropathy. Muscle Nerve 36:411–423

Berg I, Thor S, Hammarstrom P (2009) Modeling familial amyloidotic polyneuropathy (transthyretin V30M) in Drosophila melanogaster. Neurodegener Dis 6:127–138

Blake CC, Geisow MJ, Oatley SJ, Rerat B, Rerat C (1978) Structure of prealbumin: secondary, tertiary and quaternary interactions determined by Fourier refinement at 1.8 Å. J Mol Biol 121:339–356

Buxbaum JN, Ye Z, Reixach N, Friske L, Levy C, Das P, Golde T, Masliah E, Roberts AR, Bartfai T (2008) Transthyretin protects Alzheimer's mice from the behavioral and biochemical effects of Aβ toxicity. Proc Natl Acad Sci USA 105:2681–2686

Cardoso I, Saraiva MJ (2006) Doxycycline disrupts transthyretin amyloid: evidence from studies in a FAP transgenic mice model. FASEB J 20:234–239

Cardoso I, Goldsbury CS, Muller SA, Olivieri V, Wirtz S, Damas AM, Aebi U, Saraiva MJ (2002) Transthyretin fibrillogenesis entails the assembly of monomers: a molecular model for *in vitro* assembled transthyretin amyloid-like fibrils. J Mol Biol 317:683–695

Cardoso I, Merlini G, Saraiva MJ (2003) 4'-Iodo-4'-deoxydoxorubicin and tetracyclines disrupt transthyretin amyloid fibrils *in vitro* producing noncytotoxic species: screening for TTR fibril disrupters. FASEB J 17:803–809

Cardoso I, Almeida MR, Ferreira N, Arsequell G, Valencia G, Saraiva MJ (2007) Comparative *in vitro* and *ex vivo* activities of selected inhibitors of transthyretin aggregation: relevance in drug design. Biochem J 408:131–138

Cardoso I, Brito M, Saraiva MJ (2008) Extracellular matrix markers for disease progression and follow-up of therapies in familial amyloid polyneuropathy V30M TTR-related. Dis Markers 25:37–47

Cardoso I, Martins D, Ribeiro T, Merlini G, Saraiva MJ (2010) Synergy of combined doxycycline/TUDCA treatment in lowering transthyretin deposition and associated biomarkers: studies in FAP mouse models. J Transl Med 8:74

Choi SH, Leight SN, Lee VM, Li T, Wong PC, Johnson JA, Saraiva MJ, Sisodia SS (2007) Accelerated Aβ deposition in APPswe/PS1ΔE9 mice with hemizygous deletions of TTR (transthyretin). J Neurosci 27:7006–7010

Coelho T, Maia L, Da Silva AM, Waddingtion-Cruz M, Plante-Bordeneuve V, Lozeron P, Suhr OB, Campistol J, Conceicao IM, Schmidt H, Trigo P, Packman J, Grogan DR (2010) Tafamidis (Fx-1006A): a first-in-class disease-modifying therapy for transthyretin familial amyloid. Amyloid-J Protein Fold Disord 17:75–76

Coimbra A, Andrade C (1971a) Familial amyloid polyneuropathy: an electron microscope study of the peripheral nerve in five cases. I. Interstitial changes. Brain 94:199–206

Coimbra A, Andrade C (1971b) Familial amyloid polyneuropathy: an electron microscope study of the peripheral nerve in five cases. II. Nerve fibre changes. Brain 94:207–212

Connelly S, Choi S, Johnson SM, Kelly JW, Wilson IA (2010) Structure-based design of kinetic stabilizers that ameliorate the transthyretin amyloidoses. Curr Opin Struct Biol 20:54–62

Cornwell GG 3rd, Sletten K, Johansson B, Westermark P (1988) Evidence that the amyloid fibril protein in senile systemic amyloidosis is derived from normal prealbumin. Biochem Biophys Res Commun 154:648–653

Costa R, Ferreira-Da-Silva F, Saraiva MJ, Cardoso I (2008) Transthyretin protects against Aβ peptide toxicity by proteolytic cleavage of the peptide: a mechanism sensitive to the Kunitz protease inhibitor. PLoS One 3:e2899

Damas AM, Ribeiro S, Lamzin VS, Palha JA, Saraiva MJ (1996) Structure of the Val122Ile variant transthyretin—a cardiomyopathic mutant. Acta Crystallogr D Biol Crystallogr 52:966–972

Dyck PJ, Lambert EH (1969) Dissociated sensation in amyloidosis. Compound action potential, quantitative histologic and teased-fiber, and electron microscopic studies of sural nerve biopsies. Arch Neurol 20:490–507

Eneqvist T, Andersson K, Olofsson A, Lundgren E, Sauer-Eriksson AE (2000) The β-slip: a novel concept in transthyretin amyloidosis. Mol Cell 6:1207–1218

Episkopou V, Maeda S, Nishiguchi S, Shimada K, Gaitanaris GA, Gottesman ME, Robertson EJ (1993) Disruption of the transthyretin gene results in mice with depressed levels of plasma retinol and thyroid hormone. Proc Natl Acad Sci USA 90:2375–2379

Ferrão-Gonçalves AD, Souto SO, Silva JL, Foguel D (2000) The preaggregated state of an amyloidogenic protein: hydrostatic pressure converts native transthyretin into the amyloidogenic state. Proc Natl Acad Sci USA 97:6445–6450

Ferreira N, Cardoso I, Domingues MR, Vitorino R, Bastos M, Bai G, Saraiva MJ, Almeida MR (2009) Binding of epigallocatechin-3-gallate to transthyretin modulates its amyloidogenicity. FEBS Lett 583:3569–3576

Fleming CE, Saraiva MJ, Sousa MM (2007) Transthyretin enhances nerve regeneration. J Neurochem 103:831–839

Goldsteins G, Andersson K, Olofsson A, Dacklin I, Edvinsson A, Baranov V, Sandgren O, Thylen C, Hammarstrom S, Lundgren E (1997) Characterization of two highly amyloidogenic mutants of transthyretin. Biochemistry 36:5346–5352

Gorevic PD, Prelli FC, Wright J, Pras M, Frangione B (1989) Systemic senile amyloidosis. Identification of a new prealbumin (transthyretin) variant in cardiac tissue: immunologic and biochemical similarity to one form of familial amyloidotic polyneuropathy. J Clin Invest 83:836–843

Guimarães A, Viana Pinheiro A, Leite I (1988) Sural nerve biopsy in familial amyloidotic polyneuropathy: a morphological and morphometric polyneuropathy. In: Isobe T, Araki S, Uchino F, Kito S, Tsubura E (eds) Amyloid and amyloidosis. Plenum Press, New York

Hamilton JA, Steinrauf LK, Braden BC, Murrell JR, Benson MD (1996) Structural changes in transthyretin produced by the Ile84Ser mutation which result in decreased affinity for retinol binding protein. Amyloid-Int J Exp Clin Investig 3:1–12

Hammarström P, Schneider F, Kelly JW (2001) *Trans*-suppression of misfolding in an amyloid disease. Science 293:2459–2462

Hammarström P, Sekijima Y, White JT, Wiseman RL, Lim A, Costello CE, Altland K, Garzuly F, Budka H, Kelly JW (2003a) D18G transthyretin is monomeric, aggregation prone, and not detectable in plasma and cerebrospinal fluid: a prescription for central nervous system amyloidosis? Biochemistry 42:6656–6663

Hammarström P, Wiseman RL, Powers ET, Kelly JW (2003b) Prevention of transthyretin amyloid disease by changing protein misfolding energetics. Science 299:713–716

Hermansen LF, Bergman T, Jornvall H, Husby G, Ranlov I, Sletten K (1995) Purification and characterization of amyloid-related transthyretin associated with familial amyloidotic cardiomyopathy. Eur J Biochem 227:772–779

Hörnberg A, Eneqvist T, Olofsson A, Lundgren E, Sauer-Eriksson AE (2000) A comparative analysis of 23 structures of the amyloidogenic protein transthyretin. J Mol Biol 302:649–669

Hou X, Richardson SJ, Aguilar MI, Small DH (2005) Binding of amyloidogenic transthyretin to the plasma membrane alters membrane fluidity and induces neurotoxicity. Biochemistry 44:11618–11627

Hou X, Aguilar MI, Small DH (2007) Transthyretin and familial amyloidotic polyneuropathy. Recent progress in understanding the molecular mechanism of neurodegeneration. FEBS J 274:1637–1650

Inoue S, Kuroiwa M, Saraiva MJ, Guimaraes A, Kisilevsky R (1998) Ultrastructure of familial amyloid polyneuropathy amyloid fibrils: examination with high-resolution electron microscopy. J Struct Biol 124:1–12

Inouye H, Domingues FS, Damas AM, Saraiva MJ, Lundgren E, Sandgren O, Kirschner DA (1998) Analysis of X-ray diffraction patterns from amyloid of biopsied vitreous humor and kidney of transthyretin (TTR) Met30 familial amyloidotic polyneuropathy (FAP) patients: axially arrayed TTR monomers constitute the protofilament. Amyloid 5:163–174

Jacobson DR, Mcfarlin DE, Kane I, Buxbaum JN (1992) Transthyretin Pro55, a variant associated with early-onset, aggressive, diffuse amyloidosis with cardiac and neurologic involvement. Hum Genet 89:353–356

Jacobson DR, Pastore RD, Yaghoubian R, Kane I, Gallo G, Buck FS, Buxbaum JN (1997) Variant-sequence transthyretin (isoleucine 122) in late-onset cardiac amyloidosis in black Americans. N Engl J Med 336:466–473

Kakio A, Nishimoto SI, Yanagisawa K, Kozutsumi Y, Matsuzaki K (2001) Cholesterol-dependent formation of GM1 ganglioside-bound amyloid β-protein, an endogenous seed for Alzheimer amyloid. J Biol Chem 276:24985–24990

Kanda Y, Goodman DS, Canfield RE, Morgan FJ (1974) The amino acid sequence of human plasma prealbumin. J Biol Chem 249:6796–6805

Kayed R, Head E, Thompson JL, Mcintire TM, Milton SC, Cotman CW, Glabe CG (2003) Common structure of soluble amyloid oligomers implies common mechanism of pathogenesis. Science 300:486–489

Keene CD, Rodrigues CM, Eich T, Chhabra MS, Steer CJ, Low WC (2002) Tauroursodeoxycholic acid, a bile acid, is neuroprotective in a transgenic animal model of Huntington's disease. Proc Natl Acad Sci USA 99:10671–10676

Klabunde T, Petrassi HM, Oza VB, Raman P, Kelly JW, Sacchettini JC (2000) Rational design of potent human transthyretin amyloid disease inhibitors. Nat Struct Biol 7:312–321

Kohno K, Palha JA, Miyakawa K, Saraiva MJ, Ito S, Mabuchi T, Blaner WS, Iijima H, Tsukahara S, Episkopou V, Gottesman ME, Shimada K, Takahashi K, Yamamura K, Maeda S (1997)

Analysis of amyloid deposition in a transgenic mouse model of homozygous familial amyloidotic polyneuropathy. Am J Pathol 150:1497–1508
Lindgren M, Hammarström P (2010) Amyloid oligomers: spectroscopic characterization of amyloidogenic protein states. FEBS J 277:1380–1388
Liz MA, Faro CJ, Saraiva MJ, Sousa MM (2004) Transthyretin, a new cryptic protease. J Biol Chem 279:21431–21438
Longo Alves I, Hays MT, Saraiva MJ (1997) Comparative stability and clearance of [Met30]transthyretin and [Met119]transthyretin. Eur J Biochem 249:662–668
Macedo B, Batista AR, Ferreira N, Almeida MR, Saraiva MJ (2008) Anti-apoptotic treatment reduces transthyretin deposition in a transgenic mouse model of familial amyloidotic polyneuropathy. Biochim Biophys Acta 1782:517–522
Mccutchen SL, Colon W, Kelly JW (1993) Transthyretin mutation Leu-55-Pro significantly alters tetramer stability and increases amyloidogenicity. Biochemistry 32:12119–12127
Monteiro FA, Sousa MM, Cardoso I, Do Amaral JB, Guimaraes A, Saraiva MJ (2006) Activation of ERK1/2 MAP kinases in familial amyloidotic polyneuropathy. J Neurochem 97:151–161
Munarqués M, Costa PP, Saraiva MJM, Viaderfarré C, Munarbernat C, Cifuentesluna C, Fortezaalberti JF (1997) Familial amyloidotic polyneuropathy. TTR Met30 in Majorca (Spain). Amyloid Int J Exp Clin Investig 4:181–186
Nakanishi T, Yoshioka M, Moriuchi K, Yamamoto D, Tsuji M, Takubo T (2010) *S*-sulfonation of transthyretin is an important trigger step in the formation of transthyretin-related amyloid fibril. Biochim Biophys Acta 1804:1449–1456
Nakazato M, Kangawa K, Minamino N, Tawara S, Matsuo H, Araki S (1984) Revised analysis of amino acid replacement in a prealbumin variant (SKO-III) associated with familial amyloidotic polyneuropathy of Jewish origin. Biochem Biophys Res Commun 123:921–928
Nunes AF, Saraiva MJ, Sousa MM (2006) Transthyretin knockouts are a new mouse model for increased neuropeptide Y. FASEB J 20:166–168
Olofsson A, Ippel HJ, Baranov V, Horstedt P, Wijmenga S, Lundgren E (2001) Capture of a dimeric intermediate during transthyretin amyloid formation. J Biol Chem 276:39592–39599
Olofsson A, Ippel JH, Wijmenga SS, Lundgren E, Ohman A (2004) Probing solvent accessibility of transthyretin amyloid by solution NMR spectroscopy. J Biol Chem 279:5699–5707
Oza VB, Petrassi HM, Purkey HE, Kelly JW (1999) Synthesis and evaluation of anthranilic acid-based transthyretin amyloid fibril inhibitors. Bioorg Med Chem Lett 9:1–6
Pokrzywa M, Dacklin I, Hultmark D, Lundgren E (2007) Misfolded transthyretin causes behavioral changes in a *Drosophila* model for transthyretin-associated amyloidosis. Eur J Neurosci 26:913–924
Quintas A, Saraiva MJ, Brito RM (1997) The amyloidogenic potential of transthyretin variants correlates with their tendency to aggregate in solution. FEBS Lett 418:297–300
Quintas A, Saraiva MJ, Brito RM (1999) The tetrameric protein transthyretin dissociates to a non-native monomer in solution. A novel model for amyloidogenesis. J Biol Chem 274: 32943–32949
Quintas A, Vaz DC, Cardoso I, Saraiva MJ, Brito RM (2001) Tetramer dissociation and monomer partial unfolding precedes protofibril formation in amyloidogenic transthyretin variants. J Biol Chem 276:27207–27213
Raz A, Goodman DS (1969) The interaction of thyroxine with human plasma prealbumin and with the prealbumin-retinol-binding protein complex. J Biol Chem 244:3230–3237
Redondo C, Damas AM, Olofsson A, Lundgren E, Saraiva MJ (2000a) Search for intermediate structures in transthyretin fibrillogenesis: soluble tetrameric Tyr78Phe TTR expresses a specific epitope present only in amyloid fibrils. J Mol Biol 304:461–470
Redondo C, Damas AM, Saraiva MJ (2000b) Designing transthyretin mutants affecting tetrameric structure: implications in amyloidogenicity. Biochem J 348(Pt 1):167–172
Reixach N, Deechongkit S, Jiang X, Kelly JW, Buxbaum JN (2004) Tissue damage in the amyloidoses: transthyretin monomers and nonnative oligomers are the major cytotoxic species in tissue culture. Proc Natl Acad Sci USA 101:2817–2822

Rodrigues JR, Simoes CJ, Silva CG, Brito RM (2010) Potentially amyloidogenic conformational intermediates populate the unfolding landscape of transthyretin: insights from molecular dynamics simulations. Protein Sci 19:202–219

Said G, Ropert A, Faux N (1984) Length-dependent degeneration of fibers in Portuguese amyloid polyneuropathy: a clinicopathologic study. Neurology 34:1025–1032

Sánchez De Groot N, Pallarés I, Avilés FX, Vendrell J, Ventura S (2005) Prediction of "hot spots" of aggregation in disease-linked polypeptides. BMC Struct Biol 5:18

Santos SD, Cardoso I, Magalhaes J, Saraiva MJ (2007) Impairment of the ubiquitin-proteasome system associated with extracellular transthyretin aggregates in familial amyloidotic polyneuropathy. J Pathol 213:200–209

Santos SD, Costa R, Teixeira PF, Gottesman M, Cardoso I, Saraiva MJ (2008) Amyloidogenic properties of transthyretin-like protein (TLP) from *Escherichia coli*. FEBS Lett 582:2893–2898

Santos SD, Fernandes R, Saraiva MJ (2010a) The heat shock response modulates transthyretin deposition in the peripheral and autonomic nervous systems. Neurobiol Aging 31:280–289

Santos SD, Lambertsen KL, Clausen BH, Akinc A, Alvarez R, Finsen B, Saraiva MJ (2010b) CSF transthyretin neuroprotection in a mouse model of brain ischemia. J Neurochem 115:1434–1444

Saraiva MJM, Costa PP (1991) Molecular biology of the amyloidogenesis in the transthyretin related amyloidoses. In: Natvig JB, Førre Ø, Husby G, Husebekk A (eds) Amyloid and amyloidosis 1990. Kluwer Academic Publishers, Dordrecht

Saraiva MJ, Birken S, Costa PP, Goodman DS (1984) Amyloid fibril protein in familial amyloidotic polyneuropathy, Portuguese type. Definition of molecular abnormality in transthyretin (prealbumin). J Clin Invest 74:104–119

Sasaki H, Yoshioka N, Takagi Y, Sakaki Y (1985) Structure of the chromosomal gene for human serum prealbumin. Gene 37:191–197

Sebastião MP, Saraiva MJ, Damas AM (1998) The crystal structure of amyloidogenic Leu55 Pro transthyretin variant reveals a possible pathway for transthyretin polymerization into amyloid fibrils. J Biol Chem 273:24715–24722

Sebastião MP, Merlini G, Saraiva MJ, Damas AM (2000) The molecular interaction of 4'-iodo-4'-deoxydoxorubicin with Leu-55Pro transthyretin 'amyloid-like' oligomer leading to disaggregation. Biochem J 351:273–279

Sebastião MP, Lamzin V, Saraiva MJ, Damas AM (2001) Transthyretin stability as a key factor in amyloidogenesis: X-ray analysis at atomic resolution. J Mol Biol 306:733–744

Sekijima Y, Wiseman RL, Matteson J, Hammarstrom P, Miller SR, Sawkar AR, Balch WE, Kelly JW (2005) The biological and chemical basis for tissue-selective amyloid disease. Cell 121:73–85

Serag AA, Altenbach C, Gingery M, Hubbell WL, Yeates TO (2001) Identification of a subunit interface in transthyretin amyloid fibrils: evidence for self-assembly from oligomeric building blocks. Biochemistry 40:9089–9096

Sobue G, Nakao N, Murakami K, Yasuda T, Sahashi K, Mitsuma T, Sasaki H, Sakaki Y, Takahashi A (1990) Type I familial amyloid polyneuropathy. A pathological study of the peripheral nervous system. Brain 113(Pt 4):903–919

Sørensen J, Hamelberg D, Schiøtt B, Mccammon JA (2007) Comparative MD analysis of the stability of transthyretin providing insight into the fibrillation mechanism. Biopolymers 86:73–82

Sorgjerd K, Ghafouri B, Jonsson BH, Kelly JW, Blond SY, Hammarström P (2006) Retention of misfolded mutant transthyretin by the chaperone BiP/GRP78 mitigates amyloidogenesis. J Mol Biol 356:469–482

Sorgjerd K, Klingstedt T, Lindgren M, Kagedal K, Hammarström P (2008) Prefibrillar transthyretin oligomers and cold stored native tetrameric transthyretin are cytotoxic in cell culture. Biochem Biophys Res Commun 377:1072–1078

Sousa MM, Saraiva MJ (2001) Internalization of transthyretin. Evidence of a novel yet unidentified receptor-associated protein (RAP)-sensitive receptor. J Biol Chem 276:14420–14425

Sousa MM, Norden AG, Jacobsen C, Willnow TE, Christensen EI, Thakker RV, Verroust PJ, Moestrup SK, Saraiva MJ (2000) Evidence for the role of megalin in renal uptake of transthyretin. J Biol Chem 275:38176–38181

Sousa MM, Cardoso I, Fernandes R, Guimarães A, Saraiva MJ (2001a) Deposition of transthyretin in early stages of familial amyloidotic polyneuropathy: evidence for toxicity of nonfibrillar aggregates. Am J Pathol 159:1993–2000

Sousa MM, Du Yan S, Fernandes R, Guimarães A, Stern D, Saraiva MJ (2001b) Familial amyloid polyneuropathy: receptor for advanced glycation end products-dependent triggering of neuronal inflammatory and apoptotic pathways. J Neurosci 21:7576–7586

Sousa MM, Fernandes R, Palha JA, Taboada A, Vieira P, Saraiva MJ (2002) Evidence for early cytotoxic aggregates in transgenic mice for human transthyretin Leu55Pro. Am J Pathol 161:1935–1948

Sousa JC, Grandela C, Fernandez-Ruiz J, De Miguel R, De Sousa L, Magalhaes AI, Saraiva MJ, Sousa N, Palha JA (2004) Transthyretin is involved in depression-like behaviour and exploratory activity. J Neurochem 88:1052–1058

Sousa MM, Do Amaral JB, Guimaraes A, Saraiva MJ (2005) Up-regulation of the extracellular matrix remodeling genes, biglycan, neutrophil gelatinase-associated lipocalin, and matrix metalloproteinase-9 in familial amyloid polyneuropathy. FASEB J 19:124–126

Sousa JC, Marques F, Dias-Ferreira E, Cerqueira JJ, Sousa N, Palha JA (2007) Transthyretin influences spatial reference memory. Neurobiol Learn Mem 88:381–385

Tawara S, Nakazato M, Kangawa K, Matsuo H, Araki S (1983) Identification of amyloid prealbumin variant in familial amyloidotic polyneuropathy (Japanese type). Biochem Biophys Res Commun 116:880–888

Teixeira PF, Cerca F, Santos SD, Saraiva MJ (2006) Endoplasmic reticulum stress associated with extracellular aggregates. Evidence from transthyretin deposition in familial amyloid polyneuropathy. J Biol Chem 281:21998–22003

Teng MH, Yin JY, Vidal R, Ghiso J, Kumar A, Rabenou R, Shah A, Jacobson DR, Tagoe C, Gallo G, Buxbaum J (2001) Amyloid and nonfibrillar deposits in mice transgenic for wild-type human transthyretin: a possible model for senile systemic amyloidosis. Lab Invest 81:385–396

Terry CJ, Damas AM, Oliveira P, Saraiva MJ, Alves IL, Costa PP, Matias PM, Sakaki Y, Blake CC (1993) Structure of Met30 variant of transthyretin and its amyloidogenic implications. EMBO J 12:735–741

Thylén C, Wahlqvist J, Haettner E, Sandgren O, Holmgren G, Lundgren E (1993) Modifications of transthyretin in amyloid fibrils: analysis of amyloid from homozygous and heterozygous individuals with the Met30 mutation. EMBO J 12:743–748

Wahlquist J, Thylén C, Haettner E, Snadgren O, Holmgren G, Lundgren E (1991) Structure of transthyretin molecule in amyloid fibrils from the vitreous body in individuals with the Met30 mutation. In: Natvig JB, Forre O, Husby G, Husebekk A, Skogen B, Sletten K, Westermark P (eds) Amyloid and amyloidosis 1990. Kluwer Academic Publishers, Dordrecht

Westermark P, Sletten K, Johansson B, Cornwell GG 3rd (1990) Fibril in senile systemic amyloidosis is derived from normal transthyretin. Proc Natl Acad Sci USA 87:2843–2845

Yamamoto K, Hsu SP, Yoshida K, Ikeda S, Nakazato M, Shiomi K, Cheng SY, Furihata K, Ueno I, Yanagisawa N (1994) Familial amyloid polyneuropathy in Taiwan: identification of transthyretin variant (Leu55 → Pro). Muscle Nerve 17:637–641

Yan SF, Yan SD, Ramasamy R, Schmidt AM (2009) Tempering the wrath of RAGE: an emerging therapeutic strategy against diabetic complications, neurodegeneration, and inflammation. Ann Med 41:408–422

Yi S, Takahashi K, Naito M, Tashiro F, Wakasugi S, Maeda S, Shimada K, Yamamura K, Araki S (1991) Systemic amyloidosis in transgenic mice carrying the human mutant transthyretin (Met30) gene. Pathologic similarity to human familial amyloidotic polyneuropathy, type I. Am J Pathol 138:403–412

Yip CM, Darabie AA, Mclaurin J (2002) Aβ42-peptide assembly on lipid bilayers. J Mol Biol 318:97–107

Zako T, Sakono M, Hashimoto N, Ihara M, Maeda M (2009) Bovine insulin filaments induced by reducing disulfide bonds show a different morphology, secondary structure, and cell toxicity from intact insulin amyloid fibrils. Biophys J 96:3331–3340

Chapter 14
Strategies for Inhibiting Protein Aggregation: Therapeutic Approaches to Protein-Aggregation Diseases

Jennifer D. Lanning and Stephen C. Meredith

Abstract Self-aggregation of proteins and peptides is at the root of many diseases, especially neurodegenerative diseases. These conditions include Alzheimer's disease, Huntington's disease, Parkinson's disease, type-2 diabetes mellitus, and transmissible spongiform encephalopathies, which are associated with self-aggregation of amyloid β-protein (Aβ), huntingtin, α-synuclein, islet amyloid polypeptide, and the prion protein, respectively. The list of diseases for which protein/peptide aggregation is the root cause is ever expanding. There does not appear to be a single biochemical mechanism by which proteins and peptides self-associate, or a single pathogenic mechanism to explain all protein-/peptide-aggregation diseases. Nevertheless, inhibition of protein self-aggregation remains a potential target for therapeutic intervention. Beyond therapy, inhibitors of protein self-aggregation can serve as tools to help us understand the mechanisms by which aggregation occurs and harms cells. In this chapter, we examine select examples of inhibitors of protein aggregation. We have divided aggregating proteins/peptides into two types: (1) Proteins that have an unstable tertiary structure, that unfold under cellular stress, or that fail to fold correctly during biosynthesis. This instability leads to persistence of unfolded domains that can act as a nidus for self-association. (2) Peptides or proteins (or protein domains) that cannot fold at all, or fold only in the presence of a bound ligand. Examples of the first group include transthyretin, the mammalian prion protein, and certain point-mutant forms of lysozyme or α1-antitrypsin. In general, self-aggregation of these proteins results from exposure of normally buried hydrophobic residues to aqueous media. Examples of the second group include Aβ, islet amyloid polypeptide, and calcitonin. Within the second group, we also include proteins that are "peptide-like" in having domains with no unique, stable

J.D. Lanning • S.C. Meredith (✉)
Departments of Pathology, Biochemistry, and Molecular Biology, The University of Chicago, 5841 S. Maryland Ave, Chicago, IL 60637, USA
e-mail: scmeredi@uchicago.edu

F. Rahimi and G. Bitan (eds.), *Non-fibrillar Amyloidogenic Protein Assemblies—Common Cytotoxins Underlying Degenerative Diseases*, DOI 10.1007/978-94-007-2774-8_14,

tertiary fold, such as huntingtin and α-synuclein. The category of peptides and "peptide-like" proteins can be further subdivided into those containing unburied hydrophobic residues (e.g., Aβ), and those with strings of hydrogen-bonding polar residues (e.g., huntingtin and other polyglutamine proteins). To reduce the sheer volume of material in this field, we have chosen to focus mainly on one example of each type of aggregating protein or peptide, i.e., Aβ, as a peptide aggregating through the hydrophobic effect; huntingtin, as a peptide-like protein aggregating through side-chain and peptide backbone hydrogen bonding; and transthyretin, as a protein with an unstable tertiary fold. The mechanisms of self-aggregation serve as a guide for developing aggregation inhibitors. Inhibitors run the gamut of peptides, including natural and synthetic peptides, and those containing non-natural amino acids; proteins, both natural and engineered; and small molecules, including both natural and synthetic substances. The pursuit of aggregation inhibitors remains a prime goal in the search for treatments of protein aggregation diseases. In spite of some serious setbacks in clinical trials, the outlook remains bright—but the continuing task calls for equal measures of perseverance and theraeutic, as well as intellectual, modesty.

14.1 Introduction: Why Do Proteins and Peptides Aggregate?

14.1.1 Proteins vs. Peptides

14.1.1.1 Protein Amyloids

The recognition of neuritic plaques and neurofibrillary tangles as the pathognomonic lesions of Alzheimer's disease (AD) (Alzheimer 1907) was followed by the general recognition that these lesions, like other amyloids, contained insoluble β-sheet fibrils at their core (Glenner and Bladen 1966; Glenner and Wong 1984a, b; Wong et al. 1985). Over the past three decades or more, the ever-growing list of protein-aggregation diseases (PADs) seems to reflect a belated recognition of an obvious point. Self-association is so prevalent among polypeptides that it is surprising to come upon polypeptides that always remain monomeric. Physiological forms of self-association may represent a tiny segment of the conformational space that was retained from the earliest stages of protein evolution, in which the vast majority of randomly joined polypeptide chains self-associated, and of this vast majority, many or most were in the form of β-sheets. The β-sheet may be the default or generic structure of the polypeptide chain (Dobson 2001). More than 50 years ago, studies of the earliest β-sheet model peptides, poly-$_L$-Lys, showed its α-helix-to-β-sheet transition at alkaline pH and high temperatures, which was accompanied by slow precipitation of the peptide. Without using the name, they had formed amyloid fibrils in these studies (Davidson et al. 1966; Townend et al. 1966; Davidson and Fasman 1967; Greenfield et al. 1967; Greenfield and Fasman 1969).

The evolutionary pivot-point was the use of only α-amino acids, which (another surprise…) afford far greater conformational flexibility than other types of amino acids (Cheng et al. 2001; Stanger et al. 2001; Raguse et al. 2002; Allison et al. 2010).

For the purposes of this chapter, we will use the term "polypeptide" to refer to the categories, peptides and proteins, grouped together, but will consider self-association of peptides and proteins separately, to reflect differences in the causes of self-association.

Considering the proteins first: even if the β-sheet is the default conformation, it is only the second most common secondary structure found in folded globular proteins after the α-helix, which does not form insoluble self-aggregates as frequently as does the β-sheet. However, evolution also has domesticated the β-sheet, so that it, too, appears in soluble globular proteins, often within various supersecondary structures such as the β-barrel. Where β-sheets occur, *per se,* in folded globular proteins, they are often paired, and buried within the interior of the globule, surrounded by amphiphilic α-helices, which form the protein–water interface. In other cases, β-sheets stand alone (e.g., in the immunoglobulin fold), exposed to solvent.

Nevertheless, despite the domestication of the polypeptide chain, wild-type, folded globular proteins of all structural types—α-helical [cytochrome c_{552} (Pertinhez et al. 2001) and apomyoglobin (Fändrich et al. 2001)], α/β [acylphosphatase (Chiti et al. 1999) and human procarboxypeptidase A2 (Villegas et al. 2000)], α + β [α-lacto-globulin (Redfield et al. 1999) and lysozyme (Krebs et al. 2000)], as well as β [PI3-SH3 (Guijarro et al. 1998) and FNIII domain (Litvinovich et al. 1998)]—can be induced to form amyloid fibrils. Of the preceding list of proteins (in their wild-type sequences), none has been associated with disease. [In the case of lysozyme, some point-mutations are associated with human disease (Matagne and Dobson 1998; Dumoulin et al. 2006; Merlini and Bellotti 2005), as discussed below.]

In this chapter, we are concerned specifically with those proteins associated with PADs. In other words, although all proteins can (given monstrous enough treatment) form β-sheet fibrils, only some do under physiological conditions. Put another way, there is wide variation in the propensities of folded, globular proteins to form fibrils, which reflects in some way the stability of the native fold. However, what does "stability" or "instability" mean in this context?

To make a broad generalization, which undoubtedly has exceptions, proteins have a micelle-like organization, in which hydrophobic residues are contained within the protein interior while their exposure to water is minimal or close to it. At one extreme is the example of hemoglobin S, in which a point-mutation in the β-chain, Glu → Val, causes aggregation[1] of hemoglobin molecules into fibrils. However, these fibrils are not β-sheet amyloid, and the hemoglobin molecules within fibrils appear to have a completely "normal," folded structure (Eaton and Hofrichter 1990). This type of protein aggregation is to be distinguished by those

[1]Unless otherwise specified, the term "aggregation" will be used as an abbreviation for self-association or self-aggregation.

caused by conformational modification. As cogently illustrated in the case of lysozyme, amyloid fibrils form because of an alteration in the tertiary structure that mimics partial denaturation. For example, two point-mutant forms of lysozyme associated with human amyloidosis, Ile56Thr and Asp67His, were observed to have only modest alterations of enzymatic kinetic parameters (higher K_m and lower K_{cat}) and, surprisingly, little if any difference in X-ray crystallographic structure from the wild-type protein (Booth et al. 1997). However, both proteins were less stable to denaturants or heat, and at the midpoint of thermal denaturation, the proteins bound 1-anilinonaphthalene-8-sulfonic acid, indicating solvent exposure of the hydrophobic interior of the protein.

As with these mutant forms of lysozyme, the Z mutation in α1-antitrypsin is within the hinge region between one of the β-sheets (A) and the peptide loop containing the reactive center of the protein; hence, within the mobile domain of the protein. As a result of the mutation, the β-sheet opens and allows the reactive loop to insert into the top of the open space (i.e., the space vacated by the mobile part of the β-sheet), leaving the bottom half of this space free to accept the reactive loop of another α1-antitrypsin molecule (Lomas et al. 1993; Gooptu et al. 2000; Lomas and Carrell 2002). Most of the abnormally folded proteins are eliminated within the endoplasmic reticulum of the hepatocytes (where α1-antitrypsin is synthesized), but those which survive can form long protein polymers that appear as intracellular inclusion bodies by light microscopy. Similarly, immunoglobulin light-chain amyloidosis appears to arise through incorporation of destabilizing point-mutations into the light-chain sequence (Hurle et al. 1994).

In these cases, as in several point-mutant variants of transthyretin (TTR), little difference exists in the ensemble fold of the mutant and wild-type proteins under physiological pH. However, in the case of TTR, acidic conditions destabilize the native fold because they favor dissociation of the native tetrameric form of TTR into monomers, which are amyloidogenic. Even wild-type TTR is associated with (usually, clinically mild) cardiac amyloidosis, and this may occur because of the tendency of tetrameric TTR to dissociate into monomers in the absence of bound thyroid hormone or other ligands, as well as under acidic experimental conditions. Point-mutant forms of TTR are associated with either severe cardiac amyloidosis or familial amyloid polyneuropathies. Kelly and coworkers have observed that for many point-mutant forms of TTR (e.g., V30M, L55P, T119M, and V122I), the rate of tetramer dissociation to monomer is directly related to propensity of the protein to form amyloid, and thus, to disease severity. Thus, *in these cases*, tetramer dissociation to monomer is the rate-limiting step for fibril formation, while transformation of monomer into amyloid is relatively rapid (McCutchen et al. 1995; Kelly 1998; Hammarström et al. 2002; Johnson et al. 2005; Palaninathan et al. 2008). At the same time, tetramer dissociation need not be the rate-limiting step in all cases. An engineered double mutant of TTR, F87M/L110M, forms a stable monomer that is non-amyloidogenic at neutral pH, but this monomer, like that of the wild-type protein, can be induced to form amyloid, e.g., by partial acid denaturation or addition of urea (Jiang et al. 2001). Thus, once dissociated from the tetramer, tertiary structural changes within the monomer are "unmasked" and such changes are required for amyloid formation.

In summary, proteins self-associate to form amyloid when their tertiary structures are altered, but these alterations can take several forms and may occur:

1. Through mutations that leave the native fold of the polypeptide chain intact, but with a surface hydrophobic patch that mediates self-association, e.g., hemoglobin S.
2. Through mutations that alter the native fold in a way that is observable by ensemble techniques, such as X-ray crystallography. This type of aggregation appears to be a rare cause of PADs in which amyloid forms.
3. Through mutations that alter the stability of the native fold, such that a minority structure in the ensemble is partially denatured, and prone to self-aggregation (e.g., point-mutations in lysozyme and TTR). In this category, the ensemble structure of the mutant protein, as seen in X-ray crystallography, may be quite similar to that of the wild-type protein. This category can be further subdivided. In some cases (e.g., point-mutations in lysozyme), partial denaturation of the tertiary fold is rate limiting for fibril formation. In other cases (e.g., some point-mutations of TTR), quaternary structure determines whether the molecule forms amyloid. In these cases, the rate-limiting step for amyloid formation is dissociation of a native oligomer (a tetramer, in the case of TTR) into monomers, but for the monomeric subunits, partial denaturation and aggregation occur rapidly.

14.1.1.2 Amyloids Formed from Peptides, and "Peptide-Like," Intrinsically Unstructured Proteins

The popular term "protein-misfolding diseases" may not be the best term for the diseases caused by aggregation of proteins discussed above. For many amyloidogenic point-mutant variants of non-amyloidogenic (or less amyloidogenic proteins), the mutants fold normally, and ensemble structural techniques, like X-ray crystallography, generally reveal a structure that is quite similar or identical to the native one. For this reason, we have chosen to use the term, "protein-aggregation diseases" or, one might have chosen to say "protein-instability diseases," which reflects the underlying cause of the aggregation. "Protein misfolding" may apply to proteins such as the ΔF508–CFTR and other CFTR mutations (Cheung and Deber 2008), where disease results from failure of sufficient quantities of protein to fold properly and be transferred to the cell membrane; or, a fortiori, to prion diseases, in which a normally folded protein, such as mammalian PrP, is converted, by an unknown seed, into a protein with an abnormal structure that is prone to self-association (Prusiner 2001; Aguzzi et al. 2008).

The term "protein misfolding" is not merely inappropriate, however, but also inaccurate in the case of peptides, such as amyloid β-protein (Aβ), which have no native fold—no single, stable structure—as monomers. Similarly, some proteins, either entirely or in a portion of the sequence (typically ≥30 amino acids), attain a secondary or tertiary fold only in the presence of a bound ligand, e.g., α-synuclein, which acquires α-helical structure upon adsorbing to certain lipid surfaces (Zhu and Fink 2003; Uversky 2003, 2008; Uversky and Eliezer 2009; Fink 2006). In some

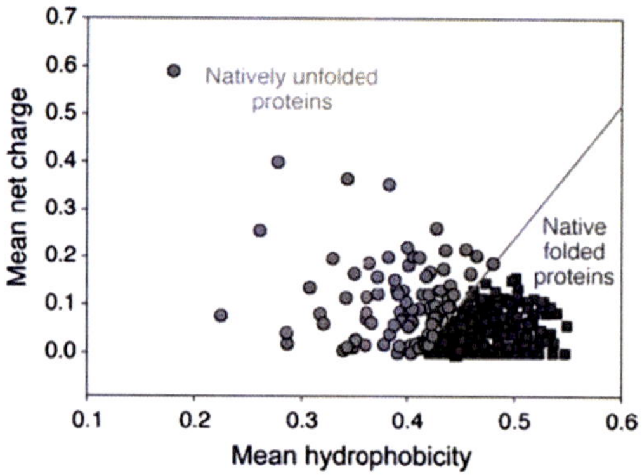

Fig. 14.1 Plot of mean net charge *versus* mean hydropathy (charge–hydropathy plot) for a set of 275 folded (*squares*) and 91 natively unfolded proteins (*circles*) [Data are from Uversky et al. (2000), and the figure is based on Dunker et al. (2008)]. Intrinsically unfolded proteins typically show a combination of low overall hydrophobicity and high net charge. Natively folded proteins typically have a micelle-like structure, requiring a core of hydrophobic amino acids; hence, they have a combination of higher hydrophobicity and lower net charge than intrinsically unfolded proteins

cases, e.g., huntingtin, only some of the ligands are known, while in other cases (e.g., other polyglutamine (polyGln)-expansion diseases), ligands remain unknown. Such proteins have been called "intrinsically unstructured," "intrinsically disordered," or "natively unfolded" (Uversky et al. 2000; Uversky and Dunker 2010; Fink 2005; Uversky 2010), and may comprise as much as 30% of the mammalian genome (Fink 2005). A unifying feature of many of these proteins is that they occupy "a unique region of charge–hydropathy phase space," low overall hydrophobicity, and high net charge compared with folded proteins (Fig. 14.1). Although this observation is surprising at first, it is completely commensurable with the micelle-like organization of folded proteins. For intrinsically unstructured proteins, the number of hydrophobic residues, which form the core of folded proteins, is too low compared with the number of charged residues, which form the surface of folded proteins. Algorithms (e.g., PONDR, among many others) now exist for predicting amino-acid sequences likely to form natively unstructured domains (Dunker et al. 2008; He et al. 2009).

In the context of PADs, we will consider such proteins to be "peptide-like" in that they lack any defined and stable tertiary structure, in whole or in part. This is the main difference between peptides or peptide-like proteins, and the proteins described in the previous section: the latter do have a well-defined tertiary structure from which a segment, usually in a small portion of the ensemble of molecules, deviates, and this leads to aggregation.

Mechanistically, it is not entirely clear why diverse, intrinsically unstructured peptides or proteins should assume a β-sheet structure over time, but it inheres to β-sheets to self-associate indefinitely, and to form insoluble fibrils. One can only speculate why intrinsically unfolded peptides and proteins came into being, aside from the dubious misery of causing neurodegenerative and other PADs. NMR spectroscopy and other biophysical techniques indicate that these proteins, though lacking extensive or temporally stable structure, often show glimmers, spatially limited and flickering formes frustes, of residual structure. Remarkably, such glimmers of structure in the midst of widespread disorder is *not* an artifact of removing the protein from its cellular environment, since such proteins or protein regions remain unstructured even within the crowded environment of the cell (Dedmon et al. 2002; McNulty et al. 2006; Selenko and Wagner 2007). However, such apparently "failed" structural elements could represent *incipient* structure, waiting to form when the protein makes contact with its ligand (Bussell and Eliezer 2001; Fuxreiter et al. 2004). Thus, intrinsically unstructured proteins, by virtue of being unstructured, provide a large surface area for interacting with ligands, while preserving *incipient* structure that may facilitate interactions with ligands. These properties would seem to apply even more to peptides like Aβ (whose presumed binding partner is unknown) and peptide hormones, such as insulin, islet amyloid polypeptide, and calcitonin. In all of these cases, the capacity for self-association and forming amyloid is the disadvantage entailed by the advantages of flexibility and facility in ligand binding.

14.1.2 Kinetics vs. Thermodynamics—The Kinetics of Fibrillization

Protein folding and denaturation of single-domain, globular proteins can usually be described by a simple two-state model:

$$U \rightleftharpoons N$$

From this equilibrium model, one expresses the equilibrium constant as

$$K = \frac{[U]}{[N]}$$

from which one obtains the free energy of the reaction, ΔG = –RT lnK. The assumption of the two-state model is that the reaction is sufficiently rapid that no intermediates can be observed. For non-aggregating proteins whose behavior follows the two-state description, the native state is at lower free energy than the unfolded state.

This description is clearly insufficient to describe the virtually irreversible formation of fibrils. For fibril-forming proteins, such as the examples of TTR or lysozyme described above, there is a relatively low kinetic barrier to a partially unfolded state:

$$U \rightleftharpoons X \rightleftharpoons N.$$

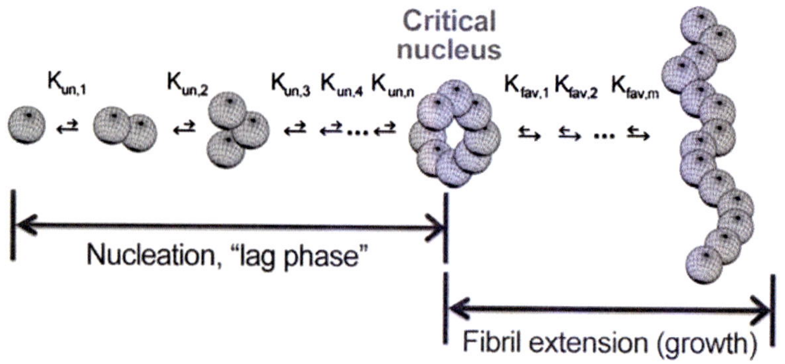

Fig. 14.2 The simplest type of nucleation–polymerization mechanism. A critical nucleus is the smallest species (in this case, the oligomer with the lowest number of subunits) for which it is energetically more favorable to add than to subtract one subunit. Monomers and oligomers are related through a series of equilibria; in the figure, $K_{un,n}$ are the equilibrium constants for thermodynamically unfavorable reactions, and $K_{fav,m}$ are the equilibrium constants for thermodynamically favorable reactions. For all steps yielding oligomers smaller than the nucleus, the equilibrium favors dissociation; for all steps yielding species larger than the nucleus, the equilibrium favors association. From this perspective, a fibril is merely a very large oligomer; in the absence of covalent modifications, such as cross-linking, it can be dissolved by dilution

In the above equation, the partially unfolded intermediate, X, is shown, for simplicity, as an intermediate in the folding pathway. This need not be so, however. An unstable form of the protein can arise through changes in physiological conditions, such as loss of a bound ligand (as in the case of TTR), through proteolysis [as in the case of serum amyloid A (SAA)], or other mechanisms.

For fibril-forming peptides, such as Aβ, the "native" form of the peptide may not exist, or, if it does, it could arise, hypothetically, by binding an unknown ligand. In any case, the unfolded form of the peptide is the only one that has been observed thus far, and can be viewed as the predominant form. Similarly, for peptide-like proteins, such as α-synuclein, a native-like form of the protein may occur when the protein is bound to neuronal vesicles, but in the lipid-free form, the unfolded protein is the predominant species.

In either case, self-association ends in fibrils, but begins as a bimolecular reaction. Most of the self-association of peptides and proteins involved in PADs can be described by a nucleation–polymerization model. The simplest example of nucleation–polymerization is shown in Fig. 14.2 (Harper and Lansbury 1997; Ferrone et al. 1985; Jarrett and Lansbury 1993), in which a solution of pure monomer nucleates homogeneously. The bimolecular reaction is followed by a similar reversible reaction, in which trimers, tetramers, and so forth, form by the addition of single molecules, culminating in the formation of a critical nucleus, which is defined as the smallest oligomer for which the addition of one molecule is energetically more favorable than the loss of one molecule.

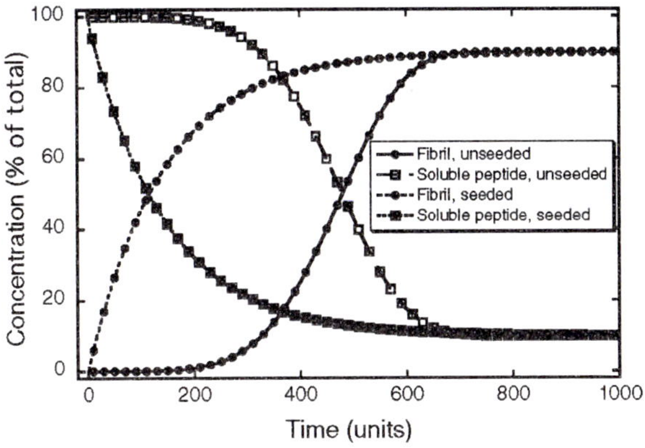

Fig. 14.3 An idealized time-course for seeded and unseeded fibril formation. The unseeded reaction can be described by a stretched exponential equation, $F = a(1 - exp(-kt^b))$, where F = concentration of peptide as fibril, k = rate constant, t = time, and a and b are parameters. In this arbitrary example, for clarity, the critical concentration (C_r) was set to 10% of the total peptide concentration, $k = 2 \times 10^{-3}$ and b = 5. As described in the chapter, typical concentrations for following fibrillization of Aβ are in the range of 1–100 μM, and the C_r is ~0.5–1 μM. The peptide concentration remaining in solution over time is $M = M_0 - F$, where M is the analytic soluble peptide concentration at any given time and M_0 is the initial monomer concentration. The terms, M and M_0 typically are measured by high-performance liquid chromatography, and expressed as analytical monomer concentration. There is no implication that all peptide in solution is monomeric. The seeded reaction eliminates the lag period, and the resulting curve can be described by a monoexponential equation, $F = a(1 - exp(-kt))$

This type of model also describes protein crystallization, and physiological types of protein aggregation, such as viral coat assembly and actin polymerization (Blow et al. 1994; Prevelige et al. 1993; Andreu and Timasheff 1986; Eaton and Hofrichter 1995; Naiki and Nakakuki 1996). The characteristic features of this type of polymerization include:

1. A lag phase, due, at least in part, to the rate-limiting process of nucleation—the formation of an ordered, multimolecular aggregate (Fig. 14.3a);
2. A critical concentration (C_r), below which the energetically unfavorable steps towards the formation of a nucleus cannot occur;
3. A growth phase, in which monomers are rapidly added to the assembled nucleus, and then to larger aggregates; and
4. A steady-state phase, in which the monomer and large aggregates are in equilibrium.

This scheme also yields the familiar kinetic pattern by which pure monomeric peptides or proteins can assemble into fibrils. (For some proteins, e.g., many point-mutant forms of TTR) no lag phase is observed, and fibrils are said to grow through "downhill polymerization" (Hurshman et al. 2004).

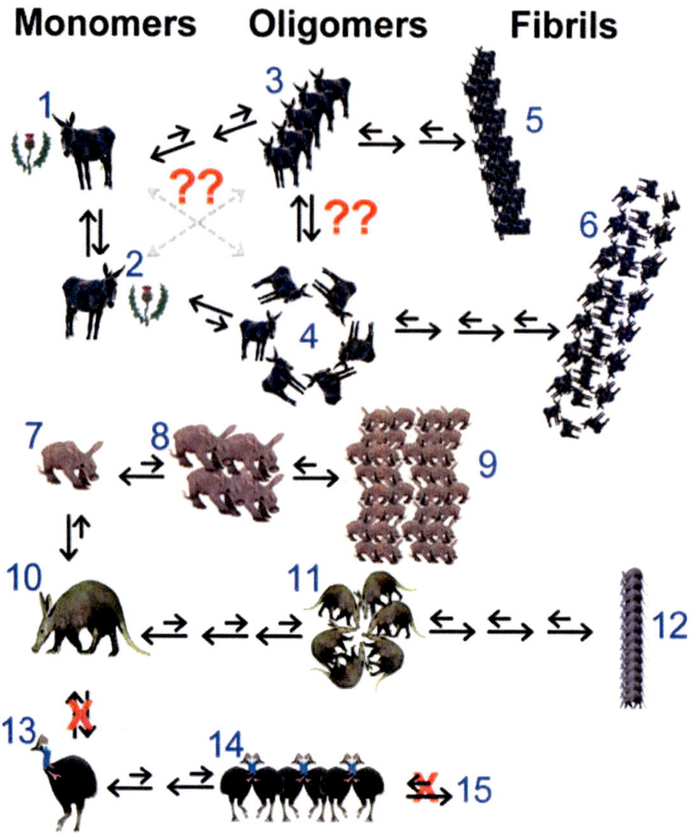

Fig. 14.4 A (highly!) schematic representation of protein self-aggregation, emphasizing oligomer and fibril polymorphism. From the viewpoint of electron microscopy, proteins can self-associate into several morphologically distinct types of insoluble aggregates, some of which have no discernible fine structural detail, and hence, are called amorphous aggregates (not represented in the figure). Some proteins conformers (*1, 2*) are easily interconvertible at the level of the monomer, like the donkey who can face east or west towards some nice thistles. Aggregation proceeds through formation of a critical nucleus, but the nuclei are polymorphic (*3, 4*); by analogy to the donkey, minor variations in aggregation conditions, e.g., forming fibrils under "quiescent" or "agitated" conditions (see Petkova et al. 2005), can lead to nuclei of different structures, and hence, to fibrils of different structures (*5, 6*). Thus, oligomers and fibrils of a single protein can have differences in fine structure, discernible by electron microscopy, antibody binding, solid-state NMR, and other techniques. This holds particularly true for intrinsically unstructured proteins and peptides, such as Aβ or the exon-1-encoded (polyGln) domain of huntingtin. It is not known whether the critical nuclei are themselves interconvertible. Fibril preparations are seldom completely uniform—there is generally some degree of polymorphism in samples. The point at which an aggregate is "committed" to a given aggregation pathway—monomer (?), dimer (?), critical nucleus (?), protofibril (?)—is not known. Fibrils can also supercoil into bundles (not represented in the figure). Folded proteins, such as TTR, can dissociate and partially unfold to form amyloid. The partially unfolded protein is like the newborn aardvark (*7*), which normally matures into an adult aardvark (*10*). Mutant TTR has a strong tendency to unfold partially—as if the mature aardvark were to regress to its immature state, indicated by the equilibrium between 7 and 10, which favors the mature aardvark/stable protein, but allows regression to the baby aardvark/partially unfolded protein. Both the newborn aardvark and the adult aardvark can form oligomers (*8* and *11*, respectively) and fibrils (*9* and *12*, respectively). Although both native and mutant TTR can form fibrils, the mutant protein has a greater tendency to do so, and does so much more rapidly than the wild-type protein. Some of the oligomeric species formed by Aβ are represented by the cassowary: monomers (*13*) can form oligomers (*14*) that are "off-pathway" for fibril formation (*15*). For any fibril-forming protein or peptide, off-pathway oligomers are possible

Experimentally, such a scheme can be demonstrated provided that pre-formed aggregates have been rigorously removed from the protein solution. Biologically, it is likely that seeded fibrillization occurs more commonly than unseeded. The addition of seeds bypasses the nucleation step, and therefore, eliminates the lag period (Fig. 14.3b). Seeds can be homogeneous—that is, identical to nuclei or larger aggregates formed by the unseeded reaction above—or heterogeneous. In the case of Aβ42, the concentration of peptide in the cerebrospinal fluid (CSF) is $\sim10^{-10}$ to 10^{-8} M (Bibl et al. 2008; Seubert et al. 1992; Galasko et al. 1998; Andreasen et al. 1999; Strozyk et al. 2003; Gustafson et al. 2007), which is considerably lower than its C_r (Wood et al. 1996; Lomakin et al. 1997; Cannon et al. 2004; O'Nuallain et al. 2004; Wetzel 2006), which is $\sim10^{-7}$ M. However, in view of these facts and much other data [e.g., the rapid formation of neuritic plaques after head trauma (Ikonomovic et al. 2004; DeKosky et al. 2010)], it seems likely that heterogeneous nucleation of Aβ fibrils could also occur, where the seeds could be lipids or other cellular debris following trauma (Matsuzaki 2007; Byström et al. 2008; Ariga et al. 2008; Matsuzaki et al. 2010; Chi et al. 2010).

The above scheme, though useful, is clearly insufficient to describe the complexity of protein aggregation leading to disease. In the past decade, the heterogeneity of both soluble oligomeric and fibrillar forms of peptides/proteins has become evident. Figure 14.4 is a highly schematic depiction of the added complexity of protein aggregation that is relevant to disease.

Several key features distinguish this scheme from the previous one, shown above, for nucleation–polymerization reactions. First, this scheme shows the multiplicity and heterogeneity of soluble oligomeric forms. Many forms of cytotoxic soluble oligomers have been described (Caughey and Lansbury 2003; Stefani 2004; Chiti and Dobson 2006; Ferreira et al. 2007; Glabe 2008; Haass and Selkoe 2007; Klein et al. 2001; Lambert et al. 1998; Lesné et al. 2006; Walsh et al. 2002; Wang et al. 2002; Bitan et al. 2003; Uversky et al. 2001; Souza et al. 2000; Conway et al. 2000a, b; Peterson et al. 2008; Sahara et al. 2007; Maeda et al. 2006), and are discussed elsewhere in this volume. It is unlikely, however, that all soluble oligomeric forms would be equally populated, or equally cytotoxic. Suffice it to say that there is not yet a universal agreement on which one or ones are most important. This lack of agreement derives in large part from the difficulty of reading data across different laboratories, where subtle differences in experimental techniques may lead to irreconcilably different experimental results. Agreement on the nature of cytotoxic oligomers is unlikely to come until structural data become available.

Second, parallel pathways are likely to exist for the formation of fibrils, and these could explain the occurrence of polymorphism of fibril structures. As first shown cogently for Aβ by Tycko and coworkers (Petkova et al. 2002, 2005, 2006; Paravastu et al. 2006, 2008), and subsequently by others for other amyloid proteins (Wetzel et al. 2007; Madine et al. 2008; Madine et al. 2009a; Andrews et al. 2009; Fändrich et al. 2009; Miller et al. 2009; Wei et al. 2010; Wegmann et al. 2010; Popova et al. 2010), even subtle variations in fibrillization conditions can yield distinct populations of fibrils, with one or another structure dominating under any given set of conditions. Advances in solid-state NMR have allowed for the development of detailed structural models of amyloid fibrils (Benzinger et al. 1998, 2000; Balbach et al. 2002; Chan et al. 2005; Ritter et al. 2005; Heise et al. 2005; Lim et al. 2008; Lührs et al. 2005; Iwata et al. 2006; Luca et al. 2007; Baxa et al. 2007; Wickner et al. 2008; Vilar et al. 2008; Bayro et al. 2010; Van Melckebeke et al. 2010). Complete NMR assignments will soon be possible in solid-state NMR, which will accelerate the pace of obtaining structural information. Detailed structural information on soluble oligomers now is starting to emerge (Chimon and Ishii 2005; Chimon et al. 2007; Walsh et al. 2010).

Since the results on fibril polymorphism were obtained with simple solutions of pure peptides, in the more complex setting of the neuron (or other cells) or CSF (or other bodily fluids), the possibilities for fibril polymorphism may be correspondingly greater. One of the important features of seeded fibril growth, however, is that progeny fibrils are structurally similar to the seeds. This could afford a much-needed glimpse into the structure of fibrils or oligomers that occur within a biological context. Paravastu et al. used authentic brain-derived Aβ amyloid fibers from patients who died with AD to seed isotopically labeled synthetic Aβ solutions, and thereby form "replicate fibrils" (Paravastu et al. 2009). Surprisingly, structural differences were noted by solid-state NMR and other techniques between these fibrils and those made from purely synthetic Aβ. Such results suggest that our understanding of even fibril structure—which is at a much more advanced state than our knowledge of oligomer structure—is only beginning to emerge.

14.2 The Diseases—What Are We Fighting Against?

14.2.1 The Inevitable List

Table 14.1 shows a current list of 43 proteins and peptides that have been associated with disease. The list is "proteocentric"—arranged by the guilty protein, rather than by the disease, extent of bodily involvement, or organ site.

Although any cell can be harmed by protein aggregation, it is obvious that the central nervous system (CNS), and neurons in particular, are especially prone to this type of damage. The anatomic selectivity of PADs is significant for neurodegenerative diseases, but has implications far beyond the confines of these diseases.

This partiality for CNS neurons remains only partly understood. Furthermore, even within the brain and among neurons, there often are no obvious reasons why *some* types of neurons are affected by protein aggregation, while others are not. In some cases, the issue is likely to be protein-expression levels (more protein leads to more protein aggregation), but in other cases (notably, the ubiquitously expressed α-synuclein and Aβ) protein expression does not explain the localization of disease entirely. α-Synuclein affects the pigmented neurons of the substantia nigra pars compacta and locus coeruleus, and to a lesser degree, the cholinergic neurons of the

Table 14.1 Self-associating proteins and protein-aggregation diseases

	Protein	Abbrev.*	Disease	Aggregation due to:	References
1	α1A voltage-dependent calcium channel subunit		Spinocerebellar ataxia type 6 (SCA6)	Polyglutamine expansion	Mantuano et al. (2003) and Kordasiewicz and Gomez (2007)
2	Androgen receptor protein		Spinal and bulbar muscular atrophy (Kennedy's disease)	Polyglutamine expansion	Palazzolo et al. (2008), Finsterer (2009), and Katsuno et al. (2010)
3	Ataxin 1		Spinocerebellar ataxia type 1 (SCA1)	Polyglutamine expansion	Limprasert et al. (1997) and Zoghbi and Orr (2009)
4	Ataxin 2		Spinocerebellar ataxia type 2 (SCA2)	Polyglutamine expansion	Figueroa and Pulst (2003) and Lastres-Becker et al. (2008)
5	Ataxin 3		Spinocerebellar ataxia type 3 (SCA3) (Machado Joseph Disease)	Polyglutamine expansion	Shehi et al. (2003), Rüb et al. (2008), and Riess et al. (2008)
6	Ataxin 7		Spinocerebellar ataxia type 7 (SCA7)	Polyglutamine expansion	Lebre and Brice (2003) and Garden and La Spada (2008)
7	β2-Microglobulin	(Aβ2M)	Peri-articular, hemodialysis-related amyloidosis	Wild-type protein	Ohashi (2001), Hasegawa et al. (2003), Ivanova et al. (2003), Jones et al. (2003), Naiki et al. (2005), Heegaard (2009), and Naiki and Nagai (2009)
8	Calcitonin	(ACal)	Medullary carcinoma of the thyroid	Wild-type protein	Reches et al. (2002) and Khurana et al. (2004)
9	α-S2C casein	(ACas)	Mammary gland corpora amylacea (from *Bos taurus*)	Wild-type protein	Niewold et al. (1999)
10	β-Amyloid precursor protein	(Aβ)	Alzheimer's disease	Wild-type protein	Glenner and Wong (1984a, b), Tycko (2006), and Murphy and LeVine (2010)

(continued)

Table 14.1 (continued)

	Protein	Abbrev.*	Disease	Aggregation due to:	References
11	ABri		Familial British dementia	Stop-codon mutation, 11-amino-acid extension	Vidal et al. (1999), Holton et al. (2001), and Gibson et al. (2005)
12	ADan		Familial Danish dementia	Stop codon mutation, 10-amio-acid extension	Holton et al. (2002) and Coomaraswamy et al. (2010)
13	APin		Calcifying epithelial odontogenic tumours (Pindborg tumours)	Residues 1–46 of FLJ3	Solomon et al. (2003)
14	Apolipoprotein AI	(AapoAI)	Familial amyloid polyneuropathy III, renal and hepatic amyloidosis	Various mutations	Nichols et al. (1988), Genschel et al. (1998), Andreola et al. (2003), Joy et al. (2003), and Hawkins (2003)
15	Apolipoprotein AII	(AapoAII)	Hereditary renal amyloidosis	Stop78G or Stop78S mutations	Hawkins (2003), Benson et al. (2001), and Yazaki et al. (2003)
16	Apolipoprotein AIV	(AapoAIV)	Systemic amyloidosis	Wild-type protein	Bergström et al. (2001, 2004)
17	Atrial natriuretic factor	(AANF)	Atrial amyloidosis	Wild-type protein	Johansson et al. (1987)
18	Atrophin-1		Dentatorubral-pallidoluysian atrophy (DRPLA)	Polyglutamine expansion	Yamada et al. (2006) and Wardle et al. (2009)
19	Cystatin C	(ACys)	Hereditary renal amyloidosis, hereditary cerebral amyloid angiopathy	L68Q mutation	Gudmundsson et al. (1972) and Ghiso et al. (1986)
20	Fibrinogen α-chain	(AFib)	Hereditary renal amyloidosis	Frame-shift mutation in fibrinogen Aα-chain	Benson et al. (1993), Uemichi et al. (1994, 1996), and Hamidi et al. (1997)
21	Gelsolin	(AGel)	Hereditary renal amyloidosis	Wild-type protein; D187N or D187Y mutant	Kiuru (1998), Fadika and Baumann (2002), Maury et al. (2003), and Liepina et al. (2004)

22	Huntingtin		Huntington's disease	Polyglutamine expansion	Hoogeveen et al. (1993), Gusella and MacDonald (1995), Trottier et al. (1995a, b), DiFiglia et al. (1997), Ross (2002), Gusella and MacDonald (2003), Young (2003), Walker (2007), Cattaneo et al. (2005), Thakur et al. (2009), and Roze et al. (2010)
23	Insulin	(AIns)	Injection site amyloidosis	Wild-type protein	Storkel et al. (1983), Swift (2002), and Shikama et al. (2010)
24	Islet amyloid polypeptide (Amylin)	(AIAPP)	Type-2 diabetes	Wild-type protein	Johnson et al. (1989), Kapurniotu (2001), Padrick and Miranker (2001), and Haataja et al. (2008)
25	Kerato-epithelin	(AKer)	Lattice corneal amyloid dystrophy	Various point mutations	Siddiqui and Afshari (2002) and Kannabiran and Klintworth (2006)
26	Lactoferrin	(ALac)	Amyloidosis in seminal vesicles, cornea and brain	Amyloidogenic fragment, or point mutations (Glu561Asp, compound Ala11Thr and Glu561Asp)	Klintworth et al. (1997), Ando et al. (2005), Nilsson MR, Dobson (2001), and Linke et al. (2005)
27	Lung surfactant protein C		Pulmonary alveolar proteinosis	Fragments of wild-type protein	Floros and Kala (1998), Trapnell et al. (2003), and Whitsett et al. (2010)
28	Lysozyme	(ALys)	Hereditary renal amyloidosis, hereditary non-neuropathic systemic amyloidosis	Various point mutations	Merlini and Belotti (2005), Röcken et al. (2006), and Trexler and Nilsson (2007)
29	Medin	(AMed)	Aortic medial amyloidosis	Resdiues 147–154 (NFGSVQFV)	Larsson et al. (2007), Peng et al. (2007), and Madine et al. (2009a, b)

(continued)

Table 14.1 (continued)

	Protein	Abbrev.*	Disease	Aggregation due to:	References
30	Monoclonal immunoglobulin heavy chain	(AH)	Immunoglobulin heavy chain amyloidosis	Fragments; sequela of monoclonal gammopathies	Eulitz et al. (1990) and Miyazaki et al. (2008)
31	Monoclonal immunoglobulin light chain or fragments	(AL)	Immunoglobulin light-chain amyloidosis, myeloma-associated amyloidosis, monoclonal plasma-cell dyscrasias	Multiple sequences, as sequela of monoglonal gammopathies	Solomon et al. (1982), Bellotti et al. (2000), Dealwis and Wall (2004), Comenzo (2006), and Sonnen et al. (2010)
32	Prion protein	(PrP^{Sc}, APrP)	Spongiform encephalopathies	Wild-type protein and point mutants	Johnson and Gibbs (1998), Hainfellner and Budka (1999), Cohen and Prusiner (1998), Prusiner (1998), Venneti (2010), and Mastrianni (2010)
33	Prolactin	(APro)	Amyloidosis of pituitary	Fragments of wild-type protein	Westermark et al. (1997)
34	Polyadenine-binding protein 2		Oculopharyngeal muscular dystrophy	Polyalanine expansion	Brais et al. (1999) and Brais (2003, 2009)
35	Semenogelin I	(ASgI)	Senile seminal vesicle amyloidosis	Fragment of wild-type protein	Linke et al. (2005)
36	Serum amyloid A protein or fragments	(AA)	Systemic secondary amyloidosis	Fragments of wild-type protein	Sipe (2000), Röcken and Shakespeare (2002), and Hatters Howlett (2002)
37	Superoxide dismutase 1		Amyotrophic lateral sclerosis	Various point mutations	Rosen et al. (1993), Hart. (2006), Rothstein (2009), and Perry et al. (2010)

38	α-Synuclein (Lewy bodies)		Parkinson's Disease, Lewy-body disease	Wild-type protein and multiple mutations	Mukaetova-Ladinska et al. (2000), Iwatsubo (2003), Jellinger (2009), Uversky and Eliezer (2009), Lees et al. (2009), Hanson and Lippa (2009), Uversky (2009), and Auluck et al. (2010)
39	TATA-box-binding protein		Spinocerebellar ataxia type 17 (SCA-17)	Polyglutamine expansion	Tsuji (2004), Friedman et al. (2008), Gao et al. (2008), and Nolte et al. (2010)
40	Tau		Frontotemporal dementias, Alzheimer's disease (neurons)	Consequence of "hyper-phosphorylation"	Goux et al. (2004), Spires-Jones et al. (2009), Meraz-Ríos et al. (2010), and Hanger and Wray (2010)
41	TDP-43 (Transactivator (Tat) RNA regulatory element (TAR) DNA-binding protein-43)		Frontotemporal dementia, amyotrophic lateral sclerosis	Hyper-phosphorylated, ubiquitinated and cleaved form of TARDBP	Neumann et al. (2006), Mackenzie and Rademakers (2008), Buratti and Baralle (2009), Geser et al. (2009), Hu and Grossman (2009), Chen-Plotkin et al. (2010), and Barmada and Finkbeiner (2010)
42	Transthyretin, mutant	(ATTR)	Familial amyloid polyneuropathy	Over 100 point mutations	Saraiva (2002), Ando et al. (2005), Hou et al. (2007), and Pepys (2009)
43	Transthyretin (wild type)	(ATTR)	Senile systemic amyloidosis, cardiac involvement	Wild-type protein	Gustavsson et al. (1991), Damas and Saraiva (2000), and Sekijima et al. (2008)

* From the Nomenclature Committee of the International Society of Amyloidosis (Westermark et al. 2005)

basal nucleus of Meynert and a few brainstem nuclei. Most other neuronal groups are spared until the latest stages of the disease or are never affected. Similarly, AD affects mainly the frontal lobes and hippocampi, and to a lesser extent, the parietal lobes. The occipital lobe has high levels of Aβ expression and develops neuritic plaques and neurofibrillary tangles, but patients with AD do not develop visual symptoms, nor are subcortical neurons affected. Huntington's disease (HD) is marked by atrophy of the caudate nucleus and the putamen (the globus pallidus atrophies secondarily). The many spinocerebellar ataxias lead to neuronal loss mainly in the cerebellum, brain stem, and spinal cord, whereas neuronal loss in amyotrophic lateral sclerosis is restricted to anterior-horn neurons of the spinal cord, and similar cells in the hypoglossal, ambiguus, and motor trigeminal cranial nerve nuclei.

While expression levels cannot account for all of the specificity of involvement, they can account for some of it, as in the cases of α-synuclein and Aβ, which are expressed more highly in brain than other tissues (see, for example, Uéda et al. 1993). However, other factors are clearly in play. Part of the brain's special susceptibility can be attributed to its high oxidative rate. The inherent leakiness of the electron-transport chain (Giasson et al. 2000; Uversky et al. 2002) places the brain under continuous oxidative stress. In addition, some neurotransmitters, e.g., the catechols, are free-radical generators and can cross-link aggregated proteins (Conway et al. 2001; Di Monte 2003). These factors seem especially important in Parkinson's disease, where exposure to toxic environmental factors such as insecticides plays an important pathogenic role.

The most striking examples of baffling localization are the prion diseases. Although it is difficult or impossible to rule out all covalent changes in the prion protein as a factor in prion diseases, the current consensus is that all human prion diseases are caused by aggregation and conformational "conversion" of the physiological form of the prion protein, PrP^{C} (for "cellular") into a pathological, aggregating form, PrP^{Sc} (for "scrapie") (Collinge 2005; Prusiner 1998; Prusiner et al. 1998; Mastrianni et al. 1999). The various types of human prion diseases arise as "strains" of the infectious prion particle, by analogy to viral strains. "Strains" have also been shown to occur in yeast prions (Bradley et al. 2002; Shorter and Lindquist 2005; Weissmann 2005; Krishnan and Lindquist 2005; Tanaka et al. 2005a; Bagriantsev and Liebman 2004; Masison et al. 1997; McGlinchey et al. 2011). In human prion diseases, each strain affects neurons in a characteristic anatomical location. For example, in fatal insomnia, whether familial or sporadic, neuronal loss is most marked in the thalamus, and the anatomic pattern survives transmission of the disease across species barriers, into mice. However, this anatomical pattern differs from that seen with another prion disease, Creutzfeldt–Jakob disease (Mastrianni et al. 1999). Again, whether familial or sporadic, spongiform change is most prominent in the cerebral and cerebellar cortices. Thus, there is an excellent clinical–anatomic correlation in both of these human prion diseases. The cerebral and cerebellar spongiform changes in Creutzfeldt–Jakob disease correlate well with the clinical findings of psychiatric symptoms, intellectual deterioration (leading to dementia), and ataxia.

The strain phenomenon of prions suggests that there may be a similar phenomenon in the amyloid diseases. As discussed above, amyloids, like prions, are structurally polymorphous, and behind or beyond structural polymorphism, could also lie pathogenic polymorphism. Historically, over the past several decades, efforts in the structural biology of amyloid fibrils have led to a unifying picture of the fibril as a β-sheet structure, most commonly parallel and in-register. In the past decade, however, attention has turned to the fine distinctions among amyloid fibrils. These differences clearly arise at the level of nucleation or early in the aggregation pathway—fibrils "breed true": progeny of different fibril types resemble their seeds (Petkova et al. 2005; Paravastu et al. 2009). Structurally, there will be similar distinctions among toxic soluble oligomers. Although the soluble oligomer may well be the main cytotoxin in these diseases, structural differences among fibrils open a window onto differences among the less temporally stable and experimentally accessible oligomers. Returning to the point raised above: why, when Aβ is expressed in frontal and occipital lobes, and plaques develop in both, do patients with AD have frontal but not occipital-lobe-related symptoms? It is important, here, to recall that a "plaque" is not a fibril; it contains fibrils, among other things, but the term is not synonymous with the term, fibrils. However, whether one considers fibrils or their precursors, are the aggregates arising in the frontal lobe structurally identical to, or different from, those that arise in the occipital lobe?

The principal distinction between an "amyloid" and a "prion" is that only the latter are infectious. The causes of prion infectiousness are beyond the scope of this article. Prions, however, are structurally related to the amyloids. Indeed, they *are* amyloids with an additional trait, infectivity, which has not been observed thus far in the other amyloids, such as Aβ. As discussed below, administration of Aβ to mice—a procedure that could have led, in the case of prions, to transmission of an infectious disease—led to immunity against Aβ, and not to infection. The immunology of amyloids and prions is at a relatively early stage of development. Nevertheless, while it seems likely that prions will remain as a special subgroup of the amyloids, strain phenomena observed for prions will also continue to be observed on the structural level in the non-infectious amyloids.

14.2.2 How Does Protein Aggregation Harm Cells?

14.2.2.1 General Principles

We will not attempt a complete answer to this large question, as it would constitute an entire book in itself, and would still remain incomplete. However, a few points need to be considered in order to understand some of the specific strategies for interfering with PADs.

There is little reason to doubt the basic premise that protein aggregates are cytotoxic, *per se*. Aggregates of proteins not associated with any disease—for example, the SH3 domain from bovine phosphatidyl-inositol-3′-kinase, and the

N-terminal domain of HypF from *Escherichia coli*—can induce cell death, showing that to damage cells, protein aggregates need not be composed of a protein that serves an important normal cellular function. In both cases, furthermore, cytotoxicity was most severe during the early stage of aggregation (Bucciantini et al. 2002; Stefani and Dobson 2003). Since the time of Alzheimer's description of neuritic plaques and neurofibrillary tangles in the disease that now bears his name, the focus began with protein aggregates visible at the level of light microscopy, and proceeded on to aggregates visible at greater levels of resolution. For the past decade and a half, there has been much work showing that soluble oligomers are more potent cytotoxins than the more visible fibrils. From the viewpoint of the physical chemist, growth and, more generally, the chemical potential and reactivity of protein aggregates should be directly related to the reactive surface area of the aggregates, which is greater in small oligomers than in fibrils.

Disrupting protein aggregation is a potential goal in combating PADs, but the question of "fibrils vs. oligomers" has sometimes generated more heat than light. This supposed opposition is a false dichotomy, however. *A priori*, the greater chemical potential of oligomers suggests that inhibitors of fibril formation will act at the level of oligomers. Of course, this need not be correct for *all* aggregation inhibitors, and one needs to be alert to the possibility that targeting fibrillization could make aggregation stall at oligomers and hence increase toxicity. In view of the lack of detailed structural information on oligomers, however, and their likely structural polymorphism, the approach to developing aggregation inhibitors must necessarily be *heuristic*, i.e., one must consider aggregation *pathways*, in which fibril structure (known) is homologous to that of oligomer and protofibril structures (unknown), whatever the main cellular villain in PADs ultimately proves to be. Thus, it is still mete and proper to study fibrillization inhibitors, even though soluble oligomers are more cytotoxic. Furthermore, there need not be a single villain in these diseases. Again, from the viewpoint of the physical chemist, the fibril represents a later stage of the reaction in which oligomers are intermediates: despite many branching and parallel pathways of the fibrillization reaction, and its great complexity, the difference between oligomers and fibrils is one of degree—of chemical potential—and not one of kind. Although structural differences between fibrils and oligomers surely exist, a heuristic approach is best because it allows us to design experimental approaches towards interfering with PADs.

14.2.2.2 How Do Protein Aggregates Harm Cells?

One can parse this question into two parts: damage done to cells by the protein aggregates *per se*, and damage done to cells and tissues *in reaction to* the initial damage done by protein aggregates. Aggregation of Aβ, for example, leads to neuroinflammation, with glial cell activation, and this specialized form of inflammation, like inflammation in general, is a two-edged sword that can damage as well as protect neurons (Mrak 2009; Schlachetzki and Hüll 2009; Eikelenboom et al. 2010; McGeer and McGeer 2010). Although neuroinflammation occurs early in AD—that is, at early stages of cognitive impairment—it is clearly a *reactive* event, and the reaction is to an injury presumably caused by Aβ aggregates. Although neuroinflammation is clearly an important potential therapeutic target (Vardy et al. 2006), our focus in this article will be on inhibiting the damage done by protein aggregates *per se*.

Cellular damage by protein and peptide aggregates can be divided into three overlapping, and non-mutually exclusive groups of mechanisms:

1. *Disruption of Cell Membranes and Cell-Membrane-Associated Structures*: With the exception of proteins and peptides with polyGln or polyGln/polyAsn domains, nearly all of the peptides and peptide-like proteins causing PADs are amphiphilic. The proteins causing PADs are globular and have the typical micelle-like organization of globular proteins, but instability of their folds leads to solvent exposure of hydrophobic groups that can interact with other such regions through the hydrophobic effect. As was first observed for the apolipoproteins (Kaiser and Kézdy 1984, 1987), and was later recognized as a general feature of intrinsically unstructured proteins (see above), amphiphilic proteins and peptides contain a higher percentage of charged (and some uncharged polar) residues than globular proteins. While the percentage of hydrophobic residues in amphiphilic proteins is approximately the same as those of globular proteins, in amphiphilic proteins the hydrophobic residues are segregated to yield a hydrophobic face or domain. In the case of Aβ, the distribution of hydrophobic residues is strikingly non-random: there are two hydrophobic stretches—residues 17–21 and 30–42, and not surprisingly, peptides from these domains can form fibrils on their own. The long *C*-terminal hydrophobic domain is derived from the (probable) transmembrane domain of the β-amyloid precursor protein. Aβ is amphiphilic, and the collapse pressure of various Aβ peptide monolayers at the air–water interface depends on the length of the *C*-terminal hydrophobic domain (Soreghan et al. 1994). The parallel, in-register β-sheet structure of Aβ peptides also derives from the amphiphilicity of this peptide. A short, non-amphiphilic segment of Aβ, Acetyl–Aβ16–22, forms antiparallel β-sheets, while a long acylation (octanoyl) increases the peptide's amphiphilicity and reverses the orientation of the β-sheet to parallel and in-register (Gordon et al. 2004). Thus Aβ, which is derived in part from a transmembrane domain of the β-APP, retains some ability to interact with lipid interfaces. In this respect, α-synuclein is even more striking. It is apolipoprotein-like, acquires α-helical structure upon binding to lipid surfaces, and has limited sequence homology to apolipoprotein A-I (Davidson et al. 1998; Perrin et al. 2000; Hatters and Howlett 2002; Bussell and Eliezer 2003; Jao et al. 2004; Bisaglia et al. 2006). Thus, it is not surprising that either of these proteins, or others like IAPP (Knight and Miranker 2004; Williamson et al. 2009), should be able to disrupt cell membranes. Much attention has come to pore-like structures (also called annular oligomers) of aggregating peptides and proteins, which have been observed not only for Aβ (Kayed et al. 2007; Wu et al. 2010; Shafrir et al. 2010), including pathogenic point-mutant forms such as E22G Aβ (Lashuel et al. 2003) and α-synuclein (Lashuel et al. 2002; Pountney et al. 2005), but also for

the prion protein (Sokolowski et al. 2003; Vendrely et al. 2005), calcitonin (Diociaiuti et al. 2006), and polyGln proteins such as huntingtin (Hirakura et al. 2000). Some of these, especially those made from Aβ, have been reported to have ion-channel properties (Hirakura et al. 2000; Arispe et al. 1993, 1996; Lin et al. 2000a, 2001; Kagan et al. 2002; Plein 2002). However, it is not clear what role, if any, such structures play in disease, nor is the structure of such peptides within membranes completely understood. Nevertheless, such proteins do interact with membranes, and there is no obvious reason for such interactions to be limited to the plasma membrane. Even without forming discrete, pore-like annular oligomers, aggregating proteins could bind to, and disrupt, normal membrane functions that are critical to cellular function and integrity.

Another aspect of membrane disruption is illustrated by the protein, tau. Hyperphosphorylation of this microtubule-associated protein is implicated in the pathogenesis of AD through its ability to disrupt microtubules. Tau is abundant in axons, and is also found in dendrites. It has six isoforms (arising through alternative splicing of a single gene), which contain three or four tubulin-binding motifs in their *C*-terminal domains, as well as other differences among isoforms. Although the isoforms have different physiological and developmental roles, the main function of these proteins is to stabilize microtubules (reviewed in Goedert and Spillantini 2006; Ballatore et al. 2007; Ittner and Götz 2011). In addition, tau proteins may engage other binding partners, including RNA (Kampers et al. 1999), presenilin 1 (Takashima et al. 1998), and src-family tyrosine kinases such as FYN5 (reviewed in Lee 2005), as well as the plasma membrane (Brandt et al. 1995; Maas et al. 2000). The physiological functions of these binding interactions are not completely understood.

Eighty-four phosphorylation sites have been identified on tau [45 serine, 35 threonine, and 4 tyrosine residues (Maas et al. 2000)]. In addition, tau is glycosylated, and other post-translational modifications have been observed, including ubiquitination (Cripps et al. 2006), sumoylation (Dorval and Fraser 2006), nitration (Nonnis et al. 2008; Reyes et al. 2011), and proteolysis (Johnson 2006; Wang et al. 2007); the physiological significance of these other modifications is not known. The physiological phosphorylation of tau is believed to modulate microtubule function: phosphorylation regulates (diminishes) the association of tau with microtubules, and through frequent cycles of phosphorylation and dephosphorylation, plays an important role in maintaining the normal asymmetry of the neuron, and with this, the processes of axonal transport. In AD, frontotemporal dementia (FTD) and other "tauopathies", changes in post-translational modification of tau occur, especially increasing phosphorylation—usually called "hyperphosphorylation" (reviewed in Mazanetz and Fischer 2007; Ballatore et al. 2007). These changes could lead to a toxic loss of microtubule function. In addition, however, as tau dissociates from microtubules, cytosolic concentrations of free tau rise, leading to aggregation and fibrillization of tau. The pathway for tau aggregation is complex. In early stages, there are non-fibrillar deposits called "pretangles," which may not contain β-sheets, as they do not bind "β-sheet-specific" dyes such as thioflavin T (Galván et al. 2001;

Maeda et al. 2006, 2007). Later, β-sheet aggregates arise, which appear on electron microscopy as "paired helical filaments;" these then form neurofibrillary tangles that are one of the hallmarks of AD. In theory, any of these aggregates could cause a so-called toxic gain-of-function.

2. *Sequestration or Inactivation of Sparse Cellular Proteins*: The main normal function of proteins is to bind ligands, *especially* other proteins. This fact is easy to forget when considering PADs. For many of the peptides and proteins involved in PADs, the functions are either unknown or poorly understood. This is the case for Aβ, the prion protein, and IAPP, for which physiological functions have been proposed and supported but remain unproven. Several physiological functions have been ascribed to huntingtin, and of these, many consist of binding other proteins, including the TATA-binding and CREB-binding proteins, which contain polyGln domains, and bind to the polyGln domain of huntingtin. In a cell-culture system, CREB-binding protein is diverted away from its normal subcellular location (nucleus to cytoplasm) by a huntingtin protein, containing an expanded polyGln domain (Chrivia et al. 1993; Trottier et al. 1995a; Nucifora et al. 2001). Furthermore, disruption of CREB-binding protein function is sufficient to cause a neurodegenerative disease: the absence of CREB-binding protein in the CNS of developing mice leads to apoptosis of post-mitotic neurons, and disruption of this gene in adult mice leads to progressive neurodegeneration in the hippocampus and in the dorsolateral striatum—a phenotype reminiscent of that occurring in HD (Mantamadiotis et al. 2002). These experiments and others (Lin et al. 2000b; Shimohata et al. 2000) suggest that in this and other polyGln-expansion diseases, sequestration of sparse and critical proteins could play a central role in pathogenesis. Huntingtin is a very large (348 kDa) protein with many other potential binding partners, as is also true for other polyGln-containing proteins that are implicated in another set of polyGln-expansion diseases, the spinocerebellar ataxias. Among the partner proteins of huntingtin is TAFII130, a regulator of CREB-mediated transcription. Although TAFII130 does not have a polyGln domain itself, polyGln expansion of SCA3 interferes with CREB-dependent transcription (Shimohata et al. 2000). This interaction on the genetic level might or might not be due to direct interaction of these two proteins. Maneuvers that increase transcriptional activity of CREB partially alleviate the defect caused by polyGln expansion (Shimohata et al. 2005). Another polyGln protein, the androgen receptor, causes a neurodegenerative disease, spinobulbar muscular atrophy, when the polyGln region is expanded. Notably, this occurs only in males, showing that polyGln expansion interferes with the normal ligand binding necessary for translocation into the nucleus where the protein normally acts as a transcription factor (Takeyama et al. 2002; Adachi et al. 2001). The ligand dependency of neurodegeneration, in this disease, is shown by the fact that androgen reduction alleviates the disease in male animals, and androgen administration induces it in female animals (Katsuno et al. 2002). As discussed below, in the case of TTR amyloidosis, ligand binding can stabilize the normal tertiary fold of mutant forms of TTR, and alleviate some of the protein deposi-

tion. Although on the surface, this effect is the "opposite" of that observed with androgens and SCA3, in fact both demonstrate the critical role played by small-molecule ligands in modulating protein structure.

14.2.2.3 "Gumming Up the Works"

It is worth re-emphasizing that the normal function of proteins is to bind ligands, especially other proteins. The difference between normally folded proteins that do not cause disease and the ones that do is partly one of degree: intrinsically unstructured peptides and proteins, or unstable globular proteins are the ones that aggregate excessively relative to the cells' capacity to deal with protein aggregates. The maintenance of *proteostasis* depends upon the combined rates of protein production and clearance in the cell (Fig. 14.5). In general, protein production is regulated by physiological needs. The clearance of proteins is regulated by a network of protein-degradation systems, the ubiquitin–proteasome and autophagy–lysosome systems. To this, one adds the "quality-control systems," which assist in the initial folding of proteins, and, in stressful situations, respond to unfolded proteins (Balch et al. 2008; Wilson et al. 2008; Muchowski and Wacker 2005). Historically, the first "stress" to be recognized as causing this type of response was heat stress, but it is now clear that a wide variety of stressors, including UV and X-radiation, oxidative and osmotic stresses, act through a common pathway that includes the chaperonins and other proteins that assist in restoring the native fold to partially unfolded proteins.

The proteostasis systems, which collectively maintain both protein quality and the balance between protein production and clearance, are the same systems that act upon, but fail to control, accumulation of the abnormal proteins of PADs. Indeed, it is debatable whether some of these proteins ought to be called abnormal, even if the effects of their accumulation are. Since the functions of many of these proteins are unknown, e.g., Aβ, it is moot whether its accumulation in AD results from a physiological response to another neuronal injury, or itself represents the pathological event initiating AD. In either case, the accumulation of aggregated proteins is pathogenic, and the cell attempts to contain the damage through the same mechanisms that maintain proteostasis throughout life. For example, many of the same genes involved in the response to polyGln protein or superoxide-dismutase-1 aggregation are also important in the cells' response to osmotic stress, an ostensibly unrelated stressor (Lamitina et al. 2006). Unbiased genetic screens for upregulation of genes in response to a protein-aggregation stress identify a large number of genes involved in protein synthesis, folding, and degradation, and RNA processing. A similar set of proteins was identified in *Drosophila* deficient in HSP70. These "networks" of responder proteins suggest a widespread link between protein aggregation and diverse stresses—a network that collectively resembles the core of proteins regulating proteostasis in both health and disease (Nollen et al. 2004; Bilen and Bonini 2007; van Ham et al. 2008; Wang et al. 2009).

Anfinsen's classical experiments on the denaturation and refolding of small proteins such as RNase, showed that the sequence of the polypeptide chain had sufficient

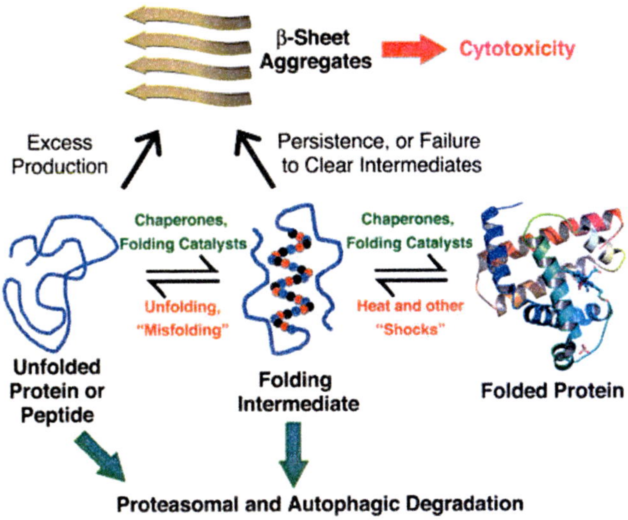

Fig. 14.5 The elements of proteostasis. Nascent proteins are unfolded; information about how to fold is intrinsic to their amino-acid sequence. Various protein-folding catalysts (chaperones, disulfide isomerases, *cis–trans* peptidyl prolyl isomerases, and others) can increase the rate of folding. Some proteins and peptides either do not fold at all, or remain unfolded until they bind to a ligand. For some peptides, such as Aβ, no folded structure has been observed; α-synuclein is an example of an intrinsically unfolded protein that adopts an α-structure upon binding to a ligand, a lipid surface. Increased production of unfolded proteins, the failure to fold a protein, the instability of the fold of a protein because of mutation, or a "shock" such as heat or oxidative stress can elicit an *unfolded-protein response*, which can include upregulation of chaperone protein expression. The ability of the cell to clear completely or partially unfolded proteins also declines with age. If the unfolded-protein response fails to correct the excess of completely or partially unfolded proteins, these can form β-sheet aggregates (oligomers and fibrils), some of which are cytotoxic

information to direct protein folding, but not at a rate sufficient to sustain life (Anfinsen 1973). Thus, there are catalysts of protein folding, including the chaperone proteins, which, in the endoplasmic reticulum, prevent "liaisons dangereuses" between nascent protein chains. As discussed, the chaperones do far more than accelerate nascent polypeptide-chain folding. There is much overlap between the chaperones and other systems, which maintain protein quality in the cell. For example, HSP70 proteins can interact with Bag-1 and CHIP, which facilitate the proteasomal degradation of damaged proteins (Imai et al. 2003). The 19S cap of the proteasome is itself a chaperone that can promote protein renaturation (Braun et al. 1999). For this reason, it is not surprising that chaperones are often found associated with protein aggregates, e.g., HSP70 chaperones in association with huntingtin and other polyGln proteins (Cummings et al. 1998; Muchowski and Wacker 2005; Jana et al. 2000) or with Lewy bodies, HSP70 in association with α-synuclein (McNaught et al. 2002), and HSP16 proteins with intracellular Aβ aggregates (Fonte et al. 2002, 2007).

Most of the neurodegenerative diseases and other PADs due to aggregation of wild-type proteins are diseases of the elderly. Early-onset, familial AD due to point-mutations within the β-amyloid precursor protein or the presenilins represent an acceleration of the course of sporadic AD—either because of overproduction of Aβ (as in the Swedish mutation of β-APP or presenilin mutations) or increased tendency towards aggregation (as in some of the point-mutations of β-APP within the Aβ sequence). It is clear that the capacity of the proteostasis systems to respond to stress declines with age. For example, the concentrations of chaperones, protein damage and cross-linking, and similar changes occur with age. Such changes in the cellular milieu help to account for the increasing prevalence of sporadic neurodegenerative diseases with age.

Several of the genes involved in proteostasis have also been implicated in control of lifespan in model organisms such as *Drosophila* and *Caenorhabditis elegans*. For example, downregulation of the heat-shock transcription factor, HSF1 opposes the effect of FOXO transcription factor DAF-16 to increase lifespan (Hsu et al. 2003; Brunet et al. 2004). Overexpression of HSF1 tends to increase lifespan. Both HSF1 and DAF-16 are regulated by SIRT1, the NAD-dependent sirtuin that regulates lifespan in *C. elegans* and other organisms by deacylating FOXO transcription factors and thereby modulating their activities (Westerheide et al. 2009). These results indicate integration of the proteostasis systems with other cellular systems that respond to stress and influence longevity of cells and, in some cases, entire organisms.

As described below, many of the potential therapeutic approaches to PADs involve blocking protein aggregation *per se*. Originally aimed at fibrillization, the question arose whether therapeutic approaches to block fibrillization could have the unwanted effect of increasing concentration of cytotoxic soluble oligomers. It has since become clear that many anti-fibrillization agents are effective more generally at blocking protein aggregation, including aggregation in pre-fibrillar aggregates such as soluble oligomers. In addition, the persistence of fibrils in or adjacent to a cell, though not necessarily harmful in itself, could put an additional burden on the protein-degradation systems. Most studies comparing the cytotoxicity of oligomers and fibrils proceed by adding each of these to cell-culture systems. In acute experiments of this type, oligomers are more toxic than fibrils. Nevertheless, the persistence of fibrils within or adjacent to cells could have direct and indirect effects on degradation of oligomers. First, although many depictions of "fibrillization pathways" show oligomers leading to fibril formation, in fact this is a partially reversible process, and fibrils can give rise to oligomers. Second, fibrils are protease-resistant, and thus persistence of fibrils could increase cellular levels of toxic oligomers over time, by tying up protein-degradation systems that would otherwise be used to degrade the more toxic soluble oligomers.

This statement can be generalized further. One of the mechanisms by which protein aggregates damage cells is by inhibiting proteasomal and autophagic systems. Neuritic plaques and Lewy bodies are large protein aggregates, containing not only the villainous aggregating proteins, Aβ and α-synuclein, respectively, but also a slew of additional proteins—a mass spectrometrist's feast—including ubiquitin and 19S and 20S proteasomal components. Thus, even if these lesions were nothing more than the cinders of a previously raging protein-aggregation fire, the presence of these proteasomal proteins suggests a failed attempt—and time/energy wasted—trying to eliminate aggregated protein. Not only fibrils, but soluble oligomers, are resistant to proteolysis. Thus, in conditions leading to the cytotoxicity of protein aggregates, there is *decreased* proteasomal activity present in cells, even in the face of increased expression of proteasomal components (Bennett et al. 1999; Cummings et al. 1998; Keller et al. 2000; Bence et al. 2001; Verhoef et al. 2002; McNaught et al. 2003; Holmberg et al. 2004; Wong et al. 2008; Emmanouilidou et al. 2010). One should recall, also, that the proteasome and autophagic pathways have roles other than the clearance of aggregated proteins: oxidized or otherwise-damaged proteins are also cleared by these systems, and their impairment could lead to the accumulation of other types of cytotoxins (Grune et al. 1997; Squier 2001; Betarbet et al. 2006; Dawson and Dawson 2003; Lev et al. 2006; Pan et al. 2008; Dasuri et al. 2010).

These observations are underscored by the fact that some heritable neurodegenerative diseases are caused by mutations in genes of the protein-degradation systems. PARK2 and PARK4 cause early-onset Parkinson's disease. PARK2 encodes a ubiquitin E3 ligase (Kitada et al. 1999; Wang et al. 2001), and PARK4 encodes a ubiquitin ligase/hydrolase (Zhang et al. 2000; Chung et al. 2001). Thus, an inherited defect in protein degradation can lead to neurodegenerative diseases, and acquired defects in these systems can contribute to them.

14.3 Specific Approaches to Inhibit Protein Aggregation—Potential Therapeutic Strategies for PADs

The remainder of this article will describe approaches to inhibit protein aggregation. This is not the sole avenue towards therapy of PADs, however. The number and diversity of PADs is ever-increasing and many forms of therapy can be directed towards improving end-organ function, apart from issues of protein aggregation. For example, several forms of therapy that target inflammation, the lipid profile, and vascular perfusion have been proposed to improve cognitive function in AD. It is wise indeed not to become *too* focused on protein aggregation.

We, however, will address only those approaches that are related to protein aggregation, *per se*, and will consider only in passing the cellular mechanisms that come into play *after* a protein or peptide aggregates, e.g., neuroinflammation by which cells in the CNS respond to Aβ aggregates. While such issues are of obvious clinical importance, they are beyond the scope of this review. In addition, we cannot consider all factors that cause expression levels of aggregating proteins to vary. These will be considered in only some instances where they are particularly germane to controlling protein aggregation, as in the case of Aβ, where the activities of the secretases play a central role in modulating production levels of Aβ peptides, as well as chain-length variants. Finally, it is beyond the scope of this review to consider all aspects of the immunology of protein aggregates. These relate centrally to the important topic of immunotherapy of PADs, for some of which (e.g., AD) there have already been extensive clinical trials. While recognizing the enormous potential of immunological approaches, this field is in great flux, and many conclusions would be premature at this time.

Finally, in view of the ever-increasing number of PADs, it would be repetitious to try to discuss them exhaustively. Instead, we have chosen to focus on three paradigmatic diseases, while alluding to other diseases as appropriate. Among the peptides and peptide-like proteins causing PADs, the mechanisms of aggregation can be grouped into two very broad types: those involving mainly the hydrophobic effect, such as Aβ, α-synuclein, the apolipoproteins, and peptide hormones; and those involving mainly polar residues, polyGln-containing proteins such as huntingtin. Without wanting to slight the other important diseases, we have chosen Aβ and huntingtin as the paradigms of each class. The third group of diseases is represented by the TTR amyloidoses. In addition to being a clinically important problem, these diseases are exemplary of protein misfolding, *per se*. In addition, they have been the subject of some of the best work on small-molecule inhibitors of aggregation. Accordingly, these diseases will be used as a paradigm of globular proteins involved in PADs.

14.3.1 Alzheimer's Disease and Aβ Peptides

14.3.1.1 Peptidic and Peptidomimetic Inhibitors of Aβ Aggregation

Recognition-Element-Containing Inhibitor Peptides

Several peptidic inhibitors of Aβ have been synthesized containing the tandem combination of a "recognition element," which binds to a putative aggregation domain of Aβ, and a "disruptive element" to disrupt aggregation. The most frequently used recognition domain is the hydrophobic domain from residues 17–21, LVFFA. Disruption of this sequence within full-length Aβ by mutation abrogates Aβ aggregation, e.g., replacement of either Phe by Thr abolishes fibrillization (Hilbich et al. 1992). Thus, while the octapeptide, Aβ16–23 can bind to Aβ1–40, it is not a fibrillation inhibitor, while the peptide with Thr substituted for Phe (QKLVTTAE) is one (Hughes et al. 1996). Phe is found in the aggregation sites of many aggregating proteins, including islet amyloid polypeptide: whereas NFGAILSS (IAPP residues 22–29) forms fibrils, peptides containing Tyr in place of Phe do not, and act as a moderate inhibitor of IAPP fibrillization (Porat et al. 2004). Indeed, even the dipeptide FF forms self-assembling nanotubes (Reches and Gazit 2003). This central hydrophobic domain of Aβ was at first proposed as a fibrillization inhibitor (Tjernberg et al. 1996, 1997; Hetényi et al. 2002); a similar strategy was employed using a hydrophobic domain from the prion protein (Chabry et al. 1998). While these peptides reduced thioflavin T fluorescence developing from Aβ solutions, they did not actually inhibit fibrillization, but rather, formed mixed fibrils with full-length Aβ peptides that were thioflavin T-fluorescence negative.

The β-breaker peptides at first modified the central hydrophobic domain using a single Pro substitution, e.g., iAβ5, NH_2–LPFFD–COOH. In other versions of these peptides, strings of Pro residues were appended to the aggregation domain (Soto et al. 1998). Some of these peptides inhibited fibril formation by several criteria, and inhibited amyloid formation in neuroblastoma cell lines. When injected into the brain after a previous injection of Aβ1–42, the β-breaker peptide also diminished cerebral Aβ deposition. Although the animal model study is remote even from what occurs in transgenic animals, these studies established the principle of balancing binding to existing or incipient aggregates, with inhibiting the propagation of β-sheet structures at that site. A subsequent study demonstrated that the iAβ5p peptide (iAβ5 with carboxamidated *C*-terminus) penetrated the blood–brain barrier and reduced Aβ deposition in two transgenic mouse AD models. By some criteria, there was increased neuronal survival in the treated mice compared with control (Permanne et al. 2002; Sigurdsson et al. 2000).

Other studies by the same group incorporated $_D$-amino acids and non-natural amino acids, and are discussed below (Soto et al. 1996; Poduslo et al. 1999). Another group used a similar strategy, by incorporating a string of three or more charged or polar amino acids (Lys, Glu, or Ser) in tandem to a related recognition sequence (KLVFF, residues 16–20 of Aβ). They observed that the two types of charged disruptive element were similarly effective while the Ser-containing peptides were ineffective as inhibitors (Lowe et al. 2001). These compounds inhibited fibrillization even at high ratios of Aβ to inhibitor (100:1, mol:mol). These studies rely almost entirely on light-scattering data, without corroborating thioflavin T or electron-microscopy data, however. In a further variation on this theme, the inhibitor peptides OR1 (RGKLVFFGR) and OR2 (RGKLVFFGR–NH2) were moderately effective inhibitors of Aβ fibril formation, but only the latter peptide also inhibited oligomerization and inhibited cytotoxicity of Aβ towards human neuroblastoma SHSY5Y cells (Austen et al. 2008).

There have been relatively few attempts to use the *C*-terminal hydrophobic domain as a recognition domain in inhibitor peptides. It is not clear why this is so, since residues ~30–42 of Aβ form a β-sheet in fibrils and oligomers, though perhaps a somewhat less stable one than the *N*-terminal β-sheet by the criterion of protection factors in hydrogen–deuterium exchange experiments (Kheterpal et al. 2000, 2003; Whittemore et al. 2005; Kheterpal and Wetzel 2006; Kodali et al. 2010). Two attempts to do so yielded modestly effective inhibitors using Aβ39–42 (Hetényi et al. 2002) or Aβ31–34 (Fülöp et al. 2004) as the recognition domain.

C- and *N*-Terminal Modifications

A related strategy has been the use of a recognition domain in tandem with a non-peptidic disruptor. Burkoth et al. (1998) synthesized an inhibitor of Aβ10–35

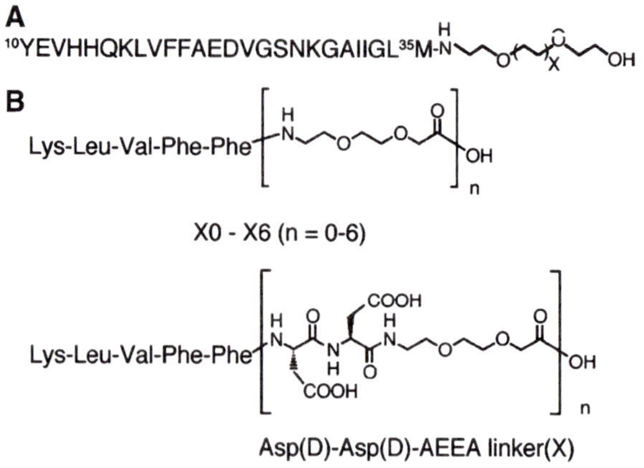

Fig. 14.6 Two types of block co-polymers with Aβ segments. (**A**) Shows Aβ10–35 with a *C*-terminal block of polyethylene glycol–3000 (From Burkoth et al. 1998). The resulting peptide slows Aβ10–35 fibrillization, and it aggregates reversibly. (**B**) Shows two inhibitors of Aβ aggregation, consisting of a "recognition element", KLVFF, and a hydrophilic "disrupting element"—either aminoethoxy ethoxy acetic acid (AEEA, which has an ethylene glycol core) or AEEA with $_D$-Asp (From Watanabe et al. 2002)

aggregation consisting of this peptide in tandem with a *C*-terminal polyethyleneglycol-3000 block (Fig. 14.6a). This is, in essence, a block copolymer, which oligomerized in a concentration-dependent manner from monomer to hexamer or heptamer. It also slowly formed fibrils that were entirely linear, i.e., without a supramolecular twist. An additional feature of the peptide was that fibrillization was pH-dependent, and could be reversed by acidification to pH ~2.5. A similar strategy was used in a peptide consisting of Aβ16–20 with a disruptor containing the hydrophilic polymer aminoethoxy ethoxy acetate, and aspartate to add charges to the block (Fig. 14.6b); the numbers of each of these elements in the disruptor were varied (Watanabe et al. 2002; Akikusa et al. 2003). These compounds showed moderate levels of efficacy in decreasing Aβ aggregation and cytotoxicity in IMR-32 neuroblastoma cells *in vitro*.

A related strategy was in the incorporation of the disaccharide trehalose (α-D-glucopyranosyl-(1 → 1)-α-$_D$-glucopyranoside) into peptides based on the β-blocker peptide, iAβ5 at various sites: *N*- and *C*-termini and the Asp side-chain (De Bona et al. 2009). The intent of incorporating trehalose was to protect the peptide from proteases while retaining the efficacy of the original peptide as a fibrillization inhibitor.

Starting with Aβ16–30, Findeis et al. (1999) surveyed numerous *N*-terminal modifications, varying polarity, charge, and size of the adducts. They determined that of all the compounds tested, cholyl–LVFFA–OH was the most effective, but was also of limited durability due to its susceptibility to proteases. This problem was addressed by substituting all-$_D$-amino acids, which yields an effective inhibitor, but with enhanced resistance to proteases in monkey CSF. In a similar strategy, Adessi et al. (2003) modified the iAβ5 (Soto et al. 1998, "β-sheet-breaker") with various *N*-terminal protectors that enhanced stability of the peptide in plasma. Several other investigators subsequently observed that *N*- or *C*-terminal Arg somewhat enhance inhibitor efficacy, and that the effect seems specific for Arg (Fülöp et al. 2004; El-Agnaf et al. 2004; Das et al. 2007), for inhibitors of both Aβ and α-synuclein aggregation. In one study, this amino acid was prominent in "hits" from a random-screening procedure for aggregation inhibitors (Kawasaki et al. 2010). The use of Arg is a more complex matter than it first appears. One study (Gibson and Murphy 2005) compared Aβ-aggregation inhibitors containing a common recognition domain—KLVFF, with various *C*-terminal-disruption domains, consisting of polyLys or polyArg. Although Lys and Arg are both cationic, as shown in Fig. 14.7a they resemble compounds at opposite ends of the Hofmeister series—e.g., amines such as betaine resembling Lys, and urea and guanidinium somewhat resembling Arg. Thus, the analysis could be put into terms of effects on solvent surface tension. Kosmotropes such as betaine increase solvent surface tension because they are excluded from the air–water (or other hydrophobe–water) interface. They are believed to stabilize protein structure by enhancing the hydrophobic effect, and thereby tending to drive hydrophobic amino-acid side-chains away from solvent and into the core of a folded protein. At the same time, they tend to decrease a protein's solubility in water (Neagu et al. 2001; Cacace et al. 1997; Lin and Timasheff 1996; Soderlund et al. 2003; Kita et al. 1994; Arakawa et al. 1990). Chaotropes have the opposite effect: they are enriched at the air–water interface, and therefore decrease surface tension; they are believed to destabilize proteins by permitting mobile hydrophobic side-chains to emerge from the core of a protein into the solvent. By the same reasoning, they also tend to enhance a protein's solubility in water. Arg is a moderate chaotrope when added to proteins as a co-solute; it destabilizes folded proteins, increases their water solubility, and therefore has been used to diminish protein aggregation (Shiraki et al. 2002, 2004; Xie et al. 2004; Qiao et al. 2001; Buchner and Rudolph 1991; Arakawa and Tsumoto 2003; Taneja and Ahmad 1994; Tsumoto et al. 2004). The effect of Arg when included in a composite peptide containing tandem recognition and disruption elements (e.g., KLVFFRRRRRR) was complex; however, contrary to the above statement (about the resemblance of Arg to urea or guanidinium), the peptide did not act like a chaotrope—it increased surface tension and accelerated fibrillization, either because it does not associate with Aβ or act as a chaotrope at all in the context of a peptide, allowing other solvent effects to dominate. The paper does, however, demonstrates convincingly a (non-linear) correlation between surface tension increase of hybrid compounds (without Aβ present) and increase in weight-average molecular weight, M_w, of Aβ plus hybrid compounds (Fig. 14.7b). Thus, the use of chaotropes as the basis for designing future inhibitors remains an intriguing possibility.

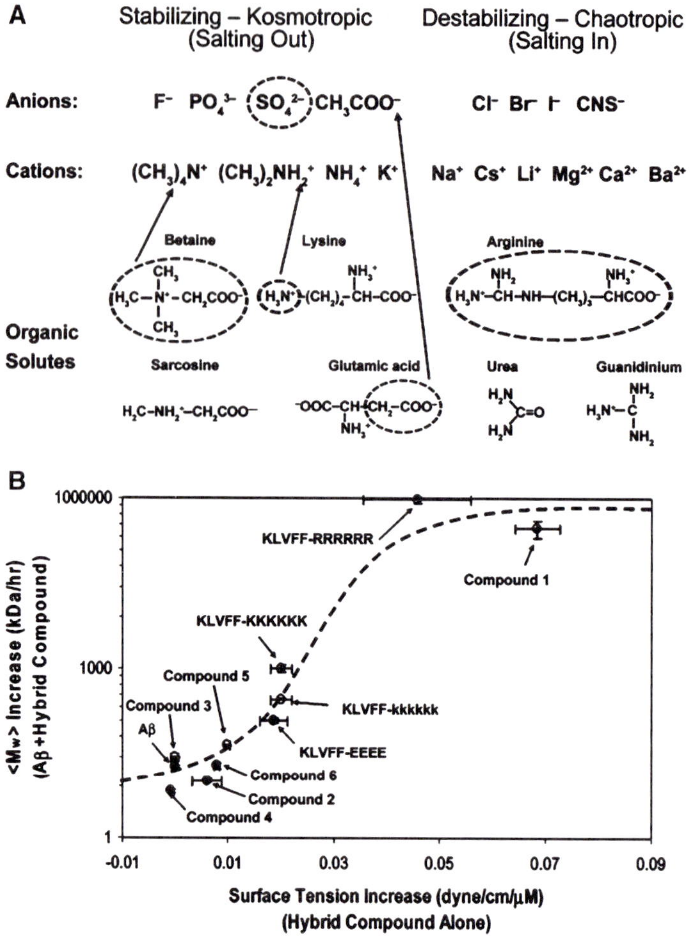

Fig. 14.7 The importance of surface tension for the aggregation of Aβ [Figure is from Gibson and Murphy (2005)]. The aggregation of Aβ occurs mainly through the hydrophobic effect. (**A**) Shows the Hofmeister series. Solutes on the *left* are kosmotropes, which stabilize protein structures, decrease their solubility, and increase solvent surface tension. Solutes on the *right* are chaotropes, which destabilize protein structures, increase their solubility, and decrease solvent surface tension. The locations of lysine, glutamic acid, and betaine side-chain groups in this series are shown. (**B**) Shows the non-linear correlation between surface tension of compounds (without Aβ) tested as aggregation inhibitors (x-axis), and the weight-average molecular weight, M_w (y-axis) attained by Aβ in the presence of inhibitor compounds (as determined by light-scattering techniques)

Peptide Backbone Modifications and Non-natural Amino Acids

N-Methyl Amino Acids

Several groups have designed protein-aggregation inhibitors based on the use of *N*-methyl amino acids (NMe-AAs, Fig. 14.8a). NMe-AAs can inhibit protein aggregation in at least three ways: (1) NMe groups eliminate a hydrogen-bond donor, the

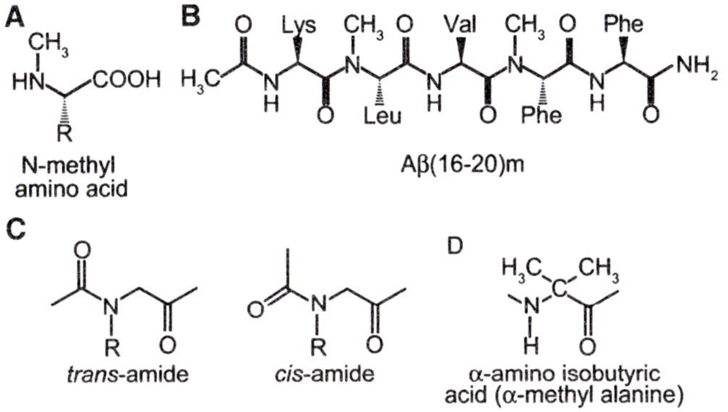

Fig. 14.8 Modified amino acids used in aggregation inhibitors. (**A**) *N*-methyl amino acid. (**B**) *N*-methyl-amino-acid-containing inhibitor of Aβ aggregation, Aβ(16–20)m, as described in Gordon et al. (2002). The peptide consists of a central hydrophobic region of the Aβ peptide, Aβ(16–20), with *N*-methylated amino acids at alternate residues—here, residues 17 and 19 (Leu and Phe, respectively). Similar inhibitory results are obtained if residues 18 and 20 (Val and Leu, respectively) are *N*-methylated. (**C**) *N*-substituted amino acids. These are the basic units of peptoids, and they can form *cis*- or *trans*- amide bonds with approximately the same propensities. (**D**) A Cα-substituted amino acid, α-amino isobutyric acid, also referred to as α-methyl alanine. This amino acid can form peptide bonds with a very narrow range of torsional angles (Φ and Ψ) at the boundary between α- and 3,10-helices, but since it is achiral, these can be either *left*- or *right*-handed helices

amide H, and thereby interfere with hydrogen bonding; (2) NMe groups are larger than the amide H, and could cause steric hindrance in the packing of β-strands into sheets; and (3) NMe groups strongly constrain the peptide bond into the *trans* conformation (Patel and Tonelli 1976; Tonelli 1971; Vitoux et al. 1981), which enables NMeAA-containing peptides to associate with fibrillizing peptides. X-NMeAA (where X = any amino acid) peptide bonds resemble X–Pro bonds, and therefore one might expect some of these to be *cis*-peptide bonds. In fact, all X–NMeAAs observed thus far have been *trans* when these occur in peptides of >3 residues, containing only $_{L}$-amino-acids. *Cis* peptide bonds have, however, been seen in the setting of linear and cyclic depsipeptides (Elseviers et al. 1988; Bersch et al. 1993), pseudopeptides (Howell et al. 1995), highly constrained cyclic peptides (Higuchi et al. 1983), or peptides containing mixtures of $_{D}$- and $_{L}$-amino acids (Vitoux et al. 1981; Penkler et al. 1993). In addition, the NMeAAs are sterically hindered and tend to be restricted in their backbone conformations to the β-sheet geometry (Tonelli 1971, 1974; Kumar et al. 1975; Manavalan and Momany 1980; Vitoux et al. 1981).

Incorporation of an NMeAA into interleukin-8 disrupted dimerization of this protein at an interface consisting of two antiparallel β-strands (Rajarathnam et al. 1994). NMeAAs were also incorporated into cyclic nanotube-forming peptides (Clark et al. 1998). These peptides are completely planar β-sheets, by virtue of the alternation of $_{L}$- and $_{D}$-amino-acids. Without NMeAAs, they form infinitely long nanotubes. When NMeAAs are incorporated at alternate residues, the NMe groups are segregated to one face of the plane; hence, stacking of the cyclic planar disks is limited to the formation of dimers.

Doig (1997) made a three-stranded β-sheet peptide called the β-meander, in which aggregation was diminished by the combined use of NMeAAs and Asp side-chains. Based on the preceding observations, they synthesized Aβ25–35 peptides congeners containing single NMeAAs (Doig et al. 2002; Hughes et al. 2000), and tested their effects on the aggregation and toxicity in PC12 cells of a segment of Aβ, Aβ25–35. The effects of the NMeAAs depended strongly on the position of the NMeAA within the peptide. In one case (NMe–Gly25–Aβ25–30), there was little or no inhibition of aggregation and toxicity; in another, there was inhibition of both aggregation and toxicity (NMe–Gly33–Aβ25–35); while in a third, the NMe only altered fibril morphology (NMe–Leu34–Aβ25–35).

A further development of the *N*-methylation strategy was to incorporate NMeAAs at alternate positions, so that the NMe groups would be segregated to one face of a β-strand (Gordon et al. 2001, .2002). In principle, this resembled the strategy that limited stacking of the cyclic planar β-sheet-like peptides described above (Clark et al. 1998). This strategy requires knowledge of a putative aggregation domain, which in these cases, was the *N*-terminal hydrophobic region of Aβ and adjacent residues, i.e., 16–22 (–KLVFFAE–). In the later paper, this domain was truncated to include only residues 16–20 (Fig. 14.8b). Only minor differences were observed between peptides containing two or three NMeAAs, and within this short peptide, little difference was seen between different placements of the alternating NMeAAs (e.g., NMe–Leu^{17} and NMe–Phe^{19} *versus* NMe–Val^{18} and NMe–Phe^{20}). A significant difference was seen, however, between peptides containing two or three alternating NMeAAs, and one containing four contiguous NMeAAs: peptides with the motif of alternating NMeAAs were clearly superior in ability to inhibit fibrillization than the peptide containing contiguous NMeAAs. The motif was also generalizable: an NMeAA-containing heptapeptide based on a putative aggregation domain in the human prion protein, residues 106–126, was effective against aggregation of the unmethylated peptide of residues 106–126. Notably, there was no cross-inhibition: although the peptide appeared to interfere with fibrillization by interrupting hydrogen bonding, side-chain homology was necessary for the inhibitor to bind to its aggregating target. A similar strategy has also been adopted for inhibiting IAPP aggregation (Yan et al. 2006). Although originally designed as "fibrillogenesis inhibitors," the peptides were notably efficient at disaggregating pre-formed Aβ fibrils. In addition, although originally designed to inhibit fibril formation, they were subsequently observed to inhibit oligomerization as well, and to bind to their target with 1:1 stoichiometry, with only moderate affinity ($K_d \sim 10^{-6}$ M). As discussed below, while this modest affinity might seem at first glance to be a problem, it is entirely consonant with their "chaperone-like" activity. In other words, they do not so much permanently block growth sites as "reset the clock" in the lengthy conformational transition to β-sheet structure.

The NMeAA-containing peptides have several intriguing physical properties not entirely related to their ability to inhibit fibrillization and disassemble fibrils (Gordon et al. 2002). To take the example of Aβ(16–20)m (Ac–K–NMeL–V–NMeF–F–$CONH_2$), it has an extraordinary stable β-strand secondary structure (presumably due to the strong tendency of NMeAAs to be constrained to β-sheet geometry) as indicated by CD (red-shifted minimum) and NMR; ellipticity remains essentially constant from 0°C to 90°C, pH 2–11, and 0–6 M guanidine-HCl. It is soluble in aqueous buffer to >30 mM, and is also highly soluble in many organic solvents, ranging in polarity from DMSO and alcohols to toluene and chloroform. The dual solubility (water and non-polar organic solvents) suggested that it might be able to diffuse through phospholipid bilayers, and this was shown to be the case for both single bilayer POPC vesicles and COS cells, without disrupting either. Presumably, the single positive charge on the peptide (Lys residue) without negative charges was responsible for successful vectorial passage into the cell. Finally, the peptide is completely protease resistant, compared with the non-methylated peptide of the same sequence, which is readily cleaved by chymotrypsin. Another NMeAA-containing inhibitor peptide, iAβ5p-A1 (Ac–LP–NMeF–FD–$CONH_2$) showed increased half-life in rat plasma and brain compared with the non-methylated counterpart (Adessi et al. 2003).

Few attempts have been made to apply the same strategy to peptides based on the *C*-terminal β-sheet domain. One group compared hexapeptides, all based on residues 32–37 of Aβ, with 0–5 NMeAAs (Pratim et al. 2009). The authors found the penta-*N*-methylated peptide (all residues in the peptide except Met) to be most effective, but the effect on fibrillization *in vitro* was modest. Their study is notable mainly for testing the inhibitor peptides in transgenic *D. melanogaster* expressing human Aβ1–42 in CNS neurons. The transgenic flies develop Aβ aggregation *in vivo*, and have reduced lifespans (Rival et al. 2009; Crowther et al. 2005, 2006; Luheshi et al. 2007). Feeding the peptide to the transgenic flies led to improvement in a locomotion (climbing) assay; one of the peptides, containing five *N*-methylations, also led to a very modest increase in lifespan.

Two studies using molecular-dynamics (MD) simulations have provided insights into the mechanisms of action of NMeAA-containing inhibitors. Soto et al. (2007) used fully atomistic MD simulations, and demonstrated that Aβ(16–20)m binds to a model protofilament of Aβ16–22 peptides at four different sites: (1), at the edge of the protofilament; (2), on the exposed face of a protofilament layer; (3), between the protofilament layers; and (4), between the protofilament strands. Binding at the different sites resulted in several modes of inhibition: (1) inhibition of monomer deposition (longitudinal growth); (2) inhibition of lateral protofilament assembly; and (3) disruption of fibril morphology, leading to disassembly of the fibril by strand removal. These observations show that Aβ(16–20)m binds to both pre-fibrillar and fibrillar targets, and help to account for the fact that the inhibitor both inhibits fibrillization and catalyzes disassembly of pre-formed fibrils.

Chebaro and Derreumaux (2009) performed coarse-grained implicit solvent MD and replica-exchange MD simulations on Aβ16–22 and Aβ16–22 monomers, and then examined a six-chain Aβ16–22 bilayer with either four copies of Aβ16–22 or four copies of mAβ16–22. They demonstrated that NMe groups reduced and blocked both β-sheet extension and lateral association of layers of Aβ16–22 and also intercalated between Aβ16–22 molecules to allow sequestration of Aβ16–22 peptides.

An important question about NMeAA-containing aggregation inhibitors is whether they act solely or mainly by removing a hydrogen-bond donor (due to replacement of the amide hydrogen by a methyl group), or also act through steric blocking of β-strands (or incipient β-strands) of the aggregating peptide. The above MD-simulation studies suggest a combination of both factors. An approach to this question was to use ester bonds in lieu of NMe groups. The ester group, like the NMe group, cannot act as a hydrogen-bond donor (Beligere and Dawson 2000; Bramson et al. 1985; Lu et al. 1997), but otherwise resembles amides in its transplanar conformation and bond lengths and angles (Ingwall and Goodman 1974). Aβ(16–20)e, an ester-containing analogue of Aβ(16–20)m, resembled the latter peptide in inhibiting fibril formation, and in other properties such as high water solubility. The main drawback of this approach for development of therapeutic agents is that ester bonds are base labile, and even at pH 7.4 were slowly hydrolyzed. Aβ(16–20)e was only slightly less effective as an inhibitor than Aβ(16–20)m, suggesting that while steric factors do play a role in the action of Aβ(16–20)m, these peptides act mainly by disrupting hydrogen bonding of the aggregating peptide.

Subsequent iterations of NMe-containing inhibitors combined this strategy with others, notably, the use of other non-natural amino acids, and $_{D}$-amino acids. These combined approaches are considered below.

$_{D}$-Amino Acids

The initial rationale for incorporating $_{D}$-amino acids into aggregation inhibitors was to limit proteolysis, which, in the case of a drug, would tend to prolong half-life of the inhibitor in plasma or tissues. In one early study, screening of a combinatorial library comprised only of $_{D}$-amino-acid-containing pentapeptides yielded ligands for Aβ with Phe in the second position and Leu in the third position (Tjernberg et al. 1997). These ligands also inhibited Aβ fibril formation. The sequence requirements for these peptides suggest binding to Aβ in an antiparallel orientation.

One might have imagined, when pondering protein self-association, that all-$_{D}$- and all-$_{L}$-amino-acid peptides cannot mix in a β-sheet. In fact, this question was tested directly with Aβ: while all-$_{L}$-Aβ in solution deposits with first-order kinetics onto immobilized all-$_{L}$-Aβ, and all-$_{D}$-Aβ deposits with first-order kinetics onto immobilized all-$_{D}$-Aβ, no binding was observed between soluble peptide and immobilized template peptides of opposite chirality, showing that this association is stereospecific (Esler et al. 1999).

This statement is not entirely generalizable, however. On the contrary, mixtures of poly-$_{D}$-Lys and poly-$_{L}$-Lys form β-sheets even more rapidly than peptides of a single chirality (Fuhrhop et al. 1987; Dzwolak et al. 2004), possibly because diastereomeric self-assembly is a more effective way of packing the chains with greater reduction of hydrating water. Diastereomeric self-assembly is not unique to poly-$_{L}$- and -$_{D}$-Lys (Zepik et al. 2002).

The Aβ sequence contains four discernible regions: a hydrophilic *N*-terminal domain (residues ~1–10), an *N*-terminal hydrophobic segment (residues ~17–21), a central hydrophilic domain (residues ~22–29) and a *C*-terminal hydrophobic domain (residues ~31–42). As discussed above, full-length Aβ peptides, and long, though truncated segments of Aβ that preserve the basic domain structure adopt a parallel, in-register β-sheet structure—with one recent exception, the Iowa mutant form of Aβ, D23N-Aβ (Tycko et al. 2009). Indeed, this appears to be true for most amyloid fibrils made of peptides of more than ~10 amino acids. Shorter Aβ peptides, such as Ac–KLVFFAE (Aβ16–22) can form antiparallel β-sheets, however (Balbach et al. 2000). Amphiphilicity is one factor determining orientation of the β-sheet: amphiphilicity favors adoption of the parallel, in-register β-sheet structure. Aβ16–22 is not amphiphilic, but adding an *N*-terminal octanoyl group makes the resulting peptide amphiphilic (as shown by collapse pressure of monolayers at the air–water interface) and reverses the β-sheet orientation from antiparallel to parallel and in-register (Gordon et al. 2004). Thus, while self-similar association of Aβ into fibrils may require all-$_L$-Aβ to bind to all-$_L$-Aβ (or all-$_D$-Aβ to bind to all-$_D$-Aβ), the same requirement does not necessarily hold for the binding of short, aggregation-inhibitor peptides to full-length Aβ. Although the studies of all-$_D$ inhibitor peptides *suggest* antiparallel β-sheet binding, this has not been proven decisively.

Chalifour et al. (2003) made the surprising observation that $_D$-enantiomers of five Aβ16–20-based peptides were more effective inhibitors or Aβ cytotoxicity and fibril formation than the corresponding $_L$-amino-acid peptides. In another study, cholyl–LVFFA–OH was found to be a better inhibitor of Aβ aggregation than the $_L$-amino-acid peptide (Findeis et al. 1999). Similarly, all-$_D$-amino acid variants of the β-breaker peptides, iAβ11 (RDLPFFPVPID) and YiAβ11 (RDLPFYPVPID) had longer half-lives in plasma than the corresponding all-$_L$-peptides (Soto et al. 1996; Poduslo et al. 1999). A modified form of one of these peptides, PUT-YiAβ11, containing *N*-terminal putrescine moiety, was found to cross the blood–brain barrier and survive in plasma longer than the all-$_L$-amino-acids form (Poduslo et al. 1999).

Two recent papers have carried the use of $_D$-amino-acid-containing peptides to the next logical step: all-$_D$-amino-acid peptides with reverse amino-acid sequences, i.e., *retro-inverso* peptides. The first of these studies (Austen et al. 2008) is based on prior work with two inhibitor peptides, RGKLVFFGR (OR1) and RGKLVFFGR–NH2 (OR2) containing the sequence KLVFF (Matharu et al. 2010). In the more recent study, these investigators tested a retro-inverso analogue of OR1, rGffvlkGr-1,5-diamino pentane. Surface plasmon resonance (SPR) studies showed that the retro-inverso peptides increase the association rate to Aβ by ~50-fold and the affinity for Aβ ~500-fold, with small changes in the dissociation rates. The authors also showed that the retro-inverso peptides afforded protection in human SHSY-5Y neuronal cells against Aβ40-induced toxicity. A second study examined the retro-inverso analogue of OR2, for which SPR showed very fast association and dissociation rates, and only moderate affinity (8.8 and 9.5 μM) for the binding of RI-OR2 to monomers and fibrils (Taylor et al. 2010). This peptide was, in fact, their most effective inhibitor, and thus demonstrates that high affinity is not the best criterion for predicting the efficacy of an aggregation inhibitor. As discussed below, similar results were also obtained by another group (Lanning et al. 2010) for an inhibitor of polyGln peptide aggregation. These peptides appear to act not by binding to and "capping" growth sites. Rather, they act in a *chaperone-like manner*, by interfering with structural transitions to the aggregative β-sheet conformation, and thereby resetting the clock, as it were, for the long process of this structural transition.

Peptoids and Modifications of the α-Carbon

Peptoids (polymers containing *N*-substituted glycine residues, Fig. 14.8c) have been used, though infrequently, to design aggregation inhibitors. Thus far, this approach has not been used for Aβ, but has been used for IAPP, as discussed below. Peptoids lack the conformational rigidity of peptides made of α-amino acids. Since side-chains are on the amide nitrogen, the peptoid backbone is achiral. By the same token, amide bonds in peptoids are tertiary, and readily isomerize between *cis* and *trans*, in contrast to most peptide bonds. Because they lack amide protons, peptoids do not form secondary structures that depend on stabilization by backbone hydrogen bonds—again, in contrast to peptides. Although these factors make peptoids structurally unconstrained compared with peptides, peptoids can form secondary structures, however, if amide side chains that restrict backbone conformation are incorporated into their sequences, e.g., branched and bulky side chains to cause steric clashes between side chains, and thus limit rotational freedom about bonds (Armand et al. 1997). Turns have also been incorporated in peptoids, either by forming macrocycles (Shin et al. 2007), or by incorporating a heterocyclic turn-inducing unit, e.g., a triazole (Pokorski et al. 2007).

With such modifications, peptoids have been used to produce bioactive molecules, with the advantage that peptoids are much more resistant to proteases than most peptides. Most of the applications of peptoids consist of mimics of α-helices, in two particular areas: antimicrobial agents, where peptoids have been used to mimic α-helical, membrane disruptive peptides (Patch and Barron 2003; Shin et al. 2007; Gorske and Blackwell 2006; Fowler et al. 2008), and as mimics of the two hydrophobic α-helical proteins in lung surfactant, SP-B and SP-C (Seurynck-Servoss et al. 2006; Brown et al. 2008). In the field of protein aggregation, the use of peptoids has been limited to analogues of amylin (islet amyloid polypeptide, IAPP) residues 21–30 (–SNNFGAILSS–). Peptoid (*N*Ser-*N*Asn-*N*Asn-*N*Phe-Gly-*N*Ala-*N*Ile-*N*Leu-*N*Ser-*N*Ser) and retropeptoid (*N*Ser-*N*Ser-*N*Leu-*N*Ile-*N*Ala-Gly-*N*Phe-*N*Asn-*N*Asn-*N*Ser) analogues of IAPP21–30 have been synthesized and examined as fibrillization inhibitors (Elgersma et al. 2007). The peptide of this part of IAPP has β-sheet structure by FTIR and CD spectroscopy, and forms fibrils. When mixed with the above peptoid or retropeptoid, the IAPP(20–29) peptide did not form fibrils (as judged by electron microscopy), and did not adopt β-sheet structure (by CD spectroscopy). The authors proposed that peptoids bind to peptides through side-chain interactions, but disrupt the hydrogen bonding that propagates the β-sheet structure of fibrils.

As discussed above, in the section on NMeAA-containing peptides, backbone hydrogen bonds are critical for Aβ aggregation, but the sequence specificity of NMeAA-containing inhibitors indicates that side-chain interactions and steric factors also play a role in inhibition. Modifications of the α-carbon cannot interfere with hydrogen bonding, of course, but several aggregation inhibitors have been designed based on the use of non-natural amino acids with modifications of the α-carbon (Fig. 14.8d).

One set of inhibitors used α-amino isobutyric acid (also referred to as α-methyl alanine, or αMeA), substituted for Ala or Leu in sequences derived from IAPP (Gilead and Gazit 2004). Ala and Leu are both strong α-helix formers, and the addition of a second α-methyl group constrains the polypeptide chain to a very narrow range of torsional angles (Φ and Ψ) at the boundary between α- and 3,10-helices. Since αMeA is achiral, it can form left- or right-handed helices. It is a naturally occurring amino acid that is incorporated into some proteins and peptides, especially peptide antibiotics produced by soil fungi. A Ramachandran plot of this amino acid occurring in 367 crystal structures shows it fairly evenly divided between α- and 3,10-helices, and between right- and left-handed helices (Venkatraman et al. 2001; Aravinda et al. 2003; Mahalakshmi and Balaram 2006). Incorporation of multiple residues of αMeA into a peptide favors 3,10-helix formation, while fewer αMeA residues in a peptide favor α-helix. Aggregation inhibitors containing this amino acid could act by trapping an α-helical intermediate, and preventing its transformation into a β-sheet.

Another modified side-chain used in aggregation inhibitors is α,β-dehydroalanine (ΔA), which has been incorporated into two peptides, P1 (K–L–V–F–ΔA–I–ΔA) and P2 (K–F–ΔA–ΔA–ΔA–F) (Abedini and Raleigh 2009). ΔA constrains the polypeptide chain to a narrow range of torsional angles, i.e., a fully extended conformation, with backbone dihedral angles, [Φ, Ψ] = [−180°, 180°] (Bhatnagar et al. 1995; Henzler et al. 2004; Crisma et al. 1999). Homopolymers of ΔA, pBrBz–$(\Delta Ala)_n$–OMe (n = 1–6) were found to form the fully extended conformation, the 2.05-helix (Crisma et al. 1999). The two inhibitor peptides, P1 and P2, both inhibited Aβ42 aggregation, but had somewhat different mechanisms. P1 maintained Aβ42 in an unstructured state, while P2 induced and maintained Aβ42 in an α-helical structure. The authors suggested that P1 and P2 bind to Aβ42 at different stages in the aggregation (earlier and later in the pathway, respectively). In addition, P1 is hypothesized to interact with the *N*-terminal β-sheet and P2 with the *C*-terminal β-sheet domain of Aβ. It is somewhat surprising that if ΔA acts by constraining its binding partners to an extended conformation, these two inhibitor peptides do not form β-sheet structures, themselves.

Other non-natural amino acids used in peptides to generate aggregation inhibitors include β-alanine and γ-aminobutyric acid (Madine et al. 2009b). Several short peptides containing β-Ala reduced fibrillization of α-synuclein and/or Aβ somewhat, but also increased conversion of monomeric peptides to pre-fibrillar aggregates. Another study examined several variants of the pentapeptide, Ac–LPFFD–NH_2 (iAβ5p) described above, but with substitutions of several non-natural amino acids (Giordano et al. 2009), including (*S*)-azetidine-2-carboxylic acid (Aze, compound 1) and its (2*R*,3*R*)-*trans*-3-phenyl substituted analogue (PhAze, compound 2) (Table 14.2). The Aze ring constrained the peptide chain into a non-β-sheet conformation, but it affords somewhat greater side-chain conformational flexibility than Pro. Compounds 3 and 4 contain β-amino-acid homologues of Pro, β-aminocarboxylic or β-aminosulfonic

Table 14.2 Sequence of iAβ5p and its synthetic analogues

Peptide	Sequence
iAβ5p	Ac–Leu–Pro–Phe–Phe–Asp(OH)–NH_2
1	Ac–Leu–**Aze**–Phe–Phe–Asp(OH)–NH_2
2	AceLeu–(*2R,3R*)–**PhAze**–Phe–Phe–Asp(OH)eNH_2
3	Ac–Leu–**β–HPro**–Phe–Phe–Asp(OH)–NH_2
4	Ac–Leu–**Pro–Ψ[CH_2SO_2]**–Phe–Phe–Asp(OH)–NH_2
5	**Me_2Tau**–Leu–Pro–Phe–Phe–Asp(OH)–NH2
6	Ac–Leu–Pro–Phe–Phe–Asp(OH)–Leu–Pro–Phe–Phe–Asp(OH)–NH_2

From Giordano et al. (2009)

acid, respectively. These compounds retain the Pro-like side-chain found in iAβ5p, but also confer protease resistance to these compounds. Compound 5 incorporates a *N,N*-dimethyl-taurine (*N,N*-dimethy-2-aminoethanesulfonic acid; Me_2Tau). Compound 6 represents a tandem repeat of iAβ5p. Although their study contains much elegant synthetic chemistry, the inhibitors are essentially of equal potency as others described above.

Combined Approaches

Several groups have combined the above approaches in order to achieve greater efficacy in their aggregation inhibitors. Grillo-Bosch et al. (2009) examined retro-enantiomers of *N*-methylated peptide inhibitors of aggregation. They compared a peptide containing a single NMeAA, NMe–Phe, in their peptide inL (NH_2–KKLVF–MeFA–$CONH_2$), to two others: an all $_D$-amino-acid peptide of the same sequence (inD), and a retro-enantiomer of inD (i.e., reverse amino-acid sequence), inrD. The group had previously shown that inL, though it does not inhibit development of thioflavin T fluorescence in Aβ42, nevertheless reduced cytotoxicity of Aβ42 (Cruz et al. 2004). This peptide differs from Aβ(16–20)m (Gordon et al. 2002) in having an additional Lys residue for greater water solubility and only one NMeAA. The inability of this peptide to prevent fibrillization is somewhat at odds with other results (Amijee et al. 2009) that one NMe–AA in a peptide based on the same "recognition sequence" of Aβ was sufficient to prevent Aβ aggregation. The dissociation between inhibiting fibrillization and inhibiting cytotoxicity suggests that a toxic species of Aβ may be off-pathway for fibril formation. In the comparison of inL, inD, and inrD, the last of these peptides was most effective in decreasing both fibrillization and cytotoxicity in cell cultures of a neuroblastoma cell line. Moderately high concentrations of these peptides (10–20 μM) were necessary to achieve robust inhibition.

The most extensive survey of peptides combining the NMeAA–, $_D$- and non-natural amino acids was carried out by Doig and co-workers (Amijee et al. 2009; Kokkoni et al. 2006). In the first of these papers, they varied peptide length, *N*-methylation sites, acetylation, and amidation of the *N*- and *C*-termini, side-chain identity, and chirality, using five compound libraries, and tested inhibition of Aβ aggregation by several methods, and cytotoxicity by MTT assay. They found an advantage in all-$_D$-amino-acid inhibitors with a free *N*-terminus and an amidated *C*-terminus, and large, branched hydrophobic side-chains at positions 1–4, while the side-chain at position 5 was less important. As stated, they also found that a single *N*-methyl group was necessary and sufficient for inhibition. Their most active compound was $_{D}$-[(chGly)–(Tyr)–(chGly)–(chGly)–(mLeu)]–NH_2 (chGly = cyclohexyl Gly). This inhibitor was quite effective at reducing cytotoxicity by their criteria, but even their most effective inhibitor did not decrease thioflavin T fluorescence to less than 40% of control values. They suggest that the inhibitor decreases cytotoxicity by diminishing the concentration of a soluble oligomer, and at the same time led to the formation of a type of Aβ fibril that does not bind thioflavin T (or one that binds thioflavin T but without causing fluorescence). In another part of these studies, they applied the same principles to a "sequence recognition element" from α-synuclein, –VAQKTV–, which was identified using solid-state NMR, and made an *N*-methylated homologue of this peptide that was able to inhibit α-synuclein aggregation.

14.3.1.2 Naturally Occurring and Engineered Proteins as Inhibitors

Transthyretin

Aside from the chaperones and other catalysts of protein folding, several proteins, notably TTR, have been identified as naturally occurring inhibitors of protein aggregation. These have served, in turn, as models for engineering novel aggregation inhibitors.

TTR is itself capable of forming amyloid, both as the wild-type protein, which causes senile systemic (especially cardiac) amyloidosis (Cornwell et al. 1998; Westermark et al. 1990), and in over 100 point-mutant forms which lead to familial amyloid polyneuropathy (FAP, Saraiva 1995) and familial cardiac amyloidosis (also called familial amyloid cardiomyopathy, FAP, Westermark et al. 1990), and CNS-selective amyloidosis (Gambetti and Russo 1998; Sekijima et al. 2005). TTR amyloidosis will be discussed in detail, below; the most successful small-molecule inhibitors of amyloidosis are directed against TTR amyloidosis. For now, discussion will be limited to a brief description of the characteristics of the wild-type protein.

TTR, formerly called prealbumin, is a 54-kDa protein composed of four identical 129 amino-acid chains, which exists mainly as a tetramer, but which can dissociate into dimers and monomers (Nilsson et al. 1975; Monaco et al. 1995). It is made in the liver and choroid plexus; it is abundant in the CSF [5–20 μg/mL (0.1–0.4 mg/mL)] and serum [170–420 μg/mL, (Giunta et al. 2005); see also (Ingenbleek et al. 1972; Vatassery et al. 1991)]. It binds and transports thyroxine (T4) and the complex of retinol and the retinol-binding protein. Although it binds T4 and many of its analogues, it is not the main thyroid-hormone-binding protein in plasma, and most of the T4-binding sites in plasma TTR are unoccupied. TTR transports 15–20% of plasma T4, but ~80% of T4 in the CNS (Hagen and Elliot 1973). An X-ray crystallographic structure of the TTR tetramer shows molecular 2,2,2 symmetry, with identical subunits around a central channel containing two T4-binding sites (Blake et al. 1978). The two T4-binding sites are structurally identical, but show negative cooperativity; the affinity for the first molecule of T4 is ~100-fold higher than for the second molecule (K_a1 ~10^8 M^{-1} and K_a2 ~10^6 M^{-1}) (Cheng et al. 1977; Wojtczak et al. 2001). Each monomer has a β-sandwich with two β-sheets, each containing four β-strands (DAGH and CBEF), and a small α-helix connecting β-strands E and F.

Recently, several lines of evidence have suggested that TTR may protect against Aβ aggregation by directly binding Aβ and inhibiting its aggregation. CSF concentrations of TTR are lower in multi-infarct dementia and in AD (Riisoen 1988; Link 1995; Serot et al. 1997) than in non-demented control subjects. In transgenic *C. elegans* expressing Aβ, co-expression of TTR abrogates cellular Aβ deposition (Castano et al. 2006). Several papers then demonstrated that Aβ binds directly to TTR (Schwarzman et al. 1994, 2004, 2005; Schwarzman and Goldgaber 1996; Tsuzuki et al. 1996). The protective effect of TTR comes about through the cleavage of the amyloid precursor protein (APP) by α-secretase. DNA microarray experiments showed that, in transgenic mice bearing the Swedish mutation of APP, the increase in Aβ is partly compensated for by an increase mRNA for TTR (Stein and Johnson 2002). Other experiments showed that the cleavage product, sAPPα, derived from the cleavage of APP by α-secretase, induces the increase in TTR expression (Stein et al. 2004).

Liu and Murphy (2006) showed that substoichiometric TTR greatly diminished the rate of Aβ aggregation, but did not decrease the concentration of Aβ in the aggregating pool. Measurement of an affinity constant of TTR for Aβ (from quenching of the intrinsic fluorescence of Trp in TTR) indicated only weak binding, K_S=2,300 M^{-1} (K_S=association constant between quencher and fluorophore from the Stern Volmer equation). Kinetic modeling (mainly from dynamic light-scattering data) and transmission–electron-microscopy data indicated that TTR reduced the average size of pre-fibrillar aggregates, reduced the length of amyloid fibrils, and reduced the rate of fibril extension (linear growth) by ~100-fold. TTR also decreased lateral association of aggregates. The authors suggest that TTR inhibits growth of aggregates, but not their initial formation. Other investigators measured the affinity of TTR for Aβ by immobilizing Aβ on a 96-well plate, and measuring competitive binding of ^{125}I–TTR (specific binding defined as bound ^{125}I–TTR alone—bound ^{125}I–TTR in the presence of 100-fold excess unlabeled TTR, Costa et al. 2008). Using this procedure, they obtained K_d=28 nM, considerably higher affinity than the previously described studies, perhaps owing to the use of different methods, and the use of ^{125}I–TTR by the latter investigators. This methodological question aside, they also compared binding of Aβ by wild-type and several point-mutant forms of TTR, and observed the following rank order of binding affinities: T119M>wild-type TTR>V30M≥Y78F>L55P. Surprisingly, one of the point-mutants (T119M) had higher affinity for Aβ than did wild-type TTR. It is unknown whether individuals with these mutants have different susceptibility to AD than the general population.

Two studies have tried to isolate the sites of interactions between Aβ and TTR, to find peptides that could be used as Aβ aggregation inhibitors (Schwarzman et al. 2005). They screened an amplified dodecapeptide FliTrxTM random peptide library (1.77×10^8 primary clones, Invitrogen), and a peptide library derived from TTR sequences to identify peptides that would inhibit Aβ aggregation. Although some consensus sequences were identified, these peptides had only moderate inhibitory activity. The same group examined 47 recombinant mutant forms of TTR that formed tetramers and were able to bind T4; only two mutant TTR molecules formed fibrils at pH 6.8 (Schwarzman et al. 2004) (wild-type TTR forms fibrils at acidic pH, but does so only very slowly at neutral pH). Most of the TTR variants also were able to bind Aβ and inhibited its aggregation, but in several of the variants (S64, A71, Q89, V107, H114, and I122), this inhibition was diminished, suggesting that TTR mutation could be a possible etiological factor in the development of AD.

Additional Natural Proteins and Peptides

Scattered reports have appeared of other naturally occurring peptides or proteins (other than chaperones and TTR) with apparent ability to inhibit aggregation of Aβ. Among these are peptides derived from α-crystallin (Santhoshkumar and Sharma 2004). α-Crystallin is an abundant, highly water-soluble protein found in the lens of the eye, with some structural similarity to small heat-shock proteins. Its main function in the eye is structural: it maintains the proper refractive index in the lens. In addition to this role, or perhaps as part of it, it also retains some chaperone activity, perhaps to prevent the formation of protein aggregates that would scatter light or form cataracts (Andley 2009). An internal peptide from α-crystallin, residues 70–88, was described as a "peptide chaperone", mini-αA-crystalline, with the ability to inhibit Aβ fibrillization *in vitro*, though the degree of inhibition appears modest.

Colostrinin is a set of Pro-rich peptides with an average molecular weight of ~6 kDa derived from colostrum (Janusz et al. 1981), the first milk produced by a mother after childbirth. A host of disparate, indeed extravagant benefits have been attributed to colostrinin, including diverse effects on innate and specific immunity, on neurological development and function, and the ability to relieve oxidative stress (Boldogha and Kruzel 2008) and the aggregation of Aβ implicated in AD (Bourhim et al. 2007). It remains to be seen which, if any, of these effects will be supported by further experimentation (Gladkevich et al. 2007).

Engineered Proteins and Other Constructs

The question remains open whether therapy should aim at eliminating Aβ fibril formation, for example, on the grounds that this would also inhibit other forms of Aβ aggregation; or at promoting Aβ fibrillization, for example, on the grounds that fibrils are more tolerable than oligomers, and whatever accelerates fibrillization will

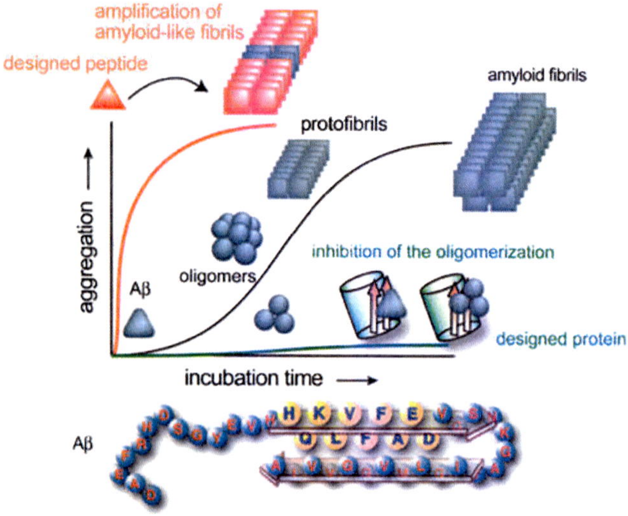

Fig. 14.9 Two strategies against Aβ aggregation (From Takahashi and Mihara 2008). A peptide, Ac–KQKLLLFLEE–NH_2 (*red triangle*), containing a modified form of the hydrophobic core sequence of Aβ (residues 16–20, –KLVFF–) does not aggregate itself, but rapidly converts soluble Aβ into fibrils. The assumption is that all cytotoxicity is due to soluble Aβ oligomers, and that fibrils are relatively innocuous. A second strategy is to incorporate Aβ-like sequences into the β-barrel of green fluorescent protein (GFP, green cylinder). This constructed protein can bind Aβ and therefore prevent it from aggregating into oligomers and fibrils

eliminate the toxic oligomers. One group developed reagents for each approach (Takahashi and Mihara 2008, Fig. 14.9). First, they developed a peptide, called LF, Ac–KQKLLLFLEE–NH_2, based on the hydrophobic stretch within the *N*-terminal β-sheet, –KLVFF–, but with a simplified sequence. The LF peptide forms fibrils rapidly itself, and also immediately transforms Aβ42 into fibrils, by co-assembling into fibrils with the latter peptide. This peptide could, perhaps, transform toxic soluble intermediates into less toxic fibrils, though one must wonder whether it could, itself, form some type of toxic intermediate. They also took the opposite approach by incorporating two Aβ-like β-strands with parallel orientation into the β-barrel of green fluorescent protein (GFP), which they refer to as a pseudo-Aβ β-sheet surface. Thus, the construct contained an Aβ-like β-strand on its surface that could bind Aβ in solution. They observed that this construct was able to bind Aβ42 and inhibit its oligomerization at substoichiometric levels. Presumably, the spacing of the Aβ-like β-strands in GFP has the wrong orientation to catalyze Aβ oligomer or fibril formation.

In a subsequent study (Takahashi et al. 2010), they showed that two GFP variants (P13H and AP93Q) with pseudo-Aβ β-sheet surfaces were able to bind Aβ with moderate affinity ($K_d = 260$ and 420 nM, respectively), and suppressed Aβ toxicity in a cell-viability assay. These two mutations were combined to generate a molecule, SFAB4, with higher affinity for Aβ ($K_d = 100$ nM). This is an ingenious approach, but it leaves several questions still open. Their methods for detecting oligomers is an ELISA assay using the monoclonal antibodies 4 G8, which preferentially recognizes Aβ oligomers, and 6E10, which recognizes Aβ monomers, oligomers, and fibrils. Using this assay, they demonstrate that when their constructs are mixed with Aβ, less Aβ oligomer remains in solution than in the absence of their constructs. They propose that their construct specifically binds Aβ oligomers, but it is also possible that the construct binds Aβ (not necessarily *as oligomer*), and then serves as a nidus upon which Aβ oligomers form, leaving less substrate for the formation of oligomers in the solution phase.

In another strategy, several groups have synthesized branched or tandem constructs of Aβ or its internal segments. One such construct was a dendrimer, resembling those originally used for immunization [(Spetzler and Tam 1995; Tam and Spetzler 1995); for review, see (Tam and Spetzler 2001), (Sadler and Tam 2002), and (Paleos et al. 2010)], and more recently, as an approach to eliminate PrP^{Sc}, the infectious form of the mammalian prion protein, from infected cells (Supattapone et al. 1999; Supattapone et al. 2001). They reported that branched polyamines, including polyamidoamide [PAMAM (Tang et al. 1996); Fig. 14.10a] dendrimers, polypropyleneimine, and polyethyleneimine, were able to eliminate PrP^{Sc} from scrapie-infected N2A cells in culture. The dendrimers were effective at non-cytotoxic concentrations, and their efficacy was related to the concentration of the branched polymer, the duration of exposure, and the density of amino groups at the surface of the dendrimers. Clearance of infectious material was attenuated by acidic pH or administration of chloroquine to cells, suggesting that the lysosome is involved in the mechanism of clearance. The authors considered the possibility that the polyamine dendrimers acted by induced expression of chaperones, but they were able to produce a similar phenomenon in a cell-free system: when a scrapie-infected mouse brain homogenate was exposed transiently to low pH (<4) followed by a polyamine (SupreFect), the scrapie material became susceptible to proteinase K, suggesting that PrP^{Sc} had been converted into a non-scrapie form of the protein. Thus, the effect is attributable to exposure of PrP^{Sc} to certain polycations, including polyamine dendrimers and SupreFect, at acidic pH, and may represent an effect on the aggregation or folding state of the protein. The authors also suggest that the dendrimers, or possibly other polyamines, could have similar effects in other PADs, including AD. This proposition was tested by another group, who tested the ability of various "generations" (of branchings) of PAMAM (polyamidoamide) dendrimers to inhibit aggregation of Aβ1–28. Again, the degree of inhibition was directly related to the "generation" number: fourth generation branched dendrimers were better inhibitors than third generation, and so forth (Klajnert et al. 2006).

Another type of dendrimer is one that includes specific sequences from an aggregating peptide or protein attached to a tree-shaped core structure. For example, Chafekar attached the KLVFF peptide on a first-generation dendrimer scaffold ((Chafekar et al. 2007); Fig. 14.10b). The first-generation scaffold was based on an amine-terminated poly(propyleneimine), with four reactive amino groups available, to each of which a Cys residue had been attached. This scaffold, the Cys dendrimer C4, reacted efficiently by native chemical ligation chemistry with four equivalents

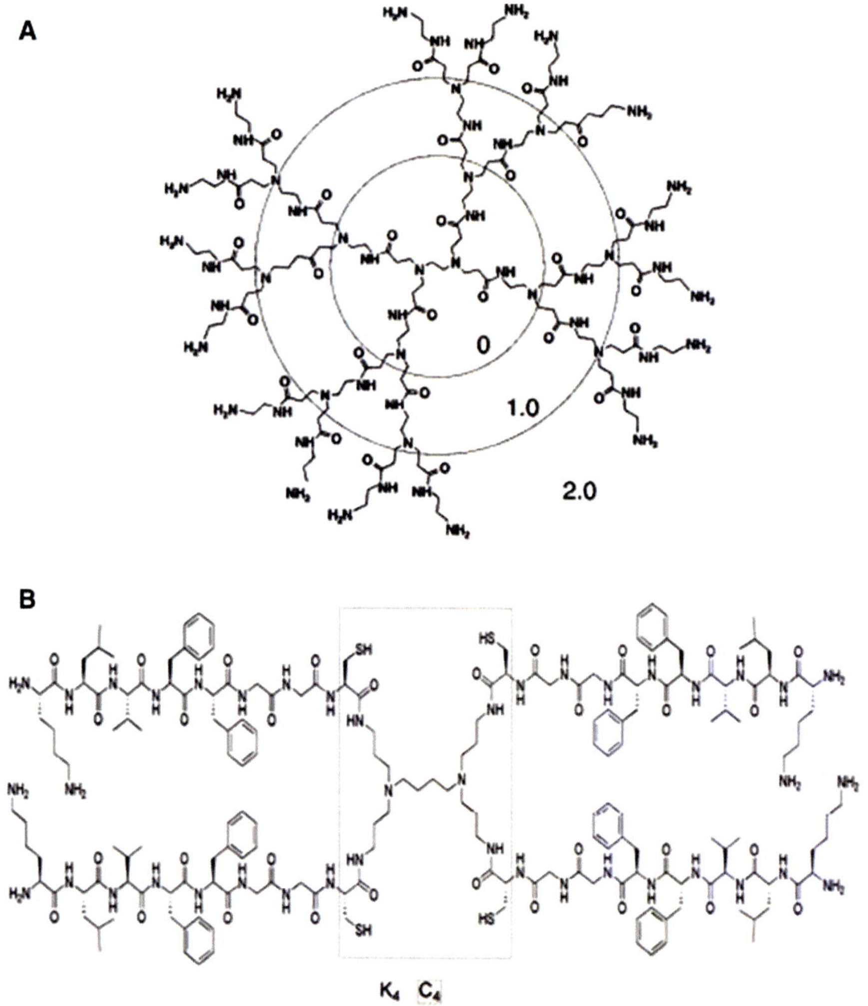

Fig. 14.10 Two types of dendrimers that inhibit protein or peptide aggregation. (**A**) Is from Supattapone et al. (1999), and shows polyamidoamide (PAMAM) dendrimers. Successive generations of branching are indicated by the *circles*. These dendrimers are one of several types tested. Their activity as inhibitors of prion proteins from infected N2A cells is related to their cationic nature. Note that these dendrimers are not derivatized. (**B**) Is from Chafekar et al. (2007), and shows a branched dendrimer bearing multiple copies of the sequence KLVFF from Aβ, linked to a cysteine dendrimer core by native chemical ligation. The core Aβ sequence has a *C*-terminal GG-thioester functionality for this purpose

of a peptide, KLVFFGG, functionalized by a *C*-terminal mercaptopropionic acid leucine (MPAL) thioester, as described by Hackeng et al. (Hackeng et al. 1999; for review, see Dawson and Kent 2000; Kent 2009). The resulting dendrimer, called K4, was more effective as an inhibitor of Aβ42 aggregation than the free peptide. In one

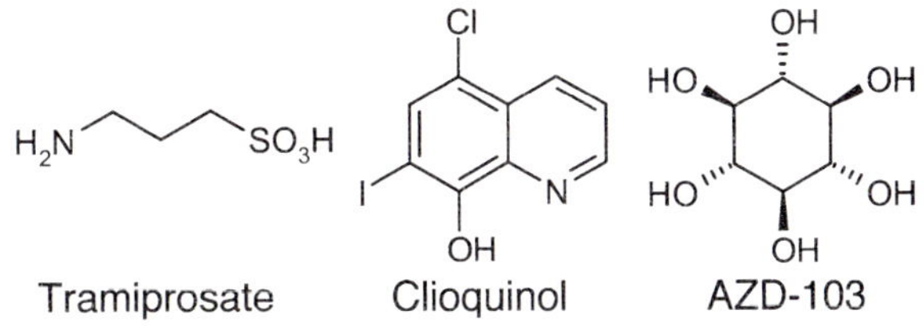

Fig. 14.11 Small molecules (other than natural products, and not including peptides) that are or have been recently in clinical trials to treat protein-aggregation diseases

way, this is not a surprising result (since free KLVFF itself forms fibrils), but the results raise an important question: how, exactly, does this dendrimer work? First, the authors note the difference between their dendrimers and the ones described above: the previous dendrimers were non-functionalized polyamines. A fourth generation dendrimer functionalized with the present Aβ would likely be unwieldy and insoluble, and unlikely to function like a polyamine dendrimer. On the contrary, the present authors limited the generation number to only one, in order to enhance solubility. Furthermore, their non-functionalized C4 dendrimer was ineffective as an inhibitor. It seems likely that their K4 dendrimer works by virtue of its ability to bind Aβ, though it is not known with what affinity, nor for how long; this binding on a branched dendrimer would seem to lack the geometry necessary for assembly of Aβ into ordered β-sheet fibrils—or, for that matter, into somewhat less ordered, partially β-sheet soluble oligomers.

A less elaborate, but still effective application of the above strategy was the use of a tandem Aβ peptide, NH_2–$A\beta_{40-1}$–$(Gly)_8$–$A\beta_{1-40}$–COOH, which is a substoichiometric inhibitor of Aβ40 aggregation (Mustafi et al. 2010). In the tandem dimer, a second Aβ molecule is attached through a flexible Gly_8 sequence to a second molecule of the reverse sequence. The authors demonstrate inhibition in the development of thioflavin T fluorescence at stoichiometric ratios as low as 1:25 (inhibitor to Aβ), and hypothesize that the construct binds a fibrillar intermediate, but the reverse sequence delays or prevents further monomer addition. The tandem arrangement of forward and reverse sequences would seem to favor binding to Aβ in a parallel, in-register orientation, but again, not necessarily with the correct fine structure to lead to the formation of ordered aggregates.

14.3.1.3 Small-Molecule Inhibitors of Aβ Aggregation

Small Molecules in Clinical Trials

From the viewpoint of most pharmaceutical companies, small molecules are vastly preferable to peptides, proteins, and other biological compounds, provided they are equally effective. Unfortunately, thus far, they are not—with the notable exceptions discussed below in the section on TTR amyloidosis. Figure 14.11 shows a collection of small molecules in clinical trials, or in consideration for clinical trials.

A case in point is tramiprosate (3-APS, NC531, Alzhemed™) a simple compound (3-amino-1-propanesulfonic acid, homotaurine), which has been in phase-III clinical trials in North America. Prior to clinical testing, tramiprosate was shown to inhibit Aβ42-induced cytotoxicity in primary neuronal cultures, but by most standards was a weak inhibitor, requiring high molar ratios to Aβ (20:1) for efficacy (Gervais et al. 2006, 2001). It reduced both soluble and insoluble loads of Aβ in transgenic (TgCRND8) mice, but again required very high concentrations (500 mg/kg). Although it is able to cross the blood–brain barrier, it decreases plasma concentrations of Aβ, suggesting that it reduces whole-body pools of Aβ that are in dynamic equilibrium with one another—perhaps by allowing peptide to partition from brain to plasma. Whatever its mechanism, a phase-II trial showed a decrease in CSF Aβ concentrations, but no improvement in cognitive tests in patients, though early data suggested some degree of improvement in individuals with early stages of AD. The results of the phase-II trial were inconclusive at best (Aisen et al. 2006, 2007), but testing proceeded to a phase-III trial involving 1,052 patients in 67 sites. The results were inconclusive because of problems with statistical models. In addition, it is difficult to know which endpoints of this, or any, trial are most meaningful. Commonly used endpoints include psychometric testing (a domain of much controversy) and measurements of hippocampal volume in MRI scans (Wong 2007; Gauthier et al. 2007); these measurements were used in the phase-III trial of tramiprosate, and failed to demonstrate any benefit of the drug. This was obviously a large and costly effort, and worth performing the autopsy upon. First, one can take the view that this trial did not prove that tramiprosate *does not* work, and certainly, one should not be discouraged from looking for other small molecules to treat AD. The trial, however, illustrates the confounding factors involved in testing any medicine, such as the use of other medications, indeterminate definitions of what a "control" group really is, and vagaries in the use of any endpoint, including the ones used in this study, as shown by the fact that *the control* group showed cognitive improvement over the test period (Wong 2007; Saumier et al. 2009). Underlying all of these problems is a lack of fundamental understanding of the mechanism of how the agent is supposed to work. What exactly does tramiprosate *do* to Aβ and to neurons affected by Aβ? This is not at all clear. It has been proposed (Stephenson and Weaver 2006) that the compound is a "glycosaminoglycan (GAG) mimetic" that could interfere with aggregation of Aβ that is stimulated by GAGs, perhaps by binding the slightly cationic sequence in Aβ, residues 13–16 (HHQK). However, does tramiprosate compete with GAGs, or mimic them? What proportion of Aβ aggregation is attributable to the effects of GAGs? How efficient or specific is tramiprosate? To the extent that it *does* inhibit Aβ aggregation (which is questionable), which part of the pathway does it affect? How could the GAG-mimetic effects lead to clearance of Aβ from brain and plasma, and do these effects account for effects *in vitro* on Aβ cytotoxicity? The pressing clinical needs may have led to a clinical trial that was on a less-than-solid ground from the viewpoint of basic biochemical data.

Another small-molecule drug now in clinical trials is PBT2, a molecule that is reportedly superior to, and replaces, its predecessor PBT1 (clioquinol, Ritchie et al. 2003). Both molecules are derivatives of 8-hydroxyquinol. Since clioquinol is an antibiotic, its bioavailability properties are known. It also acts as a divalent metal ionophore, and its use has been proposed in treatment of cancer and immunological disorders such as graft-*versus*-host disease (Ding and Lind 2009; Chen and Dou 2008). The structure of PBT2 is not available, but PBT1 contained iodine, and was problematic because of di-iodo impurities. PBT2, which does not contain iodine, has moderate affinity for divalent cations, including Cu^{2+}, and for this reason, has been reported to counteract metal-ion-induced Aβ aggregation, and the production of reduced oxygen species that may result from this type of Aβ aggregation (Ding and Lind 2009). It has also been reported to decrease Aβ synthesis in tissue culture, though the mechanism is not entirely clear, but may involve induction of matrix metalloproteinases that degrade Aβ. The production of these proteases may be due to signaling by Cu^{2+} or other divalent metal ions.

In this case, there is a more solid biochemical basis for proposing this class of compounds as potential therapeutic agents, the pros and cons of this particular one aside. Aβ binds divalent cations with high affinity (Atwood et al. 2000b). Thus, the divalent metal-binding properties of Aβ are well-documented, and this binding accelerates Aβ aggregation (Bush et al. 1994a, b; Atwood et al. 1998). Some of the bound divalent metal ions, notably Cu(II), induce the formation of reduced oxygen species such as H_2O_2 (Huang et al. 1999a; Atwood et al. 2000a), and this leads to tyrosine cross-linking and dityrosine formation (Atwood et al. 2004). In tissue-culture systems, the cytotoxicity of Aβ peptides is directly related to the amount of H_2O_2 produced in the presence of Aβ and Cu(II) (Huang et al. 1999b; Sayre et al. 1997). In the APP2576 transgenic mouse model of AD, metal ions (Zn and Fe) are enriched in areas of Aβ deposition (Smith et al. 1997). This was a sufficient basis for testing clioquinol in transgenic mice: in APP2576 transgenic mice treated orally for 9 weeks with clioquinol, there was a marked decrease in soluble Aβ in the CSF and deposited Aβ in the brain matter, while other CNS markers (APP, synaptophysin, and GFAP) were unaffected (Cherny et al. 2001). Finally, PBT2 (and clioquinol) rescued the impairment of long-term potentiation that is induced in hippocampal slices by Aβ42, and led to improvement of cognitive function in transgenic mice [Tg2576 (female Bl6/SJL), (Adlard et al. 2008)].

Thus, the basic biochemical data on metal ions and their effects on Aβ biology justified moving into clinical trials. How good were the results of these trials? The results of phase-IIa clinical trials in patients with mild AD, though reports are still very preliminary, were sufficiently encouraging (Ritchie et al. 2003) to allow continuation of the trials (Lannfelt et al. 2008). After 12 weeks of treatment, the drug appeared safe and was tolerated well, reduced CSF concentrations of Aβ peptides without altering plasma concentrations, and showed some improvement of cognitive function, as assessed by the Trail-Making Test Part B and the Category-Fluency Test (parts of the Neuropsychological Test Battery).

Another small molecule currently in clinical trials is AZD-103 (scyllo-cyclohexanehexol, scyllo-inositol), which is well-tolerated and leads to cognitive improvement in transgenic (TgCRND8) mice, as measured by the Morris water-maze test (McLaurin et al. 2000, 2006; Sun et al. 2008). It reduces Aβ deposition and astrocytic and microglial activation, two markers of neuroinflammation. The investigators also showed a moderate loss of high-molecular-weight soluble oligomers and a more substantial loss of monomers and trimers in the CSF, which they

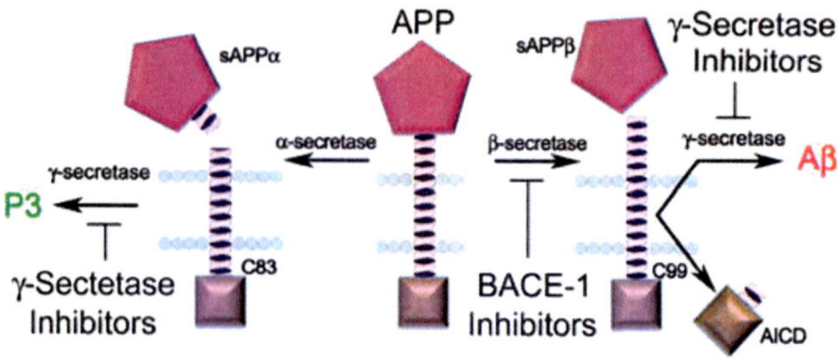

Fig. 14.12 Cleavage of APP. APP is cleaved first by either the α-secretase or β-secretase (BACE-1), after which it becomes a substrate for γ-secretase, generating either the non-pathogenic p3 peptide (which aggregates only at very high concentrations) or the pathogenic Aβ peptides. Therapeutic agents could target β-secretase, since there is only one enzyme in humans with this activity, but finding appropriate BACE-1 inhibitors has proven difficult, thus far. Several γ-secretase inhibitors have been developed and are under investigation. In theory, it might be possible to develop therapy based on up-regulating α-secretase, but as discussed in the text, this might be difficult for practical reasons. These include the fact that several enzymes have α-secretase activity, and most cleavage of APP is by α-secretase to start, before any potential therapy could be given

interpreted as being due to inhibition of high-molecular-weight oligomers (it could also be due to disaggregation of insoluble, deposited Aβ). For unknown reasons, it did not lead to a reduction in microglial activation in a mouse model of frontotemporal dementia, the transgenic Tau P301L mouse. There is some degree of stereospecificity for this compound, as epi-cyclohexanehexol is less effective. One paper reported that AZD-103 neutralizes the synaptotoxic activity of isolated small Aβ oligomers, but did not alter the stability or size distribution of these species (Townsend et al. 2006). There is little evidence from any of these studies that this reagent binds to Aβ or affects its physical properties (e.g., it has no effect on the size distribution of oligomers), or, for that matter, that it interacts physically with Aβ at all. Despite a lack of clear mechanistic data, a phase-I clinical trial took place, and indicated that the drug is tolerated well; it is not known yet if it is effective.

γ-Secretase Inhibitors

This topic is discussed briefly because it is off the central focus of this article on molecules that inhibit protein aggregation *per se*. Nevertheless, as discussed below, the failure of two large phase-III clinical trials of γ-secretase inhibitors may have important implications for the future of AD research. As discussed previously, Aβ peptides arise from the sequential cleavage of the amyloid precursor protein, first by β-secretase, followed by γ-secretase (Fig. 14.12). Normally, the cleavage of APP by α-secretase, yielding (relatively?) non-pathogenic peptides such as Aβ17–40 and Aβ17–42 (both also called p3 peptides), is vastly greater than cleavage by β-secretase, but in patients with AD, it is believed that there is a relative increase in the ratio of APP molecules cleaved by β-secretase to those cleaved by α-secretase, resulting in increased production of Aβ peptides. The action of γ-secretase is necessary to

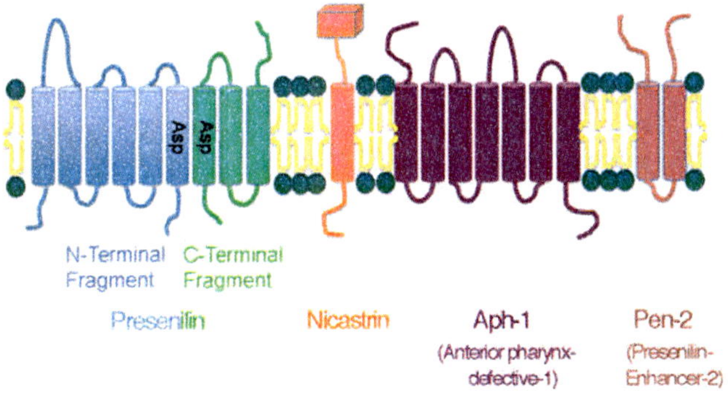

Fig. 14.13 Schematic representation of γ-secretase. The complex is an aspartyl protease, with the two critical Asp residues (indicated in the figure) located within a putative transmembrane helix. It also acts on substrates within the membrane, such as APP and Notch. It is a member of a family of proteases calls I-CLiPs (intramembranous cleaving proteases). The number and orientation of transmembrane helices are indicated in the figure. In addition to generating the Aβ and p3 peptides, the enzyme (probably with other factors) controls the length of Aβ peptides generated, e.g., the ratio of Aβ40:Aβ42. There are at least four subunits: presenilin (mutations of which give rise to forms of familial AD), nicastrin, Aph-1 (anterior pharynx-defective-1), and Pen-2 (presenilin enhancer-2)

release Aβ peptides. For these reasons, inhibition of β- or γ-secretase, and possibly activation of α-secretase represent potential ways of decreasing Aβ production, and hence, aggregation and cytotoxicity.

Although the least is known about γ-secretase activity of these three enzymatic activities, this was the first one to have been targeted for inhibition. γ-Secretase is a complex intramembranous enzyme, containing at least four protein components: presenilin, nicastrin, Aph-1A, and presenilin enhancer 2 (for review, see DeStrooper 2003, Fig. 14.13). Detailed structural information on this enzyme is still not available. In addition, the enzyme has targets other than Aβ (e.g., Notch), and it is not known what the effect will be of inhibiting all processing of targets by γ-secretase. These points may have slowed the development of inhibitors of this enzymatic activity. Nevertheless, after it was shown in the early 1990s that Aβ could be reconstituted in cells in tissue culture (Haass et al. 1992), non-toxic inhibitors of γ-secretase were developed that inhibited production of both p3 and Aβ peptides derived from APP, as would be predicted. One of these compounds, DAPT (their compound 8, *N*-[*N*-(3,5-difluorophenacetyl)-$_{L}$-alanyl]-*S*-phenylglycine t-butyl ester), was orally administered to transgenic mice that overproduce Aβ. The transgene is an alternatively spliced APP minigene, driven by the platelet-derived growth factor (PDGF-B chain) promoter and encoding (V717F)-mutated hAPP770, hAPP751, and hAPP695. These mice develop age- and brain-region-dependent Aβ deposition, dystrophic neurites, loss of presynaptic terminals, astrocytosis, and microgliosis (Games et al. 1995). A single oral dose of DAPT reduced total soluble brain cortical Aβ concentration. There was a corresponding stabilization of 10- and 12-kDa *C*-terminal fragments of APP, indicating interference with γ-secretase processing (Dovey et al. 2001).

The best studied of the γ-secretase inhibitors, Semagacestat (LY450139) had been in phase-III studies, based on previous, phase-II studies showing that the drug was safe and well-tolerated, and led to lowering of both plasma and CSF concentrations of Aβ (Fleisher et al. 2008; Bateman et al. 2009). The promise also stemmed from the fact that this was the first γ-secretase inhibitor to cross the blood–brain barrier and reduce plaque burden in the brain. Unfortunately, after much fanfare and promise, Semagacestat failed in two large phase-III clinical studies in patients with mild-to-moderate AD, because patients receiving the drug developed greater decline in cognitive function than patients receiving placebo (Samson 2010; Schor 2011; Imbimbo and Giardina 2011). The announcement by Eli Lilly in August 2010 of the premature halt to these trials came after 10 years of investigation, and clinical trials involving more than 2,600 patients, treated in 31 countries over 21 months. Another γ-secretase inhibitor, tarenflurbil (R flurbiprofen), also recently failed in two large phase-III clinical trials—this failure attributed low drug potency and inability of the drug to enter the brain efficiently (Imbimbo and Giardina 2011). Other compounds have entered phase-II clinical trials (Fleisher et al. 2008; Bateman et al. 2009, Table 14.3), but it is not clear what the fate of these trials will be with the failure of phase-III trials.

To say that this is disappointing is not merely an understatement; it is also inaccurate. This is a disaster, because such failures can spread confusion and work to the detriment and discouragement of future scientific efforts. The important issue, now, is to understand the reasons for the failures.

Several recent editorials stress the idea that γ-secretase cleaves many substrates other than β-APP, and that functional loss of these other proteins may be the root cause of the failure of γ-secretase inhibitors. Among the other substrates of γ-secretase, attention has focused mainly on Notch, which is important in embryological development, cell replication, cell lineage differentiation, angiogenesis, and perhaps most importantly, control of neurite outgrowth. In addition to Notch, cadherins are cleaved by γ-secretase, and γ-secretase is also involved in signaling through the p75NTR, CD46, CD44, and ErbB4 receptors; furthermore, these substrates may only scratch the surface, as ~60 other γ-secretase substrates have been identified (Wakabayashi and DeStrooper 2008; McCarthy et al. 2009; DeStrooper and Annaert 2010). Concerns about potential effects of γ-secretase inhibition on cell replication in the immune system and the gastrointestinal tract had been expressed long before phase-III clinical trials (Wong et al. 2004; Milano et al. 2004).

As has been known for a long time, presenilin mutations account for ~90% of the mutations associated with familial AD, and many or most of these increase production of Aβ peptides, and additionally increase the ratio of Aβ42:Aβ40, which favors amyloid production. In addition, some of the dominantly inherited missense mutations in β-APP, such as the Swedish mutation (Haass et al. 1995), increase the production of Aβ peptides. Mouse models tend to confirm the idea that overproduction of Aβ can lead to neurotoxicity, but the degree of neurotoxicity seems modest when judged by the standard of human disease: massive amyloid accumulation in mice leads to a demonstrable but only moderate memory loss, and rather modest histological evidence of neurodegeneration, e.g., one study of a transgenic mouse model using the Swedish β-APP model (Irizarry et al. 1997). In contrast, conditional

Table 14.3 γ-Secretase inhibitors in clinical development for the treatment of Alzheimer's disease (AD)

Compound	Mechanism of action	Side-effects	Development status	Company
Semagacestat (LY-450139)	Decreases newly synthesized Aβ in CSF of AD patients	No significant effects on brain plaque burden in transgenic mice Lack of data on behavioral effects in animal models of AD Gastrointestinal and skin side-effects in AD patients	Phase III—prematurely discontinued	Eli-Lilly
MK-0752	Decreases Aβ40 levels in CSF of healthy volunteers.	Inhibits Notch cleavage. Significant gastrointestinal toxicity in humans	Abandoned	Merck
E2012	Notch sparing	Lenticular opacity in rats	Phase I	Eisai
BMS-708163	Notch sparing; decreases Aβ levels in CSF of healthy volunteers.	Lack of data on brain plaque deposition in transgenic mice; lack of data on behavioral effects in animal models of AD	Phase II	Bristol Myers Squibb
PF-3084014	Notch sparing; good brain penetration; long-lasting effects on Aβ levels in animals; no rebound effect on plasma Aβ in animals	Lack of data on brain plaque deposition in transgenic mice; lack of data on behavioral effects in animal models of AD	Abandoned	Pfizer
GSI-953 (Begacestat)	Improves memory in a transgenic mouse model of AD.	Does not decrease Aβ40 levels in CSF of AD patients	Phase II	Wyeth

Source: Panza et al. (2010)

knockout of presenilins in the adult cerebral cortex lead to severe and progressive neurodegeneration, with some features (e.g., tau hyperphosphorylation) similar to that observed in human AD (Saura et al. 2004). Although this model provides clear evidence of memory loss, impairment of synaptic plasticity, and neurodegeneration, many different experimental manipulations could injure neurons, and thus, it remains unclear how completely this model replicates human disease. Obviously, one difference between this model and human disease is whether Aβ accumulates in the brain, reflecting the question of whether one considers Aβ accumulation to be an essential feature of AD or merely an epiphenomenon. Furthermore, a single presenilin mutation can lead to AD, with Aβ overproduction, even though this leaves three presenilin alleles intact, which begs an explanation for how the loss of one of four alleles could lead to a loss-of-function disease (One possibility, of course, is that the mutant gene could act in a dominant-negative manner). Nevertheless, despite possible shortcomings of the conditional presenilin-1-knockout model, these studies cast doubt upon γ-secretase inhibition as a therapeutic goal.

An additional problem with γ-secretase inhibitors is that their effect is strongly dose-dependent: at low doses, they may cause little change in Aβ40, but an *increase in* Aβ42 production. At higher doses, they decrease Aβ40 production, and at still higher doses, they also increase Aβ42 production. As Shen and Kelleher (Shen and Kelleher 2007; Kelleher and Shen 2010) pointed out, the variation in effect of γ-secretase inhibitors seems to reflect the range of effects of presenilin mutations: some mutations increase Aβ production globally, some increase mainly the longer, more amyloidogenic forms of Aβ selectively, while still others decrease Aβ production globally. Since Semagacestat did lower plasma and CSF concentrations of Aβ in patients, presumably the dosage was sufficiently high to achieve the desired goal of lowering Aβ production. Nevertheless, it is possible that in some patients, or perhaps in most patients at some times, the dose was in the range that would lead to increased Aβ production.

One further point applies not only to γ-secretase, but also to any other attempt to reduce production of Aβ, e.g., through inhibition of β-secretase (see below). The physiological function of Aβ remains unknown, and yet a function probably exists in view of the physiological regulation of Aβ production (Cirrito and Holtzman 2003; Cirrito et al. 2003; Cirrito et al. 2005; Kang et al. 2009; Bero et al. 2011). Simply suppressing Aβ production globally may not be a complex enough approach to the problem, and variations in Aβ production throughout time and anatomy of the brain may need to be taken into account. As Cirrito and Holtzman put it (2003), in thinking about Aβ production, "the devil may be in the details."

β-Secretase Inhibitors

In contrast to γ-secretase, and despite its localization to membranes, in the form of a single-pass transmembrane domain, BACE-1, the β-secretase, is a fairly ordinary aspartyl protease (Hussain et al. 1999; Vassar et al. 1999; Yan et al. 1999; Lin et al. 2000c; for review, see Cole and Vassar 2008; Citron 2004). The active site of the enzyme is on the luminal side of the membrane. β-Secretase activity is expressed ubiquitously, but the highest levels are found in the brain. Also in contrast to γ-secretase (the presenilin 1 and 2 proteins), no mutations of β-secretase have been reported to cause AD, nor any other disease. Patients with sporadic AD have higher-than-normal β-secretase, but it is not clear whether this is causative in disease or a reaction to another injury that causes the disease. BACE-1-knockout mice do not produce Aβ peptide, and therefore, there is probably no other biologically significant enzyme with β-secretase activity. The knockout mice appear to be fairly healthy and are fertile (Citron 2004). BACE-1 also cleaves a second substrate, type-III neuregulin 1, which is essential for peripheral nerve myelination in newborn mice (Willem et al. 2006). For this reason, BACE-1-knockout mice have hypomyelinated peripheral nerves, and perhaps delayed remyelination after crush injuries (Willem et al. 2006; Hu et al. 2006). There is also some contradictory evidence that BACE-1-knockout mice may have more or less anxiety-like behavior. Other than this, however, there is no evidence that BACE-1 inhibition affects adult mice.

Of the various aspartyl proteases, BACE-1 is in the structural family containing pepsin. BACE-1 has a pH optimum of ~4 and for this reason may be most active in endosomal compartments. Although BACE-1 has only modest sequence homology with pepsin, the three-dimensional structures of BACE-1 and pepsin are rather similar. Enzymes of the class to which BACE-1 belongs, including BACE-1 itself, are bilobed, each lobe containing one of the catalytic pair of Asp residues.

Because BACE-1 is a potentially attractive therapeutic target, extensive screens for small-molecule inhibitors were conducted both in silico and *in vitro*, but these yielded only low-affinity compounds, suggesting that this is a difficult enzyme to target (Villaverde et al. 2007; Hunt and Turner 2009). In this respect, it shares this unfortunate property with other aspartyl proteases, which have an extended substrate-binding site (Hong et al. 2000; Vassar 2001). Despite its structural similarity to pepsin, it also is only weakly inhibited by the transition state "statin" analogues, such as pepstatin, containing a hydroxymethylene isostere of the peptide bond. Nevertheless, one promising inhibitor was developed: OM99-2, with which the enzyme's protease ectodomain was crystallized (Hong et al. 2000). OM99-2 and its descendants (Hong et al. 2002) contain a central hydroxymethylene group in an octapeptide sequence based the β-secretase cleavage site of the Swedish mutation of β-APP. Refinements include substitution of additional hydrophobic residues to improve passage through the blood–brain barrier. The search for BACE-1 inhibitors continues, and this enzyme remains a potential target for therapeutic approaches (Vassar et al. 2009; Marks and Berg 2010; Hunt and Turner 2009; Hills and Vacca 2007).

Acetylcholinesterase Inhibitors

Proteomic and other studies have identified acetylcholinesterase (AChE) as a prominent component of the dense amyloid core of the neuritic plaque (Inestrosa et al. 1996a). Most CNS AChE exists in a tetrameric form bound to neuronal membranes (Inestrosa et al. 1982), but when it is secreted it can bind to extracellular structures such as the synaptic basement membrane and the neuromuscular junction (Geula and Mesulam 1989). The enzymatic and histochemical properties of AChE found in plaques differs from those observed in other settings such as the neurons (Alvarez et al. 1998)—perhaps because of the enzyme's association with Aβ (Inestrosa et al. 1996b).

AChE inhibitors have long been used to improve cognitive symptoms in patients with AD (Pepeu and Giovannini 2009; Villarroya et al. 2007), but it was surprising to learn that this treatment was based on more than its cholinergic effects. AChE accelerates Aβ fibril formation, forming a high-molecular-weight complex with Aβ (Reyes et al. 1997), and some, but not all anti-AChE drugs also block amyloid formation. Inhibitors that bind to the active site of AChE, such as edrophonium or tacrine, were *ineffective*, while drugs that bind to a peripheral anionic site on the enzyme, such as propidium and fasciculin, or monoclonal antibodies that were able to bind to the same part of the enzyme, were able to block amyloid formation (Reyes et al. 1997; Muñoz et al. 1999; De Ferrari et al. 2001; Bartolini et al. 2003). Molecular Dynamic simulations, and studies using peptides from the AChE sequence, identified a membrane-interacting region containing a conserved tryptophan was the likely site that catalyzed Aβ aggregation (Shin et al. 1996; Bartolini et al. 2003). A recombinant form of human AChE has been shown to induce Aβ aggregation by a direct mechanism, i.e., by forming stable AChE–Aβ complexes that catalyze Aβ aggregation (Inestrosa et al. 1996a; Inestrosa et al. 1996b; Muñoz-Ruiz et al. 2005).

This site on AChE is "druggable," and has led to the search for either old AChE inhibitors that also bind to the peripheral anionic site, or new drugs that do so without blocking the active site of the enzyme. As shown in Fig. 14.14, many of the currently commercialized drugs to treat AD are directed against AChE. Several tacrine derivatives, and variants of rivastigmine and xanthostigmine were found to block AChE catalysis of Aβ aggregation (Bolognesi et al. 2004; Belluti et al. 2005; Piazzi et al. 2003), and the donepezil derivative, AP2238 (Ezoulin et al. 2006). Concerns about hepatotoxicity, due to oxidative damage from metabolism of tacrine (Osseni et al. 1999; Dogterom et al. 1988), prompted the search for newer compounds that would have the same or better effects on Aβ aggregation, without toxicity. Among the newer agents are tacrine–antioxidant hybrid molecules, such as lipocrine, which contains a lipoic acid moiety (Rosini et al. 2005), and tacrine hybrids that contain NO donors (Fang et al. 2008). Another source of concern was inhibition of butyrylcholinesterase (BuChE) activity outside of the CNS by AChE inhibitors. The *bis*(7)-tacrine compounds contain two copies of tacrine linked by methylenic linkers of various lengths; they have ~150-fold increased potency compared to tacrine, and selectivity towards AChE rather than BuChE (Muñoz-Torrero 2008; Pang et al. 1996). BuChE normally plays a secondary role in acetylcholine cleavage in the CNS, hydrolyzing only ~20% of this compound, while AChE carries out the remainder. As AD progresses, however, the balance shifts, and thus it may be important to target BuChE as well, which many of these drugs do. The *bis*(7)-tacrines appear to act by multiple mechanisms: they block Aβ aggregation, as stated, but also inhibit BACE-1, and in N2A cells expressing the Swedish APP mutation,

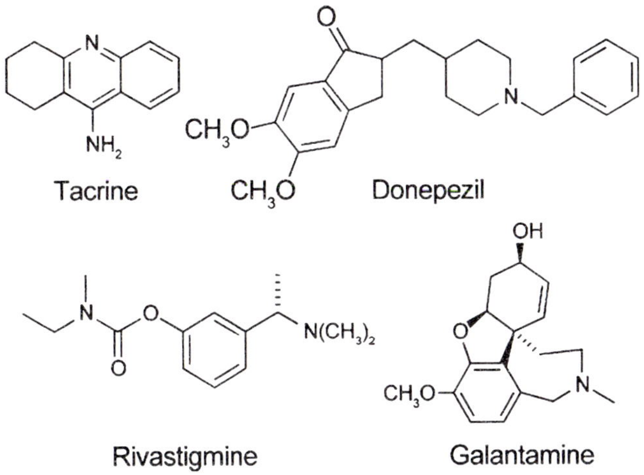

Fig. 14.14 Four commercially developed anticholinesterases currently in use for treating AD. These compounds have neuroprotective effects against glutamate- and NMDA-induced toxicity, and have been observed more recently to decrease Aβ aggregation

slightly activate α-secretase (Bolognesi et al. 2007). A number of newer compounds have been synthesized that have been reported to have similar effects on both Aβ aggregation and production by BACE-1 (Fu et al. 2008). Development of tacrine derivatives also continues (Camps et al. 2009; Rosini et al. 2005; Marco-Contelles et al. 2009; Fernández-Bachiller et al. 2010; García-Palomero et al. 2008). It is too early to tell which, if any, of these compounds will prove ultimately to affect the clinical course of AD.

And a Cast of Thousands!

Aromatic residues are critical for practically all protein–protein interactions, and it is therefore not surprising that diverse aromatic compounds have been found to bind to Aβ and block aggregation. For example, benzofuran is a moderately effective inhibitor of Aβ fibril formation (Howlett et al. 1999). The concept of "bioisosterism" was employed to direct synthesis of several new compounds, but the elegant chemistry in this study did not produce strong inhibitors (Lee and Jeon 2005, Fig. 14.15). Nevertheless, the use of an indole scaffold, including indanes and tryptophan itself in addition to indole, has led to the synthesis of other compounds with greater efficacy as Aβ aggregation inhibitors *in vitro* (Catto et al. 2010). Thus far, the best IC_{50} achieved with these compounds is in the low μM range, which is encouraging but not yet sufficient for therapeutic testing.

In the previous study, their most effective inhibitor (their 6c, R=4-$CH(CH_3)_2$, R′=H) was a hybrid molecule linking an indole with a moiety containing a diazo group and aromatic ring out of the central portion of the amyloid-staining dye,

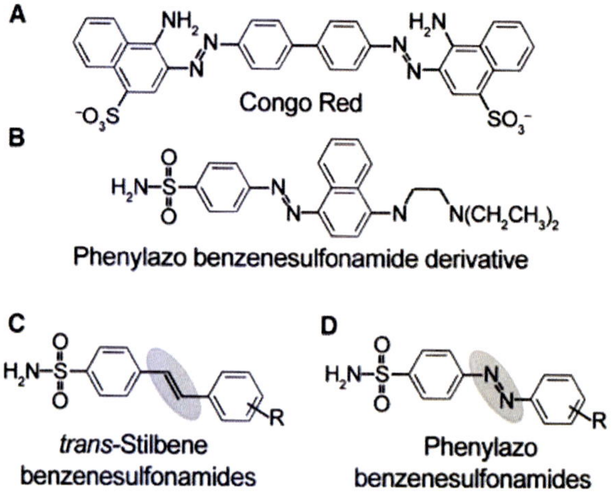

Fig. 14.15 A comparison of the amyloid-binding dye, *Congo red* (**A**), and a lead compound used for derivatization to discover aggregation inhibitors (Lin et al. 2004). The lead compounds in this study (**B**) use bioisosterism by substituting the simpler phenylazo group for the naphthylazo moiety. Another type of bioisosterism is shown in (**C**) and (**D**) the *trans*-stilbene is isosteric with the phenylazo group

Congo red. Congo red itself has served as a scaffold for derivatization to synthesize potential inhibitors of Aβ aggregation. Congo red is a symmetrical phenylazo benzenesulfonamide compound. In one study (Török et al. 2006), two forms of bioisosterism were examined: substitution of the C–C double bond in the *trans*-stilbene benzenesulfonamides for the N–N double bond in phenylazo benzenesulfonamide derivatives, and substitution of a phenylazo group for the naphthylazo group. While the *trans*-stilbene-containing compounds were inactive, simple phenylazo-like structures had possessed moderate activity. In a similar study, a series of *N*-phenyl anthranilic acid analogs were synthesized and found to have moderate activity as Aβ aggregation inhibitors (Lin et al. 2004, Fig. 14.16).

Another strategy for derivatizing the indole scaffold has been the use of organofluorines (Simons et al. 2009, Fig. 14.17). For reasons that are only partially understood, organofluorine compounds often promote the formation of α-helical structure of peptides, including Aβ. This is true not only for the familiar solvents 2,2,2-trifluorethanol and 1,1,1,3,3,3-hexafluoro-2-propanol (Vieira et al. 2003), but also solid organofluorine such poly(tetrafluoroethylene) surfaces (Giacomelli and Norde 2003). Many of these compounds had some degree of inhibitory activity, and a few of them (e.g., their compound 12, $R_1=R_2=H$, $R_3=F$) completely inhibited Aβ aggregation at high stoichiometric ratios (e.g., 10:1 = inhibitor to Aβ), and had IC_{50} at substoichiometric ratios to Aβ below unity.

A similar strategy used dopamine-related compounds as scaffolds for derivatization (Fig. 14.18). One study compared naturally occurring catechol-containing compounds (dopamine, pyrogallol, gallic acid, caffeic acid, and quercetin), and two nitrocatechols, entacapone and tolcapone. Some of these compounds are used as

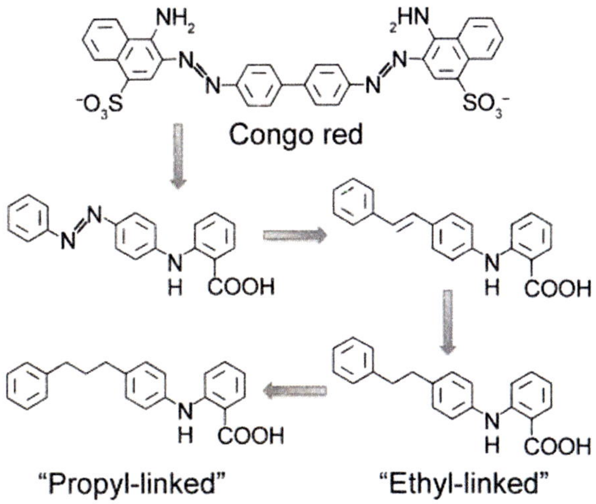

Fig. 14.16 "Evolution" of *N*-phenyl anthranilic acid analogs as amyloid aggregation inhibitors. Figure is adapted from Simons et al. (2009). Starting with *Congo red*, the structure was refined from a diazo-linked anthranilic acid derivative to an ethyl-linked moiety. The addition of a single methylene unit in the linker generated compounds with increased potency *in vitro*

Fig. 14.17 Core structure of a proposed class of organofluorine Aβ aggregation inhibitors: trifluorohydroxylindolyl-propionic acid esters, which inhibit Aβ fibrillization at substoichiometric inhibitor-to-Aβ ratios (e.g., 1:10, mol:mol) (Figure is adapted from Török et al. 2006)

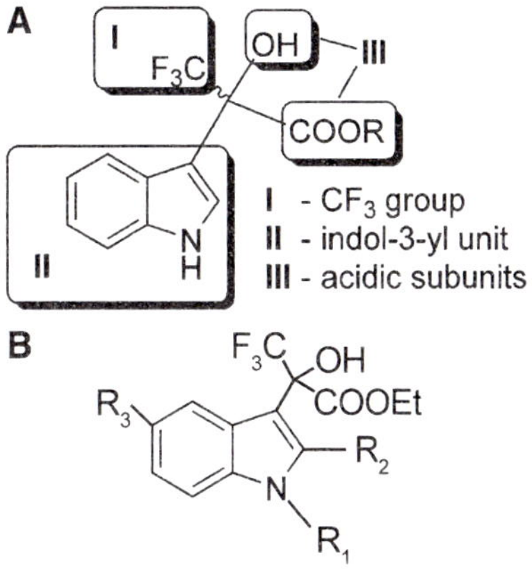

drugs against Parkinsonism because they are dopamine agonists, but they were found to have the added benefit of inhibiting not only α-synuclein, but also Aβ oligomerization and fibrillogenesis, and to protect against extracellular toxicity induced by the aggregation of both proteins (Di Giovanni et al. 2010).

The pyrones and their derivatives are yet another set of aromatic compounds with putative ability to block Aβ aggregation and cytotoxicity (Maezawa et al. 2006; Hong et al. 2007). One compound, CP2, inhibits Aβ aggregation, and blocks binding of Aβ to an immobilized monoclonal antibody, A11 (Maezawa et al. 2006), a commercially available rabbit polyclonal antibody with specificity for oligomers (Kayed et al. 2003). Another study utilized microwave synthetic methods to

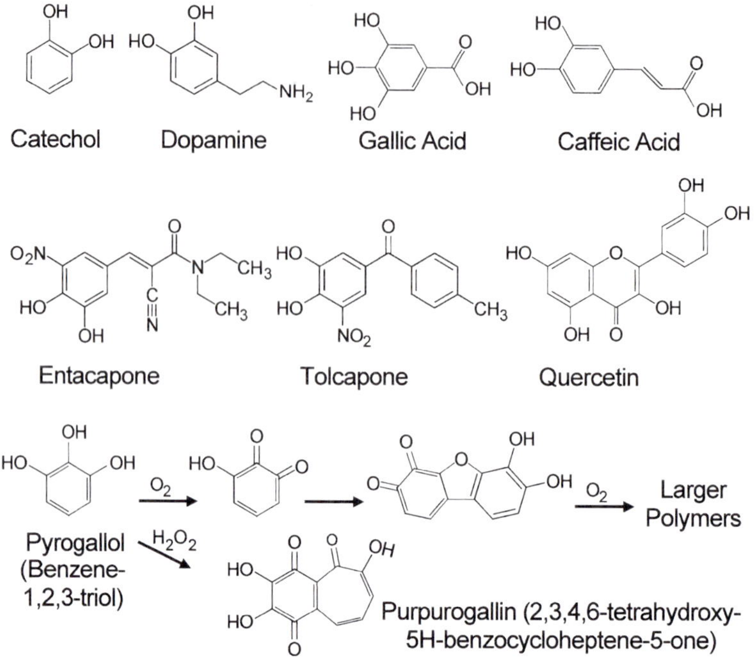

Fig. 14.18 Catechols tested for their ability to inhibit aggregation of Aβ and α-synuclein (Figure is from Di Giovanni et al. 2010). These compounds include dopamine and weak-to-moderate dopamine agonists, such as pyrogallol and gallic acid. Chemically, these compounds are catechols, which are also chemical homologues of the polyphenols, discussed in a subsequent section. The catechol homologues also include two nitrocatechols, entacapone and tolcapone, which are catechol-*O*-methyl transferase inhibitors, which are used to treat Parkinson's disease because they prevent degradation of L-DOPA into 3-methoxy-4-hydroxy-L-phenylalanine (which passes through the blood–brain barrier poorly). Querecetin in a flavonoid (discussed in a subsequent section). Caffeic acid is a hydroxycinnamic acid, for which a variety of health benefits have been claimed, though not strongly substantiated. Two other catechol derivatives tested in the studies described in the text are gallic acid and pyrogallol, which arises from decarboxylation of gallic acid. Pyrogallol is prone to oxidation and forms large polymers and compounds such as purpurogallin, both of which may be carcinogenic

decarboxylate bicyclic 2-pyridone scaffolds for derivatization, and some of these were moderately effective inhibitors of Aβ aggregation (Åberg et al. 2005).

There are literally too many compounds to count, and it is difficult to hone in on those that will be most beneficial. Furthermore, most of the above inhibitors have only moderate efficacy, and require high concentrations that could be toxic. The use of hybrid molecules is potentially a powerful approach to solving both of these problems. Bose et al. (2005) have described a novel approach to this problem (Fig. 14.19). They synthesized a bifunctional molecule consisting of a synthetic ligand for the FK506-binding protein (SLF) and Congo red, with nanomolar IC_{50} for its effects on Aβ aggregation and cytotoxicity. Congo red has weak inhibitory activity, but the hybrid molecule, in effect, "borrows" the surface and steric bulk of

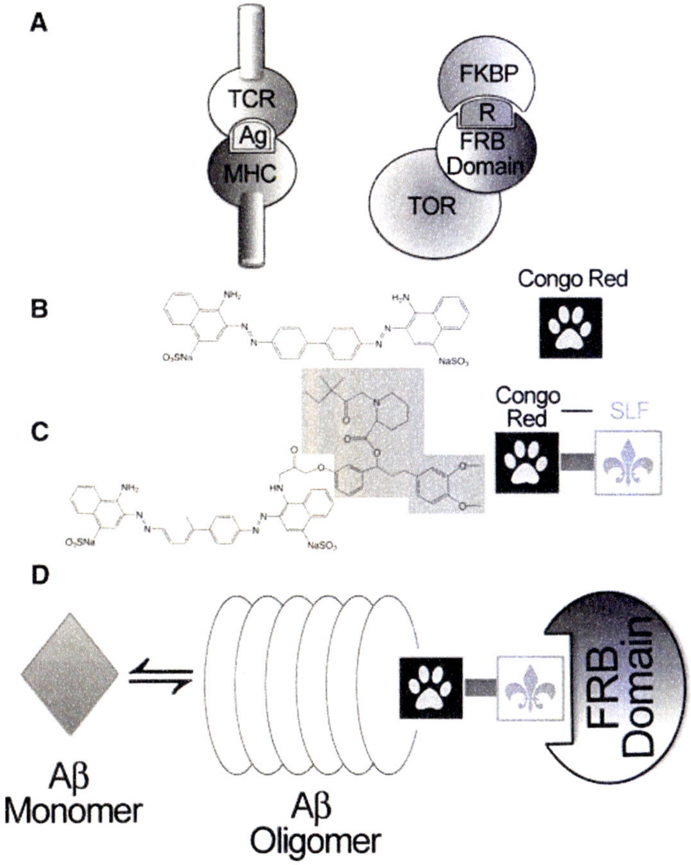

Fig. 14.19 Bifunctional molecules as a strategy for inhibiting Aβ aggregation. Ternary interactions are widely used in nature: a small molecule binds to one protein first, and then presents a composite binding surface to a second protein. (**A**) The examples from nature, used in Bose et al. (2005), include ternary complexes of: (1) MHC (major histocompatibility complex molecule), TCR (T-cell receptor protein), and Ag (peptide antigen); and (2) TOR (mammalian target of rapamycin), FRB domain [the rapamycin-binding domain of FKBP (FK506-binding protein)], and R (the macrolide, rapamycin). In this strategy, a bifunctional molecule, composed of (**B**) Congo red (represented by a Fleur-de-Lys) tethered to a synthetic ligand for FKBP, called SLF ((**C**) represented by a *paw print*), binds to both Aβ aggregates and the FRB (**D**). In (**D**), an Aβ monomer is represented by a *diamond*, and an Aβ aggregate (type unspecified) is represented by *ovals*; in this depiction, the aggregates bind to *Congo red.* The bound FRB provides "steric bulk" to inhibit further Aβ aggregation (Figure is adapted from Bose et al. 2005)

a cellular chaperone. For biochemists, understanding a "three-bodied problem," such as the complex of MHC molecules, peptide antigens, and T-cell-receptor molecules, poses great difficulties. Nature, however, has utilized this strategy often. In the case of the hybrid molecules discussed by Bose et al. (2005), the SLF moiety recruits FK506, which increases the effective molecular mass of the hybrid compound.

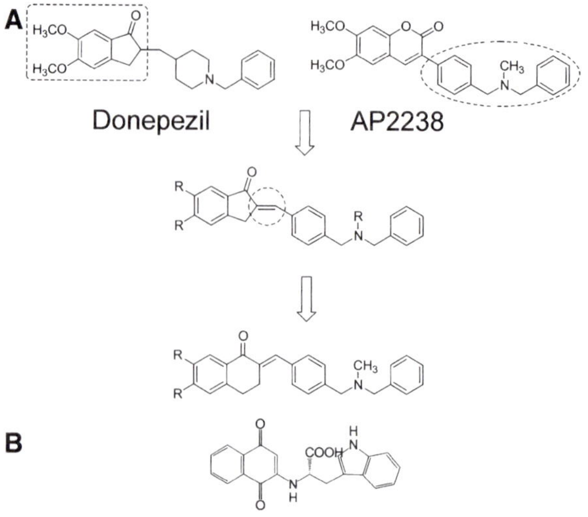

Fig. 14.20 Hybrid molecules combining indole compounds and inhibitors of AChE. (**A**) Donepezil is an AChE inhibitor that has been used to treat AD; AP2238 is also an AChE inhibitor of potency comparable to that of donepezil, which also has even better potency as an inhibitor of Aβ aggregation. In the hybrid molecule (Figure from Rizzo et al. 2010), the indanone core of donepezil is linked to the phenyl-*N*-methylbenzylamino moiety of AP2238, which maintains a key interaction of the drug with Tyr124 of the enzyme. (**B**) The molecule shown in the figure (From Scherzer-Attali et al. 2010) is 1,4-naphthoquinon-2-yl-$_{L}$-tryptophan (NQTrp), a hybrid of tryptophan and a quinone. Each of the components binds with modest affinity to Aβ, but in combination strongly inhibits Aβ aggregation and cytotoxicity in cell culture

Congo red then binds to Aβ, and by steric effects, blocks further aggregation. The hybrid blocks fibril formation, which, as discussed earlier, could be either a positive or negative effect. In this case, the hybrid compound also decreased Aβ-induced cytotoxicity, though several mechanisms appear possible. The compound could bind even to monomeric Aβ and block oligomer formation; it could bind to small oligomers, and prevent their toxic effects, e.g., by preventing their interaction with cell membranes or other targets; or it could block the enlargement of small aggregates into larger ones with cytotoxic effects.

Two other recent hybrid molecules are the following. The first combines a derivative of donepezil and phenyl-N-methylbenzylamino moiety from AP2238 (Fig. 14.20a). Both of these drugs block the AChE-associated stimulation of Aβ aggregation, and derivatives of a hybrid compound show improved potency (Stefano et al. 2010). The other is a hybrid of tryptophan and quinones, in particular anthraquinones, 1,4-naphthoquinon-2-yl-L-tryptophan (Scherzer-Attali et al. 2010, Fig. 14.20b). The two moieties, alone, bind to Aβ, albeit not strongly. In combination, they strongly inhibit Aβ oligomerization and fibrillization, and block Aβ-induced cytotoxicity in tissue culture towards the neuronal cell line, PC12. Feeding this compound to transgenic *Drosophila* expressing Aβ improved their locomotion defect, and prolonged their lifespan.

Natural Products—Polyphenols and Others

Epidemiological studies suggest that diet may have some ability to slow the onset of neurological diseases, including AD. There is cause for much skepticism on this point, which seems to involve as much (if not more) romanticism as science, but claims have been made that diets that include regular intake of red wine, fruits, vegetables and tea may delay the onset of AD in the elderly population (Luchsinger et al. 2004; Orgogozo et al. 1997; Pan et al. 2003; Dauchet et al. 2005; Dai et al. 2006; Scarmeas et al. 2006). Similar lists of dietary wonders have been touted as delaying or preventing other diseases, including atherosclerotic cardiovascular disease, cancer, diabetes, and viral infections. What could possibly unite this disparate list of dietary items and this disparate list of diseases?

It has been hypothesized that polyphenols are present in high concentrations in these foods and drinks, and that they affect basic pathophysiology underlying diseases of different etiologies, including AD (Commenges et al. 2000; Rezai-Zadeh et al. 2005; Levites et al. 2003; Bastianetto et al. 2000; Yao et al. 2001; Stackman et al. 2003; Ono et al. 2003, 2004; Rivière et al. 2006). The inhibition of amyloid aggregation by polyphenols has been reviewed well by Porat et al. (2006).

Polyphenols can be defined broadly as any compound with multiple ring structures, of which at least one is a phenolic ring. In addition to the foods named above, polyphenols can be found also in nuts, berries, cocoa, and a variety of other plant species, such as *Ginkgo Biloba*. The polyphenols include tannins (gallic acid esters of sugars) and phenylpropanoids, including lignins and flavinoids (Fig. 14.21). Polyphenols are synthesized in plants from simpler phenolic units derived from the shikimate pathway. The flavinoids are the largest group of polyphenols, and these include the flavenols, flavenones, catechins, anthocyanidins, and isoflavinoids. Plants use polyphenols as part of their defense mechanisms against UV light, oxidative damage, and pathogens, and for diverse other functions such as mating (e.g., many flower colors are polyphenols).

This section of the chapter will touch on three promising anti-amyloidogenic polyphenols: curcumin, (–)-epigallocatechin-3-gallate (EGCG), and resveratrol. Some of the relevant structures are shown in Fig. 14.21.

Curcumin

Curcumin is the main curcuminoid of turmeric, a traditional food spice and medicinal ingredient in China and India. Turmeric is a member of the ginger family (Zingiberaceae), and the curcuminoids are responsible for the yellow color of turmeric. Curcumin undergoes a keto-enol tautomerization, of which the enol form is more stable, both in solution and in the solid state.

Curcumin has been hypothesized to have anti-inflammatory, antioxidant, chemopreventative, and chemotherapeutic benefits. Several studies have described the anti-amyloidogenic properties of curcumin *in vitro* and *in vivo* (reviewed by Hamaguchi et al. 2010). Curcumin was reported to inhibit fibril formation of Aβ,

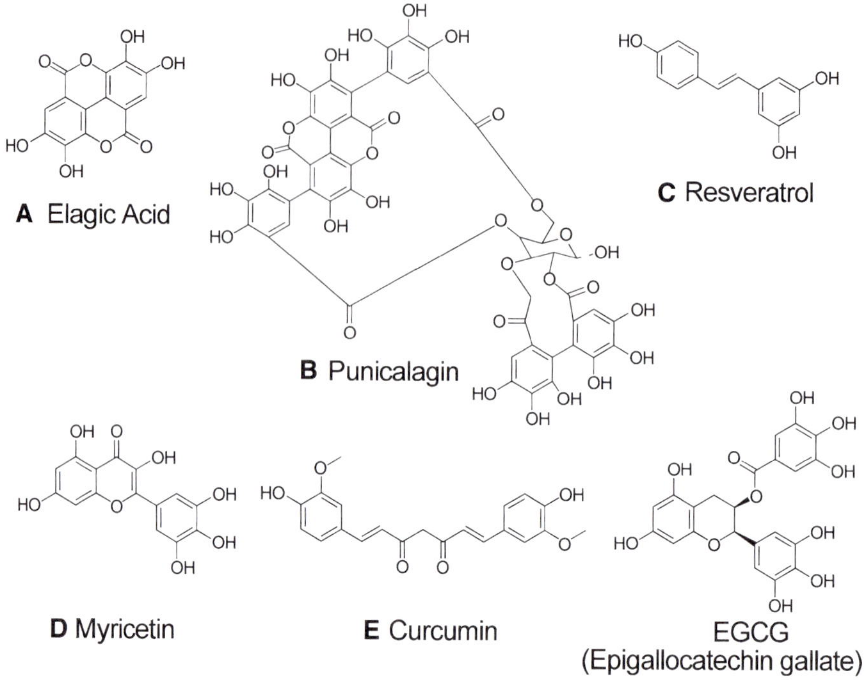

Fig. 14.21 Polyphenols that have been tested for their ability to inhibit Aβ aggregation and/or cytotoxicity. Compounds shown are (**A**) ellagic acid, (**B**) punicalagin, (**C**) resveratrol, (**D**) myricetin, (**E**) curcumin, and (**F**) epigallocatechin gallate (EGCG). Resveratrol, curcumin, and EGCG are discussed in the text. Ellagic acid and punicalagin are much-publicized polyphenols found in high quantities in pomegranates. Ellagic acid is also found in many other fruits and vegetables, and especially in berries and nuts. Punicalagin is a tannin and is found in other plants related to the pomegranate. Myrecetin is a polyphenol (a flavinol) found in grapes (including those used to make Cabernet Sauvignon, Bastianetto et al. 2000), berries, fruits, many vegetables, walnuts, herbs, and other plants

and disaggregate preformed β-amyloid fibrils *in vitro* (Yang et al. 2005). The same group also observed reductions of soluble and insoluble β-amyloid burden in aged Tg2576 mice along with suppressed inflammation and oxidative damage in the brain when treated with low doses of oral curcumin (Lim et al. 2001). Although curcumin inhibits Aβ aggregation, it is not clear in these studies whether curcumin acts directly by means of its anti-aggregative activity, or dampens a neuroinflammatory reaction to Aβ aggregation and deposition. Two phase-II trials of curcumin have begun in patients with AD. These are very small trials, and thus far, have demonstrated no significant benefit in humans (Baum et al. 2008; Hamaguchi et al. 2010 [in meeting Abstract form only, cited in Ringman et al. 2008)]. It seems likely that the quantities of curcumin needed to attain any clinical effect are quite large. Nevertheless, the idea of a natural food item preventing AD is appealing, and thus further research is underway to pursue this compound as a potential preventative or treatment for AD.

(−)-Epigallocatechin-3-Gallate (EGCG)

The beneficial medicinal qualities of green tea have been ascribed to its high levels of catechins, one type of polyphenolic flavinoid. EGCG is the major constituent of tea catechins, proposed to be an antioxidant. It is thought to act at multiple CNS targets as a neuroprotective agent, and to protect against AD in particular (reviewed in Mandel et al. 2008).

EGCG may decrease cellular production of Aβ by directing cleavage of APP towards the α-secretase pathway. The effects of this compound were studied in cell lines expressing the double mutation in APP responsible for the Swedish form of familial AD, i.e., codons 670 and 671 (LysMet→AsnLeu). This mutation results in increased production of Aβ, by increasing the proportion of APP that is cleaved by β- as opposed to α-secretase (Suzuki et al. 1994; Haass et al. 1995). In cell lines (N2A) transfected with this mutant form of human APP, and in primary neuron cultures from $TgAPP_{SWE}$ transgenic mice, EGCG was found to decrease production of Aβ, which the authors hypothesized may be due to increased cleavage of APP by α-secretase (Rezai-Zadeh et al. 2005), without change in β-secretase activity. In a follow-up study, they examined three candidate enzymes of the ADAM (a-disintegrin and metalloproteases) family of zinc metalloproteinases for α-secretase activity, ADAM9, 10, and 17. Treatment of N2A cells with EGCG led to increased activation of ADAM10, and increased release of the soluble product of α-secretase cleavage of APP (sAPP-α, Obregon et al. 2006).

Another study examined release of sAPP-α into conditioned media of human SHSY5Y neuroblastoma and rat pheochromocytoma PC12 cells, and suggested that upregulation of α-secretase proteolysis was signaled by a protein-kinase-C-dependent pathway (Levites et al. 2003). As stated earlier, upregulation of α-secretase activity is a possible strategy for preventing Aβ; however, in most patients with sporadic AD, and even in patients with the Swedish mutation of APP, APP is cleaved far more by α- than β-secretase. Since the above studies found that EGCG did not decrease β-secretase activity, an increase in α-secretase activity seems unlikely to provide much benefit.

Resveratrol

Moderate consumption of wine is associated with a decreased risk of AD (Lindsay et al. 2002; Luchsinger et al. 2004). Even in a transgenic mouse model of AD (Tg2576 mice, expressing neuronal APP695 with the Swedish mutation, double mutations at KM670/671NL), drinking Cabernet Sauvignon was found to improve spatial memory and decrease Aβ neuropathology, compared to a similar quantity of ethanol, or water alone (Wang et al. 2006). Resveratrol is *trans*-3,4,5-trihydroxystilbene, a natural polyphenol found in grapes and red wine. It has neuroprotective effects against the toxicity of Aβ in cell-culture models, although these effects occur by more than one mechanism. In cell lines expressing either the wild-type APP or the Swedish mutant APP_{695}, resveratrol reduced the amount of intracellular and secreted Aβ in several cell lines, by promoting its intracellular degradation by the proteasome (Marambaud et al. 2005). Another group posited a role for SIRT1, a class-III histone deacetylase, in microglial NFκB-mediated neurotoxicity (Chen et al. 2005). In tissue culture, Aβ caused acetylation of RelA/p65 at Lys310, which is required for full transcriptional activity of the pro-inflammatory regulator, NF-κB. This acetylation is reversible, and resveratrol was found to inhibit NF-κB signaling by promoting deacetylation of Lys310 of RelA/p65 (Yeung et al. 2004). It is important to note that in both of these studies, effects were observed only at high concentrations of resveratrol, approximately 10-fold higher than that found in red wine.

Another possible mechanism for the reported beneficial effects of resveratrol involves the AMP-activated protein kinase (AMPK) signaling pathway (Vingtdeux et al. 2010a, b; Kwon et al. 2010). AMPK targets proteins involved in cellular energy-balance regulation and inhibits mammalian target of rapamycin (mTOR), which is a repressor of autophagy. These studies suggest that resveratrol lowers Aβ levels by activating AMPK and therefore promoting autophagy and lysosomal clearance of Aβ. Experiments in an AD mouse model (APP/PS1), also demonstrated that resveratrol is orally bioavailable and able to reach concentrations of bioactivity in the mammalian brain (Vingtdeux et al. 2010b). Screening of a library of small molecules of similar chemical structure to resveratrol revealed two compounds (RSVA314 and RSVA405) that inhibited Aβ deposition, but with a potency 40 times that of resveratrol (Vingtdeux et al. 2010a). Piceatannol (*trans*-3,4,3′,5′-tetrahydroxystilbene) is another structural homologue for resveratrol, and has been reported to have greater potency in preventing oxidative stress and apoptosis in PC12 treated with high concentrations of Aβ25–35 (Kim et al. 2007). Resveratrol may have its own efficacy, or may lead to the discovery of other small molecules for preventing or treating AD.

Several studies have also suggested that resveratrol could inhibit Aβ aggregation, though the results have been inconsistent. One study, using thioflavin T fluorescence, transmission-electron microscopy, western blotting, and dot blotting, suggested that resveratrol decreases Aβ42 fibril formation, while increasing oligomerization, and yet attenuated the cytotoxicity of Aβ42 (Feng et al. 2009). They suggested that resveratrol induced formation of a non-toxic Aβ oligomer, perhaps by binding to oligomers and altering their conformation. Another study suggested that resveratrol remodels soluble pre-fibrillar oligomers, fibrillar intermediates, and fibrils but does not do so to non-toxic oligomers and monomers (Ladiwala et al. 2010). In this study, conformers were formed by dissolving Aβ42 in a succession of solvents (50% acetonitrile, followed by neat 1,1,1,3,3,3-hexafluoro-2-propanol, followed by dilute base, and finally, phosphate buffer), followed by centrifugation, followed by sonication, and incubation (25°C) either with or without agitation. In this procedure for assembling pre-fibrillar intermediates, agitation produced a toxic oligomeric species, while quiescence produced a non-toxic oligomeric species. The detailed structures of these species are not known, however, nor is it clear in what specific way resveratrol might modify their structures.

Other Natural Compounds

Aside from polyphenolic structures, many other natural extracts and compounds have been explored as potential sources for AD therapies. These include garlic extract, nicotine, inositols, phosphatidylinositol lipids and derivatives, galantamine, colostrinin, salvianolic acid B, and α-d-mannosylglycerate (Gupta et al. 2009; Moore et al. 2004; Nitz et al. 2008; McLaurin et al. 2006; Matharu et al. 2009; Durairajan et al. 2008; Ryu et al. 2008; Salomon et al. 1996; Zeng et al. 2001).

14.3.1.4 Immunotherapy for Alzheimer's Disease?

The Basic Phenomenon

This is a question, indeed, but one of the greatest importance. As stated before, it will be discussed briefly, but the reader is referred to many excellent recent review articles on this subject (Weksler et al. 2005; Arbel and Solomon 2007; Boche and Nicoll 2008; Popovich and Longbrake 2008; Shah et al. 2008; Villoslada et al. 2008; Brody and Holtzman 2008; Foster et al. 2009; Yamin et al. 2008; Weksler et al. 2009; Morgan 2009; Jicha 2009; Tarditi et al. 2009; Wisniewski 2009; Federoff 2009; Neugroschl and Sano 2009; Deane et al. 2009; Bednar 2009; Röskam et al. 2010; von Bernhardi 2010; Lemere and Masliah 2010; Fu et al. 2010; Grill and Cummings 2010; Citron 2010).

The difference between a prion and an amyloid, as discussed earlier, is that the former is an infectious subtype of amyloids. Another way to state this is to say that non-infectious amyloids are those that induce immunity, whereas the prions, as infectious agents, have mechanisms for evading the immune system, broadly defined. The immune system includes not only the T and B cells of specific (or adaptive) immunity, but also all of the cells of innate (or natural) immunity, including the phagocytic cells, the plasma defense mechanisms such as the complement system, and the physical barriers to infection. Considering the case of Kuru or atypical Creutzfeldt–Jakob disease, through endocannibalism, an infectious prion breaches some of the normal physical barriers, but the infectious particle must breach additional boundaries that allow it, first, to enter the brain, and subsequently, to enter the neuron.

Whether or not this was predicted in advance, synthetic human Aβ1–42 elicits a brisk immune reaction when injected into PDAPP transgenic mice (Schenk et al. 1999). The mice developed anti-Aβ antibodies, and this prevented deposition of Aβ within the brain substance and reversed previous deposits of Aβ. Immunization also reduced the gliosis and neuritic dystrophy that is part of this animal model of AD. Since that time, this fundamental finding has been confirmed with a variety of Aβ immunogens, adjuvants, and AD model mice (Lemere et al. 2000; Weiner et al. 2000; Das et al. 2001; Sigurdsson et al. 2001; Maier et al. 2006). Immunization reduced cerebral Aβ load by several measurements, and more importantly, resulted in improvement of the cognitive impairment associated with Aβ deposits (Janus et al. 2000; Morgan et al. 2000). Not surprisingly, the best results were when immunization was performed before the development of advanced pathology, while there were more mixed results when immunization was given after more advanced pathology developed.

The B-cell epitopes were localized mainly to the *N*-terminal residues 1–15, which, in the fibril, includes the unstructured *N*-terminal domain, and the approximate start of the *N*-terminal β-sheet (Lemere et al. 2001; Town et al. 2001; McLaurin et al. 2002; Gardberg et al. 2007). Most of the T-cell epitopes have been localized to the middle or *C*-terminal domains in the molecule, i.e., the parts that are within β-sheets in fibrils (Cribbs et al. 2003; Monsonego et al. 2001).

The above studies represent active immunization; passive immunization (immunization with antibodies raised in another animal) also was able to prevent Aβ deposition, when given to young PDAPP mice. In an instructive study, DeMattos et al. (2001) administered a monoclonal antibody (m266) to young mice. This antibody, which recognizes the central domain (residues 13–28) of Aβ, was able to sequester all plasma Aβ. Although Aβ is generated solely in the CNS in this model, and the antibody remained mainly or entirely in the plasma and did not bind to CNS deposits of Aβ, the antibody led to a rapid, 1,000-fold increase in plasma concentration of Aβ, and ultimately, to increased clearance of brain Aβ deposits. These results suggest that the antibody acts by shifting the equilibrium between brain and blood Aβ, and thereby leads to increased Aβ clearance. In older mice, with established Aβ deposition and plaque-like pathology, however, this antibody was not effective, whereas an antibody directed against the *N*-terminal domain was able to reduce cerebral Aβ load (DeMattos et al. 2001; Bard et al. 2003). In addition to clearing Aβ deposits from the brain, this treatment also succeeded in reversing cognitive deficits (Kotilinek et al. 2002), but a serious drawback was that it caused cerebral hemorrhages in areas of vascular Aβ deposition. This is worrisome, since some degree of cerebral amyloid angiopathy is present in as much as 80% of patients with sporadic AD, and is prominent in ~30% (Castellani et al. 2010, 2004). Cerebral hemorrhage appears to be a common concomitant of passive immunization with anti-Aβ antibodies (Pfeifer et al. 2002; Wilcock et al. 2004; Racke et al. 2005), perhaps because it causes a rapid influx of Aβ peptides into the blood vessel wall.

Clinical Trials of Aβ Vaccines

In the 1990s, Elan and Wyeth entered into phase-I clinical trials of an anti-Aβ vaccine, AN1792, consisting of synthetic Aβ1–42 peptide and a saponin adjuvant, QS21. Control groups consisted of a vaccine with the adjuvant, adjuvant alone, or a placebo. This vaccine was the first, and only, clinical trial of human immunization for which extensive information is available. A phase-I clinical study in 80 Alzheimer's patients with mild-to-moderate cognitive impairment showed no obvious adverse effects (Bayer et al. 2005). The phase-IIa trial for preliminary evaluation of clinical efficacy began in late 2001 on AN1792 with QS21, in 372 patients with mild-to-moderate cognitive impairment of AD. However, the trial was stopped in 2002 when ~6% of patients developed meningoencephalitis and leukoencephalopathy. The adjuvant used in this immunization trial, QS21, typically induces a strong T_H1 response, which was necessary to achieve a strong antibody response (Fox et al. 2005). Even with this adjuvant, however, in the phase-I studies, it had been noted that the antibody response was generally poor: less than a quarter of patients had antibody titers greater that 1:1,000. In fact, the usual response in these patients was a T_H2 response. Reasoning that the antigen failed to immunize because of poor solubility properties, they added polysorbate 80, which did increase the number of strong antibody responders to ~60% of patients—though not necessarily with good antibody titers—but also induced a strong pro-inflammatory response, which is also common characteristic of strong T_H1 responses (Pride et al. 2008). Despite receiving between one and three vaccinations, only ~20% showed strong antibody titers (>1:2,200). These antibodies were directed against the *N*-terminus (Lee et al. 2005).

The conclusion from these studies was that this vaccine caused activation of autoimmune T cells, probably recognizing a T-cell epitope in the central or *C*-terminal domain of Aβ, which led to a pro-inflammatory, T_H1 response. These studies illustrate the problems in developing robust active immunization vaccines. Aβ is, after all, a self-protein, and antibodies directed against it are autoantibodies, by definition. It is not unusual to develop autoantibodies, but this response is typically low affinity and low titer, except, of course, in autoimmune diseases. Thus, the reasoning seemed to be that a strong antibody response to Aβ requires strong T-cell help, and for the types of antibodies required, this meant a T_H1 response. The risk is precisely for what happened: the development of inflammation as a part of what is, in essence, an autoimmunization.

This result was obviously a major disappointment, but there is a positive aspect of a sort. A single-center study of 30 patients suggested that there was cognitive improvement in six patients with high antibody titers (Hock et al. 2003). Obviously, these numbers are too small to make much out of, but the results suggest that antibodies directed against Aβ can, indeed, slow the cognitive decline in AD and may lead to useful therapy, if the immunization conundrum can be solved. Even the most positive of these results, however, are dubious: antibody responders lost *more* total brain volume than controls in MRI studies (Schott et al. 2005; Fox et al. 2005). The authors explained these observations as indicating a "dissociation between brain volume loss and cognitive function," possibly due to amyloid removal and associated cerebral fluid shifts. This highly optimistic, though plausible explanation raises another issue, however: what is a suitable endpoint for evaluating success of therapy? If their explanation for the loss of brain volume is correct, this clearly is not a satisfactory way to evaluate response to therapy.

Eight of the patients enrolled in the phase-I study died before or during the follow-up phase, and had autopsies with neuropathology. Although the numbers are small, they had reduction in the area of neuropil covered by Aβ compared with controls, and the decrease was proportionate to the antibody titer (Holmes et al. 2009). Unfortunately, these particular patients also died with severe dementia. In a sense, this is hardly surprising, since they already had advanced disease at the start of the trial; there is no reason to expect removal of Aβ to reverse neuronal death!

The meningoencephalopathy was accompanied by leukoencephalopathy, and showed extensive T lymphocyte and macrophage infiltrates (Ferrer et al. 2004; Nicoll et al. 2006), i.e., it had features of an autoimmune meningoencephalitis. Although the vaccine succeeded in removing Aβ, there was persistence of cerebral amyloid angiopathy, and no difference from controls in the appearance of neurofibrillary tangles—again, not surprising in view of the aims of the vaccine.

A survey of the current status of vaccination trials indicates that the results of the previous trials have not led drug companies, including Elan/Wyeth, to abandon the effort. There are currently at least seven ongoing clinical trials of passive immunotherapy for patients with mild-to-moderate AD. Thus far, few results are available. In one of these studies, preliminary results indicate efficacy only in patients with an apolipoprotein-E isoform other than E4 (Grundman and Black 2008), which, unfortunately, is the apolipoprotein-E isoform that confers the greatest risk for developing AD.

Beyond passive immunization, new vaccines for active immunization are in development. These include Elan and Wyeth's new vaccine, ACC-001, which uses an *N*-terminal Aβ immunoconjugate. The vaccine appears geared to avoid a repetition of the previous results. In phase-I and -II trials in ~360 patients with mild-to-moderate AD, the vaccine has been tolerated well except for the possible development of skin lesions; in particular there have been no reports of meningoencephalitis (Strobel 2009). Novartis also has a vaccine in phase-II clinical trial. It is a Qb virus-like particle containing multiple copies of Aβ1–6. Thus far, there seems to be little antibody production among the enrolled patients, and correspondingly little lowering of CSF Aβ concentrations, cognitive improvement, or changes in brain volume as assessed by MRI studies.

So, then: what are the current prospects for vaccination against AD? At present, one can conclude only that Aβ immunotherapy is promising—a statement that is meant to be taken both ways: as a present participle and as a gerund. There is a lot of hype about Aβ immunotherapy—promises, promises, promises, but on the other hand, also a lot of well-justified anticipation.

14.3.2 Huntington's Disease and Other Polyglutamine Diseases

The polyGln-expansion diseases include at least nine neurodegenerative diseases and are caused by mutations that increase a CAG nucleotide repeat beyond a pathogenic threshold. This threshold is context dependent, i.e., it varies among proteins, probably because of the effects of flanking sequences. In most of the polyGln diseases, including HD, this threshold is ~35–40 Gln residues. Like other amyloidogenic proteins, expanded polyGln domains aggregate by a nucleation / polymerization model and form fibrils with a cross-β motif. Very little is known about the detailed

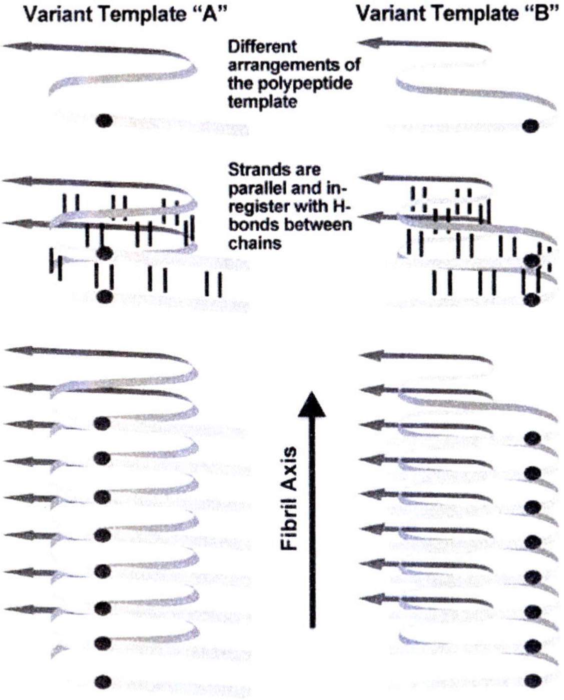

Fig. 14.22 Model of the in-register, parallel β-sheet structure of the yeast prions [PSI+], [URE3], and [PIN+] (Figure is from Shewmaker et al. 2009). These prion variants differ in the arrangements of the polypeptide template onto which additional molecules add

structure of PolyGln proteins, however, because of their low solubility and sequence redundancy of the Gln stretch, which makes structure determination problematic. Several lines of evidence support the idea that PolyGln peptides contain β-sheet structure, including data from circular dichroism, X-ray diffraction, Fourier-transform infrared spectroscopy, and computer modeling (reviewed in Ross et al. 2003). PolyGln proteins also exhibit binding of a monoclonal antibody with high selectivity for a generic conformational amyloid fibril epitope (Chen et al. 2002a).

Although not polyGln proteins in the strict sense, the yeast prions contain Gln-/Asn-rich domains with some structural similarities to polyGlns. The PSI(+) prion of *Saccharomyces cerevisiae* is a self-propagating amyloid form of Sup35p, a subunit of the translation-termination factor. Solid-state NMR of amyloid fibrils formed *in vitro* from purified recombinant Sup35(1–253), consisting of the glutamine- and asparagine-rich *N*-terminal 123-residue prion domain (N) and the adjacent 130-residue highly charged M domain indicate an in-register, parallel β-sheet structure (Shewmaker et al. 2006). More recently, amyloid fibrils of two prion variants of Sup35NM were compared by solid-state NMR, and despite subtle differences, both showed the in-register, parallel β-sheet structure (Shewmaker et al. 2009, Fig. 14.22). A similar structure had been reported earlier for the residues 10–39 of the yeast prion protein Ure2p (Chan et al. 2005) and amyloid of the Rnq1, the basis of the [PIN+] prion (Wickner et al. 2008).

PolyGln stretches are made entirely of polar residues, and as described earlier, aggregate by mechanisms different from those by which other amyloid proteins, such as Aβ, aggregate. Aβ contains clusters of sequential hydrophobic residues; in the case of α-synuclein, hydrophobic side-chains are spaced periodically, so that in certain conformations (α-helix), they can bind to lipids, while in other conformations (β-sheet) they self-associate. In all of these cases, self-association occurs mainly through the hydrophobic effect, i.e., the shielding of hydrophobic groups from the aqueous medium. In contrast, PolyGln peptides interact through hydrogen bonds, not only between peptide backbone groups, as in all amyloids, but also between side-chain amides (Starikov et al. 1999; Esposito et al. 2008; Lanning et al. 2010; Masino 2004). In solution, short PolyGln peptides adopt a polyPro-II-helix-like structure, with formation of oligomers that also have polyPro-II-helix-like structure (Darnell et al. 2007, 2009), and thus, the conversion to fibrils may require a transition from this structure to β-sheet.

As is the case for all progressive neurodegenerative diseases, there is no cure, and little in the way of treatment for the polyGln diseases. Even symptomatic treatments are of limited benefit, and all of these diseases progress inexorably to death. Many attempts have been made to test potential therapeutic agents (reviewed in Herbst and Wanker 2006; Scatena et al. 2007), with studies in both the laboratory and clinic.

HD is a uniformly fatal autosomal-dominant neurodegenerative disorder with virtually 100% penetrance. The brain shows progressive neuronal loss in select regions, notably, the GABAergic medium spiny striatal neurons. The most prominently affected areas are in the neostriatum, i.e., the caudate nucleus and (secondarily) the putamen. Damage progresses to other regions, including the cerebral cortex, substantia nigra, hippocampus, and portions of the cerebellum, thalamus, and hypothalamus. Huntingtin is a 348-kDa protein, which has been found to interact with ~20 proteins directly, and scores more indirectly. The interacting proteins have roles in transcription regulation, intracellular transport functions, and much else; indeed, it is difficult to summarize the vast number of potential roles this protein plays. The expansion of the PolyGln region in the exon-1-encoded portion of the protein is considered to be a toxic gain of function, with effects on gene transcription, proteasomal function, axonal transport, endocytosis, synaptic transmission and Ca^{2+} signaling—essentially every function that a neuron performs (Tobin and Signer 2000; Bezprozvanny 2009; Cha 2007; Li and Li 2004; Ross 2002; Rubinsztein 2002; Truant et al. 2008). It is not known exactly what accounts for the specific cellular targeting in this disease.

As with the other proteins causing PolyGln-expansion diseases, little is known about the structure of huntingtin. Recently, an X-ray crystallographic study was published of the exon-1-encoded region of huntingtin (with 17 Gln residues), as part of a fusion protein with maltose-binding protein (MBP) (Kim et al. 2009). This structure includes an *N*-terminal α-helix, the PolyGln region and adjacent polyPro region, which forms a polyPro helix (Fig. 14.23). The PolyGln region itself can adopt numerous conformations, including an α-helix, random coil, and extended loop. Thus, this region shows great conformational flexibility, and is influenced by the context of neighboring domains in huntingtin, and its many binding partners.

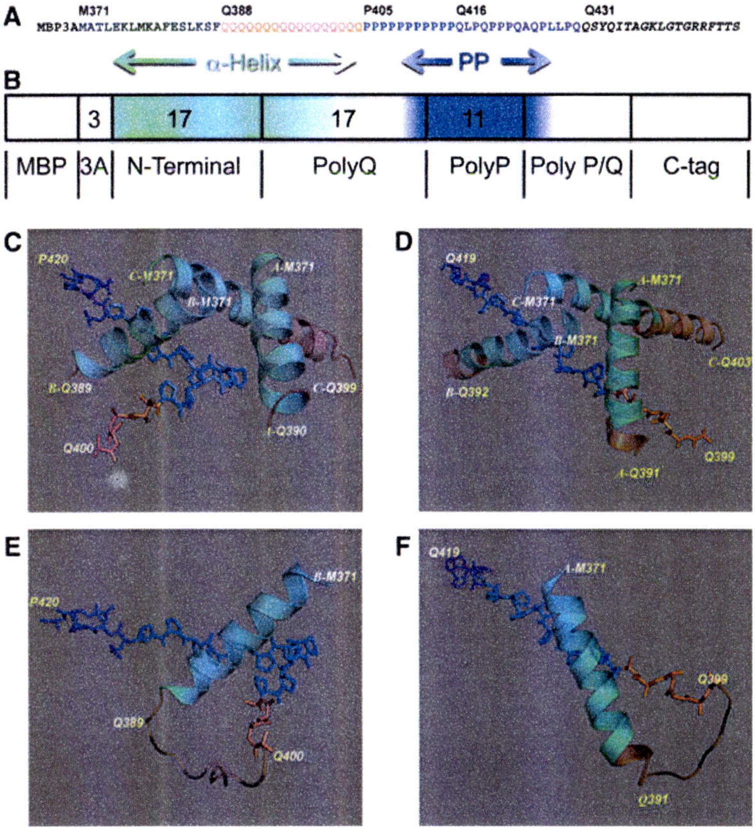

Fig. 14.23 Polymorphic structure of the exon-1-encoded portion of huntingtin (Figures are from (or adapted from) Kim et al. 2009). The top line (**A**) shows the sequence of the fusion protein composed of the MBP and a 17Q huntingtin-exon-1 protein, MBP-Htt17Q-EX1, used for these studies. The fusion protein contains (from *N*- to *C*-terminus) the MBP (not to scale), a three-alanine linker (3A), the 17 residue *N*-terminal region of Htt (green), a poly region with 17 Gln residues (poly17Q, orange), a polyPro region with 11 Pro residues (poly11P, blue), a 15-residue Gln/Pro-rich region, and a 19-residue *C*-terminal tag. The bar diagram (**B**) shows a schematic summary of structural information obtained from analysis of data from seven crystals of Htt17Q-EX1. The secondary structural elements are indicated; the shaded green box indicates the transition from α-helix to unstructured region; the blue region shows an area of polyPro helix, with a transition to an unstructured area. (**C**), (**D**), (**E**), and (**F**) represent structures of HttQ17-EX1 from different crystals. The MBP and 3Ala linker are removed for clarity. The protein forms a trimer, and the structure varies from crystal to crystal. In (**C**), from c95 crystal, for example, there is an *N*-terminal α-helix extending from Met371 to Phe 387. The *N*-terminal portion of the PolyGln region extends the helix in one chain of the trimer, but its *C*-terminal part is unstructured. In another chain of the trimer, the polyGln region is unstructured. The polyPro region forms a polyPro helix, as does the *N*-terminal part of the polyPro/PolyGln region. Further details are described in the paper

Even its mechanism of self-association is quite complex, and simple PolyGln peptides, while informative, do not convey the full complexity of the process. Several recent studies have shown that the 17-amino-acid domain at the *N*-terminus of the protein—i.e., also *N*-terminal to the PolyGln region—are important for triggering self-association of the exon-1-encoded domain (Thakur et al. 2009).

In this section, we give a synopsis of the literature concerning small-molecule and peptide inhibitors of PolyGln aggregation. This field is not yet as developed as that of inhibitors directed against β-amyloid aggregation, but several important screening methods and inhibitors have emerged, mostly directed towards huntingtin and HD.

14.3.2.1 Screening for Small-Molecule Inhibitors of PolyGln Aggregation

Screening procedures have dominated the search for inhibitors of PolyGln aggregation and the cytotoxicity that results from it. This field has had a few—strangely few—attempts at rational design of inhibitors; in this respect, this field differs from the ones described in the previous section on Aβ. Nevertheless, several small-molecule inhibitors of PolyGln expansions have been identified by these high-throughput screening methods. A challenge for developing or screening for PolyGln aggregation inhibitors is the extreme insolubility of these peptides. Nevertheless, several screening studies, both *in vitro* and *in vivo*, have yielded small molecules that could lead eventually to forms of therapy. For example, a filter retardation assay was used to demonstrate that the antibody 1C2 (as well as the dyes Congo red, thioflavin S, chrysamine G, and direct fast yellow) suppressed the aggregation *in vitro* of huntingtin exon-1 protein (Heiser et al. 2000). This antibody had previously been shown to recognize PolyGln expansions specifically in their soluble form, but not to bind to insoluble, high-molecular-weight PolyGln protein aggregates (Trottier et al. 1995b). The same group developed an automated high-throughput version of the *in vitro* filter-retardation assay and used it to screen a library of ~184,000 small molecules, in which they identified 25 benzothiazole derivatives that inhibit huntingtin fibrillogenesis in a dose-dependent manner. Most, however, were found to be cytotoxic in cell assays (Heiser et al. 2002). The benzothiazole, riluzole, had already been studied as a possible therapeutic for amyotrophic lateral sclerosis and HD (Bensimon et al. 1994; Lacomblez et al. 1996; Rosas et al. 1999; Schiefer et al. 2002).

An improved method for identifying effective inhibitors used an *ex vivo* organotypic slice-culture assay (Smith et al. 2001). Using R6/2 mice, which express an *N*-terminal huntingtin fragment with a 140 Gln expansion driven by the human huntingtin promoter, the number of inclusion bodies in hippocampal slice culture was evaluated by immunohistochemistry. The inhibitors identified using this assay, however, have failed to yield positive results *in vivo*, in mice. Although minocycline (a tetracycline antibiotic) and riluzole (a benzothiazole, described above) were potent inhibitors in the *ex vivo* organotypic slice-culture assay, there was no clear improvement in behavioral abnormalities or postmortem aggregate burden in the treated R6/2 mice (Smith et al. 2003; Hockly et al. 2006). This underscores the importance of animal model trials after identification of potential aggregation inhibitors by screenings *in vitro* or even *ex vivo*. In a review of eight human clinical trials testing possible interventions (vitamin E, Idebenone, Baclofen, Lamotrigine, creatine, coenzyme Q10+Remacemide, ethyl-eicosapentanoic acid and Riluzole), none proved effective as a disease-modifying therapy for HD (Mestre et al. 2009). The choice of which inhibitor compounds to advance to clinical trials must be made carefully to avoid wasting valuable time, money, and patient goodwill.

Another approach to screen for PolyGln aggregation inhibitors uses myoglobin as a host protein (Tanaka et al. 2001, 2002, 2004, 2005b). These investigators found that disaccharides reduce the aggregation of Mb-Gln_{35}. Trehalose was the most effective inhibitor and was shown to increase the stability of Mb-Gln_{35} in guanidine-HCl-induced denaturing experiments. In a cellular model of HD, exogenous and endogenous trehalose also decreased aggregation and enhanced cell viability. As mentioned earlier, similar observations were made for Aβ aggregation (De Bona et al. 2009). Oral administration of trehalose in the R6/2 transgenic mouse model of HD led to decreased aggregate burden in cerebrum and liver, improved motor dysfunction and extended lifespan. The authors of these studies proposed that trehalose acts as a "chemical chaperone", by interacting with and stabilizing the PolyGln protein in a non-aggregative state. Although the term "chemical chaperone" is not precisely defined, a stabilized PolyGln protein might be less prone to trigger caspase cleavage, or might be less scissile to this enzyme, and thus could avoid nuclear translocation of the toxic *N*-terminal fragment. "Chemical chaperones" could also prevent cytotoxicity by relieving the burden of "misfolded" proteins on the proteasome system.

A high-throughput fluorescence–resonance-energy-transfer (FRET)-based cellular assay has been developed to screen for small molecules that inhibit intracellular aggregation of PolyGln peptides fused to GFP or YFP. A first screen by this method in 2003 identified Y-27632, a small-molecule inhibitor of the Rho-associated kinase p160ROCK, which was subsequently shown to decrease neurodegeneration in an HD *Drosophila* model (Pollitt et al. 2003). This cellular pathway had not been identified previously as directly involved in PolyGln aggregation, demonstrating the power of a cell-based-screening assays over a simpler *in vitro* screening method. A later paper detailed a more comprehensive screen and demonstrated a high predictive value (~50%) of the primary FRET-based assay to identify compounds that rescue the disease phenotype in the HD *Drosophila* model (Desai et al. 2006). Aside from validating the FRET assay as a useful screening method, this result confirms that aggregation is an important therapeutic target.

A screen of the NINDS Custom Collection of 1,040 FDA-approved drugs and bioactive compounds for their ability to prevent aggregation of the *N*-terminal fragment of huntingtin with 58 Gln repeats *in vitro* (Wang et al. 2005). Gossypol, gambogic acid, juglone, celastrol, sanguinarine, and anthralin were among the compounds that inhibited aggregation with $IC_{50} < 15\ \mu M$. Of these, juglone and celastrol were effective in reversing the abnormal cellular localization of PolyGln-expanded huntingtin observed in mutant $Hdh^{Q111/Q111}$ striatal cell culture. Further research has revealed that celastrol exerts its neuroprotective effects by upregulating the expression of heat-shock proteins (Zhang and Sarge 2007).

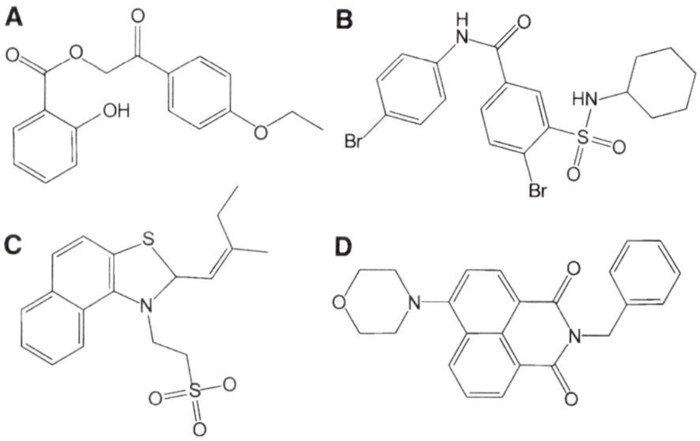

Fig. 14.24 Results of the yeast screening procedure described in Zhang et al. (2005). The figure (from Zhang et al. 2005) shows four "hits"; these compounds inhibited PolyGln aggregation in PC12 cells

Finally, a yeast-based high-throughput screen of a chemical library was used to identify chemical compounds that inhibit aggregation without significant cytotoxicity. From this assay, four compounds (Fig. 14.24) were identified as inhibiting aggregation and, in cultured brain slices from a HD model transgenic mice, were found to be non-toxic and efficacious in decreasing aggregate load (Zhang et al. 2005). A similar high-throughput fluorescence cell-based assay screened a library of ~10,000 compounds, and this yielded quinazoline as a "hit" that then served as the basis for structure–activity studies, which yielded four quinazoline derivatives with greater potency (Rinderspacher et al. 2009).

14.3.2.2 Screening for Peptide Inhibitors of PolyGln Aggregation

In comparison with other parts of the larger field of PADs, there has been relatively little work on peptide-based inhibitors of PolyGln aggregation. A screen of a combinatorial peptide library by phage display found six tryptophan-rich peptides that preferentially bound to expanded polyGln domains (Nagai et al. 2000). PolyGln-binding peptide 1 (QBP1; sequence SNWKWWPGIFD) inhibited thioredoxin–PolyGln aggregation in a turbidity assay *in vitro*, decreased aggregation of PolyGln-YFP in transfected COS-7 cells, and also reduced PolyGln-induced cytotoxicity. A subsequent paper demonstrated the efficacy of QBP1 *in vivo*, by genetically expressing the peptide inhibitor in a *Drosophila* model of HD (Nagai et al. 2003). Also in *Drosophila*, QBP1 was fused with cationic protein transduction domains (PTDs), which deliver covalently bound small molecules into cells, to show that PTD-QBP1 suppresses PolyGln-induced neurodegeneration when delivered exogenously (Popiel et al. 2007). Most recently, this group has successfully detected delivery of PTD-QBP1 into mouse brain cells upon intracerebroventricular injection (Popiel et al. 2009). Long-term administration of PTD-QBP1 to R6/2 mice improved their weight-loss phenotype, suggesting a possible therapeutic effect. Finally, SPR was used to characterize the binding specificities and affinities of aggregation inhibitors to expanded PolyGln domains (Okamoto et al. 2009). QBP1 was shown to bind specifically to thio-Q62 peptide and not to thio-Q19, suggesting that this inhibitor specifically recognizes a toxic, amyloidogenic PolyGln conformer. Congo red, conversely, is a nonspecific binder and shows no preference to a long or short PolyGln segment. Clearly, binding specificity is a desirable trait in a therapeutic agent, to help avoid side effects associated with binding to non-therapeutic targets.

14.3.2.3 Rational Design of Inhibitors of PolyGln Aggregation

High-throughput screens of random molecules have succeeded in identifying some PolyGln aggregation inhibitors. While this approach has led to discovery of several promising molecules, its obvious disadvantage is that it is not mechanism-based, and consequently general conclusions about mechanisms of inhibition—mechanisms which, if understood, could lead to more effective or comprehensive modes of treatment—may not be readily apparent. As stated earlier, the field of PolyGln diseases has been curiously sparse in rational design of aggregation inhibitors. Rationally designed inhibitors might also help in the design of therapeutic agents, but more than this, rationally designed inhibitors are valuable as structural and mechanistic probes that can help address questions about aggregate structure, the kinetics of aggregation, and the pathological basis of aggregation diseases.

Wetzel and colleagues first demonstrated that inserting Pro residues into certain positions in Aβ greatly reduced the ability of the mutated peptide to form fibrils. The authors inferred that these positions were those occurring within the β-sheet segments of the peptide (Wood et al. 1995; Williams et al. 2004). These observations led to the development of a rationally designed PolyGln aggregation inhibitor (Thakur et al. 2004). The peptides PGQ_9P^2 (sequence: K_2–Q_9–PG–Q_4PQ_4–PG–Q_9–PG–Q_9–K_2) and $PGQ_9P^{2,3}$ (sequence: K_2–Q_9–PG–Q_4PQ_4–PG–Q_4PQ_4–PG–Q_9–K_2) do not make fibrils themselves, and are also effective inhibitors of the aggregation of other PolyGln peptides. They appear to bind to the growth site of fibrils and block further propagation because of the insertion of Pro residues, which prevents β-strand formation. These inhibitor peptides were also shown to be cyto-protective when added in conjunction with toxic, preformed PolyGln aggregates. This suggests that these inhibitor peptides act as elongation inhibitors, thus supporting the elongation/sequestration theory of PolyGln neurotoxicity.

One of the outstanding questions in the field of PolyGln aggregation is the nature of PolyGln oligomers. It is clear that PolyGln peptides can make oligomers (Ossato et al. 2010; Hands and Wyttenbach 2010), but less is known about the structure of these oligomers than about those formed by Aβ, and most other proteins involved in PADs. It is not known whether these oligomers are cytotoxic, whether they are heterogeneous (one would imagine, *a priori*, that they would be), or what their relationship to fibrils (e.g., on- or off-pathway) might be. Indeed, it is not even clear that the nucleus for PolyGln aggregation is an oligomer at all. One surprising proposal (Fig. 14.25), based on careful and rigorous analysis of aggregation kinetics, is that the nucleus in the PolyGln aggregation pathway is an alternatively folded, high-energy state of the monomer. One kinetic parameter, Kn*—the equilibrium constant

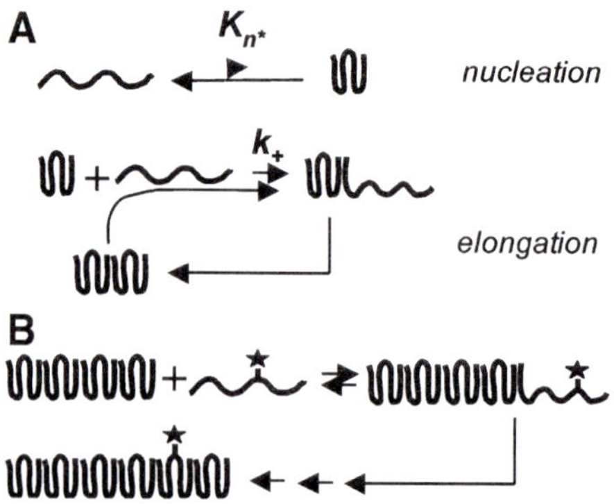

Fig. 14.25 Proposal of nucleated growth of polyGln aggregates by addition of monomers. Detailed analysis of the kinetics of PolyGln aggregation yielded the surprising result that nucleation consists of an unfavorable folding event within the monomeric protein or peptide (Chen et al. 2002a). A monomer adds to the growth site of the PolyGln aggregate, but is unstable and prone to dissociate unless it is followed by subsequent rounds of monomer addition (Figure is from Bhattacharyya et al. 2005) In the figure, K_{n*} = nucleation equilibrium constant, and k_+ = second-order rate constant for elongation of the aggregate, and the asterisk indicates biotinylated PolyGln (Q29) peptide. The authors propose a multiphase growth mechanism including an initial reversible binding step ("docking"), and subsequent, rate-limiting, rearrangements ("locking") to complete the elongation cycle. Under conditions that slowed elongation, a reversible binding step can be observed. Towards this end, they incubated labeled (*, biotinylated) Q29 with unlabeled PolyGln aggregates; loosely bound biotinylated Q29 was trapped on the growth sites by adding an excess of unlabeled PolyGln peptide

describing the monomer–nucleus equilibrium—was related well to PolyGln repeat length, and could be used to predict aggregation lag time. The relationship between PolyGln repeat length and predicted lag time was inversely correlated with, and could help to explain, age-of-onset of HD (Scherzinger et al. 1999; Chen et al. 2001, 2002b; Bhattacharyya et al. 2005). It is clear that many questions remain to be answered about PolyGln aggregation and PADs with PolyGln expansion.

One attempt at rational design of PolyGln inhibitor examined eight permutations of *N*-methylation of short PolyGln peptides as potential PolyGln aggregation inhibitors (Lanning et al. 2010). Since PolyGln peptides contain both backbone and side-chain amides, it is not clear, *a priori*, which amide should be methylated to inhibit aggregation, and which should be retained to allow binding to the target PolyGln peptide. Surprisingly, the most effective inhibitor, called 5QMe$_2$ (sequence: Anth–K–Q–Q(Me$_2$)–Q–Q(Me$_2$)–Q–CONH$_2$, where Anth is *N*-methylanthranilic acid and Q(Me$_2$) is side-chain *N*-methyl Gln), includes only side-chain methylations at alternate residues. Although somewhat similar to the *N*-methylated inhibitors of Aβ aggregation (Gordon et al. 2001, 2002), there are also important differences. The Aβ aggregation inhibitors, such as Aβ(16–20)m, are highly soluble, monomeric β-strands, and both inhibit aggregation and disassemble pre-formed fibrils. They bind to oligomers and fibrils, i.e., to Aβ peptides with some degree of β-sheet structure. In contrast, 5QMe$_2$ has a PolyPro-II-helix-like structure, and binds to PolyGln peptides when they are also in this conformation. While $5QMe_2$ is an effective aggregation inhibitor, it does not disassemble pre-formed PolyGln fibrils, and by inference, does not recognize β-sheet forms of this peptide. Furthermore, although Aβ(16–20)m inhibits Aβ aggregation by blocking backbone hydrogen-bond formation, it binds to Aβ mainly through side-chain interactions. This is shown by the fact that Aβ(16–20)m inhibits aggregation of Aβ peptides, but not other peptides, even those that aggregate through the hydrophobic effect, such as the human prion protein residues 106–129. In contrast, $5QMe_2$ binds to PolyGln through its remaining backbone hydrogen bonds, while blocking aggregation through side-chain hydrogen bonding. The $5QMe_2$ inhibitor highlights the importance of side-chain interactions in PolyGln fibrillogenesis. Subsequent experiments showed that $5QMe_2$ makes transient, 1:1 complexes with its target aggregation-prone peptides. Affinity for the target is moderate, but the kinetics of binding and desorption are very rapid, suggesting that this inhibitor acts through a mechanism reminiscent of chaperone proteins—rapidly binding and unbinding their targets, "resetting the clock" in the complex, multistep process, which includes structural transformation, of PolyGln aggregation. The development of aggregation inhibitors, especially through random screening, depends largely on identifying peptides that bind to their target. The results with $5QMe_2$ suggest that binding affinity, *per se*, may not be the only important factor: kinetics of binding and desorption may even be more important in some cases. These results again suggest a scheme for the polyGln aggregation pathway more complex than previously appreciated.

While it is apparent that much work has been accomplished in developing screening methods and identifying inhibitors of PolyGln aggregation, the field has not yet produced a safe and effective therapy for these devastating disorders. Discovery of such molecules will go hand-in-hand with a better understanding of the pathogenesis of these diseases.

14.3.3 Transthyretin (TTR) Amyloidosis

As discussed above, wild-type and mutant TTRs can form amyloid, and the mutant peptides do so because of instability of the native tetrameric protein. The precise causes of this instability vary from point-mutant to point-mutant, but as a generalization, one can say that any point-mutation (or set of them) that perturbs the monomer–dimer–tetramer equilibrium in favor of monomer formation is likely to favor TTR aggregation. The monomer, especially in the case of many of the point-mutants, is prone to partial denaturation, which renders this subunit amyloidogenic (Hurshman et al. 2008).

The key concept in TTR self-association is that it is kinetically controlled. The rate-limiting step in fibril formation, in most, if not all TTR point-mutants, is dissociation of tetramer into monomer. In most cases, the TTR point-mutant does not cause major, or any, disruption of the native fold in the monomer. If mutations did disrupt the monomer fold, the resulting protein would not be likely to be exported by

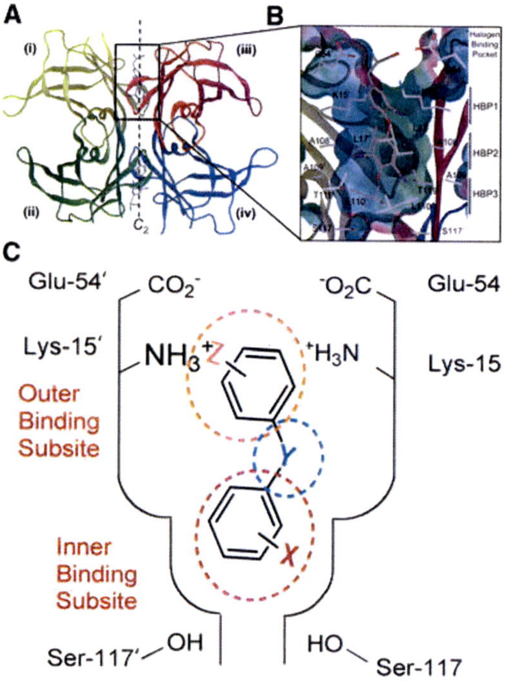

Fig. 14.26 Kinetic stabilization of TTR structure by bound ligands. (**A**) Shows the crystal structure of transthyretin (TTR), containing two molecules of bound T4 within hydrophobic binding pockets (**B**, HBPs). (**C**) Is a schematic of one of the two binding sites for T4, with a bound ligand. X and Z are substituents on the aryl ring, including alkyl, carboxyl, halide, trifluoromethyl, or hydroxyl groups; Y is a flexible linker joining the two aryl rings. Figure is from Connelly et al. (2010)

the cells that synthesize it, nor would it circulate and be able to cause amyloidosis. The dimer–dimer interface, related by a crystallographic twofold axis (C_2) of rotational symmetry, has two binding sites for thyroxin (T4). The binding of T4 stabilizes the tetrameric form of the protein. Binding of T4 shows strong negative cooperativity, however; in human blood, the vast majority of T4 binding sites are unoccupied (Ong and Kelly 2010). Theoretically, therefore, one could prevent TTR aggregation by loading the binding sites with T4. This is not possible, since to do so would require toxic concentrations of this hormone. The strategy against TTR aggregation, therefore, has been based on a search for non-toxic ligands that can bind to the T4-binding site with high affinity, and by occupying this site, stabilize the tetrameric form of TTR.

A closer look at the T4-binding sites of TTR shows that they are comprised of a set of subsites: an outer and inner binding site, with an intervening interface. These sites are made from symmetrical depressions adjacent to hydrophobic amino-acid side-chains, which form the *halogen-binding pockets* where the iodine atoms from T4 bind. Figure 14.26 shows the structure of one of the iodine-binding pockets, occupied by a "stabilizer ligand". The negative cooperativity, which is observed not only for T4 binding but also for other ligands, indicates that the binding of one ligand molecule is sufficient to induce conformational changes in the tetramer. Thus, it is not necessary to bind to two ligand molecules in order to achieve stabilization of the tetramer.

To date, over 1,000 small molecules have been synthesized that bind to the T4-binding site of TTR (Oza et al. 1999, 2002; Petrassi et al. 2000; Klabunde et al. 2000; Razavi et al. 2003; Adamski-Werner et al. 2004; Purkey et al. 2004; Petrassi et al. 2005; Johnson et al. 2005, 2008a, b, 2009). The vast majority of these compounds contain two aromatic rings, like T4, and these rings occupy the inner and outer T4-binding subsites. There are many variations on both the linkers between the two aromatic rings (e.g., linked directly as biphenyls, links through short hydrophobic chains, etc.), and on the ring substituents. The substituents enable these compounds to bind to the T4 sites through both the hydrophobic effect and electrostatic interactions with Glu and Lys residues at the periphery of the binding site. Under physiological conditions, binding effectively blocks dissociation of tetramers into monomers (Hammarström et al. 2003; Wiseman et al. 2005).

It is hardly surprising, of course, that aromatic compounds would bind to a protein. In fact, aromatic compounds bind rather promiscuously to proteins, and aromatic amino acids are involved in most protein–protein interactions (Kossiakoff and Koide 2008; Koide et al. 2007; Fellouse et al. 2007). For any stabilizer compound to qualify as a therapeutic agent, it must also be specific and selective in its binding. For example, it must not bind to other sites where T4 binds, such as the thyroid-hormone receptor, where it could act as an agonist or antagonist. Similarly, although several non-steroidal anti-inflammatory agents qualify as "stabilizer ligands," they are often contraindicated or ill-advised in patients with renal disease, and for this reason, cannot be used in patients with TTR amyloidosis, which causes renal disease (Harirforoosh and Jamali 2009; John and Herzenberg 2009). Thus, it is important to attain selective binding of these agents to TTR. Although fallible, the most common screen for selectivity is to assess affinity of an agent for TTR within the context of human plasma (Ong and Kelly 2010; Almeida et al. 2004).

Using structure-based design and screens of affinity and selectivity, many and widely various TTR stabilizers have been identified. These have included many variants of natural products, especially flavinoids and xanthone derivatives. Among synthetic compounds, five families of compounds are prominent in the long list: bisaryloxime ethers, biphenyls, 1-aryl-4,6-biscarboxydibenzofurans, 2-phenylbenzoxazoles and biphenylamines (Oza et al. 1999, 2002; Hornberg et al. 2000; Petrassi et al. 2000; Klabunde et al. 2000; Razavi et al. 2003; Adamski-Werner et al. 2004; Purkey et al. 2004; Petrassi et al. 2005; Johnson et al. 2005, 2008a, b). Many of the non-steroidal anti-inflammatory drugs (NSAIDs), including diflunisal, and flufenamic and salicylic acids, have been the basis of numerous halogenated variants (Almeida et al. 2004; Baures et al. 1998, 1999; Lueprasitsakul et al. 1990; Maia et al. 2005; Adamski-Werner et al. 2004; Miller et al. 2004; Dolado et al. 2005; Gales et al. 2005; Mairal et al. 2009). Until recently, the most potent and selective agents were the biphenyls, 2-phenylbenzoxazoles and dibenzofurans. Most recently, additional agents, including isatin and β-aminoxypropionic acid linked aryl or fluorenyl derivatives, have been identified and characterized (Gonzalez et al. 2009; Palaninathan et al. 2009).

Aside from biophysical measurements of TTR aggregation, there are several cell-based systems for evaluating the efficacy of inhibitors in mitigating cytotoxicity. For example, the V30M TTR mutant is cytotoxic in the human neuroblastoma cell line, IMR-32. This toxicity, which was attributed to oligomers but not fibrils of TTR, was inhibited by compounds that stabilize the tetrameric structure of this mutant TTR (Reixach et al. 2004). Cytotoxicity occurs in patients who later develop familial amyloid polyneuropathy, through activation of NF-κB, leading to cytokine expression in peripheral nerves; this toxicity begins before the appearance of amyloid deposits (Sousa et al. 2000, 2001). This same cell line, IMR-32 transfected with and expressing V30M TTR, was then used to test the ability of many compounds, especially NSAID (diflunisal) derivatives and a few polyphenols, to inhibit cytotoxicity. The ability of the compounds to inhibit cytotoxicity was compared with their ability to inhibit W30M TTR fibril formation *in vitro*. In general, the correlation between these two measurements was good, but a few compounds that were effective in the cell-based assays were not very active in the fibrillization-inhibition assays (Reixach et al. 2006). The authors attribute these deviations to the fact that the assays *in vitro* are performed at pH 4.4, at which fibril formation of TTR occurs more rapidly than at physiological pH. Resveratrol, a somewhat distant structural relative of T4 or diflunisal, was quite active in both assays. A similar cell-based assay system used a rat Schwannoma cell line transfected with wild-type TTR, or V30M or L55P point-mutant TTR. The occurrence of TTR aggregates was shown by a dot-blot filter assays followed by immunodetection. Using this assay, 12 compounds, previously found to inhibit TTR fibrillization *in vitro*, were assessed for their ability to do so in the cell-based assays. Again, there was a general but not uniform correlation between the results *in vitro* and *in vivo* (Cardoso et al. 2007).

As mentioned, resveratrol was active in the cell-based assay of TTR aggregate cytotoxicity, as well as in the assay of fibrillization inhibition *in vitro*. Resveratrol also bears some structural homology with diethylstilbestrol (DES), since chemically, resveratrol is *trans*-3,4′,5-trihydroxystilbene, i.e., a stilbene derivative (Fig. 14.27a). Crystallographic, and subsequent NMR structures of TTR with bound resveratrol showed that this compound fits well in the T4-binding site while maintaining its own minimal energy conformation (Klabunde et al. 2000; Commodari et al. 2005). Resveratrol, furthermore, bound in two modes related by a 180° rotation about the T4-binding channel. The main contacts were between the aromatic stilbene moiety and hydrophobic side-chains in the pocket. Resveratrol is a phytoestrogen, and the structurally related DES is actually a somewhat better inhibitor of TTR fibrillization *in vitro*. A crystallographic study of TTR with bound DES shows that it, too, binds in two different modes, deep within the T4-binding pocket (Morais-de-Sá et al. 2004, Fig. 14.27b). The two ethyl groups of DES insert snugly through hydrophobic interactions with the protein's halogen-binding pocket. Although DES itself could not be used therapeutically for TTR amyloidosis because of its strong estrogenic activity, it could serve as a basis for the design of other drugs.

Most of the inhibitors of TTR aggregation described above are designed to stabilize the tetramer and prevent its dissociation into the aggregative monomers. An alternative strategy for preventing TTR amyloidosis is the trapping of monomers

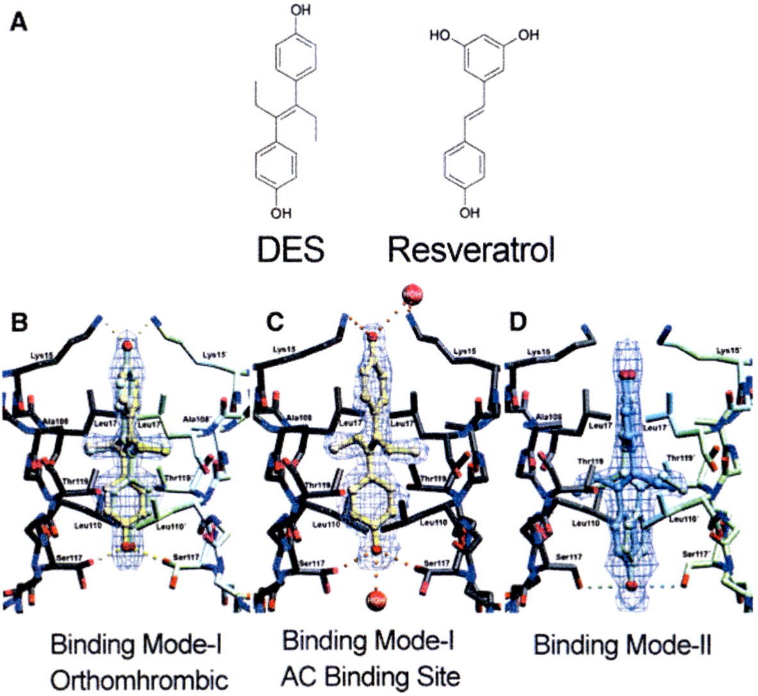

Fig. 14.27 Binding of diethylstilbesterol (DES) and resveratrol to TTR. (**A**) Shows the similarity of the synthetic nonsteroidal estrogen, DES, and the phytoestrogen, resveratrol. (**B–D**) Show the structure of DES in two crystal structures of TTR (Figure from Morais-de-Sá et al. 2004). (**B**) Is binding mode-I for the orthorhombic crystal. (**C**) Is binding mode-I present in the AC binding site of the monoclinic crystal. (**D**) Is binding mode-II, where a shift of about 2 Å toward the center of the channel is observed for the DES position

into a form that does not aggregate. This approach takes advantage of an unusual TTR mutant, T119M, which dissociates 40 times more slowly, and reassembles 90–200 times more slowly than wild-type TTR (Palhano et al. 2009). T119M TTR can be dissociated into denatured monomers by the combination of high pressure and urea concentrations. Upon removal of urea and release of high pressure, the monomers refold, but are long-lived and structurally stable as monomers, only slowly re-forming tetramers. Thus, the monomers can be incorporated with pathogenic mutant forms of TTR, such as L55P and V30M, to form mixed tetramers that are more stable than the mutant tetramers (Hammarström et al. 2002). The resulting mixed tetramers are also less prone to form amyloid than the mutant tetramers (Fig. 14.28).

Finally, the chemistry of TTR aggregation inhibitors can range even to the exotic, as is the case for a new class of inhibitors based on carboranes (dicarba-closo-dodecaboranes). These compounds are icosahedral carbon-containing boron clusters that resist catabolism, and are strongly hydrophobic and inert to many reagents. Their regioselectivity and ease of derivatization allows for facile syntheses of a wide

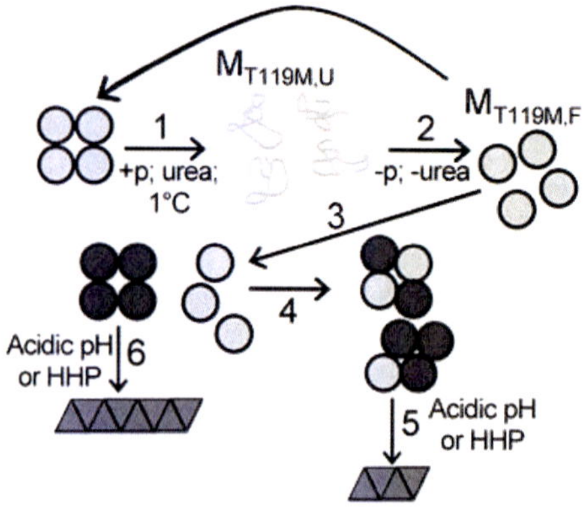

Fig. 14.28 Stabilization of unstable mutant forms of TTR by stable monomers of the T119M mutant. The T119M mutant is very thermostable. It forms unfolded monomers ($M_{T119M,U}$) by the addition of urea at high pressure (+p), and then, upon release of the high pressure (–p) and removal of urea, refolds into stable monomers ($M_{T119M,F}$) that only very slowly associate into tetramers. If the temporally stable T119M monomers (*light circles*) are mixed with other mutant TTR molecules (*dark circles*), mixed tetramers result, containing one or more monomeric units of T119M TTR. The mixed tetramers are less amyloidogenic than many mutant forms of TTR (i.e., other than T119M TTR) (Figure is from Palhano et al. 2009). In the figure, *HHT* high hydrostatic pressure; either HHT or mildly acidic conditions partially denature the protein and induce fibrillization

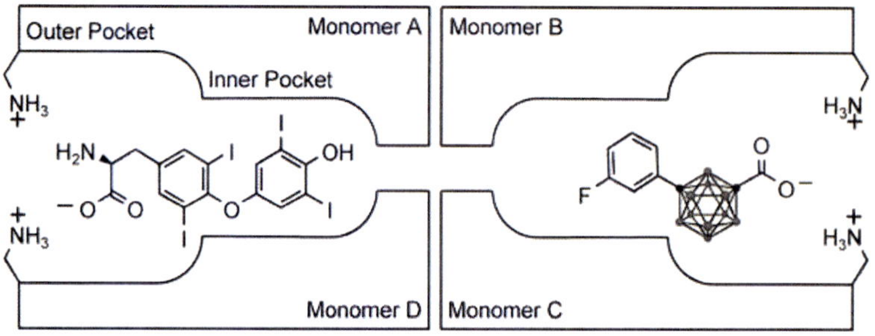

Fig. 14.29 Schematic of TTR tetramer, showing binding of TTR (*left*) and a putative binding of a carborane compound (*right*) (Figure is from Green et al. 2005); the boranes are shown as *grey circles*. The schematic representation of ligand binding sites of TTR is adapted from Green et al. (2005)

variety of novel structures. Carboranes were recently used to synthesize NSAID analogues, but lacking the cyclooxygenase-inhibiting activity of NSAIDs (Julius et al. 2007). Several carboranes were synthesized, and one of these, 1-carboxylic acid-7-[3-fluorophenyl]-1,7-dicarba-closo-dodecaborane, bound to TTR and stabilized its tetrameric form, while showing effectively no COX-1 or COX-2 inhibition at concentrations ~10-fold higher than those needed to inhibit TTR dissociation to monomer (Fig. 14.29).

The above small-molecule agents have been extensively and systematically optimized, and can serve as a model of structure-based design of aggregation inhibitors. As shown by the recent example of the carboranes, this process still has much room for future development.

14.4 Where Do We Go from Here? A Plea for Therapeutic (and Intellectual) Modesty

"In 2006, the worldwide prevalence of AD was 26.6 million. By 2050, the prevalence will quadruple, by which time 1 in 85 persons worldwide will be living with the disease." Between one-third and one-half of these individuals will require intensive care. "If interventions could delay both disease onset and progression by a modest 1 year, there would be nearly 9.2 million fewer cases of the disease in 2050, with nearly the entire decline attributable to decreases in persons needing a high level of care" (Brookmeyer et al. 2007).

Even before the "graying of America," which is occurring now that the baby boomers have started to turn 65, AD is one of the leading causes of death and disability in the US, as it is elsewhere in the world. However, the huge numbers may obscure the individual tragedies. As one writer put it (Post 2000), "A deep fault in the dreams of expanding the human lifespan is that so often dementia steals away all the plans and hopes of retirement and creates unanticipated problems that can break the human spirit." Even putting the numbers aside, finding effective and morally appropriate treatments of AD remains a top healthcare priority.

The history of our attempts to find treatments, however, has not been glorious. Indeed, by some standards, it has been dismal. Thus far, attempts at immunization have failed, or worse, have hastened death in some patients; and treatments to prevent production of Aβ peptides by inhibiting β- or γ-secretase have been disappointing, to say the least. Worse: these failures have cast unwarranted doubt on solid experimental evidence relating Aβ and the pathogenesis of AD. To be clear, we are *not* saying there is no reason to doubt the "amyloid cascade hypothesis" of AD (Hardy and Selkoe 2002). There always have been reasons to question this hypothesis, and the reasons have increased, if anything. However, we are warning against a premature rejection of this hypothesis in favor of other, even less-substantiated hypotheses, which would be as wrong as premature total acceptance of the amyloid cascade hypothesis.

Similarly, attempts to connect high serum cholesterol concentrations to AD have not led to effective treatments. Treating with statin drugs has also been a major disappointment. The many small-molecule compounds, including resveratrol and curcumin, among hundreds or thousands of others, have not failed yet, but one senses, in the public media at least, much hyperbole.

The amyloid cascade hypothesis remains exactly that: a hypothesis. Questioning it is appropriate. Nevertheless, we might start with this as a working hypothesis, and, after reviewing the many proposed treatments for AD, divide them into four broad categories:

1. Preventing the production of Aβ, e.g., through inhibition of β- or γ-secretase inhibitors.
2. Removal of Aβ peptide, including deposited Aβ, e.g., through immunization.
3. Prevention of Aβ aggregation, e.g., through peptides or small molecules that interfere with aggregation of Aβ.
4. Prevention of the effects of Aβ aggregates, e.g., through statins, curcummin, resveratrol, or other putative inhibitors of neuroinflammation.

As discussed earlier in this chapter, one possible flaw (among others) in (1) and (2) is that both of these approaches attempt to target Aβ globally, as an unconditional "enemy" of neuronal health. The physiological regulation of Aβ (Cirrito and Holtzman 2003) implies that some product of β-APP cleavage, likely Aβ itself, has a physiological function, in *some* parts of the brain, at *some* times, at least; and even if this were not true, the enzymes and the rest of the cellular machinery by which Aβ is produced, especially γ-secretase, probably serves some function(s). As for immunization, if given early in life, it runs similar risks as inhibiting the enzymes that produce Aβ. If given late, on the other hand, it might be able to clear existing Aβ deposits, but could also entail complications, such as cerebral hemorrhage. Furthermore, the existence of deposits follows long after many neurons have been damaged, and clearing Aβ deposits may be the neuronal analogue of putting out the cold ashes of a previously raging fire.

All of which does not mean that the above approaches are invalid, only that they *might* be invalid if we knew enough about the functions of Aβ and the cellular machinery for its production. As for (3) and (4), the issues are different. There seems little reason to doubt that Aβ aggregates are capable of harming or killing neurons, although there is still abundant room to debate *which* type or types of aggregates are most culpable. However, in contrast to an approach that targets any and all Aβ, inhibition of protein aggregates *per se* still appears a rational goal. However, will any of the existing reagents be effective treatments for disease? There is a long distance between *in vitro* effects and efficacy in animal models, and perhaps even a longer distance between animal models and effectiveness in humans, and no reagents that block self-association of Aβ or the downstream effects of Aβ aggregates have yet travelled these distances. The appropriate response to these failures is to be chastened; they call for scientific soul-searching, and intellectual modesty. Intellectual modesty is an awareness of the limits of one's knowledge. The opposite of intellectual modesty is *not* audacity, which is a virtue, but arrogance. In the words of Karl Popper (The Open Society and lits Enemies):

> Moreover, indeed, our intellectual as well as our ethical education is corrupt. It is perverted by the admiration of brilliance, of the way that things are said, which takes the place of critical appreciation of the things that are said (and the things that are done). It is perverted by the romantic idea of the splendor of the stage of History on which we are the actors. We are educated to act with an eye to the gallery.

A public that continues to hear over-hyped promises will not continue to support valid research into the causes of AD, and this is exactly what is needed now, perhaps more than ever. We wish to end this chapter not with the usual statement that there is much work to be done—for this is obvious —but rather, with a plea for intellectual and therapeutic modesty. In contrasting the nature of divine and human knowledge, Thomas Aquinas wrote (quoting and commenting on Augustine):

> For the human intellect is measured by *things*, so that a human concept is not true by reason of itself, but by reason of its being consonant with things, since "an opinion is true or false according as it answers to the reality."

References

Abedini A, Raleigh DP (2009) A critical assessment of the role of helical intermediates in amyloid formation by natively unfolded proteins and polypeptides. Protein Eng Des Sel 22:453–459

Åberg V, Norman F, Chorell E, Westermark A, Olofsson A, Sauer-Eriksson AE, Almqvist F (2005) Microwave-assisted decarboxylation of bicyclic 2-pyridone scaffolds and identification of Aβ-peptide aggregation inhibitors. Org Biomol Chem 3:2817–2823

Adachi H, Kume A, Li M, Nakagomi Y, Niwa H, Do J, Sang C, Kobayashi Y, Doyu M, Sobue G (2001) Transgenic mice with an expanded CAG repeat controlled by the human AR promoter show polyglutamine nuclear inclusions and neuronal dysfunction without neuronal cell death. Hum Mol Genet 10:1039–1048

Adamski-Werner SL, Palaninathan SK, Sacchettini JC, Kelly JW (2004) Diflunisal analogues stabilize the native state of transthyretin. Potent inhibition of amyloidogenesis. J Med Chem 47:355–374

Adessi C, Frossard M, Boissard C, Fraga S, Bieler S, Ruckle T, Vilbois F, Robinson SM, Mutter M, Banks WA, Soto C (2003) Pharmacological profiles of peptide drug candidates for the treatment of Alzheimer's disease. J Biol Chem 278:13905–13911

Adlard PA, Cherny RA, Finkelstein DI, Gautier E, Robb E, Cortes M, Volitakis I, Liu X, Smith JP, Perez K, Laughton K, Li Q-X, Charman SA, Nicolazzo JA, Wilkins S, Deleva K, Lynch T, Kok G, Ritchie CW, Tanzi RE, Cappai R, Masters CL, Barnham KJ, Bush AI (2008) Rapid restoration of cognition in Alzheimer's transgenic mice with 8-hydroxyquinoline analogs is associated with decreased interstitial Aβ. Neuron 59:43–55

Aguzzi A, Baumann F, Bremer J (2008) The prion's elusive reason for being. Annu Rev Neurosci 31:439–477

Aisen PS, Saumier D, Briand R, Laurin J, Gervais F, Tremblay P, Garceau D (2006) A phase II study targeting amyloid-β with 3-APS in mild-to-moderate Alzheimer disease. Neurology 67:1757–1763

Aisen PS, Gauthier S, Vellas B, Briand R, Saumier D, Laurin J, Garceau D (2007) Alzhemed: a potential treatment for Alzheimer's disease. Curr Alzheimer Res 4:473–478

Akikusa S, Nakamura K, Watanabe K-I, Horikawa E, Konakahara T, Kodaka M, Okuno H (2003) Practical assay and molecular mechanism of aggregation inhibitors of β-amyloid. J Pept Res 61:1–6

Allison JR, Müller M, van Gunsteren WF (2010) A comparison of the different helices adopted by α- and β-peptides suggests different reasons for their stability. Protein Sci 19:2186–2195

Almeida MR, Macedo B, Cardoso I, Alves I, Valencia G, Arsequell G, Planas A, Saraiva MJ (2004) Selective binding to transthyretin and tetramer stabilization in serum from patients with familial amyloidotic polyneuropathy by an iodinated diflunisal derivative. Biochem J 381:351–356

Alvarez A, Alarcón R, Opazo C, Campos EO, Muñoz FJ, Calderón FH, Dajas F, Gentry MK, Doctor BP, De Mello FG, Inestrosa NC (1998) Stable complexes involving acetylcholinesterase and amyloid-β peptide change the biochemical properties of the enzyme and increase the neurotoxicity of Alzheimer's fibrils. J Neurosci 18:3213–3223

Alzheimer A (1907) Über eine eigenartige Erkrankung der Hirnrinde. Allgemeine Zeitschrift für Psychiatrie 64:146–148

Amijee H, Madine J, Middleton DA, Doig AJ (2009) Inhibitors of protein aggregation and toxicity. Biochem Soc Trans 37:692–696

Andley UP (2009) Effects of α-crystallin on lens cell function and cataract pathology. Curr Mol Med 9:887–892

Ando Y, Nakamura M, Kai H, Katsuragi S, Terazaki H, Nozawa T, Okuda T, Misumi S, Matsunaga N, Hata K, Tajiri T, Shoji S, Yamashita T, Haraoka K, Obayashi K, Matsumoto K, Ando M, Uchino M (2002) A novel localized amyloidosis associated with lactoferrin in the cornea. Lab Invest 82:757–766

Ando Y, Nakamura M, Araki S (2005) Transthyretin-related familial amyloidotic polyneuropathy. Arch Neurol 62:1057–1062

Andreasen N, Hesse C, Davidsson P, Minthon L, Wallin A, Winblad B, Vanderstichele H, Vanmechelen E, Blennow K (1999) Cerebrospinal fluid β-amyloid(1–42) in Alzheimer disease: differences between early- and late-onset Alzheimer disease and stability during the course of disease. Arch Neurol 56:673–680

Andreola A, Bellotti V, Giorgetti S, Mangione P, Obici L, Stoppini M, Torres J, Monzani E, Merlini G, Sunde M (2003) Conformational switching and fibrillogenesis in the amyloidogenic fragment of apolipoprotein A-I. J Biol Chem 278:2444–2451

Andreu JM, Timasheff SN (1986) The measurement of cooperative protein self-assembly by turbidity and other techniques. Methods Enzymol 130:47–59

Andrews ME, Inayathullah NM, Jayakumar R, Malar EJ (2009) Conformational polymorphism and cellular toxicity of IAPP and βAP domains. J Struct Biol 166:116–125

Anfinsen CB (1973) Principles that govern the folding of protein chains. Science 181:223–230

Aquinas T (1265–1274) Summa Theologica I-II, Q93, a1, ad 1

Arakawa T, Tsumoto K (2003) The effects of arginine on refolding of aggregated proteins: not facilitate refolding, but suppress aggregation. Biochem Biophys Res Commun 304:148–152

Arakawa T, Bhat R, Timasheff SN (1990) Why preferential hydration does not always stabilize the native structure of globular proteins. Biochemistry 29:1924–1931

Aravinda S, Shamala N, Roy RS, Balaram P (2003) Non-protein amino acids in peptide design. Proc Indian Acad Sci (Chem Sci) 115:373–400

Arbel M, Solomon B (2007) Immunotherapy for Alzheimer's disease: attacking amyloid-β from the inside. Trends Immunol 28:511–513

Ariga T, McDonald MP, Yu RK (2008) Role of ganglioside metabolism in the pathogenesis of Alzheimer's disease—a review. J Lipid Res 49:1157–1175

Arispe N, Pollard HB, Rojas E (1993) Giant multilevel cation channels formed by Alzheimer disease amyloid β-protein in bilayer membranes. Proc Natl Acad Sci USA 90:10573–10577

Arispe N, Pollard HB, Rojas E (1996) Zn^{2+} interaction with Alzheimer amyloid-β protein calcium channels. Proc Natl Acad Sci USA 93:1710–1715

Armand P, Kirshenbaum K, Falicov A, Dunbrack RL Jr, Dill KA, Zuckermann RN, Cohen FE (1997) Chiral *N*-substituted glycines can form stable helical conformations. Fold Des 2:369–375

Arnold CS, Johnson GV, Cole RN, Dong DL, Lee M, Hart GW (1996) The microtubule-associated protein tau is extensively modified with *O*-linked *N*-acetylglucosamine. J Biol Chem 271:28741–28744

Atwood CS, Moir RD, Huang X, Scarpa RC, Bacarra NM, Romano DM, Hartshorn MA, Tanzi RE, Bush AI (1998) Dramatic aggregation of Alzheimer Aβ by Cu(II) is induced by conditions representing physiological acidosis. J Biol Chem 273:12817–12826

Atwood CS, Huang X, Khatri A, Scarpa RC, Kim YS, Moir RD, Tanzi RE, Roher AE, Bush AI (2000a) Copper catalyzed oxidation of Alzheimer Aβ. Cell Mol Biol (Noisy-le-Grand) 46:777–783

Atwood CS, Scarpa RC, Huang X, Moir RD, Jones WD, Fairlie DP, Tanzi RE, Bush AI (2000b) Characterization of copper interactions with Alzheimer amyloid β peptides: identification of an attomolar-affinity copper binding site on amyloid β1–42. J Neurochem 75:1219–1233

Atwood CS, Perry G, Zeng H, Kato Y, Jones WD, Ling KQ, Huang X, Moir RD, Wang D, Sayre LM, Smith MA, Chen SG, Bush AI (2004) Copper mediates dityrosine cross-linking of Alzheimer's amyloid-β. Biochemistry 43:560–568

Auluck PK, Caraveo G, Lindquist S (2010) α-Synuclein: membrane interactions and toxicity in Parkinson's disease. Annu Rev Cell Dev Biol 26:211–233

Austen BM, Paleologou KE, Ali SAE, Qureshi MM, Allsop D, El-Agna OMA (2008) Designing peptide inhibitors for oligomerization and toxicity of Alzheimer's β-amyloid peptide. Biochemistry 47:1984–1992

Bagriantsev S, Liebman SW (2004) Specificity of prion assembly *in vivo*. [*PSI*+] and [*PIN*+] form separate structures in yeast. J Biol Chem 279:51042–51048

Balbach JJ, Ishii Y, Antzutkin ON, Leapman RD, Rizzo NW, Dyda F, Reed J, Tycko R (2000) Amyloid fibril formation by Aβ16–22, a seven-residue fragment of the Alzheimer's β-amyloid peptide, and structural characterization by solid state NMR. Biochemistry 39:13748–13759

Balbach JJ, Petkova AT, Oyler NA, Antzutkin ON, Gordon DJ, Meredith SC, Tycko R (2002) Supramolecular structure in full-length Alzheimer's β-amyloid fibrils: evidence for a parallel β-sheet organization from solid-state nuclear magnetic resonance. Biophys J 83:1205–1219

Balch WE, Morimoto RI, Dillin A, Kelly JW (2008) Adapting proteostasis for disease intervention. Science 319:916–919

Ballatore C, Lee VM, Trojanowski JQ (2007) Tau-mediated neurodegeneration in Alzheimer's disease and related disorders. Nat Rev Neurosci 8:663–672

Bard F, Barbour R, Cannon C, Carretto R, Fox M, Games D, Guido T, Hoenow K, Hu K, Johnson-Wood K, Khan K, Kholodenko D, Lee C, Lee M, Motter R, Nguyen M, Reed A, Schenk D, Tang P, Vasquez N, Seubert P, Yednock T (2003) Epitope and isotype specificities of antibodies to β-amyloid for protection against Alzheimer's disease-like neuropathology. Proc Natl Acad Sci USA 100:2023–2028

Barmada SJ, Finkbeiner S (2010) Pathogenic TARDBP mutations in amyotrophic lateral sclerosis and frontotemporal dementia: disease-associated pathways. Rev Neurosci 21:251–272

Bartolini M, Bertucci C, Cavrini V, Andrisano V (2003) β-Amyloid aggregation induced by human acetylcholinesterase: inhibition studies. Biochem Pharmacol 65:407–416

Bastianetto S, Ramassamy C, Doré S, Christen Y, Poirier J, Quirion R (2000) The *Ginkgo biloba* extract (EGb 761) protects hippocampal neurons against cell death induced by β-amyloid. Eur J Neurosci 12:1882–1890

Bateman RJ, Siemers ER, Mawuenyega KG, Wen G, Browning KR, Sigurdson WC, Yarasheski KE, Friedrich SW, Demattos RB, May PC, Paul SM, Holtzman DM (2009) A γ-secretase inhibitor decreases amyloid-β production in the central nervous system. Ann Neurol 66:48–54

Baum L, Lam CW, Cheung SK, Kwok T, Lui V, Tsoh J, Lam L, Leung V, Hui E, Ng C, Woo J, Chiu HF, Goggins WB, Zee BC, Cheng KF, Fong CY, Wong A, Mok H, Chow MS, Ho PC, Ip SP, Ho CS, Yu XW, Lai CY, Chan MH, Szeto S, Chan IH, Mok V (2008) Six-month randomized, placebo-controlled, double-blind, pilot clinical trial of curcumin in patients with Alzheimer disease. J Clin Psychopharmacol 28:110–113

Baures PW, Peterson SA, Kelly JW (1998) Discovering transthyretin amyloid fibril inhibitors by limited screening. Bioorg Med Chem 6:1389–1401

Baures PW, Oza VB, Peterson SA, Kelly JW (1999) Synthesis and evaluation of inhibitors of transthyretin amyloid formation based on the nonsteroidal anti-inflammatory drug flufenamic acid. Bioorg Med Chem 7:1339–1347

Baxa U, Wickner RB, Steven AC, Anderson DE, Marekov LN, Yau WM, Tycko R (2007) Characterization of β-sheet structure in Ure2p1–89 yeast prion fibrils by solid-state nuclear magnetic resonance. Biochemistry 46:13149–13162

Bayer AJ, Bullock R, Jones RW, Wilkinson D, Paterson KR, Jenkins L, Millais SB, Donoghue S (2005) Evaluation of the safety and immunogenicity of synthetic Aβ42 (AN1792) in patients with AD. Neurology 64:94–101

Bayro MJ, Maly T, Birkett NR, Macphee CE, Dobson CM, Griffin RG (2010) High-resolution MAS NMR analysis of PI3-SH3 amyloid fibrils: backbone conformation and implications for protofilament assembly and structure. Biochemistry 49:7474–7484

Bednar MM (2009) Anti-amyloid antibody drugs in clinical testing for Alzheimer's disease. IDrugs 12:566–575

Beligere GS, Dawson PE (2000) Design, synthesis and characterization of 4-ester CI2, a model for backbone hydrogen bonding in protein α-helices. J Am Chem Soc 122:12079–12082

Bellotti V, Mangione P, Merlini G (2000) Immunoglobulin light chain amyloidosis—the archetype of structural and pathogenic variability. J Struct Biol 130:280–289

Belluti F, Rampa A, Piazzi L, Bisi A, Gobbi S, Bartolini M, Andrisano V, Cavalli A, Recanatini M, Valenti P (2005) Cholinesterase inhibitors: xanthostigmine derivatives blocking the acetylcholinesterase-induced β-amyloid aggregation. J Med Chem 48:4444–4456

Bence NF, Sampat RM, Kopito RR (2001) Impairment of the ubiquitin–proteasome system by protein aggregation. Science 292:1552–1555

Bennett MC, Bishop JF, Leng Y, Chock PB, Chase TN, Mouradian MM (1999) Degradation of α-synuclein by proteasome. J Biol Chem 274:33855–33858

Bensimon G, Lacomblez L, Meininger V (1994) A controlled trial of riluzole in amyotrophic lateral sclerosis. ALS/Riluzole Study Group. N Engl J Med 330:585–591

Benson MD, Liepnieks J, Uemichi T, Wheeler G, Correa R (1993) Hereditary renal amyloidosis associated with a mutant fibrinogen α-chain. Nat Genet 3:252–255

Benson MD, Liepnieks JJ, Yazaki M, Yamashita T, Hamidi AK, Guenther B, Kluve-Beckerman B (2001) A new human hereditary amyloidosis: the result of a stop-codon mutation in the apolipoprotein AII gene. Genomics 72:272–277

Benzinger TL, Gregory DM, Burkoth TS, Miller-Auer H, Lynn DG, Botto RE, Meredith SC (1998) Propagating structure of Alzheimer's β-amyloid(10–35) is parallel β-sheet with residues in exact register. Proc Natl Acad Sci USA 95:13407–13412

Benzinger TL, Gregory DM, Burkoth TS, Miller-Auer H, Lynn DG, Botto RE, Meredith SC (2000) Two-dimensional structure of β-amyloid(10–35) fibrils. Biochemistry 39:3491–3499

Bergström J, Murphy C, Eulitz M, Weiss DT, Westermark GT, Solomon A, Westermark P (2001) Codeposition of apolipoprotein A-IV and transthyretin in senile systemic (ATTR) amyloidosis. Biochem Biophys Res Commun 285:903–908

Bergström J, Murphy CL, Weiss DT, Solomon A, Sletten K, Hellman U, Westermark P (2004) Two different types of amyloid deposits—apolipoprotein A-IV and transthyretin—in a patient with systemic amyloidosis. Lab Invest 84:981–988

Bero AW, Yan P, Roh JH, Cirrito JR, Stewart FR, Raichle ME, Lee JM, Holtzman DM (2011) Neuronal activity regulates the regional vulnerability to amyloid-β deposition. Nat Neurosci 14:750–756

Bersch B, Koehl P, Nakatani Y, Ourisson G, Milon A (1993) ^{1}H nuclear magnetic resonance determination of the membrane-bound conformation of senktide, a highly selective neurokinin B agonist. J Biomol NMR 3:443–461

Betarbet R, Canet-Aviles RM, Sherer TB, Mastroberardino PG, McLendon C, Kim JH, Lund S, Na HM, Taylor G, Bence NF, Kopito R, Seo BB, Yagi T, Yagi A, Klinefelter G, Cookson MR, Greenamyre JT (2006) Intersecting pathways to neurodegeneration in Parkinson's disease: effects of the pesticide rotenone on DJ-1, α-synuclein, and the ubiquitin–proteasome system. Neurobiol Dis 22:404–420

Bezprozvanny I (2009) Calcium signaling and neurodegenerative diseases. Trends Mol Med 15:89–100

Bhatnagar S, Rao GS, Singh TP (1995) The role of dehydro-alanine in the design of peptides. Biosystems 34:143–148

Bhattacharyya AM, Thakur AK, Wetzel R (2005) Polyglutamine aggregation nucleation: thermodynamics of a highly unfavorable protein folding reaction. Proc Natl Acad Sci USA 102:15400–15405

Bibl M, Mollenhauer B, Esselmann H, Schneider M, Lewczuk P, Welge V, Gross M, Falkai P, Kornhuber J, Wiltfang J (2008) Cerebrospinal fluid neurochemical phenotypes in vascular dementias: original data and mini-review. Dement Geriatr Cogn Disord 25:256–265

Bilen J, Bonini NM (2007) Genome-wide screen for modifiers of ataxin-3 neurodegeneration in *Drosophila*. PLoS Genet 3:1950–1964

Bisaglia M, Schievano E, Caporale A, Peggion E, Mammi S (2006) The 11-mer repeats of human α-synuclein in vesicle interactions and lipid composition discrimination: a cooperative role. Biopolymers 84:310–316

Bitan G, Kirkitadze MD, Lomakin A, Vollers SS, Benedek GB, Teplow DB (2003) Amyloid β-protein (Aβ) assembly: Aβ40 and Aβ42 oligomerize through distinct pathways. Proc Natl Acad Sci USA 100:330–335

Blake CC, Geisow MJ, Oatley SJ, Rerat B, Rerat C (1978) Structure of prealbumin: secondary, tertiary and quaternary interactions determined by Fourier refinement at 1.8 Å. J Mol Biol 121:339–356

Blow DM, Chayen NE, Lloyd LF, Saridakis E (1994) Control of nucleation of protein crystals. Protein Sci 3:1638–1643

Boche D, Nicoll JA (2008) The role of the immune system in clearance of Aβ from the brain. Brain Pathol 18:267–278

Boldogha I, Kruzel ML (2008) Colostrinin: an oxidative stress modulator for prevention and treatment of age-related disorders. J Alzheimers Dis 13:303–321

Bolognesi ML, Bartolini M, Cavalli A, Andrisano V, Rosini M, Minarini A, Melchiorre C (2004) Design, synthesis, and biological evaluation of conformationally restricted rivastigmine analogues. J Med Chem 47:5945–5952

Bolognesi ML, Cavalli A, Valgimigli L, Bartolini M, Rosini M, Andrisano V, Recanatini M, Melchiorre C (2007) Multi-target-directed drug design strategy: from a dual binding site acetylcholinesterase inhibitor to a trifunctional compound against Alzheimer's disease. J Med Chem 50:6446–6449

Booth DR, Sunde M, Bellotti V, Robinson CV, Hutchinson WL, Fraser PE, Hawkins PN, Dobson CM, Radford SE, Blake CC, Pepys MB (1997) Instability, unfolding and aggregation of human lysozyme variants underlying amyloid fibrillogenesis. Nature 385:787–793

Bose M, Gestwicki JE, Devasthali V, Crabtree GR, Graef IA (2005) Molecular mechanisms of neurodegeneration. Biochem Soc Trans 33:543–547

Bourhim M, Kruzel M, Srikrishnan T, Nicotera T (2007) Linear quantitation of Aβ aggregation using Thioflavin T: reduction in fibril formation by colostrinin. J Neurosci Methods 160:264–268

Bradley ME, Edskes HK, Hong JY, Wickner RB, Liebman SW (2002) Interactions among prions and prion "strains" in yeast. Proc Natl Acad Sci USA 99(Suppl 4):16392–16399

Brais B (2003) Oculopharyngeal muscular dystrophy: a late-onset polyalanine disease. Cytogenet Genome Res 100:252–260

Brais B (2009) Oculopharyngeal muscular dystrophy: a polyalanine myopathy. Curr Neurol Neurosci Rep 9:76–82

Brais B, Rouleau GA, Bouchard JP, Fardeau M, Tomé FM (1999) Oculopharyngeal muscular dystrophy. Semin Neurol 19:59–66

Bramson HN, Thomas NE, Kaiser ET (1985) The use of *N*-methylated peptides and depsipeptides to probe the binding of heptapeptide substrates to cAMP-dependent protein kinase. J Biol Chem 260:15452–15457

Brandt R, Leger J, Lee G (1995) Interaction of tau with the neural plasma membrane mediated by tau's amino-terminal projection domain. J Cell Biol 131:1327–1340

Braun BC, Glickman M, Kraft R, Dahlmann B, Kloetzel PM, Finley D, Schmidt M (1999) The base of the proteasome regulatory particle exhibits chaperone-like activity. Nat Cell Biol 1:221–226

Brody DL, Holtzman DM (2008) Active and passive immunotherapy for neurodegenerative disorders. Annu Rev Neurosci 31:175–193

Brookmeyer R, Johnson E, Ziegler-Graham K, Arrighi HM (2007) Forecasting the global burden of Alzheimer's disease. Alzheimers Dement 3:186–191

Brown NJ, Wu CW, Seurynck-Servoss SL, Barron AE (2008) Effects of hydrophobic helix length and side chain chemistry on biomimicry in peptoid analogues of SP-C. Biochemistry 47:1808–1818

Brunet A, Sweeney LB, Sturgill JF, Chua KF, Greer PL, Lin Y, Tran H, Ross SE, Mostoslavsky R, Cohen HY, Hu LS, Cheng HL, Jedrychowski MP, Gygi SP, Sinclair DA, Alt FW, Greenberg ME (2004) Stress-dependent regulation of FOXO transcription factors by the SIRT1 deacetylase. Science 303:2011–2015

Bucciantini M, Giannoni E, Chiti F, Baroni F, Formigli L, Zurdo J, Taddei N, Ramponi G, Dobson CM, Stefani M (2002) Inherent toxicity of aggregates implies a common mechanism for protein misfolding diseases. Nature 416:507–511

Buchner J, Rudolph R (1991) Renaturation, purification and characterization of recombinant Fab-fragments produced in *Escherichia coli*. Bio/Technology 9:157–162

Buratti E, Baralle FE (2009) The molecular links between TDP-43 dysfunction and neurodegeneration. Adv Genet 66:1–34

Burkoth TS, Benzinger TLS, Jones DNM, Hallenga K, Meredith SC, Lynn DG (1998) *C*-terminal PEG blocks the irreversible step in β-amyloid(10–35) fibrillogenesis. J Am Chem Soc 120:7655

Bush AI, Pettingell WH, Multhaup G, d Paradis M, Vonsattel JP, Gusella JF, Beyreuther K, Masters CL, Tanzi RE (1994a) Rapid induction of Alzheimer Aβ amyloid formation by zinc. Science 265:1464–1467

Bush AI, Pettingell WH Jr, Paradis MD, Tanzi RE (1994b) Modulation of Aβ adhesiveness and secretase site cleavage by zinc. J Biol Chem 269:12152–12158

Bussell R Jr, Eliezer D (2001) Residual structure and dynamics in Parkinson's disease-associated mutants of α-synuclein. J Biol Chem 276:45996–46003

Bussell R Jr, Eliezer D (2003) A structural and functional role for 11-mer repeats in α-synuclein and other exchangeable lipid binding proteins. J Mol Biol 329:763–778

Byström R, Aisenbrey C, Borowik T, Bokvist M, Lindström F, Sani MA, Olofsson A, Gröbner G (2008) Disordered proteins: biological membranes as two-dimensional aggregation matrices. Cell Biochem Biophys 52:175–189

Cacace MG, Landau EM, Ramsden JJ (1997) The Hofmeister series: salt and solvent effects on interfacial phenomena. Q Rev Biophys 30:241–277

Camps P, Formosa X, Galdeano C, Muñoz-Torrero D, Ramírez L, Gómez E, Isambert N, Lavilla R, Badia A, Clos MV, Bartolini M, Mancini F, Andrisano V, Arce MP, Rodríguez-Franco MI, Huertas O, Dafni T, Luque FJ (2009) Pyrano[3,2-c]quinoline-6-chlorotacrine hybrids as a novel family of acetylcholinesterase- and β-amyloid-directed anti-Alzheimer compounds. J Med Chem 52:5365–5379

Cannon MJ, Williams AD, Wetzel R, Myszka DG (2004) Kinetic analysis of β-amyloid fibril elongation. Anal Biochem 328:67–75

Cardoso I, Almeida MR, Ferreira N, Arsequell G, Valencia G, Saraiva MJ (2007) Comparative *in vitro* and *ex vivo* activities of selected inhibitors of transthyretin aggregation: relevance in drug design. Biochem J 408:131–138

Castano EM, Roher AE, Esh CL, Kokjohn TA, Beach T (2006) Comparative proteomics of cerebrospinal fluid in neuropathologically-confirmed Alzheimer's disease and nondemented elderly subjects. Neurol Res 28:155–163

Castellani RJ, Smith MA, Perry G, Friedland RP (2004) Cerebral amyloid angiopathy: major contributor or decorative response to Alzheimer's disease pathogenesis. Neurobiol Aging 25:599–602

Castellani RJ, Rolston RK, Smith MA (2010) Alzheimer disease. Dis Mon 56:484–546

Cattaneo E, Zuccato C, Tartari M (2005) Normal huntingtin function: an alternative approach to Huntington's disease. Nat Rev Neurosci 6:919–930

Catto M, Aliano R, Carotti A, Cellamare S, Palluotto S, Purgatorio R, De Stradis A, Campagna F (2010) Design, synthesis and biological evaluation of indane-2-arylhydrazinylmethylene-1,3-diones and indol-2-aryldiazenylmethylene-3-ones as β-amyloid aggregation inhibitors. Eur J Med Chem 45:1359–1366

Caughey B, Lansbury PT (2003) Protofibrils, pores, fibrils, and neurodegeneration: separating the responsible protein aggregates from the innocent bystanders. Annu Rev Neurosci 26:267–298
Cha JH (2007) Transcriptional signatures in Huntington's disease. Prog Neurobiol 83:228–248
Chabry J, Caughey B, Chesebro B (1998) Specific inhibition of *in vitro* formation of protease-resistant prion protein by synthetic peptides. J Biol Chem 273:13203–13207
Chafekar SM, Malda H, Merkx M, Meijer EW, Viertl D, Lashuel HA, Baas F, Scheper W (2007) Branched KLVFF tetramers strongly potentiate inhibition of β-amyloid aggregation. Chembiochem 8:1857–1864
Chalifour RJ, McLaughlin RW, Lavoie L, Morissette C, Tremblay N, Boulé M, Sarazin P, Stéa D, Lacombe D, Tremblay P, Gervais F (2003) Stereoselective interactions of peptide inhibitors with the β-amyloid peptide. J Biol Chem 278:34874–34881
Chan JC, Oyler NA, Yau WM, Tycko R (2005) Parallel β-sheets and polar zippers in amyloid fibrils formed by residues 10–39 of the yeast prion protein Ure2p. Biochemistry 44:10669–10680
Chebaro Y, Derreumaux P (2009) Targeting the early steps of Aβ16–22 protofibril disassembly by *N*-methylated inhibitors: a numerical study. Proteins 75:442–452
Chen D, Dou QP (2008) New uses for old copper-binding drugs: converting the pro-angiogenic copper to a specific cancer cell death inducer. Expert Opin Ther Targets 12:739–748
Chen S, Berthelier V, Yang W, Wetzel R (2001) Polyglutamine aggregation behavior *in vitro* supports a recruitment mechanism of cytotoxicity. J Mol Biol 311:173–182
Chen S, Berthelier V, Hamilton JB, O'Nuallain B, Wetzel R (2002a) Amyloid-like features of polyglutamine aggregates and their assembly kinetics. Biochemistry 41:7391–7399
Chen S, Ferrone FA, Wetzel R (2002b) Huntington's disease age-of-onset linked to polyglutamine aggregation nucleation. Proc Natl Acad Sci USA 99:11884–11889
Chen J, Zhou Y, Mueller-Steiner S, Chen LF, Kwon H, Yi S, Mucke L, Gan L (2005) SIRT1 protects against microglia-dependent amyloid-β toxicity through inhibiting NF-κB signaling. J Biol Chem 280:40364–40374
Cheng SY, Pages RA, Saroff HA, Edelhoch H, Robbins J (1977) Analysis of thyroid hormone binding to human serum prealbumin by 8-anilinonaphthalene-1-sulfonate fluorescence. Biochemistry 16:3707–3713
Cheng RP, Gellman SH, DeGrado WF (2001) β-Peptides: from structure to function. Chem Rev 101:3219–3232
Chen-Plotkin AS, Lee VM, Trojanowski JQ (2010) TAR DNA-binding protein 43 in neurodegenerative disease. Nat Rev Neurol 6:211–220
Cherny RA, Atwood CS, Xilinas ME, Gray DN, Jones WD, McLean CA, Barnham KJ, Volitakis I, Fraser FW, Kim Y, Huang X, Goldstein LE, Moir RD, Lim JT, Beyreuther K, Zheng H, Tanzl RE, Masters CL, Bush AI (2001) Treatment with a copper-zinc chelator markedly and rapidly inhibits β-amyloid accumulation in Alzheimer's disease transgenic mice. Neuron 30:665–676
Cheung JC, Deber CM (2008) Misfolding of the cystic fibrosis transmembrane conductance regulator and disease. Biochemistry 47:1465–1473
Chi EY, Frey SL, Winans A, Lam KL, Kjaer K, Majewski J, Lee KY (2010) Amyloid-β fibrillogenesis seeded by interface-induced peptide misfolding and self-assembly. Biophys J 98:2299–2308
Chimon S, Ishii Y (2005) Capturing intermediate structures of Alzheimer's β-amyloid, Aβ(1–40), by solid-state NMR spectroscopy. J Am Chem Soc 127:13472–13473
Chimon S, Shaibat MA, Jones CR, Calero DC, Aizezi B, Ishii Y (2007) Evidence of fibril-like β-sheet structures in a neurotoxic amyloid intermediate of Alzheimer's β-amyloid. Nat Struct Mol Biol 14:1157–1164
Chiti F, Dobson CM (2006) Protein misfolding, functional amyloid, and human disease. Annu Rev Neurosci 75:333–366
Chiti F, Taddei N, Webster P, Hamada D, Fiaschi T, Ramponi G, Dobson CM (1999) Acceleration of the folding of acylphosphatase by stabilization of local secondary structure. Nat Struct Biol 6:380–387
Chrivia JC, Kwok RP, Lamb N, Hagiwara M, Montminy MR, Goodman RH (1993) Phosphorylated CREB binds specifically to the nuclear protein CBP. Nature 365:855–859

Chung KK, Zhang Y, Lim KL, Tanaka Y, Huang H, Gao J, Ross CA, Dawson VL, Dawson TM (2001) Parkin ubiquitinates the α-synuclein-interacting protein, synphilin-1: implications for Lewy-body formation in Parkinson disease. Nat Med 7:1144–1150
Cirrito JR, Holtzman DM (2003) Amyloid β and Alzheimer disease therapeutics: the devil may be in the details. J Clin Invest 112:321–323
Cirrito JR, May PC, O'Dell MA, Taylor JW, Parsadanian M, Cramer JW, Audia JE, Nissen JS, Bales KR, Paul SM, DeMattos RB, Holtzman DM (2003) *In vivo* assessment of brain interstitial fluid with microdialysis reveals plaque-associated changes in amyloid-β metabolism and half-life. J Neurosci 23:8844–8853
Cirrito JR, Yamada KA, Finn MB, Sloviter RS, Bales KR, May PC, Schoepp DD, Paul SM, Mennerick S, Holtzman DM (2005) Synaptic activity regulates interstitial fluid amyloid-β levels *in vivo*. Neuron 48:913–922
Citron M (2004) β-Secretase inhibition for the treatment of Alzheimer's disease—promise and challenge. Trends Pharmacol Sci 25:59–112
Citron M (2010) Alzheimer's disease: strategies for disease modification. Nat Rev Drug Discov 9:387–398
Clark TD, Buriak JM, Kobayashi K, Isler MP, McRee DE, Ghadiri MR (1998) Cylindrical β-sheet peptide assemblies. J Am Chem Soc 120:8949–8962
Cohen FE, Prusiner SB (1998) Pathologic conformations of prion proteins. Annu Rev Biochem 67:793–819
Cole SL, Vassar R (2008) The role of amyloid precursor protein processing by BACE-1, the β-secretase, in Alzheimer disease pathophysiology. J Biol Chem 283:29621–29625
Collinge J (2005) Molecular neurology of prion disease. J Neurol Neurosurg Psychiatry 76:906–919
Comenzo RL (2006) Systemic immunoglobulin light-chain amyloidosis. Clin Lymphoma Myeloma 7:182–185
Commenges D, Scotet V, Renaud S, Jacqmin-Gadda H, Barberger-Gateau P, Dartigues JF (2000) Intake of flavonoids and risk of dementia. Eur J Epidemiol 16:357–363
Commodari F, Khiat A, Ibrahimi S, Brizius AR, Kalkstein N (2005) Comparison of the phytoestrogen *trans*-resveratrol (3,4′,5-trihydroxystilbene) structures from X-ray diffraction and solution NMR. Magn Reson Chem 43:567–672
Connelly S, Choi S, Johnson SM, Kelly JW, Wilson IA (2010) Structure-based design of kinetic stabilizers that ameliorate the transthyretin amyloidoses. Curr Opin Struct Biol 20:54–62
Conway KA, Lee SJ, Rochet JC, Ding TT, Harper JD, Williamson RE, Lansbury PT Jr (2000a) Accelerated oligomerization by Parkinson's disease linked α-synuclein mutants. Ann NY Acad Sci 920:42–45
Conway KA, Lee SJ, Rochet JC, Ding TT, Williamson RE, Lansbury PT Jr (2000b) Acceleration of oligomerization, not fibrillization, is a shared property of both α-synuclein mutations linked to early-onset Parkinson's disease: implications for pathogenesis and therapy. Proc Natl Acad Sci USA 97:571–576
Conway KA, Rochet JC, Bieganski RM, Lansbury PT Jr (2001) Kinetic stabilization of the α-synuclein protofibril by a dopamine–α-synuclein adduct. Science 294:1346–1349
Coomaraswamy J, Kilger E, Wölfing H, Schäfer C, Kaeser SA, Wegenast-Braun BM, Hefendehl JK, Wolburg H, Mazzella M, Ghiso J, Goedert M, Akiyama H, Garcia-Sierra F, Wolfer DP, Mathews PM, Jucker M (2010) Modeling familial Danish dementia in mice supports the concept of the amyloid hypothesis of Alzheimer's disease. Proc Natl Acad Sci USA 107:7969–7974
Cornwell GG 3rd, Sletten K, Johansson B, Westermark P (1998) Evidence that the amyloid fibril protein in senile systemic amyloidosis is derived from normal prealbumin. Biochem Biophys Res Commun 154:648–653
Costa R, Gonçalves A, Saraiva MJ, Cardoso I (2008) Transthyretin binding to A-Beta peptide—Impact on A-Beta fibrillogenesis and toxicity. FEBS Lett 582:936–942
Cribbs DH, Ghochikyan A, Vasilevko V, Tran M, Petrushina I, Sadzikava N, Babikyan D, Kesslak P, Kieber-Emmons T, Cotman CW, Agadjanyan MG (2003) Adjuvant-dependent modulation of T_h1 and T_h2 responses to immunization with β-amyloid. Int Immunol 15:505–514

Cripps D, Thomas SN, Jeng Y, Yang F, Davies P, Yang AJ (2006) Alzheimer disease-specific conformation of hyperphosphorylated paired helical filament-Tau is polyubiquitinated through Lys-48, Lys-11, and Lys-6 ubiquitin conjugation. J Biol Chem 281:10825–10838
Crisma M, Formaggio F, Toniolo C, Yoshikawa T, Wakamiya WJ (1999) Flat peptides. J Am Chem Soc 121:3272–3278
Crowther DC, Kinghorn KJ, Miranda E, Page R, Curry JA, Duthie FA, Gubb DC, Lomas DA (2005) Intraneuronal Aβ, non-amyloid aggregates and neurodegeneration in a *Drosophila* model of Alzheimer's disease. Neuroscience 132:123–135
Crowther DC, Page R, Chandraratna D, Lomas DA (2006) A *Drosophila* model of Alzheimer's disease. Methods Enzymol 412:234–255
Cruz M, Tusell JM, Grillo-Bosch D, Albericio F, Serratosa J, Rabanal F, Giralt E (2004) Inhibition of β-amyloid toxicity by short peptides containing *N*-methyl amino acids. J Pept Res 63:324–328
Cummings CJ, Mancini MA, Antalffy B, DeFranco DB, Orr HT, Zoghbi HY (1998) Chaperone suppression of aggregation and altered subcellular proteasome localization imply protein misfolding in SCA1. Nat Genet 19:148–154
Cummings CJ, Reinstein E, Sun Y, Antalffy B, Jiang Y, Ciechanover A, Orr HT, Beaudet AL, Zoghbi HY (1999) Mutation of the E6-AP ubiquitin ligase reduces nuclear inclusion frequency while accelerating polyglutamine-induced pathology in SCA1 mice. Neuron 24:879–892
Dai Q, Borenstein AR, Wu Y, Jackson JC, Larson EB (2006) Fruit and vegetable juices and Alzheimer's disease: the Kame Project. Am J Med 119:751–759
Damas AM, Saraiva MJ (2000) TTR amyloidosis—structural features leading to protein aggregation and their implications on therapeutic strategies. J Struct Biol 130:290–299
Darnell G, Orgel JPRO, Pahl R, Meredith SC (2007) Flanking polyproline sequences inhibit β-sheet structure in polyglutamine segments by inducing PPII-like helix structure. J Mol Biol 374:688–704
Darnell GD, Derryberry JM, Kurutz JW, Meredith SC (2009) Mechanism of *cis*-inhibition of polyQ fibrillation by polyP: PPII oligomers and the hydrophobic effect. Biophys J 97:2295–2305
Das P, Murphy MP, Younkin LH, Younkin SG, Golde TE (2001) Reduced effectiveness of Aβ1–42 immunization in APP transgenic mice with significant amyloid deposition. Neurobiol Aging 22:721–727
Das U, Hariprasad G, Ethayathulla AS, Manral P, Das TK, Pasha S, Mann A, Ganguli M, Verma AK, Bhat R, Chandrayan SK, Ahmed S, Sharma S, Kaur P, Singh TP, Srinivasan A (2007) Inhibition of protein aggregation: supramolecular assemblies of arginine hold the key. PLoS One 2:e1176
Dasuri K, Ebenezer P, Zhang L, Fernandez-Kim SO, Bruce-Keller AJ, Markesbery WR, Keller JN (2010) Increased protein hydrophobicity in response to aging and Alzheimer disease. Free Radic Biol Med 48:1330–1337
Dauchet L, Amouyel P, Dallongeville J (2005) Fruit and vegetable consumption and risk of stroke: a meta-analysis of cohort studies. Neurology 65:1193–1197
Davidson B, Fasman GD (1967) The conformational transitions of uncharged poly-$_{L}$-lysine. α helix-random coil-β structure. Biochemistry 6:1616–1629
Davidson B, Tooney N, Fasman GD (1966) The optical rotatory dispersion of the β structure of poly-$_{L}$-lysine and poly-$_{L}$-serine. Biochem Biophys Res Commun 23:156–162
Davidson WS, Jonas A, Clayton DF, George JM (1998) Stabilization of α-synuclein secondary structure upon binding to synthetic membranes. J Biol Chem 273:9443–9449
Dawson TM, Dawson VL (2003) Molecular pathways of neurodegeneration in Parkinson's disease. Science 302:819–822
Dawson PE, Kent SB (2000) Synthesis of native proteins by chemical ligation. Annu Rev Biochem 69:923–960
De Bona P, Giuffrida ML, Caraci F, Copani A, Pignataro B, Attanasio F, Cataldo S, Pappalardo G, Rizzarelli E (2009) Design and synthesis of new trehalose-conjugated pentapeptides as inhibitors of Aβ(1–42) fibrillogenesis and toxicity. J Pept Sci 15:220–228

De Ferrari GV, Canales MA, Shin I, Weiner LM, Silman I, Inestrosa NC (2001) A structural motif of acetylcholinesterase that promotes amyloid β-peptide fibril formation. Biochemistry 40:10447–10457

Dealwis C, Wall J (2004) Towards understanding the structure–function relationship of human amyloid disease. Curr Drug Targets 5:159–171

Deane R, Bell RD, Sagare A, Zlokovic BV (2009) Clearance of amyloid-β peptide across the blood–brain barrier: implication for therapies in Alzheimer's disease. CNS Neurol Disord Drug Targets 8:16–30

Dedmon MM, Patel CN, Young GB, Pielak GJ (2002) FlgM gains structure in living cells. Proc Natl Acad Sci USA 99:12681–12684

DeKosky ST, Ikonomovic MD, Gandy S (2010) Traumatic brain injury—football, warfare, and long-term effects. N Engl J Med 363:1293–1296

DeMattos RB, Bales KR, Cummins DJ, Dodart JC, Paul SM, Holtzman DM (2001) Peripheral anti-Aβ antibody alters CNS and plasma Aβ clearance and decreases brain Aβ burden in a mouse model of Alzheimer's disease. Proc Natl Acad Sci USA 98:8850–8855

Desai UA, Pallos J, Ma AA, Stockwell BR, Thompson LM, Marsh JL, Diamond MI (2006) Biologically active molecules that reduce polyglutamine aggregation and toxicity. Hum Mol Genet 15:2114–2124

DeStrooper B (2003) Aph-1, Pen-2, and Nicastrin with Presenilin generate an active γ-secretase complex. Neuron 38:9–12

DeStrooper B, Annaert W (2010) Novel research horizons for presenilins and γ-secretases in cell biology and disease. Annu Rev Cell Dev Biol 26:235–260

Di Giovanni S, Eleuteri S, Paleologou KE, Yin G, Zweckstetter M, Carrupt P-A, Lashuel HA (2010) Entacapone and tolcapone, two catechol *O*-methyltransferase inhibitors, block fibril formation of α-synuclein and β-amyloid and protect against amyloid-induced toxicity. J Biol Chem 285:14941–14954

Di Monte DA (2003) The environment and Parkinson's disease: is the nigrostriatal system preferentially targeted by neurotoxins? Lancet Neurol 2:531–538

DiFiglia M, Sapp E, Chase KO, Davies SW, Bates GP, Vonsattel JP, Aronin N (1997) Aggregation of huntingtin in neuronal intranuclear inclusions and dystrophic neurites in brain. Science 277:1990–1993

Ding WQ, Lind SE (2009) Metal ionophores—an emerging class of anticancer drugs. IUBMB Life 61:1013–1018

Diociaiuti M, Polzi LZ, Valvo L, Malchiodi-Albedi F, Bombelli C, Gaudiano MC (2006) Calcitonin forms oligomeric pore-like structures in lipid membranes. Biophys J 91:2275–2281

Dobson CM (2001) The structural basis of protein folding and its links with human disease. Philos Trans R Soc Lond B 356:133–145

Dogterom P, Nagelkerke JF, Mulder GJ (1988) Hepatotoxicity of tetrahydroaminoacridine in isolated rat hepatocytes: effect of glutathione and vitamin E. Biochem Pharmacol 37:2311–2313

Doig AJ (1997) A three stranded β-sheet peptide in aqueous solution containing *N*-methyl amino acids to prevent aggregation. J Chem Soc Chem Commun 22:2153–2154

Doig AJ, Hughes E, Burke RM, Su TJ, Heenan RK, Lu J (2002) Inhibition of toxicity and protofibril formation in the amyloid-β peptide β(25–35) using *N*-methylated derivatives. Biochem Soc Trans 30:537–542

Dolado I, Nieto J, Saraiva MJ, Arsequell G, Valencia G, Planas A (2005) Kinetic assay for high-throughput screening of *in vitro* transthyretin amyloid fibrillogenesis inhibitors. J Comb Chem 7:246–252

Dorval V, Fraser PE (2006) Small ubiquitin-like modifier (SUMO) modification of natively unfolded proteins tau and α-synuclein. J Biol Chem 281:9919–9924

Dovey HF, John V, Anderson JP, Chen LZ, de Saint Andrieu P, Fang LY, Freedman SB, Folmer B, Goldbach E, Holsztynska EJ, Hu KL, Johnson-Wood KL, Kennedy SL, Kholodenko D, Knops JE, Latimer LH, Lee M, Liao Z, Lieberburg IM, Motter RN, Mutter LC, Nietz J, Quinn KP, Sacchi KL, Seubert PA, Shopp GM, Thorsett ED, Tung JS, Wu J, Yang S, Yin CT, Schenk DB, May PC, Altstiel LD, Bender MH, Boggs LN, Britton TC, Clemens JC, Czilli DL,

Dieckman-McGinty DK, Droste JJ, Fuson KS, Gitter BD, Hyslop PA, Johnstone EM, Li WY, Little SP, Mabry TE, Miller FD, Audia JE (2001) Functional γ-secretase inhibitors reduce β-amyloid peptide levels in brain. J Neurochem 76:173–181

Dumoulin M, Kumita JR, Dobson CM (2006) Normal and aberrant biological self-assembly: insights from studies of human lysozyme and its amyloidogenic variants. Acc Chem Res 39:603–610

Dunker AK, Oldfield CJ, Meng J, Romero P, Yang JY, Chen JW, Vacic V, Obradovic Z, Uversky VN (2008) The unfoldomics decade: an update on intrinsically disordered proteins. BMC Genomics 9(Suppl 2):S1

Durairajan SS, Yuan Q, Xie L, Chan WS, Kum WF, Koo I, Liu C, Song Y, Huang JD, Klein WL, Li M (2008) Salvianolic acid B inhibits Aβ fibril formation and disaggregates preformed fibrils and protects against Aβ-induced cytotoxicity. Neurochem Int 52:741–750

Dzwolak W, Ravindra R, Nicolini C, Jansen R, Winter R (2004) The diastereomeric assembly of polylysine is the low-volume pathway for preferential formation of β-sheet aggregates. J Am Chem Soc 126:3762–3768

Eaton WA, Hofrichter J (1990) Sickle cell hemoglobin polymerization. Adv Protein Chem 40:63–279

Eaton WA, Hofrichter J (1995) The biophysics of sickle cell hydroxyurea therapy. Science 268:1142–1143

Eikelenboom P, van Exel E, Hoozemans JJ, Veerhuis R, Rozemuller AJ, van Gool WA (2010) Neuroinflammation—an early event in both the history and pathogenesis of Alzheimer's disease. Neurodegener Dis 7:38–41

El-Agnaf OM, Paleologou KE, Greer B, Abogrein AM, King JE, Salem SA, Fullwood NJ, Benson FE, Hewitt R, Ford KJ, Martin FL, Harriott P, Cookson MR, Allsop D (2004) A strategy for designing inhibitors of α-synuclein aggregation and toxicity as a novel treatment for Parkinson's disease and related disorders. FASEB J 18:1315–1317

Elgersma RC, Mulder GE, Kruijtzer JA, Posthuma G, Rijkers DT, Liskamp RM (2007) Transformation of the amyloidogenic peptide amylin(20–29) into its corresponding peptoid and retropeptoid: access to both an amyloid inhibitor and template for self-assembled supramolecular tapes. Bioorg Med Chem Lett 17:1837–1842

Elseviers M, Van der Auwera L, Pepermans H, Tourwe D, Van Binst G (1988) Evidence for the bioactive conformation in a cyclic hexapeptide analogue of somatostatin containing a *cis*-peptide bond mimic. Biochem Biophys Res Commun 154:515–521

Emmanouilidou E, Stefanis L, Vekrellis K (2010) Cell-produced α-synuclein oligomers are targeted to, and impair, the 26S proteasome. Neurobiol Aging 31:953–968

Esler WP, Stimson ER, Fishman JB, Ghilardi JR, Vinters HV, Mantyh PW, Maggio JE (1999) Stereochemical specificity of Alzheimer's disease-peptide assembly. Biopolymers 49: 505–514

Esposito L, Paladino A, Pedone C, Vitagliano L (2008) Insights into structure, stability, and toxicity of monomeric and aggregated polyglutamine models from molecular dynamics simulations. Biophys J 94:4031–4040

Eulitz M, Weiss DT, Solomon A (1990) Immunoglobulin heavy-chain-associated amyloidosis. Proc Natl Acad Sci USA 87:6542–6546

Ezoulin MJ, Dong CZ, Liu Z, Li J, Chen HZ, Heymans F, Lelièvre L, Ombetta JE, Massicot F (2006) Study of PMS777, a new type of acetylcholinesterase inhibitor, in human HepG2 cells. Comparison with tacrine and galanthamine on oxidative stress and mitochondrial impairment. Toxicol *In Vitro* 20:824–831

Fadika GO, Baumann M (2002) Peptides corresponding to gelsolin derived amyloid of the Finnish type (AGelFIN) adopt two distinct forms in solution of which only one can polymerize into amyloid fibrils and form complexes with apoE. Amyloid 9:75–82

Fändrich M, Fletcher MA, Dobson CM (2001) Amyloid fibrils from muscle myoglobin. Nature 410:165–166

Fändrich M, Meinhardt J, Grigorieff N (2009) Structural polymorphism of Alzheimer Aβ and other amyloid fibrils. Prion 3:89–93

Fang L, Appenroth D, Decker M, Kiehntopf M, Roegler C, Deufel T, Fleck C, Peng S, Zhang Y, Lehmann J (2008) Synthesis and biological evaluation of NO-donor-tacrine hybrids as hepatoprotective anti-Alzheimer drug candidates. J Med Chem 51:713–716

Federoff HJ (2009) Development of vaccination approaches for the treatment of neurological diseases. J Comp Neurol 515:4–14

Fellouse FA, Esaki K, Birtalan S, Raptis D, Cancasci VJ, Koide A, Jhurani P, Vasser M, Wiesmann C, Kossiakoff AA, Koide S, Sidhu SS (2007) High-throughput generation of synthetic antibodies from highly functional minimalist phage-displayed libraries. J Mol Biol 373:924–940

Feng Y, Wang XP, Yang SG, Wang YJ, Zhang X, Du XT, Sun XX, Zhao M, Huang L, Liu RT (2009) Resveratrol inhibits β-amyloid oligomeric cytotoxicity but does not prevent oligomer formation. Neurotoxicology 30:986–995

Fernández-Bachiller MI, Pérez C, González-Muñoz GC, Conde S, López MG, Villarroya M, García AG, Rodríguez-Franco MI (2010) Novel tacrine-8-hydroxyquinoline hybrids as multifunctional agents for the treatment of Alzheimer's disease, with neuroprotective, cholinergic, antioxidant, and copper complexing properties. J Med Chem 53:4927–4937

Ferreira ST, Vieira MN, De Felice FG (2007) Soluble protein oligomers as emerging toxins in Alzheimer's and other amyloid diseases. IUBMB Life 59:332–345

Ferrer I, Boada Rovira M, Sánchez Guerra ML, Rey MJ, Costa-Jussá F (2004) Neuropathology and pathogenesis of encephalitis following amyloid-β immunization in Alzheimer's disease. Brain Pathol 14:11–20

Ferrone FA, Hofrichter J, Eaton WA (1985) Kinetics of sickle hemoglobin polymerization. II. A double nucleation mechanism. J Mol Biol 183:611–631

Figueroa KP, Pulst SM (2003) Identification and expression of the gene for human ataxin-2-related protein on chromosome 16. Exp Neurol 184:669–678

Findeis MA, Musso GM, Arico-Muendel CC, Benjamin HW, Hundal AM, Lee JJ, Chin J, Kelley M, Wakefield J, Hayward NJ, Molineaux SM (1999) Modified peptide inhibitors of amyloid β-peptide polymerization. Biochemistry 38:6791–6800

Fink AL (2005) Natively unfolded proteins. Curr Opin Struct Biol 15:35–41

Fink AL (2006) The aggregation and fibrillation of α-synuclein. Acc Chem Res 39:628–634

Finsterer J (2009) Bulbar and spinal muscular atrophy (Kennedy's disease): a review. Eur J Neurol 16:556–561

Fleisher AS, Raman R, Siemers ER, Becerra L, Clark CM, Dean RA, Farlow MR, Galvin JE, Peskind ER, Quinn JF, Sherzai A, Sowell BB, Aisen PS, Thal LJ (2008) Phase 2 safety trial targeting amyloid β production with a γ-secretase inhibitor in Alzheimer disease. Arch Neurol 65:1031–1038

Floros J, Kala P (1998) Surfactant proteins: molecular genetics of neonatal pulmonary diseases. Annu Rev Physiol 60:365–384

Fonte V, Kapulkin V, Taft A, Fluet A, Friedman D, Link CD (2002) Interaction of intracellular β-amyloid peptide with chaperone proteins. Proc Natl Acad Sci USA 99:9439–9444

Fonte V, Kipp DR, Yerg J 3rd, Merin D, Forrestal M, Wagner E, Roberts CM, Link CD (2007) Suppression of *in vivo* β-amyloid peptide toxicity by overexpression of the HSP-16.2 small chaperone protein. J Biol Chem 283:784–791

Foster JK, Verdile G, Bates KA, Martins RN (2009) Immunization in Alzheimer's disease: naïve hope or realistic clinical potential? Mol Psychiatry 14:239–251

Fowler SA, Blackwell HE (2009) Structure–function relationships in peptoids: recent advances toward deciphering the structural requirements for biological function. Org Biomol Chem 7:1508–1524

Fowler SA, Stacy DM, Blackwell HE (2008) Design and synthesis of macrocyclic peptomers as mimics of a quorum sensing signal from Staphylococcus aureus. Org Lett 10:2329–2332

Fox NC, Black RS, Gilman S, Rossor MN, Griffith SG, Jenkins L, Koller M (2005) AN1792(QS-21)-201 Study. Effects of Aβ immunization (AN1792) on MRI measures of cerebral volume in Alzheimer disease. Neurology 64:1563–1572

Friedman MJ, Wang CE, Li XJ, Li S (2008) Polyglutamine expansion reduces the association of TATA-binding protein with DNA and induces DNA binding-independent neurotoxicity. J Biol Chem 283:8283–8290

Fu H, Li W, Luo J, Lee NTK, Li M, Tsim KWK, Pang Y, Youdim MBH, Han Y (2008) Promising anti-Alzheimer's dimer *bis*(7)-tacrine reduces β-amyloid generation by directly inhibiting BACE-1 activity. Biochem Biophys Res Commun 366:631–636

Fu HJ, Liu B, Frost JL, Lemere CA (2010) Amyloid-β immunotherapy for Alzheimer's disease. CNS Neurol Disord Drug Targets 9:197–206

Fuhrhop JH, Krull M, Büldt G (1987) Precipitates with β-pleated sheet structure by mixing aqueous solutions of helical poly($_D$-lysine) and Poly($_L$-lysine). Angew Chem Int Ed Engl 26:699–700

Fülöp L, Zarándi M, Datki Z, Soós K, Penke B (2004) β-Amyloid-derived pentapeptide RIIGLa inhibits Aβ1–42 aggregation and toxicity. Biochem Biophys Res Commun 324:64–69

Fuxreiter M, Simon I, Friedrich P, Tompa P (2004) Preformed structural elements feature in partner recognition by intrinsically unstructured proteins. J Mol Biol 338:1015–1026

Galasko D, Chang L, Motter R, Clark CM, Kaye J, Knopman D, Thomas R, Kholodenko D, Schenk D, Lieberburg I, Miller B, Green R, Basherad R, Kertiles L, Boss MA, Seubert P (1998) High cerebrospinal fluid tau and low amyloid β42 levels in the clinical diagnosis of Alzheimer disease and relation to apolipoprotein E genotype. Arch Neurol 55:937–945

Gales L, Macedo-Ribeiro S, Arsequell G, Valencia G, Saraiva MJ, Damas AM (2005) Human transthyretin in complex with iododiflunisal—structural features associated with a potent amyloid inhibitor. Biochem J 388:615–621

Galván M, David JP, Delacourte A, Luna J, Mena R (2001) Sequence of neurofibrillary changes in aging and Alzheimer's disease: a confocal study with phospho-tau antibody, AD2. J Alzheimers Dis 3:417–425

Gambetti P, Russo C (1998) Human brain amyloidoses. Nephrol Dial Transplant 13(Suppl 7):33–40

Games D, Adams D, Alessandrini R, Barbour R, Berthelette P, Blackwell C, Carr T, Clemens L, Donaldson T, Gillespie E, Guido T, Hagopian S, Johnson-Wood K, Khan K, Lee M, Leibowitz P, Lieberburg I, Little S, Masliah E, McConlogue L, Montoya-Zavala M, Mucke L, Paganini L, Penniman E, Power M, Schenk D, Seubert P, Snyder B, Soriano F, Tan H, Vitale J, Wadworth S, Wolozin B, Zhao J (1995) Alzheimer-type neuropathology in transgenic mice overexpressing V717F β-amyloid precursor protein. Nature 373:523–527

Gao R, Matsuura T, Coolbaugh M, Zühlke C, Nakamura K, Rasmussen A, Siciliano MJ, Ashizawa T, Lin X (2008) Instability of expanded CAG/CAA repeats in spinocerebellar ataxia type 17. Eur J Hum Genet 16:215–222

García-Palomero E, Muñoz P, Usan P, Garcia P, Delgado E, De Austria C, Valenzuela R, Rubio L, Medina M, Martínez A (2008) Potent β-amyloid modulators. Neurodegener Dis 5:153–156

Gardberg AS, Dice LT, Ou S, Rich RL, Helmbrecht E, Ko J, Wetzel R, Myszka DG, Patterson PH, Dealwis C (2007) Molecular basis for passive immunotherapy of Alzheimer's disease. Proc Natl Acad Sci USA 104:15659–15664

Garden GA, La Spada AR (2008) Molecular pathogenesis and cellular pathology of spinocerebellar ataxia type 7 neurodegeneration. Cerebellum 7:138–149

Gauthier S, Aisen PS, Ferris SH, Saumier D, Duong A, Haine D, Garceau D, Suhy J, Oh J, Lau W, Sampalis J (2007) Effect of tramiprosate in patients with mild-to-moderate Alzheimer's disease: exploratory analyses of the MRI sub-group of the Alphase study. J Nutr Health Aging 13:550–557

Genschel J, Haas R, Propsting MJ, Schmidt HH (1998) Apolipoprotein A-I induced amyloidosis. FEBS Lett 430:145–149

Gervais F, Chailfour R, Garceau D, Kong X, Laurin J, McLaughlin R, Morissette C, Paquette J (2001) Glycosaminoglycan mimetics: a therapeutic approach to cerebral amyloid angiopathy. Amyloid 8(Suppl 1):28–35

Gervais F, Paquette J, Morissette C, Krzywkowski P, Yu M, Azzi M, Lacombe D, Kong X, Aman A, Laurin J, Szarek WA, Tremblay P (2006) Targeting soluble Aβ peptide with Tramiprosate for the treatment of brain amyloidosis. Neurobiol Aging 28:537–547

Geser F, Martinez-Lage M, Kwong LK, Lee VM, Trojanowski JQ (2009) Amyotrophic lateral sclerosis, frontotemporal dementia and beyond: the TDP-43 diseases. J Neurol 256:1205–1214

Geula C, Mesulam M (1989) Special properties of cholinesterases in the cerebral cortex of Alzheimer's disease. Brain Res 498:185–189

Ghiso J, Jensson O, Frangione B (1986) Amyloid fibrils in hereditary cerebral hemorrhage with amyloidosis of Icelandic type is a variant of α-trace basic protein (cystatin C). Proc Natl Acad Sci USA 83:2974–2978

Giacomelli CE, Norde W (2003) Influence of hydrophobic teflon particles on the structure of amyloid β-peptide. Biomacromolecules 4:1719–1726

Giasson BI, Duda JE, Murray IV, Chen Q, Souza JM, Hurtig HI, Ischiropoulos H, Trojanowski JQ, Lee VM (2000) Oxidative damage linked to neurodegeneration by selective α-synuclein nitration in synucleinopathy lesions. Science 290:985–989

Gibson TJ, Murphy RM (2005) Design of peptidyl compounds that affect β-amyloid aggregation: importance of surface tension and context. Biochemistry 44:8898–8907

Gibson G, El-Agnaf OM, Anwar Z, Sidera C, Isbister A, Austen BM (2005) Structure and neurotoxicity of novel amyloids derived from the BRI gene. Biochem Soc Trans 33:1111–1112

Gilead S, Gazit E (2004) Inhibition of amyloid fibril formation by peptide analogues modified with α-aminoisobutyric acid. Angew Chem Int Ed Engl 43:4041–4044

Giordano C, Masi A, Pizzini A, Sansone A, Consalvi V, Chiaraluce R, Lucente G (2009) Synthesis and activity of fibrillogenesis peptide inhibitors related to the 17–21 β amyloid sequence. Eur J Med Chem 44:179–189

Giunta S, Valli MB, Galeazzi R, Fattoretti P, Corder EH, Galeazzi L (2005) Transthyretin inhibition of amyloid β aggregation and toxicity. Clin Biochem 38:1112–1119

Glabe CG (2008) Structural classification of toxic amyloid oligomers. J Biol Chem 283:29639–29643

Gladkevich A, Bosker F, Korf J, Yenkoyan K, Vahradyan H, Aghajanov M (2007) Proline-rich polypeptides in Alzheimer's disease and neurodegenerative disorders—Therapeutic potential or a mirage? Prog Neuropsychopharmacol Biol Psychiatry 31:1347–1355

Glenner GG, Bladen HA (1966) purification and reconstitution of the periodic fibril and unit structure of human amyloid. Science 154:271–272

Glenner GG, Wong CW (1984a) Alzheimer's disease: initial report of the purification and characterization of a novel cerebrovascular amyloid protein. Biochem Biophys Res Commun 120:885–890

Glenner GG, Wong CW (1984b) Alzheimer's disease and Down's syndrome: sharing of a unique cerebrovascular amyloid fibril protein. Biochem Biophys Res Commun 122:1131–1135

Goedert M, Spillantini MG (2006) A century of Alzheimer's disease. Science 314:777–781

Gonzalez A, Quirante J, Nieto J, Almeida MR, Saraiva MJ, Planas A, Arsequell G, Valencia G (2009) Isatin derivatives, a novel class of transthyretin fibrillogenesis inhibitors. Bioorg Med Chem Lett 19:5270–5273

Gooptu B, Hazes B, Chang WS, Dafforn TR, Carrell RW, Read RJ, Lomas DA (2000) Inactive conformation of the serpin α(1)-antichymotrypsin indicates two-stage insertion of the reactive loop: implications for inhibitory function and conformational disease. Proc Natl Acad Sci USA 97:67–72

Gordon DJ, Sciarretta KL, Meredith SC (2001) Inhibition of β-amyloid(40) fibrillogenesis and disassembly of β-amyloid(40) fibrils by short β-amyloid congeners containing *N*-methyl amino acids at alternate residues. Biochemistry 40:8237–8245

Gordon DJ, Tappe R, Meredith SC (2002) Design and characterization of a membrane permeable *N*-methyl amino acid containing peptide that inhibits Aβ(1–40) fibrillogenesis. J Pept Res 60:37–55

Gordon DJ, Balbach JJ, Tycko R, Meredith SC (2004) Increasing the amphiphilicity of an amyloidogenic peptide changes the β-sheet structure in the fibrils from antiparallel to parallel. Biophys J 86:428–434

Gorske BC, Blackwell HE (2006) Interception of quorum sensing in *Staphylococcus aureus*: a new niche for peptidomimetics. Org Biomol Chem 4:1441–1445

Goux WJ, Kopplin L, Nguyen AD, Leak K, Rutkofsky M, Shanmuganandam VD, Sharma D, Inouye H, Kirschner DA (2004) The formation of straight and twisted filaments from short Tau peptides. J Biol Chem 279:26868–26875

Green NS, Foss TR, Kelly JW (2005) Genistein, a natural product from soy, is a potent inhibitor of transthyretin amyloidosis. Proc Natl Acad Sci USA 104:14545–14550
Greenfield N, Fasman GD (1969) Computed circular dichroism spectra for the evaluation of protein conformation. Biochemistry 8:4108–4116
Greenfield N, Davidson B, Fasman GD (1967) The use of computed optical rotatory dispersion curves for the evaluation of protein conformation. Biochemistry 6:1630–1637
Grill JD, Cummings JL (2010) Current therapeutic targets for the treatment of Alzheimer's disease. Expert Rev Neurother 10:711–728
Grillo-Bosch D, Carulla N, Cruz M, Sánchez L, Pujol-Pina R, Madurga S, Rabanal F, Giralt E (2009) Retro-enantio *N*-methylated peptides as β-amyloid aggregation inhibitors. ChemMedChem 4:1488–1494
Grundman M, Black R (2008) Clinical trials of bapineuzumab, a β-amyloid-targeted immunotherapy in patients with mild to moderate Alzheimer's disease [abstract O3-04-05]. Alzheimers Dement 4:T166
Grune T, Reinheckel T, Davies KJ (1997) Degradation of oxidized proteins in mammalian cells. FASEB J 11:526–534
Gudmundsson G, Hallgrimsson J, Jonasson TA, Bjarnason O (1972) Hereditary cerebral haemorrhage with amyloidosis. Brain 95:387–404
Guijarro JI, Sunde M, Jones JA, Campbell ID, Dobson CM (1998) Amyloid fibril formation by an SH3 domain. Proc Natl Acad Sci USA 95:4224–4228
Gupta VB, Indi SS, Rao KS (2009) Garlic extract exhibits antiamyloidogenic activity on amyloid-β fibrillogenesis: relevance to Alzheimer's disease. Phytother Res 23:111–115
Gusella JF, MacDonald ME (1995) Huntington's disease. Semin Cell Biol 6:21–28
Gusella JF, MacDonald ME (2003) Huntingtin: a single bait hooks many species. Curr Opin Neurobiol 8:425–430
Gustafson DR, Skoog I, Rosengren L, Zetterberg H, Blennow K (2007) Cerebrospinal fluid β-amyloid 1–42 concentration may predict cognitive decline in older women. J Neurol Neurosurg Psychiatry 78:461–464
Gustavsson A, Engstrom U, Westermark P (1991) Normal transthyretin and synthetic transthyretin fragments form amyloid-like fibrils *in vitro*. Biochem Biophys Res Commun 175:1159–1164
Haass C, Selkoe DJ (2007) Soluble protein oligomers in neurodegeneration: lessons from the Alzheimer's amyloid β-peptide. Nat Rev Mol Cell Biol 8:101–112
Haass C, Schlossmacher MG, Hung AY, Vigo-Pelfrey C, Mellon A, Ostaszewski BL, Lieberburg I, Koo EH, Schenk D, Teplow DB, Selkoe DJ (1992) Amyloid β-peptide is produced by cultured cells during normal metabolism. Nature 359:322–325
Haass C, Lemere CA, Capell A, Citron M, Seubert P, Schenk D, Lannfelt L, Selkoe DJ (1995) The Swedish mutation causes early-onset Alzheimer's disease by β-secretase cleavage within the secretory pathway. Nat Med 1:1291–1296
Haataja L, Gurlo T, Huang CJ, Butler PC (2008) Islet amyloid in type 2 diabetes, and the toxic oligomer hypothesis. Endocr Rev 29:303–316
Hackeng TM, Griffin JH, Dawson PE (1999) Protein synthesis by native chemical ligation: expanded scope by using straightforward methodology. Proc Natl Acad Sci USA 96:10068 10073
Hagen GA, Elliot WJ (1973) Transport of thyroid hormones in serum and cerebrospinal fluid. J Clin Endocrinol 37:415–422
Hainfellner JA, Budka H (1999) Disease associated prion protein may deposit in the peripheral nervous system in human transmissible spongiform encephalopathies. Acta Neuropathol (Berl) 98:458–460
Hamaguchi T, Ono K, Yamada M (2010) Curcumin and Alzheimer's disease. CNS Neurosci Ther 16:285–297
Hamidi AL, Liepnieks JJ, Uemichi T, Rebibou JM, Justrabo E, Droz D, Mousson C, Chalopin JM, Benson MD, Delpech M, Grateau G (1997) Renal amyloidosis with a frame shift mutation in fibrinogen α-chain gene producing a novel amyloid protein. Blood 90:4799–4805
Hammarström P, Jiang X, Hurshman AR, Powers ET, Kelly JW (2002) Sequence-dependent denaturation energetics: a major determinant in amyloid disease diversity. Proc Natl Acad Sci USA 99:16427–16432

Hammarström P, Wiseman RL, Powers ET, Kelly JW (2003) Prevention of transthyretin amyloid disease by changing protein misfolding energetics. Science 299:713–716

Hands SL, Wyttenbach A (2010) Neurotoxic protein oligomerization associated with polyglutamine diseases. Acta Neuropathol 120:419–437

Hanger DP, Wray S (2010) Tau cleavage and tau aggregation in neurodegenerative disease. Biochem Soc Trans 38:1016–1020

Hanson JC, Lippa CF (2009) Lewy body dementia. Int Rev Neurobiol 84:215–228

Hardy J, Selkoe DJ (2002) The amyloid hypothesis of Alzheimer's disease: progress and problems on the road to therapeutics. Science 297:353–356

Harirforoosh S, Jamali F (2009) Renal adverse effects of nonsteroidal anti-inflammatory drugs. Expert Opin Drug Saf 8:669–681

Harper JD, Lansbury PT Jr (1997) Models of amyloid seeding in Alzheimer's disease and scrapie: mechanistic truths and physiological consequences of the time-dependent solubility of amyloid proteins. Annu Rev Biochem 66:385–407

Hart PJ (2006) Pathogenic superoxide dismutase structure, folding, aggregation and turnover. Curr Opin Chem Biol 10:131–138

Hasegawa K, Ohhashi Y, Yamaguchi I, Takahashi N, Tsutsumi S, Goto Y, Gejyo F, Naiki H (2003) Amyloidogenic synthetic peptides of β2-microglobulin—a role of the disulfide bond. Biochem Biophys Res Commun 304:101–106

Hatters DM, Howlett GJ (2002) The structural basis for amyloid formation by plasma apolipoproteins: a review. Eur Biophys J 31:2–8

Hawkins PN (2003) Hereditary systemic amyloidosis with renal involvement. J Nephrol 16:443–448

He B, Wang K, Liu Y, Xue B, Uversky VN, Dunker AK (2009) Predicting intrinsic disorder in proteins: an overview. Cell Res 19:929–949

Heegaard NH (2009) β2-microglobulin: from physiology to amyloidosis. Amyloid 16:151–173

Heise H, Hoyer W, Becker S, Andronesi OC, Riedel D, Baldus M (2005) Molecular-level secondary structure, polymorphism, and dynamics of full-length α-synuclein fibrils studied by solid-state NMR. Proc Natl Acad Sci USA 102:15871–15876

Heiser V, Scherzinger E, Boeddrich A, Nordhoff E, Lurz R, Schugardt N, Lehrach H, Wanker EE (2000) Inhibition of huntingtin fibrillogenesis by specific antibodies and small molecules: implications for Huntington's disease therapy. Proc Natl Acad Sci USA 97:6739–6744

Heiser V, Engemann S, Bröcker W, Dunkel I, Boeddrich A, Waelter S, Nordhoff E, Lurz R, Schugardt N, Rautenberg S, Herhaus C, Barnickel G, Böttcher H, Lehrach H, Wanker EE (2002) Identification of benzothiazoles as potential polyglutamine aggregation inhibitors of Huntington's disease by using an automated filter retardation assay. Proc Natl Acad Sci USA 99(Suppl 4):16400–16406

Henzler Wildman KA, Ramamoorthy A, Wakamiya T, Yoshikawa T, Crisma M, Toniolo C, Formaggio F (2004) A study of a Cα, β-didehydroalanine homo-oligopeptide series in the solid-state by 13C cross-polarization magic angle spinning NMR. J Pept Sci 10:336–341

Herbst M, Wanker EE (2006) Therapeutic approaches to polyglutamine diseases: combating protein misfolding and aggregation. Curr Pharm Des 12:2543–2555

Hetényi C, Szabo Z, Klement E, Datki Z, Kortvelyesi T, Zarandi M, Penke B (2002) Pentapeptide amides interfere with disaggregation of β-amyloid peptide of Alzheimer's disease. Biochem Biophys Res Commun 292:931–936

Higuchi N, Kyogoku Y, Shin M, Inouye K (1983) Origin of slow conformer conversion of triostin A and interaction ability with nucleic acid bases. Int J Pept Protein Res 21:541–545

Hilbich C, Kisters-Woike B, Reed J, Masters CL, Beyreuther K (1992) Substitutions of hydrophobic amino acids reduce the amyloidogenicity of Alzheimer's disease βA4 peptides. J Mol Biol 228:460–473

Hills ID, Vacca JP (2007) Progress toward a practical BACE-1 inhibitor. Curr Opin Drug Discov Devel 10:383–391

Hirakura Y, Azimov R, Azimova R, Kagan BL (2000) Polyglutamine-induced ion channels: a possible mechanism for the neurotoxicity of Huntington and other CAG repeat diseases. J Neurosci Res 60:490–494

Hock C, Konietzko U, Streffer JR, Tracy J, Signorell A, Müller-Tillmanns B, Lemke U, Henke K, Moritz E, Garcia E, Wollmer MA, Umbricht D, de Quervain DJ, Hofmann M, Maddalena A, Papassotiropoulos A, Nitsch RM (2003) Antibodies against β-amyloid slow cognitive decline in Alzheimer's disease. Neuron 38:547–554

Hockly E, Tse J, Barker AL, Moolman DL, Beunard JL, Revington AP, Holt K, Sunshine S, Moffitt H, Sathasivam K, Woodman B, Wanker EE, Lowden PA, Bates GP (2006) Evaluation of the benzothiazole aggregation inhibitors riluzole and PGL-135 as therapeutics for Huntington's disease. Neurobiol Dis 21:228–236

Holmberg CI, Staniszewski KE, Mensah KN, Matouschek A, Morimoto RI (2004) Inefficient degradation of truncated polyglutamine proteins by the proteasome. EMBO J 23:4307–4318

Holmes C, Boche D, Wilkinson D, Yadegarfar G, Hopkins V, Bayer A, Jones RW, Bullock R, Love S, Neal JW, Zotova E, Nicoll JA (2009) Long-term effects of Aβ42 immunisation in Alzheimer's disease: follow-up of a randomised, placebo-controlled phase I trial. Lancet 372:216–223

Holton JL, Ghiso J, Lashley T, Rostagno A, Guerin CJ, Gibb G, Houlden H, Ayling H, Martinian L, Anderton BH, Wood NW, Vidal R, Plant G, Frangione B, Revesz T (2001) Regional distribution of amyloid-Bri deposition and its association with neurofibrillary degeneration in familial British dementia. Am J Pathol 158:515–526

Holton JL, Lashley T, Ghiso J, Braendgaard H, Vidal R, Guerin CJ, Gibb G, Hanger DP, Rostagno A, Anderton BH, Strand C, Ayling H, Plant G, Frangione B, Bojsen-Moller M, Revesz T (2002) Familial Danish dementia: a novel form of cerebral amyloidosis associated with deposition of both amyloid-Dan and amyloid-β. J Neuropathol Exp Neurol 61:254–267

Hong L, Koelsch G, Lin X, Wu S, Terzyan S, Ghosh AK, Zhang XC, Tang J (2000) Structure of the protease domain of memapsin 2 (β-secretase) complexed with inhibitor. Science 290:150–153

Hong L, Turner RT, Koelsch G, Shin D, Ghosh AK, Tang J (2002) Crystal structure of memapsin 2 (β-secretase) in complex with an inhibitor OM00-3. Biochemistry 41:10963–10967

Hong H-S, Maezawa I, Yao N, Diaz-Avalos R, Rana S, Hua DH, Cheng RH, Lam KS, Jin L-W (2007) Combining the rapid MTT formazan exocytosis assay and the MC65 protection assay led to the discovery of carbazole analogs as small molecule inhibitors of A-β oligomer-induced cytotoxicity. Brain Res 1130:223–234

Hoogeveen AT, Willemsen R, Meyer N, de Rooij KE, Roos RA, van Ommen GJ, Galjaard H (1993) Characterization and localization of the Huntington disease gene product. Hum Mol Genet 2:2069–2073

Hornberg A, Eneqvist T, Olofsson A, Lundgren E, Sauer-Eriksson AE (2000) A comparative analysis of 23 structures of the amyloidogenic protein transthyretin. J Mol Biol 302:649–669

Hou X, Aguilar MI, Small DH (2007) Transthyretin and familial amyloidotic polyneuropathy. Recent progress in understanding the molecular mechanism of neurodegeneration. FEBS J 274:1637–1650

Howell PL, Pangborn WA, Marshall GR, Zabrocki J, Smith GD (1995) A thyrotropin-releasing hormone analogue: pGlu–Phe–$_D$–Pro–Ψ[CN_4]–NMe at 293 and 107 K. Acta Crystallogr C 51:2575–2579

Howlett DR, Perry AE, Godfrey F, Swatton JE, Jennings KH, Spitzfaden C, Wadsworth H, Wood SJ, Markwell RE (1999) Inhibition of fibril formation in β-amyloid peptide by a novel series of benzofurans. Biochem J 340:283–289

Hsu AL, Murphy CT, Kenyon C (2003) Regulation of aging and age-related disease by DAF-16 and heat-shock factor. Science 300:1142–1145

Hu WT, Grossman M (2009) TDP-43 and frontotemporal dementia. Curr Neurol Neurosci Rep 9:353–358

Hu X, Hicks CW, He W, Wong P, Macklin WB, Trapp BD, Yan R (2006) BACE-1 modulates myelination in the central and peripheral nervous system. Nat Neurosci 9:1520–1525

Huang X, Atwood CS, Hartshorn MA, Multhaup G, Goldstein LE, Scarpa RC, Cuajungco MP, Gray DN, Lim J, Moir RD, Tanzi RE, Bush AI (1999a) The Aβ peptide of Alzheimer's disease directly produces hydrogen peroxide through metal ion reduction. Biochemistry 38:7609–7616

Huang X, Cuajungco MP, Atwood CS, Hartshorn MA, Tyndall JD, Hanson GR, Stokes KC, Leopold M, Multhaup G, Goldstein LE, Scarpa RC, Saunders AJ, Lim J, Moir RD, Glabe C, Bowden EF, Masters CL, Fairlie DP, Tanzi RE, Bush AI (1999b) Cu(II) potentiation of Alzheimer Aβ neurotoxicity. Correlation with cell-free hydrogen peroxide production and metal reduction. J Biol Chem 274:37111–37116

Hughes SR, Goyal S, Sun JE, Gonzalez-Dewhitt P, Fortes M, Riedel NG, Sahasrabudhe SR (1996) Two hybrid system as a model to study the interaction of β-amyloid peptide monomers. Proc Natl Acad Sci USA 93:2065–2070

Hughes E, Burke RM, Doig AJ (2000) Inhibition of toxicity in the β-amyloid peptide fragment β-(25–35) using *N*-methylated derivatives—a general strategy to prevent amyloid formation. J Biol Chem 275:25109–25115

Hunt CE, Turner AJ (2009) Cell biology, regulation and inhibition of β-secretase (BACE-1). FEBS J 276:1845–1859

Hurle MR, Helms LR, Li L, Chan W, Wetzel R (1994) A role for destabilizing amino acid replacements in light-chain amyloidosis. Proc Natl Acad Sci USA 91:5446–5450

Hurshman Babbes AR, Powers ET, Kelly JW (2008) Quantification of the thermodynamically linked quaternary and tertiary structural stabilities of transthyretin and its disease-associated variants: the relationship between stability and amyloidosis. Biochemistry 47:6969–6984

Hurshman AR, White JT, Powers ET, Kelly JW (2004) Transthyretin aggregation under partially denaturing conditions is a downhill polymerization. Biochemistry 43:7365–7381

Hussain I, Powell D, Howlett DR, Tew DG, Meek TD, Chapman C, Gloger IS, Murphy KE, Southan CD, Ryan DM, Smith TS, Simmons DL, Walsh FS, Dingwall C, Christie G (1999) Identification of a novel aspartic protease (Asp 2) as β-secretase. Mol Cell Neurosci 14:419–427

Ikonomovic MD, Uryu K, Abrahamson EE, Ciallella JR, Trojanowski JQ, Lee VM, Clark RS, Marion DW, Wisniewski SR, DeKosky ST (2004) Alzheimer's pathology in human temporal cortex surgically excised after severe brain injury. Exp Neurol 190:192–203

Imai J, Yashiroda H, Maruya M, Yahara I, Tanaka K (2003) Proteasomes and molecular chaperones: cellular machinery responsible for folding and destruction of unfolded proteins. Cell Cycle 2:585–590

Imbimbo BP, Giardina GA (2011) γ-Secretase inhibitors and modulators for the treatment of Alzheimer's disease: disappointments and hopes. Curr Top Med Chem 11:1555–70 (Epub ahead of print)

Inestrosa NC, Silberstein L, Hall ZW (1982) Association of the synaptic form of acetylcholinesterase with extracellular matrix in cultured mouse muscle cells. Cell 29:71–79

Inestrosa NC, Alvarez A, Calderon F (1996a) Acetylcholinesterase is a senile plaque component that promotes assembly of amyloid β-peptide into Alzheimer's filaments. Mol Psychiatry 1:359–361

Inestrosa NC, Alvarez A, Perez CA, Moreno RD, Vicente M, Linker C, Casanueva OI, Soto C, Garrido J (1996b) Acetylcholinesterase accelerates assembly of amyloid-β-peptides into Alzheimer's fibrils: possible role of the peripheral site of the enzyme. Neuron 16:881–891

Ingenbleek Y, De Visscher M, De Nayer P (1972) Measurement of prealbumin as index of protein-calorie malnutrition. Lancet 2:106–109

Ingwall RT, Goodman M (1974) Polydepsipeptides. III. Theoretical conformational analysis of randomly coiling and ordered depsipeptide chains. Macromolecules 7:598–605

Irizarry MC, McNamara M, Fedorchak K, Hsiao K, Hyman BT (1997) APPSw transgenic mice develop age-related Aβ deposits and neuropil abnormalities, but no neuronal loss in CA1. J Neuropathol Exp Neurol 56:965–973

Ittner LM, Götz J (2011) Amyloid-β and tau—a toxic pas de deux in Alzheimer's disease. Nat Rev Neurosci 12:65–72

Ivanova MI, Gingery M, Whitson LJ, Eisenberg D (2003) Role of the *C*-terminal 28 residues of β2-microglobulin in amyloid fibril formation. Biochemistry 42:13536–13540

Iwata K, Fujiwara T, Matsuki Y, Akutsu H, Takahashi S, Naiki H, Goto Y (2006) 3D structure of amyloid protofilaments of β2-microglobulin fragment probed by solid-state NMR. Proc Natl Acad Sci USA 103:18119–18124

Iwatsubo T (2003) Aggregation of α-synuclein in the pathogenesis of Parkinson's disease. J Neurol 250(suppl 3):III 11–III 14

Jana NR, Tanaka M, Wang G, Nukina N (2000) Polyglutamine length-dependent interaction of Hsp40 and Hsp70 family chaperones with truncated *N*-terminal huntingtin: their role in suppression of aggregation and cellular toxicity. Hum Mol Genet 9:2009–2018

Janus C, Pearson J, McLaurin J, Mathews PM, Jiang Y, Schmidt SD, Chishti MA, Horne P, Heslin D, French J, Mount HT, Nixon RA, Mercken M, Bergeron C, Fraser PE, St George-Hyslop P, Westaway D (2000) Aβ peptide immunization reduces behavioral impairment and plaques in a model of Alzheimer's disease. Nature 408:979–982

Janusz M, Staroscik K, Zimecki M, Wieczorek Z, Lisowski J (1981) Chemical and physical characterization of a proline-rich polypeptide from sheep colostrum. Biochem J 199:9–15

Jao CC, Der-Sarkissian A, Chen J, Langen R (2004) Structure of membrane-bound α-synuclein studied by site-directed spin labeling. Proc Natl Acad Sci USA 101:8331–8336

Jarrett JT, Lansbury PT Jr (1993) Seeding "one-dimensional crystallization" of amyloid: a pathogenic mechanism in Alzheimer's disease and scrapie? Cell 73:1055–1058

Jellinger KA (2009) Formation and development of Lewy pathology: a critical update. J Neurol 256(Suppl 3):270–279

Jiang X, Smith CS, Petrassi HM, Hammarström P, White JT, Sacchettini JC, Kelly JW (2001) An engineered transthyretin monomer that is nonamyloidogenic, unless it is partially denatured. Biochemistry 40:11442–11452

Jicha GA (2009) Is passive immunization for Alzheimer's disease 'alive and well' or 'dead and buried'? Expert Opin Biol Ther 9:481–491

Johansson B, Wernstedt C, Westermark P (1987) Atrial natriuretic peptide deposited as atrial amyloid fibrils. Biochem Biophys Res Commun 148:1087–1092

John R, Herzenberg AM (2009) Renal toxicity of therapeutic drugs. J Clin Pathol 62:505–515

Johnson G (2006) Tau phosphorylation and proteolysis: insights and perspectives. J Alzheimers Dis 9:243–250

Johnson RT, Gibbs CJ Jr (1998) Creutzfeldt–Jakob disease and related transmissible spongiform encephalopathies. N Engl J Med 339:1994–2004

Johnson KH, O'Brien TD, Betsholtz C, Westermark P (1989) Islet amyloid, islet-amyloid polypeptide, and diabetes mellitus. N Engl J Med 321:513–518

Johnson SM, Wiseman RL, Sekijima Y, Green NS, Adamski-Werner SL, Kelly JW (2005) Native state kinetic stabilization as a strategy to ameliorate protein misfolding diseases: a focus on the transthyretin amyloidoses. Acc Chem Res 38:911–921

Johnson SM, Connelly S, Wilson IA, Kelly JW (2008a) Biochemical and structural evaluation of highly selective 2-arylbenzoxazolebased transthyretin amyloidogenesis inhibitors. J Med Chem 51:260–270

Johnson SM, Connelly S, Wilson IA, Kelly JW (2008b) Toward optimization of the linker substructure common to transthyretin amyloidogenesis inhibitors using biochemical and structural studies. J Med Chem 51:6348–6358

Johnson SM, Connelly S, Wilson IA, Kelly JW (2009) Toward optimization of the second aryl substructure common to transthyretin amyloidogenesis. Inhibitors using biochemical and structural studies. J Med Chem 52:1115–1125

Jones S, Manning J, Kad NM, Radford SE (2003) Amyloid-forming peptides from β2-microglobuli—insights into the mechanism of fibril formation *in vitro*. J Mol Biol 325:249–257

Joy T, Wang J, Hahn A, Hegele RA (2003) APOA1 related amyloidosis: a case report and literature review. Clin Biochem 36:641–645

Julius RL, Farha OK, Chiang J, Perry LJ, Hawthorne MF (2007) Synthesis and evaluation of transthyretin amyloidosis inhibitors containing carborane pharmacophores. Proc Natl Acad Sci USA 104:4808–4813

Kagan BL, Hirakura Y, Azimov R, Azimova R, Lin MC (2002) The channel hypothesis of Alzheimer's disease: current status. Peptides 23:1311–1315

Kaiser ET, Kézdy FJ (1984) Amphiphilic secondary structure: design of peptide hormones. Science 223:249–255

Kaiser ET, Kézdy FJ (1987) Peptides with affinity for membranes. Annu Rev Biophys Biophys Chem 16:561–581
Kampers T, Pangalos M, Geerts H, Wiech H, Mandelkow E (1999) Assembly of paired helical filaments from mouse tau: implications for the neurofibrillary pathology in transgenic mouse models for Alzheimer's disease. FEBS Lett 451:39–44
Kang JE, Lim MM, Bateman RJ, Lee JJ, Smyth LP, Cirrito JR, Fujiki N, Nishino S, Holtzman DM (2009) Amyloid-β dynamics are regulated by orexin and the sleep-wake cycle. Science 326:1005–1007
Kannabiran C, Klintworth GK (2006) TGFβI gene mutations in corneal dystrophies. Hum Mutat 27:615–625
Kapurniotu A (2001) Amyloidogenicity and cytotoxicity of islet amyloid polypeptide. Biopolymers 60:438–459
Katsuno M, Adachi H, Kume A, Li M, Nakagomi Y, Niwa H, Sang C, Kobayashi Y, Doyu M, Sobue G (2002) Testosterone reduction prevents phenotypic expression in a transgenic mouse model of spinal and bulbar muscular atrophy. Neuron 35:843–854
Katsuno M, Banno H, Suzuki K, Adachi H, Tanaka F, Sobue G (2010) Clinical features and molecular mechanisms of spinal and bulbar muscular atrophy (SBMA). Adv Exp Med Biol 685:64–74
Kawasaki T, Onodera K, Kamijo S (2010) Selection of peptide inhibitors of soluble Aβ(1–42) oligomer formation by phage display. Biosci Biotechnol Biochem 74:2214–2219
Kayed R, Head E, Thompson JL, McIntire TM, Milton SC, Cotman CW, Glabe CG (2003) Common structure of soluble amyloid oligomers implies common mechanism of pathogenesis. Science 300:486–489
Kayed R, Head E, Sarsoza F, Saing T, Cotman CW, Necula M, Margol L, Wu J, Breydo L, Thompson JL, Rasool S, Gurlo T, Butler P, Glabe CG (2007) Fibril specific, conformation dependent antibodies recognize a generic epitope common to amyloid fibrils and fibrillar oligomers that is absent in prefibrillar oligomers. Mol Neurodegener 2:18
Kelleher RJ 3rd, Shen J (2010) γ-Secretase and human disease. Science 330:1055–1056
Keller JN, Hanni KB, Markesbery WR (2000) Impaired proteasome function in Alzheimer's disease. J Neurochem 75:436–439
Kelly JW (1998) The alternative conformations of amyloidogenic proteins and their multi-step assembly pathways. Curr Opin Struct Biol 8:101–106
Kent SB (2009) Total chemical synthesis of proteins. Chem Soc Rev 38:338–351
Kheterpal I, Wetzel R (2006) Hydrogen/deuterium exchange mass spectrometry—a window into amyloid structure. Acc Chem Res 39:584–593
Kheterpal I, Zhou S, Cook KD, Wetzel R (2000) Aβ amyloid fibrils possess a core structure highly resistant to hydrogen exchange. Proc Natl Acad Sci USA 97:13597–13601
Kheterpal I, Lashuel HA, Hartley DM, Walz T, Lansbury PT Jr, Wetzel R (2003) Aβ protofibrils possess a stable core structure resistant to hydrogen exchange. Biochemistry 42:14092–14098
Khurana R, Agarwal A, Bajpai VK, Verma N, Sharma AK, Gupta RP, Madhusudan KP (2004) Unraveling the amyloid associated with human medullary thyroid carcinoma. Endocrinology 145:5465–5470
Kim HJ, Lee KW, Lee HJ (2007) Protective effects of piceatannol against β-amyloid-induced neuronal cell death. Ann N Y Acad Sci 1095:473–482
Kim MW, Chelliah Y, Kim SW, Otwinowski Z, Bezprozvanny I (2009) Secondary structure of huntingtin amino-terminal region. Structure 17:1205–1212
Kita Y, Arakawa T, Lin TY, Timasheff SN (1994) Contribution of the surface free energy perturbation to protein-solvent interactions. Biochemistry 33:15178–15189
Kitada T, Asakawa S, Hattori N, Matsumine H, Yamamura Y, Minoshima S, Yokochi M, Mizuno Y, Shimizu N (1999) Mutations in the parkin gene cause autosomal recessive juvenile parkinsonism. Nature 392:605–608
Kiuru S (1998) Gelsolin-related familial amyloidosis, Finnish type (FAF), and its variants found worldwide. Amyloid 5:55–66
Klabunde T, Petrassi HM, Oza VB, Raman P, Kelly JW, Sacchettini JC (2000) Rational design of potent human transthyretin amyloid disease inhibitors. Nat Struct Biol 7:312–321

Klajnert B, Cortijo-Arellano M, Cladera J, Bryszewska M (2006) Influence of dendrimer's structure on its activity against amyloid fibril formation. Biochem Biophys Res Commun 345:21–28

Klein WL, Krafft GA, Finch CE (2001) Targeting small Aβ oligomers: the solution to an Alzheimer's disease conundrum? Trends Neurosci 24:219–224

Klintworth GK, Valnickova Z, Kielar RA, Baratz KH, Campbell RJ, Enghild JJ (1997) Familial subepithelial corneal amyloidosis—a lactoferrin-related amyloidosis. Invest Ophthalmol Vis Sci 38:2756–2763

Knight JD, Miranker AD (2004) Phospholipid catalysis of diabetic amyloid assembly. J Mol Biol 341:1175–1187

Kodali R, Williams AD, Chemuru S, Wetzel R (2010) Aβ(1–40) forms five distinct amyloid structures whose β-sheet contents and fibril stabilities are correlated. J Mol Biol 401:503–517

Koide A, Tereshko V, Uysal S, Margalef K, Kossiakoff AA, Koide S (2007) Exploring the capacity of minimalist protein interfaces: interface energetics and affinity maturation to picomolar Kd of a single-domain antibody with a flat paratope. J Mol Biol 373:941–953

Kokkoni N, Stott K, Amijee H, Mason JM, Doig AJ (2006) *N*-Methylated peptide inhibitors of β-amyloid aggregation and toxicity. Optimization of the inhibitor structure. Biochemistry 45:9906–9918

Kordasiewicz HB, Gomez CM (2007) Molecular pathogenesis of spinocerebellar ataxia type 6. Neurotherapeutics 4:285–294

Kossiakoff AA, Koide S (2008) Understanding mechanisms governing protein–protein interactions from synthetic binding interfaces. Curr Opin Struct Biol 18:499–506

Kotilinek LA, Bacskai B, Westerman M, Kawarabayashi T, Younkin L, Hyman BT, Younkin S, Ashe KH (2002) Reversible memory loss in a mouse transgenic model of Alzheimer's disease. J Neurosci 22:6331–6335

Krebs MR, Wilkins DK, Chung EW, Pitkeathly MC, Chamberlain AK, Zurdo J, Robinson CV, Dobson CM (2000) Formation and seeding of amyloid fibrils from wild-type hen lysozyme and a peptide fragment from the β-domain. J Mol Biol 300:541–549

Krishnan R, Lindquist S (2005) Structural insights into a yeast prion illuminate nucleation and strain diversity. Nature 435:765–772

Kumar NG, Izumiya N, Miyoshi M, Sugano H, Urry DW (1975) Conformational and spectral analysis of the polypeptide antibiotic *N*-methylleucine gramicidin *S* dihydrochloride by nuclear magnetic resonance. Biochemistry 14:2197–2207

Kwon KJ, Kim HJ, Shin CY, Han SH (2010) Melatonin potentiates the neuroprotective properties of resveratrol against β-amyloid-induced neurodegeneration by modulating AMP-activated protein kinase pathways. J Clin Neurol 6:127–137

Lacomblez L, Bensimon G, Leigh PN, Guillet P, Meininger V (1996) Dose-ranging study of riluzole in amyotrophic lateral sclerosis. Amyotrophic Lateral Sclerosis/Riluzole Study Group II. Lancet 347:1425–1431

Ladiwala AR, Lin JC, Bale SS, Marcelino-Cruz AM, Bhattacharya M, Dordick JS, Tessier PM (2010) Resveratrol selectively remodels soluble oligomers and fibrils of amyloid Aβ into off pathway conformers. J Biol Chem 285:24228–24237

Lambert MP, Barlow AK, Chromy BA, Edwards C, Freed R, Liosatos M, Morgan TE, Rozovsky I, Trommer B, Viola KL, Wals P, Zhang C, Finch CE, Krafft GA, Klein WL (1998) Diffusible, nonfibrillar ligands derived from Aβ1–42 are potent central nervous system neurotoxins. Proc Natl Acad Sci USA 95:6448–6453

Lamitina T, Huang CG, Strange K (2006) Genome-wide RNAi screening identifies protein damage as a regulator of osmoprotective gene expression. Proc Natl Acad Sci USA 103:12173–12178

Lannfelt L, Blennow K, Zetterberg H, Båtsman S, Ames D, Harrison J, Masters CL, Targum S, Bush AI, Murdoch R, Wilson J, Ritchie CW (2008) Safety, efficacy, and biomarker findings of PBT2 in targeting Aβ as a modifying therapy for Alzheimer's disease: a phase IIa, double-blind, randomised, placebo-controlled trial. Lancet Neurol 7:779–786

Lanning JD, Hawk AJ, Derryberry J, Meredith SC (2010) Chaperone-like *N*-methyl peptide inhibitors of polyglutamine aggregation. Biochemistry 49:7108–7118

Larsson A, Söderberg L, Westermark GT, Sletten K, Engström U, Tjernberg LO, Näslund J, Westermark P (2007) Unwinding fibril formation of medin, the peptide of the most common form of human amyloid. Biochem Biophys Res Commun 361:822–828

Lashuel HA, Petre BM, Wall J, Simon M, Nowak RJ, Walz T, Lansbury PT Jr (2002) α-Synuclein, especially the Parkinson's disease-associated mutants, forms pore-like annular and tubular protofibrils. J Mol Biol 322:1089–1102

Lashuel HA, Hartley DM, Petre BM, Wall JS, Simon MN, Walz T, Lansbury PT Jr (2003) Mixtures of wild-type and a pathogenic (E22G) form of Aβ40 *in vitro* accumulate protofibrils, including amyloid pores. J Mol Biol 332:795–808

Lastres-Becker I, Rüb U, Auburger G (2008) Spinocerebellar ataxia 2 (SCA2). Cerebellum 7:115–124

Lebre AS, Brice A (2003) Spinocerebellar ataxia 7 (SCA7). Cytogenet Genome Res 100:154–163

Lee G (2005) Tau and src family tyrosine kinases. Biochim Biophys Acta 1739:323–330

Lee SM, Jeon R (2005) Synthesis of 6-[2-(benzoxazol-2-ylmethylamino)ethoxy]-1-alkyl-1*H*-indole-2-carboxylic acid and inhibitory activity on β-amyloid aggregation. Arch Pharm Res 28:1219–1223

Lee M, Bard F, Johnson-Wood K, Lee C, Hu K, Griffith SG, Black RS, Schenk D, Seubert P (2005) Aβ42 immunization in Alzheimer's disease generates Aβ *N*-terminal antibodies. Ann Neurol 58:430–435

Lees AJ, Hardy J, Revesz T (2009) Parkinson's disease. Lancet 373:2055–2066

Lemere CA, Masliah E (2010) Can Alzheimer disease be prevented by amyloid-β immunotherapy? Nat Rev Neurol 6:108–119

Lemere CA, Maron R, Spooner ET, Grenfell TJ, Mori C, Desai R, Hancock WW, Weiner HL, Selkoe DJ (2000) Nasal Aβ treatment induces anti-Aβ antibody production and decreases cerebral amyloid burden in PD-APP mice. Ann N Y Acad Sci 920:328–331

Lemere CA, Maron R, Selkoe DJ, Weiner HL (2001) Nasal vaccination with β-amyloid peptide for the treatment of Alzheimer's disease. DNA Cell Biol 20:705–711

Lesné S, Koh MT, Kotilinek L, Kayed R, Glabe CG, Yang A, Gallagher M, Ashe KH (2006) A specific amyloid-β protein assembly in the brain impairs memory. Nature 440:352–357

Lev N, Melamed E, Offen D (2006) Proteasomal inhibition hypersensitizes differentiated neuroblastoma cells to oxidative damage. Neurosci Lett 399:27–32

Levites Y, Amit T, Mandel S, Youdim MB (2003) Neuroprotection and neurorescue against Aβ toxicity and PKC-dependent release of nonamyloidogenic soluble precursor protein by green tea polyphenol (–)-epigallocatechin-3-gallate. FASEB J 17:952–954

Li SH, Li XJ (2004) Huntingtin-protein interactions and the pathogenesis of Huntington's disease. Trends Genet 20:146–154

Li X, Lu F, Wang JZ, Gong CX (2006) Concurrent alterations of *O*-GlcNAcylation and phosphorylation of tau in mouse brains during fasting. Eur J Neurosci 23:2078–2086

Liepina I, Janmey P, Czaplewski C, Liwo A (2004) Towards gelsolin amyloid formation. Biopolymers 76:543–548

Lim GP, Chu T, Yang F, Beech W, Frautschy SA, Cole GM (2001) The curry spice curcumin reduces oxidative damage and amyloid pathology in an Alzheimer transgenic mouse. J Neurosci 21:8370–8377

Lim KH, Nguyen TN, Damo SM, Mazur T, Ball HL, Prusiner SB, Pines A, Wemmer DE (2008) Solid-state NMR structural studies of the fibril form of a mutant mouse prion peptide PrP89–143(P101L). Solid State Nucl Magn Reson 29:183–190

Limprasert P, Nouri N, Nopparatana C, Deininger PL, Keats BJ (1997) Comparative studies of the CAG repeats in the spinocerebellar ataxia type 1 (SCA1) gene. Am J Med Genet 74:488–493

Lin TY, Timasheff SN (1996) On the role of surface tension in the stabilization of globular proteins. Protein Sci 5:372–381

Lin H, Bhatia R, Lal R (2000a) Fresh and globular amyloid-β protein (1–42) induces rapid cellular degeneration: evidence for AβP channel-mediated cellular toxicity. FASEB J 14:1233–1243

Lin X, Antalffy B, Kang D, Orr HT, Zoghbi HY (2000b) Polyglutamine expansion down-regulates specific neuronal genes before pathologic changes in SCA1. Nat Neurosci 3:157–163

Lin X, Koelsch G, Wu S, Downs D, Dashti A, Tang J (2000c) Human aspartic protease memapsin 2 cleaves the β-secretase site of β-amyloid precursor protein. Proc Natl Acad Sci USA 97:1456–1460

Lin H, Bhatia R, Lal R (2001) Amyloid β-protein forms ion channels: implications for Alzheimer's disease pathophysiology. FASEB J 15:2433–2444

Lin S-J, Shiao Y-J, Chi CW, Yang L-M (2004) Aβ Aggregation inhibitors. Part 1: synthesis and biological activity of phenylazo benzenesulfonamides. Bioorg Med Chem Lett 14:1173–1176

Lindsay J, Laurin D, Verreault R, Hébert R, Helliwell B, Hill GB, McDowell I (2002) Risk factors for Alzheimer's disease: a prospective analysis from the Canadian study of health and aging. Am J Epidemiol 156:445–453

Link CD (1995) Expression of human β-amyloid peptide in transgenic *Caenorhabditis elegans*. Proc Natl Acad Sci USA 92:9368–9372

Linke RP, Joswig R, Murphy CL, Wang S, Zhou H, Gross U, Rocken C, Westermark P, Weiss DT, Solomon A (2005) Senile seminal vesicle amyloid is derived from seminogelin I. J Lab Clin Med 145:187–193

Litvinovich SV, Brew SA, Aota S, Akiyama SK, Haudenschild C, Ingham KC (1998) Formation of amyloid-like fibrils by self-association of a partially unfolded fibronectin type III module. J Mol Biol 280:245–258

Liu L, Murphy RM (2006) Kinetics of inhibition of β-amyloid aggregation by transthyretin. Biochemistry 45:15702–15709

Liu F, Iqbal K, Grundke-Iqbal I, Hart GW, Gong CX (2004) *O*-GlcNAcylation regulates phosphorylation of tau: a mechanism involved in Alzheimer's disease. Proc Natl Acad Sci USA 101:10804–10809

Lomakin A, Teplow DB, Kirschner DA, Benedek GB (1997) Kinetic theory of fibrillogenesis of amyloid β-protein. Proc Natl Acad Sci USA 94:7942–7947

Lomas DA, Carrell RW (2002) Serpinopathies and the conformational dementias. Nat Rev Gen 3:759–768

Lomas DA, Evans DL, Stone SR, Chang WS, Carrell RW (1993) Effect of the Z mutation on the physical and inhibitory properties of α1-antitrypsin. Biochemistry 32:500–508

Lowe TL, Strzelec A, Kiessling LL, Murphy RM (2001) Structure–function relationships for inhibitors of β-amyloid toxicity containing the recognition sequence KLVFF. Biochemistry 40:7882–7889

Lu W, Qasim MA, Laskowski MJ, Kent SBH (1997) Probing intermolecular main chain hydrogen bonding in serine proteinase–protein inhibitor complexes: chemical synthesis of backbone-engineered turkey ovomucoid third domain. Biochemistry 36:673–679

Luca S, Yau WM, Leapman R, Tycko R (2007) Peptide conformation and supramolecular organization in amylin fibrils: constraints from solid-state NMR. Biochemistry 46:13505–13522

Luchsinger JA, Tang MX, Siddiqui M, Shea S, Mayeux R (2004) Alcohol intake and risk of dementia. J Am Geriatr Soc 52:540–546

Lueprasitsakul W, Alex S, Fang SL, Pino S, Irmscher K, Kohrle J, Braverman LE (1990) Flavonoid administration immediately displaces thyroxine (T4) from serum transthyretin, increases serum free T4, and decreases serum Thyrotropin in the rat. Endocrinology 126:2890–2895

Luheshi LM, Tartaglia GG, Brorsson AC, Pawar AP, Watson IE, Chiti F, Vendruscolo M, Lomas DA, Dobson CM, Crowther DC (2007) Systematic *in vivo* analysis of the intrinsic determinants of amyloid β pathogenicity. PLoS Biol 5:e290

Lührs T, Ritter C, Adrian M, Riek-Loher D, Bohrmann B, Döbeli H, Schubert D, Riek R (2005) 3D structure of Alzheimer's amyloid-β(1–42) fibrils. Proc Natl Acad Sci USA 102:17342–17347

Maas T, Eidenmuller J, Brandt R (2000) Interaction of tau with the neural membrane cortex is regulated by phosphorylation at sites that are modified in paired helical filaments. J Biol Chem 275:15733–15740

Mackenzie IR, Rademakers R (2008) The role of transactive response DNA-binding protein-43 in amyotrophic lateral sclerosis and frontotemporal dementia. Curr Opin Neurol 21:693–700

Madine J, Jack E, Stockley PG, Radford SE, Serpell LC, Middleton DA (2008) Structural insights into the polymorphism of amyloid-like fibrils formed by region 20–29 of amylin revealed by solid-state NMR and X-ray fiber diffraction. J Am Chem Soc 130:14990–15001

Madine J, Copland A, Serpell LC, Middleton DA (2009a) Cross-β spine architecture of fibrils formed by the amyloidogenic segment NFGSVQFV of medin from solid-state NMR and X-ray fiber diffraction measurements. Biochemistry 48:3089–3099

Madine J, Wang X, Brown DR, Middleton DA (2009b) Evaluation of β-alanine- and GABA-substituted peptides as inhibitors of disease-linked protein aggregation. Chembiochem 10:1982–1987

Maeda S, Sahara N, Saito Y, Murayama S, Ikai A, Takashima A (2006) Increased levels of granular tau oligomers: an early sign of brain aging and Alzheimer's disease. Neurosci Res 54:197–201

Maeda S, Sahara N, Saito Y, Murayama M, Yoshiike Y, Kim H, Miyasaka T, Murayama S, Ikai A, Takashima A (2007) Granular tau oligomers as intermediates of tau filaments. Biochemistry 46:3856–3861

Maezawa I, Hong HS, Wu HC, Battina SK, Rana S, Iwamoto T, Radke GA, Pettersson E, Martin GM, Hua DH, Jin LW (2006) A novel tricyclic pyrone compound ameliorates cell death associated with intracellular amyloid-β oligomeric complexes. J Neurochem 98:57–67

Mahalakshmi R, Balaram P (2006) Non-protein amino acids in the design of secondary structure scaffolds. Methods Mol Biol 340:71–94

Maia F, Almeida Mdo R, Gales L, Kijjoa A, Pinto MM, Saraiva MJ, Damas AM (2005) The binding of xanthone derivatives to transthyretin. Biochem Pharmacol 70:1861–1869

Maier M, Seabrook TJ, Lazo ND, Jiang L, Das P, Janus C, Lemere CA (2006) Short amyloid-β (Aβ) immunogens reduce cerebral Aβ load and learning deficits in an Alzheimer's disease mouse model in the absence of an Aβ-specific cellular immune response. J Neurosci 26:4717–4728

Mairal T, Nieto J, Pinto M, Almeida MR, Gales L, Ballesteros A, Barluenga J, Pérez JJ, Vázquez JT, Centeno NB, Saraiva MJ, Damas AM, Planas A, Arsequell G, Valencia G (2009) Iodine atoms: a new molecular feature for the design of potent transthyretin fibrillogenesis inhibitors. PLoS One 4:e4124

Manavalan P, Momany FA (1980) Conformational energy studies on *N*-methylated analogs of thyrotropin releasing hormone, enkephalin, and luteinizing hormone-releasing hormone. Biopolymers 19:1943–1973

Mandel SA, Amit T, Weinreb O, Reznichenko L, Youdim MB (2008) Simultaneous manipulation of multiple brain targets by green tea catechins: a potential neuroprotective strategy for Alzheimer and Parkinson diseases. CNS Neurosci Ther 14:352–365

Mantamadiotis T, Lemberger T, Bleckmann SC, Kern H, Kretz O, Martin Villalba A, Tronche F, Kellendonk C, Gau D, Kapfhammer J, Otto C, Schmid W, Schütz G (2002) Disruption of CREB function in brain leads to neurodegeneration. Nat Genet 31:47–54

Mantuano E, Veneziano L, Jodice C, Frontali M (2003) Spinocerebellar ataxia type 6 and episodic ataxia type 2: differences and similarities between two allelic disorders. Cytogenet Genome Res 100:147–153

Marambaud P, Zhao H, Davies P (2005) Resveratrol promotes clearance of Alzheimer's disease amyloid-β peptides. J Biol Chem 280:37377–37382

Marco-Contelles J, León R, de los Ríos C, Samadi A, Bartolini M, Andrisano V, Huertas O, Barril OX, Luque FJ, Rodríguez-Franco MI, López B, López MG, García AG, do Carmo Carreiras M, Villarroya M (2009) Tacripyrines, the first tacrine-dihydropyridine hybrids, as multitarget-directed ligands for the treatment of Alzheimer's disease. J Med Chem 52:2724–2732

Marks N, Berg MJ (2010) BACE and γ-secretase characterization and their sorting as therapeutic targets to reduce amyloidogenesis. Neurochem Res 35:181–210

Martone RL, Zhou H, Atchison K, Comery T, Xu JZ, Huang X, Gong X, Jin M, Kreft A, Harrison B, Mayer SC, Aschmies S, Gonzales C, Zaleska MM, Riddell DR, Wagner E, Lu P, Sun SC, Sonnenberg-Reines J, Oganesian A, Adkins K, Leach MW, Clarke DW, Huryn D, Abou-Gharbia M, Magolda R, Bard J, Frick G, Raje S, Forlow SB, Balliet C, Burczynski ME, Reinhart PH, Wan HI, Pangalos MN, Jacobsen JS (2009) Begacestat (GSI-953): a novel, selective

thiophene sulfonamide inhibitor of amyloid precursor protein γ-secretase for the treatment of Alzheimer's disease. J Pharmacol Exp Ther 331:598–608
Masino L (2004) Polyglutamine and neurodegeneration: structural aspects. Protein Pept Lett 11:239–248
Masison DC, Maddelein ML, Wickner RB (1997) The prion model for [URE3] of yeast: spontaneous generation and requirements for propagation. Proc Natl Acad Sci USA 94:12503–12508
Mastrianni JA (2010) The genetics of prion diseases. Genet Med 12:187–195
Mastrianni JA, Nixon R, Layzer R, Telling GC, Han D, DeArmond SJ, Prusiner SB (1999) Prion protein conformation in a patient with sporadic fatal insomnia. N Engl J Med 340:1630–1638
Matagne A, Dobson CM (1998) The folding process of hen lysozyme: a perspective from the 'new view'. Cell Mol Life Sci 54:363–371
Matharu B, Gibson G, Parsons R, Huckerby TN, Moore SA, Cooper LJ, Millichamp R, Allsop D, Austen B (2009) Galantamine inhibits β-amyloid aggregation and cytotoxicity. J Neurol Sci 280:49–58
Matharu B, El-Agnaf O, Razvi A, Austen BM (2010) Development of retro-inverso peptides as anti-aggregation drugs for β-amyloid in Alzheimer's disease. Peptides 31:1866–1872
Matsuzaki K (2007) Physicochemical interactions of amyloid β-peptide with lipid bilayers. Biochim Biophys Acta 1768:1935–1942
Matsuzaki K, Kato K, Yanagisawa K (2010) Aβ polymerization through interaction with membrane gangliosides. Biochim Biophys Acta 1801:868–877
Maury CP, Nurmiaho-Lassila EL, Boysen G, Liljestrom M (2003) Fibrillogenesis in gelsolin-related familial amyloidosis. Amyloid 10(Suppl 1):21–25
Mazanetz MP, Fischer PM (2007) Untangling tau hyperphosphorylation in drug design for neurodegenerative diseases. Nat Rev Drug Discov 6:464–479
McCarthy JV, Twomey C, Wujek P (2009) Presenilin-dependent regulated intramembrane proteolysis and γ-secretase activity. Cell Mol Life Sci 66:1534–1555
McCutchen SL, Lai Z, Miroy G, Kelly JW, Colon W (1995) Comparison of lethal and nonlethal transthyretin variants and their relationship to amyloid disease. Biochemistry 34:13527–13536
McGeer EG, McGeer PL (2010) Neuroinflammation in Alzheimer's disease and mild cognitive impairment: a field in its infancy. J Alzheimers Dis 19:355–361
McGlinchey RP, Kryndushkin D, Wickner RB (2011) Suicidal [PSI+] is a lethal yeast prion. Proc Natl Acad Sci USA 108:5337–5341
McLaurin J, Golomb R, Jurewicz A, Antel JP, Fraser PE (2000) Inositol stereoisomers stabilise an oligomeric aggregate of Alzheimer amyloid β peptide and inhibit Aβ-induced toxicity. J Biol Chem 275:18495–18502
McLaurin J, Cecal R, Kierstead ME, Tian X, Phinney AL, Manea M, French JE, Lambermon MH, Darabie AA, Brown ME, Janus C, Chishti MA, Horne P, Westaway D, Fraser PE, Mount HT, Przybylski M, St George-Hyslop P (2002) Therapeutically effective antibodies against amyloid-β peptide target amyloid-β residues 4–10 and inhibit cytotoxicity and fibrillogenesis. Nat Med 8:1263–1269
McLaurin J, Kierstead ME, Brown ME, Hawkes CA, Lambermon MHL, Phinney AL, Darabie AA, Cousins JE, French JE, Lan MF, Chen F, Wong SSN, Mount HTJ, Fraser PE, Westaway D, St George-Hyslop P (2006) Cyclohexanehexol inhibitors of Aβ aggregation prevent and reverse Alzheimer phenotype in a mouse model. Nat Med 12:801–808
McNaught KS, Shashidharan P, Perl DP, Jenner P, Olanow CW (2002) Aggresome-related biogenesis of Lewy bodies. Eur J Neurosci 16:2136–2148
McNaught KS, Belizaire R, Isacson O, Jenner P, Olanow CW (2003) Altered proteasomal function in sporadic Parkinson's disease. Exp Neurol 179:38–46
McNulty BC, Young GB, Pielak GJ (2006) Macromolecular crowding in the *Escherichia coli* periplasm maintains α-synuclein disorder. J Mol Biol 355:893–897
Meraz-Ríos MA, Lira-De León KI, Campos-Peña V, De Anda-Hernández MA, Mena-López R (2010) Tau oligomers and aggregation in Alzheimer's disease. J Neurochem 112:1353–1367
Merlini G, Bellotti V (2005) Lysozyme: a paradigmatic molecule for the investigation of protein structure, function and misfolding. Clin Chim Acta 357:168–172

Mestre T, Ferreira J, Coelho MM, Rosa M, Sampaio C (2009) Therapeutic interventions for disease progression in Huntington's disease. Cochrane Database Syst Rev 3:CD006455
Milano J, McKay J, Dagenais C, Foster-Brown L, Pognan F, Gadient R, Jacobs RT, Zacco A, Greenberg B, Ciaccio PJ (2004) Modulation of notch processing by γ-secretase inhibitors causes intestinal goblet cell metaplasia and induction of genes known to specify gut secretory lineage differentiation. Toxicol Sci 82:341–358
Miller SR, Sekijima Y, Kelly JW (2004) Native state stabilization by NSAIDs inhibits transthyretin amyloidogenesis from the most common familial disease variants. Lab Invest 84:545–552
Miller Y, Ma B, Nussinov R (2009) Polymorphism of Alzheimer's Aβ17–42 (p3) oligomers: the importance of the turn location and its conformation. Biophys J 97:1168–1177
Miyazaki D, Yazaki M, Gono T, Kametani F, Tsuchiya A, Matsuda M, Takenaka Y, Hosh Y 2nd, Ikeda S (2008) AH amyloidosis associated with an immunoglobulin heavy chain variable region (VH1) fragment: a case report. Amyloid 15:125–128
Monaco HL, Rizzi M, Coda A (1995) Structure of a complex of two plasma proteins: transthyretin and retinol-binding protein. Science 268:1039–1041
Monsonego A, Maron R, Zota V, Selkoe D, Weiner H (2001) Immune hyporesponsiveness to amyloid-β peptide in amyloid precursor protein transgenic mice: implications for the pathogenesis and treatment of Alzheimer's disease. Proc Natl Acad Sci USA 98:10273–10278
Moore SA, Huckerby TN, Gibson GL, Fullwood NJ, Turnbull S, Tabner BJ, El-Agnaf OM, Allsop D (2004) Both the $_D$-(+) and $_L$-(−) enantiomers of nicotine inhibit Aβ aggregation and cytotoxicity. Biochemistry 43:819–826
Morais-de-Sá E, Pereira PJ, Saraiva MJ, Damas AM (2004) The crystal structure of transthyretin in complex with diethylstilbestrol: a promising template for the design of amyloid inhibitors. J Biol Chem 279:53483–53490
Morgan D (2009) The role of microglia in antibody-mediated clearance of amyloid-β from the brain. CNS Neurol Disord Drug Targets 8:7–15
Morgan D, Diamond DM, Gottschall PE, Ugen KE, Dickey C, Hardy J, Duff K, Jantzen P, DiCarlo G, Wilcock D, Connor K, Hatcher J, Hope C, Gordon M, Arendash GW (2000) Aβ peptide vaccination prevents memory loss in an animal model of Alzheimer's disease. Nature 408:982–985
Mrak RE (2009) Neuropathology and the neuroinflammation idea. J Alzheimers Dis 18:473–481
Muchowski PJ, Wacker JL (2005) Modulation of neurodegeneration by molecular chaperones. Nat Rev Neurosci 6:11–22
Mukaetova-Ladinska EB, Hurt J, Jakes R, Xuereb J, Honer WG, Wischik CM (2000) α-Synuclein inclusions in Alzheimer and Lewy body diseases. J Neuropathol Exp Neurol 59:408–417
Muñoz FJ, Aldunate R, Inestrosa NC (1999) Peripheral binding site is involved in the neurotrophic activity of acetylcholinesterase. Neuroreport 10:3621–3625
Muñoz-Ruiz P, Rubio L, García-Palomero E, Dorronsoro I, del Monte-Millán M, Valenzuela R, Usán P, de Austria C, Bartolini M, Andrisano V, Bidon-Chanal A, Orozco M, Luque FJ, Medina M, Martínez A (2005) Design, synthesis, and biological evaluation of dual binding site acetylcholinesterase inhibitors: new disease-modifying agents for Alzheimer's disease. J Med Chem 48:7223–7233
Muñoz-Torrero D (2008) Acetylcholinesterase inhibitors as disease-modifying therapies for Alzheimer's disease. Curr Med Chem 15:2433–2455
Murphy MP, LeVine H 3rd (2010) Alzheimer's disease and the amyloid-β peptide. J Alzheimers Dis 19:311–323
Mustafi SM, Garai K, Crick SL, Baban B, Frieden C (2010) Substoichiometric inhibition of Aβ1–40 aggregation by a tandem $Aβ_{40-1}$–Gly_8–$Aβ_{1-40}$ peptide. Biochem Biophys Res Commun 397:509–512
Nagai Y, Tucker T, Ren H, Kenan DJ, Henderson BS, Keene JD, Strittmatter WJ, Burke JR (2000) Inhibition of polyglutamine protein aggregation and cell death by novel peptides identified by phage display screening. J Biol Chem 275:10437–10442
Nagai Y, Fujikake N, Ohno K, Higashiyama H, Popiel HA, Rahadian J, Yamaguchi M, Strittmatter WJ, Burke JR, Toda T (2003) Prevention of polyglutamine oligomerization and neurodegeneration by the peptide inhibitor QBP1 in *Drosophila*. Hum Mol Genet 12:1253–1259

Naiki H, Nagai Y (2009) Molecular pathogenesis of protein misfolding diseases: pathological molecular environments *versus* quality control systems against misfolded proteins. J Biochem 146:751–756

Naiki H, Nakakuki K (1996) First-order kinetic model of Alzheimer's β-amyloid fibril extension *in vitro*. Lab Invest 74:374–383

Naiki H, Yamamoto S, Hasegawa K, Yamaguchi I, Goto Y, Gejyo F (2005) Molecular interactions in the formation and deposition of β2-microglobulin-related amyloid fibrils. Amyloid 12:15–25

Neagu A, Neagu M, Der A (2001) Fluctuations and the Hofmeister effect. Biophys J 81:1285–1294

Neugroschl J, Sano M (2009) An update on treatment and prevention strategies for Alzheimer's disease. Curr Neurol Neurosci Rep 9:368–376

Neumann M, Sampathu DM, Kwong LK, Truax AC, Micsenyi MC, Chou TT, Bruce J, Schuck T, Grossman M, Clark CM, McCluskey LF, Miller BL, Masliah E, Mackenzie IR, Feldman H, Feiden W, Kretzschmar HA, Trojanowski JQ, Lee VM (2006) Ubiquitinated TDP-43 in frontotemporal lobar degeneration and amyotrophic lateral sclerosis. Science 314:130–133

Nichols WC, Dwulet FE, Liepnieks J, Benson MD (1988) Variant apolipoprotein AI as a major constituent of a human hereditary amyloid. Biochem Biophys Res Commun 156:762–768

Nicoll JA, Barton E, Boche D, Neal JW, Ferrer I, Thompson P, Vlachouli C, Wilkinson D, Bayer A, Games D, Seubert P, Schenk D, Holmes C (2006) Aβ species removal after Aβ42 immunization. J Neuropathol Exp Neurol 65:1040–1048

Niewold TA, Murphy CL, Hulskamp-Koch CA, Tooten PC, Gruys E (1999) Casein related amyloid, characterization of a new and unique amyloid protein isolated from bovine corpora amylacea. Amyloid 6:244–249

Nilsson MR, Dobson CM (2003) *In vitro* characterization of lactoferrin aggregation and amyloid formation. Biochemistry 42:375–382

Nilsson SF, Rask L, Peterson PA (1975) Studies on thyroid hormone-binding proteins. II. Binding of thyroid hormones, retinol-binding protein, and fluorescent probes to prealbumin and effects of thyroxine on prealbumin subunit self-association. J Biol Chem 250:8554–8563

Nitz M, Fenili D, Darabie AA, Wu L, Cousins JE, McLaurin J (2008) Modulation of amyloid-β aggregation and toxicity by inosose stereoisomers. FEBS J 275:1663–1674

Nollen EA, Garcia SM, van Haaften G, Kim S, Chavez A, Morimoto RI, Plasterk RH (2004) Genome-wide RNA interference screen identifies previously undescribed regulators of polyglutamine aggregation. Proc Natl Acad Sci USA 101:6403–6408

Nolte D, Sobanski E, Wissen A, Regula JU, Lichy C, Müller U (2010) Spinocerebellar ataxia type 17 associated with an expansion of 42 glutamine residues in TATA-box binding protein gene. J Neurol Neurosurg Psychiatry 81:1396–1399

Nonnis S, Cappelletti G, Taverna F, Ronchi C, Ronchi S, Negri A, Grassi E, Tedeschi G (2008) Tau is endogenously nitrated in mouse brain: identification of a tyrosine residue modified *in vivo* by NO. Neurochem Res 33:518–525

Nucifora FC Jr, Sasaki M, Peters MF, Huang H, Cooper JK, Yamada M, Takahashi H, Tsuji S, Troncoso J, Dawson VL, Dawson TM, Ross CA (2001) Interference by huntingtin and atrophin-1 with CBP-mediated transcription leading to cellular toxicity. Science 291:2423–2428

O'Nuallain B, Williams AD, Westermark P, Wetzel R (2004) Seeding specificity in amyloid growth induced by heterologous fibrils. J Biol Chem 279:17490–17499

Obregon DF, Rezai-Zadeh K, Bai Y, Sun N, Hou H, Ehrhart J, Zeng J, Mori T, Arendash GW, Shytle D, Town T, Tan J (2006) ADAM10 activation is required for green tea (−)-epigallocatechin-3-gallate-induced α-secretase cleavage of amyloid precursor protein. J Biol Chem 281:16419–16427

Ohashi K (2001) Pathogenesis of β2-microglobulin amyloidosis. Pathol Int 51:1–10

Okamoto Y, Nagai Y, Fujikake N, Akiko Popiel H, Yoshioka T, Toda T, Inui T (2009) Surface plasmon resonance characterization of specific binding of polyglutamine aggregation inhibitors of the expanded polyglutamine stretch. Biochem Biophys Res Commun 378:634–639

Ong DS, Kelly JW (2010) Chemical and/or biological therapeutic strategies to ameliorate protein misfolding diseases. Curr Opin Cell Biol 23:231–238

Ono K, Yoshiike Y, Takashima A, Hasegawa K, Naiki H, Yamada M (2003) Potent anti-amyloidogenic and fibril-destabilizing effects of polyphenols *in vitro*: implications for the prevention and therapeutics of Alzheimer's disease. J Neurochem 87:172–181
Ono K, Hasegawa K, Naiki H, Yamada M (2004) Anti-amyloidogenic activity of tannic acid and its activity to destabilize Alzheimer's β-amyloid fibrils *in vitro*. Biochim Biophys Acta 1690:193–202
Orgogozo JM, Dartigues JF, Lafont S, Letenneur L, Commenges D, Salamon R, Renaud S, Breteler MB (1997) Wine consumption and dementia in the elderly: a prospective community study in the Bordeaux area. Rev Neurol (Paris) 153:185–192
Ossato G, Digman MA, Aiken C, Lukacsovich T, Marsh JL, Gratton E (2010) A two-step path to inclusion formation of huntingtin peptides revealed by number and brightness analysis. Biophys J 98:3078–3085
Osseni RA, Debbasch C, Christen M-O, Rat P, Warnet J-M (1999) Tacrine-induced reactive oxygen species in a human liver cell line: the role of anethole dithiolethione as a scavenger. Toxicol *In Vitro* 13:683–688
Oza VB, Petrassi HM, Purkey HE, Kelly JW (1999) Synthesis and evaluation of anthranilic acid-based transthyretin amyloid fibril inhibitors. Bioorg Med Chem Lett 9:1–6
Oza VB, Smith C, Raman P, Koepf EK, Lashuel HA, Petrassi HM, Chiang KP, Powers ET, Sachettinni J, Kelly JW (2002) Synthesis, structure, and activity of diclofenac analogues as transthyretin amyloid fibril formation inhibitors. J Med Chem 45:321–332
Padrick SB, Miranker AD (2001) Islet amyloid polypeptide: identification of long-range contacts and local order on the fibrillogenesis pathway. J Mol Biol 308:783–794
Palaninathan SK, Mohamedmohaideen NN, Snee WC, Kelly JW, Sacchettini JC (2008) Structural insight into pH-induced conformational changes within the native human transthyretin tetramer. J Mol Biol 382:1157–1167
Palaninathan SK, Mohamedmohaideen NN, Orlandini E, Ortore G, Nencetti S, Lapucci A, Rossello A, Freundlich JS, Sacchettini JC (2009) Novel transthyretin amyloid fibril formation inhibitors: synthesis, biological evaluation, and X-ray structural analysis. PLoS One 4:e6290
Palazzolo I, Gliozzi A, Rusmini P, Sau D, Crippa V, Simonini F, Onesto E, Bolzoni E, Poletti A (2008) The role of the polyglutamine tract in androgen receptor. J Steroid Biochem Mol Biol 108:245–253
Paleos CM, Tsiourvas D, Sideratou Z, Tziveleka LA (2010) Drug delivery using multifunctional dendrimers and hyperbranched polymers. Expert Opin Drug Deliv 7:1387–1398
Palhano FL, Leme LP, Busnardo RG, Foguel D (2009) Trapping the monomer of a non-amyloidogenic variant of transthyretin: exploring its possible use as a therapeutic strategy against transthyretin amyloidogenic diseases. J Biol Chem 284:1443–1453
Pan T, Jankovic J, Le W (2003) Potential therapeutic properties of green tea polyphenols in Parkinson's disease. Drugs Aging 20:711–721
Pan M, Maitin V, Parathath S, Andreo U, Lin SX, St Germain C, Yao Z, Maxfield FR, Williams KJ, Fisher EA (2008) Presecretory oxidation, aggregation, and autophagic destruction of apoprotein-B: a pathway for late-stage quality control. Proc Natl Acad Sci USA 105:5862–5867
Pang Y-P, Quiram P, Jelacic T, Hong F, Brimijoin S (1996) Highly potent, selective, and low cost *bis*-tetrahydroaminacrine inhibitors of acetylcholinesterase. J Biol Chem 271:23646–23649
Panza F, Frisardi V, Imbimbo BP, Capurso C, Logroscino G, Sancarlo D, Seripa D, Vendemiale G, Pilotto A, Solfrizzi V (2010) γ-Secretase inhibitors for the treatment of Alzheimer's disease: the current state. CNS Neurosci Ther 16:272–284
Paravastu AK, Petkova AT, Tycko R (2006) Polymorphic fibril formation by residues 10–40 of the Alzheimer's β-amyloid peptide. Biophys J 90:4618–4629
Paravastu AK, Leapman RD, Yau WM, Tycko R (2008) Molecular structural basis for polymorphism in Alzheimer's β-amyloid fibrils. Proc Natl Acad Sci USA 105:18349–18354
Paravastu AK, Qahwash I, Leapman RD, Meredith SC, Tycko R (2009) Seeded growth of β-amyloid fibrils from Alzheimer's brain-derived fibrils produces a distinct fibril structure. Proc Natl Acad Sci USA 106:7443–7448

Patch JA, Barron AE (2003) Helical peptoid mimics of magainin-2 amide. J Am Chem Soc 125:12092–12093

Patel DJ, Tonelli AE (1976) *N*-methylleucine gramicidin-*S* and (di-*N*-methylleucine) gramicidin-*S* conformations with *cis* $_L$-Orn–$_L$-*N*–MeLeu peptide bonds. Biopolymers 15:1623–1635

Peng S, Larsson A, Wassberg E, Gerwins P, Thelin S, Fu X, Westermark P (2007) Role of aggregated medin in the pathogenesis of thoracic aortic aneurysm and dissection. Lab Invest 87:1195–1205

Penkler LJ, Van Rooyen PH, Wessels PL (1993) Conformational analysis of μ-selective [D-Ala2, MePhe4]enkephalins. Int J Pept Protein Res 41:261–274

Pepeu G, Giovannini MG (2009) Cholinesterase inhibitors and beyond. Curr Alzheimer Res 6:86–96

Pepys MB (2009) A molecular correlate of clinicopathology in transthyretin amyloidosis. J Pathol 217:1–3

Permanne B, Adessi C, Saborio GP, Fraga S, Frossard MJ, Van Dorpe J, Dewachter I, Banks WA, Van Leuven F, Soto C (2002) Reduction of amyloid load and cerebral damage in a transgenic mouse model of Alzheimer's disease by treatment with a β-sheet breaker peptide. FASEB J 16:860–862

Perrin RJ, Woods WS, Clayton DF, George JM (2000) Interaction of human α-synuclein and Parkinson's disease variants with phospholipids. Structural analysis using site-directed mutagenesis. J Biol Chem 275:34393–34398

Perry JJ, Shin DS, Tainer JA (2010) Amyotrophic lateral sclerosis. Adv Exp Med Biol 685:9–20

Pertinhez TA, Bouchard M, Tomlinson EJ, Wain R, Ferguson SJ, Dobson CM, Smith LJ (2001) Amyloid fibril formation by a helical cytochrome. FEBS Lett 495:184–186

Peterson DW, Zhou H, Dahlquist FW, Lew J (2008) A soluble oligomer of tau associated with fiber formation analyzed by NMR. Biochemistry 47:7393–7404

Petkova AT, Ishii Y, Balbach JJ, Antzutkin ON, Leapman RD, Delaglio F, Tycko R (2002) A structural model for Alzheimer's β-amyloid fibrils based on experimental constraints from solid state NMR. Proc Natl Acad Sci USA 99:16742–16747

Petkova AT, Leapman RD, Guo Z, Yau WM, Mattson MP, Tycko R (2005) Self-propagating, molecular-level polymorphism in Alzheimer's β-amyloid fibrils. Science 307:262–265

Petkova AT, Yau WM, Tycko R (2006) Experimental constraints on quaternary structure in Alzheimer's β-amyloid fibrils. Biochemistry 45:498–512

Petrassi HM, Klabunde T, Sacchettini J, Kelly JW (2000) Structure-based design of *N*-phenyl phenoxazine transthyretin amyloid fibril inhibitors. J Am Chem Soc 122:2178–2192

Petrassi HM, Johnson SM, Purkey HE, Chiang KP, Walkup T, Jiang X, Powers ET, Kelly JW (2005) Potent and selective structure-based dibenzofuran inhibitors of transthyretin amyloidogenesis: kinetic stabilization of the native state. J Am Chem Soc 127:6662–6671

Pfeifer M, Boncristiano S, Bondolfi L, Stalder A, Deller T, Staufenbiel M, Mathews PM, Jucker M (2002) Cerebral hemorrhage after passive anti-Aβ immunotherapy. Science 298:1379

Piazzi L, Rampa A, Bisi A, Gobbi S, Belluti F, Cavalli A, Bartolini M, Andrisano V, Valenti P, Recanatini M (2003) 3-(4-[[Benzyl(methyl)amino]methyl]phenyl)- 6,7-dimethoxy-2H-2-chromenone (AP2238) inhibits both acetylcholinesterase and acetylcholinesterase induced β-amyloid aggregation: a dual function lead for Alzheimer's disease therapy. J Med Chem 46:2279–2282

Plein H (2002) Amyloid β-protein forms ion channels. Trends Neurosci 25:137

Poduslo JF, Curran GL, Kumar A, Frangione B, Soto C (1999) β-Sheet breaker peptide inhibitor of Alzheimer's amyloidogenesis with increased blood–brain barrier permeability and resistance to proteolytic degradation in plasma. J Neurobiol 39:371–382

Pokorski JK, Jenkins LM, Feng H, Durell SR, Bai Y, Appella DH (2007) Introduction of a triazole amino acid into a peptoid oligomer induces turn formation in aqueous solution. Org Lett 9:2381–2383

Pollitt SK, Pallos J, Shao J, Desai UA, Ma AA, Thompson LM, Marsh JL, Diamond MI (2003) A rapid cellular FRET assay of polyglutamine aggregation identifies a novel inhibitor. Neuron 40:685–694

Popiel HA, Nagai Y, Fujikake N, Toda T (2007) Protein transduction domain mediated delivery of QBP1 suppresses polyglutamine-induced neurodegeneration *in vivo*. Mol Ther 15:303–309
Popiel HA, Nagai Y, Fujikake N, Toda T (2009) Delivery of the aggregate inhibitor peptide QBP1 into the mouse brain using PTDs and its therapeutic effect on polyglutamine disease mice. Neurosci Lett 449:87–92
Popova LA, Kodali R, Wetzel R, Lednev IK (2010) Structural variations in the cross-β core of amyloid β fibrils revealed by deep UV resonance Raman spectroscopy. J Am Chem Soc 132:6324–6328
Popovich PG, Longbrake EE (2008) Can the immune system be harnessed to repair the CNS? Nat Rev Neurosci 9:481–493
Popper K (1945) The open society and its enemies. Routledge. p 276. http://en.wikipedia.org/wiki/The_Open_Society_and_Its_Enemies
Porat Y, Mazor Y, Efrat S, Gazit E (2004) Inhibition of islet amyloid polypeptide fibril formation: a potential role for heteroaromatic interactions. Biochemistry 43:14454–14462
Porat Y, Abramowitz A, Gazit E (2006) Inhibition of amyloid fibril formation by polyphenols: structural similarity and aromatic interactions as a common inhibition mechanism. Chem Biol Drug Des 67:27–37
Post SG (2000) Defining the task. In: The moral challenge of Alzheimer's disease. Johns Hopkins University Press, Baltimore, p 3
Pountney DL, Voelcker NH, Gai WP (2005) Annular α-synuclein oligomers are potentially toxic agents in α-synucleinopathy. Neurotox Res 7:59–67
Pratim Bose P, Chatterjee U, Nerelius C, Govender T, Norström T, Gogoll A, Sandegren A, Göthelid E, Johansson J, Arvidsson PI (2009) Poly-*N*-methylated amyloid β-peptide (Aβ) *C*-terminal fragments reduce Aβ toxicity *in vitro* and in *Drosophila melanogaster*. J Med Chem 52:8002–8009
Prevelige PE, Thomas D, King J (1993) Nucleation and growth phases in the polymerization of coat and scaffolding subunits into icosahedral procapsid shells. Biophys J 64:824–835
Pride M, Seubert P, Grundman M, Hagen M, Eldridge J, Black RS (2008) Progress in the active immunotherapeutic approach to Alzheimer's disease: clinical investigations into AN1792-associated meningoencephalitis. Neurodegener Dis 5:194–196
Prusiner SB (1998) Prions. Proc Natl Acad Sci USA 95:13363–13383
Prusiner SB (2001) Shattuck lecture—neurodegenerative diseases and prions. N Engl J Med 344:1516–1526
Prusiner SB, Scott MR, DeArmond SJ, Cohen FE (1998) Prion protein biology. Cell 93:337–348
Purkey HE, Palaninathan SK, Kent KC, Smith C, Safe SH, Sacchettini JC, Kelly JW (2004) Hydroxylated polychlorinated biphenyls selectively bind transthyretin in blood and inhibit amyloidogenesis: rationalizing rodent PCB toxicity. Chem Biol 11:1719–1728
Qiao ZS, Guo ZY, Feng YM (2001) Putative disulfide-forming pathway of porcine insulin precursor during its refolding *in vitro*. Biochemistry 40:2662–2668
Racke MM, Boone LI, Hepburn DL, Parsadainian M, Bryan MT, Ness DK, Piroozi KS, Jordan WH, Brown DD, Hoffman WP, Holtzman DM, Bales KR, Gitter BD, May PC, Paul SM, DeMattos RB (2005) Exacerbation of cerebral amyloid angiopathy-associated microhemorrhage in amyloid precursor protein transgenic mice by immunotherapy is dependent on antibody recognition of deposited forms of amyloid β. J Neurosci 25:629–636
Raguse TL, Porter EA, Weisblum B, Gellman SH (2002) Structure–activity studies of 14-helical antimicrobial β-peptides: probing the relationship between conformational stability and antimicrobial potency. J Am Chem Soc 124:12774–12785
Rajarathnam K, Clark-Lewis I, Sykes BD (1994) ^{1}H NMR studies of interleukin 8 analogs: characterization of the domains essential for function. Biochemistry 33:6623–6630
Razavi H, Palaninathan SK, Powers ET, Wiseman RL, Purkey HE, Mohamedmohaideen NN, Deechongkit S, Chiang KP, Dendle MT, Sacchettini JC, Kelly JW (2003) Benzoxazoles as transthyretin amyloid fibril inhibitors. Synthesis, evaluation, and mechanism of action. Angew Chem Int Ed Engl 42:2758–2761
Reches M, Gazit E (2003) Casting metal nanowires within discrete self-assembled peptide nanotubes. Science 300:625–627

Reches M, Porat Y, Gazit E (2002) Amyloid fibril formation by pentapeptide and tetrapeptide fragments of human calcitonin. J Biol Chem 277:35475–35480

Redfield C, Schulman BA, Milhollen MA, Kim PS, Dobson CM (1999) α-Lactalbumin forms a compact molten globule in the absence of disulfide bonds. Nat Struct Biol 6:948–952

Reixach N, Deechongkit S, Jiang X, Kelly JW, Buxbaum JN (2004) Tissue damage in the amyloidoses: transthyretin monomers and nonnative oligomers are the major cytotoxic species in tissue culture. Proc Natl Acad Sci USA 101:2817–2822

Reixach N, Adamski-Werner SL, Kelly JW, Koziol J, Buxbaum JN (2006) Cell based screening of inhibitors of transthyretin aggregation. Biochem Biophys Res Commun 348:889–897

Reyes AE, Perez DR, Alvarez A, Garrido J, Gentry MK, Doctor BP, Inestrosa NC (1997) A monoclonal antibody against acetylcholinesterase inhibits the formation of amyloid fibrils induced by the enzyme. Biochem Biophys Res Commun 232:652–655

Reyes JF, Fu Y, Vana L, Kanaan NM, Binder LI (2011) Tyrosine nitration within the proline-rich region of Tau in Alzheimer's disease. Am J Pathol 178:2275–2285

Rezai-Zadeh K, Shytle D, Sun N, Mori T, Hou H, Jeanniton D, Ehrhart J, Townsend K, Zeng J, Morgan D, Hardy J, Town T, Tan J (2005) Green tea epigallocatechin-3-gallate (EGCG) modulates amyloid precursor protein cleavage and reduces cerebral amyloidosis in Alzheimer transgenic mice. J Neurosci 25:8807–8814

Riess O, Rüb U, Pastore A, Bauer P, Schöls L (2008) SCA3: neurological features, pathogenesis and animal models. Cerebellum 7:125–137

Riisoen H (1988) Reduced prealbumin (transthyretin) in CSF of severely demented patients with Alzheimer's disease. Acta Neurol Scand 78:455–459

Rinderspacher A, Cremona ML, Liu Y, Deng SX, Xie Y, Gong G, Aulner N, Többen U, Myers K, Chung C, Andersen M, Vidović D, Schürer S, Brandén L, Yamamoto A, Landry DW (2009) Potent inhibitors of Huntingtin protein aggregation in a cell-based assay. Bioorg Med Chem Lett 19:1715–1717

Ringman JM, Cole GM, Teng E, Badmaev V, Bardens J, Frautschy S, Rosario E, Fein J, Porter V, Vanek Z, Sugar C, Yau A, Cummings JL (2008) Oral curcumin for the treatment of mild-to-moderate Alzheimer's disease: tolerability and clinical and biomarker efficacy results of a placebo-controlled 24-week study. In: Proceedings of the abstract of international conference on Alzheimer's disease, Chicago, USA, 26–31 July 2008, p T774

Ritchie CW, Bush AI, Mackinnon A, Macfarlane S, Mastwyk M, MacGregor L, Kiers L, Cherny R, Li QX, Tammer A, Carrington D, Mavros C, Volitakis I, Xilinas M, Ames D, Davis S, Beyreuther K, Tanzi RE, Masters CL (2003) Metalprotein attenuation with iodochlorhydroxyquin (clioquinol) targeting Aβ amyloid deposition and toxicity in Alzheimer disease: a pilot phase 2 clinical trial. Arch Neurol 60:1685–1691

Ritter C, Maddelein ML, Siemer AB, Lührs T, Ernst M, Meier BH, Saupe SJ, Riek R (2005) Correlation of structural elements and infectivity of the HET-s prion. Nature 435:844–848

Rival T, Page RM, Chandraratna DS, Sendall TJ, Ryder E, Liu B, Lewis H, Rosahl T, Hider R, Camargo LM, Shearman MS, Crowther DC, Lomas DA (2009) Fenton chemistry and oxidative stress mediate the toxicity of the β-amyloid peptide in a *Drosophila* model of Alzheimer's disease. Eur J Neurosci 29:1335–1347

Rivière C, Richard T, Quentin L, Krisa S, Mérillon JM, Monti JP (2006) Inhibitory activity of stilbenes on Alzheimer's β-amyloid fibrils *in vitro*. Bioorg Med Chem 15:1160–1167

Rizzo S, Bartolini M, Ceccarini L, Piazzi L, Gobbi S, Cavalli A, Recanatini M, Andrisano V, Rampa A (2010) Targeting Alzheimer's disease: Novel indanone hybrids bearing a pharmacophoric fragment of AP2238. Bioorg Med Chem 18(5):1749–60. Epub 2010 Feb 4. PMID: 20171894 [PubMed - indexed for MEDLINE]: http://www.ncbi.nlm.nih.gov/pubmed/20171894

Röcken C, Shakespeare A (2002) Pathology, diagnosis and pathogenesis of AA amyloidosis. Virchows Arch 440:111–122

Röcken C, Becker K, Fandrich M, Schroeckh V, Stix B, Rath T, Kahne T, Dierkes J, Roessner A, Albert FW (2006) ALys amyloidosis caused by compound heterozygosity in exon 2 (Thr70Asn) and exon 4 (Trp112Arg) of the lysozyme gene. Hum Mutat 27:119–120

Rosas HD, Koroshetz WJ, Jenkins BG, Chen YI, Hayden DL, Beal MF, Cudkowicz ME (1999) Riluzole therapy in Huntington's disease. Mov Disord 14:326–330

Rosen DR, Siddique T, Patterson D, Figlewicz DA, Sapp P, Hentati A, Donaldson D, Goto J, O'Regan JP, Deng HX, Rahmani Z, Krizus A, McKenna-Yasek D, Cayabyab A, Gaston SM, Berger R, Tanzi RE, Halperin JJ, Herzfeldt B, Van den Bergh R, Hung W-Y, Bird T, Deng G, Mulder DW, Smyth C, Laing NG, Soriano E, Pericak–Vance MA, Haines J, Rouleau GA, Gusella JS, Horvitz HR, Brown RH Jr (1993) Mutations in Cu/Zn superoxide dismutase gene are associated with familial amyotrophic lateral sclerosis. Nature 362:59–62

Rosini M, Andrisano V, Bartolini M, Bolognesi ML, Rehfla P, Minarini A, Tarozzi A, Melchiorre C (2005) Rational approach to discover multipotent anti-Alzheimer drugs. J Med Chem 48:360–363

Röskam S, Neff F, Schwarting R, Bacher M, Dodel R (2010) APP transgenic mice: the effect of active and passive immunotherapy in cognitive tasks. Neurosci Biobehav Rev 34:487–499

Ross CA (2002) Polyglutamine pathogenesis: emergence of unifying mechanisms for Huntington's disease and related disorders. Neuron 35:819–822

Ross CA, Poirier MA, Wanker EE, Amzel M (2003) PolyQ fibrillogenesis: the pathway unfolds. Proc Natl Acad Sci USA 100:1–3

Rothstein JD (2009) Current hypotheses for the underlying biology of amyotrophic lateral sclerosis. Ann Neurol 65(Suppl 1):S3–S9

Roze E, Bonnet C, Betuing S, Caboche J (2010) Huntington's disease. Adv Exp Med Biol 685:45–63

Rüb U, Brunt ER, Deller T (2008) New insights into the pathoanatomy of spinocerebellar ataxia type 3 (Machado-Joseph disease). Curr Opin Neurol 21:111–116

Rubinsztein DC (2002) Lessons from animal models of Huntington's disease. Trends Genet 18:202–209

Ryu J, Kanapathipillai M, Lentzen G, Park CB (2008) Peptide inhibition of β-amyloid peptide aggregation and neurotoxicity by α-$_D$-mannosylglycerate, a natural extremolyte. Peptides 29:578–584

Sadler K, Tam JP (2002) Peptide dendrimers: applications and synthesis. J Biotechnol 90:195–229

Sahara N, Maeda S, Murayama M, Suzuki T, Dohmae N, Yen SH, Takashima A (2007) Assembly of two distinct dimers and higher-order oligomers from full length tau. Eur J Neurosci 25:3020–3029

Salomon AR, Marcinowski KJ, Friedland RP, Zagorski MG (1996) Nicotine inhibits amyloid formation by the β-peptide. Biochemistry 35:13568–13578

Samson K (2010) NerveCenter: phase III Alzheimer trial halted: search for therapeutic biomarkers continues. Ann Neurol 68:A9–A12

Santhoshkumar P, Sharma KK (2004) Inhibition of amyloid fibrillogenesis and toxicity by a peptide chaperone. Mol Cell Biochem 267:147–155

Saraiva MJ (1995) Transthyretin mutations in health and disease. Hum Mutat 5:191–196

Saraiva MJ (2002) Hereditary transthyretin amyloidosis: molecular basis and therapeutical strategies. Expert Rev Mol Med 4:1–11

Saumier D, Aisen PS, Gauthier S, Vellas B, Ferris SH, Duong A, Suhy J, Oh J, Lau W, Garceau D, Haine D, Sampalis J (2009) Lessons learned in the use of volumetric MRI in therapeutic trials in Alzheimer's disease: the ALZHEMED (Tramiprosate) experience. J Nutr Health Aging 13:370–372

Saura CA, Choi SY, Beglopoulos V, Malkani S, Zhang D, Shankaranarayana Rao BS, Chattarji S, Kelleher RJ 3rd, Kandel ER, Duff K, Kirkwood A, Shen J (2004) Loss of presenilin function causes impairments of memory and synaptic plasticity followed by age-dependent neurodegeneration. Neuron 42:23–36

Sayre LM, Zelasko DA, Harris PL, Perry G, Salomon RG, Smith MA (1997) 4-Hydroxynonenal-derived advanced lipid peroxidation end products are increased in Alzheimer's disease. J Neurochem 68:2092–2097

Scarmeas N, Stern Y, Mayeux R, Luchsinger JA (2006) Mediterranean diet, Alzheimer disease, and vascular mediation. Arch Neurol 63:1709–1717

Scatena R, Martorana GE, Bottoni P, Botta G, Pastore P, Giardina B (2007) An update on pharmacological approaches to neurodegenerative diseases. Expert Opin Investig Drugs 16:59–72

Schenk D, Barbour R, Dunn W, Gordon G, Grajeda H, Guido T, Hu K, Huang J, Johnson-Wood K, Khan K, Kholodenko D, Lee M, Liao Z, Lieberburg I, Motter R, Mutter L, Soriano F, Shopp G, Vasquez N, Vandevert C, Walker S, Wogulis M, Yednock T, Games D, Seubert P (1999) Immunization with amyloid-β attenuates Alzheimer-disease-like pathology in the PDAPP mouse. Nature 400:173–177

Scherzer-Attali R, Pellarin R, Convertino M, Frydman-Marom A, Egoz-Matia N, Peled S, Levy-Sakin M, Shalev DE, Caflisch A, Gazit E, Segal D (2010) Complete phenotypic recovery of an Alzheimer's disease model by a quinone-tryptophan hybrid aggregation inhibitor. PLoS One 5:e11101

Scherzinger E, Sittler A, Schweiger K, Heiser V, Lurz R, Hasenbank R, Bates GP, Lehrach H, Wanker EE (1999) Self-assembly of polyglutamine-containing huntingtin fragments into amyloid-like fibrils: implications for Huntington's disease pathology. Proc Natl Acad Sci USA 96:4604–4609

Schiefer J, Landwehrmeyer GB, Lüesse HG, Sprünken A, Puls C, Milkereit A, Milkereit E, Kosinski CM (2002) Riluzole prolongs survival time and alters nuclear inclusion formation in a transgenic mouse model of Huntington's disease. Mov Disord 17:748–757

Schlachetzki JC, Hüll M (2009) Microglial activation in Alzheimer's disease. Curr Alzheimer Res 6:554–563

Schor NF (2011) What the halted phase III γ-secretase inhibitor trial may (or may not) be telling us. Ann Neurol 69:237–239

Schott JM, Price SL, Frost C, Whitwell JL, Rossor MN, Fox NC (2005) Measuring atrophy in Alzheimer disease: a serial MRI study over 6 and 12 months. Neurology 65:119–124

Schwarzman AL, Goldgaber D (1996) Interaction of transthyretin with amyloid β-protein: binding and inhibition of amyloid formation. Ciba Found Symp 199:146–160, discussion 160–4

Schwarzman AL, Gregori L, Vitek MP, Lyubski S, Strittmatter WJ, Enghilde JJ, Bhasin R, Silverman J, Weisgraber KH, Coyle PK, Zagorski MG, Talafous J, Eisenberg M, Saunders AM, Roses AD, Goldgaber D (1994) Transthyretin sequesters amyloid β protein and prevents amyloid formation. Proc Natl Acad Sci USA 91:8368–8372

Schwarzman AL, Tsiper M, Wente H, Wang A, Vitek MP, Vasiliev V, Goldgaber D (2004) Amyloidogenic and anti-amyloidogenic properties of recombinant transthyretin variants. Amyloid 11:1–9

Schwarzman AL, Tsiper M, Gregori L, Goldgaber D, Frakowiak J, Mazur-Kolecka B, Taraskina A, Pcheina S, Vitek MP (2005) Selection of peptides binding to the amyloid β protein reveals potential inhibitors of amyloid formation. Amyloid 12:199–209

Sekijima Y, Wiseman RL, Matteson J, Hammarström P, Miller SR, Sawkar AR, Balch WE, Kelly JW (2005) The biological and chemical basis for tissue-selective amyloid disease. Cell 121:73–85

Sekijima Y, Kelly JW, Ikeda S (2008) Pathogenesis of and therapeutic strategies to ameliorate the transthyretin amyloidoses. Curr Pharm Des 14:3219–3230

Selenko P, Wagner G (2007) Looking into live cells with in-cell NMR spectroscopy. J Struct Biol 158:244–253

Serot JM, Christmann D, Dubost T, Couturier M (1997) Cerebrospinal fluid transthyretin: aging and late onset Alzheimer's disease. J Neurol Neurosurg Psychiatry 63:506–508

Seubert P, Vigo-Pelfrey C, Esch F, Lee M, Dovey H, Davis D, Sinha S, Schlossmacher M, Whaley J, Swindlehurst C, McCormack R, Wolfert R, Selkoe D, Lieberburg I, Schenk D (1992) Isolation and quantification of soluble Alzheimer's β-peptide from biological fluids. Nature 359:325–327

Seurynck-Servoss SL, Dohm MT, Barron AE (2006) Effects of including an *N*-terminal insertion region and arginine-mimetic side chains in helical peptoid analogues of lung surfactant protein B. Biochemistry 45:11809–11818

Shafrir Y, Durell SR, Anishkin A, Guy HR (2010) β-Barrel models of soluble amyloid-β oligomers and annular protofibrils. Proteins 78:3458–3472
Shah RS, Lee HG, Xiongwei Z, Perry G, Smith MA, Castellani RJ (2008) Current approaches in the treatment of Alzheimer's disease. Biomed Pharmacother 62:199–207
Shehi E, Fusi P, Secundo F, Pozzuolo S, Bairati A, Tortora P (2003) Temperature-dependent, irreversible formation of amyloid fibrils by a soluble human ataxin-3 carrying a moderately expanded polyglutamine stretch (Q36). Biochemistry 42:14626–14632
Shen J, Kelleher RJ 3rd (2007) The presenilin hypothesis of Alzheimer's disease: evidence for a loss-of-function pathogenic mechanism. Proc Natl Acad Sci USA 104:403–409
Shewmaker F, Wickner RB, Tycko R (2006) Amyloid of the prion domain of Sup35p has an in-register parallel β-sheet structure. Proc Natl Acad Sci USA 103:19754–19759
Shewmaker F, Kryndushkin D, Chen B, Tycko R, Wickner RB (2009) Two prion variants of Sup35p have in-register parallel β-sheet structures, independent of hydration. Biochemistry 48:5074–5082
Shikama Y, Kitazawa J, Yagihashi N, Uehara O, Murata Y, Yajima N, Wada R, Yagihashi S (2010) Localized amyloidosis at the site of repeated insulin injection in a diabetic patient. Intern Med 49:397–401
Shimohata T, Nakajima T, Yamada M, Uchida C, Onodera O, Naruse S, Kimura T, Koide R, Nozaki K, Sano Y, Ishiguro H, Sakoe K, Ooshima T, Sato A, Ikeuchi T, Oyake M, Sato T, Aoyagi Y, Hozumi I, Nagatsu T, Takiyama Y, Nishizawa M, Goto J, Kanazawa I, Davidson I, Tanese N, Takahashi H, Tsuji S (2000) Expanded polyglutamine stretches interact with TAFII130, interfering with CREB-dependent transcription. Nat Genet 26:29–36
Shimohata M, Shimohata T, Igarashi S, Naruse S, Tsuji S (2005) Interference of CREB-dependent transcriptional activation by expanded polyglutamine stretches—augmentation of transcriptional activation as a potential therapeutic strategy for polyglutamine diseases. J Neurochem 93:654–663
Shin I, Silman I, Weiner LM (1996) Interaction of partially unfolded forms of Torpedo acetylcholinesterase with liposomes. Protein Sci 5:42–51
Shin SB, Yoo B, Todaro LJ, Kirshenbaum K (2007) Cyclic peptoids. J Am Chem Soc 129:3218–3225
Shiraki K, Kudou M, Fujiwara S, Imanaka T, Takagi M (2002) Biophysical effect of amino acids on the prevention of protein aggregation. J Biochem 132:591–595
Shiraki K, Kudou M, Nishikori S, Kitagawa H, Imanaka T, Takagi M (2004) Arginine ethylester prevents thermal inactivation and aggregation of lysozyme. Eur J Biochem 271:3242–3247
Shorter J, Lindquist S (2005) Prions as adaptive conduits of memory and inheritance. Nat Rev Genet 6:435–450
Siddiqui N, Afshari NA (2002) The changing face of the genetics of corneal dystrophies. Curr Opin Ophthalmol 13:199–203
Sigurdsson EM, Permanne B, Soto C, Wisniewski T, Frangione B (2000) *In vivo* reversal of amyloid-β lesions in rat brain. J Neuropathol Exp Neurol 59:11–17
Sigurdsson EM, Scholtzova H, Mehta PD, Frangione B, Wisniewski T (2001) Immunization with a nontoxic/nonfibrillar amyloid-β homologous peptide reduces Alzheimer's disease-associated pathology in transgenic mice. Am J Pathol 159:439–447
Simons LJ, Caprathe BW, Callahan M, Graham JM, Kimura T, Lai Y, LeVine H 3rd, Lipinski W, Sakkab AT, Tasaki Y, Walker LC, Yasunaga T, Ye Y, Zhuang N, Augelli-Szafran CE (2009) The synthesis and structure–activity relationship of substituted *N*-phenyl anthranilic acid analogs as amyloid aggregation inhibitors. Bioorg Med Chem Lett 19:654–657
Sipe JD (2000) Serum amyloid A: from fibril to function. Current status. Amyloid 7:10–12
Smith MA, Harris PL, Sayre LM, Perry G (1997) Iron accumulation in Alzheimer's disease is a source of redox-generated free radicals. Proc Natl Acad Sci USA 94:9866–9868
Smith DL, Portier R, Woodman B, Hockly E, Mahal A, Klunk WE, Li XJ, Wanker E, Murray KD, Bates GP (2001) Inhibition of polyglutamine aggregation in R6/2 HD brain slices-complex dose-response profiles. Neurobiol Dis 8:1017–1026

Smith DL, Woodman B, Mahal A, Sathasivam K, Ghazi-Noori S, Lowden PA, Bates GP, Hockly E (2003) Minocycline and doxycycline are not beneficial in a model of Huntington's disease. Ann Neurol 54:186–196

Soderlund T, Alakoskela JM, Pakkanen AL, Kinnunen PK (2003) Comparison of the effects of surface tension and osmotic pressure on the interfacial hydration of a fluid phospholipid bilayer. Biophys J 85:2333–2341

Sokolowski F, Modler AJ, Masuch R, Zirwer D, Baier M, Lutsch G, Moss DA, Gast K, Naumann D (2003) Formation of critical oligomers is a key event during conformational transition of recombinant syrian hamster prion protein. J Biol Chem 278:40481–40492

Solomon A, Frangione B, Franklin EC (1982) Bence Jones proteins and light chain of immunoglobulins Preferential association of the $V\lambda_{VI}$ subgroup of human light chains with amyloidosis AL(λ). J Clin Invest 70:453–460

Solomon A, Murphy CL, Weaver K, Weiss DT, Hrncic R, Eulitz M, Donnell RL, Sletten K, Westermark G, Westermark P (2003) Calcifying epithelial odontogenic (Pindborg) tumor-associated amyloid consists of a novel human protein. J Lab Clin Med 142:348–355

Sonnen AF, Yu C, Evans EJ, Stuart DI, Davis SJ, Gilbert RJ (2010) Domain metastability: a molecular basis for immunoglobulin deposition? J Mol Biol 399:207–213

Soreghan B, Kosmoski J, Glabe C (1994) Surfactant properties of Alzheimer's Aβ peptides and the mechanism of amyloid aggregation. J Biol Chem 269:28551–28554

Soto C, Kindy MS, Baumann M, Frangione B (1996) Inhibition of Alzheimer's amyloidosis by peptides that prevent β-sheet conformation. Biochem Biophys Res Commun 226:672–680

Soto C, Sigurdsson EM, Morelli L, Kumar RA, Castano EM, Frangione B (1998) β-sheet breaker peptides inhibit fibrillogenesis in a rat brain model of amyloidosis: Implications for Alzheimer's therapy. Nat Med 4:822–826

Soto P, Griffin MA, Shea J-E (2007) New insights into the mechanism of Alzheimer amyloid-β fibrillogenesis inhibition by *N*-methylated peptides. Biophys J 93:3015–3025

Sousa MM, Yan SD, Stern D, Saraiva MJ (2000) Interaction of the receptor for advanced glycation end products (RAGE) with transthyretin triggers nuclear transcription factor κB (NF-κB) activation. Lab Invest 80:1101–1110

Sousa MM, Cardoso I, Fernandes R, Guimaraes R, Saraiva MJ (2001) Deposition of transthyretin in early stages of familial amyloidotic polyneuropathy: evidence for toxicity of nonfibrillar aggregates. Am J Pathol 159:1993–2000

Souza JM, Giasson BI, Chen Q, Lee VM, Ischiropoulos H (2000) Dityrosine cross-linking promotes formation of stable α-synuclein polymers. Implication of nitrative and oxidative stress in the pathogenesis of neurodegenerative synucleinopathies. J Biol Chem 275:18344–18349

Spetzler JC, Tam JP (1995) Unprotected peptides as building blocks for branched peptides and peptide dendrimers. Int J Pept Protein Res 45:78–85

Spires-Jones TL, Stoothoff WH, de Calignon A, Jones PB, Hyman BT (2009) Tau pathophysiology in neurodegeneration: a tangled issue. Trends Neurosci 32:150–159

Squier TC (2001) Oxidative stress and protein aggregation during biological aging. Exp Gerontol 36:1539–1550

Stackman RW, Eckenstein F, Frei B, Kulhanek D, Nowlin J, Quinn JF (2003) Prevention of age-related spatial memory deficits in a transgenic mouse model of Alzheimer's disease by chronic *Ginkgo biloba* treatment. Exp Neurol 184:510–520

Stanger HE, Syud FA, Espinosa JF, Giriat I, Muir T, Gellman SH (2001) Length-dependent stability and strand length limits in antiparallel β-sheet secondary structure. Proc Natl Acad Sci USA 98:12015–12020

Starikov EB, Lehrach H, Wanker EE (1999) Folding of oligoglutamines: a theoretical approach based on thermodynamics and molecular mechanics. J Biomol Struct Dyn 17:409–427

Stefani M (2004) Protein misfolding and aggregation: new examples in medicine and biology of the dark side of the protein world. Biochim Biophys Acta 1739:5–25

Stefani M, Dobson CM (2003) Protein aggregation and aggregate toxicity: new insights into protein folding, misfolding diseases and biological evolution. J Mol Med 81:678–699

Stefano Rizzo S, Bartolini M, Ceccarini L, Piazzi L, Gobbi S, Cavalli A, Recanatini M, Andrisano V, Rampa A (2010) Targeting Alzheimer's disease: novel indanone hybrids bearing a pharmacophoric fragment of AP2238. Bioorg Med Chem 18:1749–1760

Stein TD, Johnson JA (2002) Lack of neurodegeneration in transgenic mice overexpressing mutant amyloid precursor protein is associated with increased levels of transthyretin and the activation of cell survival pathways. J Neurosci 22:7380–7388

Stein TD, Anders NJ, DeCarli C, Chan SL, Mattson MP, Johnson JA (2004) Neutralization of transthyretin reverses the neuroprotective effects of secreted amyloid precursor protein (APP) in APPSw mice resulting in tau phosphorylation and loss of hippocampal neurons: support for the amyloid hypothesis. J Neurosci 24:7707–7717

Stephenson V, Weaver DF (2006) Mechanism of action of the anti-Alzheimer's drug 3-APS. Alzheimers Dement 2(Suppl 1):P4–P436

Storkel S, Schneider HM, Muntefering H, Kashiwagi S (1983) Iatrogenic, insulin-dependent, local amyloidosis. Lab Invest 48:108–111

Strobel G (2009) An eFAD prevention trial—one man's view. Alzheimer Research Forum [online]. http:// www.alzforum.org/new/detail.asp?id=2273

Strozyk D, Blennow K, White LR, Launer LJ (2003) CSF Aβ42 levels correlate with amyloid-neuropathology in a population-based autopsy study. Neurology 60:652–656

Sun Y, Zhang G, Hawkes CA, Shaw JE, McLaurin J, Nitz M (2008) Synthesis of scyllo-inositol derivatives and their effects on amyloid β peptide aggregation. Bioorg Med Chem 16:7177–7184

Supattapone S, Nguyen HO, Cohen FE, Prusiner SB, Scott MR (1999) Elimination of prions by branched polyamines and implications for therapeutics. Proc Natl Acad Sci USA 96:14529–14534

Supattapone S, Wille H, Uyechi L, Safar J, Tremblay P, Szoka FC, Cohen FE, Prusiner SB, Scott MR (2001) Branched polyamines cure prion-infected neuroblastoma cells. J Virol 75:3453–3461

Suzuki N, Cheung TT, Cai XD, Odaka A, Otvos L Jr, Eckman C, Golde TE, Younkin SG (1994) An increased percentage of long amyloid β protein secreted by familial amyloid β protein precursor (βAPP717) mutants. Science 264:1336–1340

Swift B (2002) Examination of insulin injection sites: an unexpected finding of localized amyloidosis. Diabet Med 19:881–882

Takahashi T, Mihara H (2008) Peptide and protein mimetics inhibiting amyloid-β peptide aggregation. Acc Chem Res 41:1309–1318

Takahashi T, Ohta K, Mihara H (2010) Rational design of amyloid β peptide-binding proteins: pseudo-Aβ β-sheet surface presented in green fluorescent protein binds tightly and preferentially to structured Aβ. Proteins 78:336–347

Takashima A, Murayama M, Murayama O, Kohno T, Honda T, Yasutake K, Nihonmatsu N, Mercken M, Yamaguchi H, Sugihara S, Wolozin B (1998) Presenilin 1 associates with glycogen synthase kinase-3β and its substrate tau. Proc Natl Acad Sci USA 95:9637–9641

Takeyama K, Ito S, Yamamoto A, Tanimoto H, Furutani T, Kanuka H, Miura M, Tabata T, Kato S (2002) Androgen-dependent neurodegeneration by polyglutamine-expanded human androgen receptor in *Drosophila*. Neuron 35:855–864

Tam JP, Spetzler JC (1995) Chemoselective approaches to the preparation of peptide dendrimers and branched artificial proteins using unprotected peptides as building blocks. Biomed Pept Proteins Nucleic Acids 1:123–132

Tam JP, Spetzler JC (2001) Synthesis and application of peptide dendrimers as protein mimetics. Curr Protoc Immunol, Chapter 9, Unit 9.6

Tanaka M, Morishima I, Akagi T, Hashikawa T, Nukina N (2001) Intra-and intermolecular β-pleated sheet formation in glutamine-repeat inserted myoglobin as a model for polyglutamine diseases. J Biol Chem 276:45470–45475

Tanaka M, Machida Y, Nishikawa Y, Akagi T, Morishima I, Hashikawa T, Fujisawa T, Nukina N (2002) The effects of aggregation-inducing motifs on amyloid formation of model proteins related to neurodegenerative diseases. Biochemistry 41:10277–10286

Tanaka M, Machida Y, Niu S, Ikeda T, Jana NR, Doi H, Kurosawa M, Nekooki M, Nukina N (2004) Trehalose alleviates polyglutamine-mediated pathology in a mouse model of Huntington disease. Nat Med 10:148–154

Tanaka M, Chien P, Yonekura K, Weissman JS (2005a) Mechanism of cross-species prion transmission: an infectious conformation compatible with two highly divergent yeast prion proteins. Cell 121:49–62

Tanaka M, Machida Y, Nukina N (2005b) A novel therapeutic strategy for polyglutamine diseases by stabilizing aggregation-prone proteins with small molecules. J Mol Med 83:343–352

Taneja S, Ahmad F (1994) Increased thermal stability of proteins in the presence of amino acids. Biochem J 303:147–153

Tang MX, Redemann CT, Szoka FC Jr (1996) *In vitro* gene delivery by degraded polyamidoamine dendrimers. Bioconjug Chem 7:703–714

Tarditi A, Caricasole A, Terstappen G (2009) Therapeutic targets for Alzheimer's disease. Expert Opin Ther Targets 13:551–567

Taylor M, Moore S, Mayes J, Parkin E, Beeg M, Canovi M, Gobbi M, Mann DMA, Allsop D (2010) Development of a proteolytically stable retro-inverso peptide inhibitor of β-amyloid oligomerization as a potential novel treatment for Alzheimer's disease. Biochemistry 49:3261–3272

Thakur AK, Yang W, Wetzel R (2004) Inhibition of polyglutamine aggregate cytotoxicity by a structure-based elongation inhibitor. FASEB J 18:923–925

Thakur AK, Jayaraman M, Mishra R, Thakur M, Chellgren VM, Byeon IJ, Anjum DH, Kodali R, Creamer TP, Conway JF, Gronenborn AM, Wetzel R (2009) Polyglutamine disruption of the huntingtin exon 1 *N*-terminus triggers a complex aggregation mechanism. Nat Struct Mol Biol 16:380–389

Tjernberg LO, Naslund J, Lindqvist F, Johansson J, Karlstrom AR, Thyberg J, Terenius L, Nordstedt C (1996) Arrest of β-amyloid fibril formation by a pentapeptide ligand. J Biol Chem 271:8545–8548

Tjernberg LO, Lilliiehook C, Callaway DJE, Naslund J, Hahne S, Thyberg J, Terenius L, Nordstedt C (1997) Controlling amyloid β-peptide fibril formation with protease stable ligands. J Biol Chem 272:12601–12605

Tobin AJ, Signer ER (2000) Huntington's disease: the challenge for cell biologists. Trends Cell Biol 10:531–536

Tonelli AE (1971) On the stability of *cis* and *trans* amide bond conformations in polypeptides. J Am Chem Soc 93:7153–7155

Tonelli AE (1974) Conformational characteristics of polypeptides containing isolated $_{L}$-proline residues with cis peptide bonds. J Mol Biol 86:627–635

Tong Y, Yamaguchi H, Giaime E, Boyle S, Kopan R, Kelleher RJ 3rd, Shen J (2010) Loss of leucine-rich repeat kinase 2 causes impairment of protein degradation pathways, accumulation of α-synuclein, and apoptotic cell death in aged mice. Proc Natl Acad Sci USA 107:9879–9884

Török M, Abid M, Mhadgut SC, Török B (2006) Organofluorine inhibitors of amyloid fibrillogenesis. Biochemistry 45:5377–5383

Town T, Tan J, Sansone N, Obregon D, Klein T, Mullan M (2001) Characterization of murine immunoglobulin G antibodies against human amyloid-β1–42. Neurosci Lett 307:101–104

Townend R, Kumosinski TF, Timasheff SN (1966) The circular dichroism of the β structure of poly-$_{L}$-lysine. Biochem Biophys Res Commun 23:163–169

Townsend M, Cleary JP, Mehta T, Hofmeister J, Lesné S, O'Hare E, Walsh DM, Selkoe DJ (2006) Orally available compound prevents deficits in memory caused by the Alzheimer amyloid-β oligomers. Ann Neurol 60:668–676

Trapnell BC, Whitsett JA, Nakata K (2003) Pulmonary alveolar proteinosis. N Engl J Med 349:2527–2539

Trexler AJ, Nilsson MR (2007) The formation of amyloid fibrils from proteins in the lysozyme family. Curr Protein Pept Sci 8:537–557

Trottier Y, Devys D, Imbert G, Saudou F, An I, Lutz Y, Weber C, Agid Y, Hirsch EC, Mandel JL (1995a) Cellular localization of the Huntington's disease protein and discrimination of the normal and mutated form. Nat Genet 10:104–110

Trottier Y, Lutz Y, Stevanin G, Imbert G, Devys D, Cancel G, Saudou F, Weber C, David G, Tora L, Agid Y, Brice A, Mandel J-L (1995b) Polyglutamine expansion as a pathological epitope in Huntington's disease and four dominant cerebellar ataxias. Nature 378:403–406

Truant R, Atwal RS, Desmond C, Munsie L, Tran T (2008) Huntington's disease: revisiting the aggregation hypothesis in polyglutamine neurodegenerative diseases. FEBS J 275:4252–4262

Tsuji S (2004) Spinocerebellar ataxia type 17: latest member of polyglutamine disease group highlights unanswered questions. Arch Neurol 61:183–184

Tsumoto K, Umetsu M, Kumagai I, Ejima D, Philo JS, Arakawa T (2004) Role of arginine in protein refolding, solubilization, and purification. Biotechnol Prog 20:1301–1308

Tsuzuki F, Fukatsu R, Hayashi Y, Yoshida T, Sasaki N, Takamaru Y, Yamaguchi H, Tateno M, Fujii N, Takahata N (1996) Amyloid β-protein and transthyretin, sequestrating protein colocalize in normal human kidney. Neurosci Lett 222:163–166

Tycko R (2006) Molecular structure of amyloid fibrils: insights from solid-state NMR. Q Rev Biophys 39:1–55

Tycko R, Sciarretta KL, Orgel JP, Meredith SC (2009) Evidence for novel β-sheet structures in Iowa mutant β-amyloid fibrils. Biochemistry 48:6072–6084

Uéda K, Fukushima H, Masliah E, Xia Y, Iwai A, Yoshimoto M, Otero DA, Kondo J, Ihara Y, Saitoh T (1993) Molecular cloning of cDNA encoding an unrecognized component of amyloid in Alzheimer disease. Proc Natl Acad Sci USA 90:11282–11286

Uemichi T, Liepnieks JJ, Benson MD (1994) Hereditary renal amyloidosis with a novel variant fibrinogen. J Clin Invest 93:731–736

Uemichi T, Liepnieks JJ, Yamada T, Gertz MA, Bang N, Benson MD (1996) A frame shift mutation in the fibrinogen Aα chain gene in a kindred with renal amyloidosis. Blood 87:4197–4203

Uversky VN (2003) A protein-chameleon: conformational plasticity of α-synuclein, a disordered protein involved in neurodegenerative disorders. J Biomol Struct Dyn 21:211–234

Uversky VN (2008) α-Synuclein misfolding and neurodegenerative diseases. Curr Protein Pept Sci 9:507–540

Uversky VN (2009) Intrinsic disorder in proteins associated with neurodegenerative diseases. Front Biosci 14:5188–5238

Uversky VN (2010) The mysterious unfoldome: structureless, underappreciated, yet vital part of any given proteome. J Biomed Biotechnol 2010:568068

Uversky VN, Dunker AK (2010) Understanding protein non-folding. Biochim Biophys Acta 1804:1231–1264

Uversky VN, Eliezer D (2009) Biophysics of Parkinson's disease: structure and aggregation of α-synuclein. Curr Protein Pept Sci 10:483–499

Uversky VN, Gillespie JR, Fink AL (2000) Why are "natively unfolded" proteins unstructured under physiologic conditions? Proteins 41:415–427

Uversky VN, Lee HJ, Li J, Fink AL, Lee SJ (2001) Stabilization of partially folded conformation during α-synuclein oligomerization in both purified and cytosolic preparations. J Biol Chem 276:43495–43498

Uversky VN, Yamin G, Souillac PO, Goers J, Glaser CB, Fink AL (2002) Methionine oxidation inhibits fibrillation of human α-synuclein *in vitro*. FEBS Lett 517:239–244

van Ham TJ, Thijssen KL, Breitling R, Hofstra RM, Plasterk RH, Nollen EA (2008) *C. elegans* model identifies genetic modifiers of α-synuclein inclusion formation during aging. PLoS Genet 4:e1000027

Van Melckebeke H, Wasmer C, Lange A, Ab E, Loquet A, Böckmann A, Meier BH (2010) Atomic-resolution three-dimensional structure of HET-s(218–289) amyloid fibrils by solid-state NMR spectroscopy. J Am Chem Soc 132:13765–13775

Vardy ER, Hussain I, Hooper NM (2006) Emerging therapeutics for Alzheimer's disease. Expert Rev Neurother 6:695–704

Vassar R (2001) The β-secretase, BACE: a prime drug target for Alzheimer's disease. J Mol Neurosci 17:157–170

Vassar R, Bennett BD, Babu-Khan S, Kahn S, Mendiaz EA, Denis P, Teplow DB, Ross S, Amarante P, Loeloff R, Luo Y, Fisher S, Fuller J, Edenson S, Lile J, Jarosinski MA, Biere AL, Curran E,

Burgess T, Louis JC, Collins F, Treanor J, Rogers G, Citron M (1999) β-Secretase cleavage of Alzheimer's amyloid precursor protein by the transmembrane aspartic protease BACE. Science 286:735–741

Vassar R, Kovacs DM, Yan R, Wong PC (2009) The β-secretase enzyme BACE in health and Alzheimer's disease: regulation, cell biology, function, and therapeutic potential. J Neurosci 29:12787–12794

Vatassery GT, Quach HT, Smith WE, Benson BA, Eckfeld JH (1991) A sensitive assay of transthyretin (prealbumin) in human cerebrospinal fluid in nanogram amount by ELISA. Clin Chim Acta 197:19–25

Vendrely C, Valadié H, Bednarova L, Cardin L, Pasdeloup M, Cappadoro J, Bednar J, Rinaudo M, Jamin M (2005) Assembly of the full-length recombinant mouse prion protein I. Formation of soluble oligomers. Biochim Biophys Acta 1724:355–366

Venkatraman J, Shankaramma SC, Balaram P (2001) Design of folded peptides. Chem Rev 101:3131–3152

Venneti S (2010) Prion diseases. Clin Lab Med 30:293–309

Verhoef LG, Lindsten K, Masucci MG, Dantuma NP (2002) Aggregate formation inhibits proteasomal degradation of polyglutamine proteins. Hum Mol Genet 11:2689–2700

Vidal R, Frangione B, Rostagno A, Mead S, Révész T, Plant G, Ghiso J (1999) A stop-codon mutation in the BRI gene associated with familial British dementia. Nature 399:776–781

Vieira EP, Hermel H, Möhwald H (2003) Change and stabilization of the amyloid-β(1–40) secondary structure by fluorocompounds. Biochim Biophys Acta 1645:6–14

Vilar M, Chou HT, Lührs T, Maji SK, Riek-Loher D, Verel R, Manning G, Stahlberg H, Riek R (2008) The fold of α-synuclein fibrils. Proc Natl Acad Sci USA 105:8637–8642

Villarroya M, García AG, Marco-Contelles J, López MG (2007) An update on the pharmacology of galantamine. Expert Opin Investig Drugs 16:1987–1998

Villaverde MC, Gonzalez-Louro L, Sussman F (2007) The search for drug leads targeted to the β-secretase: an example of the roles of computer assisted approaches in drug discovery. Curr Top Med Chem 7:980–990

Villegas V, Zurdo J, Filimonov VV, Avilés FX, Dobson CM, Serrano L (2000) Protein engineering as a strategy to avoid formation of amyloid fibrils. Protein Sci 9:1700–1708

Villoslada P, Moreno B, Melero I, Pablos JL, Martino G, Uccelli A, Montalban X, Avila J, Rivest S, Acarin L, Appel S, Khoury SJ, McGeer P, Ferrer I, Delgado M, Obeso J, Schwartz M (2008) Immunotherapy for neurological diseases. Clin Immunol 128:294–305

Vingtdeux V, Chandakkar P, Zhao H, d'Abramo C, Davies P, Marambaud P (2010a) Novel synthetic small-molecule activators of AMPK as enhancers of autophagy and amyloid-β peptide degradation. FASEB J 25:219–231

Vingtdeux V, Giliberto L, Zhao H, Chandakkar P, Wu Q, Simon JE, Janle EM, Lobo J, Ferruzzi MG, Davies P, Marambaud P (2010b) AMP-activated protein kinase signaling activation by resveratrol modulates amyloid-β peptide metabolism. J Biol Chem 285:9100–9113

Vitoux B, Aubry A, Cung MT, Boussard G, Marraud M (1981) *N*-methyl peptides. III. Solution conformational study and crystal structure of N-pivaloyl-L-prolyl-N-methyl-N′-isopropyl-L-alaninamid. Int J Pept Protein Res 17:469–479

von Bernhardi R (2010) Immunotherapy in Alzheimer's disease: where do we stand? Where should we go? J Alzheimers Dis 19:405–421

Wakabayashi T, DeStrooper B (2008) Presenilins: members of the γ-secretase quartets, but part-time soloists too. Physiology 23:194–204

Walker FO (2007) Huntington's disease. Lancet 369:218–228

Walsh DM, Klyubin I, Fadeeva JV, Cullen WK, Anwyl R, Wolfe MS, Rowan MJ, Selkoe DJ (2002) Naturally secreted oligomers of amyloid β protein potently inhibit hippocampal long-term potentiation *in vivo*. Nature 416:535–539

Walsh P, Neudecker P, Sharpe S (2010) Structural properties and dynamic behavior of nonfibrillar oligomers formed by PrP(106–126). J Am Chem Soc 132:7684–7695

Wang M, Suzuki T, Kitada T, Asakawa S, Minoshima S, Shimizu N, Tanaka K, Mizuno Y, Hattori N (2001) Developmental changes in the expression of parkin and UbcR7, a parkin-interacting and ubiquitin-conjugating enzyme, in rat brain. J Neurochem 77:1561–1568

Wang HW, Pasternak JF, Kuo H, Ristic H, Lambert MP, Chromy B, Viola KL, Klein WL, Stine WB, Krafft GA, Trommer BL (2002) Soluble oligomers of β amyloid (1–42) inhibit long-term potentiation but not long-term depression in rat dentate gyrus. Brain Res 924:133–140
Wang J, Gines S, MacDonald ME, Gusella JF (2005) Reversal of a full-length mutant huntingtin neuronal cell phenotype by chemical inhibitors of polyglutamine-mediated aggregation. BMC Neurosci 6:1
Wang J, Ho L, Zhao Z, Seror I, Humala N, Dickstein DL, Thiyagarajan M, Percival SS, Talcott ST, Pasinetti GM (2006) Moderate consumption of Cabernet Sauvignon attenuates Aβ neuropathology in a mouse model of Alzheimer's disease. FASEB J 20:2313–2320
Wang YP, Biernat J, Pickhardt M, Mandelkow E, Mandelkow EM (2007) Stepwise proteolysis liberates tau fragments that nucleate the Alzheimer-like aggregation of full-length tau in a neuronal cell model. Proc Natl Acad Sci USA 104:10252–10257
Wang J, Farr GW, Hall DH, Li F, Furtak K, Dreier L, Horwich AL (2009) An ALS-linked mutant SOD1 produces a locomotor defect associated with aggregation and synaptic dysfunction when expressed in neurons of *Caenorhabditis elegans*. PLoS Genet 5:e1000350
Wardle M, Morris HR, Robertson NP (2009) Clinical and genetic characteristics of non-Asian dentatorubral-pallidoluysian atrophy: A systematic review. Mov Disord 24:1636–1640
Watanabe K-I, Nakamura K, Akikusa S, Okada T, Kodaka M, Konakahara T, Okuno H (2002) Inhibitors of fibril formation and cytotoxicity of β-amyloid peptide composed of KLVFF recognition element and flexible hydrophilic disrupting element. Biochem Biophys Res Commun 290:121–124
Wegmann S, Jung YJ, Chinnathambi S, Mandelkow EM, Mandelkow E, Muller DJ (2010) Human Tau isoforms assemble into ribbon-like fibrils that display polymorphic structure and stability. J Biol Chem 285:27302–27313
Wei G, Jewett AI, Shea JE (2010) Structural diversity of dimers of the Alzheimer amyloid-β(25–35) peptide and polymorphism of the resulting fibrils. Phys Chem Chem Phys 12:3622–3629
Weiner HL, Lemere CA, Maron R, Spooner ET, Grenfell TJ, Mori C, Issazadeh S, Hancock WW, Selkoe DJ (2000) Nasal administration of amyloid-β peptide decreases cerebral amyloid burden in a mouse model of Alzheimer's disease. Ann Neurol 48:567–579
Weissmann JS (2005) Birth of a prion: spontaneous generation revisited. Cell 122:165–168
Weksler ME, Gouras G, Relkin NR, Szabo P (2005) The immune system, amyloid-β peptide, and Alzheimer's disease. Immunol Rev 205:244–256
Weksler ME, Pawelec G, Franceschi C (2009) Immune therapy for age-related diseases. Trends Immunol 30:344–350
Westerheide SD, Anckar J, Stevens SM Jr, Sistonen L, Morimoto RI (2009) Stress-inducible regulation of heat shock factor 1 by the deacetylase SIRT1. Science 323:1063–1066
Westermark P, Sletten K, Johansson B, Cornwell GG 3rd (1990) Fibril in senile systemic amyloidosis is derived from normal transthyretin. Proc Natl Acad Sci USA 87:2843–2845
Westermark P, Eriksson L, Engstrom U, Enestrom S, Sletten K (1997) Prolactin-derived amyloid in the aging pituitary gland. Am J Pathol 150:67–73
Westermark P, Benson MD, Buxbaum JN, Cohen AS, Frangione B, Ikeda S, Masters CL, Merlini G, Saraiva MJ, Sipe JD (2005) Nomenclature Committee of the International Society of Amyloidosis. Amyloid: toward terminology clarification. Report from the Nomenclature Committee of the International Society of Amyloidosis. Amyloid 12:1–4
Wetzel R (2006) Kinetics and thermodynamics of amyloid fibril assembly. Acc Chem Res 39:671–679
Wetzel R, Shivaprasad S, Williams AD (2007) Plasticity of amyloid fibrils. Biochemistry 46:1–10
Whitsett JA, Wert SE, Weaver TE (2010) Alveolar surfactant homeostasis and the pathogenesis of pulmonary disease. Annu Rev Med 61:105–119
Whittemore NA, Mishra R, Kheterpal I, Williams AD, Wetzel R, Serpersu EH (2005) Hydrogen–deuterium (H/D) exchange mapping of Aβ1–40 amyloid fibril secondary structure using nuclear magnetic resonance spectroscopy. Biochemistry 44:4434–4441

Wickner RB, Dyda F, Tycko R (2008) Amyloid of Rnq1p, the basis of the [*PIN*[+]] prion, has a parallel in-register β-sheet structure. Proc Natl Acad Sci USA 105:2403–2408
Wilcock DM, Rojiani A, Rosenthal A, Subbarao S, Freeman MJ, Gordon MN, Morgan D (2004) Passive immunotherapy against Aβ in aged APP-transgenic mice reverses cognitive deficits and depletes parenchymal amyloid deposits in spite of increased vascular amyloid and microhemorrhage. J Neuroinflammation 1:24
Willem M, Garratt AN, Novak B, Citron M, Kaufmann S, Rittger A, DeStrooper B, Saftig P, Birchmeier C, Haass C (2006) Control of peripheral nerve myelination by the β-secretase BACE-1. Science 314:664–666
Williams AD, Portelius E, Kheterpal I, Guo JT, Cook KD, Xu Y, Wetzel R (2004) Mapping Aβ amyloid fibril secondary structure using scanning proline mutagenesis. J Mol Biol 335:833–842
Williamson JA, Loria JP, Miranker AD (2009) Helix stabilization precedes aqueous and bilayer-catalyzed fiber formation in islet amyloid polypeptide. J Mol Biol 393:383–396
Wilson MR, Yerbury JJ, Poon S (2008) Potential roles of abundant extracellular chaperones in the control of amyloid formation and toxicity. Mol Biosyst 4:42–52
Wiseman RL, Johnson SM, Kelker MS, Foss T, Wilson IA, Kelly JW (2005) Kinetic stabilization of an oligomeric protein by a single ligand binding event. J Am Chem Soc 127:5540–5551
Wisniewski T (2009) AD vaccines: conclusions and future directions. CNS Neurol Disord Drug Targets 8:160–166
Wojtczak A, Neumann P, Cody V (2001) Structure of a new polymorphic monoclinic form of human transthyretin at 3 A resolution reveals a mixed complex between unliganded and T4-bound tetramers of TTR. Acta Crystallogr D Biol Crystallogr 57:957–967
Wong GT (2007) FDA deems U.S. Alzhemed trial results inconclusive, Alzheimer Research Forum, http://www.alzforum.org/new/detail.asp?id=1647, Posted 28 Aug 2007
Wong CW, Quaranta V, Glenner GG (1985) Neuritic plaques and cerebrovascular amyloid in Alzheimer disease are antigenically related. Proc Natl Acad Sci USA 82:8729–8732
Wong GT, Manfra D, Poulet FM, Zhang Q, Josien H, Bara T, Engstrom L, Pinzon-Ortiz M, Fine JS, Lee HJ, Zhang L, Higgins GA, Parker EM (2004) Chronic treatment with the γ-secretase inhibitor LY-411,575 inhibits β-amyloid peptide production and alters lymphopoiesis and intestinal cell differentiation. J Biol Chem 279:12876–12882
Wong HK, Bauer PO, Kurosawa M, Goswami A, Washizu C, Machida Y, Tosaki A, Yamada M, Knöpfel T, Nakamura T, Nukina N (2008) Blocking acid-sensing ion channel 1 alleviates Huntington's disease pathology via an ubiquitin–proteasome system-dependent mechanism. Hum Mol Genet 17:3223–3235
Wood SJ, Wetzel R, Martin JD, Hurle MR (1995) Prolines and amyloidogenicity in fragments of the Alzheimer's peptide β/A4. Biochemistry 34:724–730
Wood SJ, MacKenzie L, Maleeff B, Hurle MR, Wetzel R (1996) Selective inhibition of Aβ fibril formation. J Biol Chem 271:4086–4092
Wu JW, Breydo L, Isas JM, Lee J, Kuznetsov YG, Langen R, Glabe C (2010) Fibrillar oligomers nucleate the oligomerization of monomeric amyloid-β but do not seed fibril formation. J Biol Chem 285:6071–6079
Xie Q, Guo T, Lu J, Zhou HM (2004) The guanidine like effects of arginine on aminoacylase and salt-induced molten globule state. Int J Biochem Cell Biol 36:296–306
Yamada M, Shimohata M, Sato T, Tsuji S, Takahashi H (2006) Polyglutamine disease: recent advances in the neuropathology of dentatorubral-pallidoluysian atrophy. Neuropathology 26:346–351
Yamin G, Ono K, Inayathullah M, Teplow DB (2008) Amyloid β-protein assembly as a therapeutic target of Alzheimer's disease. Curr Pharm Des 14:3231–3246
Yan R, Bienkowski MJ, Shuck ME, Miao H, Tory MC, Pauley AM, Brashier JR, Stratman NC, Mathews WR, Buhl AE, Carter DB, Tomasselli AG, Parodi LA, Heinrikson RL, Gurney ME (1999) Membrane-anchored aspartyl protease with Alzheimer's disease β-secretase activity. Nature 402:533–537

Yan L-M, Tatrek-Nossol M, Velkova A, Kazantzis A, Kapurniotu A (2006) Design of a mimic of nonamyloidogenic and bioactive human islet amyloid polypeptide (IAPP) as nanomolar inhibitor of IAPP cytotoxic fibrillogenesis. Proc Natl Acad Sci USA 103:2046–2051

Yang F, Lim GP, Begum AN, Ubeda OJ, Simmons MR, Ambegaokar SS, Chen PP, Kayed R, Glabe CG, Frautschy SA, Cole GM (2005) Curcumin inhibits formation of amyloid β oligomers and fibrils, binds plaques, and reduces amyloid *in vivo*. J Biol Chem 280:5892–5901

Yao Z, Drieu K, Papadopoulos V (2001) The *Ginkgo biloba* extract EGb 761 rescues the PC12 neuronal cells from β-amyloid-induced cell death by inhibiting the formation of β-amyloid-derived diffusible neurotoxic ligands. Brain Res 889:181–190

Yazaki M, Liepnieks JJ, Barats MS, Cohen AH, Benson MD (2003) Hereditary systemic amyloidosis associated with a new apolipoprotein AII stop codon mutation Stop78Arg. Kidney Int 64:11–16

Yeung F, Hoberg JE, Ramsey CS, Keller MD, Jones DR, Frye RA, Mayo MW (2004) Modulation of NF-κB-dependent transcription and cell survival by the SIRT1 deacetylase. EMBO J 23:2369–2380

Young AB (2003) Huntingtin in health and disease. J Clin Invest 111:299–302

Zeng H, Zhang Y, Peng L, Shao H, Menon NK, Yang J, Salomon AR, Freidland RP, Zagorski MG (2001) Nicotine and amyloid formation. Biol Psychiatry 49:248–257

Zepik H, Shavit E, Tang M, Jensen TR, Kjaer K, Bolbach G, Leiserowitz L, Weissbuch I, Lahav M (2002) Chiral amplification of oligopeptides in two-dimensional crystalline self-assemblies on water. Science 295:1266–1269

Zhang YQ, Sarge KD (2007) Celastrol inhibits polyglutamine aggregation and toxicity though induction of the heat shock response. J Mol Med 85:1421–1428

Zhang Y, Gao J, Chung KK, Huang H, Dawson VL, Dawson TM (2000) Parkin functions as an E2-dependent ubiquitin- protein ligase and promotes the degradation of the synaptic vesicle-associated protein, CDCrel-1. Proc Natl Acad Sci USA 97:13354–13359

Zhang X, Smith DL, Meriin AB, Engemann S, Russel DE, Roark M, Washington SL, Maxwell MM, Marsh JL, Thompson LM, Wanker EE, Young AB, Housman DE, Bates GP, Sherman MY, Kazantsev AG (2005) A potent small molecule inhibits polyglutamine aggregation in Huntington's disease neurons and suppresses neurodegeneration *in vivo*. Proc Natl Acad Sci USA 102:892–897

Zhu M, Fink AL (2003) Lipid binding inhibits α-synuclein fibril formation. J Biol Chem 278:16873–16877

Zoghbi HY, Orr HT (2009) Pathogenic mechanisms of a polyglutamine-mediated neurodegenerative disease, spinocerebellar ataxia type 1. J Biol Chem 284:7425–7429

Index

F. Rahimi and G. Bitan (eds.), *Non-fibrillar Amyloidogenic Protein Assemblies—Common Cytotoxins Underlying Degenerative Diseases*, DOI 10.1007/978-94-007-2774-8,